Estuaries and Nutrients

Contemporary Issues in Science and Society

Estuaries and Nutrients, edited by ***Bruce J. Neilson*** and ***L. Eugene Cronin,*** 1981

Pesticides: *Contemporary Roles in Agriculture, Health, and the Environment,* edited by ***T. J. Sheets*** and ***David Pimentel,*** 1979

Solar Energy: *Chemical Conversion and Storage,* edited by ***Richard R. Hautala, R. Bruce King,*** and ***Charles Kutal,*** 1979

Estuaries and Nutrients

Edited by

Bruce J. Neilson
Virginia Institute of Marine Science
and
L. Eugene Cronin
Chesapeake Research Consortium

Humana Press · Clifton, New Jersey

The Humana Press Inc.
Crescent Manor
PO Box 2148
Clifton, New Jersey 07015

Library of Congress Cataloging in Publication Data

International Symposium on the Effects of Nutrient Enrichment in Estuaries (1979: Williamsburg, Va.)
Estuaries and Nutrients.

(Contemporary issues in science and society) (Chesapeake Research Consortium publication; no. 90)

"Proceedings of an International Symposium on the Effects of Nutrient Enrichment in Estuaries, Williamsburg, Virginia, 29–31, May, 1979, conducted by the Chesapeake Research Consortium for the Chesapeake Bay Program, United States Environmental Protection Agency"—Verso t.p.

Bibliography: p.

Includes Index.

1. Estuarine ecology—Congresses. 2. Eutrophication—Congresses. 3. Estuarine pollution—Congresses. I. Neilson, Bruce J. II. Cronin, L. Eugene (Lewis Eugene), 1917- . III. Chesapeake Research Consortium. IV. United States. Environmental Protection Agency. Chesapeake Bay Program. V. Series. VII. Series: Chesapeake Research Consortium publication; no. 90.

QH541.5.E8I57 1979 574.5'26365 81-83901
ISBN 0-89603-035-0 AACR2

Printed in the United States of America

CONTENTS

Case Studies

Contributed Papers

Contributed Complementary Studies in the York River

PREFACE

Estuaries are eternally enriched. Their positions at the foot of watersheds and their convenience as receiving bodies for the wastes of cites, towns and farms results in continuous addition of nutrients - those elements and compounds which are essential for organic production. Such materials must be added to these complex bodies of water to sustain production, since there is a net loss of water and its contents to the oceans. Enrichment from land and the ocean and the subsequent cycling of the original chemicals or their derivatives contribute to the extraordinarily high values of estuaries for human purposes. Many estuaries are able to assimilate large quantities of nutrients despite the great fluctuations which occur with variations in the flow from tributaries. The nutrients can be stored, incorporated in standing crops of plants, released, cycled and exported - and the system frequently achieves high production of plants and and animals without creation of any undesirable results of enrichment.

Excessive enrichment with the same elements and compounds can, however, be highly detrimental to estuaries and their uses. Coastal cities are usually located on the estuaries which provided a harbor for them - and which now receive partially treated sewage and other wastes from the expanding population and industrial activity. Conversion of woodlands to agricultural use and the extensive application of fertilizers have resulted in the flow of large quantities of nutrients down the hill or slopes and eventually into the estuary.

The results of excessive enrichment can be dramatic and serious. As phytoplankton populations increase, they can change qualitatively so that a rich system feeding many components in the food web becomes a jungle, with massive production of species of little value in the energy systems.

Recognition that nutrient loading into the Chesapeake Bay system had been and was the source of serious and extensive damage to parts of the estuary - and that the resident population is projected to double within a few decades - resulted in the identification of excessive enrichment as one of the three areas of principal attention in the 5-year Chesapeake Bay Program conducted by the U.S. Environmental Protection Agency in cooperation with the State of Maryland and the Commonwealths of Pennsylvania and Virginia. Under a grant to the Chesapeake Research Consortium, Inc., to develop the background for our present state of knowledge about enrichment of estuaries, an **International Symposium on the Effects of Nutrient Enrichment in Estuaries** was convened at Williamsburg, Virginia on May 29-31 of 1979. Forty-four papers were presented and discussed. Of these, thirty-three survived the rigors of review and editing, and the demands on the time of the authors, and are presented in this volume.

Papers are arranged in four sets. *Invited Review Papers* were solicited from recognized experts in various aspects of estuarine enrichment to provide the overview and summary so necessary for the education of all scientists and managers concerned with nutrient questions. The series is presented as it was at the Symposium, with broad background papers followed by reviews in specific areas. Six *Case Studies* are reported to illustrate problems and experiences in several parts of the world. *Contributed Papers* included a variety of reports on recent research, with a group of four papers derived from an interesting cooperative approach to a single estuarine area.

We hope that these papers will be of value. That value may appear in many forms - in improved management in the Chesapeake Bay area and in other estuaries, in stimulation of further advance in comprehension of estuaries and the questions of adequate and excessive enrichment, in guidance for the expenditure of sufficients funds for prevention of serious problems without wastage because of ignorance, and in improved knowledge among managers, scientists and citizens concerned with achievement and maintenance of the wisest balances in our environment.

We are pleased to acknowledge many essential contributions to this volume. The authors are the most important source, followed closely by the Chesapeake Bay Program of EPA for its financial support through Grant R 806 189010 and for professional support of the effort. The Planning Committee - L. Eugene Cronin, Donald R. Heinle, Andrew J. McErlean, Bruce J. Neilson and Thomas H. Pheiffer - labored productively in planning and arranging the Symposium and in assuring excellent peer review and quality control without intended injury to the freedom of the authors. Peers from many fields of science and management provided constructive review and comments on manuscripts.

Mr. Eric Rellihan worked hard and well as manager of the Symposium. Ms. Maxine Smith has been highly conscientious in preparing camera-ready copy. Ms. Linda Kilch provided intelligent assistance in preparing a Symposium brochure, advertisements of the Symposium and other useful materials. Wright Robinson performed the tedious and difficult tasks of preparing the Index. Finally, Humana Press has been most patient and courteous during completion of our material.

To these and to all of the others who participated in the Symposium, we express our sincere appreciation.

Bruce J. Neilson

L. Eugene Cronin

INVITED REVIEW PAPERS

SPECIAL CHARACTERISTICS OF ESTUARIES

Robert B. Biggs* and L. Eugene Cronin**
**College of Marine Studies*
University of Delaware
Newark, Delaware 19711

***Chesapeake Research Consortium, Inc.*
Suite 297, 1414 Forest Drive
Annapolis, Maryland 21403

ABSTRACT: Estuaries are characterized by the gradient of salinity in a semi-enclosed coastal system. A working classification for drowned river estuaries has been developed and is based on the dominance of certain terms in the salt balance equation. Tidal and wind energy, as well as freshwater flow and density gradients, are responsible for mixing. Suspended and bottom sediment distribution may be characterized for each estuarine type. These sediments are of exceptional importance to the routes and fates of nutrients and other chemical materials. Although particular estuaries are transient in geological or evolutionary time spans, the estuarine environment has been common through a long geologic time, resulting in the evolution of a group of organisms uniquely capable of using the salinity gradient to a competitive advantage. These species, well adapted to the rigorous estuarine environment, frequently produce high standing crops and biomass from a small number of species. The range of nutrient inputs to the estuarine system is reviewed, along with the major nutrient processes and pathways which operate internally. Particularly successful models of nutrient movement are discussed.

INTRODUCTION

There are three definitive characteristics which, together, distinguish the estuary from all other bodies of water. The first two are geomorphic; the body of water must be semi-enclosed and coastal, and with a free connection to the open sea; the third is chemical; there must be a salinity gradient caused by the dilution of seawater with freshwater from upland drainage. Each of the other components or processes which occur in estuaries can occur in aquatic or marine environments. The same physical, chemical, biological, and geological materials and processes are parts of other systems. The quantities of certain components and the rates of some processes are exceptional (high or low) in estuaries and the net product is distinctly different from fresh water lakes, streams, and rivers, and from the open ocean.

Geological Occurrence of the Estuarine Environment

There have been estuaries for at least the last 200 million years. The record of their occurrence has been firmly established on the basis of paleontological and/or or geochemical evidence. Coastal plain estuaries have been formed and destroyed principally by the rise and fall of sea level. During the last million years there have been at least four (and perhaps as many as ten) stands of sea level at least as high as present and there have been as many lowerings, some to elevations of -100 meters (33).

FIGURE 1. A Holocene sea level rise curved based on the carbon-14 age of peats and other materials which can be inferred to have been deposited at or very near to sea level. (2)

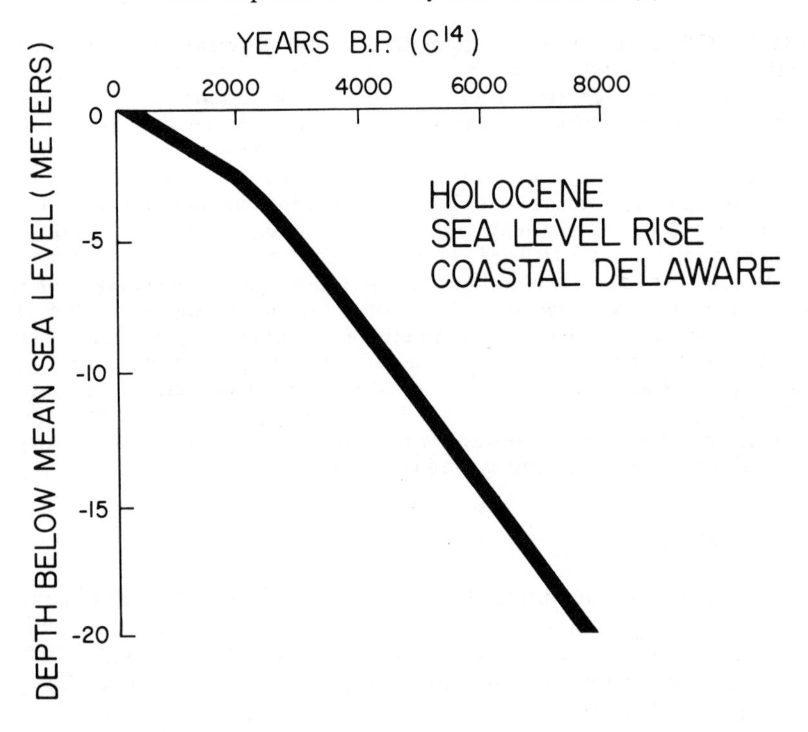

Figure 1 illustrates the most recent rise of sea level derived by age dating salt marsh peats found below present sea level on the Delmarva Peninsula. Other evidence (from oystem bioherms and peats found offshore, submarine canyons) indicates that, about 18,000 years before present, sea level stood about 100 meters below its present level and that the coast in the mid-Atlantic region was about 70 kilometers seaward of its present position.

Modern (Holocene) drowned river valley estuaries have been formed by the flooding of preexisting drainage systems. The development and migration of coastal barrier systems across an older topography has produced bar-build estuaries. Test borings for bridge crossings of the submerged valley with a maximum depth of -60 meters beneath present sea level. Clearly these valleys were carved by one or more low stands of sea level when the estuarine environment was located on the present continental shelf. Other deposits of sand, mud and peat located inland of the present coast and up to 10 meters above present sea level are interpreted as estuarine deposits and indicate that, during the present sea level are interpreted as estuarine deposits and indicate that, during the present transgression, we have not yet reached the maximum possible rise of sea level.

Estuaries move in time and space in response to changes in sea level and changes in the rate of sedimentation. Thus, a particular estuary, as defined by its location, is ephemeral. We wish to make the point, though, that the geologic evidence supports the contention that the estuarine environment is permanant so long as there is freshwater runoff from an upland source.

Classification of Estuarine Types

If we define the geomorphic characteristics of an estuarine basin and then examine the variation of salinity within the basin as a function of freshwater discharge and tidal velocity, a scheme for classifying estuaries emerges. The classification scheme was first developed by Pritchard (27) and is based on the advection - diffusion equation for salt. This salt balance equation states that the time rate of change observed in the salinity at a fixed point in the estuary is caused by two different physical processes, diffusion and advection. Diffusive processes are defined as a flux of salt, advective processes as a flux of salt and a flux of water. Both advective and diffusive processes may occur along the longitudinal, lateral, and vertical axes of the estuary. In particular estuaries, under defined freshwater discharge, tidal velocity and wind conditions, these terms may be of the same order of magnitude.

Type A Estuaries

Let us start with a coastal river valley whose bottom is below sea level, with very small or no measurable tides, and with freshwater discharge at the landward end. The freshwater pressure head will push the salt water out of the upper reaches of the estuary until some point downstream where the valley widens. At that point, the freshwater will ride over the more dense saltwater, spreading out as a thinner and thinner surface layer in the seaward direction. The landwardmost penetration of the bottom salty water is, therefore, a function of river discharge. As river discharge increases, the tip of the salt wedge moves seaward. When there is a velocity difference between two fluids in contact, shear forces at the interface create waves. The observed phenomena of wind moving over water is somewhat analogous to the situation here, where surface freshwater is moving relative to deeper saltwater. Internal waves are formed at the interface and these waves break upward. As a result, masses of saltwater mix upward with the seaward-flowing freshwater, increasing its salinity in a seaward direction. At the same time, for every volume of deep seawater which becomes entrained in the surface water, an equal volume of seawater must enter the mouth of the estuary along the bottom and flow upstream. As freshwater discharge increases, the interfacial waves increase and the bottom, landward-directed, saltwater flow increases. These breaking internal waves transport both salt and water to the upper seaward-flowing layer by the process of vertical advection. Pritchard terms this type of estuary a salt wedge or Type A estuary and points to the Southwest Pass of the Mississippi River as an example. A schematic illustration is provided in Figure 2. Schubel (31) has defined a mixing index for estuaries (ratio of volume of freshwater entering during a half-tidal period to the tidal prism) and has stated that, where this ratio is less than unity and where the ratio of width to depth is small, the estuary will be highly stratified.

FIGURE 2. The Type A (salt wedge) estuary is characterized by the dominance of river flow. Mixing occurs as the result of upward breaking internal waves. A prounced salt wedge is present, though the position of its upstream tip varies with freshwater discharge. From Prtichard, 1955 (27).

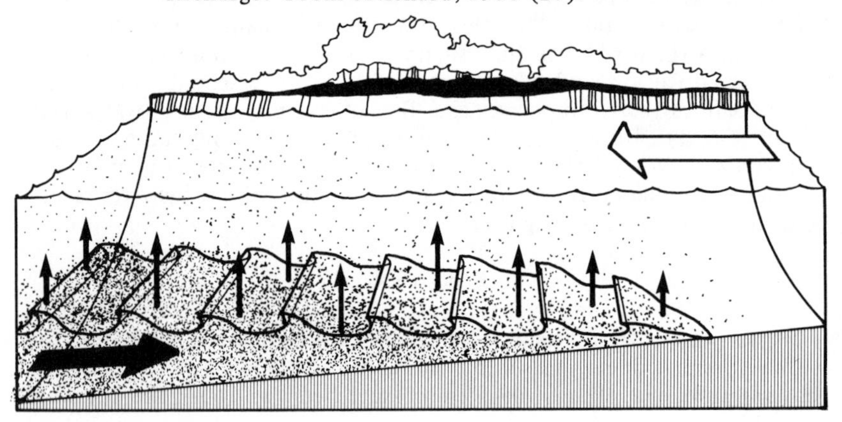

Sand-sized particles, moving in the river bedload, will be deposited somewhat upstream from the tip of the salt wedge as the unidirectional downstream current velocity decreases. Smaller, suspended particles transported in the upper layers may move downstream of the salt wedge, may be modified by flocculation or biodeposition and settle by gravitational forces. As these suspended particles settle through the lower, salty layer, they may be transported back upstream by the net upstream flow of the lower layer to be deposited just seaward of the tip of the salt wedge. Thus, the tip of the salt wedge is a region of intense sedimentation, and as the salt wedge moves up and down the length of the basin in response to changing river flow, so does the zone of most active sedimentation.

In the Southwest Pass of the Mississippi, the tip of the salt wedge may be located between the jetties at the seaward end of the channel during high river stages and may migrate 300 m upstream at very low flow period. Suspended sediment concentrations may exceed 400 mg ℓ^{-1}, the euphotic zone is very shallow and the salinity at any one point may vary from $0^{o}/oo$ to $30^{o}/oo$. Benthic populations are low because of the high suspended sediment load, the salinity range, and the instability of the bottom. Both natural and anthropogenic dissolved materials entering the head of the estuary are carried seaward and continuously diluted by seawater in the upper layer. Materials entering the estuary attached to or becoming attached to suspended matter may be sedimented out in the vicinity of the salt wedge or may be transported through the estuary and onto the shelf.

Type B Estuaries

Let us now introduce a tidal force which is sufficiently strong to prevent the river flow from dominating the circulation of the estuary. As described in the mixing index, suppose we make the half-tidal river flow small compared to the intertidal volume, say 1/10. The added turbulence is sufficient to make the vertical diffusion term significant in the overall salt balance equation. Fresh (or low salinity) water mixes downward and salt (or higher salinity) water mixes upward, and the salt wedge is replaced by a halocline where the difference in salinity between the surface and bottom waters is relatively constant regardless of position in the estuary. Prtichard has termed such an estuary a Type B estuary (Figure 3) and cites the James River as an example. The ratio of width to depth is larger in a Type B estuary than in a Type A estuary.

In addition to the longitudinal and vertical salinity gradients, a modest lateral salinity gradient exists with more salinie water on the left side (looking downstream in the northern hemisphere) and fresher water on the right side. This lateral salinity variation is caused by the earth's rotation (Coriolis force) beneath the moving waters. In the northern hemisphere, flowing water will be deflected to the right. In a Type B estuary, this lateral gradient is modest, certainly no larger than the vertical gradient.

FIGURE 3. Type B estuaries have significant tidal mixing in addition to freshwater discharge. Both advective and diffusive processes mix salt water upward and freshwater downward. In addition there is a measurable lateral salinity gradient due to the Coriolis effect. From Pritchard, 1955 (27).

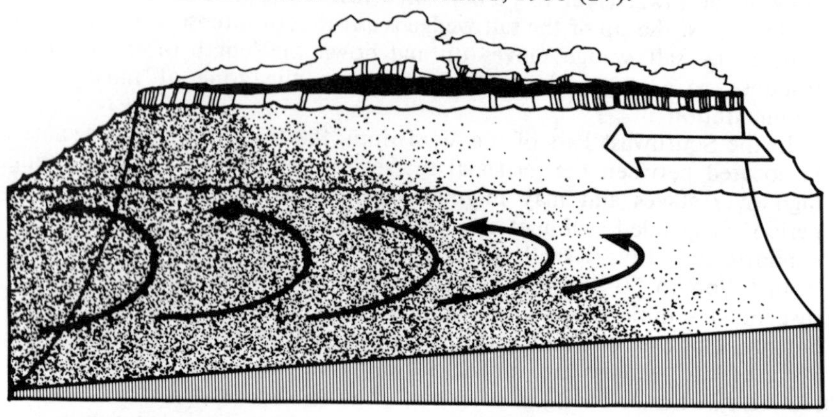

The seaward flow in the surface layer may be an order of magnitude greater than the freshwater river discharge (R). If the total seaward flow is 10 R in the surface layer at the mouth of the estuary, then there must be a landward flow 9 R in the bottom waters to preserve volume continuity. The mixing process in a Type B estuary is illustrated in Figure 4. Although

FIGURE 4. Schematic flow pattern, flow volume and salinity distribution in a Type B (partially mixed) estuary. Volume rate of flow is expressed in terms of the freshwater discharge (R). From Schubel, 1971 (31).

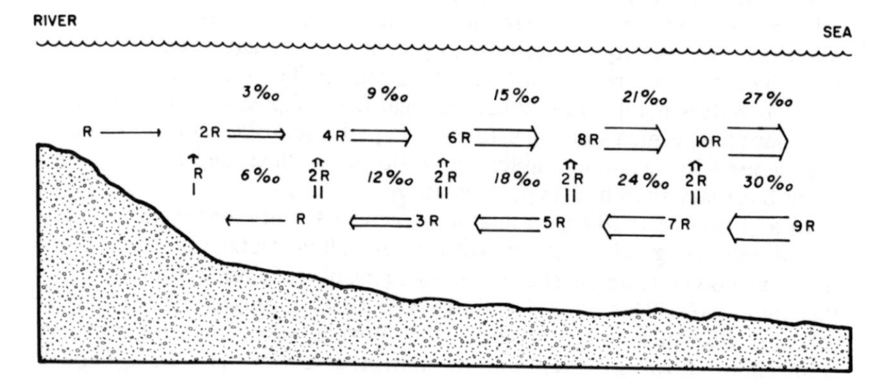

both the surface and bottom layer experience reversing tidal currents, there is a net seaward flow in the surface layer and a net landward flow in the bottom layer (Figure 5). The upstream limit of salt water intrusion moves landward or seaward in response to both the tide and to river flow.

FIGURE 5. This figure illustrates the concept of "net flow" in an estuarine system. Salty, near-bottom waters and the materials suspended in them have a net movement toward the head of the estuary while fresher near-surface waters experience a net seaward motion. From Cronin and Mansueti 1971 (11).

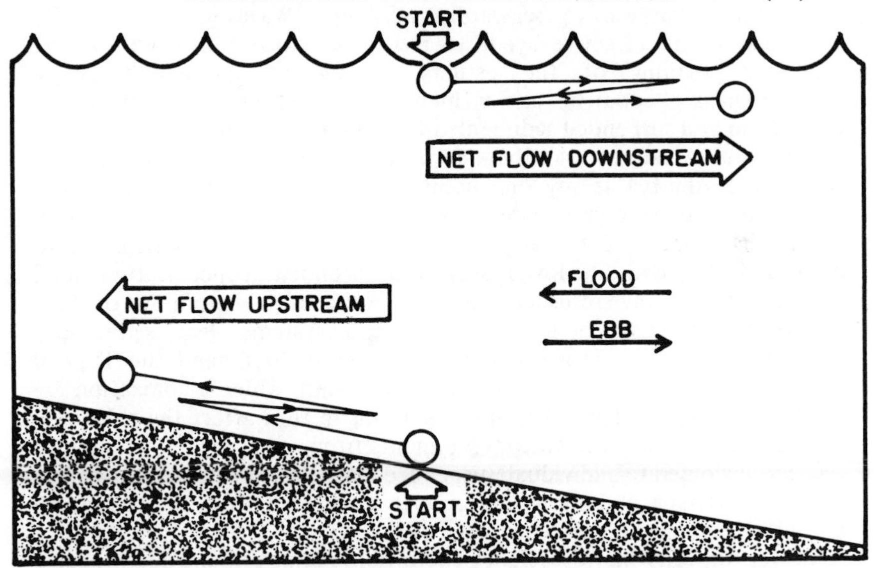

Bedload material moving seaward with the freshwater discharge is deposited where reversing tidal currents are first observed. Suspended material may be transported into the estuary in the upper layer and either mix or settle downward to be carried towards the head of the estuary in the lower layer. In the suite of various sized suspended materials (which may be composed of more than one individual particle) which enter the head of the estuary, the coarsest may settle by gravitation, while others may form agglomerates (either by flocculation or the creation of fecal pellets) before settling. Some particles, though, escape immediate sedimentation, move downstream, settle into the lower layer and are carried back upstream to the limit of salt intrusion by the net flow. Because of the higher level of mixing between the upper and lower layers in a Type B estuary, and because new deposits are easily resuspended by tidal and wind driven currents, there will probably be an accumulation of suspended sediments in the water column, producing a turbidity maximum in which the turbidity levels exceed those of the proximal river or estuary. Maximum sedimentation in a Type B estuary will occur between the upstreams and downstream limit of salt water intrusion. In a Type B estuary, very little of the suspended sediment introduced from upland sources escapes through the estuary to the sea (5) and the average concentration of suspended sediment in the water column is lower than in a Type B estuary by a factor of 5 to 10.

As previously mentioned the Type B estuary is wide relative to its depth. This phenomenon allows normal winds to play an important role in the circulation pattern of the estuary and to disturb the long-term average direction and rate of flow. In the short-term, (one-to-three days), net flows may be landward or seaward at all depths. Waves generated by the wind can be an effective agent to resuspend bottom sediments in the shallower portions of the estuary throughout its length. Thus, meterological phenomena can influence the short-term behavior of the circulation and suspended sediments of the Type B estuary.

The Type A estuary was characterized by its range salinity of (from 0 to oceanic salinity) at any one point, by the instability of the bottom sediments, and by high suspended sediment concentrations. The Type B estuary has a much lower range of salinity at a particular location, up to an order of magnitude lower suspended sediment concentration, and, except for wind-wave resuspension, a relatively stable bottom. All of these phenomena affect the biology of the estuarine systems. Decreased salinity range may permit a larger group of organisms to inhabit the Type B estuary. Increased bottom stability can permit wider colonization by benthic organisms. These benthic organisms in turn affect the substrate. Epifauna may armor the bottom sediments from resuspension, while filter feeders agglomerate individual suspended particles, creating fecal pellets which are larger and more difficult to resuspend than the individual particles of which they are composed (18).

Lower turbidity increases light penetration, permitting the development of algal mats and rooted aquatic plants as well and enhances primary productivity by phytoplankton. Some organisms have evolved elaborate life cycles to take maximum advantage of the Type B estuarine circulation pattern. Type B estuaries frequently contain exceptionally large populations of a relatively small number of species which have successfully adapted to the rigorous and highly variable environment. The biota has large components in phytoplankton, zooplankton, submerged and emergent aquatic vegetation, dense benthic communities and a variety of fish. The fish include ocean-source migrants entering the estuary to spawn or feed, permanent residents and intruders from fresh water. High concentrations of nutrients, phytoplankton and zooplankton in the low salinity region are associated with the intense use of many Type B estuaries as hatching and nursery sites for fish species. Cronin and Mansueti (11) have contrasted the cycle of the croaker (*Micropogon udulatus*) and the shad (*Alosa saphidissima*). Both of these organisms exhibit the importance of the Type B estuary as a nursery ares (Figures 6, 7).

FIGURE 6. The adults of anadromous species utilize the estuary as a route to freshwater spawning grounds while larval and young use the estuary as a nursery. From Cronin and Mansueti, 1971 (11).

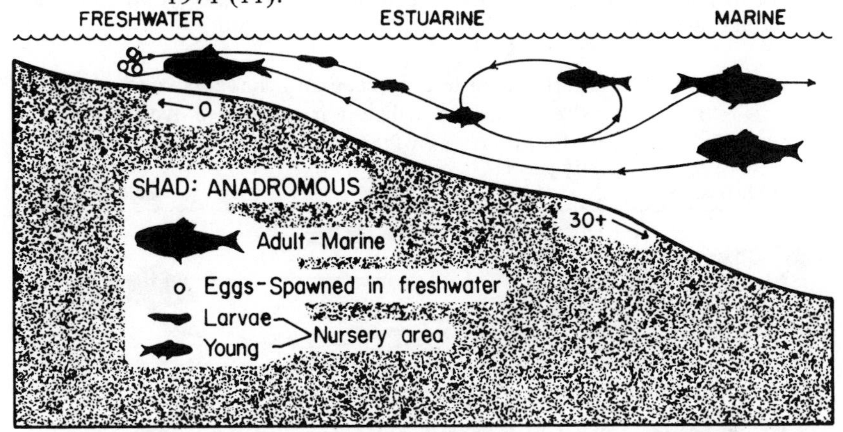

The intensity of the biological activity in a Type B estuary can affect the chemistry of the system in a number of ways. One of the most dramatic involves the dissolved oxygen concentration in the lower layer. In the main stem Chesapeake Bay (a Type B estuary) during the summer months, the combination of high water temperatures and intense biological activity results in a high dissolved oxygen demand. Dissolved oxygen can reach the lower layer (refer to Figure 3) by vertical diffusion and advection. These processes are not adequate to satisfy the oxygen demand of the deep waters and the bottom, and, as the lower waters move towards the head of the estuary, their dissolved oxygen concentration frequently decreases to undetectable levels. This situation does not occur during the remaining months of the year when water temperature and biological activity are lower and river flow and vertical mixing processes are higher.

Type C Estuaries

Suppose now that the tidal currents are increased to provide sufficient energy to completely erase the vertical salinity gradient that was present in a Type B estuary. Now there will be a longitudinal salinity gradient but no vertical gradient. The Coriolis force will be responsible for a lateral salinity gradient with higher salinity on the left side of the estuary, looking seaward (Figure 8). Pritchard has termed such an estuary a Type C estuary or a vertically homogenous estuary. The dominant terms in the salt balance equation become longitudinal and lateral advection, and lateral diffusion. To maintain volume continuity, there will be a net upstream flow at all depths on the left side of the estuary and a net downstream flow at all depths on the right side of the estuary. Schubel's mixing index will be much less than unity (at most 0.05). Such estuaries will have a large width to depth ratio and the wind may be an important short-term perturbation. The lower reach of Delaware Bay is an example of a Type C estuary.

FIGURE 7. Adult croakers spawn in the ocean and the larvae and young move into the estuary to utilize it as a nursery. Adults seasonally use the estuary as a feeding ground. From Cronin Mansueti 1971 (11).

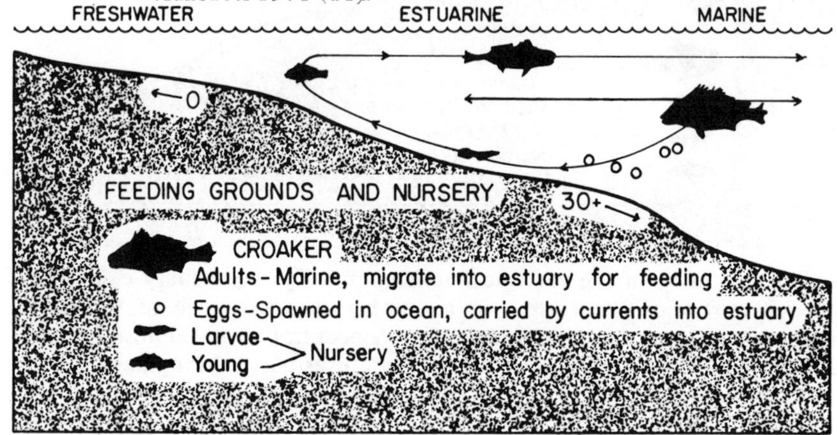

Suspended materials brought to the estuary by freshwater discharge move seaward with the net flow which occurs on the right side of the estuary. Oceanic source materials move landward along the left side of the estuary (looking downstream). Fine sediments are deposited near the head of salt intrusion and in areas where islands, shoals, or tributaries cause a modification in the net flow pattern. The high tidal velocities at all depths cause considerable resuspension and redistribution of bottom materials. Mobile bedforms like megaripples and sand waves occur frequently in a Type C estuary.

Pollutants associated with suspended materials derived from riverine sources are deposited on the right side of the estuary and materials injected into the system along the right side of the estuary move seaward with the net flow along the right side. Pollutants from oceanic sources or those which are injected into the left side of the estuary move towards the head of the estuary along the left side. The organisms which inhabit a Type C estuary are similar to those of a Type B estuary except that their geographic distribution is different. Because of the strong lateral as well as longitudinal salinity gradient, organisms which tolerate high salinity will favor the left side of the estuary while those which tolerate lower salinity will favor the right side.

FIGURE 8. In a Type C (vertically homogenous) estuary, tidal energy dominates the mixing process. There is a lateral and longitudinal salinity gradient but no vertical gradient. From Pritchard, 1955 (27).

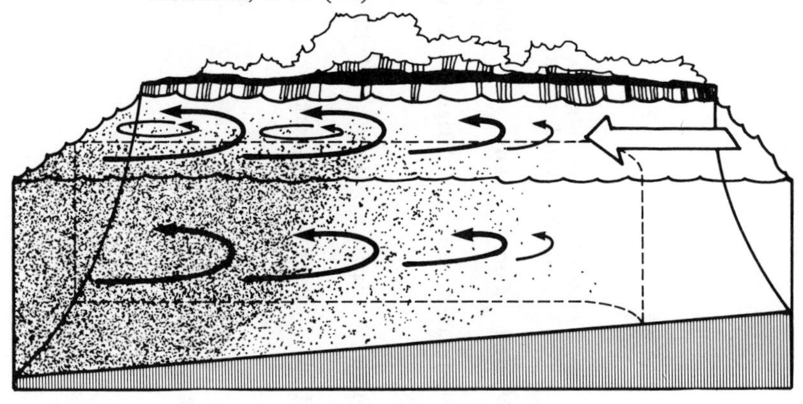

Type D Estuaries

Consider now an estuary where the tidal mixing forces are so great compared to river discharge that there is neither a vertical or lateral salinity gradient, that is to say, the only gradient is a longitudinal gradient. This is Pritchard's Type D or sectionally homogenous estuary. The dominant terms in the salt balance equation are horizontal advection and horizontal diffusion. Pritchard does not identify a natural estuary with these characteristics and it is probable that none exist.

Summary, Estuarine Types

In Table 1, we have attempted to summarize the general characteristic of Pritchard's estuarine types. The reader is cautioned to recognize that natural estuaries represent a continuum from Types A through C and that, under varying river flows, a particular estuary may shift from one type to another.

Low-frequency, high intensity storms can affect estuaries in a number of ways. Freshwater runoff from extreme precipitation changes circulation patterns, reduces salinity, and transports to the estuary natural and anthropogenic materials which have long accumulated along the course of river bottoms. Gross et al. (17) have measured the suspended sediment discharge from the Susquehanna River to Chesapeake Bay. In the absence of major floods, the Susquehanna discharged $0.9 \pm 0.3 \times 10^6$ metric tons of sediment per year. Over half of the annual contribution occurred during the spring freshet, which usually lasted about one month. During the ten years of measurement (1966-1976), two hurricanes (Agnes,

TABLE 1. General drowned river valley estuarine characteristics.

Estuarine Type[1]	Dominant Mixing Force	Mixing Energy	Width/Depth Ratio	Salinity Gradient	Mixing Index[2]	Turbidity	Bottom Stability	Biological Productivity	Example
A	River Flow	Low	Low	Longitudinal Vertical	≥ 1	V. High	Poor	Low	Southwest Pass Mississippi River
B	River Flow, Tide	Moderate	Moderate	Longitudinal Vertical Lateral	$< \frac{1}{10}$	Moderate	Good	V. High	Chesapeake Bay
C	Tide, Wind	High	High	Longitudinal Lateral	$< \frac{1}{20}$	High	Fair	High	Delaware Bay
D	Tide, Wind	V. High	V. High	Longitudinal	?	High	Poor	Moderate	?

[1]Follows Pritchard's advection-diffusion classification scheme (27).

[2]Follows Schubel's definition: $MI = \frac{\text{vol. freshwater discharge on 1/2 tidal period}}{\text{vol. tidal prism}}$ (31).

June 24-30, 1972; Eloise, September 26-30, 1975) caused very high runoff from the Susquehanna (Agnes, 26 x 10^3 m^3 sec^{-1}; Eloise, 17 x 10^3 m^3 sec^{-1}). Based on peak discharge Agnes was a 200 year storm and Eloise was a 25 year storm. Runoff associated with Agnes deposited 30 x 10^6 tons of sediment to the Bay while Eloise deposited 10 x 10^6 tons. Gross et al. concluded that about 50 x 10^6 tons of suspended sediment were discharged from the Susquehanna during the 1966-1976 interval. Eighty percent of that total (40 x 10^6 tons) was discharged in the two extreme events (ten days) while the remaining 20 percent was contributed during the remainder of the decade (3,640 days). Detailed data on the broad effects of Agnes on the Chesapeake may be found in Laird (22).

NUTRIENTS IN ESTUARINE SYSTEMS

Inputs

The composition of the river waters that mix with seawater in estuaries varies with rate of freshwater discharge from the drainage basin and with the geological/geochemical character of each drainage basin. Within-basin discharge variations approach the between-basin watershed variations for major elements (Carpenter et al., 1969). The major dissolved components (Na, K, Ca, Mg, Cl, SO_4) from an estuarine sample can be calculated with knowledge of the concentration of the components in the river and ocean end members, that is, within accuracies approaching our analytical capability to measure these components, they behave linearly.

Freshwater and seawater composition of the micronutrient components show extreme variations (over several orders of magnitude) as illustrated in Table 2. In addition, estuarine systems can show strong flow-related (frequently seasonal) variations. Table 3 illustrates the variation in nutrient concentration in several Indian estuaries where freshwater discharge is strongly seasonal. Both nitrate and silica show a strong pulse during the monsoon. High concentrations persist through the post-monsoon period. Phosphate does not show such a strong seasonal pattern. During extreme runoff events in temperate estuaries, the same pattern can develop. Tropical storm Agnes, which produced the 200-year flood in the Susquehanna River basin, added nitrate sufficient to increase the ambient concentration in the upper Chesapeake to two-to-three times normal concentration for the summer season. High levels of nitrate persisted for about a month after the flood event (32). In both the annual monsoon pulse and the extreme event pulse, estuarine phosphate concentrations did not change markedly from pre- or post-event levels.

Anthropogenic inputs of nutrients to estuarine systems are a cause of overenrichment. The inputs from these sources can dominate the sum of all of the inputs of nutrients to estuarine systems - (see, for example, Nixon - this volume; Carpenter et al., 1969 (9); Ryther and Dunstan, 1971

TABLE 2. Nutrient concentrations in various estuaries.

		μm			
Estuary-Sample	References	NO_3^-	NH_4^+	$Si(OH)_4$	PO_4
Delaware	1				
0°/oo		150	50	110	2
30°/oo		1	1	7	0.3
Susquehanna-Chesapeake	2,3				
0°/oo		100	10	-	-
30°/oo		1	1	-	-
Zaire (Africa)	4				
0°/oo		8	0.5	160	1.2
35°/oo		0	0	0	0
Magdalena (S. America)	5				
0°/oo		17	-	225	3
35°/oo		0	-	0	0.2
Scheldt (Belgium	6,7				
0°/oo		0	600	230	40
30°/oo		30	40	10	2
Potomac-Chesapeake	8				
0°/oo		110	200	-	32
10°/oo		1	1	-	0.2
Hudson	9,10				
0°/oo		40	30	100	5
30°/oo		5	5	40	1.5

References
1. Sharp, Church and Culberson, personal communication
2. Carpenter et .al., 1969 (9)
3. McCarthy et al., 1975 (24)
4. Van Bennekan et al., 1978
5. Fanning and Maynard, 1978 (13)
6. Wollast and Debroen, 1971 (36)
7. Van Bennekan et al., 1975
8. Jaworski et al., 1972 (8)
9. Garside et al., 1976 (16)
10. Simpson, 1975 (35)

(30). One of the more spectacular increases of nutrients to a major estuarine system has been reported by Vander Eijk (1979) for the Rhine River-Wadden Sea-North Sea. In a ten-year period (1959-1968) the annual phosphate load of the Rhine increased from 4,000 to 15,000 tons of phosphorus and ammonia plus nitrate increased from 190,000 to 340,000 tons (as nitrogen).

TABLE 3. Range in nutrient concentrations in the estuaries of Goa. (Values in µg-at/l) From Quasim, 1979 (28).

		Pre-Monsoon	Monsoon	Post-Monsoon
Mandovi	Phosphate (P)	0.11 - 1.78	0.04 - 1.75	0.001 - 1.52
	Nitrate (N)	0.08 - 1.72	0.25 - 4.11	0.21 - 2.79
	Silicate (Si)	—	17.84 -134.30	—
Zuari	Phosphate (P)	0.29 - 2.04	0.26 - 1.17	0.10 - 1.42
	Nitrate (N)	0.25 - 1.47	0.36 - 4.56	0.35 - 8.03
	Silicate (Si)	—	12.43 - 89.98	—
Cumbarjua	Phosphate (P)	0.39 - 1.91	0.13 - 1.26	0.004 - 1.45
	Nitrate (N)	0.13 - 1.43	0.42 - 3.87	0.17 - 5.93
	Silicate (Si)	—	20.15 - 114.15	—

Internal Reactions

As documented in Pritchard and Schubel (this volume), Ianviello (19), and Boon (6), it is technically difficult and very costly to attempt to measure the flux of mass through a given cross-section of an estuary. In addition, nutrients may not behave conservatively within the estuary. Particular nutrient forms may be gained or lost by chemical reactions, by biological uptake or metabolic activity and/or adsorption on or desorption from suspended matter (see Boyle (8); Carpenter et al., (9); Garside, (16); Liss, (23)).

Estuarine scientists have sometimes attempted to avoid the mass-balance problems in estuarine nutrient dynamic studies by relating nonconservative parameters to salinity or chlorinity (see Carpenter et al., (9); Peterson et al., (26); Sdholkovitz et al., (34)). In the absence of biogeochemical processes occurring within an estuary or export to the atmosphere, the composition of a particular element within that estuary will be a linear function of its concentration in the river and the ocean. Chloride (or salinity) is the usual reference or conservative property upon which variations of other elements are measured. Figure 9 illustrates the hypothetical behavior of substances in the estuarine environment plotted against salinity. Deviations of the parameters from linearity in Figures 9b, 9c, and 9d indicate nonconservative behavior in that there is a net loss or gain in the estuary. These curves do not provide insight into dynamic estuarine processes which may be occurring. A linear response of a nutrient along the salinity gradient does not necessarily mean that the substance is not participating in biogeochemical processes within the system; it means only that losses and gains are balanced and the estuary is in a "steady-state" with respect to that particular substance. Peterson et al. (26) measured dissolved silica distribution in San Francisco Bay, concluding that, during high freshwater discharge, the silica-salinity distribution was linear (Figure 9a) and that during low discharge and high

FIGURE 9. Idealized micronutrient-salinity relations showing concentration and mixing of nutrient-rich river water with nutrient-poor seawater. From Peterson et al., 1975 (26). a. Expected concentration-salinity distribution of a substance behaving in a conservative manner (for example, chloride) in an estuary. b. Expected concentration-salinity distribution of a substance for which the estuary is a source (for example, particulate carbon). c. Expected concentration-salinity distribution of a substance for which the estuary is a sink (for example, phosphorus). d. Expected concentration-salinity distribution of a substance for which the estuary is a pronounced sink that is, where the concentration of the substance in the estuary is lower than the river and the ocean (for example Si).

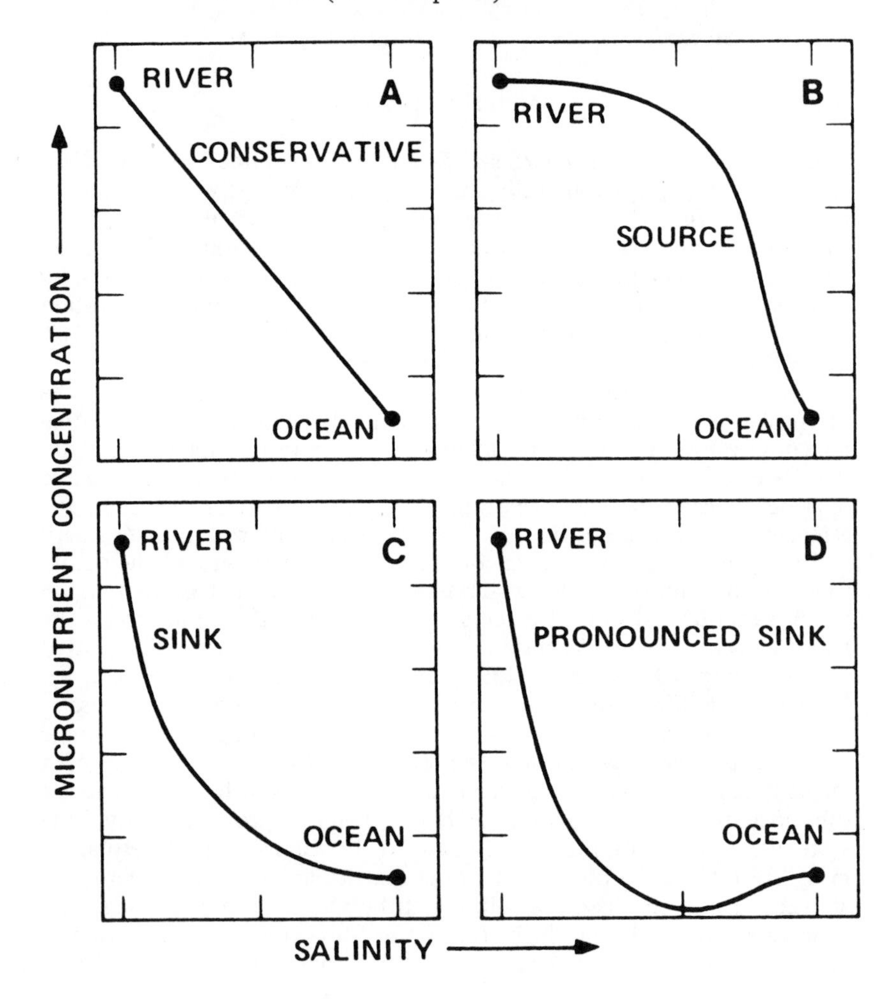

primary productivity, the dissolved silica followed distributions indicating that the estuary served as a sink (Figures 9c, 9d). Similar plots for nitrite plus nitrate, and ammonia showed that the estuary could serve as a source (Figure 9b) or a sink (Figures 9c, 9d) for these nutrients. Recently, Rattray and Officer (29) have developed a relatively simple model of the Peterson et al. (26) silica data for San Francisco Bay (Figure 10). This model avoids the necessity of numerical integration (and knowledge of the eddy diffusivity term) of the conservation equation. As Rattray and Officer point out, it should be possible to use the model to determine estuarine uptake rates from the relation between salinity and the distribution of other non-conservative substances.

As pointed out previously, salinity-nutrient plots provide a "net" or "apparent" rate of nutrient utilization, representing the balance of gains or losses at points along the estuary. Individual process gains or losses may be orders of magnitude greater and may change during the year.

The largest portion of the freshwater phosphorus entering the estuary is adsorbed on ferric oxides and hydroxides which themselves are adsorbed onto or flocculated with suspended sediment (Kramer et al., (21)). The most important controls determining the adsorption-desorption reaction of phosphorus in sediment appears to be redox potential, calcium concentrations (in the freshest part of the estuary), pH, and all of those factors (permeability, water content, bioturbation, and physical mixing) which can cause interstitial waters to mix with overlying waters. Correll et al. (10) have produced a model of the major pathways of estuarine phosphorus cycling within the bacteria, phytoplankton-zooplankton regime. Bricker and Troup (1975) examined the apparent maximum flux of phosphate from the sediment to the overlying water and found it to be about 5 percent of the total amount present in the water of the Chesapeake. Liss (23) has proposed that phosphate release from sediment serves as a geochemical buffer to the phosphate in the overlying waters of the estuary. Webb, O'Connor, Monbet et al., and McCarthy (this volume) examine pathways and mechanisms for nutrient translocation in estuarine systems, while Nixon (this volume) has provided a perspective on the proportion of internal remineralization to external supply. Our point in this discussion is to demonstrate that we are now developing sufficient data on a number of estuaries to measure both "net" rate of nutrient utilization and internal rates and processes.

Management of Nutrient Effects in Estuarine systems

The health of an estuarine ecosystem is that state in which the components and processes remain well within specified limits of system integrity selected to assure that there is no diminution in the capacity of the system to render its basic services to society throughout the indefinite future (Darnell and Soniat, this volume). In general, the response of a

FIGURE 10. Model silica-salinity relationships calculated for San Francisco Bay. From Rattray and Officer, 1979 (29).

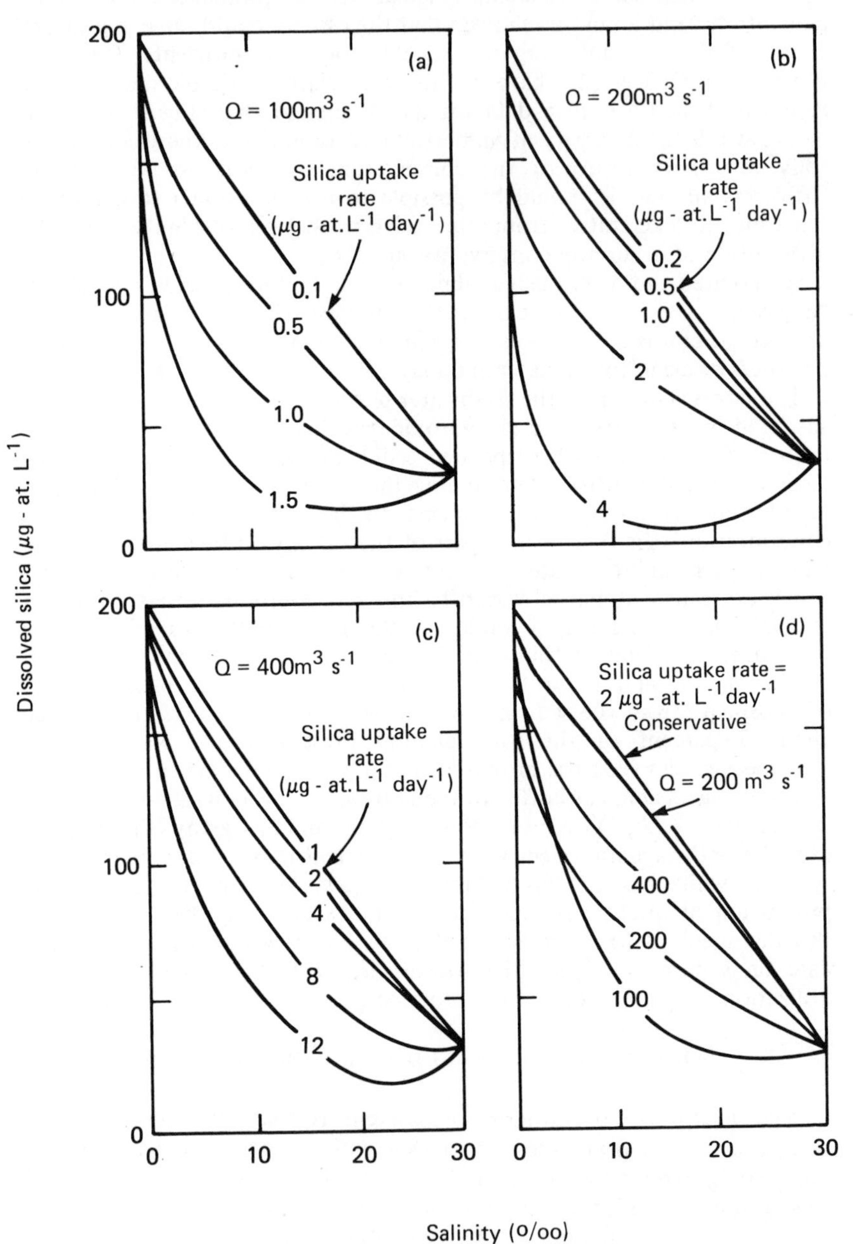

particular estuary to a particular input of nutrients cannot be predicted a priori, nor can the effects of removal of a nutrient source be quantified a priori. We do not yet have the predictive capability that appears to be present and widely applicable in freshwater systems (see Schindler, this volume), although much recent progress has been made (see Lee and Jones; Darnell and Soniat; Jaworski and Villa, this volume). In freshwater systems, simple models using phosphorus inputs and residence times have proven to be of significant utility in the prediction of algal standing crops. The occurrence of high concentrations of a wide variety of materials dissolved in the estuarine system associated with enrichment may exert a strong influence on the quantity and quality of primary production in estuarine systems (Ryther, this volume).

ACKNOWLEDGEMENT

This paper is Publication No. 89 of the Chesapeake Research Consortium, Inc.

REFERENCES

1. Basan, P.B. and R.W. Frey. 1977. Actual paleontology and neoichology of salt marshes near Sapelo Island, Georgia. *In*: Crimos, P.T. and J.C. Harper, (eds.)., *Trace Fossils*, Seal House Press, Liverpool. p. 41-70.
2. Belknap, D.F. and J.C. Kraft. 1977. Holocene relative sea level change and coastal stratigraphic units on the northwest flank of the Baltimore Canyon geosyncline. J. Sed. Petrol. 47:610-629.
3. Bennekom, A.J. Van, W.W.C. Giekes, and S.B. Tijssen. 1975. Eutrophication of Dutch coastal waters. Proc. Roy. Soc. London Series B, 169:359-374.
4. Bennekom, A.J. Van, G.W. Berger, W. Helder, and R.P.V. DeVries. 1978. Nutrient distribution in the Zaire Estuary and River Plume. Neth. Jour. Sea. Res. 12:296-323.
5. Biggs, R.B. 1970. Sources and distribution of suspended sediments in northern Chesapeake Bay. Marine Geology 9:187-201.
6. Boon, J.D. 1978. Suspended solids transport in a salt marsh creek - an analysis of errors. *In* B. Kjerfve (ed.), *Estuarine Transport Processes.* U. of S. Carolina Press, p. 147-160.
7. Bowden, K.F. 1978. Mixing processes in estuaries. *In* Kjerfe, B. (ed.), *Estuarine Transport Processes*, U. South Carolina Press, p. 11-36.
8. Boyle, E.A. 1974. On the chemical mass balance in estuaries. Geochim, Cosmochim. Acta. 38:1719-1728.
9. Bricker, D.P., III. and B.N. Troup. 1975. Sediment water exchange in Chesapeake Bay. *In* Cronin, L.E. (ed.), *Estuarine Research* 1:3-27. Academic Press, New York.

10. Carpenter, J.H., W.L. Bradford, and V. Grant. 1975. Processes affecting the composition of estuarine waters. *In* Cronin, L.E. (ed.), *Estuarine Research* 1:188-214. Academic Press, New York.
11. Carpenter, J.H., D.W. Prtichard, and R.C. Whaley. 1969. Observations of eutrophication and nutrient cycles in some coastal plain estuaries. *In Eutrophication: Causes, Consequences, and Correctives*, p. 210-221. Nat. Acad. Sci., Washington, D.C.
12. Correll, D.A., M.A. Faust, and D.J. Severn. 1975. Phosphorus flux and cycling in estuaries. *In* Cronin, L.E. (ed.), *Estuarine Research*, 1:108-136. Academic Press, New York.
13. Cronin, L.E. and A.J. Mansueti. 1971. The biology of the estuary. *In* Douglas, P.A. and R.H. Stroud (eds.), *A Symposium on the Biological Significance of Estuaries*. Sport Fishing Inst., Washington, D.C. p. 14-39.
14. Duncan, W.H. 1974. Vascular halophytes of the Atlantic and Gulf coasts of North America north of Mexico. *In* Reimold, R.J. and W.H. Queen (eds.), *Ecology of Halophytes*. Academic Press, New York. p. 23-50.
15. Fanning, K.A. and V.I. Maynard, 1978. Dissolved boron and nutrient in the mixing plumes of major tropical rivers. Neth. Jour. Sea Res. 12:345-354.
16. Frey, R.W. and P.B. Basan. 1978. Coastal salt marshes. *In* Davis, R.A. (ed.), *Coastal Sedimentary Environments*. Springer-Verlag, New York. p. 101-170.
17. Gardner, L.R. and W. Kitchens. 1978. Sediment and chemical exchanges between salt marshes and coastal waters. *In* Kjerfve, B. (ed.), *Estuarine Transport Processes*. U. of S. Carolina Press, Columbia, S.C., p. 191-208.
18. Garside, C., T.C. Malone, O.A. Roels, and B.A. Sharfstein. 1976. An evaluation of sewage-derived nutrients and their influence on the Hudson estuary and New York Bight. Est. Coast Mar. Sci. 4:281-289.
19. Gross, M.G., M. Karweit, W.B. Cronin, and J.R. Schubel. 1978. Suspended sediment discharge from the Susquehanna River to Northern Chesapeake Bay, 1966-1976. Estuaries 162:106-110.
20. Haven, D.S. and R. Morales-Alamo. 1966. Aspects of biodeposition by oysters and other invertebrate filter feeders. Limnol. and Oceanogr. 11:487-498.
21. Ianviello, J.P. 1977. Tidally induced residual currents of estuaries of constant breadth and depth. J. Mar. Res. 35(4):755-786.
22. Jaworski, N.A., D.W. Lear and O. Villa, 1972. Nutrient management in the Potomac Estuary. *In* Likens, G.E. (ed.), *Nutrients and Eutrophication*. Amer. Soc. Limnol. and Oceanogr., p. 246-273.
23. Kramer, J.R., S.E. Herbes, and H.E. Allen. 1972. Phosphorus, analysis of water, biomass, and sediment. *In Nutrients in Natural Waters*. p. 51-100. Wiley, New York, 457 p.
24. Laird, B. (ed.) 1976. *The effects of tropical storm Agnes on the Chesapeake Bay estuarine system*. Chesapeake Research Consortium, Pub. 54. Johns Hopkins Press, Baltimore, MD. 639 p.
25. Liss, P.S. 1976. Conservative and nonconservative behavior of dissolved constituents during estuarine mixing. *In* Burton, J.D. and P.S. Liss (eds.), *Estuarine Chemistry*, Academic Press, p. 93-130.

26. McCarthy, J.J., W.R. Taylor and J.L. Taft. 1975. The dynamics of nitrogen and phosphorus cycling in the open waters of Chesapeake Bay. *In* Church, T.M. (ed.), *Marine Chemistry of the Coastal Environment*, p. 664-681, Amer. Chem. Soc. Symposium Series, Vol. 18.
27. MacDonald, K.B. 1977. Plant and animal communities of Pacific North American salt marshes. *In* Chapman, V.J. (ed.), *Wet Coastal Ecosystems*. Elsevier, Amsterdam, p. 167-191.
28. Peterson, D.H., T.J. Conomos, W.W. Broenkow, and E.P. Sceivani. 1975. Processes controlling the dissolved silica distribution in San Francisco Bay. *In* Cronin, L.E. (ed.), *Estuarine Research*, 1:153-187. Academic Press, New York.
29. Prtichard, D.W. 1955. Estuarine circulation patterns. Proc. Am. Soc. Civil Eng. 81, Separate No. 717.
30. Qasim, S.Z. 1979. Production in some tropical environments. *In* Dunbar, M.J. (ed.), *Marine Production Mechanics*, p. 31-70, Cambridge Univ. Press, London, 338 p.
31. Rattray, M., Jr., and C.B. Officer. 1979. Distribution of a nonconservative constituent in an estuary with application to the numerical simulation of dissolved silica in the San Francisco Bay. Est. and Coastal Mar. Sci. 8:489-494.
32. Ryther, J.H. and W.M. Dunstan. 1971. Nitrogen, phosphorus and eutrophication in the coastal marine environment. Science 171:1008-1013.
33. Schubel, J.R. 1971. Estuarine circulation and sedimentation. *In The Estuarine Environment*, published by Amer. Geol. Inst., Washington, D.C.
34. Schubel, J.R., W.R. Taylor, V.E. Grant, W.B. Cronin, and M. Glendening. 1976. Effects of Agnes on the distribution of nutrients in upper Chesapeake Bay. *In* Laird, B.E. (ed.), *The Effects of Tropical Storm Agnes on the Chesapeake Bay* Estuarine System. Chesapeake Research Consortium, Pub. 54, p. 311-319, Johns Hopkins Press, Baltimore, MD.
35. Shakelton, N.J. and N.D. Opdyke. 1973. Oxygen isotope and paleomagnetic stratigraphy of equitorial core V25-238. Quaternary Res. 3:39-55.
36. Sholkovitz, E.R., E.A. Boyle, and N.B. Price. 1978. Removal of dissolved humic acids and iron during estuarine mixing. Earth and Planet. Sci. Newsletter 40:130-136.
37. Simpson, H.J., D.E. Hammond, B.L. Deck, and S.C. Williams. 1975. Nutrient budgets in the Hudson Estuary. *In* Church, T.M. (ed.), *Marine Chemistry in the Coastal Environment*. Amer. Chem. Society Series, Vol. 18.
38. Van der Eijk, E.M. 1979. Nutrient concentrations in the Rhine River, Wadden Sea, and North Sea. *In Marine Pollution* Mechanics, Dunbar, M.J. (ed.), p. 31-70, Cambridge Univ. Press, London 338p.
39. Wollast, R. and F. DeBroen. 1971. Study of the behavior of dissolved silica in the estuary of the Scheldt. Geochim. Cosmochim. Acta. 35:613-620.

CONCEPTUAL MODELS AND PROCESSES OF NUTRIENT CYCLING IN ESTUARIES

Kenneth L. Webb
School of Marine Science and
Virginia Institute of Marine Science
College of William and Mary
Gloucester Point, Virginia 23062

INTRODUCTION

This paper is primarily concerned with conceptual models of nutrient cycling in estuaries. A conceptual model is largely a theoretical construction containing the essential attributes of the system. It should be no more complicated than need be to serve the intended purpose (31). Conceptual models are often represented by diagrams of boxes representing components and arrows between boxes representing relationships or transfer of materials or energy. These models should be connected to the real world if they are to be useful. This is most often accomplished by measurements of concentrations and rates. In this paper I will not present a great number of rates for exchanges taking place in any estuary. It is my intention to present a point of departure for understanding what may happen in the Chesapeake or other estuary when it is enriched with nutrients. In other papers in this volume, Nixon (46) and Smith (63) present rate as well as concentration data from Narragansett Bay and Kaneohe Bay, respectively.

This paper occasionally digresses into speculations. It is my belief that this is profitable to stimulate discussion and research, even if the speculations ultimately prove to be incorrect.

SIMPLE CONCEPTUAL MODELS

Conceptual models can exist in a hierarchy of complexity. My first model, of one compartment in size, represents the entire estuary and contains bi-directional arrows indicating the possible fluxes of nutrients between the estuary and other systems (Figure 1). This estuary is

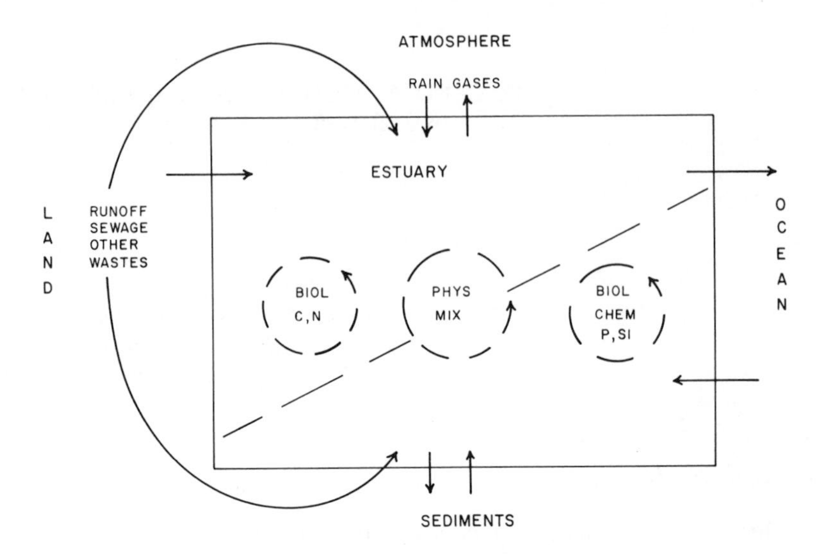

FIGURE 1. Model of moderately stratified estuary indicating major routes of inputs and outputs of nutrients and the kinds of cycles and processes considered.

considered moderately stratified as contrasted to highly stratified or vertically homogeneous (52) and is intended to be a useful model for Chesapeake Bay and its subtributaries. Estuaries are interesting and unique aquatic environments in that they have inputs from both terrigenous and oceanic sources, as well as from the atmosphere and sediments. As indicated in Figure 1, freshwater or land inputs such as runoff, point and nonpoint sources, etc. are likely to be in the surface water of the estuary. These land derived inputs can also take the indirect routes of ground water seepage through the sediment or wind driven dust from agricultural lands. Inputs from the ocean are usually in the deeper water as indicated on the right of the box (Figure 1).

This simple box model of Figure 1 becomes a mass balance if all inputs and outputs of a given substance are added up. This kind of treatment can be very useful as indicated by the nutrient budget work of Jaworski et al. (21) on the Potomac system and a number of other works abstracted in our eutrophication bibliography (75).

The one compartment model becomes inadequate when we wish to consider processes and transfers within the compartment. Some of these processes and nutrient cycles we wish to consider are indicated within the box of Figure 1. Processes which affect nutrients can be either biological, chemical or physical. The circles indicating nutrient cycles within the box (Figure 1) are meant to show that transformations of carbon and nitrogen in temperate estuaries are largely biological, while those of phosphorus and silica may be both biological and chemical. Silica uptake and incorporation by diatoms is considered biological, while its regeneration is largely chemical through dissolution. Phosphorus cycles biologically, but its exchange with sediments is probably chemical and possibly controlled by oxygen concentration (45, 51, 76).

The rate of physical mixing, indicated by the center circle of Figure 1, varies greatly within estuaries and can greatly affect the rates of biological and chemical transformations of the nutrients, e.g. in an estuary in which light does not reach the bottom, sediment released phosphate is unlikely to be incorporated into phytoplankton until it reaches the lighted surface zone. Physical models are usually designed for very different purposes than are biological models and the need for better communication between physical and biological oceanographers is apparent (8). We must also have appropriate or suitable levels of sophistication in our knowledge of the physical, chemical and biological processes within the estuary in order to understand the effects of nutrient enrichment within the estuary.

MORE COMPLEX CONCEPTUAL MODELS

Traditional physical two layered models depict more adequately moderately stratified estuaries and have been used to construct long term mass balances, especially of conservative properties. Such models tend to be inadequate when dealing with short term phenomena of interest to biologists or nutrient chemists; this is especially true in situations where either or both the degree of stratification and rate of vertical mixing vary on the short term (15, 16). There is a need to appreciate the appropriateness of time scales of sampling related to the time scale of the process being studied (8, 14, 61).

Biological models can be trophic models, nutrient transformation models, etc. Nixon (46) presents, in his Figure 1, a very nice depiction of the evolution, knowledge and complexity of conceptual models incorporating both trophic and nutrient concepts. For our purposes we will use a phytoplankton-herbivore-carnivore conceptual model to which we have added compartments for bacteria, dead organic matter and inorganic nutrients (Figure 2). It works equally well for C, N, and P as elements of flux and less well for elements such as silica. My concern is with the inputs indicated on the left of Figure 2 which include inorganic nutrients and dead organic matter such as contained in sewage and various allochthanous sources, and how they affect the system. All of the compartments contain arrows indicating that those materials can be removed from the system largely along the pathways indicated in Figure

FIGURE 2. Model of some major compartments of the estuary related to nutrient processes. Adapted from papers by Caperon (2, 3).

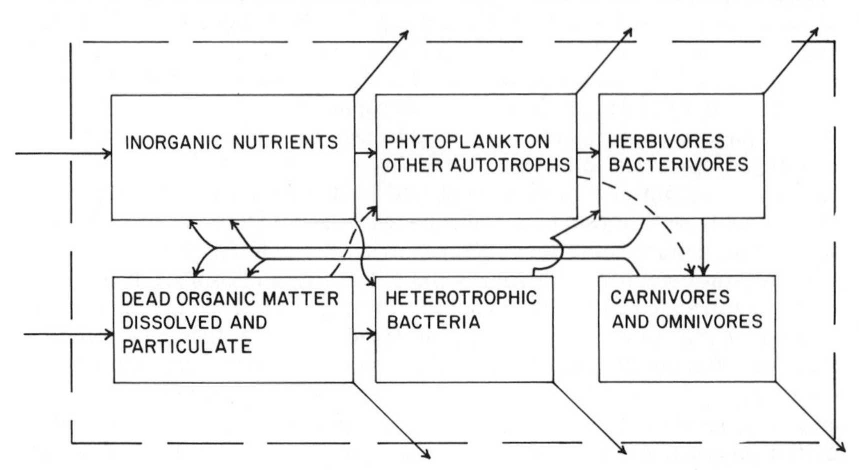

1. Conversely, inputs (not shown) can occur along these same pathways for each of the indicated compartments. This model allows us to deal with another level of hierarchy within the system and relationships between components on that level which is not possible in the model of Figure 1. The most important features of this model include the composition and nature of the inputs, the composition and functioning of the organisms or material within the compartments, the connectivity and rates of transfer between compartments, and the controls of the processes. Much of this kind of information we possess, much of it we know very little about, and some of which we think we know is in fact dogma, which is not an adequate reflection of reality.

A major concern of scientists and lay people alike is the consequences of increased inorganic nutrient loading to estuaries. A priori, many scientists and nonscientists alike would expect the concentration in the first or second compartment, i.e. the inorganic nutrients or phytoplankton, to increase. Indeed this often appears to be the assumption upon which water quality models are based. However, both theoretical modelling approaches (2, 3, 55, 56) and experimental observations (2, 37, 63) indicate that the effects of increased nutrient loading may be reflected in other compartments. For example, in a controlled enrichment of a freshwater system (37), "enrichment yields an increase in the concentration of zooplankton but no change in the steady-state concentration of primary producers" (56). Concentration of materials in any compartment is the result of the relative balance between input and output processes. The rates of transfer between compartments, in fact, may be more important than concentrations within them, although knowledge of both is needed. Up to now, far more emphasis has been directed to collecting concentration data than to obtaining transfer rate data.

Many of the classical questions of plankton investigations, e.g. how can so many species of phytoplankton coexist (20), what are the underlying causes of species succession in the phytoplankton (36), and does patchiness (7) bear a functional relationship to these phenomena, are also of significance to the question of system response to nutrient enrichment. Population models (77, 78), Monod models (70) and other theoretical models simulating either resource competition or predator-prey relationships are presently used to test hypotheses related to the above questions. Multispecies continuous culture (19, 43) techniques are also available for experimental testing of the same hypotheses. The joining of the theoretical and experimental approaches appears promising (44, 70) and may be very useful in furthering our knowledge of ecosystem response to nutrient enrichment.

The Phytoplankton Compartment

The kinds of organisms most often thought of as being the first processors of added nutrients to the estuarine environment are the photoautotrophs: phytoplankton and macrophytes. Although there is little known about the relative importance of their role, the autotrophic and heterotrophic bacteria may also utilize these materials as indicated in Figure 2. Until the 1970s, phytoplankton were often thought of as being diatoms large enough to be caught in nets. As early as the first decade of this century (32, 33), the observation was made that plankton nets retained only a small part of the plankton algae. For nutrient utilization as well as primary productivity, the large-sized group of phytoplankton may be of trivial importance although it plays other roles. The dominance of the small-sized phytoplankton has been observed numerous times and includes observations from Chesapeake Bay (39, 72).

As pointed out earlier, we need to know the composition of the compartments of the system as well as rates of transfers and control mechanisms. This discussion suggests that we know much more about species composition, abundance, chemical composition, etc. of the large size photoautotrophs than we do of the small forms. This statement is reinforced by recently published observations (22, 74) that the most abundant photoautotrophic organism in the marine environment is a chroococcoid cyanobacteria (blue green algae) of about 0.5 to 1 μm in diameter. Unpublished observations of my colleague L.W. Haas indicate that these organisms are of significant abundance in the Chesapeake Bay and adjacent waters. This discussion is not intended to imply that large phytoplankton are unimportant and that work on them is over-rated or insignificant; it is intended to point out by comparison that we know next to nothing about the roles of organisms responsible for most of the primary production in the marine and estuarine environments.

Phytoplankton cell size as a controlling factor (49) and its implications in trophic relationships (57, 58) has received much attention in the past decade. Surface area to volume considerations indicate that small-sized phytoplankton should have a competitive advantage over large-sized phytoplankton when nutrients are in short supply (12). As indicated by Ryther and Officer (58) this hypothesis has been controversial. The earlier implication related to nutrient enrichment is that added nutrients should favor growth of larger-sized phytoplankton (49). Much evidence from the environment supports this hypothesis; not only is the phytoplankton community generally dominated by small-size forms as discussed above, but patterns from areas subject to coastal upwelling suggest that nanoplankton dominate between upwelling events and that larger forms dominate during upwellings (13, 34). Likewise blooms of large-celled dinoflagellates in Chesapeake Bay appear to be related to circulation patterns (71) or tidally related destratification events (17) and thus possibly directly to nutrient enrichment. A small amount of work indicates that other pollutants may favor small phytoplankton either by direct enhancement of growth, e.g. petroleum hydrocarbons (50), or by selective toxicity to larger forms, e.g. chlorinated hydrocarbons (47). A consideration of these relationships is largely speculative at the moment (10).

Work of Ryther (reviewed in 58) in two Long Island, New York bays indicated that restricted ocean exchange and nutrient enrichment changed the phytoplankton community from a mixed diatom-small flagellate to a small *Nannochloris atomus* dominated community. These results do not necessarily contradict the above hypothesis but do emphasize that reality is complicated. Added nutrients are apt to be very low in silica compared to the natural environmental nutrient cycles and, as correctly pointed out by Ryther and Officer (58), the response in the New York bays may have been influenced by severe silica depletion.

RATIOS OF NUTRIENTS UTILIZED FOR PLANT GROWTH

Although most algae adhere to the Redfield (53) model and require C, N, and P in an atomic ratio of 106:16:1, these ratios are only approximate and there are species to species variations (Table 1). Diatoms and some chrysophytes also require silica and many blue green bacteria can fix atmospheric nitrogen, thereby not requiring the usual nitrogen nutrient in the form of nitrate, ammonia, etc. An obvious implication of nutrient enrichment to estuaries is that if nutrient proportionality is changed, species composition of the phytoplankton may also be changed. A possibility previously mentioned is a decline in silica concentrations, resulting in a reduction of the diatom component of the community. The silica aspect is treated in the literature (58, 59, 60). Excessive phosphorus relative to nitrogen can be hypothesized to favor blue green bacteria,

TABLE 1. Inorganic nutrient requirements of some major categories of phytoplankton based in part on the hypothesis of Redfield (53).

	Nutrient Requirements (Excluding N_2) Atomic Proportions Relative to P			
	Si	C	N	P
Diatoms Some Chrysophytes	20	106	16	1
Most Algae		106	16	1
N-Fixing		106		1

which can fix nitrogen, although this only appears to occur in the extremely low salinity portions of the estuary.

Changes in both species composition and chemical composition of the phytoplankton population will have effects on the next trophic level. This subject is treated in part in the following section.

The Herbivore Compartment

Obviously, an estuarine organism must have an adequate supply of food (proper size and quality) or it will perish. As pointed out earlier, relative importance of different groups of photoautotrophs in the estuarine environment is in question and likewise in question is their relative food value for primary consumers, including the economically important species (58). Nonetheless, marine and estuarine science students are still taught the dogma of what could be called the diatom-dinoflagellate-copepod syndrome. This simple food chain model of diatom-copepod-fish appears entrenched as a central tenet in marine science although there is good reason to believe that it has outlived its usefulness.

An increasing number of organisms have been observed to survive and grow better on a non-diatom diet. Lasker (28) indicated that new anchovy larvae need algae about 40 μm in diameter in order for their first feeding to be successful. These algae are usually the naked dinoflagellate *Gynmodinum splendens*, which is an excellent food for larvae of the northern anchovy *Engraulis mordax*. Diatoms also seems to be poor food for both larval and adult American oysters (*Crassostrea virginica*). Work by Mann and Ryther (35) indicated that although several bivalve molluscs grew well on a diet predominantly of the diatom *Phaeodactylum tricornulum*, several others including *C. virginica* grew very little and exhibited a decrease in a condition index. The experience of the Virginia

Institute of Marine Science's larval culture laboratory of Dr. John Dupuy also indicates that *P. tricornulum* is a poor conditioning agent for adult oysters and that it does not satisfactorily contribute to a diet for larval *C. virginica*. In fact, a good larval diet consists of representatives of the Chlorophyceae, Haptophyceae and Prasinophyceae-all algae which might be predicted to increase in dominance in estuaries if silica became limiting.

Adequacy of food for marine organisms is more complicated than just its species composition. Relative proportions of protein, carbohydrate and lipid, in addition to food quality, are also of significance. Environmental factors which change a food species composition may also affect its suitability for food. Flaak and Epifanio (11) munipulated the growth conditions of the diatom *Thalassiosira pseudonana* to produce six diets of differing protein content and carbon:nitrogen ratios. Juvenile oysters, *C. virginica*, of about 2 cm shell height grew much better with the low protein, high C:N diets although the food species was identical. Since nutrient availability is known to affect algal C:N (41), it is anticipated that one effect of nutrient enrichment may be a change of food quality, even if species composition is not changed. Good food for oysters, however, will not necessarily be good food for other organisms.

NUTRIENT CYCLES

Numerous attempts have been made to depict conceptual models of inorganic and organic nutrient cycles in the aquatic environment. In considering the impact of nutrient enrichment on estuaries, we must remember that phytoplankton, macrophytes and other photoautotrophic organisms are not the only users of the inorganic inputs. Heterotrophic bacteria and fungi, if oxidizing organic material poor in an element essential for their growth, will utilize the same sources as do the photoautotrophs. Two elements, nitrogen and sulfur, exist in a number of oxidation states and the incorporation of nitrogen and much of the sulfur into biological material is in the most reduced form. Likewise the nitrogen and much of the sulfur liberated or mineralized from biological materials is originally in its most reduced form. These reduced forms can be utilized as energy sources by chemoautotrophic microorganisms. Thus if inorganic nutrient enrichments are in the reduced form, both heterotrophic and chemoautoautrophic microorganisms potentially compete with phytoplankton and macrophytes for these materials.

The Nitrogen Cycle

The nitrogen cycle is both interesting and complex due to its number of oxidation states (-5 to +3), which are biologically active (Figure 3). Organic nitrogen is in the reduced (¯3) state in both cell protoplasm and in

FIGURE 3. The nitrogen cycle including oxidation state, biologically possible significant form, name of compound, and biological transformations which are aerobic or anaerobic. Letter designations for processes are named in Table 2. Processes A, B and C are reductions and move down; D is an oxidation and moves up. Arrows to the right indicate where material is known to enter the process and arrows to the left indicate where material is known to leave the process.

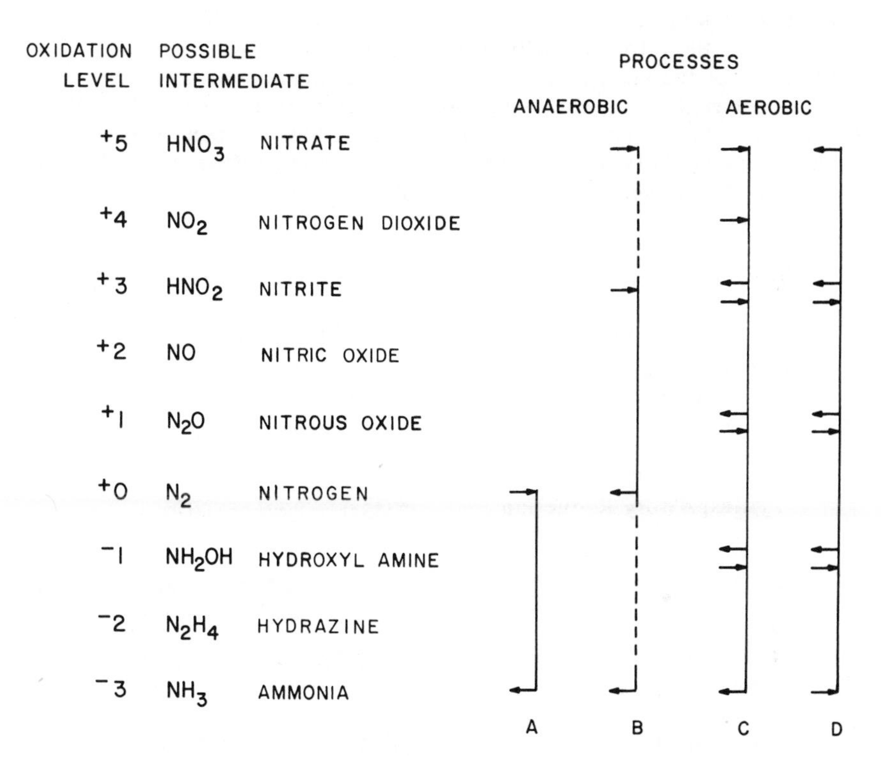

such compounds as urea, uric acid and amino acids which may occur in the environment. Ammonia, a major input to estuaries both from sewage and from release by the sediments, is also the most reduced form of nitrogen and can be used directly by phytoplankton and other microorganisms in assimilatory processes without further reduction. Uptake and utilization of these numerous nitrogenous compounds by plants is treated in this volume by McCarthy (38).

A number of the transformations of the nitrogen cycle are given specific names, some of which are logical and for some of which the derivation is not apparent to the inexperienced. A list of terms and processes is given in Table 2. Some confusion exists related to the chemical notation for nitrogen forms as well as units for concentration terms. Unfortunately the symbology for nitrogen dioxide (NO_2), a major form of atmospheric nitrogen pollution, is the same as many investigators use for the nitrite ion which should properly be indicated by NO_2^-. Confusion also results when authors do not specify clearly whether weights are for the nitrate ion, as often indicated in the text, or for nitrate-nitrogen which is the more useful and commonly used unit. The processes and some of the specific intermediates will be briefly discussed in the following section.

Most of the nitrogen within the cycle we have shown in Figure 3 is in the form of the elemental gas, N_2; the remaining non-elemental forms are often referred to as combined nitrogen. It is often assumed that the biosphere, in the absence of major modifications by man, is in a steady state balance between elemental and combined nitrogen through the process of nitrogen fixation (Table 2, A) and denitrification (Table 2, B).

TABLE 2. Nomenclature of some of the processes of the nitrogen cycle.

NAME	PROCESS
A. Nitrogen Fixation (oxygen sensitive)	$N_2 \longrightarrow NH_3$
B. Dissimilatory Reduction (oxygen sensitive)	
1. Respiratory Reduction (nitrate respiration)	$NO_3^- \longrightarrow NO_2^-$
2. Denitrification	$NO_3^-, NO_2^-, N_2O \rightarrow$ gaseous products (N_2O, N_2, NH_3)
C. Assimilatory Reduction (ammonia sensitive)	NO_3^-, etc $\longrightarrow NH_3$
D. Nitrification	$NH_3 \longrightarrow NO_3^-$
E. Ammonification	"R-NH_2" $\longrightarrow NH_3$

Nitrogen fixation is a subject of intense interest to agriculturists and there is a plethora of reviews on the subject; consequently it will be treated here only briefly. Nitrogen fixation in the estuarine environment apparently has not been reported from the water column but may be of considerable significance at interfaces between the water and either plants or sediments including adjacent splash zones as well as within sediments.

Ammonification, process E, Table 2, is simply the release of ammonia during the catabolism of nitrogen containing biological materials. It is often referred to as mineralization or remineralization in the marine sciences. In the estuarine water column, animals release the majority of their nitrogenous waste products as ammonia although approximately 20 percent of the total may be urea and to a lesser extent amino acids and other compounds. There is still considerable disagreement concerning whether animals or bacteria and fungi are more important mineralizers of nitrogen. I believe that as more becomes known of the role of protozoa and heterotrophic flagellates in the estuarine water column, the relative role of bacteria will be found less significant in nitrogen mineralization than currently believed. Bacteria, however, may play a more significant role in the breakdown of detritus in sediments, although this too may be questioned.

Nitrate and intermediates between nitrate and ammonia can serve as terminal electron acceptors in anaerobic respiration for a number of microorganisms; the process is generally termed dissimilatory reduction. Some organisms can carry out only the beginning steps from nitrate to nitrite and the more restricted terms of nitrate respiration or respiratory reduction are applied to the process in these organisms. Denitrification is the more extensive dissimilatory process starting with NO_3^- or NO_2^- (or N_2O) and ending in gaseous products, N_2O or N_2. Recent work (e.g. 27) indicates that ammonia may also be an end product. Thus dissimilatory reduction may be the source of some of the ammonia in anoxic sediments.

Assimilatory reduction is the reduction of oxidized nitrogen to ammonia which is incorporated into amino acids, proteins and polynucleotides. This process occurs in microbial heterotrophs as well as photoautotrophs. As indicated in Figure 3, any number of intermediates between nitrate and ammonia may be the starting material. Under some conditions, such as low light or an increase in ammonia concentration when phytoplankton are utilizing nitrate, intermediates such as nitrite may be released by the organism (18).

When there are compounds in the environment which contain energy, there will undoubtedly evolve microorganisms which can live by oxidizing these compounds. Ammonia is such a compound and its biological oxidation is called nitrification. There is also heterotrophic nitrification (73) and it has been shown to be significant in sewage-impacted freshwater environments; perhaps it will be so shown in estuarine environments. Autotrophic nitrification is classically a two organism process for the oxidation of ammonia to nitrate, *Nitrosomonas* oxidizing ammonia to nitrite and *Nitrobacter* oxidizing nitrite to nitrate. Biochemical intermediates can be starting materials but usually none accumulates including nitrite.

Ultimately intermediates at all the oxidation states of nitrogen will be found that are biologically active, perhaps even in the natural unmodified nitrogen cycle. Nitrogen dioxide, a major form of atmospheric nitrogen pollution not yet attracting attention in the marine environment, has recently been shown to be metabolizable by terrestrial plants. Nitrogen from NO_2 becomes reduced and appears in amino acids (54, 81).

Nitrous oxide, N_2O, has become the subject of increasing interest in the past decade and has been the subject of at least two recent Ph.D. dissertations (4, 9). Nitrous oxide plays a major role in stratospheric chemistry and helps regulate the ozone concentration of the stratosphere. It is an intermediate in both the oxidation and reduction pathways in the nitrogen cycle. It is thus apparent that both increased loadings of inorganic nutrients providing additional nitrogen and increased loading of organic materials increasing anoxic conditions may increase natural production of nitrous oxide from estuaries. Evidence of this nature has been provided for the upper Potomac where highest concentrations of N_2O were associated with sewage inputs (24). Our understanding of the sources, sinks and the processes affecting N_2O inputs into the atmosphere are deficient and more research effort should be directed toward understanding associated critical processes.

Nitrite has been of interest for decades, perhaps because the analytical technique is reliable and was developed early. Primary and secondary nitrite peaks have long been observed in the oceans with the primary maxima often observed below the oceanic euphotic zone (25); the secondary nitrite maxima is associated with the oxygen minimum layer usually well below the thermocline (69). These observations have been used to support the hypothesis that the upper maximum is derived from phytoplankton while the lower maxima are derived as intermediates from dissimilatory reduction although numerous suggestions have been made as to the nitrite origin.

Nitrite was of interest in estuaries in the 1930s but the recent tendency has often been to report only combined values for nitrate and nitrite. Seasonal distributions for two estuaries are shown in Figure 4 and the early work was discussed in the classic text by Sverdrup et al. (64) *The Oceans*. The general estuarine seasonal distribution of nitrite includes a peak in the fall, September to October, and a lesser peak in late winter or early spring and was reported for Friday Harbor, Washington and the Gulf of Maine in the 1930s (64). Nitrite concentrations during that time period, as they usually are for most estuarine locations today, were much lower than the nitrate concentrations. McCarthy (40) first called my attention to high nitrite concentrations in Chesapeake Bay in the August to October period of the early 1970s. At this time nitrite occurred in the virtual absence of nitrate, an observation repeated in the York River subestuary of the bay in August, 1978 (76). Since our 1978 observation, I have looked at many data sets for seasonal nitrite distribution and my evaluation of the trend is that the absolute concentration of the nitrite peak in the fall is increasing with time, is occurring earlier in the season, and is more frequently in excess of nitrate concentrations. These historical trends are suggestive of a possible link to nutrient enrichment and may serve as an indicator thereof.

Both the biological and spatial source of the nitrite is of considerable interest. The seasonal distribution indicates that the peaks are observed just after the hottest and coldest part of the year. The Arrhenius' law plots of ammonia and nitrite oxidation rate constants at estuarine pHs (80) indicate that nitrite oxidation may be greatly reduced compared to

FIGURE 4. Seasonal nitrite distributions in the English Channel in 1931 and in the York River, Virginia in 1969 and 1972 (1, 5, 23).

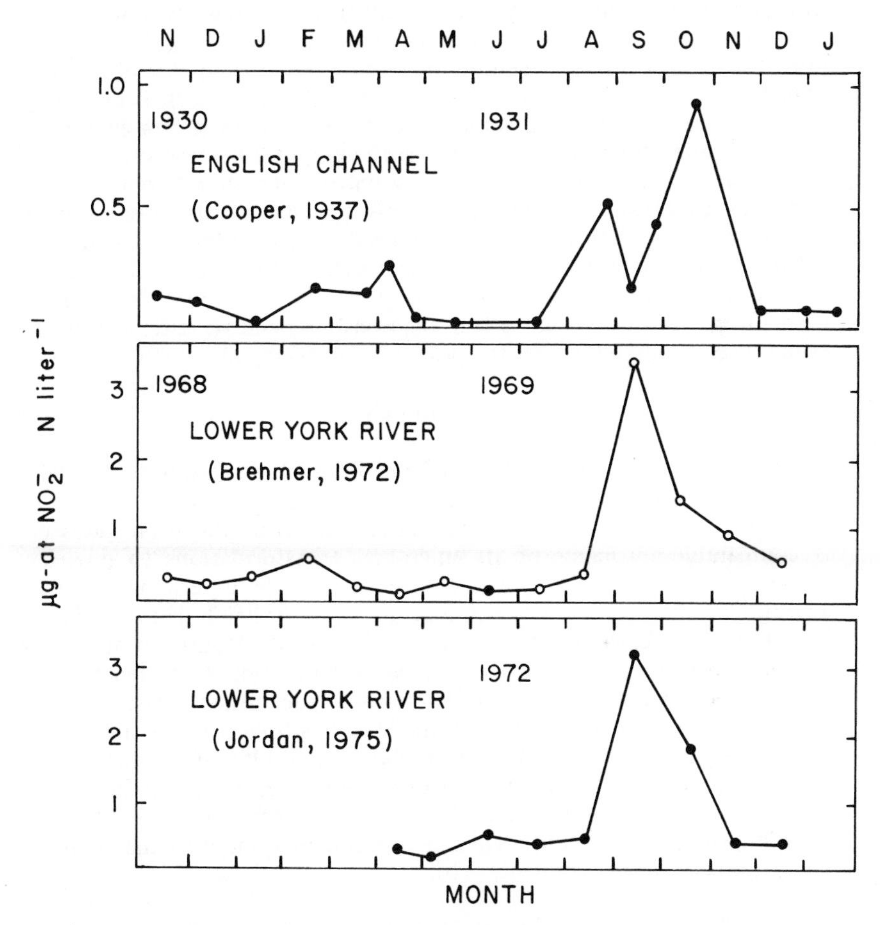

ammonia oxidation at the extremes of the biological temperature range. The model of Laudelout et al. (29) similarly shows nitrite accumulation at 30°C but not at lower temperatures (see also 26). The Chesapeake Bay and York River observations are most often in the surface waters compatible with an earlier upstream source. These relationships may help explain the seasonal distribution of nitrite but does little to explain the historic trends.

The observations from the ocean, indicating nitrite maxima in oxygen minimum, suggests that oxygen concentration may have an affect on nitrite concentration and reduced oxygen levels are associated with increased organic loading often accompanying nutrient enrichment. The work of Laudelout et al. (30) and the data cited by Painter (48) both indicate that the affinity for oxygen, expressed as the half saturation constant, is greater for *Nitrosomonas* oxidizing ammonia than it is for *Nitrobacter* oxidizing nitrite. Thus under conditions of low oxygen concentrations, one should expect an accumulation of nitrite if nitrification is occurring. The hypothesis that the nitrite peaks observed in estuaries are from nitrification, rather than other processes, is appealing. Historic trends mentioned above may be a result of increased organic loading of the estuaries which increase the extent and duration of anoxic conditions and thus the boundary between oxic and anoxic conditions. In fact, some aspect of nitrite concentrations, such as a nitrite:nitrate ratio as related to temperature, might be considered an additional index of eutrophication of estuaries. Of course much of the above is speculative and it remains for future investigations to determine the control mechanisms, the cause and effect relationships, and the sites within the estuarine environment where the processes are taking place.

The Sulfur Cycle

The sulfur cycle will only briefly be mentioned. Since sulfate is a major anion in seawater, additions are not likely to directly affect growth of organisms. Sulfur, like nitrogen, exists in a number of oxidation states and is comparable in many of its microbial transformations to nitrogen. Sulfate, like nitrate, can be used by bacteria as a terminal electron acceptor in anoxic environments, and may be a principle pathway in the oxidation of organic matter in anoxic sediments. Hydrogen sulfide, a potential end product of dissimilative sulfate reduction, reacts with iron-forming iron sulfides. Such reactions impact other nutrient cycles, especially in sediments, in that if the iron was formerly associated with phosphate, the phosphate becomes mobile as a result and can diffuse into the overlying oxic sediment or water column. A potential impact of increased organic loading to an estuary is to increase the extent and duration of anoxic conditions, and consequently increase the input of phosphorus from the sediments (76).

The Phosphorus Cycle

Compared to the nitrogen and sulfur cycles, the phosphorus cycle is rather uncomplicated although by no means completely understood. Phosphorus occurs almost exclusively in an oxidation state of +5 although one sees occasional references to possible oxidation state changes (62). In the phosphorus cycle, phosphate is taken up by both autotrophs (12, 65,

66) and heterotrophic microbes (12) and incorporated into cellular constituents. When these organisms are eaten, the phosphorus is mineralized as phosphate and some cycling as organic phosphorus undoubtedly occurs. Some phosphorus-containing detritus makes its way to the sediment through sedimentation. Dissolved phosphate may be in equilibrium with sediment phosphate, the major estuarine reservoir, through chemical processes mediated by oxygen concentrations indirectly. These relationships were briefly discussed earlier. It is also probable that plants, such as sea grasses which have leaves in the water environment and roots and rhizomes in the sediments, may serve to speed up the equilibration process between the two environments (42). The works of Taft et al. (65, 66, 67, 68) and Correll et al. (6, 12) provide contemporary information on phosphorus concentration and dynamics in the Chesapeake Region.

The Silica Cycle

The cycle of silica is considerably different than that of carbon, nitrogen, phosphorus or sulfur from a biological point of view in that only the utilization is biological and the mineralization is largely a chemical solution process. On one hand this may give the sediments a greater role in regeneration than it has even for nitrogen and phosphorus, and on the other hand silica may be responsible for wide swings in species composition throughout the year because it is required only by diatoms. The species composition and food web implications of nutrient enrichments which are poor in silica have been mentioned earlier and are covered in detail in the paper by Ryther and Officer (58). Note, however, that very little is known about silica concentrations or about the silica cycle in Chesapeake Bay or its subestuaries. For detailed consideration of the marine chemistry of silica the reader is referred to a review paper (79).

SCENARIO AND RECOMMENDATIONS

A reasonable scenario for nutrient enrichment of a classic oligotrophic, nitrogen-limited temperate zone estuary might be as follows: small additions of nutrient inputs will probably increase overall production, the increased biomass possibly showing up at any trophic level from primary producer on up but probably in secondary or tertiary levels. Additional steps of nutrient additions will undoubtedly produce changes in species composition at all trophic levels within the water column and result in greater input to the environment of fecal pellets or dying and senecent individuals of species which are not grazed. Such settling to the bottom in small increments will increase the productivity of benthos; additional inputs will remain for the microbiol populations to metabolize with the result that the oxic-anoxic boundary will rise within the sediments to the sediment surface and possibly into the water column during the warmer portions of the year.

A completely anoxic sediment has obvious implications for aerobic organisms:most will die. Breakdown of nitrogenous compounds will proceed with ammonium as the most abundant nitrogen end product of either aerobic or anaerobic metabolism. Under conditions where the oxic:anoxic interface is within or at the surface of the sediments populations of nitrifying bacteria will oxidize some of the ammonia to nitrate. We have little idea of any trophic relationships which might depend upon nitrifying bacteria; however, they provide the substrate, nitrate and nitrite, for the denitrification process. Thus if the sediment is completely anoxic the insitu supply of nitrite and nitrate (nitrification) will be depleted and consequently denitrification, the natural process which returns combined nitrogen to the atmosphere as N_2, will cease.

With a completely anoxic sediment the sulfur cycle will take on increased significance, producing sulfide which frees bound phosphate which can then diffuse into the water column. The flux of phosphate from the sediment under these conditions should exceed the usual ratio of N:P under oxic mineralization conditions. Factors which change the proportions of available nutrients should change the species composition of the phytoplankton populations when these nutrients make their way to the surface waters.

Based on the above review and other papers from the symposium, I view the critical research needs related to nutrient input into Chesapeake Bay as follows:

1. Since the sediments are the major nutrient pool within the estuary, we need to know more of the processes within the sediments which affect the storage capacity, release rates and transformation of materials within the sediments. Nutrient regeneration rates must be related to oxygen concentrations and the rates of organic inputs; the rates of silica flux must be added to our knowledge. The sediments as the major reservoir of nutrients may have a limit to their storage capacity.
2. Since physical movement of water transports both the chemicals and organisms of interest, e.g. sediment released nutrients to the surface water where they are utilized by phytoplankton, we need to know much more about rates of vertical mixing within the water column as well as other local water movements. These questions require increased collaboration and communication among biologists, hydrographers and physical oceanographers, and should be carried out with awareness of appropriate time scales for sampling, etc.
3. Our understanding of the food webs of economically important organisms is inadequate to manage the system successfully. What are the natural relationships and what are the probable responses of species, communities and the total system to nutrient enrichment?

4. The availability, cycling and significance of silica in Chesapeake Bay has inadvertently been ignored by most previous investigators. This needs to be corrected.
5. Two intermediates in the nitrogen cycle appear to play a more significant role than previously considered. The role of N_2O in the nitrogen cycle of the Chesapeake Bay should be investigated; we have only recently become aware of the possible significance of this intermediate. On the other hand, although NO_2^- has been investigated for decades, its potential importance has been frequently ignored of late. Since it currently appears to be increasing in concentration at certain times of the year, more attention is due it. Even though it is a toxic material, it would seem unlikely that it will reach toxic concentrations in the estuary. Processes controlling the nitrite phenomenon and its significance should be investigated. Other intermediates may become significant as nutrient enrichment increases.

ACKNOWLEDGEMENTS

This paper is Publication No. 87 of the Chesapeake Research Consortium, Inc. and No. 941 of the Virginia Institute of Marine Science. Development of ideas expressed in this manuscript would not have been possible without the continued support of the Oceanography Section, National Science Foundation, under grants GA-35866, OCE75-20241, OCE77-20228 and OCE78-20433, the Virginia Institute of Marine Science, and the Office of Sea Grant, National Oceanic and Atmospheric Administration, U.S. Department of Commerce. Special thanks are extended to my friends and colleagues for their intellectual exchanges and support and to R.G. Wiegert and C.F. D'Elia who reviewed the manuscript and pointed out several pertinent literature citations. Grant No. R806189010 from the U.S. Environmental Protection Agency to the Chesapeake Research Consortium provided the opportunity to interact more extensively with my Chesapeake Bay colleagues and to participate in this symposium.

REFERENCES

1. Brehmer, M.L. 1972. Biological study of Virginia's estuaries. VIMS manuscript, 88 pp, plus data report.
2. Caperon, J., S.A. Cattell and G. Kransnick. 1971. Phytoplankton kinetics in a subtropical estuary: eutrophication. Limnol. Oceanogr. 16:599-607.
3. Caperon, J. 1975. A trophic level ecosystem model analysis of the plankton community in a shallow water subtropical estuarine embayment, pages 691-709. *In* L.E. Cronin (ed.), Estuarine Research

Vol. 1, Chemistry, Biology, and the Estuarine System. Academic Press, Inc., New York.

4. Cohen, Y. 1978. Studies on the marine chemistry of nitrous oxide. Ph.D. Thesis, Oregon State University. 121 pp.
5. Cooper, L.H.N. 1937. The nitrogen cycle in the sea. J. Mar. Biol. Assoc. U.K. 22:183-204.
6. Correll, D.L., M.A. Faust and D.J. Severn. 1975. Phosphorus flux in estuaries, pages 108-136. *In* L.E. Cronin (ed.), Estuarine Research, Vol. I. Academic Press, New York.
7. Cushing, D.H. 1962. Patchiness. Rappt. Process-Verbaux Reunions, Conseil Perm. Intern. Exploration Mer. 153:152-163.
8. D'Elia, C.E., K.L. Webb and R.L. Wetzel. 1980. Time Vary Hydrodynamics and Water Quality in an estuary, pp. 597-606. *In* B.J. Neilson and L.E. Cronin (eds.), International symposium on nutrient enrichment in estuaries. Humana Press, Inc., Clifton, N.J.
9. Elkins, J.W. 1978. Aquatic sources and sinks for nitrous oxide. Ph.D. Thesis, Harvard University. 158 pp.
10. Eppley, R.W. and C.S. Weiler. 1979. The dominance of nanoplankton as an indicator of marine pollution: a critique. Oceanologica Acta 2:241-245.
11. Flaak, A.R. and C.E. Epifanio. 1978. Dietary protein levels and growth of the oyster *Crassostrea virginica*. Mar. Biol. 45:157-163.
12. Friebele, D.S., D.L. Correll and M.A. Faust. 1978. Relationship between phytoplankton cell size and the rate of orthophosphate uptake: in situ observations of an estuarine population. Mar. Biol. 45:39-52.
13. Garrison, D.L. 1976. Contribution of the netplankton and nanoplankton to the standing stock and primary productivity in Monterey Bay, California during the upwelling season. Fish. Bull. 74:183-194.
14. Gunnerson, C.G. 1966. Optimizing sampling intervals in tidal estuaries. J. Sanit. Eng. Div. Am. Soc. Civil Engrs. 92:103-123.
15. Haas, L.W. 1977. The effect of the spring-neap tidal cycle on the vertical salinity structure of the James, York and Rappahannock rivers, Virginia, U.S.A. Estuarine Coastal Mar. Sci. 5:485-496.
16. Haas, L.W., F.J. Holden and C.S. Welch. 1980. Short term changes in the vertical salinity distribution of the York River estuary associated with the neap-spring tidal cycle, pp. 586-595. *In* B.J. Neilson and L.E. Cronin (eds.), International symposium on nutrient enrichment in estuaries. Humana Press, Inc., Clifton, New Jersey.
17. Haas, L.W., S.J. Hastings and K.L. Webb. 1980. Phytoplankton response to a tidally induced cycle of stratification and mixing, pp. 619-636. *In* B.J. Neilson and L.E. Cronin (eds.), International symposium on nutrient enrichment in estuaries. Humana Press, Inc., Clifton, New Jersey.
18. Harrison, P.J. and C.O. Davis. 1977. Use of the perturbation technique to measure nutrient uptake rates of natural phytoplankton populations. Deep-Sea Res. 24:247-255.
19. Harrison, P.J. and C.O. Davis. 1979. The use of outdoor phytoplankton continuous cultures to analyze factors influencing species selection. J. Exp. Mar. Biol. Ecol. 41:9-23.

20. Hutchinson, G.E. 1961. The paradox of the plankton. Amer. Nat. 45:137-145.
21. Jaworski, N.A., D.W. Lear and O. Villa. 1972. Nutrient management in the Potomac estuary, pages 246-273. *In* G.E. Likens (ed.), Nutrients and Eutrophication, Special Symposia Vol. 1, Amer. Soc. Limnol. Oceanogr.
22. Johnson, P.W. and J. McN. Sieburth. 1979. Chroococcoid cyanobacteria in the sea: A ubiquitous and diverse phototrophic biomass. Limnol. Oceanogr. 24:928-935.
23. Jordan R.A., R.W. Virnstein, J.E. Illowsky and J. Colvocoresses. 1975. Yorktown Power Station ecological study, Phase II final technical report. Study conducted by VIMS for Virginia Electric and Power Company. 462 pp.
24. Kaplan, W.A., J.W. Elkins, C.E. Kolb, M.B. McElroy, S.C. Wofsy and A.P. Duran. 1978. Nitrous oxide in fresh water systems: An estimate for the yield of atmospheric N_2O associated with disposal of human waste. Pure Appl. Geophys. 116:424-438.
25. Kiefer, D.A., R.J. Olsen and O. Holm-Hansen. 1976. Another look at the nitrite maximum in the central North Pacific Deep-Sea Res. 23:1199-1208.
26. Knowles, G., Al. L. Downing and M.J. Barrett. 1965. Determination of kinetic constants for nitrifying bacteria in mixed culture, with the aid of an electronic computer. J. Gen. Microbiol. 38:263-278.
27. Koike, I. and A. Hattori. 1978. Denitrification and ammonia formation in anaerobic coastal sediments. Appl. Environ. Microbiol. 35:278-282.
28. Lasker, R. 1978. The relation between oceanographic conditions and larval anchovy food in the California current identification of factors contributing to recruitment failure. Rapp. P.-V. Reun. Cons. Int. Explor. Mer. 173:212-230.
29. Laudelout, H., R. Lambert, J.L. Fripiat and M.L. Pham. 1974. Effet de la temperature sur la vitesse d'oxydation de l'ammonium en nitrate par des cultures mixtes de nitrifiants. Ann. Microbiol. (Paris) 125B:75-84.
30. Laudelout, H., R. Lambert, and M.L. Pham. 1976. Effect of pH and partial pressure of oxygen on nitrification. Ann. Microbiol. (Paris) 127A:367-382.
31. Levins, R. 1966. The strategy of model building in population biology. Amer. Scientist 54:421-431.
32. Lohmann, H. 1903. Neu Untersuchungen uber den Richtum des Meers an Plankton and uber die Brauchbarkeit der verschiedenen Fangmethoden. Wiss. Meeresunters. 7:1-86.
33. Lohmann, H. 1908. Untersuchugen zur Fesstellund des vollstandigen Gehaltes des Meeres an Plankton. Wiss. Meeresunters. 10:131-370.
34. Malone, T.C. 1971. The relative importance of nanoplankton and netplankton as primary producers in the California Current System. Fish. Bull. 69:799-820.
35. Mann, R., and J.H. Ryther. 1977. Growth of six species of bivalve molluscs in a waste recycling - aquaculture system. Aquaculture 11:231-245.

36. Margalef, R. 1962. Succession in marine populations. Advg. Frontiers Pl. Sci., New Delhi 2:137-188.
37. McAllister, C.D., R.J. LeBrasseau and T.R. Parsons. 1972. Stability of enriched aquatic ecosystems. Science 175:562-564.
38. McCarthy, J.J. 1980. Uptake of major nutrients by estuarine plants, pp. 139-163. *In* B.J. Neilson and L.E. Cronin (eds.), International symposium on nutrient enrichment in estuaries. Humana Press, Inc., Clifton, N.J.
39. McCarthy, J.J., W.R. Taylor and M.E. Loftus. 1974. Significance of nanoplankton in the Chesapeake Bay estuary and problems associated with the measurement of nanoplankton productivity. Mar. Biol. 24:7-16.
40. McCarthy, J.J., W.R. Taylor and J.L. Taft. 1977. Nitrogenous nutrition of the plankton in the Chesapeake Bay. I. Nutrient availability and phytoplankton preferences. Limnol. Oceanogr. 22:996-1011.
41. McCarthy, J. J. and J. C. Goldman. 1979. Nitrogenous nutrition of marine phytoplankton in nutrient-depleted waters. Science 203:670-672.
42. McRoy, C. P., R. J. Barsdate and M. Nebert. 1972. Phosphorus cycling in an eelgrass (*Zostera marina* L.) ecosystem. Limnol. Oceanogr. 17:58-67.
43. Menzel, D. W. and J. Case. 1977. Concept and design: controlled ecosystem pollution experiment. Bull. Mar. Sci. 27:1-7.
44. Mickelson, M. J., H. Maske and R. C. Dugdale. 1979. Nutrient-determined dominance in multispecies chemostat cultures of diatoms. Limnol. Oceanogr. 24:298-315.
45. Mortimer, C. H. 1971. Chemical exchanges between sediments and water in the Great Lakes-Speculations on probable regulatory mechanisms. Limnol. Oceanogr. 16:387-404.
46. Nixon, S.W. 1980. Remineralization and nutrient cycling in coastal marine ecosystems, pp. 111-138. *In* B.J. Neilson and L.E. Cronin (eds.), International symposium on nutrient enrichment in estuaries. Humana Press, Inc., Clifton, New Jersey.
47. O'Connors, H. B., C. F. Wurster, C. D. Powers, D. C. Biggs and R. G. Rowland. 1978. Polychlorinated biphenyls may alter marine trophic pathways by reducing phytoplankton size and production. Science 20:373-379.
48. Painter, H. W. 1970. A review of literature on inorganic nitrogen metabolism in microorganisms. Water Res. 4:393-450.
49. Parsons, T. R. and M. Takahashi. 1973. Environmental control of phytoplankton cell size. Limnol. Oceanogr. 18:511-515.
50. Parsons, T. R., W. K. W. Li, and R. Waters. 1976. Some preliminary observations on the enhancement of phytoplankton growth by low levels of mineral hydrocarbon. Hydrobiology 51:85-89.
51. Patrick, W. H., Jr. and R. A. Khalid. 1974. Phosphate release and sorption by soils, sediments: effect of aerobic and anaerobic conditions. Science 186:53-55.
52. Pritchard, D. W. 1967. Observations of circulation in coastal plain estuaries, pages 37-44 *In* G. H. Lauff (ed.), Estuaries, American Association for the Advancement of Science, Publication 83.

53. Redfield, A. C. 1958. The biological control of chemical factors in the environment. Amer. Scientist 46:1-221.
54. Rogers, H. H., J. C. Campbell and R. J. Volk. 1979. Nitrogen-15 dioxide uptake and incorporation by *Phaseolus vulgaris* (L.) Science 206:333-335.
55. Rosenzweig, M. L. 1971. Paradox of enrichment: destabilization of exploitation ecosystems in ecological time. Science 171:385-387.
56. Rosenzweig, M. L. 1972. Stability of enriched aquatic ecosystems - reply. Science 175:564-565.
57. Ryther, J. H. 1969. Photosynthesis and fish production in the sea. The production of organic matter and its conversion to higher trophic forms of life vary throughout the world ocean. Science 166:72-76.
58. Ryther, J.H. and C.B. Officer. 1980. Impact of nutrient enrichment on water uses, pp. 247-261. *In* B.J. Neilson and L.E. Cronin (eds.), International symposium on nutrient enrichment in estuaries. Humana Press, Inc., Clifton, New Jersey.
59. Schelske, C. L. and E. F. Stoermer. 1972. Phosphorus, silica and eutrophication of Lake Michigan, pages 157-171 *In* G. E. Lukens (ed.), Nutrients and Eutrophication, Special Symposia, Vol. 1, Amer. Soc. Limnol. Oceanogr.
60. Schelske, C. L., M. S. Simmons and L. E. Feldt. 1975. Phytoplankton responses to silica enrichments in Lake Michigan. Verh. Internat. Verein. Limnol. 19:911-921.
61. Shugart, H. H., Jr. (Ed.) 1978. Time series and ecological processes. Society for Industrial and Applied Mathematics, Philadelphia. 303 pp.
62. Silverman, M. P. and H. L. Ehrlich. 1964. Microbial formation and degradation of minerals. Adv. Appl. Microbiol. 6:153-206.
63. Smith, S.V. 1980. Responses of Kaneohe Bay, Hawaii, to relaxation of sewage stress, pp. 247-261. *In* B.J. Neilson and L.E. Cronin (eds.) International symposium on nutrient enrichment in estuaries. Humana Press, Inc., Clifton, New Jersey.
64. Sverdrup, H. U., M. W. Johnson and R. H. Fleming. 1942. The Oceans, their physics, chemistry and general biology. Prentice-Hall, Inc. New York. 1087 pp.
65. Taft, J. L., W. R. Taylor and J. J. McCarthy. 1975. Uptake and release of phosphorus by phytoplankton in the Chesapeake Bay Estuary, USA. Mar. Biol. 33:21-32.
66. Taft, J. L. and W. R. Taylor. 1976. Phosphorus dynamics in some coastal plain estuaries, pages 79-89. *In* M. Wiley (ed.), Estuarine Processes, Vol. I, Academic Press, New York.
67. Taft, J. L. and W. R. Taylor. 1976. Phosphorus distribution in the Chesapeake Bay. Chesapeake Sci. 17:67-73.
68. Taft, J. L., M. E. Loftus and W. R. Taylor. 1977. Phosphate uptake from phosphomonoesters by phytoplankton in the Chesapeake Bay. Limnol. Oceanogr. 22:1012-1021.
69. Thomas, W. H. 1966. On denitrification in the northeastern tropical Pacific Ocean. Deep-Sea Res. 13:1109-1114.
70. Tilman, D. 1977. Resource competition between planktonic algae: an experimental and theoretical approach. Ecology 58:338-348.

71. Tyler, M. A. and H. M. Seliger. 1978. Annual subsurface transport of a red tide dinoflagellate to its bloom area: water circulation patterns and organism distributions in the Chesapeake Bay. Limnol. Oceanogr. 23:227-246.
72. Van Valkenburg, S.D. and D.A. Flemer. 1974. The distribution and productivity of nanoplankton in a temperate estuarine area. Estuarine Coastal Mar. Sci. 2:311-322.
73. Verstraete, W. and M. Alexander. 1972. Heterotrophic nitrification in samples from natural environments. Naturwissenschaften 59:79-80.
74. Waterbury, J. B., S. W. Waston, R. R. Guillard and L. E. Brand. 1979. Widespread occurrence of a unicellular, marine, planktonic cyanobacterium. Nature 277:293-294.
75. Webb, K.L., D.M. Hayward, J.M. Baker and B. Murray. 1979. Estuarine response to nutrient enrichment, a counterpart of eutrophication: A bibliography. Special Scientific Report No. 95, Virginia Institute of Marine Science, Gloucester Point, Virginia.
76. Webb, K.L. and C.F. D'Elia. 1980. Nutrient and oxygen redistribution during a spring-neap tidal cycle in a temperate estuary. Science 207:983-985.
77. Wiegert, R.G. 1974. Competition: a theory based on realistic, general equations of population. Science 185:539-542.
78. Wiegert, R.G. 1979. Population models: experimental tools for analysis of ecosystems, pages 233-279. *In* D.J. Horn, G.R. Stairs and R.D. Mitchell (eds.), Analysis of ecological systems. Ohio State University Press, Columbus.
79. Wollast, R. 1974. The silica problem, pages 359-392. *In* E. D. Goldgerb (ed.), The Sea, Volume 5, marine chemistry. J. Wiley & Sons, New York.
80. Wong-Chong, G. M. and R. C. Loehr. 1975. The kinetics of microbial nitrification. Water Res. 9:1099-1106.
81. Yoneyama, T. and H. Sasakawa. 1979. Transformation of atmospheric NO_2 absorbed in spinach leaves. Plant Cell Physiol. 263-266.

PHYSICAL AND GEOLOGICAL PROCESSES CONTROLLING NUTRIENT LEVELS IN ESTUARIES[1]

Donald W. Pritchard and Jerry R. Schubel
Marine Sciences Research Center
State University of New York at Stony Brook

As described in detail in other papers given in this symposium, sources of nutrients include the surface fresh water drainage from land, ground water inflow, direct waste discharges, rainfall, and fallout of particulates from the atmosphere. In some cases nutrients are supplied to the estuary from adjacent coastal waters, though more often nutrients are exported from the estuary to these coastal waters. Fixation of gaseous nitrogen from the atmosphere also contributes an input of nitrogen-based nutrients to some parts (generally lower salinity regions) of some estuaries. Within the estuary there are proximate sources of nutrients, such as regeneration from bottom sediments and oxidation of organic material within the water column.

This paper is concerned with the physical processes of transport and dispersion of nutrients within and through the estuary; with the manner in which suspended inorganic sediments modify these transport and dispersion processes; and with the suspended sediment and bottom sediment as short-term and long-term storage reservoirs for nutrients in the estuary.

AN ACCOUNTING OF THE PROCESSES AFFECTING THE NUTRIENT LEVELS IN AN ESTUARY

Once within the estuary, the distribution of concentration of any given chemical species of a particular nutrient depends upon various physical, chemical, biological and geological processes. In order to give an accounting of these processes, we consider a small (elemental) fixed volume in the estuary. The time rate of change of concentration of the dissolved form of a particular nutrient species within this elemental volume results from:

[1]Contribution No. 265 of the Marine Sciences Research Center.

(a) The difference between the advective flux into and out of the volume through the boundaries of the volume;

(b) The difference between the diffusive flux into and out of the volume through the boundaries of the volume;

(c) The rate of loss of this nutrient species by chemical conversion (oxidation or reduction) to another chemical form within the elemental volume;

(d) The rate of gain of this nutrient species by chemical conversion (oxidation or reduction) of another chemical form of the nutrient within the elemental volume;

(e) Rate of biological uptake of the nutrient species by phytoplankton and/or fixed plants within the elemental volume;

(f) Rate of gain of this nutrient species by metabolic activity within the elemental volume;

(g) Rate of biochemical production of this nutrient species by processing of organic material by bacteria, or by the fixation of the elemental or molecular form by bacterial or other microbial activity;

(h) Rate of loss of this nutrient species by such processes as denitrification by bacteria within this elemental volume;

(i) Rate of loss of this nutrient species by absorption onto suspended material within the elemental volume;

(j) Rate of gain of this nutrient species by desorption from suspended material within the elemental volume.

Thus, designating the concentration of the dissolved form of nutrient species i by $(C_d)i$, then the time rate of change of this concentration within the elemental volume can be written as:

$$\text{(1)} \qquad \partial (C_d)_i / \partial t = (a)+(b)-(c)+(d)-(e)+(f)+(g)-(h)-(i)+(j)$$

If the elemental volume has the bottom of the estuary as one boundary, then the diffusive and/or advective exchanges of the nutrient across the water/sediment interface may be treated as a boundary source or sink. Additional terms on the RHS of Eq. (1) would then be required to represent this source or sink. Similarly, if the air/water interface forms one boundary of the elemental volume, then additional terms will be required to represent the sources (i.e., rainfall, particulate fallout) and sinks (i.e., loss in spray) at this boundary. Gaseous exchanges of molecular oxygen and nitrogen are also important processes at the air/water interface which contribute indirectly to the control of nutrient levels in natural water bodies.

Each of the terms on the RHS of Eq. (1) can be expressed in mathematical form, and the equation is conceptually soluble, at least numerically. However, a number of the terms in the equation involve coefficients, such as diffusion coefficients or reaction rate constants, which are not well known, or which must be determined for the particular situation, that is, the particular physical and chemical environment in which the process is occurring.

Equation (1) cannot be treated as a single equation. It represents one of a set of equations, each of which treats the time rate of change of concentration of one of a sequence of chemical species of a particular nutrient. Each member of the sequence may be formed by oxidation and reduction reactions from adjacent members, and hence source terms in one equation of the set become sink terms in another, and vice versa.

Also, the equation set represented by Eq. (1) cannot be treated as an isolated set. The time rate of change of the concentration of a particular nutrient species which is absorbed on particles, and hence in suspended form, would be expressed by similar equation, and hence there would be an equation set for the suspended forms of the nutrient sequence. Some of the terms in Eq. (1) would not have counter-parts in the corresponding equation for the suspended form of the nutrient, and others will change signs; that is, absorption onto particles becomes a source term in this second equation set, while desorption becomes a sink term. Also, an added term would be required involving the net veritcal flux of suspended nutrient due to sinking (sedimentation) through the elemental volume.

The two sets of equations should be solved simultaneously, certainly a formidable task. Fortunately our task in this paper is less formidable, that of describing the processes represented by just a few of the terms which constitute each of the equations in the two equation sets. Specifically we are concerned with the physical processes of advective and diffusive transport of dissolved and suspended material, and with what we will classify here as geological processes; that is, absorption onto and desorption from particulate matter, the sinking of suspended material through the water column, sedimentation onto the bottom, resuspension from the bottom, and exchange of the dissolved form of the nutrient between the bottom sediments and the water column just over the bottom.

A SCENARIO OF THE CLASSICAL CONCEPT OF MOVEMENT AND DISPERSION OF A DISSOLVED AND OF A SUSPENDED SUBSTANCE IN AN ESTUARY

Consider an estuary of simple geometeric form, occupying an elongated indenture in a coastline, with a river entering at the upper end as the major fresh water source. A single V-shaped channel is assumed to extend longitudinally through the estuary, with gradually increasing depths from the head of the estuary to the mouth.

A dissolved or suspended material, such as a nutrient, in such an estuary would be subjected to an oscillatory movement directed generally along the axis of the estuary as a result of the ebb and flood tidal flow. The material would also be advected by a longer term, non-tidal circulation pattern associated with the distribution of density (as primarily controlled by salinity) and with the stress of wind. Prior to the last decade most attention was given to the density driven circulation. The classical picture of the non-tidal flow in an estuary was then that of a seaward moving upper layer and an up-estuary flowing lower layer. These flows could often be much larger, individually, than the volume rate of inflow of fresh water, but the difference between these flows, that is, the net non-tidal flow through any section, was considered to be just sufficient to move the inflowing fresh water through the estuary into the adjacent coastal ocean. In order to satisfy continuity, there must be a vertical flow from the lower layer into the upper layer over most of the length of the estuary.

Thus dissolved and suspended material in the surface layers would have a net non-tidal motion towards the ocean, while material in the lower layers would be moved toward the head of the estuary. Material would also be advected vertically from deeper layers into the surface layers. Turbulent motions in the internal field of flow, augmented by eddies caused by the tidal flows interacting with bottom roughness and shoreline irregularities, and by wind induced stirring, will result in a non-advective diffusion of material from regions of higher concentration towards regions of lower concentration.

Material in suspended form will also be subjected to a downward flux due to settling of the particles, with this effect being dependent on particle size and density. Near the head of the estuary there is a reach where the non-tidal up-estuary flow in the bottom layers weakens and finally ceases, to be replaced by a tidal averaged flow directed seaward at all depths. This region thus represents a convergence area for suspended material. The combination of the sinking of particles from the seaward flowing upper layer into the up-estuary flowing lower layer, the transport in this layer of the suspended material up to the region where the up-estuary flow weakens and then ceases, results in a concentration of suspended material in this reach of the estuary. It is primarily this physical mechanism, rather than any chemical flocculation of material, which results in this upper reach of the estuary being visually recognizable as a region of maximum turbidity. It is also a region of high sedimentation rate. Nutrients absorbed onto the particles are also concentrated in this part of the estuary, and the flux of nutrients onto the bottom with the settling particles is correspondingly high. Under suitable conditions, this reach can also be a region of high release of nutrients from the bottom to the overlying water column.

While this classical description of the fate of dissolved suspended material in an estuary has served to explain many of the gross features of the concentration distribution of these substances in a number of estuaries, studies over the past decade have increasingly indicated that, particularly at times scales of less than four weeks or so, other processes contribute significantly to the material balance in the estuary. This new evidence primarily concerns the advective processes.

A CLOSER LOOK AT THE ADVECTIVE PROCESS IN ESTUARIES

The advective process appears in the equations for the time rate of change of nutrient concentration in the form of a spatial gradient of the advective flux, which in turn is the product of the velocity and the concentration at a given point and time. The velocity field which enters the advective flux encompasses a wide range of temporal and spatial scales, including the oscillatory motion of tidal currents, the non-tidal, long-term density driven flows, and the intermediate time scales of motions associated with meteorological phenomena, primarily the wind. Because of the complex coupling of the velocity field with the concentration distribution of a dissolved or suspended substance via the advection-diffusion equation, the concentration of a dissolved or suspended material also exhibits variations encompassing the same range of time scales. Now consider the significance of these various scales of motion, and how well they can be measured, or computed.

The oscillatory tidal velocities are important in respect to the time varying concentration distribution in the oscillatory plume from a point source of nutrients, such as a sewage plant or other waste discharge. In addition to this importance of the tidal velocities in contributing to the advective fluxes at time scales of a tidal cycle and less, and at spatial scales of a tidal excursion and less, they are also important at longer time scales and larger spatial scales in two very important ways. First, the convariance at the surface of the tidal rise and fall of water and the longitudinal oscillatory tidal currents, and in the internal field of motion, the co-variance of the longitudinal, lateral and vertical oscillatory tidal currents with each other, can lead to residual (non-tidal) currents which vary in time only with the longer-term tidal components (i.e., the fortnightly tide), but which can have spatial variations across the estuary and along the estuary at scales both smaller than and larger than a tidal excursion. The velocities associated with these residual currents are identical with the Stokes velocities of classical wave theory. The second way in which tidal velocities are important to longer-term processes is in providing the turbulent energy for mixing, or eddy diffusion.

Records from *in situ* continuous recording current meters can be processed by suitable filtering to separate the signal associated with the astronomical tide from the signal resulting from short period turbulent motions and instruments noise, and also from the longer period non-tidal motions. The signal of tidal period produced from processing a record having a duration of at least two weeks obtained from a well calibrated current meter of suitable design is a very good representation of the actual horizontal component of the tidal current. Harmonic analysis of such records will provide values of the amplitude and phase of the major astronomical components of the tidal current. It can be shown that the equation of continuity can be applied independently to the instantaneous field of motion, the non-tidal or tidally averaged field of motion, and to the field of motion of tidal period. Thus integration of the equation of continuity can be used to obtain the vertical component of the tidal currents from the spatial variations in the horizontal components. Given a suitable initial base of current observations, the purely oscillatory, tidal part of the field of motion can then be predicted for any other time.

Except for the effects of bottom friction, the tidal currents are essentially barotrophic, and independent of the density distribution. Vertically averaged two-dimensional hydrodynamic numerical models have been very successful in computing the field of vertically averaged tidal currents from tide gauge records at the boundaries of the modelled area, after adjustment of the bottom frictional coefficients based on tidal gauge data and tidal current data obtained from an adequate number of locations in the estuary. Empirical data on the vertical variation of the tidal velocity and in the phase of the time variation of the tidal velocity can be used to give a reasonable estimate of the three-dimensional tidal velocity field from the computed vertical averaged velocities.

The classical density driven flow pattern of a seaward directed flow in the upper layers and an up-estuary directed flow in the lower layers in moderately strong estuaries with channel depths of at least 3 m represents the important advective field on time scales of several weeks and longer. Lateral variations in the flow pattern, particularly in relatively shallow, broad estuaries, have been shown to be important at these longer time scales. These lateral variations arise from the effects of variations in depth on bottom friction, from spatial variations in the Stokes Velocity, and from the effects of coriolis force.

A recent study by Pritchard and Rives (7) can be used to examine how well motion at these time scales can be measured. During the 20-day period, October 13 through November 2, 1977, 22 current meters were deployed on two cross-sections in the Chesapeake Bay in the vicinity of Calvert Cliffs. One section, containing 13 meters arrayed on five vertical moorings, extended eastward across the bay from Kenwood Beach. The other, containing nine meters on three vertical moorings, extended eastward from Cove Point. The two sections were separated by about 15 km, with the northern section having a width of about 14 km and the southern section having a width of about 10 km. Two current meters on the northern section failed to function, and three meters on the southern section failed to produce usable records over a significant portion of the deployment period. Eight other meters were arrayed on three vertical moorings along an oblique section interior to the two cross-sections. On the basis of an observed similarity in the data from stations at about the same distance from the western shore, and having the same depth, the records from the three interior stations were employed to fill in some of the data gaps on the two cross-sections. Tide gauges were also installed at the four corners of the study area.

Both the current meter data and the tide gauge data were processed by filtering to separate the variations in water surface elevation and current speed having tidal period from the longer-term tidal averaged fluctuations.

The purely harmonic tidal variation in surface elevation had a range of about 50 cm, while the amplitude of the longitudinal component of the tidal current ranged from ± 15 cm/sec to ± 40 cm/sec, depending on position in the cross-section. The tidally averaged water surface elevations showed aperiodic variations with the time interval between major events ranging from about two days to about four days, with a maximum range exceeding the range of the purely harmonic tide. The tidally averaged current meter records also showed aperiodic variations in the longitudinal component with the time interval between major events ranging from two to four days, and with a maximum amplitude equal to or slightly exceeding the amplitude of the purely harmonic tidal current. Recent studies by Wang and Elliot (10), Elliot (2), Elliot and Wang (3), and Elliot et al (4) suggest that such non-tidal variations in water surface elevations and in current velocities are characteristic of estuaries.

Leaving for the moment the time variations of the tidal mean velocities, consider now the average values of the longitudinal component of the tidal mean velocity over the period of record. The spatial distribution in each cross-section of these long-term (20-day) averaged tidal mean velocities is clearly in agreement with the classical two-layered estuarine flow pattern. The flows in the upper layer of the estuary were directed seaward while the flows in the lower layer were directed landward (i.e., up the estuary). As is consistant with the effects of the earth's rotation, the boundary between the seaward flowing upper layer and the landward flowing lower layer is deeper on the western side of the Chesapeake Bay where it intersects the sloping bottom of the bay at a depth of about 12 m, than on the eastern side of the bay, where the up-estuary flowing lower layer extends to within 7.5 m of the water surface. The magnitude of the non-tidal velocities in the upper layer is greater in the center of the estuary than over the shallow side reaches. The lower layer shows a core of maximum landward directed velocities centered on the deepest part of the cross-section, but well above the actual bottom. This distribution of velocity magnitudes is consistent with the expected effects of bottom friction.

One measure of how well the advective flux can be determined directly from measurements is to see how well the measured mean volume rate of flow through each of these two sections satisfied continuity. Certainly the advective flux of some water borne dissolved or suspended component cannot be determined any better than the advective flux of the water itself!

We proceed by developing the relationship for continuity of volume for a two-layered estuary. Let V_X be the instantaneous volume of the estuary from the head of tide to a cross-section at longitudinal position $\underline{X}$ within the estuary. The x-axis is directed longitudinally down the estuary from an origin at the head of the estuary. Designate the volume rate of inflow of fresh water to the estuary up-stream of the cross-section at $\underline{X}$ by Q_R;

the area of the upper layer of the cross-section in which the long-term non-tidal flow is directed down the estuary by σ_u; and the area of the lower layer of the cross-section in which the long-term non-tidal flow is directed up the estuary by σ_ℓ. The time rate of change of the volume V_X is then given by

$$\partial V_x/\partial t = Q_R - \iint_{\sigma_u} u d\sigma - \iint_{\sigma_\ell} u d\sigma \tag{2}$$

where u is the longitudinal component of the instantaneous velocity at a point y, z in the cross-section at longitudinal position X.

Averaging Eq. (2) over a tidal cycle of period τ gives

$$\partial \bar{V}_x/\partial t = \bar{Q}_R - \frac{1}{\tau}\int_\tau \{\iint_{\sigma_u} u d\sigma\} dt - \frac{1}{\tau}\int_\tau \{\iint_{\sigma_\ell} u d\sigma\} dt \tag{3}$$

where

$$\bar{V}_x = \frac{1}{\tau}\int_\tau V_x dt \text{ and } \bar{Q}_R = \frac{1}{\tau}\int_\tau Q_R dt \tag{4}$$

are the average values of V_X and Q_R, respectively, over the tidal cycle. There is a tendency to exchange the order of integration of the second and third terms on the RHS of Eq. (2), thus giving the integration over the cross-sectional area of the tidal averaged velocity field. Such an exchange in the order of integration is permissible in the third term on the RHS, since σ_ℓ, the area of the lower layer of the cross-section, is considered to be constant with time. It is not permissible in the second term, however, since the area of the upper layer of the cross-section, σ_u, does vary with time over the tidal cycle. The upper boundary of σ_u is the water surface itself, which moves up and down with the tidal wave and also as a result of meteorological forcing.

The second term on the RHS of Eq. (3) is the tidal mean value of the advective flux of water through the upper layer of the cross-section. This term can be treated in parts by considering the instantaneous area σ_u to be composed of the tidal average value of σ_u, designated by $\bar{\sigma}_u$, plus a zero-centered time varying area of tidal period designated by σ_η. Since $\bar{\sigma}_u$ is not a function of time with respect to integrations over the tidal cycle, Eq. (3) can be written:

$$\sigma \bar{V}_x/\sigma t = \bar{Q}_R - \iint_{\bar{\sigma}_u} \{\frac{1}{\tau}\int_\tau u dt\} d\sigma - \iint_{\sigma_\ell} \{\frac{1}{\tau}\int_\tau u dt\} d\sigma$$

$$- \frac{1}{\tau}\int_\tau \{\iint_{\sigma_\eta} u d\sigma\} dt \tag{5}$$

Now, in the second and third terms on the RHS of Eq. (5), designate $u = \bar{u} + U + u'_{\tau}$ where $\bar{u} = \frac{1}{\tau}\int_{\tau} u\,dt$, and where U is the longitudinal component of the purely oscillatory (zero-centered) tidal velocity and u'_{τ} is the zero-centered turbulent velocity. Also, in the fourth term on the RHS of (5) designate $u = u_{\bar{\sigma}_u} + u'_{\sigma_u}$ where $u_{\bar{\sigma}_u} = \frac{1}{\sigma_u}\iint_{\bar{\sigma}_u} u\,d\sigma$, that is. $u_{\bar{\sigma}_u}$ is the average value of the instantaneous velocity over the tidal mean cross-sectional area. It is therefore constant with respect to any spatial integration. Consequently, Eq. (5) becomes:

$$\partial\bar{V}_x/\partial t = \bar{Q}_R - \iint_{\bar{\sigma}_u} \bar{u}\,d\sigma - \iint_{\sigma_\ell} \bar{u}\,d\sigma - \frac{1}{\tau}\int_{\tau} \sigma_\eta u_{\bar{\sigma}_u}\,dt \quad (6)$$

where it has been assumed that the time variations in the velocity deviation term, $u'_{\bar{\sigma}u}$, are not correlated with the time variations in σ_η.

Eq. (6) can be further simplified by designating

$$\bar{Q}_{\bar{\sigma}_u} = \iint_{\bar{\sigma}u} \bar{u}\,d\sigma; \qquad \bar{Q}_{\sigma_\ell} = \iint \bar{u}\,d\sigma \quad (7)$$

and

$$u_{\bar{\sigma}_u} = \bar{u}_{\bar{\sigma}_u} + U_{\bar{\sigma}_u} + u_{\bar{\sigma}_{u,\tau}} \quad (8)$$

where

$$\bar{u}_{\bar{\sigma}_u} = \frac{1}{\bar{\sigma}_u}\int_{\tau} u_{\bar{\sigma}_u}\,dt$$

and $U_{\bar{\sigma}_u}$ is the purely oscillatory (zero-centered) tidal velocity averaged over the tidal mean area of the upper layer cross-section. Eq. (6) then can be written

$$\partial\bar{V}_x/\partial t = \bar{Q}_R - \bar{Q}_{\bar{\sigma}_u} - \bar{Q}_{\sigma_\ell} - \langle\sigma_{\eta *}U_{\bar{\sigma}_u}\rangle_{\tau} \quad (9)$$

where

$$\langle\sigma_\eta U_{\bar{\sigma}_u}\rangle_{\tau} = \frac{1}{\tau}\int_{\tau} \sigma_\eta U_{\bar{\sigma}_u}\,dt$$

now note that the term $\bar{Q}_{\bar{\sigma}_u}$ is the integral over the tidal mean upper layer cross-sectional area of the tidal mean velocity. It is not the same as $\bar{Q}_{\sigma_u}$, which would represent the tidal mean value of the integral over the instantaneous upper layer cross-sectional area of the instantaneous velocity. The difference between $\overline{Q}_{\sigma_u}$ and $\overline{Q}_{\overline{\sigma}_u}$ is given by the last term in Eq. (9). This term, which is the tidal mean value of the product of the time varying intertidal area, σ_η and the time varying areal mean value of the tidal velocity, $U_{\overline{\sigma}_u}$ is in fact the Stokes Transport, familar to students of classical wave theory. Interest in Stokes-type transports and residual velocities in estuaries was stimulated by a paper written by Longuet-Higgins (6), who treated the general subject of residual currents arising from the co-variation of the various components of purely oscillatory (zero-centered) motions. Tee (9) and Ianuiello (5) have recently treated somewhat different aspects of the dynamics of Stokes-type velocities in estuaries.

The Stokes Transport term in Eq. (9) can be approximated by

$$\langle \sigma_\eta * U_{\overline{\sigma}_u} \rangle_\tau = \tfrac{1}{2}\, \overline{b}_\eta\, \eta_o U_o \cos(\varepsilon_\eta - \varepsilon_U) \qquad (10)$$

where $\overline{b}_\eta$ is the tidal mean width of the estuary at the surface, η_0 is the amplitude of the oscillatory tidal variation in the water surface (i.e., the "tide") averaged across the estuary, U_0 is the amplitude of the oscillatory tidal current averaged over the upper layer, and ε_η-ε_U is the difference in the phase of the tide and the tidal current. For a progressive wave, maximum flood current occurs at the time of High Water and maximum ebb current occurs at the time of Low Water. Since in our coordinate system ebb flow is positive and flood flow is negative, then for a progressive wave ε_η-ε_U = π, and cos (ε_η-ε_U) = -1.0. For a standing wave slack water before ebb occurs at High Water, and slack water before flood occurs at Low Water. Maximum flood flow then occurs 3 hours and 6 minutes, or one-quarter of a tidal cycle prior to Low Water. Consequently, ε_η-ε_U = $\pi/2$, and cos (ε_η-ε_U) = 0.

Thus the Stokes Transport is a maximum if the tidal wave in the estuary is a pure progressive wave, and is zero if the tidal wave is a pure standing wave. This should be obvious, since this longitudinal component of the Stokes Transport arises from the fact that in a progressive wave the velocity is in the direction of wave motion under the crest and in the opposite direction under the trough; that is, the current is flooding during the period when the water level is above mean tide level, and is ebbing during the period when the water level is below mean tide level. Therefore, with a purely oscillatory zero-centered tidal velocity, more water moves up the estuary through a cross-section of the estuary during the flood period than moves downstream during the ebb period. This net up-estuary transport due to the co-variance of the rise and fall of the water level and the flood and ebb of the tide must be compensated for by a residual (non-tidal) flow directed down the estuary, in order to satisfy continuity.

In most estuaries the tidal characteristics are intermediate between those associated with a progressive tidal wave and those associated with a standing wave; that is, the maximum flood current occurs prior to High Water by a time interval less than 3.1 hours, and a similar relationship occurs between the time of maximum ebb current and the time of Low Water. Over a major segment of the Chesapeake Bay, the time differences between maximum flood and High Water, and between maximum ebb and Low Water, are less than one hour, and hence the Stokes Transport for this estuary is within a factor of 0.875 of that which would occur if the tidal wave were purely progressive.

Each of the terms on the RHS of Eq. (9) were evaluated for the 20-day period of study for each of the two cross-sections described above. The result of this evaluation is given in Table 1. The last column gives the value of $\partial \bar{V}_X / \partial t$ obtained for evaluating the RHS of Eq. (9), and not from direct measurement of the change in the tidal mean volume of the estuary above the two sections. Our best estimate from available tidal data is that $\partial \bar{V}_X / \partial t$ for the 20 days of the study had a value close to zero. However, note that the surface area of the estuary above the Kenwood Beach section is 2.72×10^9 m^2. A difference of just over 6 cm in the tidal mean water level averaged over this surface area, between the start and end of the study period, would correspond to a value for the magnitude of $\partial \bar{V}_X / \partial t$ of 100 $m^3 \cdot s^{-1}$. Variations in the tidal mean water surface elevation of as much as 1 m have occurred over this reach of the Chesapeake Bay. Over a 20-day period, this would correspond to a value of $\partial \bar{V}_X / \partial t$ of 1572 m^3/sec. However, such a change in tidal mean water surface elevation has been observed over periods as short as 4 days, which would correspond to a value of $\partial \bar{V}_X / \partial t$ of 7859 m^3/sec. It is evident that the storage term cannot in general be neglected in considering tidal mean continuity in the Chesapeake Bay.

TABLE 1. Values of the various advective transport terms in Equation (9) evaluated for two cross-sections in the Chesapeake Bay for the period October 13 through November 2, 1977. Values are in cubic meters per second.

Section	$\bar{Q}_R$	$\bar{Q}_{\bar{\sigma}_u}$	$\bar{Q}_{\tilde{\sigma}_u}$	$\langle \sigma_\eta U_{\bar{\sigma}_u} \rangle_\tau$	$\partial \bar{V}_x / \partial t$
Kenwood Beach	3028	6235	-2539	-270	-398
Cove Point	3036	7736	-4706	-320	+326

In any case, Eq. (9) evaluates to -398 $m^3 \cdot s^{-1}$ for the Kenwood Beach section and +326 $m^3 \cdot s^{-1}$ for the Cove Point section. This difference of 724 $m^3 \cdot s^{-1}$ is impossible, since it would imply that the tidal mean water surface elevation over the surface area of the bay between the two cross-sections increased by about 7 m between the start and end of the 20-day study period. The maximum credible difference in values of $\partial \overline{V}_x / \partial t$ for the two sections in this study is about 20 $m^3 \cdot s^{-1}$.

There thus appears to be an error in the estimates for the advective flux terms of about -400 $m^3 \cdot s^{-1}$ for the Kenwood Beach section and +325 $m^3 \cdot s^{-1}$ for the Cove Point section. The error cannot be in the estimate of $\overline{Q}_R$. Uncertainties in the calibration of river gauges, together with the fact that a significant part of the inflow of fresh water is from ungauged areas and the run-off from these areas must be estimated using average run-off factors, may result in uncertainties in $\overline{Q}_R$ of some 10 percent. However, the value of $\overline{Q}_R$ for the Kenwood Beach Section would have to be increased by some 400 $m^3 \cdot s^{-1}$, and the value for the Cove Point would have to be decreased by some 325 $m^3 \cdot s^{-1}$ to account for the failure of the data to satisfy continuity at the two sections. The value of the mean fresh water inflow to the Bay above the Cove Point Section must be greater than for the Kenwood Beach Section, but by no more than about 10 $m^3 \cdot s^{-1}$.

The errors cannot be in the Stokes Transport term, either. Differences in this term between the two sections can only result from differences in the width of the sections and the amplitudes of the tide and tidal current at the two sections, all of which are reasonably well known.

The indicated errors must therefore be in the computations of the mean advective flux through the upper and lower layers of the two sections, which is based on the processing of the current meter records to obtain values of the tidal mean velocity at each meter, and on the numerical integration of these tidal mean velocity values over the tidal mean cross-sectional area. This is not surprising. Each current meter is sampling a large portion of the cross-section, amounting in this study to about 1.5×10^4 m^2. Each current meter is subject to calibration and least count or record reading errors, the effects of which are however minimized by the processes of time averaging and spatial integration. Biological fouling can lead to changes in calibration coefficients and sometimes to failure of the current meters. In this study fouling did not, however, appear to be an important factor.

In any case, note that if the tidal mean speed in the upper layer of the Kenwood Beach section were over-estimated by 0.2 $cm \cdot s^{-1}$, and the mean speed in the lower layer of that section were underestimated by the same amount, while this same error but with opposite signs occurred in the Cove Point section, the difference in the evaluation of Eq. (9) for the two sections would be explained. Statistical analysis of the various possible error terms suggests that the uncertainty in the sectional averaged tidal mean speed is probably on the order of $\pm$ 0.5 $cm \cdot sec^{-1}$, and hence the agreement between the results of the evaluation of the continuity relationship between the two sections is in fact better than should be expected.

As shown in Table 1, the Stokes Transport appears to have a magnitude of only about one-tenth of the inflow of fresh water. However, the average fresh water inflow for the period of study of October 13 to November 2, 1977 was about seven times the long-term average for this part of the year. In a normal year there are significant periods of time, on the order of 3 months, during which the weekly averaged fresh water inflow is 300 $m^3 \cdot s^{-1}$ (10,600 cfs) or less. The Stokes Transport has a relatively small variation in magnitude with time at any given section, perhaps on the order of $\pm$ 50 percent over the interval between neap tide and spring tide. Thus the Stokes Transport at a given cross-section can readily equal or exceed the fresh water inflow to the estuary up-stream from the section for significant periods of time.

The uncertainty in the measured advective flux of water through a cross-section as indicated above may appear to be sufficiently small for many purposes. What does this uncertainty in the measurements of the velocity field imply in terms of the uncertainty that might be expected in the direct measurement of the advective flux of a particular dissolved or suspended substance?

It is doubtful that the weighted mean tidally averaged sectionally integrated concentration of dissolved or suspended nutrient concentration can be determined with any greater accuracy than can the net non-tidal flux of water. This is true even though measurements of concentration in an individual sample might be made with very high precision. The problem is that the measurements of concentration should be made at the same frequency and locations as the current measurements. Since no in situ devices equivalent to current meters exist for measurements of nutrient concentrations or suspended sediment concentrations, the amount of research vessel time and manpower requirements to attain samples for concentration measurement at the same time intervals and locations as for current measurements makes such an attempt impractical.

It is conceivable that an effort could be mounted to obtain measurements of the concentration of a dissolved or suspended material at a sufficient frequency over some integral number of tidal cycles and at a sufficient number of points in a cross-section, such that the weighted mean, tidally averaged sectionally integrated concentration of the subject substance could be determined with an uncertainty of the same order as is possible for the longitudinal component of the velocity. Assuming this is so, then what kind of uncertainty might be expected in the computed, long-term mean advective flux of the substance through the cross-section based on a measurement set over a period of some integral number of tidal cycles?

The instantaneous flux at any point is given by the product $u \cdot c$ where c is the instantaneous concentration of the subject material at a point in the cross-section, and as before, u is the instantaneous velocity. The tidal averaged flux through the cross section of area σ can then be expressed by

$$\overline{(F_c)_\sigma} = \frac{1}{\tau}\int_\tau \{\iint_{\sigma_u} (uc)d\sigma\}dt + \frac{1}{\tau}\int_\tau \{\iint_{\sigma_\ell} (uc)d\sigma\}dt \quad (11)$$

where the cross-section is divided into an upper layer in which the tidal mean velocity is directed seaward and in which the cross-sectional area is time depended, and a lower layer in which the tidal mean velocity is directed up the estuary and in which the cross-sectional area is constant with time. Note that this division is useful only for averaging periods sufficiently long so that such a classical estuarine circulation pattern is satisfied by the data. As will be discussed more fully later, the averaging period necessary to ensure that the averaged flow distribution will fit this pattern is on the order of ten days, at least in the Chesapeake Bay and its tributary estuaries. In the upper, low salinity reaches of most estuaries, and over the entire reach of some estuaries having low fresh water inflows and high tidal mixing, a lower layer of reverse tidal mean flow does not occur. In such cases, $\sigma_\ell = 0$ and the last term in Eq. (11) would be dropped. The important features of the development which follows will apply whether or not a lower layer of up-estuary tidal mean flow exists.

We proceed in a manner anologous to that used in the development of Eq. (9). First, the cross-sectional area of the upper layer is sub-divided into a tidal mean value $\overline{\sigma}_u$, which is independent of time with respect to integration over a tidal period, and a zero-centered, time varying area, σ_η, which oscillates with tidal period. For any spatial or time varying function f, we designate

$$\langle f\rangle_\tau = \bar{f} = \frac{1}{\tau}\int_\tau f dt \quad (12)$$

and

$$\langle f\rangle_\sigma = \frac{1}{\sigma}\iint_\sigma f d\sigma \quad (13)$$

Now, we utilize the fact that for spatial integration over areas which are independent of time over a tidal cycle, the order of time and spatial integration can be interchanged, and we make substitutions of the type

$$u = \bar{u} + U + u'_\tau \; ; \qquad c = \bar{c} + C + c'_\tau \quad (14)$$

or of the type

$$u = u_\sigma + u'_\sigma \; ; \qquad c = c_\sigma + c'_\sigma \quad (15)$$

depending on the type of integral being evaluated. Eq. (11) then becomes, for the tidal averaged advective flux through the cross-section $\sigma(x,t)$ of a dissolved or suspended substance having a concentration $c(x,y,z,t)$,

$$\overline{(F_c)_\sigma} = \frac{1}{\tau}\int_\tau \{\iint_\sigma u c\, d\sigma\}dt = \bar{Q}_{\bar{\sigma}_u}\bar{c}_{\bar{\sigma}_u} + \bar{Q}_{\sigma_\ell}\bar{c}_{\sigma_\ell} + \bar{\sigma}_u \langle U_{\bar{\sigma}_u} C_{\bar{\sigma}_u}\rangle_\tau$$

$$+ \sigma_\ell \langle U_{\sigma_\ell} C_{\sigma_\ell}\rangle_\tau + \bar{u}_{\bar{\sigma}_u}\langle \sigma_\eta C_{\bar{\sigma}_u}\rangle_\tau + \bar{c}_{\bar{\sigma}_u}\langle \sigma_\eta U_{\bar{\sigma}_u}\rangle_\tau$$

$$+ \langle \sigma_\eta U_{\bar{\sigma}_u} C_{\bar{\sigma}_u}\rangle_\tau + [\bar{\sigma}_u \langle \bar{u}'_{\bar{\sigma}_u}\bar{c}'_{\bar{\sigma}_u} + \langle U'_{\bar{\sigma}_u} C'_{\bar{\sigma}_u} + u'_\tau c'_\tau\rangle_\tau\rangle_{\bar{\sigma}_u}$$

$$+ \bar{\sigma}_\ell \langle \bar{u}'_{\sigma_\ell} c'_{\sigma_\ell} + \langle U'_{\sigma_\ell} C'_{\sigma_\ell} + u'_\tau c'_\tau\rangle_\tau\rangle_{\sigma_\ell}] \qquad (16)$$

where

$$\bar{Q}_{\bar{\sigma}_u} = \bar{\sigma}_u \bar{u}_{\bar{\sigma}_u}; \qquad \bar{Q}_{\sigma_\ell} = \sigma_\ell \bar{u}_{\sigma_\ell}$$

$$\bar{u}_{\bar{\sigma}_u} = \frac{1}{\bar{\sigma}_u}\iint_{\bar{\sigma}u}\{\frac{1}{\tau}\int_\tau u\,dt\}d\sigma; \qquad \bar{c}_{\bar{\sigma}_u} = \frac{1}{\bar{\sigma}_u}\iint_{\bar{\sigma}_u}\{\frac{1}{\tau}\int_\tau u\,dt\}d\sigma$$

$$U_{\bar{\sigma}_u} = \frac{1}{\bar{\sigma}_u}\iint_{\bar{\sigma}_u} U\,d\sigma; \qquad C_{\bar{\sigma}_u} = \frac{1}{\bar{\sigma}_u}\iint_{\bar{\sigma}_u} C\,d\sigma$$

$$\bar{u}'_{\bar{\sigma}_u} = \frac{1}{\tau}\int_\tau u\,dt - \frac{1}{\bar{\sigma}_u}\iint_{\bar{\sigma}_u}\{\frac{1}{\tau}\int_\tau u\,dt\}d\sigma$$

$$\bar{c}'_{\sigma_u} = \frac{1}{\tau}\int_\tau c\,dt - \frac{1}{\bar{\sigma}_u}\iint_{\bar{\sigma}_u}\{\frac{1}{\tau}\int_\tau c\,dt\}d\sigma$$

$$U'_{\bar{\sigma}_u} = U - \frac{1}{\bar{\sigma}_u}\iint_{\bar{\sigma}_u} U\,d\sigma; \qquad C'_{\bar{\sigma}_u} = C - \frac{1}{\bar{\sigma}_u}\iint C\,d\sigma$$

$$u'_\tau = u - U; \qquad c'_\tau = c - C$$

and where the terms with subscript σ_ℓ in Eq (16) are defined by an anologous equation set to the above. The symbol U represents the zero-centered tidal harmonic oscillation of the longitudinal component of the velocity, and C represents the zero-centered tidal harmonic oscillation in the concentration of the subject substance. The third, fourth, fifth, sixth and seventh terms on the RHS of Eq. (16) thus represent Stokes-type flux terms. Given information on the amplitude and phase of the harmonic fluctuations in σ_η, U and C, these terms can be evaluated.

The terms containing parameters with a superscript "prime", which are grouped together within the square brackets "()", represent tidal mean and areal mean cross-products of velocity and concentration deviation terms which arise out of the averaging process. These terms might be considered to represent an effective longitudinal diffusive flux of the subject dissolved or suspended substance.

If simultaneous measurements of velocity and concentration could be obtained at a sufficient number of points in the cross-section, at closely spaced time intervals over an adequate time period, the double integration (over the cross-sectional area and then over time) as represented by the term contained between the equal signs in Eq. (16) could be evaluated. Why, then, have we bothered to present the somewhat complex expansion represented by the RHS of Eq. (16)? As discussed earlier, the resources required to undertake such a sampling effort, particularly in large estuaries such as the Chesapeake Bay or in its major tributary estuaries, would be large and the effort would be costly. Secondly, even if such a measurement effort were to be made, the accuracy to which the tidal mean sectionally integrated advective flux of the subject substance could be determined would have to be sufficiently high so that the effort would be worthwhile. It is not a priori evident that such would be the case. Evaluation of several of the terms on the RHS of Eq. (16) can provide an estimate of the uncertainty in the direct determination of such a flux term. Finally, some investigators have argued that the long-term average flux of a dissolved or suspended substance through a given cross-section can be estimated with adequate accuracy from the product of the long term average flux of water through the section (usually taken to be equal to the average fresh water inflow rate) times the long term average sectional mean concentration of the subject material. The development given above is indicative of the actual complexity of the process. While not all of the terms on the RHS of Eq. (16) are necessarily important in contribute significantly in most estuaries, and there is no a priori reason to neglect the cross-products of the velocity and concentration deviation terms, that is, the group of terms contained within the square brackets "{}" in Eq. (16).

The sum of the first two terms on the RHS of Eq. (16) certainly contributes an important part of the tidal mean advective flux of a given substance through a cross-section of an estuary. Using the results of the evaluation of Eq. (9) discussed earlier, together with a simulated but realistic estimate of the tidal mean, areally averaged concentration of suspended sediment in the upper and lower layers of the Chesapeake Bay in the reach between Kenwood Beach and Cove Point, we can evaluate the possible range of values of the sum of these first two flux terms for the case of the flux of suspended sediment.

The average values of the volume transport terms $\bar{Q}_{\bar{\sigma}_u}$ and $\bar{Q}_{\sigma_\ell}$, and the range of uncertainties in these values, as obtained from the evaluation summarized in Table 1, for measurements made over the 20-day period October 13 through November 2, 1977 in the reach of the Chesapeake Bay between Kenwood Beach and Cove Point are:

$$\bar{Q}_{\bar{\sigma}_u} = 6950 \pm 250 \ m^3 \cdot s^{-1}; \qquad \bar{Q}_{\sigma_\ell} = -3620 \pm 120 \ m^3 \cdot s^{-1}$$

Though measurements of the suspended sediment concentration were not made during this field study, characteristic values for this reach of the bay are 4.0 ppm for the upper layer and 7.5 ppm for the lower layer. Assuming the same relative uncertainty in the tidal mean, sectional averaged values of suspended sediment concentration as for the corresponding volume flux terms, the range of possible values for the sum of the first two terms in Eq. (16) is given by

$$\{\bar{Q}_{\bar{\sigma}_u} \bar{c}_{\bar{\sigma}_u} + \bar{Q}_{\sigma_\ell} \bar{c}_{\sigma_\ell}\}_{max} = +209 \text{ tonnes/tidal cycle}$$

and

$$\bar{Q}_{\bar{\sigma}_u} c_{\bar{\sigma}_u} + \bar{Q}_{\sigma_\ell} c_{\sigma_\ell}\}_{min} = -151 \text{ tonnes/tidal cycle}$$

Thus, from this data set (part measured, part simulated) we conclude that we cannot be certain whether the advective flux of sediment is directed down the estuary or up the estuary. This uncertainty would apply equally to nutrients absorbed into the suspended sediment.

That the same uncertainty could exist for estimates of the flux of a dissolved substance is indicated by a similar analysis for the flux of salt. Salinity measurements were made during the field study in the Chesapeake Bay we have been describing. The mean salinity for the upper layer for the period of study was 9.88 ± 0.18 $^o/oo$, and for the lower layer 16.69 ± 0.30 $^o/oo$. Note that the uncertainty in these values as indicated is about half the relative uncertainty assigned to the suspended sediment estimates above, because of the higher precision possible for salinity measurements, together with the large number of measurements it is possible to obtain using in situ temperature and conductivity sensors. This uncertainty arises primarily from the large observed variance of the concentration, both in time and space, and not from uncertainties in individual measurements. The range in that part of the advective flux of salt which is given by the sum of the first two terms on the RHS of Eq. (16), is then for this data set,

$$\{\overline{Q}_{\overline{\sigma}_u}\overline{s}_{\overline{\sigma}_u} + \overline{Q}_{\sigma_\ell}\overline{s}_{\sigma_\ell}\}_{max} = 6.74 \times 10^5 \text{ tonnes/tidal cycle}$$

$$\{\overline{Q}_{\overline{\sigma}_u}\overline{s}_{\overline{\sigma}_u} + \overline{Q}_{\sigma_\ell}\overline{s}_{\sigma_\ell}\}_{min} = 6.47 \times 10^4 \text{ tonnes/tidal cycle}$$

Although both the upper and lower estimates of the advective transport of salt indicate a seaward flux due to the contribution of these two terms, there is an order of magnitude difference in the two estimates.

It is our opinion that even with the best possible set of simultaneous measurements of velocity and concentration of a particular dissolved or suspended substance, evaluation of the tidal mean, sectionally integrated advective flux of the substance could not be made with any greater accuracy than indicated by these examples.

For averaging periods less than the 20-day interval attained in the study described here, even larger uncertainties than those indicated above are likely to occur. The time record of the tidal mean velocity for any of the 26 current meters in the described study showed variations about the 20-day mean value of the same order as the amplitude of the tidal velocity. Even though the flow distribution as indicated by the 20-day mean values was characteristic of the classical two-layered estuarine flow pattern, this pattern was evident for only about half of the individual tidal cycle averages. Change in the pattern of flow was relatively rapid, with the tidal mean flow during one 24-hour period being directed down the estuary at all depths, to be replaced during the following 24-hour period with tidal mean flow over the entire cross-section directed up the estuary, only to return to down-estuary flow over the entire cross-section during the following 24-hour period. Two layered flow occurred prior to and following this period of rapid fluctuation in the non-tidal flows. As shown in the referenced study, these changes were highly correlated with changes in wind speed and direction over the bay.

Most of the discussion so far has dealt with the advective flux of material, and there has been little treatment of the other important physical process, that of turbulent diffusion. Fluorescent dye tracer studies can be used to characterize small and intermediate scale diffusion processes in estuaries. However, it is doubtful that direct measurements can provide adequate estimates of the diffusive flux of nutrients.

In view of the questions raised above concerning the direct measurements of both the advective and the diffusive flux dissolved or suspended substances in estuaries, how then, can we hope to quantify the advective and diffusive flux of nutrients in estuaries? One possible approach is through the use of numerical models of the hydrodynamics and kinematics of estuaries.

NUMERICAL MODELING OF THE ADVECTIVE AND DIFFUSIVE PROCESSES IN ESTUARIES

It is not our intent to discuss in detail the set of hydrodynamic equations for which solutions are sought by numerical simulation in numerical models of estuaries, nor to describe the numerical methods of solution. We do want to describe briefly the potential use of numerical modeling, and to indicate the present status of numerical modeling.

The equation of mass (and volume) continuity, the momentum equations, and the advection-diffusion equations for salt and for other conservative components can all be expressed in a format suitable for numerical solution. Conceptually, it should be possible - given the time varying boundary conditions of fresh water inflow into the estuary, water surface elevation and bottom salinity at the mouth the estuary, and wind stress on the surface of the estuary - to obtain as output from the numerical simulation the distribution of velocity and salinity in time and space within the estuary. Fully three-dimensional, transient state numerical models of estuaries have been described. However, full implementation and application to any real estuary has not been accomplished, to our knowledge. There are two primary reasons for this. One involves cost. Stability constraints require that the greater the spatial resolution desired, the shorter must be the time step for computation. The practical consequence of this requirement is high cost of running these models for any significant simulated time interval. These costs will ultimately decrease, both because of the development of more efficient techniques for numerical solution (for example, the use of implicit rather than explicit methods), and because computer costs for doing a particular numerical operation, and for data storage, will certainly continue to decrease.

The second reason that multi-layered, three-dimensional numerical models of estuaries have not been fully implemented is that there are certain terms in the equations which are not a priori determined by physical principles. Specifically, these are bottom frictional coefficients, coefficients of eddy viscosity, and coefficients of eddy diffusion. In some formulations of the eddy stress terms and eddy diffusion terms, other names are given to the various coefficients that arise, but there are still undetermined parameters. In some cases these parameters may depend on the scale of temporal and spatial resolution in the model. In order to obtain values of these parameters applicable to the particular estuary and the particular model formulation, model output must be compared to an extensive set of actual data from the prototype estuary, and the coefficients adjusted by trial and error to attain the best simulation of the field data. Model output for another prototype time period should then be compared to actual data for that period, to verify that the process of model adjustment was not unique to the one data set.

Although existing data for some estuaries may be suitable for such adjustment and verification, in most cases a much more intensive and extensive program than had been undertaken in the estuary must be accomplished to provide the data necessary for model adjustment and verification. Such costly field program have not been carried out for any of the larger estuarine systems.

The point to be made, however, is that, while data of the type described earlier are generally not adequate for direct determination of the advective and diffusive flux of a dissolved or suspended substance, such data, if extended sufficiently in time and space, are adequate for purposes of adjustment and verification of numerical models. The reason for this is that the appropriate values of the various coefficients are determined by adjustment of these coefficients so that the model reproduces the amplitude and phase of the tidal variations in velocity and salinity, and also the spatial variations in the non-tidal velocity and salinity distributions to within the degree of accuracy of the data. The values of the coefficients are not allowed to vary abruptly in time or space. With an adequate data base there is sufficient redundancy such that uncertainty of the data discussed earlier in the evaluation of the direct computation of advective flux do not significantly affect the deduced values of the coefficients.

Once adjusted and verified, the model is constrained, by its very formulation, to satisfy continuity. Anomalies of the type discussed earlier, where computations from velocity measurements of the flux of water through two adjacent cross-sections could not be made to balance with the observed time rate of changes in volume of the segment of the estuary between the two sections, cannot occur with a properly formulated model.

Therefore continued development and improvement of three-dimensional models should be encouraged, as should the collection of data suitable for model adjustment and verification. In the meantime, much can be learned from models which have been simplified by averaging over one or more spatial dimensions. In relatively strong estuaries such as the Chesapeake Bay and its tributary estuaries, laterally averaged two-dimensional models which treat the lateral and vertical variation in the time dependent distribution of velocity and of a water borne conservative property can describe the major advective and diffusive processes which affect the distribution of nutrients. In shallow estuaries with relatively low fresh water inflow and strong tidal or wind induced mixing, such that the waterway is very nearly homogeneous with depth, a vertically averaged two-dimensional model is most appropriate. These models now exist and are being exercised in estuaries.

Another advantage of the use of numerical modeling of the physical hydrodynamics and kinematics of estuaries is that the advection-diffusion equations for a dissolved substance can be generalized to include such nonconservative processes as absorption and desorption onto and from particles, settling of particles, and chemical and biological sources and sinks. As more is learned of these processes so that they can be simulated mathematically and so that the various exchange coefficients and rate constants can be determined, a complete simulation of the fate of nutrients in estuaries is conceptually possible. At present, however, we probably know more about the manner in which the physical processes of movement and dispersion affect the levels of nutrients in estuaries than we do about the non-conservative biological, chemical and geological processes which in many cases are of equal or greater importance.

GEOLOGICAL PROCESSES AFFECTING NUTRIENT LEVELS IN ESTUARIES

Fine-grained sediments are an important reservoir for nutrients in many estuaries, acting at various times as both source and sink. Absorption of nutrients onto suspended particulate matter and agglomeration of inorganic particles with phytoplankton by filter-feeding zooplankton are two mechanisms that accelerate the incorporation of nutrients into bottom sediments. Once deposited, nutrients may be permanently lost from the system through burial or may be regenerated and released by a variety of processes. Regeneration is controlled primarily by microbially-mediated degradation of organic materials through a series of aerobic and anaerobic respiration processes. The release of these regenerated nutrients to the overlying waters is controlled by a combination of biological processes, particularly bioturbation, and physical processes, such as irrigation, molecular diffusion, and turbulent diffusion and mixing.

Regeneration of dissolved nitrogen and phosphorus from sedimentary organic materials is greatest below the sediment surface, where anaerobic processes become important. As a result of the fact that new sediment is continually being added to the surface, together with the fact that there is a finite amount of readily degradable organic matter in any sediment layer, a maximum in the concentration of regenerated inorganic dissolved nutrients in the interstitial waters will occur quite close to but below the sediment/water interface. There will therefore be a gradient of decreasing concentration upwards and downwards from this maximum. Since the concentration in interstitial waters rapidly builds up to levels well above those occurring in the overlying waters, the upward gradient of decreasing concentration is usually quite large. A relatively large decrease with depth in concentration of inorganic nutrients may also sometimes occur below the maximum. However, the effects of compaction tend to restrict the rate of downward diffusion as compared to upward diffusion.

In undisturbed, non-bioturbated sediments this vertical transport can be described by a modification of Fick's first law of diffusion (1). That is

$$J = \phi D_{sed} (\partial c/\partial z)_{z=0}$$

where ϕ is the porosity (expressed as cm^3 of pore water per cm^3 of wet sediment), D_{sed} is the sediment diffusivity ($cm^2 \cdot s^{-1}$) and $(\partial c/\partial z)_{z=0}$ is the concentration gradient in the interstitial waters at the sediment water interface (expressed as moles per cm^3 of pore water per cm of depth). In some bioturbated system, D_{sed} should be replaced with a pseudo eddy diffusivity because the sediments are so intensely reworked that irrigation of animal burrows effectively purges the sediments of accumulated nutrients.

In some restricted environments, gases, such as methane and hydrogen, are produced as degradation by-products. These gases can sometimes reach concentration levels sufficiently above saturation as to form bubbles in the interstitial waters which sweep upwards and effectively stir these waters. This process may also purge nutrients from the sediment.

Alternate resuspension and deposition of sediment as a result of the variation in current speed over the tidal cycle, or as a result of water motions caused by surface wind waves, can also substantially increase the flux of regenerated nutrients from bottom sediments into the overlying waters. In many estuaries characterized by fine-grained sediment, such as the Chesapeake Bay, tidal scour reworks the fine-grained sediments to a depth of at least several millimeters every ebb and flood. This action breaks up agglomerated sediments, increasing the surface area available for attack by bacteria, and purges the sediment-water transition zone of any accumulated dissolved nutrients.

The processes of regeneration and release of nutrients from sediments can be important in the overall nutrient budget of an estuary. In the Chesapeake Bay, for example, Taft et al. (8) estimate that nitrogen is released from the sediments at an average annual rate of about 100×10^6 micromoles per second, with nearly all of the released nitrogen being in the form of ammonia during the summer, and approximately one-half ammonia and one-half nitrate at other times of the year. This input is a proximate source, and is several times larger than the inputs from two of the principal ultimate sources of nitrogen to the bay. The Susquehanna River inflow is the source of an annual average input of 28×10^6 micromoles per second, while municipal sewage discharge from Baltimore provides some 10×10^6 micromoles per second to the upper Chesapeake Bay. According to Taft (personal communication, 1979), the sediments are not only the major reservoirs for nutrients in the Chesapeake Bay, but they are also, over the year, the principal source of nutrients to the water column.

REFERENCES

1. Berner, Robert A. 1971. Principles of chemical sedimentology. International Series in the Earth and Planetary Sciences. McGraw-Hill Book Company, New York.
2. Elliott, A.J. 1978. Observations of meterological induced circulation in the Potomac Estuary. Estuarine and Coastal Marine Sciences, 6:285-299.
3. Elliott, A.J., and Dong-Ping Wang. 1978. The effect of meteorological forcing on the Chesapeake Bay: The coupling between an estuarine system and its adjacent coastal waters. *In* Hydrodynamics of Estuaries and Fjords, edited by J.C.J. Nihoul. Elsevier Scientific Publishing Company, Amsterdam.
4. Elliott, A.J., D.P. Wang and D.W. Pritchard. 1978. The circulation near the head of the Chesapeake Bay. Jour. of Mar. Res., Vol 36, No. 4, pp. 643-655.
5. Ianuiello, John P. 1977. Tidally induced residual currents in estuaries of constant breadth and depth. J. of Mar. Res., Vol. 35, No. 4, 1977, pp. 755-786.
6. Longuet-Higgins, M.S. 1969. On the transport of mass by time-varying ocean currents. Deep-Sea Res., 16, pp. 431-447.
7. Pritchard, D.W. and Stephen R. Rives. 1979. Physical hydrography and dispersion in a segment of the Chesapeake Bay adjacent to the Calvert Cliffs Nuclear Power Plant. Chesapeake Bay Institute, Special Report 74, August 1979.
8. Taft, J.L., A.J. Elliott, and W.R. Taylor. 1978. Box Model analysis of Chesapeake Bay ammonium and nitrate fluxes. *In* Estuarine Interactions (p. 115-130) M.L. Wiley (ed.), Academic Press, New York, 603 pp.
9. Tee, K.T. 1976. Tide-induced residual current, a 2-D nonlinear numerical tidal model. J. of Mar. Res., Vol. 34, No. 4, pp. 603-628.
10. Wang, Don-Ping, and Alan J. Elliott. 1978. Non-tidal variability in the Chesapeake Bay and Potomac River: Evidence for non-local forcing. Journal of Physical Oceanography, Vol. 8, No. 2.

STUDIES OF EUTROPHICATION IN LAKES AND THEIR RELEVANCE TO THE ESTUARINE ENVIRONMENT

D.W. Schindler
Experimental Lakes Project
Canadian Dept. of Fisheries and Oceans
Freshwater Institute
501 University Crescent
Winnipeg, Canada
R3T 2N6

In the past ten years, there has been little interaction between scientists studying the eutrophication of freshwaters and those studying the eutrophication of estuaries. One reason for this became apparent when I began a literature search prior to writing this paper. A large amount of estuarine science is tucked away in limited-circulation gray reports and symposium volumes, which I could access only with difficulty and considerable time delay, if at all. The situation is probably similar for estuarine scientists, for limnologists, too, have been inclined to hide much excellent work on eutrophication in limited-circulation documents. Once I discovered this fact, the most fruitful course appeared to be to review the "high spots" of recent work on the eutrophication of lakes, particularly aspects which might be useful to estuarine ecologists.

RECENT PROGRESS IN THE STUDY OF THE EUTROPHICATION OF FRESHWATERS

At the eutrophication symposium in Madison in 1968 (National Academy of Sciences 1969), a very gloomy picture of the eutrophication problem emerged. Water bodies were seriously affected in populous areas throughout Europe and North America. Control of the problem in surviving water bodies was expected to cost billions. Little hope was expressed of the recovery of lakes, because everyone "knew" that once hypolimnetic anoxia developed, phosphorus in sediments was able to escape the control of ferric iron to be recycled in the lake each year -

perhaps forever. A multi-nutrient approach to controlling the problem was taken, and review papers were given at the symposium on the roles of trace elements, alkaline earths, etc., as potential controls on eutrophication. No quantitative predictions for eutrophication were expressed.

The first bold step in quantifying the eutrophication process actually took place the same year at the NAS symposium. In a several hundred page report to the Organization for Economic Cooperation and Development, Paris, Dr. R.A. Vollenweider (33) reviewed the possible roles of several critical nutrients, including studies of processes ranging from physiology to geochemistry. Two conclusions from this report were to pave the way for freshwater management guidelines and eutrophication studies for the next decade. The first was to deduce from a number of lines of evidence that phosphorus was the nutrient most often to blame for eutrophication problems. The second was to concentrate on *phosphorus supply* rather than *phosphorus concentration* as the key to eutrophication control. Vollenweider described lucidly how man-made alterations to geochemical cycles in a lake's watershed could cause serious eutrophication problems. He perceived that lakes were not microcosms, but that events occurring in the entire watershed of a lake could dramatically affect the lake's productivity, a concept which has grown very popular in recent years. Vollenweider's simple plot of phosphorus "loading" versus mean depth (Figure 1) was quickly grasped by managers and incorporated in lake management schemes around the globe. Within a few years, great effort was devoted to controlling phosphorus inputs, usually by placing restrictions on detergent chemistry or by removing phosphorus from waste waters before discharging them into lakes.

Convenient as it was, the early Vollenweider model had serious flaws. The words "eutrophic", "mesotrophic" and "oligotrophic" appeared as three magical bands on the graph, without any explanation of what the bands represented. Worry was expressed by management agencies, lest "borderline" lakes in their jurisdiction cross the magical line into the never-never land of eutrophic lakes. In many cases, increasing the phosphorus input by only a few percent would accomplish this. The spectre of irrecoverability was still in peoples' minds.

But not everyone was willing to place all his faith in this early model. A few people grumbled that some lakes with high phosphorus loadings, which would appear in the eutrophic portion of Figure 1, were really quite clear - similar to lakes in the oligotrophic sector. Something was obviously lacking.

In 1969, Vollenweider changed his model to incorporate water renewal as a predictive parameter using simple chemostat theory. Dillon (5) tested this version of the model in a number of lakes of southern Ontario with a wide variety of flushing rates and phosphorus loadings, confirming that it fit very well. Lakes which remained oligotrophic in character despite high phosphorus loadings were found to have high water renewal rates (6). In

FIGURE 1. One version of Vollenweider's first model for the prediction of eutrophication from phosphorus input. Taken from Volume I of a report to the International Joint Commission in 1969 entitled "Pollution of Lake Erie, Lake Ontario and the International Section of the St. Lawrence River" by the International Lake Erie and Lake Ontario-St. Lawrence River Water Pollution Boards.

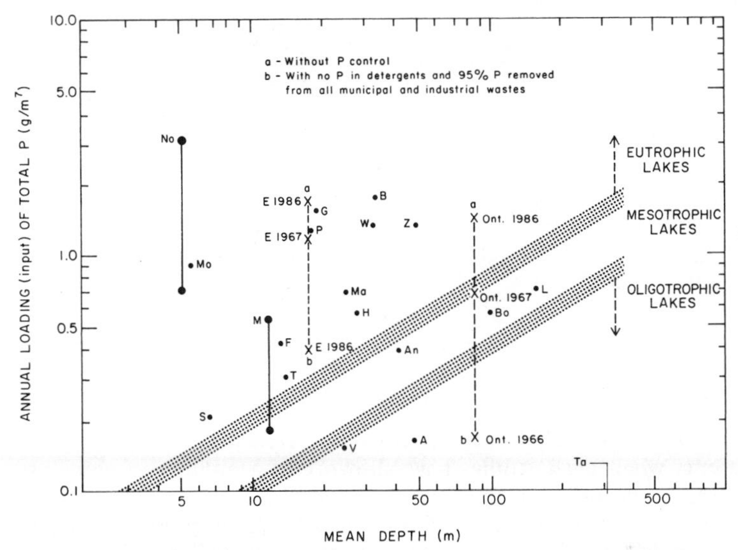

LEGEND: A - Aegrisee (Switzerland), An - Lake Annecy (France), B - Baldeggersee (Switzerland), Bo - Lake Constance (Austria, Germany, Switzerland), F - Lake Fureso (Denmark), G - Greifensee (Switzerland), H - Hallwillersee (Switzerland), L - Lake Geneva (France, Switzerland), M - Lake Mendota (U.S.A.), Ma - Lake Malaren (Sweden), Mo - Moses Lake (U.S.A.), No - Lake Norrviken (Sweden), P - Pfaffikersee (Switzerland), S - Lake Sebasticook (U.S.A.), V - Lake Vanern (Sweden), W - Lake Washington (U.S.A.), Z - Zurichsee (Switzerland), E - Lake Erie, Ont. - Lake Ontario.

later refinements of the model, Vollenweider (35, 36) was able to express eutrophication quantitatively, in μg l^{-1} of chlorophyll, as a function of phosphorus input and water renewal (Figure 2). Many slight variations of this simple model have been constructed (20, 30). Overall, this model has proven to be of enormous utility. Its predictions show that the relationship between nutrient input and phytoplankton response is a continuous one, with no sharp changes in slope. There is no "break point" in the curve above which lakes deteriorated rapidly.

FIGURE 2. The relationship between phytoplankton standing crop as chlorophyll *a* and phosphorus loading corrected for water renewal, from Vollenweider (36). Lp is the phosphorus input in g $M^{-2}d^{-1}$, $\overline{z}$ is the mean depth in M and qs is the outflow per unit of lake surface in M a^{-1}.

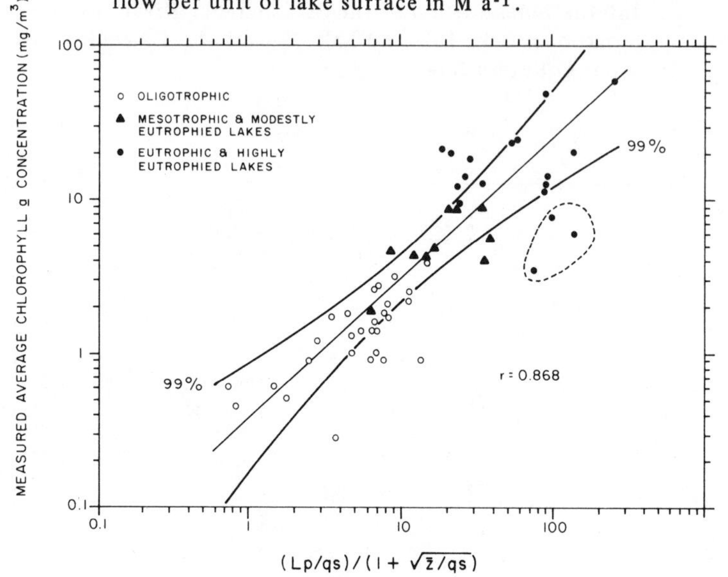

Why was it possible to predict eutrophication with models based on one nutrient, when algal studies have shown that a number of elements may limit algal growth? In this respect it is worthwhile examining possible reasons for the excellent relationship between chlorophyll and phosphorus which holds with very few exception in freshwater (Figure 3).

I believe that we can dismiss the whole suite of so-called trace elements as being of little importance to quantitative predictions of eutrophication - although there may be rare exceptions. The study of trace element deficiencies for algae requires scrupulous laboratory studies. A single fingerprint on the inside of a culture vessel can alter the course of an experiment. Lakes and their watersheds are very dirty glassware, and except for a few cases where bizarre geochemical anomalies occur, prolonged trace element deficiencies are unlikely to occur in lakes.

Even the macronutrients carbon and nitrogen proved to be of little quantitative importance as causal factors. Schindler (24, 25) showed that without phosphorus, these elements had no effect on eutrophication. On the other hand, if phosphorus were supplied in excess of algal demands relative to nitrogen and carbon, algae would call upon atmospheric and sedimentary carbon and nitrogen sources to balance their nutrient accounts, fixing enough to the elements to produce the proportions of nutrients which algae prefer (known as Redfield numbers to estuarine and marine workers) (27). The net effect of these interactions is to keep the growth of freshwater phytoplankton in most lakes proportional to the supply of phosphorus. By the early 1970s, it was obvious that studies of productivity and standing crop without regard for other nutrients were relatively meaningless. The era of "Lindemania," when productivity was measured for its own sake, began to wane under the onslaught of eutrophicationists. It was obvious that carbon was not the "dynamic" element which drove the productivity of lakes, but a rather "parasitic" element - with tens of carbon atoms frantically tagging each phosphorus atom like so many teen-age groupies after a rock and roll star.

Laboratory studies to detect the "limiting nutrient" have been shown to be of little use in developing nutrient control strategies for lakes. Such studies conducted in Lake 227 showed that the lake was chronically carbon-limited throughout the summer months, for a period of several years (28, 29). Yet years of observation have shown that all during this period the lake was in the process of slowly overcoming its carbon deficit by drawing and storing atmospheric CO_2, until finally, in 1975, after six years of fertilization, phytoplankton were no longer carbon limited (26). It is, of course, futile to tailor costly management strategies to the relatively short period of nonequilibrium when nutrient incomes to a water body are changed, when what is desired are control measures which will be effective for long periods.

Some preliminary steps have also been made in understanding what controls algal species composition. In the above studies, a low nitrogen to phosphorus ratio was shown to trigger the onset of blooms of nitrogen-fixing bluegreens, presumably because of the competitive advantage which these would have over other algae in balancing N:P requirements (27). Small as this step may seem, it allows us a mechanism to control the algal group which most objectionable in most freshwaters. Since this discovery, the converse, increasing nitrogen inputs to inhibit blooms of nitrogen-fixing bluegreens, while maintaining high productivity, has also been used to advantage (J. Barica, in prep.).

1. Lake 227 is a small (area 5 ha, mean depth 4.4 m) softwater lake in the Precambrian Shield in the Experimental lakes area of northwestern Ontario, the watershed of the lake, which is uninhabitated, is covered by virgin stands of jackpin and blackspruce. The lake has been fertilized with known doses of phosphate and nitrate since June, 1969. The original concentration of dissolved inorganic carbon (DIC) in the epilimnion in summer was less than 50 $\mu M\ l^{-1}$. In the first few years of fertilization, algal photosynthesis depleted epilimnetic DIC concentrations to nearly undetectable levels ($<5\ \mu M\ l^{-1}$), causing high rates of invasion of CO_2 from the atmosphere, and evidence of severe carbon limitation at midday. From midsummer 1974 onward, epilimnetic DIC concentrations were usually 150-200 $\mu M\ l^{-1}$, and no carbon limitation of photosynthesis was observed.

FIGURE 3. The relationship between chlorophyll concentration in mg M^{-3} (ordinate) and total phosphorus concentration in Japanese lakes and ponds. From Sakamoto (22).

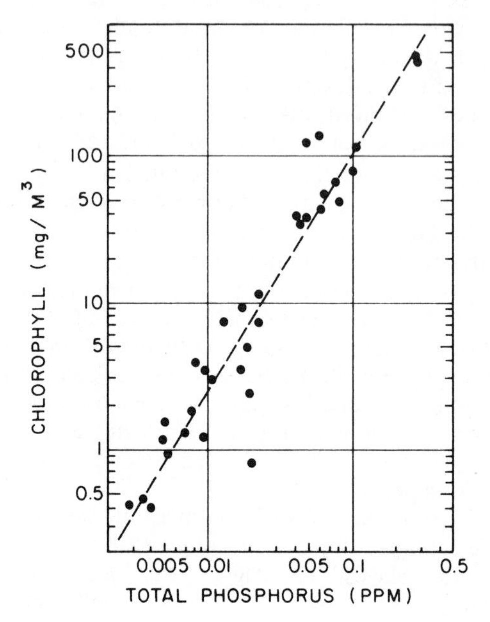

The rapid depletion of silica in summer due to increased phosphorus loading has also been shown to favor algal shifts - in this case from diatoms to bluegreens or greens which are less valuable to the planktonic food chain (23). In a second major step, Kilham and Kilham (13) have been able to predict the species of algae which will succeed one another as silica is progressively depleted.

Several detailed studies of lake recovery from eutrophication have now been completed. Two spectacularly successful recoveries have been Lake Washington in the United States (8) and Lake Norrviken in Sweden (1). In both cases, after sources of phosphorus were eliminated, depletion of the element in the lake occurred at a rate which was little different from that predicted by water renewal times (Figure 4). Why weren't lakes irrecoverable, due to return of phosphorus from anoxic sediments, as previously believed? Ahlgren (1) showed that the return of phosphorus from sediments decreases rapidly after external sources of the element are eliminated (Figure 5). Controlling external sources appears to be the key to reducing internal recycling of phosphorus. It is reasonable that the return of phosphorus from sediments must, at steady state, reflect the amount of surface loading. Unfortunately, new steady-states may develop only after long time-lags in some instances, for example when sediment

FIGURE 4. The total phosphorus concentration in Lake Norrviken, Sweden, after diversion of sewage (solid line). Annual fluctuations correspond to the onset and disappearance of hypolimnetic anoxia. Dotted and dashed lines are predicted calculations based on water renewal, as assessed by the conservative ion potassium. From Ahlgren (1).

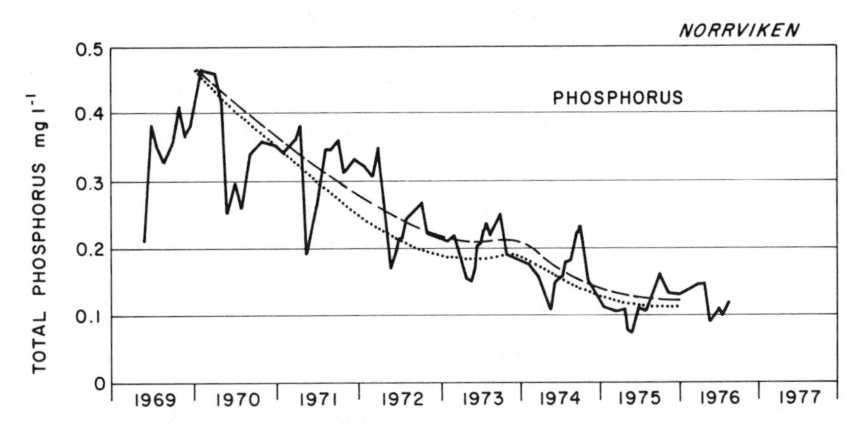

mixing and long periods of high nutrient input have enriched sediments with phosphorus to considerable depths. In some cases, the time lags in recovery may be unacceptably long, as in the case of Lake Trummen (2), and Shagawa Lake (14), so that more drastic methods must be employed to rehabilitate a lake.

FIGURE 5. Rate of release of phosphorus from the sediments of Lake Norrviken following diversion of sewage in 1969. From Ahlgren (1).

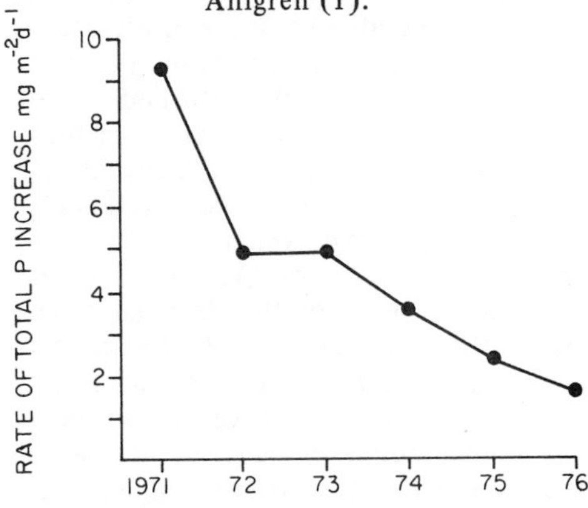

Eutrophication study in freshwater has turned into a phosphorus bandwagon. Oblivious to the large amount of variance about the phosphorus inputs-chlorophyll relationship, whatever its form, investigators are repeating studies of this same general model - it's the "in" thing to do. The situation is reminiscent of that which occurred in molecular biology following the cracking of the genetic code. My guess is that the existing model cannot be further refined without incorporation of other factors - for example, the effects of grazing or regeneration of nutrients from sediments by organisms (31).

A COMPARISON OF FRESHWATER AND ESTUARINE EUTROPHICATION

The control of eutrophication in estuaries has appeared to be much more complex than it is in lakes. Supplies of both nitrogen and phosphorus to many estuaries are far in excess of the demands of phytoplankton. Indeed, elements like phosphorus, nitrogen and silicon, which are extremely reactive in most freshwaters, often behave almost as conservatively as chloride in estuaries which have been overwhelmed with anthropogenic wastes.

Also, past studies have shown that nitrogen limitation is prevalent in major estuaries of eastern North America, if nutrient limitation is significant at all (21). Because background data do not exist for these waters, it is impossible to tell whether nitrogen limitation has developed as the result of the high ratio of P relative to N in anthropogenic pollutants, or whether nitrogen was limiting under near-pristine conditions as well. One might speculate that at least the less saline reaches of most estuaries were originally phosphorus limited, because the N:P ratio in unpolluted river water is usually high, as it is in oligotrophic lakes.

The "evolution" of optimal nutrient ratios for phytoplankton growth, as discussed above in freshwater, may not be able to occur in estuaries, due to their short water residence times and domination by physical processes. Nitrogen fixed from the atmosphere or returned from sediments is swept into the opent ocean so rapidly that it cannot accumulate to the degree which is common in freshwater lakes. While nitrogen fixation by estuarine plankton and marshlands does occur (15, 11, 7), no one appears to have assessed the long-term contribution of such processes to estuarine nutrient budgets. In light of the above, it is surprising, although heartening, that the recovery of the Potomac from eutrophication appears to follow the predictions of freshwater phosphorus-based models reasonably well (papers in this symposium by Jaworski and by Lee and Jones). Further study will be necessary to show whether this agreement is purely coincidental, or whether we have misinterpreted something fundamental in estuarine nutrient dynamics, as the importance of atmosphere-water exchange was misgauged in lakes.

Detailed mass-balance studies of nutrient cycling in estuaries, which might reveal such problems, are remarkably few. Indeed, only in Narragansett Bay, Rhode Island, have enough major processes been studied to allow some sort of coherent assessment of the relative contribution of various internal processes. For example, it appears that recycling of nitrogen from benthic organisms may be far more important in coastal waters than in freshwaters (S. Nixon, this symposium). When coupled with physical parameters, this study should allow prediction of the capacity of coastal waters for overcoming nitrogen deficiencies by internal processes.

The species composition of estuarine phytoplankton appears to respond to changes in nitrogen and silicon relative to phosphorus as they do in freshwater (12, 19, 16). There are some differences in detail. For example, in freshwaters which have been studied, seasonal silica cycles are dominated by thermal stratification and its effects on internal recycling processes. In contrast, the major source of silicon on an annual basis to estuaries appears to be an external one, the freshwater intflow. Although a slow replenishment of silicon from sediments occurs in at least some estuaries (32, 19), the process is not rapid enough to be a major source to the element.

In all the above problems, one is frustrated by the lack of "before" and "after" data. As in most eutrophied freshwaters, we have little idea of what baseline conditions really were, so that it is difficult to assess the extent to which a system can be rehabilitated. It appears that experimental enrichment of small, pristine estuaries might yield insights to estuarine processes of the same sort that experimental nutrient addition have for lakes. Because of the heavy settlement and usage of temperate east coast estuaries and their watersheds, it may be necessary to conduct such experiments in the subarctic or arctic. I believe that they would prove invaluable nevertheless..

Due to the apparent complexity of mixing processes in estuaries, the simple, one-element models which have proven useful in lakes are unlikely to be applicable, even where limitation by a single nutrient occurs. A recent extension of the Vollenweider-type model for the treatment of heavily-enriched embayments in the Great Lakes (4) is probably closer to what is required. Chapra points out that extension of the model to segments of an enriched embayment (or estuary) is easily done. In fact, similar models have been used to calculate mixing parameters from chloride concentrations in estuaries.

Finally, there are few studies of recovery and rehabilitation of eutrophic estuaries. The Thames estuary in England appears to have undergone a remarkable recovery (9, 10). Papers in this symposium appear to indicate that the Potomac, too, is improving (Jaworski; Lee and Jones; papers in this volume). When one views the enormous amounts of nutrients entering many eastern estuaries, amounts large enough to render primary nutrient elements essentially conservative, he is impressed by the enormous cost of controlling effluents sufficiently to have any effect on phytoplankton standing crop and productivity. In many cases, algal uptake of nutrients appears to be hindered by high turbidity, a situation which occurs only rarely in freshwaters (3). As Gross (10) has pointed out, the time scale between initiation and implementation of a major cleanup for a large estuary is likely to be decades, rather than years. It seems likely that a useful initial approach would be to undertake a few pilot-scale studies of small estuaries, hoping to obtain recovery results which will be useful in developing schemes for larger systems. It would be interesting to compare the results of controlling both turbidity and nutrient inputs with those controlling nutrients alone. In freshwaters, the rapid recoveries of Lake Washington, Lake Norrviken and a number of smaller water bodies from eutrophication have kept hopes alive for the recovery of the St. Lawrence Great Lakes (8, 1, 17, 25). Recovery of the latter has been delayed by their long water-residence times, and by delays in implementation of nutrient control schemes.

Only recently is there some evidence that Lake Ontario is responding to these massive cleanup efforts (S. Chapra and H. Dobson, unpublished manuscript). I suspect that estuarine managers must be equally patient. Regardless of its salinity, a water body's response to release from decades of insults cannot be instantaneous.

REFERENCES

1. Ahlgren, I. 1977. Role of sediments in the process of recovery of a eutrophicated lake. *In* H.L. Golterman (ed.). Interactions between sediments and freshwater: Proceedings of an international symposium, Amsterdam, 1976. Junk, The Hague, PUDOC, Wageningen, The Netherlands.
2. Bjork, S. 1972. Swedish lake restoration program gets results. Ambio. 1:153-165.
3. Brunskill, G. J., D.W. Schindler, S.K. Holmgren, H. Kling, P. Campbell, M.P. Stainton, F.A.J. Armstrong and B.W. Graham. 1979. Nutrients, chlorophyll, phytoplankton and primary production in Lake Winnipeg. Can. Fish. Mar. Serv. Tech. Rep. (in prep.).
4. Chapra, S.C. 1979. Applying phosphorus loading models to embayments. Limnol. Oceanogr. 24:163-168.
5. Dillon, P.J. 1974. The prediction of phosphorus and chlorophyll concentrations in lakes. Ph.D. thesis, Univ. Toronto. 328 p.
6. ——————. 1975. The phosphorus budget of Cameron Lake, Ontario: The importance of flushing rate to the degree of eutrophy of lakes. Limnol. Oceanogr. 20:28-39.
7. Dunstan, W.M. and L.P. Atkinson. 1976. Sources of new nitrogen for the South Atlantic Bight. p. 69-78 in M. Wiley (ed.). Estuarine Processes, V.I. Academic Press, N.Y.
8. Edmondson, W.T. 1972. The present condition of Lake Washington. Verh. Internt. Verein. Limnol. 18:284-291.

9. Gameson, A.L.H., M.J. Barrett, and J.S. Shewbridge. 1973. The aerobic Thames Estuary. pp 843-850. *In* S.H. Jenkins (ed.). Advances in Water Pollution Research, Sixth International Congress, Jerusalem. June 8-23, 1972. Pergamon Press, New York.
10. Gross, M. Grant. 1976. Estuarine cleanup - can it work? pp. 3-14 in M. Wiley (ed.). Estuarine Processes, V.I. Academic Press, New York.
11. Haines, E., A. Chalmers, R. Hanson and B. Sherr. 1976. Nitrogen pools and fluxes in a Georgia Salt Marsh. 241-254. *In* M. Wiley (ed.). Estuarine Processes, V.2. Academic Press, New York and London.
12. Jaworski, N..A., D.W. Lear Jr., and O. Villa, Jr. 1972. Nutrient management in the Potomac Estuary. 246-273 *In* G.E. Likens (ed.). Nutrients and Eutrophication: The Limiting Nutrient Controversy. Amer. Soc. Limnol. Oceanogr. Special Symp. No. 1. Allen Press, Lawrence, Kansas.
13. Kilham, S.S., and P. Kilham. 1978. Natural community bioassays: Prediction of results based on nutrient physiology and competition. Verh. Internat. Verein. Limnol. 20:68-74.
14. Larsen, D.P., K.W. Malueg, D.W. Schults, and R.M. Brice. 1975. Response of eutrophic Shagawa Lake, Minnesota, U.S.A., to point-source phosphorus reduction. Verh. Internat. Verein. Limnol. 19:884-892.
15. Marsho, T.V., R.P. Burchard and R. Fleming. 1975. Nitrogen fixation in the Rhode River estuary of Chesapeake Bay. Can. J. Microbiol. 21:1348-1356.
16. McCarthy, J.J., W.R. Taylor and J.L. Taft. 1977. Nitrogenous nutrition of the plankton in the Chesapeake Bay. 1. Nutrient availability and phytoplankton preferences. Limnol. Oceanogr. 22:996-1011.
17. Michalski, M.F.P., K.H. Nicholls, and M.G. Johnson. 1975. Phosphorus removal and water quality improvements in Gravenhurst Bay, Ontario. Verh. Internat. Verein. Limnol. 19:644-659.
18. National Academy of Sciences. 1969. Eutrophication: Causes, consequences, correctives. Washington, D.C. 661 p.
19. Peterson, D.H., T.J. Conomos, W.W. Broenkow and E.P. Scrivani. 1975. Processes controlling the dissolved silica distribution in San Francisco Bay. pp. 153-187. *In* L.E. Cronin (ed.). Estuarine Research. Volume 1. Chemistry, Biology, and the Estuarine System. Academic Press, New York.
20. Rast, W. and G.F. Lee. 1978. Summary analysis of the North American (U.S. portion) OECD Eutrophication Project: Nutrient loading - lake response relationships and trophic state indices. U.S.E.P.A., EPA-600/3-78-008. Corvallis Experimental Research Laboratory, Corvallis, Ore. 454 pp.
21. Ryther, J.H., and W.M. Dunstan. 1971. Nitrogen, phosphorus, and eutrophication in the coastal marine environment. Science 171:1008-1013.
22. Sakamoto, M. 1966. Primary production by phytoplankton community in some Japanese lakes and its dependence on lake depth. Arch. Hydrobiol. 62(1):1-28.

23. Schelske, C. and E. Stoermer. 1971. Eutrophication, silica depletion and predicted changes in algal quality in Lake Michigan. Science 173:423-424.
24. Schindler, D.W. 1974. Eutrophication and recovery in experimental lakes: Implications for lake management. Science 184:897-899.
25. ——————. 1975. Whole-lake eutrophication experiments with phosphorus, nitrogen and carbon. Verh. Internat. Verein. Limnol. 19:3221-3231.
26. ——————. 1976. Biogeochemical evolution of phosphorus limitation in nutrient enriched lakes of the Precambrian Shield. 647-664. *In* J.O. Nriagu (ed.). Environmental Biogeochemistry. Vol. 2. Metals transfer and Ecological Mass Balances. Ann Arbor Science Publishers, Ann Arbor, Mich.
27. ——————. 1977. The evolution of phosphorus limitation in lakes. Science 195:260-262.
28. Schindler, D.W., F.A.J. Armstrong, S.K. Holmgren and G.J. Brunskill. 1971. Eutrophication of lake 227, Experimental Lakes Area, northwestern Ontario, by addition of phosphate and nitrate. J. Fish. Res. Board Can. 28:1763-1782.
29. Schindler, D.W., and E.J. Fee. 1973. Diurnal variation of dissolved inorganic carbon and its use in estimating primary production and CO_2 invasion in lake 227. J. Fish. Res. Board Can. 30:1501-1510.
30. Schindler, D.W., E.J. Fee, and T. Ruszczynski. 1978. Phosphorus input and it consequences for phytoplankton standing crop and production in the Experimental Lakes Area and in similar lakes. J. Fish. Res. Board Can. 35:190-196.
31. Shapiro, J., V. Lamarra and M. Lynch. 1975. Biomanipulation: An ecosystem approach to lake restoration. 85-96. *In* P.L. Brezonik and J.L. Fox (eds.). Proc. Symp. on Water Quality Management through Biological Control. ENV-07-75-1. Dept. of Environmental Engineering Science. Univ. of Florida, Gainesville.
32. Simpson, H.J., D.E. Hammond, B.L. Deck and S.C. Williams. 1975. Nutrient budgets in the Hudson River Estuary. 618-635 *In* Marine Chemistry in the Coastal Environment. Amer. Chem. Soc. Symposium Series, No. 18.
33. Vollenweider, R.A. 1968. Water management research. OECD, Paris. DAS/CSI/68.27.183 p.
34. ——————. 1969. Moglichkeiten and Grenzen elementarer Modelle des Stoffbilanz von Seen. Arch. Hydrobiol. 66:1-36.
35. ——————. 1975. Input-output models. Schweiz. Z. Hydrol. 37:53-84.
36. ——————. 1976. Advances in defining critical loading levels phosphorus in lake eutrophication. Mem. Ist. Ital. Idrobiol. 33:53-83.

SOURCES OF NUTRIENTS AND THE SCALE OF EUTROPHICATION PROBLEMS IN ESTUARIES

Norbert A. Jaworski
U.S. Environmental Protection Agency
Environmental Research Laboratory
Duluth, MN 55804

ABSTRACT: A comprehensive analysis of external sources of nutrients is presented including an impact, comparison of external loadings, and the resulting scale of eutrophication. The major emphasis of the analysis is on nitrogen and phosphorus.

The relative contribution of various external sources for five major ecosystems is delineated. Discussion of seasonal and long-term trends of external sources is presented. The impact of external sources on the eutrophication process is evaluated. A scale of eutrophication for estuarine ecosystems is suggested for comparing the impact of nutrient enrichment.

Comparisons of external nutrient loadings and the scale of eutrophication for 13 estuarine and freshwater ecosystems are made. A detailed comparison of the five estuaries of the Chesapeake Bay is presented from which relationships between external nutrient loadings ($g/m^2/yr$) and eutrophic conditions are suggested.

Analysis of the relationship between external nutrient loadings, nitrogen/phosphorus ratio, and eutrophic state suggests that, for the East Coast estuarine ecosystems, if the phosphorus loading is 1.0 $g/m^2/yr$ or less, excessive eutrophic conditions can be prevented. The favorable response of the Potomac estuary to phosphorus control demonstrates that excessive eutrophic conditions may be alleviated with advanced wastewater treatment, depending on many factors including nutrient loadings.

INTRODUCTION

Nutrient budgets are necessary in understanding and in aiding control of the eutrophication process. Most nutrient budgets have been developed for total phosphorus and total nitrogen. Recently, analytical methods have been improved and their cost reduced, thus facilitating the addition of other forms of nitrogen and phosphorus in routine field studies. Many of the budgets focus on external sources, but the internal sources and sinks within a given water body have also been considered.

This paper focuses on external sources of nutrients (particularly nitrogen and phosphorus), their impact and the comparison of external loadings, and the resulting scale of eutrophication. Nutrient sources of five major ecosystems are discussed. Seasonal and long-term trends are examined for Chesapeake Bay. The impact of eutrophication is evaluated and a scale of eutrophication for estuarine ecosystems is proposed. Comparisons of external nutrient loadings and the scale of eutrophication for 13 select estuarine systems and freshwater lakes are made. A more detailed analysis is presented for the Chesapeake Bay and its five major estuaries.

EXTERNAL NUTRIENT BUDGETS FOR FIVE MAJOR ECOSYSTEMS

External Budget for the Hudson River Basin

For the Hudson River basin, the external loadings of phosphorus and nitrogen have been combined into two sources, municipal/industrial and land runoff as shown in Table 1 (29, 44). The major external loading of

TABLE 1. Annual External Nutrient Budget - Hudson River Basin - (1968).

Source	Phosphorus (kg/day)	Nitrogen (kg/day)
Municipal/Industrial	8,000	36,000
Land Runoff	3,000	21,000
Total	11,000	57,000

phosphorus was from municipal and industrial sources, which contributed about 73 percent of the total. Likewise, for nitrogen the major source in the external budget was municipal and industrial sources with a contribution of 63 percent of the total.

External Nutrient Budget for San Joaquin Delta

The external nutrient budget for the San Joaquin delta which flows into San Francisco Bay is shown in Table 2 (3). The major source of phosphorus was municipal wastewater discharges, about 60 percent of the total input. For nitrogen, the major source was industrial discharges, about 47 percent of the total. While land runoff currently is not a major

TABLE 2. Annual External Nutrient Budget - San Joaquin Delta - (1968).

Source	Phosphorus (kg/day)	Nitrogen (kg/day)
Municipal	4,000	11,000
Industrial	200	15,000
Land Runoff	2,500	6,000
Total	6,700	32,000

part of the total external nutrient budget, it is anticipated that by the year 2000, nitrogen loading from this source will be the largest nitrogen contribution to the delta, primarily from irrigation return systems. A plan to manage agricultural drainage is currently being developed (47).

External Nutrient Budget for Lake Superior

Although Lake Superior is not an estuarine system, the nutrient budget shown in Table 3 introduces an interesting aspect of atmospheric nutrient

TABLE 3. Annual External Nutrient Budget - Lake Superior - (1975).

Source	Phosphorus (kg/day)	Nitrogen (kg/day)
Municipal/Industrial	500	3,000
Land Runoff	6,300	100,000
Air	3,000	153,000
Shoreline Erosion	700	4,000
Total	10,500	260,000

input (19). The major source of phosphorus was land runoff, with a contribution of approximately 60 percent. The second largest phosphorus source was the air. For nitrogen, however, the major contribution (59 percent) was from the air. Land runoff was the second largest contributor, providing 38 percent of the total nitrogen.

External Nutrient Budget for the Potomac Estuary

Table 4 shows the contribution of nutrients to the Potomac estuary from the upper Potomac basin and from the wastewater discharges of the Washington, D.C. metropolitan area (27, 28). As can be seen in Table 4, the major contribution of total phosphorus (84 percent) was from

TABLE 4. Annual External Nutrient Budget - Potomac Estuary (1969-1970).

Parameter	Upper Potomac River Basin (kg/day)	Wastewater Discharges (kg/day)	Total (kg/day)
Total Phosphorus	2,080	11,300	13,380
Inorganic Phosphorus	1,200	8,600	9,800
Organic Phosphorus	880	2,700	3,580
Total Nitrogen	26,880	27,300	54,180
NH_3 Nitrogen	2,080	24,100	26,180
NO_2 and NO_3 Nitrogen	16,700	500	17,200
Organic Nitrogen	8,100	2,700	10,800
Total Organic Carbon	182,000	36,400	218,400
Total Carbon	795,000	102,300	897,300

wastewater discharges. The major component of total phosphorus was in the inorganic form. The total nitrogen input to the estuary was 54,180 kg/day in 1969-1970, of which approximately 50 percent was from wastewater discharges in the Washington area and the remainder from the upper Potomac basin, which includes both point and nonpoint sources. The major form of nitrogen from wastewater discharges was ammonia, while the predominant form from the upper Potomac basin was nitrate. Table 4 also shows the total organic carbon and total carbon loadings to the estuary. The major contribution of carbon, both inorganic and organic, was from the upper basin.

External Nutrient Budget for Chesapeake Bay

The external nutrient budget for Chesapeake Bay, as shown in Table 5, is presented in two systems, the entire Bay and the Bay proper (9, 12, 17, 27, 28, 41, 46). The first system includes the nutrient budget for the entire Bay, which includes the Patuxent, Potomac, Rappahannock, York, and James estuarine systems. The budget also is presented for only those discharges into the Bay proper, thus not including the five above-mentioned estuaries.

The major source of phosphorus entering the entire Chesapeake Bay was wastewater discharges within the tidal ecosystem, about 69 percent of the total. The major nitrogen contribution, approximately 68 percent, was from the upper basin land runoff. The term "upper basin land runoff" includes both point and nonpoint contributions. For the Bay proper, the major source of phosphorus (72 percent) was wastewater discharges. The largest nitrogen loading to the Bay proper was from runoff (71 percent).

TABLE 5. Annual Nutrient Budget - Chesapeake Bay - (1971).

Source	Phosphorus (kg/day)	Nitrogen (kg/day)
Entire Chesapeake Bay Including Estuaries		
Municipal/Industrial	28,700	87,700
Upper Basin Land Runoff	10,200	195,400
Air	2,500	14,800
Total	41,400	297,900
Chesapeake Bay Proper Excluding Estuaries		
Municipal/Industrial	16,900	45,900
Upper Basin Land Runoff	5,200	131,500
Air	1,400	8,200
Total	23,500	185,600

Discussion

For the five systems, the largest phosphorus and nitrogen sources are:

	Ecosystem Watershed Description	Largest Phosphorus Source	Largest Nitrogen Source
Hudson	Forested/ Industrial/ Urban	Wastewater Discharges (73%)	Wastewater Discharges (63%)
San Joaquin	Agricultural/ Industrial/ Urban	Wastewater Discharges (60%)	Wastewater Discharges (47%)
Lake Superior	Forested/ Rural	Land Runoff (60%)	Air (59%)
Potomac Estuary	Forested/ Urban	Wastewater Discharges (84%)	Wastewater/Land Runoff (50%) (50%)
Chesapeake Bay	Forested/ Industrial/ Urban	Wastewater Discharges (69%)	Land Runoff (68%)

As can be seen in the above tabulation, the major source of phosphorus or nitrogen can vary from ecosystem to ecosystem. Likewise, the relative percent contribution of a given source can also vary significantly depending on the geographical location, industrial/urban development, and agricultural practices.

SEASONAL AND LONG-TERM TRENDS OF EXTERNAL NUTRIENT INPUTS

Seasonal Nutrient Input Variations

For the upper Potomac basin, which is above the fall line near the nation's capital, nutrient inputs are presented in Tables 6 (1966) and 7 (1969-70) (26, 35, 48).

TABLE 6. Seasonal Nutrient Inputs - Upper Potomac River Basin (1966).

	Flow (m^3/s)	Phosphorus (kg/day)	Nitrogen (kg/day)
January	450	4,000	44,700
February	550	5,100	60,300
March	700	6,600	85,900
April	670	6,300	79,300
May	480	4,300	48,900
June	290	2,300	23,300
July	180	1,300	11,900
August	200	1,500	14,000
September	160	1,200	9,800
October	210	1,600	14,700
November	220	1,700	15,500
December	330	2,700	28,100
Average	370	3,200	36,400

To illustrate variations in seasonal contributions, the hydrologic flow pattern for 1966 (Table 6) was typical for the Potomac system with 1969-1970 (Table 7) being a period of low spring flows. As can be seen in Tables 6 and 7, the monthly contributions of phosphorus and nitrogen usually follow the monthly flow hydrograph with significant variations in monthly contribution of nutrients, such as for the month of February. For the upper Potomac River basin, streamflows are usually high in the spring months and low in the summer months. However, in comparing August,1966 and August, 1969, a considerably larger contribution of nutrients was observed in 1969.

A flow versus nutrient-input relationship has been developed as can be seen in Table 8 for the Susquehanna River basin above Conowingo Dam (8). When the flow is at 335 m^3/s, the phosphorus contribution is

TABLE 7. Seasonal Nutrient Inputs - Upper Potomac Basin (February 1969 - February 1970).

Month	Flow (m^3/s)	T. Phosphorus (kg/day)	Inorganic Phosphorus (kg/day)	TKN-Nitrogen (kg/day)	NH_3-Nitrogen (kg/day)	NO_2+ NO_3 (kg/day)	Total Nitrogen (kg/day)
Feb	220	2,040	950	10,100	2,150	10,500	20,600
Mar	280	2,500	1,500	12,000	2,400	20,600	32,600
Apr	230	2,150	1,000	10,300	2,300	9,900	20,200
May	130	1,250	400	5,300	1,700	3,600	8,900
Jun	70	700	200	3,500	950	1,300	4,800
Jul	80	800	250	3,850	1,000	2,250	6,100
Aug	270	2,400	1,250	12,000	2,400	14,500	26,500
Sep	145	1,300	500	6,700	1,550	4,600	11,300
Oct	75	800	200	3,750	1,050	1,450	5,200
Nov	120	1,150	400	5,400	1,400	3,200	8,600
Dec	265	2,900	1,350	11,800	2,350	17,900	29,700
Jan	395	3,550	2,550	17,700	3,150	3,950	21,650
Feb	715	7,000	5,250	30,300	4,850	87,250	117,550
Average	230	2,100	1,200	10,200	2,100	14,000	24,000

TABLE 8. Flow/Nutrient - Input Relationships - Susquehanna River - (1969-1970).

Flow (m^3/s)	Total Phosphorus (kg/day)	Inorganic Phosphorus (kg/day)	Inorg-P Total-P (%)	Total Nitrogen (kg/day)	Inorganic Nitrogen (kg/day)	Inorg-N Total-N (%)
335	1,100	500	45	36,400	26,400	73
1,670	7,000	4,150	59	181,000	134,600	74
3,350	16,500	10,500	64	363,600	272,800	75

estimated to be 1100 kg/day, with nitrogen 36,400 kg/day. However, when the flow is at 3,350 m^3/s, the estimated phosphorus contribution increases about 15-fold to 16,500 kg/day, and the nitrogen increases about 10 fold to 363,600 kg/day. The percentage of inorganic phosphorus increases substantially as the flow increases, while the percentage of inorganic nitrogen increases only slightly as the flow increases.

Long-term Nutrient-Input Variations

To demonstrate the annual variability in land runoff contributions of nutrients, data obtained from analysis of the raw water intake of the District of Columbia water treatment plant were utilized (27). The data in Table 9 show the annual variation of flow and of nitrate nitrogen input.

TABLE 9. Annual Variation in Nitrate Input - Potomac River Basin.

Year	Mean Flow (m^3/s)	Annual Nitrate Nitrogen Input (kg/day)
1960	360	20,000
1961	400	19,900
1962	355	16,400
1963	265	12,100
1964	310	15,900
1965	270	12,100
1966	225	17,400
1967	360	35,000

As can be seen in Table 9, there can be a significant yearly input of nitrate nitrogen which is the major form of nitrogen input into the estuary. The annual nitrate input increased over 2.0-fold during the 1966-1967 period, with a corresponding stream flow increase of 1.6-fold.

Long-term trends of nutrient contributions from wastewater discharges in the Washington, D.C. metropolitan area to the Potomac estuary are shown in Table 10 (27). In 1913, phosphorus and nitrogen loadings were

TABLE 10. Wastewater Loading Input Trends - Potomac Estuary (Washington, D.C., Metropolitan Area).

Year	Population	Wastewater Flow (m^3/sec)	Phosphorus (kg/day)	Nitrogen (kg/day)
1913	320,000	1.8	500	2,900
1932	575,000	3.2	900	5,200
1954	1,590,000	8.3	2,500	14,400
1960	1,860,000	9.5	4,550	16,900
1965	2,100,000	12.2	8,500	19,100
1970	2,600,000	13.9	11,000	27,300
1977	2,880,000	15.4	3,409	23,000
1978	3,140,000	16.8	2,600	24,900
1979	3,290,000	17.6	2,400	22,700

about 500 and 2,900 kg/day, respectively. These nutrients increased to a maximum of about 11,000 kg/day of phosphorus and 27,300 kg/day of nitrogen in 1970. However, with the initiation of advanced waste treatment at the wastewater facilities, the total phosphorus contribution to the estuary in 1979 had been reduced significantly to 2,400 kg/day, with a lesser reduction in nitrogen loadings to about 22,700 kg/day.

Discussion

The data for the Potomac estuary of the Chesapeake Bay ecosystem, as presented in Tables 6 and 7, demonstrate vividly that the seasonal contribution from land runoff can vary from month to month in a given year, and from year to year as shown in Table 8. For example, for the months of July and August, 1969, there was three-fold increase in phosphorus loading from the upper basin. Moreover, when the nitrogen contribution in February, 1969 is compared to February, 1970, a 5.7-fold increase can be seen.

For estuaries with rather short hydraulic retention times and large contributions from land runoff, it may be necessary to express external nutrient loading on either a seasonal or monthly basis.

In developing and evaluating control strategies, it is necessary to determine these loadings on a long-term statistical basis. As part of a nutrient input study of the Chesapeake Bay in 1969-1971 (8), a detailed analysis of stream flow and nutrient contributions from the major rivers was made. For seven major river systems, a high correlation between monthly flow and monthly phosphorus and nitrogen loading was obtained.

Utilizing such relationships and historical streamflow records, it is possible, with proper use of statistical techniques, to determine estimates of contributions from land runoff. These techniques can add significantly in determining the contribution from land runoff on a more reliable statistical basis and in analyzing historical responses of an ecosystem. In addition, an analysis of population trends and wastewater flows, as demonstrated for the Potomac estuary, can be used in determining historical contributions from municipal discharges.

IMPACT AND SCALE OF THE EUTROPHICATION PROCESSES

To illustrate the impact of external nutrient loadings on the eutrophication process, data from the Potomac estuary (see Figure 1) collected in a survey on July 8, 1977, are shown in Figures 2, 3, and 4 (6). Figure 1 shows the Potomac estuary divided into three major reaches. The upper reach, 64 kilometers (40 miles), is tidal and mainly fresh water. Freshwater blue-green algal blooms are dominant in the summer. Over 15 m^3/s of wastewater is discharged into the upper reach from 12 major treatment facilities serving the Washington, D.C. metropolitan area. The middle reach, where salt water intrusion occurs, is often referred to as the transition zone. The lower reach begins 107 kilometers (67 miles) below Chain Bridge where salt water intrusion from Chesapeake Bay is large, often reaching 16,000 mg/l. The bottom waters of the lower reach frequently become anoxic during the summer.

The physiological data of the upper and middle reaches of the Potomac estuary are shown in Figure 2, including pH, temperature, transparency, and salinity. The pH remained relatively constant, while temperature decreased slightly in the middle reach of the estuary. In the upper reach, the transparency was generally low, as a result of high sediment loadings from the upper river basin and from high turbidity loadings from the wastewater discharges, but increased below Milepoint 40. The salinity intrusion began about Milepoint 40.

Figure 3 shows the major forms of phosphorus. The large increase in phosphorus at Milepoint 8 was due to contributions from wastewater discharges. The total phosphorus remained relatively constant after Milepoint 8 in the upper reach, with a slight decrease around Milepoint 40. Although the major form of phosphorus from the wastewater plants was inorganic, the major form at Milepoint 10 was organic. It appears that the inorganic form was utilized bacterially and converted into the organic form during assimilation of the carbon and nitrogen from the wastewater discharges. Farther down the estuary, the major form was still organic and can be attributed mainly to phosphorus in the algal cells. Below Milepoint 40, when saltwater intrusion began, there was remineralization, and inorganic phosphorus became the predominant form at Milepoint 44.

Figure 4 shows the concentration of various forms of nitrogen along the estuary. At Milepoint 8, the large increase in nitrogen was due to

FIGURE 1. The Potomac estuary.

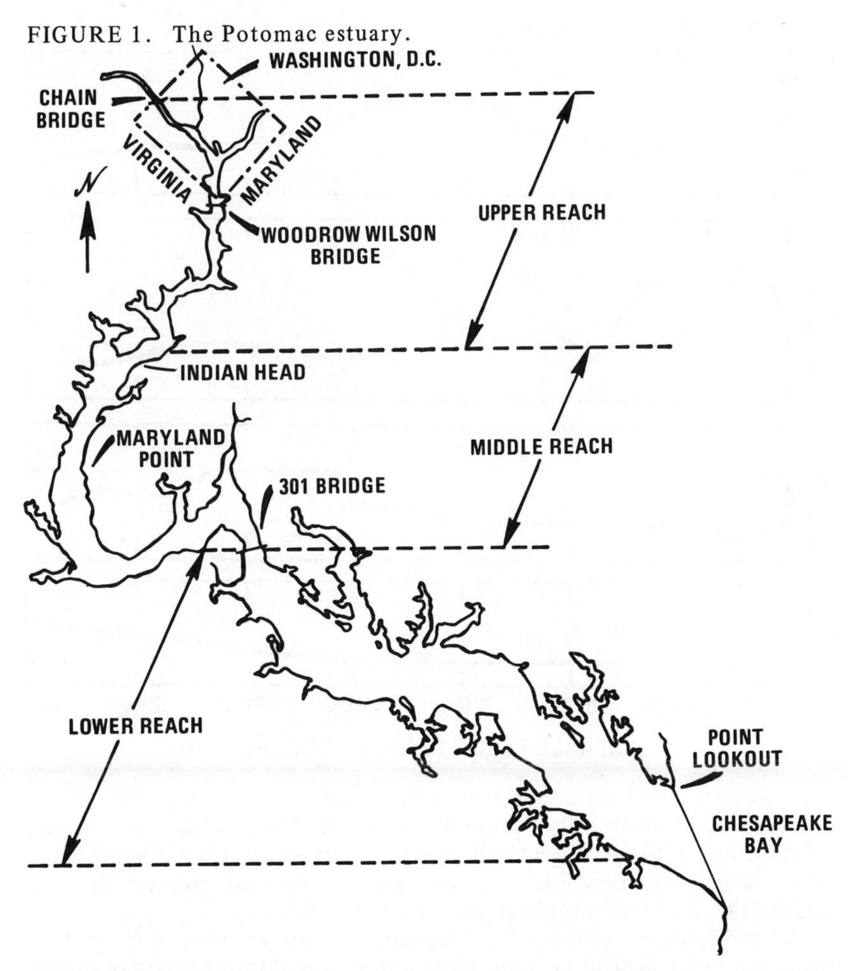

wastewater discharges. However, the total concentration decreased sharply, in contrast to phosphorus, in the lower part of the upper reach. Although the major form of nitrogen from the wastewater plants was as ammonia, it was quickly oxidized to the nitrate form, as indicated in Figure 4. A large portion of the nitrate nitrogen was incorporated into algal nitrogen. Below Milepoint 40, remineralization began to take place and ammonia increased. The predominant form below Milepoint 40 was organic nitrogen.

Figure 5 presents the carbon and chlorophyll concentrations along the estuary for the same time period. The chlorophyll level reached a maximum of about 120 μg/l below Milepoint 30. Total carbon concentrations remained essentially constant throughout the length of the

FIGURE 2. Potomac estuary physiology, July 27, 1977.

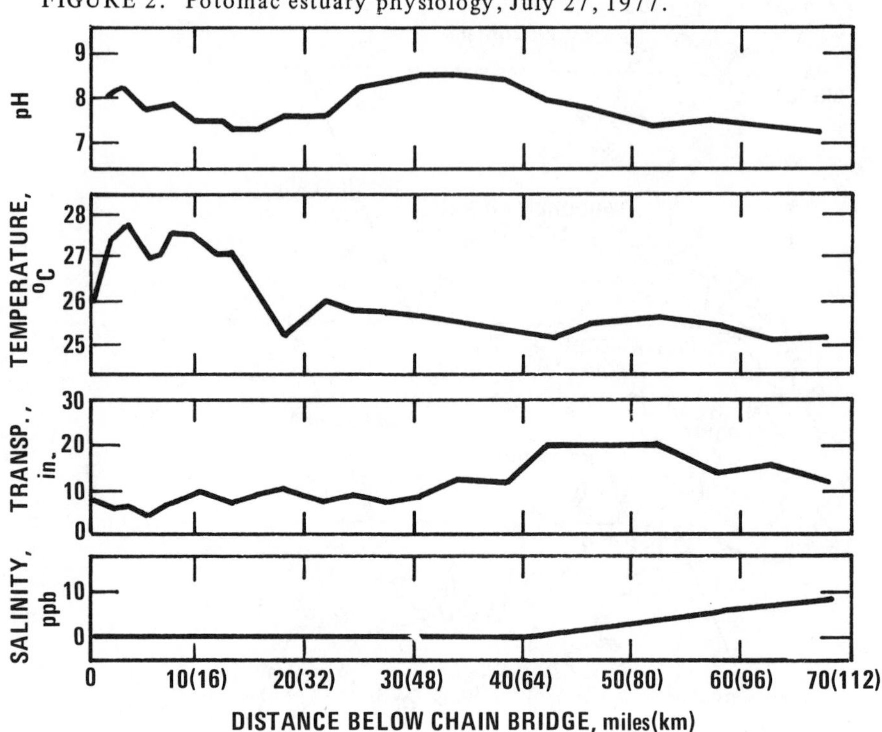

upper estuary. However, as can be seen in Figure 5, organic carbon began to increase significantly around Milepoint 30, with the largest contribution to the increase attributed to the die-off of algal blooms. High total carbon contribution and resulting biochemical oxygen demand reduced the dissolved oxygen in the lower reach (5, 27).

The phosphorus, nitrogen, and carbon data presented in Figures 3, 4, and 5 represent typical nutrient dynamics of the Potomac estuary during summer conditions. Of the approximately 25,000 kg/day (see Table 4) of ammonia nitrogen discharged into the upper reach, almost all is oxidized to nitrate nitrogen. More dissolved oxygen resources are needed to oxidize the ammonia than are required by the carbonaceous material discharged from the wastewater facilities into the upper reach (5, 27).

The nitrate nitrogen is rapidly utilized by the planktonic community in the upper Potomac estuary. The concomitant conversion of inorganic carbon to organic carbon is over 100,000 kg/day or about 300 percent more than from the total organic carbon loading from wastewater discharges (27). Similar calculations for the Hudson estuary and New York Bight by Garside, et al. (15) indicate that the eutrophication process contributes about 43 percent of the total organic carbon to these

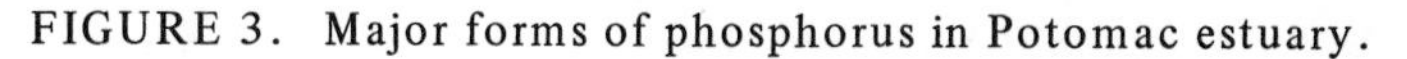

FIGURE 3. Major forms of phosphorus in Potomac estuary.

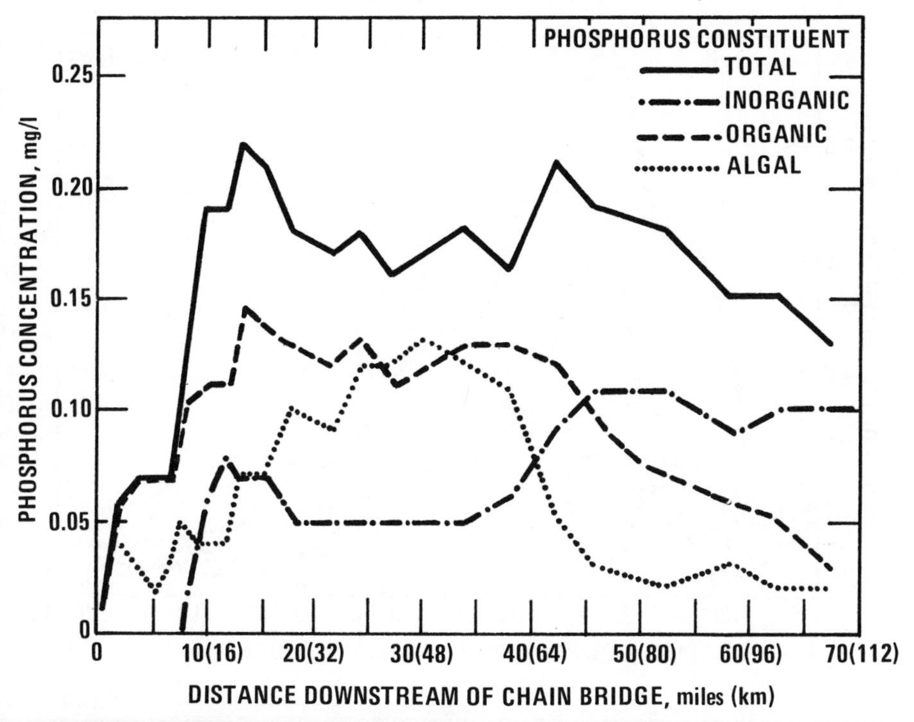

estuarine systems.

Studies of the Pamlico River estuary (10), Albermarle Sound (4, 52), and Oslo Fjord (2) have also attributed the high organic content in these estuarine systems mainly to planktonic growths. Anoxic conditions in the lower depths of Oslo Fjord are also attributed mainly to the eutrophication process.

Field and mathematical simulation studies of the Potomac estuary by Clark and Jaworski (5) have indicated that algal photosynthesis and respiration are a major component in the dissolved oxygen budget. Sensitivity analyses have shown that algal photosynthesis and respiration rates are of paramount importance in developing the dissolved oxygen budget of the Potomac estuary. At times, the dissolved oxygen condition can be depressed from 1.0 to 4.0 mg/l, attributed to algal respiration.

Discussion

High external nutrient loadings and the resulting eutrophication process can have three directly measurable impacts on the dissolved oxygen of an estuarine system: 1) a biochemical oxygen demand exerted by the oxidation of ammonia nitrogen, 2) a biochemical oxygen demand in the

FIGURE 4. Concentration of various nitrogen forms along Potomac estuary.

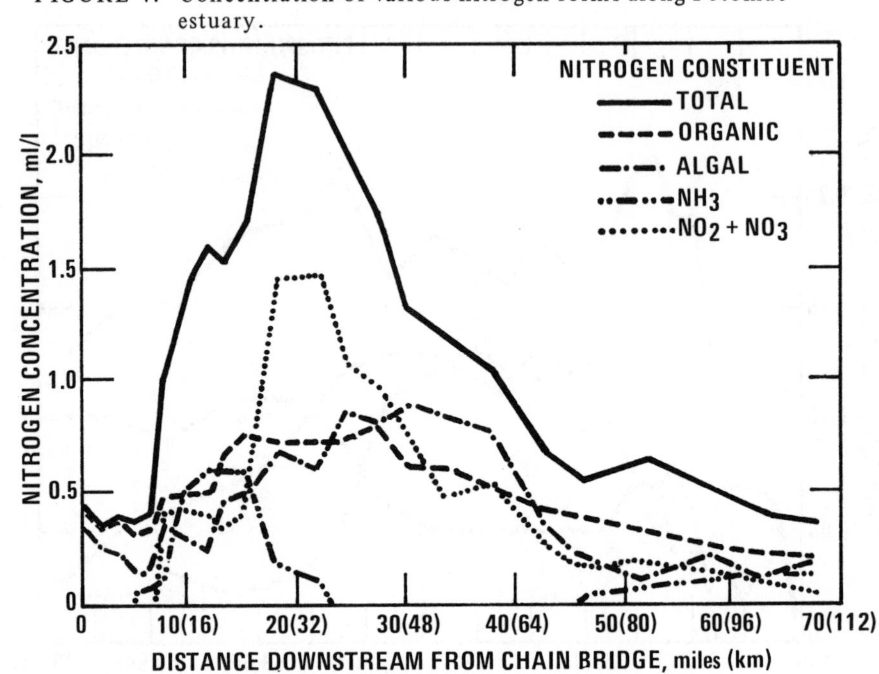

water column created by the conversion of inorganic carbon to organic carbon, and 3) a significant increase in total organic carbon of the system and its secondary impact, such as the benthic oxygen demand. There also can be an impact on the dissolved oxygen demand due to photosynthesis and respiration.

For freshwater lakes, general descriptive terms of the scale of eutrophication (i.e.. oligotrophic, mesotrophic, and eutrophic) have been successfully used by Vollenweider (51) and others. Depending on the authors, the descriptive terms included both an indication of the degree of nutrient enrichment and the impact of the enrichment on the ecosystem.

Attempts have been made to develop both scientific and management indexes by McErlean and Reed (36) and water quality criteria by Jaworski and Villa (30) to aid in providing a scale of eutrophication for estuary ecosystems. However, due to the lack of definitive and comparable data for a large number of estuarine ecosystems, a simpler approach was utilized for a general scale similar to the descriptive term used for freshwater lakes. The subjective description scale as proposed below focuses mainly on the impact of nutrient enrichment on the ecosystem rather than on the degree of nutrient enrichment; i.e., the concentration of phosphorus or nitrogen. This approach was taken in that portions of many estuaries which are highly enriched can be light limited, thus having

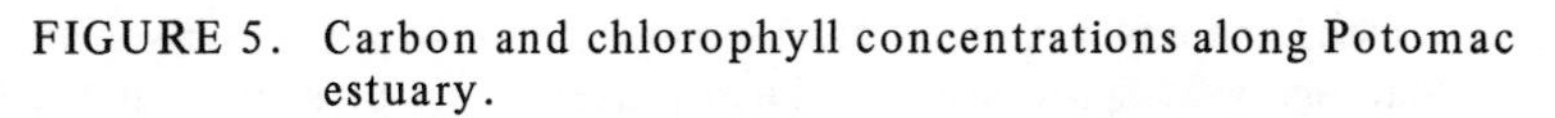

FIGURE 5. Carbon and chlorophyll concentrations along Potomac estuary.

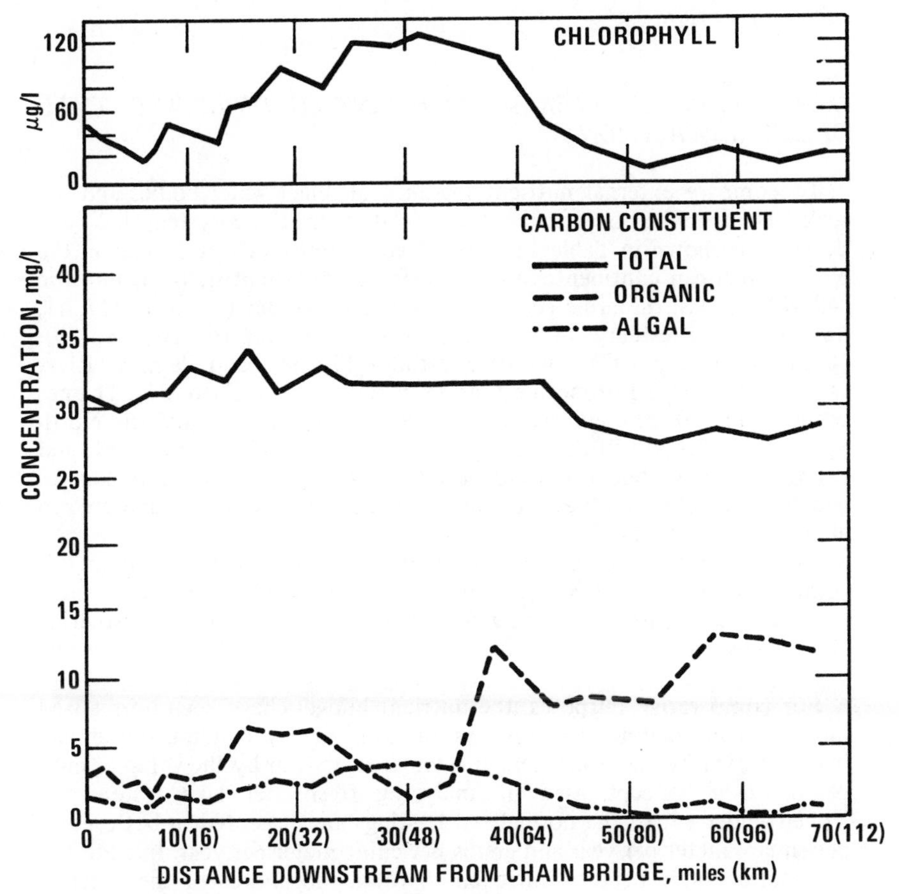

no major eutrophication problems.

For relative comparative purposes, the subjective descriptions of the scale of eutrophication for estuarine ecosystems proposed are as follows:

Hyper-eutrophic - Indicative of very excessive nuisance conditions, anoxic conditions, and "undesirable" biological communities.

Eutrophic - Indicative of excessive nuisance conditions, low dissolved oxygen concentrations, and "undesirable" biological communities.

Non-Eutrophic - Indicative of biologically healthy and productive ecosystems with "desirable" biological communities.

The above estuarine environments are incorporated into a comparison of external loadings versus the scale of eutrophication as presented in the following section.

COMPARISON OF EXTERNAL NUTRIENT LOADINGS AND SCALE OF EUTROPHICATION

To compare external nutrient loadings of select water bodies and the scale of eutrophication, a tabulation of data from 13 ecosystems has been developed, shown in Table 11. The 13 ecosystems include 3 lakes in the North American continent: Superior (19, 43, 49), Baptiste (45), and Erie (43, 49); 6 U.S. estuarine ecosystems: Chesapeake Bay (9, 13, 17, 27, 41, 46), Potomac estuary (17, 27), Albemarle Sound (4, 16, 18, 39), Narragansett Bay (37), Delaware estuary (7, 38), and Pamlico River estuary (32, 33); 3 European estuarine systems: Archipelago (31), Thames estuary (11, 20, 21), and River-on-Tyne estuary (22, 23); and the North Sea (24). Data used in these comparisons, including the hydrographic and external nutrient budgets, were obtained from various literature sources and from unpublished files. In certain instances, data were extrapolated to aid in the comparison.

Except for the North Sea, the estuarine ecosystems utilized in this comparison have relatively shallow average depths compared to most freshwater lakes, usually ranging from about 4 to 9 meters. The estuarine ecosystems have relatively short retention times, usually less than one year.

For comparative purposes, the nutrient budgets have been normalized into nutrient loadings (in $g/m^2/yr$ or $g/m^3/yr$) by dividing the annual budget (g/yr) by the ecosystem surface area (m^2) or by the volume (m^3), based on a concept used in analyzing freshwater lake systems by Vollenweider (50). The normalized loadings are presented in both grams per square meter per year and grams per cubic meter per year. In addition, the atomic ratio of total nitrogen to total phosphorus of the external loadings and a subjective description including the scale of the eutrophic condition of the ecosystem are presented. The scale of eutrophication for a given ecosystem was obtained directly, or interpreted from, the data sources referenced and/or in some cases from personal knowledge of the author.

As can be seen from Table 11, the external phosphorus loadings on the surface area basis varied from 0.04 $g/m^2/yr$ for Lake Superior to over 190 $g/m^2/yr$ for the Tyne estuary. For nitrogen, the external loading on a surface area basis varied from 1.0 $g/m^2/yr$ for the North Sea to 750 $g/m^2/yr$ for the Tyne estuary. Using the North Sea and Lake Superior for comparison purposes, estuarine ecosystems had a significantly higher phosphorus and nitrogen loading. The highest loadings were for the Delaware, Thames, and Tyne estuaries. All three have short retention times, receive a high contribution of industrial waste, and have no major algal bloom problems.

For the Narragansett, Albemarle, and Chesapeake estuarine ecosystems, nutrient loadings were very comparable. These three ecosystems can be considered to be healthy with no "excessive" eutrophication problems. However, some of the estuaries of Chesapeake Bay do have excessive eutrophic conditions. The Potomac estuary, an estuary of the Chesapeake, had considerably higher external loading and was considered to be hyper-eutrophic. However, there has been a significant decrease in phosphorus loading due to the recent institution of advanced waste treatment.

Phosphorus loadings to the estuarine ecosystems presented in Table 11 were all higher than for Lake Erie. In general, it appears that the higher the phosphorus loading, the more excessive were the eutrophic conditions. Nitrogen loadings into the estuarine systems were also generally higher than Lake Erie except Albermarle Sound and Narragansett Bay. Albemarle Sound appeared to be phosphorus limited (4), while Narragansett Bay appeared to be nitrogen limited (37).

Using the concept of the nitrogen to phosphorus ratio as developed by Redfield (40), insight can be gained into the possible limiting nutrient based on the nitrogen to phosphorus (N/P) ratio of external nutrient loadings. The ratios presented in Table 11 are for total nitrogen and total phosphorus. The nitrogen to phosphorus ratios presented in Table 11 suggest that most of the estuarine systems were nitrogen limited (an N/P below 16 is assumed to be nitrogen limited). When compared to Lake Superior and Lake Erie, which had N/P ratios of 60 and 27, respectively, and which were considered to be phosphorus limited, Albemarle Sound and Chesapeake Bay also appeared to be phosphorus limited.

For the ecosystem presented in Table 11, a rigorous comparison of external nutrient loadings and the scale of eutrophication is difficult to make unless physical factors such as depth and retention time as utilized by Vollenweider (50), Jaworski (25), and Lee (34) are included. To facilitate a more rigorous comparison, a detailed analysis was made of six estuarine ecosystems as presented in Table 12 with similar hydrodynamic configurations and a fairly consistent historical data base. The five estuaries of the Chesapeake Bay are generally long and narrow with the major wastewater discharges at the headwaters of the estuarine ecosystem. Average depth and retention times are comparable and the upper reaches, although tidal, are mainly freshwater systems. The watersheds of the estuaries have generally similar land use patterns, streamflow characteristics, and climatological conditions. The large differences in nutrient loadings among the estuaries and the institution of phosphorus removal at major wastewater treatment facilities discharging into the Upper Potomac estuary greatly enhance these detailed comparisons.

For the 1969-1971 time frame, a comparison of the external loadings for nitrogen and phosphorus indicates that the Potomac, James, and

TABLE 11. External Nutrient Loadings for Selected Ecosystems.

Ecosystem	Description	Surface Area (m^2)	Volume (m^3)	Average Depth (m)	Average Retention Time (yr)
North Sea (1972)	Non-eutrophic	575,000	54,000,000	93.3	--
Lake Superior (1976)	Oligotrophic lake	81,450	1,900,000	233.7	177.0
Lake Erie (1976)	Eutrophic lake	25,750	470,000	18.2	2.5
Chesapeake Bay with tribs. (1969-1971)	Localized eutrophic conditions	11,500	74,000	6.5	1.16
Potomac Estuary (1969-1971)	Hyper-eutrophic estuary	1,250	7,150	5.8	1.07
Albemarle Sound (1974)	Non-eutrophic	1,250	6,500	4.3	0.42
Narragansett Bay (1977)	Non-eutrophic	450	2,200	8.8	0.14
Delaware Estuary (1976)	Highly Industrial	300	1,330	4.3	0.22
Pamlico River Estuary (1976)	Eutrophic	230	980	4.2	0.71
Archipelago-Stockholm (1971) with nitrogen fixation	Eutrophic estuary	210	--	--	--
Thames Estuary (1971)	Highly Industrial	112	850	6.7	0.10
Baptiste Lake (Canada) (1977)	Naturally eutrophic lake	9	85	9.3	4.75
Tyne Estuary (1972)	Dissolved Oxygen problems	6	55	9.2	0.04

TABLE 11. (cont'd)

	External Phosphorus Loading			External Nitrogen Loading			Atomic N/P Ratio of Loading
	(g/yr) (10^6)	($g/m^2/yr$)	($g/m^3/yr$)	(g/yr) (10^6)	($g/m^2/yr$)	($g/m^3/yr$)	
North Sea	93,440	0.16	0.0017	565,000	1.0	0.01	14
Lake Superior	3,600	0.04	0.0003	95,200	1.2	0.01	60
Lake Erie	20,000	0.77	0.042	238,700	9.3	0.5	27
Chesapeake Bay	51,100	1.31	0.20	109,100	9.5	1.5	16
Potomac Estuary	5,300	4.30	0.74	25,200	20.2	3.5	11
Albemarle Sound	1,050	0.80	0.16	8,900	7.1	1.4	20
Naragansett Bay	820	1.82	0.29	2,200	4.9	0.8	6
Delaware Estuary	5,670	18.90	4.21	30,000	100.0	22.5	12
Pamilico River Estuary	905	3.93	0.92	2,840	12.3	2.89	7
Archipelago-Stockholm	720	3.48	--	5,990	28.5	--	19
Thames Estuary	12,300	109.8	15.40	51,800	462.5	64.8	10
Baptiste Lake	3.5	0.38	0.04	38	4.2	0.4	24
Tyne Estuary	1,142	190.3	20.70	4,500	750.0	81.8	9

TABLE 12. External Nutrient Loadings for the Chesapeake Bay and its Estuaries.

Ecosystem		Ecological Description	Surface Area	Volume	Average Depth	Average Retention Time
			(m^2) (10^6)	(m^3) (10^6)	(m)	(yrs)
Patuxent	1963	Non-eutrophic	137	660	4.8	1.70
	1969-71	Eutrophic	137	660	4.8	1.70
	1978	Eutrophic	137	660	4.8	1.70
York	1969-71	Non-eutrophic	210	910	4.3	0.72
Rappahannock	1969-71	Non-eutrophic	400	1,780	4.5	1.27
James	1969-71	Eutrophic	600	2,400	3.6	0.39
Potomac	1913	Non-eutrophic	1,250	7,150	5.8	1.07
	1954	Eutrophic	1,250	7,150	5.8	1.07
	1969-71	Hyper-eutrophic	1,250	7,150	5.8	1.07
	1977-78	Eutrophic	1,250	7,150	5.8	1.07
Chesapeake Bay (Including Tribs)	1969-71	Localized eutrophic conditions	11,500	74,000	6.5	1.16
Chesapeake Bay (Excluding Tribs)	1969-71		6,500	52,000	8.4	1.32

TABLE 12. (cont'd)

		External Phosphorus			External Nitrogen			Atomic N/P
		(g/yr)	($g/m^2/yr$)	($g/m^3/yr$)	(g/yr)	($g/m^2/yr$)	($g/m^3/yr$)	Ratio of Loading
		(10^6)			(10^6)			
Patuxent	1963	170	1.24	0.26	930	6.7	1.4	12
	1969-71	250	1.82	0.38	1,110	8.1	1.7	10
	1978	420	3.06	0.64	1,500	11.4	2.4	8
York	1969-71	160	0.76	0.18	1,190	5.6	1.3	17
Rappahannock	1969-71	180	0.45	0.10	1,500	3.8	0.8	19
James	1969-71	1,780	2.70	0.70	10,300	15.6	4.2	13
Potomac	1913	910	0.73	0.13	18,600	14.8	2.6	46
	1954	2,000	1.63	0.28	22,600	18.1	3.1	26
	1969-71	5,380	4.30	0.80	25,200	20.2	3.5	11
	1977-78	2,520	2.01	0.35	32,800	26.2	4.6	30
Chesapeake Bay (Including Tribs)	1969-71	15,000	1.30	0.20	109,100	9.5	1.5	16
(Excluding Tribs)	1969-71	7,350	1.10	0.10	70,160	10.8	1.3	22

Patuxent had the highest values. For the non-eutrophic York and Rappahannock systems, the phosphorus loadings were 0.76 and 0.45 $g/m^2/yr$, respectively. The loadings for the entire Chesapeake Bay, including estuaries, were 1.3 and 9.5 $g/m^2/yr$ for phosphorus and nitrogen, respectively, indicating that the system, while biologically productive, was not being stressed as much as the Potomac, Patuxent, and James.

In comparing the nitrogen to phosphorus ratios for the 1969-1971 period, the Patuxent, James, and Potomac appeared to have been nitrogen limited. The York, Rappahannock, and Chesapeake Bay appeared to have been phosphorus-limited systems.

The external data loadings for the Potomac in 1913 (26) and for the Patuxent in 1963 (14) in conjunction with the 1969-1971 loadings for the York and Rappahannock suggest that when the external phosphorus loading was 1.2 $g/m^2/yr$ or less, non-eutrophic conditions existed. Data for the Potomac vividly demonstrate that when phosphorus removal was instituted at the wastewater treatment facilities discharging into the upper estuary, the ecosystem reverted from a hyper-eutrophic condition in 1969-1971 to a eutrophic condition in 1977-1978. In 1977-1978, the nitrogen loading increased, mainly due to a larger contribution from the upper basin. The external total carbon loading in 1977-1978 was not significantly changed from the loading in 1969-1971 (1).

The nitrogen/phosphorus ratio of the external loadings in 1977-1978 was 30, indicating that phosphorus was limiting. Analysis of water quality data (1) of the upper Potomac estuary in 1977-1978 indicates N/P ratios of 30 to 40, also suggesting that phosphorus appeared to have been limiting. In addition, chlorophyll levels have been reduced by 20-30 percent since 1969-1971 and the transparency has increased by 20-30 percent. Total organic carbon has also been reduced by 30-40 percent, attributed primarily to a reduction in eutrophic conditions.

Discussion

Data from the two tables suggest that for estuarine ecosystems on the east coast of the United States, if the N/P ratio is above 16 and external phosphorus loadings are less than 1.0 $g/m^2/yr$, "excessive" eutrophic conditions will not prevail. These data further suggest that the "permissible" phosphorus loading is 0.75 $g/m^2/yr$ or less.

For estuarine ecosystems that are nitrogen limited (N/P ratio less than 16), the data in Tables 11 and 12 are somewhat conflicting with regard to suggesting a permissible loading for nitrogen. If a phosphorus loading of 0.75 is "permissible," it suggests, based on the N/P ratio, that the "permissible" nitrogen loading should be about 5.4 $g/m^2/yr$. The nitrogen loading for the Patuxent in 1963 was 6.7 $g/m^2/yr$, which is above the "permissible" state. The 1969-1971 nitrogen loading for the Pamlico River estuary was 12.8 $g/m^2/yr$, and was also nitrogen limited. However, its eutrophic scale appears to have been similar to that of the Patuxent in 1963.

This difference may possibly be explained by 1) the existence of nitrogen fixation, 2) the occurrence of denitrification, 3) a possibly greater percentage of phosphorus loss to the sediment when compared to nitrogen, or 4) a difference in nutrient dynamics, including recycling. Further analysis of ambient water quality, including dissolved oxygen resources, may also explain this difference.

Except for the Potomac, the data in Table 12 reflect wastewater discharges with no phosphorus removal. If phosphorus removal (90 percent) were incorporated into all other wastewater treatment facilities that discharge directly into the Chesapeake Bay and its estuaries, phosphorus loading would be less than 1.0 $g/m^2/yr$ and the N/P ratio would be greater than 16. These loadings and ratios suggest that excessive eutrophic conditions would be averted. The response of the Potomac estuary to phosphorus control strongly supports this hypothesis.

Data for Albemarle Sound, the Potomac, York, and Rappahannock estuaries, and projections for the Chesapeake Bay are somewhat contrary to conclusions of Ryther and Dunstan (42). They suggest that for coastal waters, nitrogen is usually the critical limiting factor to algal growth and that phosphorus is usually in excess. They further suggest that removal of phosphorus in detergents and substituting nitrilotriacetic acid (NTA) would be counterproductive in controlling eutrophic conditions in coastal waters. Data developed in this report indicate that phosphorus control by advanced wastewater treatment appears to be a feasible method of alleviating excessive eutrophic conditions in some estuarine systems.

The comparison presented in this section has been on an annual basis. For estuarine systems with short hydrologic retention times and highly varying external loading, seasonal or monthly analyses may be required. In addition, an analysis of the N/P ratios, based on the reactive forms, may also be required.

SUMMARY

External Nutrient Budget Considerations

Data from detailed analysis of five ecosystems are summarized as follows:

1. The relative contribution of nutrients from air, land runoff, and wastewater sources can vary significantly from one ecosystem to another. For example, the relative contribution from wastewater discharges of phosphorus varied from 84 percent for the Potomac estuary to less than 5 percent for Lake Superior.

2. On a seasonal basis, the contribution of nutrients from land runoff exhibits the greatest variability of the three major nutrient sources. The contribution of nutrients is greatly influenced by the seasonal hydrologic cycle.
3. The contribution of nutrients from land runoff on an annual basis can vary over two-fold from year to year, again depending on the annual hydrologic cycle.
4. Since there appears to be a good correlation between stream flows and nutrient loadings from land runoff, statistical significance can be added to short-term nutrient budgets by the proper use of statistical techniques and the longer historical stream-flow records.
5. Historical external nutrient budgets can be developed from population data from municipal wastewater sources and from hydrologic data correlations for land runoff sources.

Impact and Scale of Eutrophication

A study of the impact and scale of eutrophication of Albemarle Sound, Pamlico River estuary, Potomac estuary, Hudson estuary, Oslo Fjord, and the New York Bight estuary suggests the following:

1. Oxidation of ammonia nitrogen can have a significant impact on the oxygen resources of an estuarine ecosystem. For the Potomac estuary, this oxidation process has about three times the impact on the oxygen resources than the oxidation of carbonaceous material from wastewater discharges.
2. The amount of organic carbon fixed due to the eutrophication process can be one of the major sources of organic carbon to estuarine ecosystems.
3. Data from the New York Bight, Oslo Fjord, and other estuarine ecosystems suggest that from 40 percent to over 75 percent of the organic carbon in these estuaries is attributed to the eutrophication process.
4. Anoxic conditions in the lower depths of many estuarine ecosystems can be attributed to the eutrophication process.
5. A scale of eutrophication for estuarine ecosystems of hyper-trophic, eutrophic, and non-eutrophic conditions is proposed.

Comparison of Nutrient Loading Versus Scale of Eutrophication

The study of 13 selected ecosystems and a detailed analysis of the Chesapeake Bay and its 5 major estuaries is summarized as follows:

1. Phosphorus loading varies from 0.04 for Lake Superior to over 190 $g/m^2/yr$ for the Tyne estuary. Nitrogen loading varies from 1.2 for the North Sea to over 750 $g/m^2/yr$ for the Tyne estuary.

2. Except for three estuarine ecosystems which receive large amounts of apparently toxic industrial wastewater, phosphorus and nitrogen loadings ($g/m^2/yr$) are generally indicative of the eutrophic condition of a given estuary.
3. Based on nitrogen/phosphorus ratio anlayses, six of the selected 13 estuarine ecosystems were phosphorus limited and seven were nitrogen limited.
4. From an analysis of the Chesapeake Bay and its five major estuaries, three estuaries appear to be nitrogen limited and two appear to be phosphorus limited. The two apparently phosphorus-limited estuaries were also considered to be non-eutrophic.
5. Data from 13 water bodies and the Chesapeake Bay and its 5 major estuaries suggest that if phosphorus loading is less than 1.0 $g/m^2/yr$, excessive eutrophic conditions can be prevented. A "permissible" phosphorus loading of 0.75 $g/m^2/yr$ is suggested.
6. The favorable response of the Potomac estuary to phosphorus control demonstrates that excessive eutrophic conditions may be alleviated in estuarine ecosystems depending on many factors, including nutrient loadings.
7. The studies suggest that use of the nutrient loading concept can yield insights as to the success of advanced wastewater treatment to prevent excessive eutrophic conditions in many estuaries. Utilization of advanced wastewater treatment to reduce phosphorus loads may be appropriate for estuaries where the major source of phosphorus is wastewater discharges.
8. For estuarine ecosystems with short hydrologic retention times and highly varying external loadings, seasonal or monthly analysis may be required.

REFERENCES

1. Annapolis Field Station, Environmental Protection Agency. Data Files, 1979.
2. Balmer, P., et al. 1977. Management of urban runoff and waste water in the Oslo Fjord area. *Nordic Hydrology* 8(4):237-248.
3. Bain, R., et al. 1968. Effects of the San Joaquin master drain on water quality of the San Francisco Bay and delta. Appendix, Part C.: Nutrients and biological response. U.S. Department of the Interior. San Francisco, Cal.
4. Bowden, W. B. and J.E. Hobbie. 1977. Nutrients in Albemarle Sound, North Carolina. University of North Carolina Sea Grant College Publication, UNC-SG-75-25.
5. Clark, L.J., and N.A. Jaworski. 1972. Nutrient transport and dissolved oxygen budget studies in the Potomac estuary. U.S. Environmental Protection Agency Technical Report No. 37. Annapolis Field Office, Region III.

6. Clark, L.J., and S.E. Roesch. 1978. Assessment of 1977 water quality conditions in the Upper Potomac estuary. U.S. Environmental Protection Agency, Annapolis Field Office, Region III.
7. Clark L.J., et al. 1978. A water quality modelling study of the Delaware estuary. U.S. Environmental Protection Agency Technical Report No. 62. Annapolis Field Office, Region III.
8. Clark, L.J., et al. 1973. Nutrient enrichment and control requirements in the upper Chesapeake Bay. Summary conclusions. U.S. Environmental Protection Agency Technical Report No. 56. Annapolis Field Office, Region III.
9. Chesapeake Bay waste load allocation study. 1975. Hydroscience, Inc. Westwood, N.J.
10. Copeland, B.J., and J.E. Hobbie. 1972. Phosphorus and eutrophication in the Pamlico River estuary, N.C., 1966-1969 - a summary. Water Resources Research Institute of the University of North Carolina Report No. 65.
11. Cremer, H.W., Chairman of Survey Committee, et al. 1964. Effects of polluting discharges on the Thames estuary. Chapters 1 and 4. Report of the Thames survey committee and the water pollution research laboratory. Water Pollution Research Technical Paper No. 11. Department of Scientific and Industrial Research. Her Majesty's Stationery Office. London, England.
12. Correll, D.L., et al. 1978. Rural nonpoint pollution studies in Maryland (nonpoint pollution studies on agricultural land use types prevalent in the coastal plain zone of Maryland). U.S. Environmental Protection Agency Report No. EPA-904/9-78-002. Athens, Ga.
13. Cronin, W.B. 1971. Volumetric, areal, and tidal statistics of the Chesapeake Bay estuary and its tributaries, Special Report No. 20. Chesapeake Bay Institute of the The Johns Hopkins University. Baltimore, Md.
14. Federal Water Pollution Control Administration. 1969. The Patuxent River: water quality management technical evaluation. U.S. Department of the Interior, Charlottesville, Va.
15. Garside, C., et al. 1976. *Estuarine and Coastal Marine Science* 4(3):281-289.
16. Giese, G.L., et al. 1979. Hydrology of major estuaries and sounds of North Carolina. U.S. Geological Survey, Water Resources Investigations 79-46. Table 4.2--Monthly and annual gross water budget for Albemarle Sound.
17. Guide, V., and O. Villa, Jr. 1972. Chesapeake Bay nutrient input study. U.S. Environmental Protection Agency Technical Report No. 47. Annapolis Field Office, Region III.
18. Harrison, W.H. and J.E. Hobbie. 1974. Nitrogen budget of a North Carolina estuary. Water Resources Research Institute of the University of North Carolina Report No. 86.
19. International Joint Commission. Introduction and material balance. Chapter 3 of The waters of Lake Huron and Lake Superior. Volume III (Part A) Lake Superior.
20. James, A. 1971. Marine pollution from estuaries. Proceedings of the Challenge Society, IV(79).

21. James, A. 1971. Personal communication. July, 1979. Department of Civil Engineering, University of Newcastle-upon-Tyne. Newcastle, England.
22. James, A. 1976. Pollution of the River Tyne estuary - the use of mathematical models. Water Pollution Control, 75(3):322-340.
23. James, A. (ed.). 1972. Pollution of the River Tyne Estuary. University of Newcastle-upon-Tyne, Department of Civil Engineering, Bulletin No. 42.
24. James, A., and P.C. Head. 1970. The discharge of nutrients from estuaries and their effect on primary productivity. United Nations Food and Agriculture Organization. FIR:MP/70/E-19. Rome, Italy.
25. Jaworski, N.A. 1979. Section VIII - Multiple-state lakes and special topics limnological characteristics of the Potomac Estuary, *In* L. Seyb and K. Randolph, North American Project - A study of U.S. Water Bodies, EPA-600/3-77-086 (NTIS No. PB 275674), US EPA/Corvallis.
26. Jaworski, N.A. 1969. Nutrients in the Upper Potomac River Basin. U.S. Department of the Interior Technical Report No. 15.
27. Jaworski, N.A., et al. 1971. A water resource - water supply study of the Potomac Estuary. U.S. Environmental Protection Agency Technical Report No. 35.
28. Jaworski, N.A., et al. 1969. Nutrients in the Potomac River Basin. U.S. Department of the Interior Technical Report No. 9.
29. Jaworski, N.A., and L.J. Hetling. 1970. Relative contribution of nutrients to the Potomac River basin from various sources. U.S. Department of the Interior, Federal Water Pollution Control Administration, Middle Atlantic Region, Technical Report No. 31.
30. Jaworski, N.A., and O. Villa, Jr. 1979. A suggested approach for developing estuarine water quality criteria for management of eutrophication. Presented at an International Symposium on the Effects of Nutrient Enrichment in Estuaries. Williamsburg, Va.
31. Kalgren, L., and K. Ljungstrom. 1975. Nutrient budgets for the inner Archipelago of Stockholm. *Journal of Water Pollution Control Federation* 47(4):823-833.
32. Kuenzler, E.J., et al. 1979. Nutrient kinetics of phytoplankton in the Pamlico River, North Carolina. Water Resources Research Institute of the University of North Carolina, Report No. 139.
33. Lauria, D. Personal communication. July, 1979. Environmental Science and Engineering Department, University of North Carolina. Chapel Hill, N.C.
34. Lee, G.F., and R.A. Jones. 1979. Application of the OECD eutrophication modeling approach to estuaries. Presented at an International Symposium on Nutrient Enrichment in Estuaries. Williamsburg, Va.
35. Marks, J.W., et al. 1969-1970. Water quality of the Potomac estuary transport survey. U.S. Environmental Protection Agency, Annapolis Field Office, Region III.
36. McErlean, A.J., and G.J. Reed. 1979. On the application of water quality indices to the detection, measurement and assessment of nutrient enrichment in estuaries. Paper presented at the International

Symposium on the Effects of Nutrient Enrichment in Estuaries. Williamsburg, Va.

37. Olsen, S., and V. Lee. 1979. A summary and preliminary evaluation of data pertaining to the water quality of Upper Narragansett Bay. Coastal Resources Center, University of Rhode Island.
38. Peters, J.G. Personal communication. July, 1979. Delaware River Basin Commission. Trenton, N.J.
39. Quarterly Report. 1979. Investigation of the Chowan River estuary algal bloom. North Carolina Department of Natural Resources and Community Development.
40. Redfield, A.C. 1958. The biological control of chemical factors in the environment. *American Scientist* 46:205-221 .
41. Robertson, P.G. Personal communication. July, 1979. Water Resources Administration, Maryland Department of Natural Resources. Annapolis, Md.
42. Ryther, J.H., and W.M. Dunstan. 1971. Nitrogen, phosphorus, and eutrophication in the coastal marine environment. *Science* 170 (1008-1013).
43. Thomas, N. Personal communication. July, 1979. U.S. Environmental Protection Agency, Grosse Ile Field Station, Mich.
44. Tofflemire, T.J., and L.J. Hetling. 1979. Pollution sources and loads in the lower Hudson River. Second Annual Symposium on Hudson River Ecology. Institute of Environmental Medicine, New York University Medical Center. Sterling Forest, N.Y.
45. Trew, D.O., et al. 1978. The Baptiste Lake study. Summary report. Alberta Environment Pollution Control Division, Water Quality Control Branch.
46. Trexler, P.L. Personal communication. July, 1979. State Water Control Board, Commonwealth of Virginia.
47. U.S. Bureau of Reclamation. 1979. Agricultural drainage and salt management in the San Joaquin valley. San Joaquin Interagency Drainage Program. Preliminary edition: A special report including recommended plan and draft environmental impact report.
48. U.S. Environmental Protection Agency. 1969-1970-1971. Water quality survey of the upper Chesapeake Bay. Data report by the Annapolis Field Office staff. Technical Report No. 24. Annapolis Field Office, Region III.
49. Vallentyne, J.R., and N.A. Thomas, Canadian and U.S. Co-Chairmen. 1978. Report of Task Group III, a technical group to review phosphorus loadings. Fifth year review of Canada-United States Great Lakes Water Quality agreement.
50. Vollenweider, R.A. 1975. Input-output models, with special reference to the phosphorus loading concept in limnology. *Schweiz. Z. Hydrol.*, 37:53-84.
51. Vollenweider, R.A. 1968. Scientific fundamentals of the eutrophication of lakes and flowing waters, with particular reference to nitrogen and phosphorus as factors in eutrophication. OECD Technical Report DAS/CSI 68(27). Paris, France.
52. Witherspoon, A.M., et al. 1979. Response of phytoplankton to water quality in the Chowan River system. Water Resources Research Institute of the University of North Carolina.

REMINERALIZATION AND NUTRIENT CYCLING IN COASTAL MARINE ECOSYSTEMS

Scott W. Nixon
Graduate School of Oceanography
University of Rhode Island
Kingston, Rhode Island 02881

ABSTRACT: Our views of remineralization and nutrient cycling in coastal marine ecosystems have changed considerably over the last 30 years. The major trend has been an increasing appreciation for the complexity of processes involved, including some marked changes in our assessment of the importance of bacteria with respect to smaller animals and in our perception of the association between bacteria and particulate matter in the sea. Among the more recent developments in this area is a growing awareness of the importance of the coupling between benthic and pelagic communities in coastal waters. There appears to be a strong linear correlation between the organic matter produced in the overlying water and the amount of organic matter consumed on the bottom in almost all of the coastal environments for which annual data are available. The large amount of organic matter consumed by the benthos (perhaps 25-50 percent of that produced) is associated with a large flux of inorganic nutrients from the sediments to the overlying water. The stoichiometry of net benthic nutrient regeneration differs from that of pelagic regeneration, however, and simple Redfield type models probably cannot be applied. The amount of fixed inorganic nitrogen returned to the water across the sediment-water interface appears to be about half of that expected on the basis of the flux of phosphorus. This behavior, along with the fact that an appreciable amount of organic matter in coastal waters gets remineralized on the bottom, contributes to the low N/P ratio that is characteristic of these areas and may be responsible for the observation that nitrogen is commonly the nutrient most limiting for primary production. Recent direct measurements of the flux of dissolved N_2 across the sediment-water interface indicate that denitrification is probably responsible for the loss of fixed nitrogen during decomposition in the sediments. If this is a widespread phenomenon, estuaries, bays, and other coastal waters may be major sinks in the marine nitrogen cycle and important terms in the global nitrogen budget. However, the fact that eutrophication appears to be an increasing problem in many estuaries is dramatic warning that anthropogenic nutrient inputs can overwhelm the recycling and remineralization processes in coastal waters.

CHANGING VIEWS OF DECOMPOSITION AND REMINERALIZATION IN COASTAL MARINE ECOSYSTEMS

The seasonal cycles of major nutrients in coastal sea water have been described and studied for at least 50-60 years (5, 11, 77, 13). As a result, it is often assumed that we have a much clearer picture of the processes regulating estuarine and coastal marine nutrient dynamics than is actually the case. One of the major advances in our understanding of these processes, however, has been the realization that seasonal cycles in abundance of nutrients do not reflect a simple seasonal cycle in their uptake and regeneration. Applications of radioactive and stable isotope tracer techniques, as well as indirect evidence from budget studies and nutrient stoichiometry, have shown that the large seasonal changes in nutrients from inorganic to organic form and back again result from shifts in the relative rates of very much more rapid uptake and regeneration processes which occur on the order of hours or days rather than months (74). While studies of decomposition in the sea were begun at least as early as the 1930s (104), our increasing awareness of the dynamic nature of nutrient cycling has led to a greater appreciation for the importance of processes influencing the decomposition of organic matter and the regeneration of nutrients in marine and fresh water systems (3).

It is humbling to realize, however, that in spite of this awareness and a substantial research effort, our knowledge of these processes is far from complete. Our perceptions are changing rapidly, and some very basic questions are still unresolved. For example, in his review of nutrient cycling written just ten years ago, Johannes (49) concluded that: "...It has been demonstrated beyond reasonable doubt that most of the nitrogen and phosphorus incorporated into aquatic plants and animals is usually regenerated by processes other than direct bacterial action."

More recent research has provided conflicting evidence, and Fenchel and Harrison (24) interpreted the results of their studies of macrophyte decomposition with the observation that: "Bacteria have a high rate of excretion of dissolved phosphorus both in organic and in inorganic forms. Grazers played a modest role compared with bacteria in the regeneration of inorganic phosphorus and pure bacterial systems showed a high rate of phosphorus-cycling."

And in a discussion of his studies of pelagic remineralization off the coast of California, Harrison (38) wrote that: "In terms of nitrogen fluxes, microzooplankton could not, on the average, account for most of the ammonium remineralization in the <183-μm size fraction. The conclusion...would seem to be that bacteria account for most of the nitrogen remineralized."

I am not anxious to use this review for a detailed examination of the various experiments which have been used to support these different conclusions. I am not even sure the data are really available to resolve the question of the relative importance of bacterial action compared to other processes. It is evident, however, that it no longer seems as clear as it once did which major group of organisms is responsible for most of the organic matter decomposition and nutrient regeneration in coastal waters.

In spite of this kind of basic uncertainty, I think there has been a progression in our understanding of these processes, at least on a qualitative level. After reviewing a substantial (but by no means complete) sample of the literature dealing with this subject, it seemed to me that an historical summary such as that shown in Figure 1 might be accurate enough to be of some use in providing perspective on the present state of our knowledge. Prior to about 1950, the central concept of nutrient dynamics was the "great cycle,"in which rather simple food chains ended in dead organic matter which was decomposed by bacteria and fungi, thus

FIGURE 1. Conceptual models of remineralization and nutrient cycling in coastal marine ecosystems as they were understood at various times in the past. The solid lines represent flows that were thought to be important or which at least seemed to receive a lot of attention in the literature. It is an informal history derived from a determined but by no means exhaustive review of the literature, and the beginnings and endings of various views are somewhat fuzzier than they appear here.

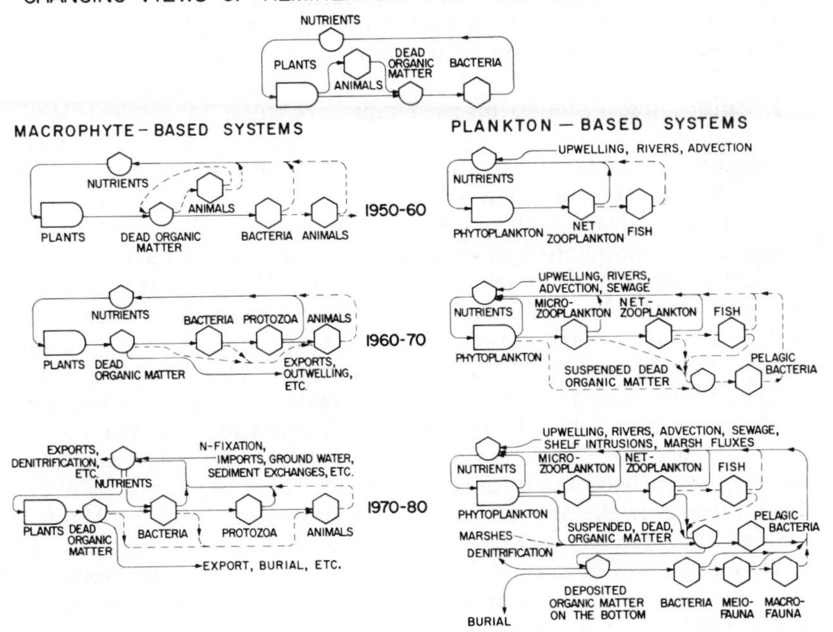

releasing nutrients which were used again by plants to produce new organic matter. As ecological research began on a larger scale during the period between 1950-1960, different views began to develop among those working in macrophyte-based systems such as sea grass meadows, salt marshes, and kelp beds and those working in plankton-based systems such as Long Island Sound.

In the macrophyte systems it became clear that Petersen's (70) early observations on the importance of detritus seemed to apply as a general principle, and that little of the plant material in these systems was eaten live. Bacteria were still seen as the major agents of nutrient regeneration, though studies such as Hutchinson's (44) survey of vertebrate excretions and Kuenzler's (54) work on marsh mussels began to evaluate larger animals as "nutrient pumps."

The prevailing view of pelagic environments seemed to remain simple for somewhat longer. A straightforward grazing food chain led from phytoplankton (with the emphasis usually on diatoms) to the net zooplankton (usually thought of as copepods) to fish. While it was recognized that in particular areas physical processes such as upwelling, advection of offshore water, or river flow were important in bringing nutrients into the coastal waters, the emphasis in nutrient cycling was on the role of copepod excretion. It is worth remembering that Harris's (35) pioneering study of the role of zooplankton excretion in the nitrogen budget of Long Island Sound was first published just 20 years ago.

The next ten years of research throughout the 1960s emphasized the role of smaller animals (microzooplankton, meiofauna) in the nutrient cycles of both detrital systems and plankton communities (42, 48, 27, 73). Isotope tracer studies revealed rapid exchanges of nutrients during excretion and decomposition and confirmed the importance of autolysis in releasing inorganic nutrients from cells (113, 49). Basic considerations of ecological energetics minimized the importance of fish and other conspicuous animals, and the nutritional qualities (high carbon, low nitrogen and phosphorus) of detritus suggested that bacteria themselves should take up rather than release nutrients (64, 101, 100). It was unclear, however, whether the meiofauna and microzooplankton were "dead ends" in the food chain and served mainly as nutrient pumps through their excretion, or if they were also important in the higher trophic dynamics of the system. The concept of detrital organic matter as an important food source and substrate for nutrient regeneration was extended to pelagic systems (79) and the axiom that pelagic bacteria were largely confined to particles began to be challenged (115, 114). It may have been the influence of the activist environmental movement of the 1960s, as well as the dramatic results of earlier studies such as Ryther's (83) revelations about Long Island duck farms and phytoplankton blooms, that made sewage and eutrophication prominent terms in papers dealing with coastal marine nutrient dynamics.

The radioactive and stable isotope tracer studies of excretion and organic matter decomposition begun in marine systems during the 1960s were continued and expanded throughout the 1970s. As a result, it became clear that it was "necessary to distinguish between the rates of nutrient uptake and excretion by the decomposer microorganisms and the actual rate of mineralization, i.e., the release of nutrients from the organic substrate" (24). Thus, while bacteria growing on nutritionally poor macrophyte detritus took up nitrogen and phosphorus, in the longer run they also excreted or released through autolysis more of the nutrients than they took up. An atom of inorganic nitrogen or phosphorus in the water might pass through the bacteria many times before the nutrients in the substrate could be added to the inorganic pool (8). As mentioned earlier, these studies, as well as nutrient budget estimates in which excretion rates were measured directly (38) have led us to reassess the importance of bacteria in nutrient regeneration. While the meiofauna and microzooplankton that received so much attention during the 1960s probably are major sources of regenerated nutrients, it now appears that the early concept of the importance of bacterial regeneration may also be correct in many marine systems. An intriguing result of a number of laboratory decomposition studies, however, is the demonstration that protozoa accelerate the bacterial breakdown of detritus through mechanisms that remain unknown (25).

In addition to this reassessment, data collected during recent years have shown that most of the bacterial activity in the sea is due to free-living bacterioplankton, and that pelagic decomposition appears not to be restricted to particle surfaces. Again, the change in perspective is dramatic and should encourage us to be modest in assessing the present state of certainty with regard to many of the questions now under study. In their 1972 review of the literature dealing with the association of microorganisms and particles in the sea, Wiebe and Pomeroy commented that "it is almost axiomatic to state that the vast majority of bacteria in the sea are associated with detritus...Yet one is hard-pressed indeed to supply support for this idea when the actual data are examined." Some feel for the extent of the change in our perceptions can be gotten by comparing the opening of Seke and Kennedy's (90) paper written in 1969 "...it has been shown that most of the biomass of microorganisms is absorbed upon particulate matter in the sea..." with the conclusion in Azam and Hudson's (6) recent study that "...roughly 90 percent of microbial heterotrophic activity is due to free-living organisms, presumably bacteria.."

The last ten years have brought other changes in our view of coastal marine nutrient cycling besides those dealing directly with animals, bacteria, and particles. Among the most significant have been reassessments of the interactions of marshes and coastal waters and an increasing awareness of the importance of benthic-pelagic coupling in nearshore areas. The role of marshes in coastal marine nutrient cycling is a complex problem that is the subject of a separate review and analysis (63), and I don't think it is necessary to go into the question of marsh-estuarine

FIGURE 2. The relationship between the amount of organic matter fixed and imported and the amount of organic matter metabolized on the bottom over an annual cycle in a variety of coastal marine systems.

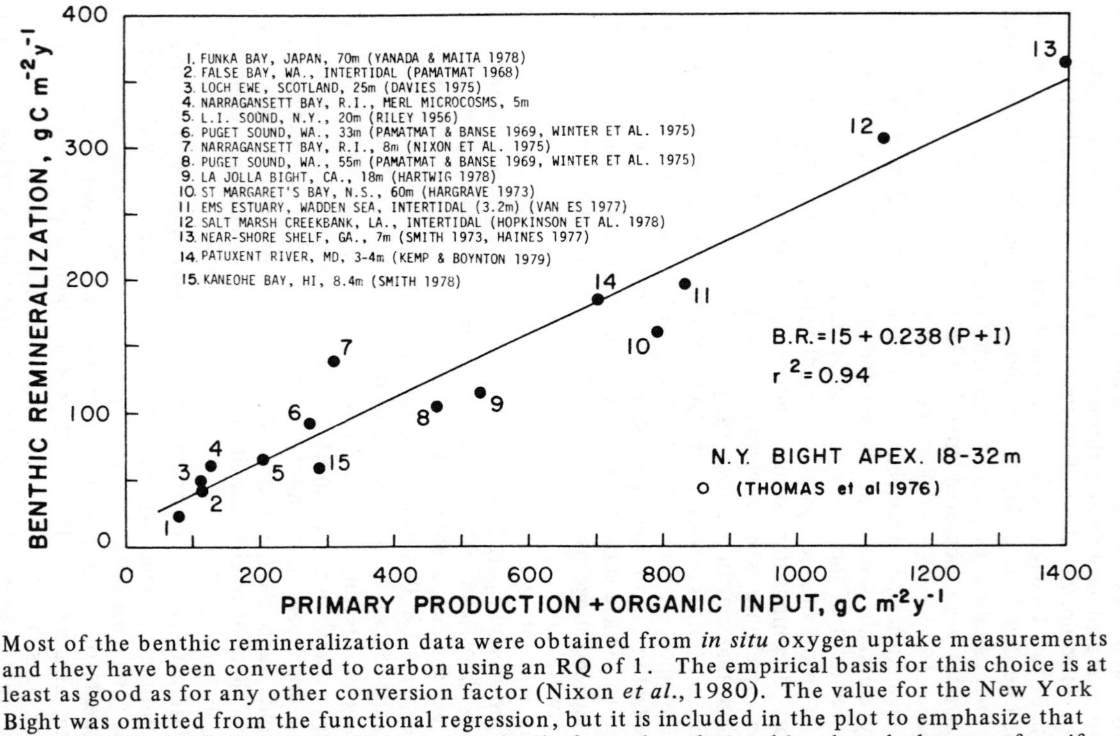

Most of the benthic remineralization data were obtained from *in situ* oxygen uptake measurements and they have been converted to carbon using an RQ of 1. The empirical basis for this choice is at least as good as for any other conversion factor (Nixon *et al.*, 1980). The value for the New York Bight was omitted from the functional regression, but it is included in the plot to emphasize that there are likely to be areas which depart markedly from the relationship, though there are few, if any, others yet reported in the literature. Over half of the organic matter assigned to the budget for the New York Bight by Thomas *et al.* (1976) was from inputs which may not really influence the entire area.

interactions as part of this paper. The traditional view of marshes as "buffering" nutrient cycles by exporting and importing nutrients at different times of year may have some merit, at least in a qualitative sense, but the importance of this term in the nutrient balance of any larger estuary appears very questionable.

In contrast to nutrient fluxes from the marshes, the importance of benthic remineralization in estuarine and perhaps coastal marine nutrient cycling seems to have been reasonably well established during recent years. As early as 1973, Hargrave (33) was able to use data from a number of lakes and some marine systems to show that benthic oxygen uptake could be related to a complex parameter which included primary production and the depth of the mixed layer in the water column. However, a variety of different types of coastal marine systems have now been studied over an annual cycle, including intertidal flats, deep bays, and salt marshes, and it is remarkable to find that there is a strong linear relationship between the amount of organic matter produced and/or imported and the amount of organic matter consumed on the bottom (Figure 2). Within the range of organic loading and production found in most coastal marine systems, between one quarter and one half of all the organic matter appears to be mineralized by the benthos. The connection between organic matter consumed (as measured by oxygen uptake or carbon dioxide release) and nutrients released, however, is not well documented for marine bottom communities. In extensive in situ studies of bottom communities in Narragansett Bay, Rhode Island, over an annual cycle, we have found that while there is a good correlation between oxygen uptake and inorganic nitrogen and phosphorus release, the stoichiometry of the relationship does not necessarily agree well with a simple Redfield model (59, 60, 62). In a number of other environments, on the other hand, there does not appear to be any consistent relationship between oxygen uptake and nutrient regeneration over various portions of the seasonal cycle (26, 50). But after examining the growing amount of literature on this subject, it appears to me that there is usually at least some correlation between the amount of inorganic nitrogen released and the oxygen taken up by a variety of different coastal marine sediments during summer, the period for which most data are available (Figure 3). In spite of the uncertainty in the exact stoichiometry of benthic nutrient regeneration, sufficient data have now been collected to give some feel for the magnitude and direction of the inorganic nitrogen (primarily ammonia) and phosphorus exchange at the sediment-water interface in coastal marine systems (Table 1). Unfortunately, it is difficult to evaluate the importance of benthic flux numbers such as these without referencing them to other terms in the overall nutrient budget of the area.

FIGURE 3. The relationship between oxygen uptake and ammonia release by the sediments in a variety of coastal marine systems during summer. The positive intercept for oxygen uptake makes the stoichiometric interpretation of the slope of the functional regression difficult. The bottom of Kaneohe Bay consists of well oxygenated, highly bioturbated $CaCO_3$ sediments, and this system was not included in the functional regression. Additional data from the New York Bight have shown higher ammonia fluxes for stations in Raritan Bay and the Bight Apex which receive urban sewage, while deeper stations on the shelf fall near the regression line (O'Reilly and Smith, personal communication).

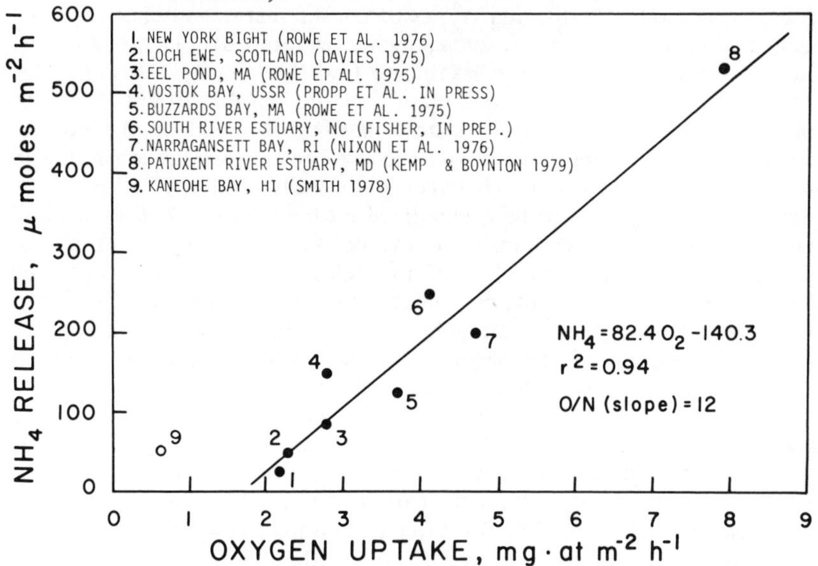

THE RELATIVE IMPORTANCE OF "NEW" AND RECYCLED NUTRIENTS IN COASTAL MARINE ECOSYSTEMS

The conceptual models of remineralization and nutrient cycling in Figure 1 are largely derived from a host of independent studies in different locations, each of which emphasized a particular aspect of the system of concern. It is probably safe to say that there is no environment for which reliable quantitative measurements of all, or even most, of the major flows shown in Figure 1 have been made. With the exception of perhaps one salt marsh (102), there are few, if any, coastal marine systems for which well defined and constrained annual budgets for any nutrient have been published. This is understandable, given the great difficulty of making such budgets, but it is also very unfortunate in light of the great power of the nutrient budget as a tool in ecosystem analysis and management. The budget provides a rigorous quantitative framework which tests our understanding of the system and provides a broad perspective from which we can derive a more objective assessment of the importance of various processes. If all we know is the amount of nitrogen entering an estuary from sewage or any other source, we have not learned anything that is very useful. In order to assess the importance of nutrient inputs in sewage, it is necessary to compare that term to others in the nutrient budget to see how much of the primary production it might sustain, how the input compares with recycled nutrients that are already turning over in the system, and how compares with other inputs and losses.

Since there are so few estuaries or other coastal waters for which the necessary data are available, I have tried to summarize some of the large amount of information on nutrients in Narragansett Bay to use as an example of the potential importance of nutrient cycling relative to nutrient inputs in coastal systems. With the exception of the net zooplankton excretion numbers, the measurements have all been made by my laboratory, and I am aware that some people may object to my including previously unpublished data in what is largely a review paper. Nevertheless, with the exception of the budgets for Kaneohe Bay, Hawaii (96), the analysis of Narragansett Bay is more complete than that of any other system I have found described in the literature, and the details of the individual measurements will be published soon. The major point I want to make with the budget is the great importance of nutrient cycling in the Bay (Table 2). However, there is a considerable range in the relative importance of "new" and "recycled" terms in the budgets for the three major nutrients and in the various forms of each nutrient. For example, while the input of total nitrogen is 1.4 times larger than the measured recycling of nitrogen, the input of total phosphorus is only 80 percent of that recycled and silica inputs are less than 14 percent of that recycled. The importance of recycling processes is clear, even though a major term, the excretion of nutrients by microzooplankton and pelagic bacteria, is missing from the budget. This is particularly true for ammonia, phosphate, and silicate, the dissolved inorganic nutrient forms that are most important in regulating the primary production in the Bay (52) (Table 3).

If the estimate of annual primary production in Narragansett Bay (310 g C m^{-2} y^{-1}, (27)) is converted to a nutrient demand, it becomes apparent that nutrient inputs (without being recycled) could support, at the most, only some 24-50 percent of the annual production, depending on the nutrient considered (Table 4). The critical role of recycling is emphasized even more if particulate and dissolved organic forms of nitrogen and phosphorus are excluded from the budget. The estimates in Table 4 also suggest that the role of bacteria and microzooplankton in the water

TABLE 1. The net flux of oxygen, ammonia, and phosphate measured at the sediment-water interface in various coastal marine systems during summer.

	O_2 mmoles m^{-2} h^{-1}	NH_4 µmoles m^{-2} h^{-1}	PO_4	T °C
Lock Ewe, Scotland (Davies 1975)[1]	0.8-1.55	20-80	--	9-12
Buzzard's Bay, MA (Rowe et al. 1975)[2]	1.85	125	-15	16
Eel Pond, MA (Rowe et al. 1975)[3]	1.40	85	16	20
Narragansett Bay, R.I. (Nixon et al. 1976)[4]	2.35	200	30-50	18-21
Long Island Sound, Conn. (Aller 1977)[5]	---	50-200	5-20	22
New York Bight, N.Y. (Rowe et al. 1976)[6]	1.10	25	2	8
Patuxent River Estuary, Md. (Kemp & Boynton 1979, Kemp, pers. comm.)[7]	3.95	710	48	25-30
Pamlico River Estuary, N.C. (Harrison & Hobbie 1975)[8]		45	--	25-30
South River Estuary, N.C. (Fisher in prep.)[9]	2.05	250	17	21-24
Off Cap Blanc, West Africa (Rowe et al. 1977)[10]	---	235	50	
Vostok Bay, USSR (Propp et al. in press)[11]	1.40	150	20	20
Maizuru Bay, Japan (Yoshida & Kimata 1969)[12]	---	13-32	--	22
Kaneohe Bay, Hawaii (Harrison, in prep.; Smith 1978)[13]	0.60	54	3	25
La Jolla Bight, Cal (Hartwig 1976)[14]		40	6	11-15

[1] June & July, 2 measurements, depth = 20-30 m
[2] June, 1 measurement, depth = 17m
[3] July, 3 measurements, depth = 2 m
[4] June, July, Aug. for 3 stations, 30 measurements, depth = 9m
[5] July, 3 stations
[6] August, mean of 2 dark control chambers, depth = 35m
[7] June & August, 5 nutrient and 39 O_2 measurements, depth = 3.4m
[8] 1 measurement in a lab trough using disturbed sediment
[9] May, 5 measurements from 3 stations for NH_4 & PO_4, 3 measurements for O_2
[10] Mean for 2 stations, 1 measurement each, depth = 25 m
[11] August, dark chambers, depth = 5.7 m
[12] July, 3 measurements
[13] Annual mean, bioturbated carbonate sediment, depth = 8m
[14] June, July, Aug., 10 measurements, depth = 18m

TABLE 2. The amount of nitrogen, phosphorus and silica in various forms input to Narragansett Bay (264×10^6 m^2) and recycled wihtin the bay, units of 10^6 g-at of N, P, or Si y^{-1}.

Nutrient Inputs[1]

Source	PN	DON	NH_4	NO_2	NO_3	ΣN	PP	DOP	PO_4	ΣP	$Si(OH)_4$
Fixation	0.20					0.20					
Rain						2.8	?	?	0.09	0.09	?
Runoff						16.2	?	?	0.8	0.8	?
Rivers	19.5	68.1	62.4	5.75	79.3	235	6.02	3.71	7.53	17.3	66.2
Sewage	47.1	128	96.4	0.5	6.1	278	7.85	3.78	10.11	21.7	19.7
Total	66.6	196	159	6.25	85.4	532	13.9	7.49	18.4	39.9	85.9

$$\Sigma N/\Sigma P = 13.3$$

$$(NH_4 \ \& \ NO_2 + NO_3)/PO_4 = 13.6$$

Recycled Nutrients

	DON	NH_4	ΣN	DOP	PO_4	ΣP	$Si(OH)_4$
Menhaden[2]	--	0.75	0.75		0.09	0.09	--
Ctenophores[3]	3.70	4.35	8.05	0.15	0.60	0.75	--
Net Zooplankton[4]	35	64.0	98.5	?	7.74		--
Benthos[5]	30	234	264	<9.49	31.6	41.1	640
Total	68.7	303	371	9.64	40.0	49.7	640

$$\Sigma N/\Sigma P = 7.5$$

$$NH_4/PO_4 = 7.6$$

[1]Fixation data for bacteria in sediments (Seitzinger et al. 1978); all other inputs from Nixon et al. (in prep.)
[2]Durbin (1976)
[3]Kremer (1975). Excretion integrated over 100 days assuming 20 ml/m^3 population density.
[4]Vargo (1976). The DON measured was in the form of urea, and Vargo considered his results "of doubtful significance.
[5]Nitrogen from Nixon et al. (1976) with DON estimated as 10 percent of ΣN flux from laboratory measurements in which NH_4 was 77 percent of ΣN. In the field NO_3 was taken up or lost; phosphorus from Nixon et al. (1980); silica from Nixon et al. (in prep).

column is very important in maintaining the primary production of the Bay. If the published estimates of the potential contribution of various terms in the nitrogen budgets for different areas are brought together, it appears that the anlaysis for Narragansett Bay is not atypical, and that much of the primary production of many coastal marine systems is supported by nutrient recycling rather than by nutrient inputs alone (Table 5).

TABLE 3. Ratio of "new"/recycled nutrients in Narragansett Bay over the annual cycle.

DON	NH_4	NO_3	ΣN	DOP	PO_4	ΣP	$Si(OH)_4$
2.85	0.52		1.43	0.78	0.46	0.95	0.13

TABLE 4. The relative importance of various nutrient sources in Narragansett Bay over an annual cycle.

	Contribution as % of the amount of N, P, or Si required to sustain primary production
Inputs:	
NH_4	15
ΣN	52
PO_4	29
ΣP	62
$Si(OH)_4$	19-36
Recycled:	
NH_4	29
ΣN	36
PO_4	62
ΣP	65
$Si(OH)_4$	142-267
Total:	
NH_4	44
ΣN	88
PO_4	91
ΣP	127
$Si(OH)_4$	161-303

It is also evident from Tables 2 and 5 that bottom communities provide a substantial portion of the recycled nutrients in coastal marine waters, as expected from the metabolic data discussed earlier. While pelagic regeneration may often be a larger term in the nutrient budget, especially if bacteria and microzooplankton are included, I want to concentrate on the benthic recycling. The reason for such an emphasis is that an exciting picture is beginning to emerge which suggests that it is the passage of a significant fraction of the organic matter through benthic food chains and sediments during regeneration that makes the nutrient dynamics of coastal marine systems different from those of the open sea.

TABLE 5. Contribution of "new" and recycled nitrogen in various coastal systems as a percent of the amount required for primary production over an annual cycle.

	Inputs	Recycled	
		Pelagic	Benthic
Long Island Sound, CT[1]	305	50	53
Narragansett Bay, R.I.[2]	52	10	26
Kaneohe Bay, Hawaii[3]	20	21	15
Pamlico River Estuary, N.C.[4]			
(winter)	964		7.8
(summer)	7		1.5
North Sea Inshore[5]	3-9		62-105
Lac Des Allemands, LA[6]	>43		21
San Francisco Bay[7]	218	26	100
Georgia Continental Shelf[8]	1.3		
La Jolla Bight, Cal.[9]			5
Southern California Bight[10]		92	
Saanich Inlet, B.C.[10]		74	
Loch Ewe, Scotland[11]			21
Patuxent River Estuary, Md.[12]			
(winter)	40		0
(summer)	7		20-200
Off Cap Blanc, Africa[13]		30-40	30-40
Funka Bay, Japan[14]		21	

[1]Harris (1959); Bowman (1977)
[2]Nixon et al. (in prep.)
[3]Smith (1978 and subsequent data)
[4]Harrison and Hobbie (1974)
[5]Billen (1978)
[6]Day et al. (1977)
[7]Peterson (1979)
[8]Haines (1975)
[9]Hartwig (1975)
[10]Harrison (1978)
[11]Davies (1975)
[12]Kemp and Boynton (1979)
[13]Rowe et al. (1977)
[14]Yanada and Maita (1978)

SEDIMENT-WATER NUTRIENT EXCHANGES AND COASTAL MARINE NUTRIENT DYNAMICS

Nutrient dynamics of coastal waters differ from those offshore in a number of ways, including the presence of a summer phosphate maximum (92, 99, 61) and a remarkable abundance of phosphate relative to nitrogen (78, 51, 46). It is relatively straightforward to demonstrate that a large net flux of phosphate from the sediments to the overlying water during summer is responsible for the characteristic seasonal phosphate cycle, at least in Narragansett Bay (62) but the question of the anomalously low N/P ratio found in virtually all coastal water is more intriguing. The question is of biological as well as geochemical interest, since it is the low N/P ratio which makes nitrogen such a major limiting nutrient in coastal waters (84, 28). While the ratio of fixed inorganic nitrogen to phosphorus remains relatively constant over much of the ocean at about 16/1 by atoms (76, 22, 2), the N/P ratio of coastal water is often below 10 and approaches zero as virtually all of the nitrogen is removed during periods of intense phytoplankton activity. The N/P ratio of the phytoplankton in Narragansett Bay and other coastal waters, however, does not appear to be significantly different from the 16/1 usually found in the phytoplankton of the open sea (T.J. Smayda, unpublished; 34, 51, 29).

The nutrient budget given in Table 2 clearly shows that the mean annual N/P ratio of about 3 for Narragansett Bay (52) cannot be due to nutrient inputs from rivers, sewage, or other sources, as the N/P ratio of the inputs is greater than 13. Since the mechanism responsible for maintaining the low N/P ratio does not appear to lie in the inputs or the phytoplankton uptake, at least for Narragansett Bay, it seems reasonable to suppose that it may arise in some way during recycling of the nutrients. The fact that the amount of inorganic nitrogen and phosphorus recycled is considerably greater than the amount put into the Bay (Table 3) adds support to such a conclusion.

TABLE 6. Relative amounts of nitrogen phosphorus excreted by some marine zooplankton.

	Atom Ratios	
	NH_4-N/PO_4-P	$\Sigma N/\Sigma P$
Mixed zooplankton, L.I. Sound (Harris 1959)	7.0	
Mixed zooplankton, Narragansett Bay (Martin 1968)	9.3	
Mixed zooplankton, Narragansett Bay (Vargo 1976)	8.8	11.8
Calanus sp. (Butler et al. 1969)	10.9	12.0
Sagitta hispida (Beers 1964)	11.3	
Mnemiopsis ledyi (Kremer 1975)	7.4	11

If the recycling of nitrogen and phosphorus is responsible for producing the anomalously low N/P ratios, it is likely that the mechanism is involved with benthic rather than pelagic remineralization. While it does appear that the ratio of N/P in marine zooplankton excretion is somewhat low

TABLE 7. Relative amounts of carbon, nitrogen, and phosphorus in the organic matter of coastal marine waters.

	Atom Ratios	
Phytoplankton	C/N	N/P
Field samples, Narragansett Bay (Smayda unpublished)	7.6	12-16
Field samples, L.I. Sound (Harris & Riley 1956)	7.6	15.7
Field samples, mixed diatoms (Redfield 1934)	6.4	29.6
Captured field sample, low NO_3 (Antia et al. 1963)	17.5	15-18
high NO_3	3.6	12-15
Various diatoms in culture (Parsons et al. 1961)	4.4-7.0	5.9-15
Culture, *Ditylum brightwellii* (Strickland et al. 1969)[a] On NO_3	6.7	
On NH_4	4.7	
Zooplankton		
Mixed zooplankton (Harris & Riley 1956)		24.6
Copepods (Redfield 1934)	6.4	29.6
Copepods (Beers 1966)	5.1	26.9
Chaetognaths (Beers 1966)	4.1	27.5
Euphausids & Mysids (Beers 1966)	4.7	14.9
Ctenophore (*M.ledyi*) (Kremer 1975)	4.0	31
Fish/fish larvae (Curl 1962)	3.6-5.9	10.2-13.2
Detritus & Particulates		
Particulates off California (>1 μ) (Holm-Hansen & Mague 1973)	6-9	14-44
Particulates in Pamlico River estuary, N.C. (Harrison & Hobbie 1974)	12-13	
Particulates in Funka Bay, Japan (Yanada & Maita 1978)	10	19
Resuspended matter in Narragansett Bay, R.I. (Oviatt & Nixon 1975)	9-12	
Deposited material in Puget Sound (Stephens et al. 1967)	7-12	
Zooplankton fecal pellets (Butler et al. 1970)	13	
Seaweed detritus in Nova Scotia (Mann 1972)	13	
Eelgrass leaves, live (Harrison, in Mann 1972)	7-9	
dead	13-18	
Marsh grass, *S. alterniflora*, dead (Nixon et al. 1976)	40-60	
Exported marsh detritus, Chesapeake Bay (Heinle & Flemer 1976)	34	4

[a]as modified by Banse 1974

(Table 6), this discrepancy comes about because the animals are synthesizing tissue which generally contains a higher N/P ratio than the food they are ingesting (Table 7). Once the animals die, sink to the bottom, and become remineralized, the nutrient balance ought to be restored to the 15 or 16/1 found in the phytoplankton tissue. Of course, if there were a very large flux of high nitrogen animal tissue out of coastal

TABLE 8. Atom ratios of oxygen, inorganic nitrogen, and inorganic phosphorus measured in fluxes between marine bottom communities and the overlying water.

Location	Author	O/N	N/P
Buzzards Bay, MA[1]	Rowe et al. 1975	28-46	3-4
Eel Pond, MA[2]	Rowe et al. 1975	32	2-7
Narragansett Bay, R.I.[3]	Nixon et al. 1975	27-33	6
Narragansett Bay, R.I.[4]	McCaffrey et al. 1978		4
Narragansett Bay, R.I.[5]	Elderfield et al. in press		3-8
Rhode Island Sound[6]	Elderfield et al. in press		11
L.I. Sound, CT[7]	Aller 1977		13
N.Y. Bight[8]	Rowe et al. 1976	87	4
Patuxent River Estuary, Md.[9]	Kemp & Boynton (1979)	26	3
South River Estuary, N.C.[10]	Fisher et al. (in prep.)	31	18
Lock Ewe, Scotland[11]	Davies (1975)	55	
N.W. Atlantic[12]	Smith et al. 1978	42-51	
Off Cap Blanc, Africa[13]	Rowe et al. 1977		8
Coastal Sea of Japan[14]	Propp et al. (in press)	76	
Kaneohe Bay, HI[15]	Harrison (in prep.); Smith (1978)	20	20

[1]4 measurements, winter and summer in Buzzards Bay
[2]3 measurements in summer only
[3]annual cycle from 3 stations
[4]summer near mid bay
[5]summer and winter, 4 stations
[6]summer and winter, 1 station
[7]seasonal measurements from 3 stations using laboratory cores
[8]1 station, summer, dark unpoisoned chambers only
[9]annual mean, 2 stations (Kemp, personal communications)
[10]from regression lines through positive fluxes only. A large intercept makes the interpretation of the appropriate ratio difficult.
[11]annual cycle at one station
[12]one measurement at each of two stations
[13]2 stations, one measurement at each station
[14]6 measurements at 1 station, spring-fall
[15]the annual cycle, 3 stations on high bioturbated calcium-carbonate sediments.

waters, it would result in a low residual N/P ratio in the water. Energetic considerations alone, however, make such an explanation unlikely, since the level of secondary production must be a relatively small fraction of the primary production and the associated accumulation of nutrients in animal tissue is thus a small part of the total system nutrient budget.

One of the major ways in which coastal waters differ from the open sea is that a much larger fraction of the organic matter in shallow water is remineralized on the bottom rather than through pelagic food chains. If we calculate the N/P ratio of the net sediment-water nutrient flux in Narragansett Bay, it is markedly low in nitrogen relative to phosphorus released (Figure 4) and the oxygen taken up (59). Similar anomalies are apparent in most of the benthic flux data from other systems as well (Table 8). Since the Narragansett Bay data were collected over an annual cycle and at various locations in the Bay, the low N/P ratio of the benthic flux cannot be explained by differences in the regeneration rate of nitrogen and phosphorus or by transients in the composition of the organic matter falling on the bottom at any one time. It is also not possible to attribute the apparent deficiency in nitrogen to a preferential accumulation of this nutrient in the sediments. If the sedimentation rate calculated from ^{210}Pb profiles by Santschi et al. (85) of about 1.5 mm y^{-1} is combined with sediment porosity (68 percent) and density (2.6 g cm^{-3}) data obtained by Pilson et al. (72) and organic carbon, nitrogen, and phosphorus profiles analyzed by Hurtt (45) and Sheith (91), respectively, it appears that the reverse may be true (Table 9).

At one point, preliminary field measurements suggested that an appreciable fraction of the nitrogen might be released to the overlying water as dissolved organic nitrogen, and that this might account for the low inorganic N/P ratio (59). However, there is a large amount of background DON in coastal water, the analytical error in DON measurements is considerable, and there are often erratic changes in control bottles incubated along with benthic chambers. As a result of

TABLE 9. Amount of carbon, nitrogen, and phosphorus buried each year in Narragansett Bay sediments expressed as a percent of the amount incorporated in primary production (310 g C m^{-2} y^{-1}; Furnas et al. 1976).[1]

	% Buried
Carbon	4.5
Nitrogen	1.5
Phosphorus	
Organic	0.7
Inorganic	3.1
Total P	3.8

[1]Sedimentation $\cong$ 1.5 mm y^{-1} (Santschi et al. 1978)

FIGURE 4. The relationship between the flux of phosphate and ammonia between sediments and the overlying water over an annual cycle at stations in the upper, middle, and lower West Passage of Narragansett Bay. The Redfield model is shown with a broken line while the solid line is a functional regression through the data (from Nixon *et al.*, 1980).

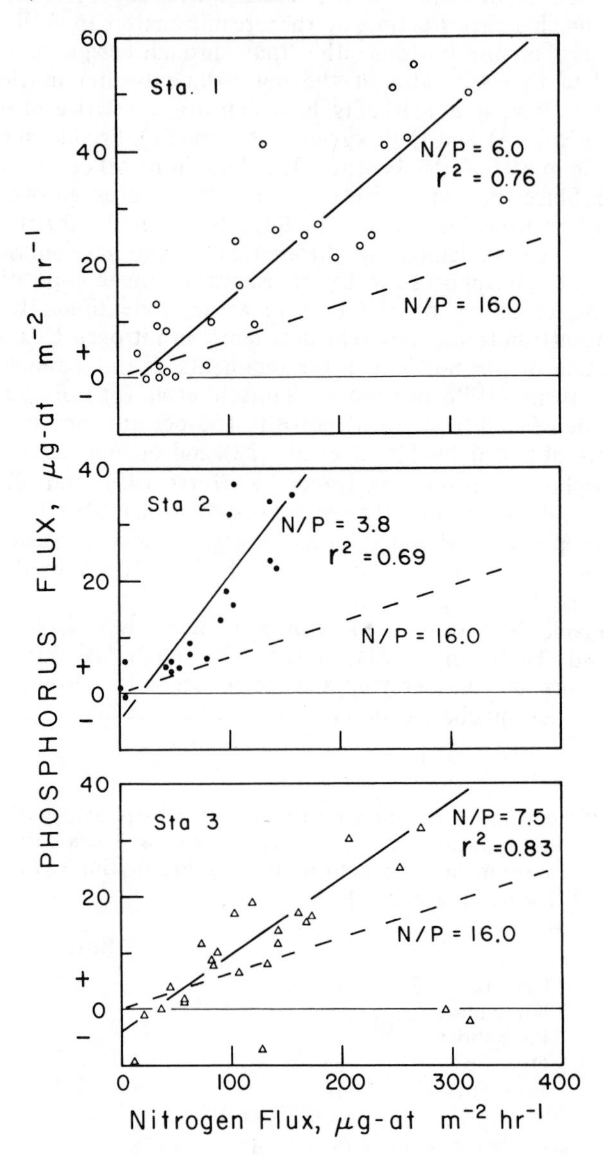

these problems, it is very difficult to measure DON fluxes from marine sediments during short-term in situ incubations. On the basis of a number of subsequent measurements using gently aerated box cores and 24 h incubations under laboratory conditions, it seems more likely that DON represents only about 10 percent of the fixed nitrogen flux from the sediments (Table 10).

The remaining possibility is that nitrogen is lost from the sediments through denitrification. This explanation is particularly attractive, since the conversion of fixed nitrogen to N_2 would actually remove nitrogen from the marine nutrient cycle and thus produce the long-term stoichiometric anomalies we have found in the Bay sediment-water fluxes. Anoxic sediments near the redox discontinuity layer would provide an ideal environment for denitrifying bacteria, since there could be a supply of nitrate from nitrification of the abundant ammonia in the aerobic pore waters above. Unfortunately, it is not easy to obtain meaningful and believable measurements of the rate of production of N_2 from subtidal marine sediments. The few estimates that have been reported have relied on assumptions about pore water nitrate profiles and models of diffusive flux in the sediment, or on measurements of the potential denitrification activity of sediment-water slurries prepared from sections of cores. Recently, however, we have developed a method which gives us the first direct area-based N_2 flux measurements from vertically intact cores of subtidal marine sediments from Narragansett Bay (88, 89). The results from one series of incubations extending over 30 days are shown in Figure 5. A vertically intact core of sediment about 5 cm deep was kept at 23°C

TABLE 10. Contribution of various forms of nitrogen to the flux from Narragansett Bay sediments to the overlying water, percent of the total fixed nitrogen flux.[1]

Temperature, °C	DON	NH_4	NO_2	NO_3
10	6.9	80.5	2.5	10.1
	--	74.0	3.4	22.6
15	14.0	72.9	3.5	9.6
	10.2	76.5	2.4	10.9
20	--	84.9	4.4	10.7
	5.3	78.1	5.2	1.4
$\bar{X}$	9.1	77.8	3.6	12.6

[1]Values for duplicate cores at each temp. DON, NO_2, NO_3 run simultaneously during 12-24 h dark incubations, NH_4 run the preceeding day during a 3 h dark incubation.

in the dark in a gas tight glass container with overlying sea water in which the N_2 had been replaced with He. The water was exchanged every one or two days with freshly sparged $He/CO_2/O_2$ water to prevent the depletion of O_2 or the accumulation of metabolic wastes. The water over the core was also stirred continuously to keep the water and a small overlying gas phase in equilibrium in the container. The gas phase was sampled periodically between water changes and analyzed by gas chromatography for N_2, N_2O, and O_2. The high initial fluxes of N_2 shown in Figure 5 resulted from the movement of N_2 out of the sediment pore water which had been in equilibrium with the high concentration of N_2 normally found in sea water ($\sim$0.8 atm.). After about 12 days, however, the supply of N_2 in the pore waters would have been exhausted, even at a flux rate of 50 μ moles m^{-2} h^{-1}. The continual release of N_2 after this time must have been from denitrification in the sediment. The resulting flux of about 125-150 μg-at N m^{-2} h^{-1} is equal to 50-100 percent of the ammonia flux found at this temperature, and could account for essentially all of the missing nitrogen needed to balance the N/P ratio of the nutrients being returned to the overlying water from benthic regeneration. While a small flux of N_2O gas was also measured, the amount of nitrogen lost in this form is insignificant in the nitrogen budget for the Bay (88).

The remarkable conclusion seems to be that much of the nitrogen limitation of primary production in coastal waters arises because of low fixed inorganic N/P ratios brought about because some 25-50 percent of the organic matter fixed in these areas is remineralized on the bottom. The burrowing and bioturbation of the sediments by macrofauna helps to mix the sedimented organic matter down through the top 5 cm or so of sediment into a microenvironment which is capable of supporting high rates of denitrification which remove some 5-25 percent of the nitrogen originally incorporated into organic matter from the coastal marine nutrient cycle. In the open ocean, where virtually all of the regeneration is pelagic and oxygen concentrations are usually high, simple stoichiometric models often apply and phosphorus may play a larger role in regulating production rates. It is not clear, however, why fresh water lakes behave differently from coastal marine systems (see the review by Schindler, this volume). While denitrification also takes place in lake sediments, it appears that nitrogen limitation is not common in lakes because blooms of nitrogen fixing blue-green algae dominate if there is low nitrogen and adequate phosphorus. In coastal marine systems, high rates of algal or bacterial nitrogen fixation seem to be limited to mud flats, salt marshes, and, perhaps, sea grass meadows. There does appear to be some bacterial nitrogen fixation in subtidal marine sediments, but ^{15}N calibrated acetylene reduction measurements with Narragansett Bay sediments have shown that the amount is very small compared with the loss of nitrogen from denitrification (87). The reasons for the apparent lack of an abundant and active pelagic nitrogen fixing blue-green algal flora in estuaries and coastal marine waters are obscure, but of great interest.

Without an input from nitrogen fixation, high rates of benthic remineralization and nutrient cycling in these areas appear to make them strong sinks in the marine nitrogen cycle and an important term in assessing the global nitrogen budget.

FIGURE 5. The flux of nitrogen gas from intact cores of Narragansett Bay sediment at 23° C. The broken line indicates the time required to remove all the pore water nitrogen at a flux rate of 50 μmoles m^{-2} h^{-1}. See text for explanation (from Setzinger and Nixon, 1979, Seitzinger *et al.*, 1980).

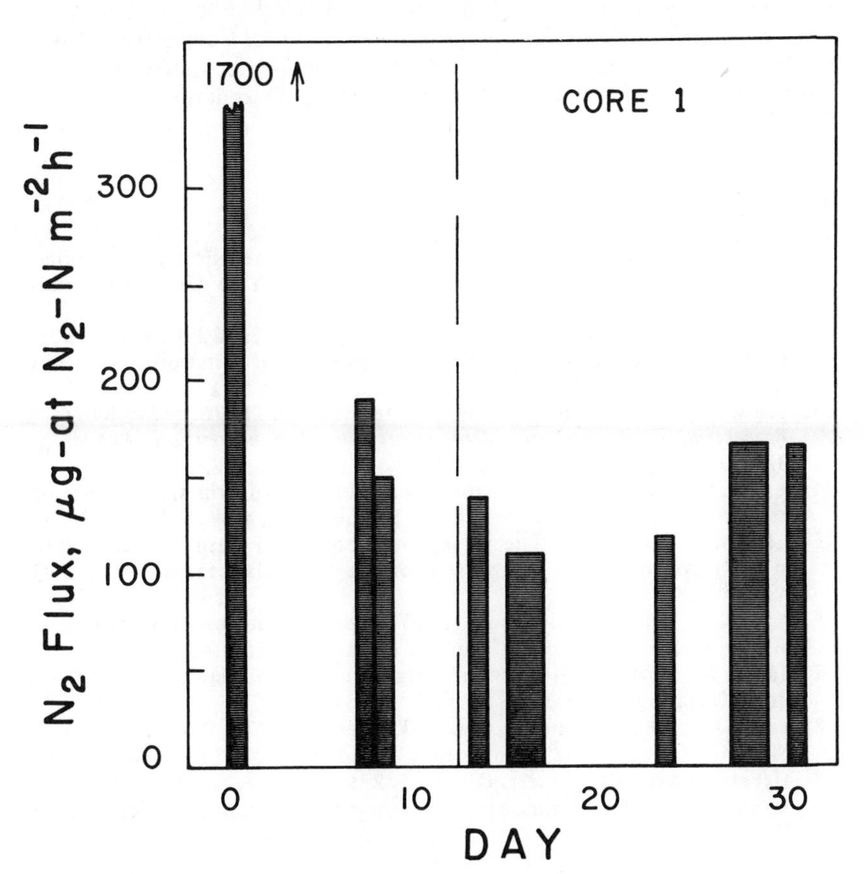

ACKNOWLEDGEMENTS

In writing a review paper, one owes a debt to all of those who have contributed to the literature in the field over the years, and I acknowledge mine gladly. A number of students, colleagues, and friends have also contributed to the development of this paper. Several people have allowed me to use their unpublished data, including Tom Fisher, Mike Kemp, and Walter Boynton at the University of Maryland; Steve Smith at the University of Hawaii; and Sybil Seitzinger, Suzanne Heffernan, Michael Pilson, Barbara Furnas, Jonathan Garber, Jim O'Reilly, Terry Smith, Jack Kelly, Gabe Vargo, Ted Smayda, and Candace Oviatt at the University of Rhode Island. Support for the Narragansett Bay work described here came from the Office of Sea Grant Programs, U.S. Dept. of Commerce, NOAA, the U.S. Environmental Protection Agency - Marine Ecosystems Research Laboratory (MERL) and the National Science Foundation, Grant No. OCE-7827124.

REFERENCES

1. Aller, R.C. 1977. The influence of macrobenthos on chemical diagenesis of marine sediments. Ph.D. Thesis, Yale University, New Haven, CT, p. 600.
2. Alvarez-Borrego, S., D. Guthrie, C.H. Culberson, and P.K. Park. 1975. Test of Redfield's model for oxygen-nutrient relationships using regression analysis. Limnol. Oceanogr. 20:795-805.
3. Anderson, J.M., and A. Macfadyen (eds.). 1976. *The Role of Terrestrial and Aquatic Organisms in Decomposition Processes.* Blackwell Scientific, London.
4. Antia, N.H. et al. 1963. Further measurements of primary production using a large-volume plastic sphere. Limnol. Oceanogr. 8:166-183.
5. Atkins, W.R.G. 1925. The phosphate content of fresh and salt waters in its relationship to the growth of the algal plankton. J. Mar. Biol. Assoc. U.K. 13:119-150.
6. Azam, F., and R.E. Hodson. 1977. Size distribution and activity of marine microheterotrophs. Limnol. Oceanogr. 22(3):492-501.
7. Banse, K. 1974. On the vertical distribution of zooplankton in the sea. Prog. Oceanogr. 2:56-125.
8. Barsdate, R.J., T. Fenchel, and R.T. Prentki. 1974. Phosphorus cycle of model ecosystems: significance for decomposer food chains and effects of bacterial grazers. Oikos 25:239-251.
9. Beers, J.R. 1964. Ammonia and inorganic phosphorus excretion by the planktonic chaetognath, *Sagitta hispida* Conant. J. Cons. Perm. Int. Explor. Mer. 29:123-129.
10. Beers, J.R. 1966. Studies on the chemical composition of the major zooplankton groups in the Sargasso Sea off Bermuda. Limnol. Oceanogr. 11:520-528.
11. Billen, G. 1978. A budget of nitrogen recycling in North Sea sediments off the Belgian Coast. Est. Coastal Mar. Sci. 7:127-146.

12. Bowman, M.J. 1977. Nutrient distributions and transport in Long Island Sound. Est. Coastal Mar. Sci. 5:531-548.
13. Brand, T. von, N.W. Rakestraw, and C.E. Renn. 1937. The experimental decomposition and regeneration of nitrogenous organic matter in sea water. Biol. Bull., Mar. Biol. Lab., Woods Hole 72:165-175.
14. Butler, E.I., E.D.S. Corner, and S.M. Marshall. 1969. On the nutrition and metabolism of zooplankton. VI. Feeding efficiency of *Calanus* in terms of nitrogen and phosphorus. J. Mar. Biol. Assn. U.K. 49:977-1001.
15. Butler, E.I., E.D.S. Corner, and S.M. Marshall. 1970. On the nutrition and metabolism of zooplankton. VII. Seasonal survey of nitrogen and phosphorus excretion by *Calanus* in the Clyde Sea area. J. Mar. Biol. Ass. U.K. 50:525-560.
16. Cooper, L.H.N. 1933. Chemical constituents of biological importance in the English Channel, Nov. 1930-Jan. 1932. J. Mar. Biol. Ass. 18:617-628.
17. Curl, H., Jr. 1962. Analyses of carbon in marine plankton organisms. J. Mar. Res. 20:181-188.
18. Davies, J.M. 1975. Energy flow through the benthos in a Scottish sea loch. Mar. Biol. 31:353-362.
19. Durbin, A.G. 1976. The role of fish migration in two coastal ecosystems: 1. The Atlantic menhaden, *Brevoortia tyrannus*, in Narragansett Bay, R.I., 2. The anadromous alewife, *Alosa pseudoharengus*, in Rhode Island ponds. Ph.D. Thesis, Univ. of Rhode Island, Kingston, Rhode Island, p. 216.
20. Elderfield, H., N. Leudtke, R.J. McCaffrey, and M. Bender. In press. Benthic flux studies in Narragansett Bay. Am. J. of Sci.
21. Es, F.B. van. 1977. A preliminary carbon budget for a part of the EMS estuary: The Dollard. Helgo. wiss. Meeresunters. 30:282-294.
22. Fleming, R.H. 1940. The composition of plankton and units for reporting populations and production. Proc. Sixth Pacific Sci. Congr. 3:535-540.
23. Fenchel, T. 1969. The ecology of marine microbenthos. Part IV. Ophelia 6:1-182.
24. Fenchel T., and P. Harrison. 1976. The significance of bacterial grazing and mineral cycling for the decomposition of particulate detritus, 285-299. *In* J.M. Anderson and A. Macfadyen (eds.), The Role of Terrestrial and Aquatic Organisms in Decomposition Processes, Blackwell Sci., London.
25. Fenchel T. 1977. The significance of bactivorous protozoa in the microbial community of detritial particles, 529-544. *In* J. Cairns, Jr. (ed.), Aquatic Microbial Communities, Garland Publishing, New York.
26. Fisher, T.R., P.R. Carlson, and R.T. Barber. Sediment nutrient fluxes in three North Carolina estuaries. Submitted to *Limnol. Oceanogr.*
27. Furnas, M.J., G.L. Hitchcock, and T.J. Smayda. 1976. Nutrient-phytoplankton relationships in Narragansett Bay during the 1974 summer bloom, 118-134. *In* M.L. Wiley (ed.), Estuarine Processes, Vol. 1, Uses, Stresses and Adaptation to the Estuary, Academic Press, New York.

28. Goldman, J.C., K.R. Tenore, and H.I. Stanley. 1973. Inorganic nitrogen removal from wastewater: effect on phytoplankton growth in coastal marine waters. Sci. 180:955-956.
29. Goldman, J.C., J.J. McCarthy, and D.G. Peavey. 1979. Growth rate influence on the chemical composition of phytoplankton in oceanic waters. Nature 279:210-214.
30. Haines, E.B. 1975. Nutrient inputs to the coastal zone: the Georgia and South Carolina shelf, 303-322. *In* L.E. Cronin (ed.), Estuarine research, Vol. 1, Academic Press, New York.
31. Haines, E.B. 1976. Stable carbon isotope ratios in the biota, soils and tidal water of a Georgia salt marsh. Est. Coastal Mar. Sci. 4:609-616.
32. Haines, E.B. 1977. The origins of detritus in Georgia salt marsh estuaries. Oikos 29:254-260.
33. Hargrave, B.T. 1973. Coupling carbon flow through some pelagic and benthic communities. J. Fish. Res. Bd. Can. 30(9):1317-1326.
34. Harris, E., and G.A. Riley. 1956. Oceanography of Long Island Sound, 1952-1954. VIII. Chemical Composition of the Plankton. Bull. Bingham oceanogr. Coll. 15:315-323.
35. Harris, E. 1959. The nitrogen cycle in Long Island Sound. Bull. Bingham oceanogr. Coll. 17:31-65.
36. Harrison, J.T. In prep. Biological mediation of benthic nutrient flux in Kaneohe Bay, Hawaii. Ph.D. Thesis, Univ. of Hawaii.
37. Harrison, W.G., and J.E. Hobbie. 1974. Nitrogen budget of a North Carolina estuary. Water Res. Res. Inst., Univ. of North Carolina, Report No. 86, p. 172.
38. Harrison, W.G. 1978. Experimental measurements of nitrogen remineralization in coastal waters. Limnol. Oceanogr. 23(4):694-694.
39. Hartwig, E.O. 1975. The impact of nitrogen and phosphorus release from a siliceous sediment on the overlying water, 103-117. In: M. Wiley (ed.), Estuarine Processes, Vol. 1. Academic Press, New York.
40. Hartwig, E.O. 1978. Factors affecting respiration and photosynthesis by the benthic community of a subtidal siliceous sediment. Mar. Biol. 46:282-293.
41. Heinle, D.R., and D.A. Flemer. 1976. Flows of materials between poorly flooded tidal marshes and an estuary. Mar. Biol. 35:359-373.
42. Holm-Hansen, O., and T.H. Mague. 1973. Chemical composition of particulate matter, 123-124. *In* Research on the maritime food chain. Progress Report 1972-73. Univ. of California, unpublished manuscript.
43. Hopkinson, C.S., J.W. Day, Jr., and B.T. Gael. 1978. Respiration studies in a Louisiana salt marsh. An. Centro. Cienc. Del Mary. Limnol. Univ. Nal. Auton. Mexico 5(1):225-238.
44. Hutchinson, G.E. 1950. Survey of contemporary knowledge of biogeochemistry. III. The biogeochemistry of vertebrate excretion. Bull. Amer. Mus. Nat. Hist. 96:544.
45. Hurtt, A. 1978. The distribution of hydrocarbons in Narragansett Bay sediment cores. M.S. thesis, Univ. of Rhode Island, Kingston, Rhode Island.
46. Jeffries, H.P. 1962. Environmental characteristics of Raritan Bay, a polluted estuary. Limnol. Oceanogr. 7:21-31.

47. Johannes, R.E. 1964. Phosphorus excretion as related to body size in marine animals: microzooplankton and nutrient regeneration. Science 146:923-924.
48. Johannes, R.E. 1965. Uptake and release of dissolved organic phosphorus by representatives of a coastal marine ecosystem.. Limnol. Oceanogr. 9:224-234.
49. Johannes, R.E. 1969. Nutrient regeneration in lakes and oceans, 203-212. *In* M.R. Droop and E.J.F. Wood (eds.), Advances in the Microbiology of the Sea, Vol. 1. Academic Press, New York.
50. Kemp, W.M., and W. Boynton. 1979. Nutrient budgets in a coastal plain estuary: Sources, sinks and internal dynamics. Amer. Soc. Limnol. Oceanogr., 42 Annual Meeting, Abstracts.
51. Ketchum, B.H., R.F. Vaccaro and Nathaniel Corwin. 1958. The annual cycle of phosphorus and nitrogen in New England coastal waters. J. Mar. Research 17:282-301.
52. Kremer, J.N., and S.W. Nixon. 1978. *A Coastal Marine Ecosystem, Simulation and Analysis*, Ecological Studies 24. Springer-Verlag, New York.
53. Kremer, P. 1975. The Ecology of the ctenophore *Mnemiopsis leidyi* in Narragansett Bay. Ph.D. Thesis, University of Rhode Island, Kingston, Rhode Island, p. 311.
54. Kuenzler, E.J. 1961. Phosphorus budget of a mussel population. Limnol. Oceanogr. 6:400-415.
55. Mann, K.H. 1972. Macrophyte production and detritus food chains in coastal waters. Mem. Ist. Ital. Idrobiol. 29:353-383.
56. Martin, J.H. 1968. Phytoplankton-zooplankton relationships in Narragansett Bay. III. Seasonal changes in zooplankton excretion rates in relation to phytoplankton abundance. Limnol. Oceanogr. 13:63-71.
57. McCaffrey, R.J., A.C. Myers, E.Davey, G. Morrison, M. Bender, N. Luedtke, D. Cullen, P. Froelich, and G. Klinkhammer. 1978. Benthic fluxes of nutrients and manganese in Narragansett Bay, Rhode Island, Limnol. Oceanogr., in prep.
58. Nixon, S.W., and C.A. Oviatt. 1973. Ecology of a New England salt marsh. Ecological Monogr. 43(4):463-498.
59. Nixon, S.W., C.A. Oviatt, and S.S. Hale. 1976. Nitrogen regeneration and the metabolism of coastal marine bottom communities, 269-283. *In*: J.M. Anderson and A. Macfadyen (eds.), The Role of Terrestrial and Aquatic Organisms in Decomposition Processes. Blackwell Scientific Pub., London.
60. Nixon, S.W., C.A. Oviatt, J. Garber, and V. Lee. 1976. Diel metabolism and nutrient dynamics in a salt marsh embayment. Ecology 57(4):740-750.
61. Nixon, Scott W. and Virginia Lee. 1980. The flux of carbon, nitrogen and phosphorus between coastal lagoons and offshore waters, 12 p. In: Unesco, 1980 (in press). Coastal Lagoons: Present and Future Research, Part II - Proceedings. (Unesco technical papers in Marine Science).
62. Nixon, S.W., J.R. Kelly, B.N. Furnas, and C.A. Oviatt. 1980. Phosphorus regeneration and the metabolism of coastal marine bottom communities, 219-242. *In* K.R. Tenore and B.C. Coull (eds.), Marine Benthic Dynamics. Univ. of South Carolina Press, Columbia.

63. Nixon, S.W. 1980. Between coastal marshes and coastal waters - a review of twenty years of speculation and research on the role of salt marshes in estuarine productivity and water chemistry, 437-525. *In* P. Hamilton and K. MacDonald (eds.), Estuarine and Wetland Processes, Plenum Publishing, N.Y.
64. Odum, E.P. and de la Cruz, A.A. 1967. Particulate organic detritus in a Georgia salt marsh-estuarine ecosystem, 383-388. *In* G. Lauff (ed.), Estuaries. Amer. Assos. Adv. Sci. Publ. 83.
65. Odum, E.P. 1968. A research challenge: Evaluating the productivity of coastal and estuarine water, 63-64. *In* Proc. 2nd Sea Grant Conf., Grad. School of Oceanography, Univ. of Rhode Island, Kingston, Rhode Island.
66. Oviatt, C.A., and S.W. Nixon. 1975. Sediment resuspension and deposition in Narragansett Bay. Est. Coastal Mar. Sci. 3:201-217.
67. Pamatmat, M.M., and K. Banse. 1969. Oxygen consumption by the seabed. 2. *In situ* measurement to a depth of 180 m. Limnol. Oceanogr. 14:250-259.
68. Pamatmat, M.M. 1968. Ecology and metabolism of a benthic community on an intertidal sandflat. Int. Revue. ges. Hydrobiol. 53(2):211-298.
69. Parsons, T.R., K. Stephens, and J.D.H. Strickland. 1961. On the chemical composition of eleven species of marine phytoplankters. J. Fish. Res. Bd. Can. 18:1001-1016.
70. Petersen, C.J.G. 1915. A preliminary result of the investigation on the valuation of the sea. Rep. Danish Biol. Sta. 23:29-33.
71. Peterson, David H. 1979. Sources and sinks of biologically reactive oxygen, carbon, nitrogen, and silica in northern San Francisco Bay, pp. 175-193. In: T. John Conomos (ed.) *San Francisco Bay: The Urbanized Estuary*. San Francisco, CA.
72. Pilson, M.E.Q., R. Beach, G. Douglas, and C. Cummings. 1978. Sediment chemistry in the MERL microcosms, 541-626. *In* Marine Ecosystem Research Laboratory Annual Report, Grad. School of Oceanography, Univ. of Rhode Island.
73. Pomeroy, L.R. 1970. The strategy of mineral cycling, 171-190. *In* R.F. Johnston (ed.), Annual Review of Ecology and Systematics, Vol. 1.
74. Pomeroy, L.R. 1974. Cycles of essential elements. *In* Benchmark Papers in Ecology, Vol. 1. Dowden, Hutchinson & Ross, Inc.
75. Propp, M.V., V.G. Tarasoff, I.I. Gherbadgi, and N.V. Lootzik. 1980. Benthic pelagic oxygen and nutrient exchange in a coastal region of the sea of Japan, 265-284. *In* K.R. Tenore and B.C. Coull (eds.), Marine Benthic Dynamics, Univ. South Carolina Press, Columbia.
76. Redfield, A.C. 1934. On the proportions of organic derivatives in sea water-their relation to the composition of the plankton, 176-192. *In* James Johnstone Memorial Volume. Liverpool. Univ. Press, Liverpool.
77. Renn, C.E. 1937. Bacteria and the phosphorus cycle in the sea. Biol. Bull. 72:190-195.
78. Riley, G.A. 1941. Plankton studies III. Long Island Sound. Bull. Bingham Oceanogr. Coll. 7(3):1-93.
79. Riley, G.A. 1970. Particulate organic matter in the sea. Adv. Mar. Biol. 8:1-118.

80. Rowe, G.T., C.H. Clifford, K.L. Smith, Jr., P.L. Hamilton. 1975. Benthic nutrient regeneration and its coupling to primary productivity in coastal waters. Nature 255:215-217.
81. Rowe, G.T., K.L. Smith, Jr., and C.H. Clifford. 1976. Benthic-pelagic coupling in the New York Bight. *In* M.G. Gross (ed.), ASLO Special Symposium, Vol. 2, 1975.
82. Rowe, G.T., C.H. Clifford, and K.L. Smith, Jr. 1977. Nutrient regeneration in sediments off Cap Blanc, Spanish Sahara. Deep-Sea Res. 24:57-63.
83. Ryther, J.H. 1954. The ecology of phytoplankton blooms in Moriches Bay and Great South Bay, Long Island, New York. Biol. Bull. 106:198-209.
84. Ryther, J.H., and W.M. Dunstan. 1971. Nitrogen, phosphorus, and eutrophication in the coastal marine environment. Science 171:1008-1013.
85. Santschi, P.H., Y.H. Li, and W.S. Broecker. 1978. Ratioactive trace metal cycling, 640-715. *In* Marine Ecosystems Research Laboratory Annual Report, Grad. School of Oceanography, Univ. of Rhode Island, Kingston, Rhode Island.
86. Schindler, D.W. Eutrophication in lakes and its relevance to the estuarine environment. Proc. Int. Symp. on Nutrient Enrichment in Estuaries, Williamsburg, VA, 1979.
87. Seitzinger, S., S. Burke, J. Garber, S. Nixon, M.E.Q. Pilson. 1978. Nitrogen fixation and denitrification measurements in Narragansett Bay sediments. 41st Annual Meeting Amer. Society of Limnol. and Oceanogr. Abstracts.
88. Seitzinger, S., S.W. Nixon. 1979. Denitrification and nitrous oxide production in Narragansett Bay sediments. 42nd Annual Meeting Amer. Society of Limnol. and Oceanogr. Abstracts.
89. Seitzinger, S., S. Nixon, M. Pilson and S. Burke. 1980. Denitrification and N_2O production in near-shore marine sediments. Geochem. Cosmochem. Acta. 44:1853-1860.
90. Seki, H., and O.D. Kennedy. 1969. Marine bacteria and other heterotrophs as food for zooplankton in the Strait of Georgia during the winter. J. Fish. Res. Bd. Can. 26:3165-3173.
91. Sheith, M.S. 1974. Nutrients in Narragansett Bay sediments. M.S. Thesis. Univ. of Rhode Island, Kingston, Rhode Island.
92. Smayda, T.J. 1957. Phytoplankton studies in lower Narragansett Bay. Limnol. Oceanogr. 2:342-359.
93. Smith, K.L., Jr. 1973. Respiration of a sublittoral community. Ecology 54:1065-1075.
94. Smith, S.L. 1978. The role of zooplankton in the nitrogen dynamics of a shallow estuary. Est. Coastal Mar. Sci. 7:555-565.
95. Smith, S.V. 1978. Kaneohe Bay sewage relaxation experiment: Pre-Diversion Report, Hawaii. Inst. Mar. Biol. Mimeo.
96. Smith, S.V. 1981. Responses of Kaneohe Bay, Hawaii to relaxation of sewage stress, Proc. Int. Symp. on Nutrient Enrichment in Estuaries, Williamsburg, VA, 1979.
97. Stephens, K., R.W. Sheldon, and T.R. Parsons. 1967. Seasonal variations in the availability of food for benthos in a coastal environment. Ecology 48:852-855.

98. Strickland, J.D.H., O. Holm-Hansen, R.W. Eppley, and R.J. Linn. 1969. The use of a deep tank in plankton ecology. 1. Studies of the growth and composition of phytoplankton at low nutrient levels. Limnol. Oceanogr. 14:23-34.
99. Taft, J.L., and W.R. Taylor. 1976. Phosphorus dynamics in in some coastal plain estuaries, 79-89. *In* M. Wiley (ed.), Estuarine Processes, Vol. 1. Academic Press, New York.
100. Thayer, G.W. 1974. Identity and regulation of nutrients limiting phytoplankton production in the shallow estuaries near Beaufort, N.C. Oecologia (Berl.) 14:75-92.
101. Ustach, J.F. 1969. The decomposition of *Spartina alterniflora*. M.S. Thesis. North Carolina State Univ. at Raleigh, N.C., p. 26.
102. Valiela, Ivan and John M. Teal. 1979. The nitrogen budget of a salt marsh ecosystem. Nature 280:652-656.
103. Vargo, G.A. 1976. The influence of grazing and nutrient excretion by zooplankton on the growth and production of the marine diatom, *Skeletonema costatum* (Greville) Cleve, in Narragansett Bay. Ph.D. Thesis, Univ. of Rhode Island, Kingston, Rhode Island.
104. Waksman, S.A., C.L. Carey, and H.W. Reuszer. 1933. Marine bacteria and their role in the cycle of life in the sea. I. Decomposition of marine plant and animal residues by bacteria. Biol. Bull. Mar. Biol. Lab., Woods Hole 65:57-79.
105. Watt, W.D., and F.R. Hayes. 1963. Tracer study of the phosphorus cycle in sea water. Limnol. Oceanogr. 8:276-285.
106. Wiebe, W.J., and L.R. Pomeroy. 1972. Microorganisms and their association with aggregates and detritus in the sea: A microscopic study. Mem. Ist. Ital. Idrobiol. 29:325-352.
107. Williams, P.J. 1970. Heterotrophic utilization of dissolved organic compounds in the sea. 1. Size distribution of population and relationship between respiration and incorporation of growth substrates. J. Mar. Biol. Assoc. U.K. 50:859-870.
108. Winter, D.F., K. Banse, and G.C. Anderson. 1975. The dynamics of phytoplankton blooms in Puget Sound, a Fjord in the Northwestern United States. Mar. Biol. 29:139-176.
109. Woodwell, G.M., D.E. Whitney, C.A.S. Hall, and R.A. Houghton. 1977. The Flax Pond ecosystem study: Exchanges of carbon in water between a salt marsh and Long Island Sound. Limnol. Oceanogr. 22:833-838.
110. Yanaha, M., and M. Yoshiaki. 1978. Production and decomposition of particulate organic matter in Funka Bay, Japan. Est. Coastal Mar. Sci. 6:523-533.
111. Yoshida, Y. and M. Kimata. 1969. Studies on the marine microorganisms utilizing inorganic nitrogen compounds-IV. On the liberation rates of inorganic nitrogen compounds from bottom muds to sea water. Bull. of Jap. Soc. Sci. Fish. 35(3):303-306.

UPTAKE OF MAJOR NUTRIENTS BY ESTUARINE PLANTS

James J. McCarthy
Museum of Comparative Zoology
Harvard University
Cambridge, Massachusetts 02138

INTRODUCTION

Within the literature that treats the utilization of major nutrients by aquatic plants there is no evidence that the estuarine habitat is unique with regard to plant nutrition. In fact, observed patterns for the relationship between phytoplankton utilization of inorganic nitrogen and phosphorus and the quantity or quality of these nutrients seem to apply generally for organisms of marine, estuarine, and fresh water origin.

Much of the impetus for the laboratory physiological and autecological investigations that have addressed problems in aquatic plant nutrition during the last decade has arisen from field observations. For aquatic habitats in which the concentrations of major nutrients are frequently near the limits of analytical detection, it is tempting to conclude that autotroph production is nutrient limited. Such limitation is, however, difficult to demonstrate, and in addressing this issue one must often use inference derived from laboratory studies with clonal isolates.

Experimentation designed to investigate aquatic plant nutrition in a natural setting is complicated by the experimental need to perturb the immediate environment of the organism. Standard means of measuring the physiological activity of an aquatic plant population require capturing and containing the population in isolation from herbivores. In an environment free from grazing pressure, the plant demand for nutrient per unit volume of contained medium increases with time, and moreover, in nature the herbivores may serve as an important source of recycled nutrients. There are other potentially serious container effects which can compromise the quality of physiological measurements made during the incubation of captured samples (41). For example, it has been demonstrated that both with and without selective removal of different size classes of organisms, captured natural samples can undergo substantial population changes (decreases as well as increases) in periods of a day or less (106).

In many respects the estuaries provide a good setting for field experiments in the nutrition of aquatic plants. First, biomass levels which are above those common in neritic and oceanic waters permit the experimentalist to correspondingly reduce the duration of the incubation period in nutritional studies. Second, the great range in estuarine nutrient concentrations particularly facilitates experimentation on nutrient preferences. Third, in many estuaries the spatial and temporal hydrographic features are conducive to an experimental design which treats temperature and salinity as variables.

The following review will consider results derived from laboratory biochemical and physiological experiments as well as those from field studies. In the laboratory investigation the population is identifiable and the environment is both definable and controllable, whereas in field studies it is often difficult or impossible to even inventory either the majority of the resident populations or the important physical and chemical aspects of the environment. Consequently, the physiological ecologist interested in interpreting field observations often must contend with an understanding from laboratory experiments that is of much greater sophistication than that based upon field investigations.

One purpose of this review is the demonstration of the great range in data quality obtainable from laboratory and field investigation. It begins with a consideration of laboratory physiological studies, proceeds to field observations relating to the effects of three environmental variables, and finally considers the problem of relating rates of nutrient uptake to growth. This last area is a good example of the use of sophisticated laboratory studies in generating the understanding necessary to interpret field observations.

BIOCHEMICAL CONSIDERATIONS

Recent treatises on algal physiology and biochemistry have reviewed current knowledge of the pathways and energetic requirements associated with the assimilation of major nutrients (16, 59, 76). The capacity to utilize the dominant dissolved inorganic forms of nitrogen (NO_3^-, NO_2^-, and NH_4^+) and phosphorus (PO_4^{3-}) is near universal among the algae, while for dissolved organic nitrogen, clones of a single genus may vary dramatically in their potential to meet nutritional needs with specific organic compounds (46). The enzymes responsible for catalyzing reaction steps in assimilatory NO_3^- reduction {nitrate reductase (NR),, nitrite reductase (NiR), and glutamate dehydrogenase (GDH)} in marine phytoplankton have been characterized (29), and they appear to function similarly to enzymes that have long been subjects of biochemical investigations in fungi and higher plants. More recently, glutamine synthetase (GS) and glutamate synthase (GOGAT) have been identified as alternatives to the glutamate dehydrogenase pathway for ammonium assimilation (35, 98). It is also now recognized that there are at least two paths of urea utilization: some algae have urease whereas others have a urea amidolyase system (61).

In contrast to assimilatory pathways, we know relatively little regarding the mechanism of uptake, or transport from the growth medium across the plasmalemma to the cytoplasm. Particularly for nitrogen, it can be assumed that this step is one of mediated transport. Intracellular concentrations for NO_3^-, NO_2^-, and NH_4^+ must be several orders of magnitude higher than typical natural water concentrations in order for the assimilatory enzymes to function efficiently (29, 40). The actual pool of substrate within the cell may be of a volume that is physically small relative to the total cell volume, and the concentration of substrate in such a pool would then be proportionally greater than a measurable whole cell concentration. An exception to the pattern of extremely high K_m values for assimilatory enzymes relative to typical natural media concentrations is glutamine synthetase. It has been shown to have a K_m of 29 μg-atoms N·liter^{-1} in a marine diatom (35), and 12-15 μg-atoms N·liter in duck weed (91).

In algae, only one specific membrane-bound transferase for nitrogen has been identified, a nitrate and chloride-actived ATPase (34). It has a K_m of 0.9 μg-atoms N·liter^{-1}, and thus could be an effective transport system for nitrate at typical natural water concentrations. One can postulate that similar transport systems, possibly involving carrier molecules which are continuously synthesized and allowed to accumulate on the cell membrane during periods of reduced nutrient availability, are operative for other nitrogenous and phosphorus nutrients (55, 62, 79). Evidence that the enzymes involved must be subject to feedback control is seen in data from studies of nitrogen deficient phytoplankton which are exposed to a pulse of NO_3^- and NH_4^+. Both nutrients are taken up initially at high rates (18) but nitrate reductase activity is absent while NH_4^+ availability is adequate to meet the total nitrogen requirement (25). Hence, the NO_3^- which entered the cell is not reduced, and the internal pooling of NO_3^- apparently suppresses the activity of the NO_3^- carrier system. For *Chlorella* cultures in which nitrate reductase activity has been suppressed by NH_4^+ presence, NO_3^- will be taken up a few hours after exposure (77). Since nitrate reductase is synthesized during the lag period, it can be hypothesized that there is a close coupling between nitrate reductase activity and the actual uptake or transport system. The transport system may, however, simply be sensitive to feedback response from accumulation of nitrate in a pool which is too small to be conveniently measured.

Another class of enzymes associated with the cell surface, which are widely represented in algae that populate waters depleted in inorganic phosphate, are the alkaline phosphatases (2, 66, 90, 94). They facilitate the utilization of organically bound phosphomonoesters by a hydrolysis which releases PO_4^{3-} in immediate proximity to the cell surface.

ENVIRONMENTAL FACTORS

Dependence of Phytoplankton Nutrient Uptake Upon Light

The effect of light on the inorganic nutrition of aquatic plants has two distinct components: the relationship between uptake and/or assimilation of nutrients as a function of the quantity of light, and the degree to which rates of uptake and/or assimilation vary between the light and the dark portions of the day.

In a plant, any active transport across the plasmalemma or subsequent energy requiring transformation of a nutrient ion ultimately depends upon photosynthesis. Feedback mechanisms may operate at several levels, and with interspecific individuality. When the total incorporation of a nutrient by a plant is measured at the end of a dark period following prior exposure to light, and the incorporation rate is shown to be less in darkness than in light, one cannot easily discern whether the actual uptake step, a transformation process such as NO_3^- reduction, or the coupling of the nutrient material to carbon skeletons was most affected by the cessation of photosynthetic activity. *Euglena gracilis* exhibits higher light than dark uptake rates for PO_4^{3-} grown under both phosphorus-sufficient and-deficient conditions (12, 13). However, some natural populations of phytoplankton show no reduction in rates of PO_4^{3-} incorporation following transfer from natural light to total darkness (97). In studies with nitrogen-limited populations grown in continuous culture out of doors, NO_3^- (65) and NO_3^- plus NH_4^+ (45) uptake remained comparable during light and dark periods.

Furthermore, MacIsaac and Dugdale (64) demonstrated that when plotting nitrogen uptake versus light, it can be shown that natural phytoplankton assemblages in coastal upwelling regions have a half-saturation constant for both NO_3^- and NH_4^+ incorporation which is 1-14 percent of the light at the sea surface. These populations were probably not nitrogen limited, and this light level is similar to that reported by Packard (80) as necessary for maximum NR activity. Eppley et al. (31) have shown, however, that for *Emiliania (Coccolithus) huxleyi* grown in continuous culture, substantially reduced night time activities in NR, NiR, and GDH did not prevent this alga from stripping all available NO_3^- and NH_4^+ from the media during darkness. Under similar experimental conditions, *Skeletonema costatum* had markedly reduced rates of nitrogen uptake in darkness.

Discussions regarding the importance of the functionality between rates of nutrient utilization, or indeed any other light-dependent physiological process, and quantity of light must take into consideration both the vertical distribution of organisms and the temporal pattern in this distribution along the vertical light gradient. In an open ocean situation characterized by high physical stability of the mixed layer, near surface populations of organisms that lack an effective means for regulating their depth in the water column in response to light exposure may experience photosynthetic inhibition. In estuaries and coastal embayments where the mixed layer is shallow, and occasionally coincident with the euphotic zone (86), rates of mixing may be adequate to prevent photosynthetic inhibition in phytoplankton by reducing exposure to near surface light (51). By analogy, the highly mixed shallow euphotic zone would be less likely to show vertical structure in either phytoplankton distribution or phytoplankton nutritional state. Routine procedures for assessing phytoplankton physiological responses entail collecting a sample from a specified depth of light penetration and subsequently holding the samples for an incubation period at the light level represented by the depth of capture. If this period is long relative to a representative natural period of exposure to constant light (perhaps more than only a few minutes), then the physiological rates determined from incubated samples and inferences drawn concerning light inhibition can be erroneous.

Preferential Utilization of Different Forms of Nitrogen and Phosphorus

Estuarine waters can provide a unique experimental setting for determining the nutritional preferences of natural assemblages of phytoplankton. The great temporal range in the concentrations of nitrogenous nutrients plus the relatively high levels of biomass particularly facilitate the execution of nitrogen preference studies. There is a great wealth of nutrient data for estuaries, some of which has been incorporated into models of estuarine biological and chemical processes, (see, for example, 93, 99) but most such efforts have given little attention to the importance of the rapid recycling of some forms of nitrogen.

Among the nitrogenous nutrients, NO_3^- and NH_4^+ (the latter to a lesser degree in most cases) are likely to be the dominant allochthonous forms, whereas both NH_4^+ and urea are the forms most likely to be rapidly recycled in situ. Through the use of ^{15}N enriched preparations of these nutrients, the significance of each individual form of nitrogenous nutrient can be assessed relative to that of other forms. The role of dissolved organic nitrogen as a plant nutrient has, with the exception of urea, received little attention. Certain amino acids are known to be utilized by coastal phytoplankton (85, 108), but the nutritional role of this class of compounds is much more difficult to study and, furthermore, the bulk of the dissolved organic nitrogen (DON), which has not been adequately characterized, is generally of unknown nutritional significance to phytoplankton. Some direct and indirect evidence suggests, however, that dissolved free amino acids probably play a minor role (85) while DON other than amino acids and urea contributes negligibly to photoautotroph nutrition (100).

Some estuaries exhibit great temporal cycles in dissolved phosphorus availability while others do not (95). In most instances, the PO_4^{3-} ion is the dominant form of phosphorus, but at least a portion of the dissolved organic phosphorus (DOP) pool turns over very rapidly (97). There is, in fact, clear evidence that the phosphorus needs of some phytoplankton can be met by organic phosphorus (57, 94). For much of the year in some estuaries, the abundance of DOP may exceed that of PO_4^{3-} (10, 96), but the only component of the aggregate DOP category known to be nutritionally useful to phytoplankton is the monophosphoester (MPE). A portion of the DOP released by phytoplankton can be reassimilated by the same or other phytoplankton in marine waters and freshwaters (58, 60), and whereas the assimilatory path for MPE utilization, alkaline phosphatase mediated hydrolysis, is well known (2, 81, 94), the contribution of MPE to the phosphorus nutrition of populations in nature is extremely difficult to estimate (see, for example, 66).

Two field programs have examined the relative importance of different nitrogenous nutrients in meeting the nutritional needs of estuarine phytoplankton. Harvey and Caperon (53) studied the utilization of NO_3^-, NH_4^+, and urea by natural phytoplankton assemblages in the southern sector of Kaneohe Bay, Hawaii, which receives a discharge of secondary treated sewage. Although there is good evidence that the sewage input has substantially altered plankton nutrition in the southern part of the bay, (7) it is now estimated that this input of nitrogen is contributing only 3.5 percent of the phytoplankton nitrogen demand (53). Total nitrogenous nutrient availability is dominated by NH4 and urea, and it rarely exceeded a few μg-atoms N·liter^{-1}. Typical quantities of available $PO_4^{=}$ are great enough to dismiss considerations of phytoplankton production rates limited by phosphorus, but neither is there any clear indication that nitrogen is limiting. Only for urea are there estimates of both maximum and in situ rates of uptake. In 4 of 18 experiments, the maximum exceeded in situ rate by 1.5-2.5 fold, and otherwise they are indistinguishable. From the continuous culture analogy (see below) these findings would indicate that in most instances the phytoplankton were not showing the marked enhancement of V_m' characteristic of nutrient-stressed steady state populations. In the Kaneohe Bay study, the order of nutrient perference, like that found in enriched natural populations (67), proceeded from NH_4^+, to urea, to NO_3^-.

In a longer term study of phytoplankton nutrition in the main body of the Chesapeake Bay (69, 70, 97) there was also no clear indication that either nitrogenous or phosphorous nutrients limited the rate of autotroph production. In addition to NO_3^-, NH_4^+, and urea, this study also examined the significance of NO_2^- as a nitrogen source for the phytoplankton. For the Chesapeake Bay study, a consistent pattern in nitrogenous nutrient preference emerges. With 98 percent regularity, an NH_4^+ concentration in excess of 1-1.5 μg-atoms N·liter suppressed NO_3^- utilization, irrespective of

NO_3^- concentration, such that NO_3^- accounted for < 5 percent of the phytoplankton nitrogen diet. Other laboratory and field data (14, 18, 25, 77, 92) have indicated the potential of NH_4^+ to suppress NO_3^- uptake and/or assimilation, but the universality of this phenomenon at common natural concentrations of NH_4^+ is only now becoming recognized. McCarthy et al. (70) made use of a relative preference index (RPI) calculated for each form of nitrogenous nutrient in assessing the competitive interaction between suitable alternative substrates

$$RPI_{Ni} = \frac{\dfrac{V_{Ni}}{\Sigma\ V_N}}{\dfrac{(S_{Ni})}{\Sigma\ (S_N)}} \qquad (1)$$

where Ni is a particular form of nitrogenous nutrient N, V is the rate of uptake and S is the concentration. This approach has also been applied in coastal studies (27), and thus far it has demonstrated that when NH_4^+ availability is adequate to meet the entire nitrogenous demand of the aggregate populations, urea may or may not be utilized while NO_3^- is rejected. An RPI for NH_4^+ is never less than unity, indicating that it is never rejected, while those for NO_3^- never exceed unity. Moreover, the RPI values for NO_3^-, NH_4^+, urea, and NO_2^- all converge at values of unity when availability of the more preferred forms is insufficient to meet the nutritional needs of the organisms.

For macroalgae, the saturation of nutrient uptake occurs at higher concentrations than for microalgae, so the likelihood of observing simultaneous uptake of NH_4^+ and NO_3^- at concentrations typical of coastal waters increases. As with microalgae, it is, however, difficult to design uptake experiments for realistic ranges of substrate concentration. In part, this may be due to an actual difference in the nature of the relationship of the uptake response to substrate concentration, or it may simply result from a substantial depletion of substrate during the uptake experiment. This latter explanation is consistent with the elevated values noted in the low concentration range of the S/V vs S transformation used by Topinka (102) and Hanisak and Harlin (48).

Within the data for macroalgal uptake of NO_3^- and NH_4^+ there seems to be no consistent pattern in the relationship between uptake and substrate concentration. In some studies saturation kinetics have been observed (47, 48) while for others the uptake response apparently consists of a low substrate component which saturates plus a higher substrate component which increases linearly with concentration (17, 47, 102). D'Elia and DeBoer (17) have discussed the importance of avoiding preconceived notions regarding the functional dependence of nutrient uptake upon concentration.

For some species of marine macroalgae, the maximum rates of NH_4^+ uptake are much greater than the maximum rates of NO_3^- uptake (17, 47, 48), and at high concentrations of both NH_4^+ and NO_3^-, some algae, e.g. *Macrocystis pyrifera* and *Fucus spiralis*, are capable of simultaneous uptake of both substrates (47, 102). For several other algae there is evidence of such simultaneous uptake at lower, less than saturating, NH_4^+ concentrations (3, 17, 48, 19, 102). This situation is analgous to that of the convergence of RPI values towards unity for all forms of nitrogen when the concentration(s) of the preferred form(s) is less than adequate to saturate the plant requirement. Since NO_3^- uptake proceeds in the presence of NH_4^+, the NO_3^- ion can pool intracellularly without reduction to NO_2^- during NH_4^+ uptake and assimilation. Since some macroalgae are capable of substantial storage of NO_3^- (11), the results of short term nutrient disappearance studies must be interpreted with caution.

As with microalgae and macroalgae, submerged aquatic macrophytes show preference for the more reduced forms of nitrogen. In a study of the nitrogen metabolism of the duck weed, *Spirodela oligorrhiza*, Ferguson and Bollard (37) demonstrated virtually complete suppression of NO_3^- uptake in the presence of NH_4^+. The relatively small amount of NO_3^- which was taken up could be accounted for in internal pools which remained a constant fraction of plant mass throughout several days of plant growth. That NO_2^- was taken up concurrently with NH_4^+ lends support to the hypothesis that NH_4^+ or products of its assimilation, even in plants which have high levels of NR and NiR activity, prevent the reduction of NO_3^- (36). NH_4^+ at concentrations >2 μg-atoms N·liter^{-1} was shown to suppress NO_3^- assimilation in *Myriophyllum spicatuus* (78).

In studies of plant nutrient uptake, efforts should be taken to minimize bacterial interference and in studies of aquatic grass and macroalgal uptake, it is particularly important to eliminate interference due to microscopic epiphytes. McRoy and Goering (74) provide good evidence for the potential of epiphytic involvement in such experiments, and the study undertaken by Bird (3) is one of the few which reportedly involved serious efforts to minimize contamination by microalgal epiphytes.

Marsh Plant Habitats

Spatial variation in plant heights among particular salt marsh grasses such as *Spartina alterniflora* has led to the suggestion that the short forms are morphs which have been stunted by environmental conditions. The short form of *S. alterniflora* occurs in habitats which range from low to high salinity (4), and data from several areas would now support the hypothesis that nutrient scarcity, particularly nitrogen, is the agent most responsible for stunting in this plant (4, 39, 104, 105). In each instance taller plant growth was stimulated in native short stands when nitrogen was either applied to the substrate surface or injected into the substrate interstices. Increased aerial biomass remained detectable weeks to months later, whereas no response was noted following nitrogen additions to native tall stands. There was no apparent effect of phosphorus additions with or without nitrogen. For another marsh plant, *Suaeda maritima*, the seedlings in the upper marsh area do not mature to the same heights as those in the lower marsh, and they have been shown to have lower total nitrogen, lower NO_3^- content, and lower NR activity (89). Within 72 hours of an addition of NO_3^- to the high marsh habitat, the NR activity in *S. maritima* increased to 50 times its previous level.

A issue which becomes important in addressing the nutrition of submerged aquatic plants is the relative efficiency of uptake pathways operative within the roots and the shoots. From a study with three species of a freshwater plant, *Potomogeton*, which had variously submerged leaves, floating leaves, and both submerged and floating leaves, Denny (19) concluded that anatomy and morphology were important in determining the principal site of nutrient absorption. There are, however, certain apparent physiological constraints. NR activity in the leaves of salt marsh plants is commonplace, whereas only one-third of 18 species examined has significant rates of activity in root tissues (90). In the leaves, enzyme activity was shown to increase with added NO_3^- and, as with microalgae the enzyme K_m of 2.4-3.0 x 10^{-4}M suggests the need for internal pooling of substrate.

There is some evidence that aquatic macrophytes can translocate nutrients from roots to emergent tissues. With ^{15}N labeled NO_3^-, NH_4^+, and urea McRoy and Goering (74) clearly demonstrated the rapid mobility of the labeled N from the root environment to ephiphytic algae attached to the leaves of *Zostera marina*. It is known that sea grasses are largely dependent upon phosphorus obtained from the sediment (72) and that a substantial portion of the phosphate absorbed from the sediments by *Z. marina* and *S. alterniflora* is released to the water via the leaves (73, 84). For nitrogen, however, the root to epiphyte transfer provides the first evidence for physiological leakage of combined nitrogen across the leaves of sea grass with subsequent absorption by the epiphyte (74).

Two studies with freshwater plants show physiological patterns which are in contrast to those in *Z. marina*. *Carex aquatilis*, characteristic of tundra ponds, takes NH_4^+ into root tissues with a Ks of 8.4-12.5 μg-atoms N·liter^{-1}, and rapidly translocates this nitrogen through the leaves to the water (71). Nitrogen could have been lost in another chemical form but, in a mass balance study for another freshwater plant, *Myriophyllum spicatum*, less than 1 percent of the nitrogen taken up by the roots was lost to the medium through the foliage (78). These authors also found little evidence for foliage to root translocation of nitrogen.

THE RELATIONSHIP BETWEEN PHYTOPLANKTON NUTRIENT UPTAKE, POPULATION GROWTH, AND NUTRIENT CONCENTRATIONS

The dependence of algal nutrient uptake rates upon nutrient concentration has been the subject of numerous investigations in recent years. The results reported by Harvey (52) for PO_4^{3-} uptake by *Phaeodactylum tricornutum* as replotted by Dugdale (21), and those reported by Eppley and Coatsworth (24) for NO_3^- uptake by *Ditylum brightwellii* were seminal contributions to our understanding of the uptake process. They demonstrated that nutrient uptake (V) is a hyperbolic function of substrate concentration (S) with the half saturation constant (K_s) equivalent to the concentration necessary to achieve half the maximal rate of uptake.

$$V = V_{max} \frac{S}{K_s + S} \tag{2}$$

This idea provided substance for hypotheses regarding the suitability of specific nutrient regimes for selected species or perhaps even races of species. Individual patterns of uptake response to nutrient concentration could be used both to ascertain the fitness of a specific clone to a given environment and to explain the temporal succession of dominant species as related to seasonal changes in nutrient availability (21, 56). Eppley et al. (30) demonstrated that the K_s values for NO_3^- and NH_4^+ uptake were different for oceanic and neritic clones, that these values were similar to typical nutrient concentrations for the habitat or origin, and that there was a direct correlation between cell size and K_s value. MacIsaac and Dugdale (63) also offered evidence for the significance of the K_s value as a measure of environmental fitness in their demonstration that average values for natural mixed population assemblages differed between oligotrophic, oceanic, and eutrophic neritic waters. Carpenter and Guillard (9) added further support to this notion with their observation that the pattern in K_s values for clones of a single species isolated from oceanic and neritic waters reflects the nutrient regime characteristic of the waters of origin for each clone.

The hypothetical potential for competitive advantage associated with paired K_s and V_{max} values proposed by Dugdale (21) is shown in Figure 1. The separate lines can be viewed as representative of a species or as a clone-specific response pattern to a nutrient such as NO_3^-. The uptake rate is in units of mass of the element of nutritional significance, i.e. the nitrogen in NO_3^-, which is taken up per unit organism mass of the same element per unit time. The convenience of expressing uptake in this manner, one which is equivalent to a growth constant in an equation for exponential growth (23), will be seen below.

Intuitively, one can argue that there must be a non trivial cost to maintaining both a high V_{max} as a large number of active sites, carrier

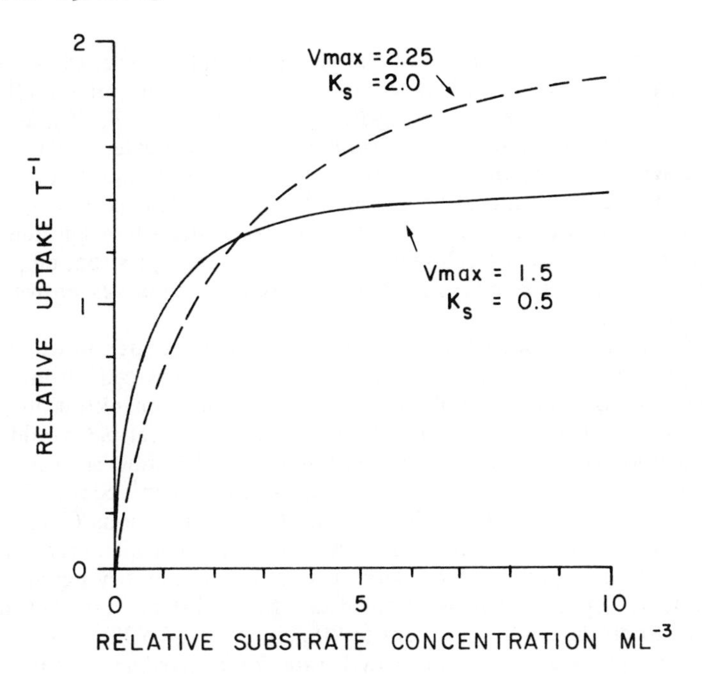

FIGURE 1. Hypothetical competitive interaction between two species or races which are characterized by different parameters for K_s and V_{max}.

molecules, etc., and a low K_s as a transport system with particularly high substrate affinity. If the cost were trivial, and if no other factors governed competitive interaction between phytoplankton, then selective pressure would lead to dominance by a species which has both a high V_{max} and a low K_s. One serious limitation to a broad application for this principle of competitive interaction is the species' specific light-dark variability in capacity for nutrient uptake,which is discussed above. If, for example, the quantity of carrier molecule or the significance of feedback from internal pools to the transport process vary markedly from light to dark periods, then competitive advantage with respect to the acquisition of a limiting nutrient may be determined primarily by the degree to which the uptake capacity for each competitor is diminished during the dark period (12, 13).

Most of the published data on K_s values for inorganic nutrient uptake by phytoplankton in culture are for nitrogen as either NH_4^+ or NO_3^-, and with broad confidence intervals few of these fall outside the range of 0.1-1.0 μg-atoms N·liter^{-1}. For much of the year, NO_3^- and NH_4^+ concentrations in the euphotic waters of major estuarine systems exceed these concentrations (50, 69, 88), whereas in the open ocean and in some estuaries during certain periods of the year (e.g. mid to lower Chesapeake Bay in summer), the nutrient concentrations approach the limits of analytical detection, i.e. <0.05 μg-atoms N·liter^{-1} (32, 68, 69). In low nutrient waters the question of phytoplankton growth rate regulation by nutrient availability becomes important. Is it possible for a phytoplankton population to achieve more than a small fraction of its maximum growth potential when nutrient concentrations are an order of magnitude less than the K_S value? In the following treatment of this question, emphasis will be placed on nitrogenous nutrients because of the more extensive data for this element.

Clearly, what is needed is an understanding of the degree to which the rate of nutrient uptake and the rate of synthesis for new cellular mass are coupled. To look only at published data sets for phytoplankton nutrient uptake by laboratory cultures as a function of concentrations would give one the impression that the maximum uptake rate for most species is not achieved at concentrations less than 3-5 μg-atoms N-liter^{-1}. Since V_{max} is approached asymptotically at high substrate concentrations (Figure 1), there is no precise threshold below which nutrient availability can be judged to limit uptake. One useful parameter in identifying nutrient limitation is K_{lim} which is defined as the substrate concentration sufficient to produce a V equivalent to 90 percent V_{max} (38).

The concept of a maximum growth rate for microalgae is subject to several interpretations. In the strictest sense, it is the rate attained under optimal environmental conditions for a certain species or race, but the multiple dimensions of the experimental matrix necessary to define these conditions is cumbersome, and furthermore, in the laboratory it is always to some degree unrealistic.

An operational definition of the maximum growth rate is that rate attained under specified environmental conditions. It is in this latter context that the term μ_{max} is usually applied in laboratory culture studies. When the composite of the populations is said to be growing at its maximal rate, the reference is usually meant to be interpreted within the context of prevailing environmental conditions. If these conditions were to change, then the same species or races constituting the assemblage would have a different μ_{max}, and if this is less than the previous μ_{max} then quite possibly some dominant constituent species would be replaced by others. To a large degree this notion is, and will remain for the foreseeable future, an abstraction. The vast majority of the species which dominate natural assemblages of phytoplankton are too small and too fragile to permit routine identification and enumeration via standard techniques for collection, preservation and microscopic examination.

In an operational sense, the maximum rate of growth is either the rate of population increase during exponential growth in batch culture, or the growth rate just prior to washout in chemostat culture. Goldman and Peavey (44) have shown that the maximum rates of growth and the maximum values for cellular nitrogen content are identical for *Dunaliella tertiolecta* grown in batch and chemostat culture.

In a study of the relationship between the parameters of nutrient uptake and nutrient regulated growth, Eppley and Thomas (33) concluded that half-saturation constants for growth from batch culture experiments were equivalent to half-saturation constants for nutrient uptake. Eppley et al. (30) observed, however, that the V_{max} for nitrogen uptake varied with time following nutrient depletion in batch cultures, and hence could not be equated with a maximum growth rate.

The coupling of nutrient uptake and cellular growth has been investigated in continuous cultures maintained at steady state in chemostats or turbidostats, which permit the investigator to select a specific growth rate by regulating the dilution rate of the culture volume. Under steady state conditions the chemical composition of both the medium and the organisms, in addition to the physiological state of the organisms, can then be related to a single growth rate.

In a recent theoretical treatment and experiments with PO_4^{3-} limited chemostats, Burmaster (5) and Goldman (42) provided evidence that the Monod model, which relates steady state growth rates to nutrient concentration, and the Droop model, which relates growth rates to cellular content of the limiting element, both adequately describe algal growth in chemostat culture. It has become clear, however, that for nutrient-limited steady state phytoplankton populations the maximum potential uptake can result in high, near maximal rates of growth at nutrient concentrations which are well below K_s values approximated from uptake curves, indeed even at concentrations which are below the current detection limits for phosphate (6, 42) and nitrogen (8, 26, 37). An example of this is seen in Figure 2, which represents *Thalassiosira pseudonana* (3H) growth as a function of NH_4^+ concentration. Uptake of NH_4^+ as a function of concentration for a population which was maintained at a specific growth rate of 1.55 day^{-1} prior to its exposure to variable nutrient concentrations for the purpose of measuring the uptake response (^{15}N incorporation) is demonstrated in Figure 3. Note, though, that at this steady state growth rate, the growth chamber concentration was below the limits of detection (Figure 2). Before considering these data further, the general relationships between steady state growth rates and uptake potential need to be examined.

Subsequent to the variable V_{max} observation with batch culture (30), numerous other investigators have observed V_{max} to vary with nutritional state (8, 15, 17, 26, 68, 81, 83). Dugdale (22) suggested V_m' be used to denote this "nutrient specific uptake rate" characteristic of a steady state population and hypothesized that

$$V_m' = \frac{\rho_m}{Q} \qquad (3)$$

FIGURE 2. Growth rate of *Thalassiosira pseudonana* (3H) as a function of substrate concentration in chemostat culture. Solid half circles represent unmeasurable NH_4^+ concentrations (<0.03 μg-atoms NH_4^+-N·liter^{-1}); * represents a value determined during culture washout. (After Goldman and McCarthy (43).

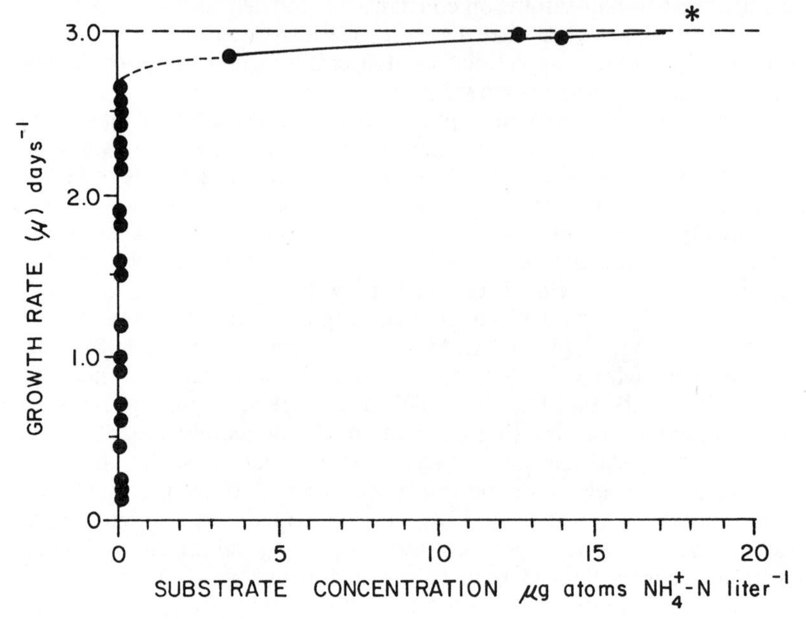

where Q is cellular content of the limiting element (M·cell^{-1}) and ρ is the rate of cellular uptake (M·cell^{-1}·T^{-1}). By analogy ρ and ρ_m (maximum ρ) can be substituted for V and V_{max} in Eq. 2. In steady state populations V_m' should then be a function of the growth rate; furthermore, Dugdale proposed that if ρ_m is constant, then V_m' should yield a hyperbola of the form $y = 1/x$. Data for nitrogen uptake by *Thalassiosira pseudonana* oceanic clone 13-1 (26) and for estuarine clone 3-H (68) in addition to phosphorus uptake by both *Scenedesmus* sp. (83) and by *Monochrysis lutheri* (6) indicate that ρ_m is not constant and hence the term ρ_m should be used to denote the maximum rate of nutrient uptake per cell for a population in steady state at a specific growth rate.

The elevation of V_m' at reduced growth rates can serve to benefit the organism inhabiting the nutrient deficient regime. It may be an effective strategy for coping with transient nutrient availability on scales which cannot be conveniently measured (1, 68, 103), by permiting the starved cell to minimize the time necessary to reestablish a maximum rate of non-nutrient limited growth following a period of nutrient limitation. It is, moreover, this potential for maintaining a V_m' well in excess of a given μ during steady state growth which permits the population to grow at near

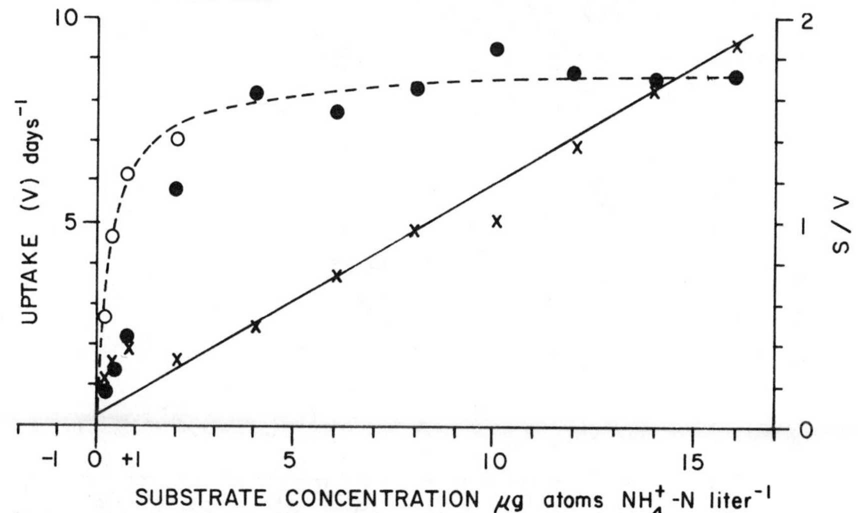

FIGURE 3. NH_4^+ uptake by *Thalassiosira pseudonana* grown in steady state chemostat culture at 1.55 day^{-1} as a function of nutrient concentration (solid circles). Uptake experiments were of five minutes duration. At S < 4 μg-atoms N·liter^{-1}, > 70 percent was utilized, and this is reflected in the elevation of these data points in the S vs S/V linear transformation (X). The degree of departure at low values for S relative to a linear extrapolation from high values for S was used to estimate the V values (open circles) which would be consistent with a true hyperbolic form (dashed line). For the culture conditions used in these experiments (43),> 25 percent of the 0.2 μg-atom NH_4^+-N·liter^{-1} addition would have been utilized in <0.5 min.

maximal rates when ambient nutrient concentrations are well below both the uptake parameters K_{lim} and K_s, and at times even below the current limits of nutrient detection. This concept was discussed by Caperon and Ziemann (8) and, by returning to Figure 3, it can be seen that for a steady state population growth rate of 1.55 day^{-1}, $\hat{V}_m$ is ~8.5 day^{-1} and K_s is 0.4 μg-atoms NH_4^+·liter^{-1}. As described in the figure legend, uptake data at low values of S are particularly difficult to obtain. A smooth curve drawn through the observed uptake values would give the impression that a concentration of approximately 0.5 μg-atoms NH_4^+·liter^{-1} would be necessary in order to maintain a V equivalent to the steady state growth rate, whereas the growth chamber concentration is at least an order of magnitude lower. If, however, we assume the V vs S relationship to be hyperbolic, then by estimating V values for low S values through extrapolation from the linear transformation, one can see the possibility of attaining a V of ~ 1.55 day^{-1} at a concentration < 0.03 μg-atoms N·liter^{-1}. What appears, therefore, as a discrepancy between data sets like those in Figures 2 and 3 can be reconciled if it is remembered that the former is a composite for multiple steady states and the latter is representative of a single steady state. Evidence suggests that for a single population, the K_S does not change as μ or V_m' change (13, 26) and from a K_S value and a relationship such as that shown in Figure 4, it is possible

FIGURE 4. Maximum rate of NH_4^+ uptake per unit plant N (V_{max}) at specific growth rates (μ) for *Thalassiosira pseudonana* (3-H) grown in an NH_4^+ limited chemostat. The circles represent rates determined 120 minutes after nutrient supply ceased. Rates of uptake adequate to meet the requirements of growth are depicted by the dotted line. (After McCarthy and Goldman (68)).

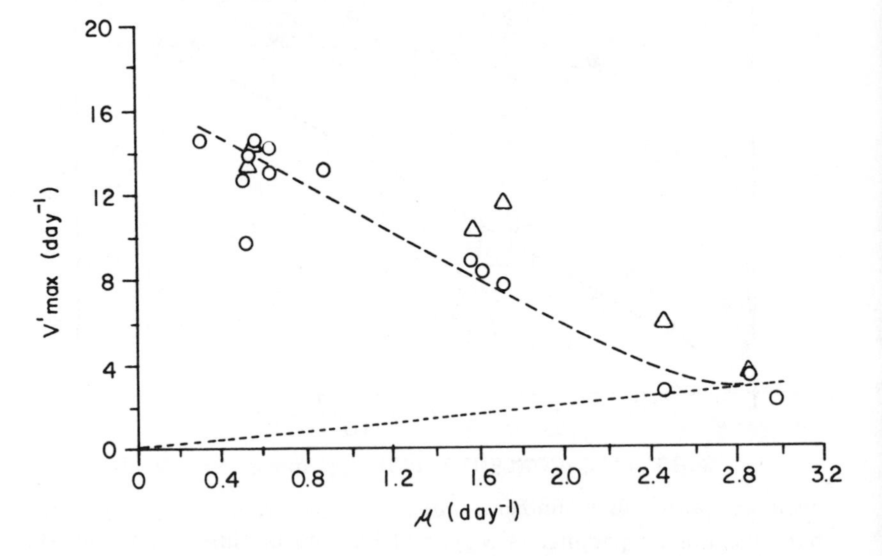

to estimate at least the lower limits for the S which corresponds to a given growth rate (8).

The potential to very V_m' in response to growth rate or nutritional state may play an important role in interspecific competition. The degree to which populations in their natural settings can be likened to those growing in continuous culture is unknown. Jannasch (54) has discussed this problem, and certainly, as shown by Eppley et al. (31) and Chisholm and Stross (13), a species' specific response to a L-D cycle may have profound implications on competitive interactions. It would seem that most species or races inhabiting a particular environment are similarly well suited to that environment, and indeed the fact that one can generate natural assemblage composite plots of hyperbolic form for nutrient kinetics (63) argues that the variability in K_S values for components of the composite must be small (109). It may be, however, that in certain natural settings, such as the central Pacific gyres (28) and the Chesapeake Bay during low nutrient conditions (70), the continuous culture is a good analogue of phytoplankton in nature, while in other habitats, such as upwelled water off Peru (107) or the Chesapeake Bay in winter, it most certainly is not.

Nevertheless, it is conceivable that different populations respond differently, or at least at different rates, to nutritional stress. The two hypothetical populations shown in Figure 1 were given K_s values more like those of large diatoms characteristic of moderate to high nutrient waters. If nutrient concentrations decrease to the point of producing a nutrient stress response in the form of an elevated V_m', then the rate at which enhanced uptake capacity develops can influence competitive interactions between two populations. In Figure 5, the curves from Figure

FIGURE 5. Variation in hypothetical competitive interaction in Figure 1 as a function of different population specific rates of V_m' enhancement. T_0 values are shown in Figure 1. At subsequent time T_1 the V_{max} values which would be observed are greater by 2 and 2.4 fold for the dashed and solid line populations respectively. After two such intervals, T_2, the intersection of the curves occurs at a concentration which is only 20 percent of that at T_0.

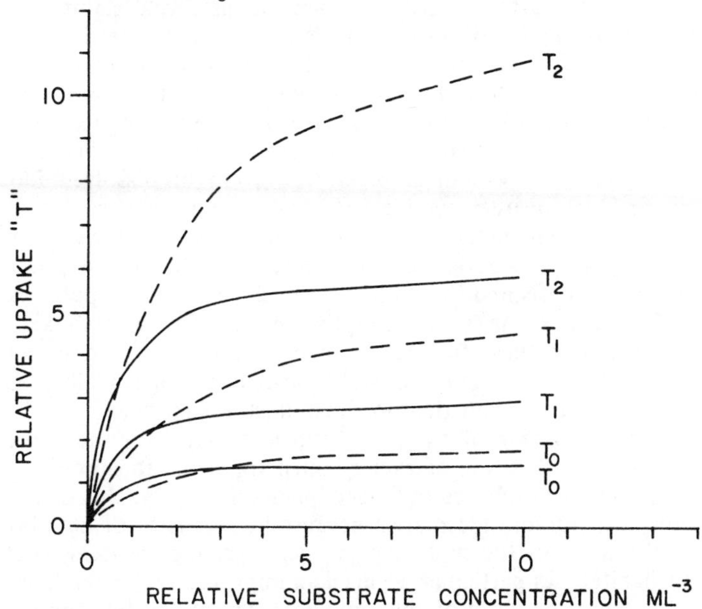

1 are plotted as T_0. If over a given time interval the population with the higher initial V_m' increases its V_m' by a factor of 2, and the other by 2.4, an increase in V_m' of 4 fold for the former and 5.7 fold for the latter after two such intervals will lead to an 80 percent reduction in the value of the

concentration for intersection of the two curves. If the population with the lower initial V_m' were permitted to develop its V_m' at the faster rate, the value of the concentration for intersection of the curves would increase. It is, of course, difficult to produce data from laboratory cultures which can be used to provide inference in analyzing shifts in dominance of populations in nature. It is clear that uptake kinetics alone cannot be used to predict the outcome of competition between two species (13, 31, 75, 101). There are numerous other factors which influence both net growth and abundance of populations, and even for cultured isolates from specific waters, it is unknown as to whether the race sampled for the clone used in initiating the laboratory culture is representative of the race or races which dominate populations of the same species in the same location at a later time.

CONCLUSIONS

The nutrition of aquatic plants is an important element in any model which describes the fluxes of organic and biologically reactive inorganic material in near surface waters. In general, estuarine habitats are no exception, but the importance of a plant metabolism term in nutrient budgets does vary greatly from estuary to estuary. In systems which have high flushing rates and heavy nutrient loading, the Hudson estuary, for example, the major plant nutrients behave as conservative properties (87). In contrast, for an estuary with a low average flushing rate, the Chesapeake Bay, for example, plant utilization of dissolved inorganic nutrients and conjugate recycling can result in multiple transformations and fates of nutrient material during its residence within the estuary (69). This capacity for utilization of nutrients, the fate of the product of autotrophic synthesis,and their interrelationships are rather poorly known features for most estuaries. A descriptive model of nutrient fluxes, which fails to incorporate these features is obviously possible for some estuaries, while for others a suitable model, particularly one which offers a predictive capability for nutrient loading, must include them.

Adequate accounting of the plant utilization term for the purpose of modeling nutrient fluxes in an estuary must apportion this process among phytoplankton, macrophytes and marsh plants. Presently, this is an almost impossible task for most estuaries because of a paucity of data both on nutritional aspects of the biota and physical exchange processes between different habitats. In particular we need an improved understanding of the rates of exchange between of the sediment interstital waters and standing waters in the nutrition of attached plants.

From recent literature, patterns which relate nutrient utilization to concentration have emerged, but there are few data which relate population growth rates to nutrient availability. It is evident from continuous culture experiments that we are prevented from drawing inference regarding the degree to which growth is nutrient limited merely because nutrient concentrations are extremely low. Furthermore, we do not know the suitability of the continuous culture as an analogue of the natural system. It is certainly less than perfect, but in conjunction with accommodation for transient responses to perturbation, it may suffice for some estuaries which have long residence times and well mixed euphotic zones.

An increased capacity to assess the nutritional sufficiency of a habitat would have great practical significance in estuarine studies. It is clear that phytoplankton can respond to nutritional stress by increasing their potential for short term nutrient uptake. We know less about this faculty in other aquatic plants, and in general its significance is difficult to assess in studies of natural systems. In this regard, more sophisticated experimental approaches in both the laboratory and the field are badly needed.

Data for particulate elemental composition can be of use in assessing the degree to which the maximum growth potential of the dominant populations of aquatic plants is being realized, but interference from detritus is difficult or impossible to either avoid or quantify in most natural waters. A larger data base from estuaries in which single or multiple populations can be selectively sampled without contamination is needed in order to justify the search for a routine methodology which eliminates the detrital problems.

From their pristine to their heavily nutrient-polluted state, estuaries constitute an attractive habitat for the study of plankton nutrition and elemental cycling. For certain types of studies the polluted estuary actually remains an ideal site for investigation. Exploitation of opportunities to study these processes in estuaries during a period of change in man's use, such as the investigations of Caperon and co-workers in Kaneohe Bay, will better permit us to relate the controlled laboratory experiment to the natural system.

REFERENCES

1. Allen, T.F.H. 1977. Scale in microscopic algal ecology: a neglected dimension. Phycologia 16:253-257.
2. Berman, T. 1970. Alkaline phosphatases and phosphorus availability in Lake Kinneret. Limnol. Oceanogr. 15:663-664.
3. Bird, K.T. 1976. Simultaneous assimilation of ammonium and nitrate by Gelidium nudifrons Gelidiales:Rhodophyla). J. Phycol. 12:238-241.
4. Broome, S.W., W.W. Woodhouse, and E.D. Seneca. 1975. The relationship of mineral nutrients to growth of *Spartina alterniflora* in North Carolina: II. The effects of N, P, and Fe fertilizers. Soil. Sci. Soc. Amer. Proc. 39:301-302.
5. Burmaster, D.E. 1979. The continuous culture of phytoplankton:

mathematical equivalence among three steady-state models. Amer. Natur. 113:123-134.

6. Burmaster, D.E. and S.W. Chisholm. 1979. A comparison of two methods for measuring phosphate uptake by *Monochrysis lutheri* grown in continuous culture. J. Exp. Mar. Biol. Ecol. 39:187-202.
7. Caperon, J., S.A. Cattell, and G. Krasnick. 1971. Phytoplankton kinetics in a subtropical estuary: eutrophication. Limnol. Oceanogr. 16:599-607.
8. Caperon, J. and D.A. Zieman. 1976. Synergistic effects of nitrate and ammonium ion on the growth and uptake kinetics of *Monochrysis lutheri* in continuous culture. Mar. Biol. 36:73-84.
9. Carpenter, E.J., and R.R.L. Guillard. 1971. Intraspecific differences in nitrate half-saturation constants for three species of marine phytoplankton. Ecology 52:183-185.
10. Carpenter, J.M., D.W. Pritchard, and R.C. Whaley. 1969. Observations of eutrophication and nutrient cycles in some coastal plain estuaries, p. 210-221. *In*: Eutrophication: Causes, consequences, correctives. Natl. Acad. Sci. Publ. 1700.
11. Chapman, A.R.O., J.W. Markham, and K. Luning. 1978. Effects of nitrate concentration on the growth and physiology of *Laminaria saccharina* (Phaeophyta) in culture. J. Phycol. 14:195-198.
12. Chisholm, S.W. and R.G. Stross. 1976. Phosphate uptake kinetics in *Euglena gracilis* (Z) (Euglenophyceae) grown on light/dark cycles. I. Synchronized batch cultures. J. Phycol. 12:210-217.
13. ———————. 1976. Phosphate uptake kinetics in *Euglena* gracilis (Z) (Euglenophyceae) grown in light/dark cycles. II. Phased PO_4-limited cultures. J. Phycol. 12:217-222.
14. Conway, H.L. 1977. Interactions of inorganic nitrogen in the uptake and assimilation by marine phytoplankton. Mar. Biol. 39:221-232.
15. Conway, H.L. and P.J. Harrison. 1977. Marine diatoms grown in chemostats under silicate or ammonium limitation. IV. Transient response of *Cheatoceros debilis, Skeletonema costatum*, and *Thalassiosira gravida* to a single addition of the limiting nutrient. Mar. Biol. 43:33-43.
16. Darley, W.M. 1974. Silicification and Calcification, p. 655-675. *In*: W.D.P. Steward (ed.), Algal physiology and biochemistry. Univ. Calif.
17. D'Elia, C. and J.A. DeBoer. 1978. Nutritional studies of two red algae. II. Kinetics of ammonium and nitrate uptake. J. Phycol. 14:266-272.
18. DeManche, J.M., H.C. Curl, Jr., D.W. Lundy, and P.L. Donaghay. 1979. The rapid response of the marine diatom *Skeletonema costatum* to changes in external and internal nutrient concentration. Mar. Biol. 53:323-334.
19. Denny, P. 1972. Sites of nutrient absorption in aquatic macrophytes. J. Ecol. 60:819-829.
20. Droop, 1973. Some thoughts on nutrient limitation in algae. J. Phycol. 9:264-272.
21. Dugdale, R.C. 1967. Nutrient limitation in the sea: dynamics, identification, and significance. Limnol. Oceanogr. 12:685-695.
22. ———————. 1977. Modeling, p. 789-806. *In*:E.D. Goldberg et al. (eds.). The sea, V. 6. Wiley-Interscience.

23. Dugdale, R.C. and J.J. Goering. 1967. Uptake of new and regenerated forms of nitrogen in primary productivity. Limnol. Oceanogr. 12:196-206.
24. Eppley, R.W., J.L. Coatsworth. 1968. Nitrate and nitrite uptake by *Ditylum brightwellii*. Kinetics and mechanisms. J. Phycol. 4:151-156.
25. Eppley, R.W., J.L. Coatsworth, and L. Solorzano. 1969. Studies of nitrate reductase in marine phytoplankton. Limnol. Oceanogr. 14:194-205.
26. Eppley, R.W. and E.H. Renger. 1974. Nitrogen assimilation of an oceanic diatom in nitrogen-limited continuous culture. J. Phycol. 10:15-23.
27. Eppley, R.W., E. H. Renger, W. G. Harrison, and J. J. Cullen. 1979. Ammonium distribution in southern California coastal waters and its role in the growth of phytoplankton. Limnol. Oceanogr. 24:495-509.
28. Eppley, R. W., E. H. Renger, E. L. Venrick, and M. M. Mullin. 1973. A study of plankton dynamics and nutrient cycling in the central gyre of the North Pacific Ocean. Limnol. Oceanogr. 18:534-551.
29. Eppley, R.W. and J.N. Rogers. 1970. Inorganic nitrogen assimilation of *Ditylum brightwellii*, a marine plankton diatom. J. Phycol. 6:344-351.
30. Eppley, R.W., J.N. Rogers, and J.J. McCarthy. 1969. Half-saturation constants for uptake of nitrate and ammonia by marine phytoplankton. Limnol. Oceanogr. 114:912-920.
31. Eppley, R.W., J.N. Rogers, J.J. McCarthy, and A. Sournia. 1971. Light/dark periodicity in nitrogen assimilation of the marine phytoplankton *Skeletonema costatum* and *Coccolithus huxleyi* in N-limited chemostat culture. J. Phycol. 7:150-154.
32. Eppley, R.W., J.H. Sharp, E.H. Renger, M.J. Perry, and W.G. Harrison. 1977. Nitrogen assimilation by phytoplankton and other microorganisms in the surface waters of the central North Pacific Ocean. Mar. Biol. 39:111-120.
33. Eppley, R.W. and W.H. Thomas. 1968. Comparison of half-saturation constants for growth and nitrate uptake of a marine phytoplankton. J. Phycol. 5:375-379.
34. Falkowski, P.G. 1975. Nitrate uptake in marine phytoplankton (nitrate, chloride) activated adenosine triphosphatase from *Skeletonema costatum* (Bacillariophycaea). J. Phycol. 11:323-326.
35. Falkowski, P.G. and R.B. Rivkin. 1976. The role of glutamine synthetase in the incorporation of ammonium in *Skeletonema costatum* (Bacillariophyceae). J. Phycol. 12:448-450.
36. Ferguson, A.R. 1969. The nitrogen metabolism of *Spirodella oligorrhiza* II. Control of the enzymes of nitrate assimilation. Planta. 88:353-363.
37. Ferguson, A.R. and E.G. Bollard. 1969. The nitrogen metabolism of *Spirodella oligorrhiza* I. Utilization of ammonium, nitrate and nitrite. Planta. 88:344-352.
38. Finenko, Z.Z. and D.K. Drupatkina-Akinina. 1974. Effect of inorganic phosphorus on the growth rate of diatoms. Mar. Biol. 26:193-201.
39. Gallagher, J.L. 1975. Effect of an ammonium nitrate pulse on the

growth and elemental composition of natural stands of *Spartina alterniflora* and *Juncus roemerianus*. Amer. J. Bot. 62:644-648.

40. Gayler, G.R. and W.R. Morgan. 1976. An NADP-dependent glutamate dehydrogenase in chloroplasts from the marine green alga *Caulerpa simpliciuscula*. Plant Physiol. 58:283-287.
41. Gieskes, W.W.C. and G.W. Kraay. 1979. Current ^{14}C methods for measuring primary production: gross underestimates in oceanic waters. Neth. J. Sea Res. 13:
42. Goldman, J.C. 1977. Steady state growth of phytoplankton in continuous culture: Comparison of internal and external nutrient equations. J. Phycol. 13:251-258.
43. Goldman, J.C. and J.J. McCarthy. 1978. Steady state growth and ammonium uptake of a fast-growing marine diatom. Limnol. Oceanogr. 23:695-703.
44. Goldman, J.L. and D.G. Peavey. 1979. Steady state growth and chemical composition of the marine chlorophyle *Dunaliella tertiolecta* in nitrogen-limited continuous cultures. Appl. Environ. Microbiol. 38:894-901.
45. Goldman, J.C. and J.H. Ryther. 1975. Nutrient transformation in mass cultures of marine algae. J. Environ. Engineer. Div. ASCE Proc. Paper 11358. 101:351-364.
46. Guillard, R.R.L. 1963. Organic sources of nitrogen for marine centric diatoms. *In*: Symposium on Marine Microbiology, pp. 93-104 (ed. C.H. Oppenheimer). Charles C. Thomas, Springfield, Illinois.
47. Haines, K.C. and P.A. Wheeler. 1978. Ammonium and nitrate uptake by the marine macrophytes *Hypnea musciformis* (Rhodophyta) and *Macrocystis pyrifera* (Phaeophyta). J. Phycol. 14:319-324.
48. Hanisak, M.D. and M.M. Harlin. 1978. Uptake of inorganic nitrogen by *Codium fragile* subsp. *tomentosoides* (Chlorophyta). J. Phycol. 14:450-454.
49. Harlin, M.M. and J.S. Cragie. 1978. Nitrate uptake by *Laminaria longicruris* (Phaeophyceae). J. Phycol. 14:464-467.
50. Harris, E. 1959. The nitrogen cycle in Long Island Sound. Bull. Bingham oceanogr. Coll. 17:31-65.
51. Harris, G. and B. Piccinin. 1977. Photosynthesis by natural phytoplankton populations. Arch. Hydrobiol. 4:405-457.
52. Harvey, H.W. 1963. The chemistry and fertility of sea waters. Cambridge Univ.
53. Harvey, W.A. and J. Caperon. 1976. The rate of utilization of urea, ammonium, and nitrate by natural populations of marine phytoplankton in a eutrophic environment. Pac. Sci. 30:329-340.
54. Jannasch, H. 1974. Steady state and the chemostat in ecology Limnol Oceanogr. 19:716-720.
55. Jeanjean, B. and G. Ducet. 1974. Carrier turnover and phosphate uptake in *Chlorella pyrenoidosa*. *In* U. Zimmerman and J. Dainty (eds.), Membrane transport in plants. Springer-Verlag.
56. Kilham, P. 1971. A hypothesis concerning silica and the freshwater planktonic diatoms. Limnol. Oceanogr. 16:10-18.
57. Kuenzler, E.J. 1965. Glucose -6- phosphate utilization by marine algae. J. Phycol. 1:156-164.

58. -------. 1970. Dissolved organic phosphorus excretion by marine phytoplankton. J. Phycol. 6:7-13.
59. Kuhl, A. 1974. Phosphorus, p. 636-654. *In* W.D.P. Steward (ed.), Algal physiology and biochemistry. Univ. Calif.
60. Lean, D.R.S. and C. Nalewajko. 1976. Phosphate exchange and organic phosphorus excretion by freshwater algae. J. Fish. Res. Bd. Can. 33:1312-1323.
61. Leftley, J.W. and P.J. Syrett. 1973. Urease and ATP: Urea amidolyase activity in unicellular algae. J. Gen. Microbiol. 77:109-115.
62. Liu, M.S. and J.A. Hellebust. 1974. Utilization of amino acids as nitrogen sources and their effects on nitrate reductase in the marine diatom *Cyclotella cryptica*. Can. J. Microbiol., 20:1119-1125.
63. MacIsaac, J.J. and R.C. Dugdale. 1969. The kinetics of nitrate and ammonia uptake by natural populations of marine phytoplankton. Deep-Sea Res. 16:45-57.
64. -------. 1972. Interactions of light and inorganic nitrogen in controlling nitrogen uptake in the sea. Deep-Sea Res. 19:209-232.
65. Malone, T.C., C. Garside, K.C. Haines, and O.A. Roels. 1975. Nitrate uptake and growth of *Chaetoceros* sp. in large outdoor continuous culture. Limnol. Oceanogr. 20:9-19.
66. McCarthy, J.J. and E.S. Carpenter. 1979. *Oscillatoria* (Trichodesmium) *theibautii* (Cyanophyta) in the central North Atlantic Ocean. J. Phycol. 15:75-82.
67. McCarthy, J.J. and R.W. Eppley. 1972. A comparison of chemical, isotopic, and enzymatic methods for measuring nitrogen assimilation of marine phytoplankton. Limnol. Oceanogr. 17:371-382.
68. McCarthy, J.J. and J.C. Goldman. 1979. Nitrogenous nutrition of marine phytoplankton in nutrient-depleted waters. Science. 203:670-672.
69. McCarthy, J.J., W.R. Taylor, and J.L. Taft. 1975. The dynamics of nitrogen and phosphorus cycling in the open waters of the Chesapeake Bay, 664-681. *In*: T.M. Church (ed.), Marine chemistry in the coastal environment. ACS Symposium Series, No. 18.
70. -------. 1977. Nitrogenous nutrition of the plankton in the Chesapeake Bay. I. Nutrient availability and phytoplankton preferences. Limnol. Oceanogr. 22:996-1010.
71. McRoy, C.P. and V. Alexander. 1975. Nitrogen kinetics in aquatic plants in artic Alaska. Aquat. Bot. 1:3-10.
72. McRoy, C.P. and R.J. Barsdate. 1970. Phosphate absorption in eelgrass. Limnol. Oceanogr. 15:6-13.
73. McRoy, C.P., R.J. Barsdate, and M. Nebert. 1972. Phosphorus cycling in an eelgrass (*Zostera marina* L,) ecosystem. Limnol. Oceanogr. 17:58-67.
74. McRoy, C.D. and J.J. Goering. 1974. Nutrient transfer between the seagrass *Zostera marina* and its epiphytes. Nature. 298:173-174.
75. Mickelson, M.J., H. Maske, and R.C. Dugdale. 1979. Nutrient-determined dominance in multispecies chemostat cultures of diatoms. Limnol. Oceanogr. 24:298-315.
76. Morris, I. 1974. Nitrogen assimilation and protein synthesis, p.

583-609. *In*: W.D.P. Stewart (ed.), Algal physiology and biochemistry. Univ. Calif.
77. Morris, I. and P.J. Syrett. 1963. The development of nitrate reductase in *Chlorella* and its repression by ammonium. Arch. Mikrobiol. 47:32-41.
78. Nichols, D.S. and D.R. Keeney. 1976. Nitrogen nutrition of *Myriophyllum spicatum*: uptake and translocation of ^{15}N by shoots and roots. Freshwat. Biol. 6:145-154.
79. North, B.B. and G.C. Stephens. 1972. Amino acid transport in *Nitzschia ovalis* Arnott. J. Phycol. 8:64-68.
80. Packard, T.T. 1973. The light dependence of nitrate reductase activity in marine phytoplankton. Limnol. Oceanogr. 18:466-469.
81. Perry, M.J. 1972. Alkaline phosphatase activity in subtropical central North Pacific waters using a sensitive fluorometric method. Mar. Biol. 15:113-119.
82. -------. 1976. Phosphate utilization by an oceanic diatom in phosphorus-limited chemostat culture and in oligotrophic waters of the central North Pacific. Limnol. Oceanogr. 21:88-107.
83. Rhee, G-Yull. 1973. A continuous culture study of phosphate uptake, growth rate and polyphosphate in *Scenedesmus sp*. J. Phycol. 9:495-506.
84. Reimold, R.J. 1972. The movement of phosphorus through the saltmarsh cord grass *Spartina alterniflora* Loisel. Limnol. Oceanogr. 17:606-611.
85. Schell, D.M. 1974. Uptake and regeneration of free amino acids in marine waters of southeast Alaska. Limnol. Oceanogr. 19:260-170.
86. Seliger, H.H. and M.E. Loftus. 1975. Dinoflagellate accumulations in Chesapeake Bay. p. 181-205. *In*: V.R. LoCicero (ed.), Proceedings of the first international conference on toxic dinoflagellate blooms. Mass. Sci. and Tech. Found.
87. Simpson, H.J., D.E. Hammond, B.L. Deck, and S.C. Williams. 1975. Nutrient budgets in the Hudson River estuary, p. 618-635. *In*: T. Church (ed.), Marine chemistry in the coastal environment. ASC Symposium Series, No. 18.
88. Stanley, D., and J.E. Hobbie. 1979. Nitrogen recycling in a North Carolina coastal river. Limnol. Oceanogr.
89. Stewart, G.R., J.A. Lee, and T.O. Orebamjo. 1972. Nitrogen metabolism of halophytes. I. nitrate reductase activity in Suaeda maritima. New Phytol. 71:263-267.
90. -------. 1973. Nitrogen metabolism of halophytes II. nitrate availability and utilization. New Phyto. 72:539-546.
91. Stewart, R.G. and D. Rhodes. 1977. A comparison of the characteristics of glutamine synthetase and glutamate dehydrogenase from New Phytol. 79-257-268.
92. Strickland, J.D.H., O. Holm-Hansen, R.W. Eppley, and R.J. Linn. 1969. The use of a deep tank in plankton ecology - I. Studies of the growth and composition of phytoplankton at low nutrient levels. Limnol. Oceanogr. 14:23-34.
93. Taft, J.L., A.J. Elliott, and W.R. Taylor. 1978. Box model analysis of Chesapeake Bay ammonium and nitrate fluxes, p. 115-130. *In*: M.

Wiley (ed.), Estuarine interactions. Academic.

94. Taft, J.L., M.E. Loftus and W.R. Taylor. 1977. Phosphate uptake from phosphomonoesters by phytoplankton in Chesapeake Bay. Limnol. Oceanogr. 22:1012-1021.
95. Taft, J.L. and W.R. Taylor. 1976a. Phosphorus dynamics in some coastal plain estuaries. p. 79-89. *In* M. Wiley (ed.), Estuarine Processes, V. 1. Academic Press.
96. ————. 1976b. Phosphorus distribution in the Chesapeake Bay. Ches. Sci. 17:67-73.
97. Taft, J.L., W.R. Taylor, and J.J. McCarthy. 1975. Uptake and release of phosphorus by phytoplankton in the Chesapeake Bay estuary. Mar. Biol. 33:21-32.
98. Tempest, D.W., J.L. Meers, and C.M. Brown. 1973. Glutamate synthetase (GOGAT): a key enzyme in the assimilation of ammonia by prokaryotic organisms, p. 167-182. *In*: S. Prusiner and E.R. Stadtman (eds.), The enzymes of glutamine metabolism. Academic Press.
99. Thomann, R.V., D.J. O'Connor, and D.M. DiTorro. 1971. Modeling of the nitrogen and algal cycles in estuaries. Proc. 5th International Water Pollution Research Conference. Pergamon Press.
100. Thomas, W.H., E.H. Renger, and A.N. Dodson. 1971. Near-surface organic nitrogen in the eastern tropical Pacific Ocean. Deep-Sea Res. 18:65-71.
101. Tilman, D. 1977. Resource competition between planktonic algae: an experimental and theoretical study. Ecology 58:338-348.
102. Topinka, J. 1978. Nitrogen uptake by *Fucus spiralis* (Phaeophyceae) J. Phycol. 14:241-247.
103. Turpin, D.H. and P.J. Harrison. 1979. Limiting nutrient patchiness and its role in phytoplankton ecology. J. exp. mar. Biol. Ecol. 39:151-166.
104. Valiela, I. and J.M. Teal. 1974. Nutrient limitation in salt marsh vegetation. p. 547-563. *In* R.J. Reimold and W.H. Queen (eds.), Ecology of halophytes. Academic.
105. Valiela, I., J.M. Teal, and W.G. Deuser. 1978. The nature of growth forms in the salt marsh grass *Spartina alterniflora*. Amer. Natur. 112:461-470.
106. Venrick, E.L., J.R. Beers, and J.F. Heinbokel. 1977. Possible consequences of containing microplankton for physiological rate measurements. J. exp. mar. Biol. Ecol. 26:55-76.
107. Walsh, J.J. and R.C. Dugdale. 1971. A simulation model of the nitrogen flow in the Peruvian upwelling system. Inv. Pesq. 35:309-330.
108. Wheeler, P.A., B.B. North, and G.C. Stephens. 1974. Amino acid uptake by marine phytoplankters. Limnol. Oceanogr. 19:249-259.

INDICATORS AND INDICES OF ESTUARINE OVERENRICHMENT[1]

A.J. McErlean* and Gale Reed
University of Maryland Center
For Environmental and Estuarine Studies
Box 775
Cambridge, Maryland 21613

**Present address:*
U.S. Environmental Protection Agency
Environmental Research Laboratory
Gulf Breeze, Florida 32561

ABSTRACT: Indicators and indices of nutrient overenrichment, which might be applied to the problems of estuarine overenrichment, are reviewed from the literature from the time of the national symposium on eutrophication (1968) to the present. While many existing methods have been applied to estuarine work and some new ones have been developed, there do not appear to be a large number of usable techniques for estuarine investigations.

INTRODUCTION

A review of estuarine eutrophication index and indicator usage is essentially a review of investigative methodology. This measurement methodology has reached a high level of sophistication for lakes, bogs, and other inland waters. Only within the last few decades, however, has estuarine and coastal eutrophication become a concern. Since eutrophication is so historically a limnetic problem, the review process is essentially that of tracing the transfer of freshwater concepts and research methods to the marine and estuarine situation. In large measure, estuarine researchers and managers have attempted to apply concepts and theories elaborated or determined in freshwater contexts to coastal situations.

[1]Contribution No. X-317 of the Environmental Research Laboratory, Gulf Breeze.

Most workers have tacitly assumed basic similarities between the two situations. Recently, however, the identity and roles of "limiting" nutrients as well as the mechanisms of eutrophication have been reevaluated. This topic has been reexamined lately (see for instance, the ASLO publication on limiting nutrients and other papers in this symposium volume).

Some enrichment indicators, such as algal blooms and fish kills, are probably as old as recorded history (see Exodus 7:19-21) while indices appear to be a more recent development. In any case, the potential of valid trophic indicators or indices is immense. While review of desirable index and indicator criteria is considered beyond the scope of this paper[2] it is important to underscore the potential utility and the practical limitation of such tools. Their value is great when specificity to cause is high and the undesirable condition has not yet occurred. Overreliance on any single tool without follow-up investigation or independent validation would seem imprudent. It is doubtful that these tools will ever replace tools such as modeling or intensive surveys. Properly considered, they are valuable adjuncts rather than replacements for more detailed scientific investigation.

Indices and indicators which reflect overenrichment have been reviewed primarily for freshwater (32, 62). Swartz (72) has reviewed Chesapeake Bay-specific indicators and measurement techniques. Only Hooper (32) has attempted what might be considered a broad review of index methodology, and that primarily for freshwater. These sources as well as the recent literature form the basis for our review of indices and indicators. Some workers, who are especially familiar with freshwater overenrichment problems, have critized transfer attempts or have cautioned about overenthusiastic application of methods or theory from fresh to saline waters. The possible lack of mechanism rigor, the hydrodynamic uniqueness of estuaries, and possible fundamental differences with respect to type and chemical form of limiting nutrients have all been suggested as strong considerations that might affect transfer. Schindler (68) expressed these concerns eloquently as follows:

> "Eutrophication of estuaries of the eastern United States has been little studied, and even its extent is poorly known. Only in recent years has any concern at all been expressed for these bodies of water. As a result, no extensive body of scientific methodology has been developed specially for estuarine studies, and the techniques and axioms employed in freshwater lakes, where eutrophication has been a matter of concern for decades, are likely to be employed by workers in the future."

[2]The reader is referred to Ott (57) and to Hooper (32) for a discussion of such criteria.

This paper attempts to review index and indicator application specifically for estuarine waters, to categorize historic methodology, and to highlight some techniques which seem more promising than others. We have attempted to restrict our review to studies that included nominal mention of trophic state or those we felt were involved with enrichment problems. Mathematical modeling and nutrient loading concepts are not considered as index or indicator tools; these topics are reviewed and discussed elsewhere in this volume.

REVIEW OF INDICES AND INDICATORS

The NAS Symposium devoted to "Eutrophication: causes, consequences correctives" viewed the detection and measurement of eutrophication as key factors in understanding and expressing the condition of (predominantly) freshwaters. Inferring from the content of the published articles, the expression of eutrophication was seen chiefly as a biological phenomenon (five of seven papers dealt with biological topics, one with physical factors, and one reviewed the state-of-the art of index application). Hooper (32), after establishing desirable index criteria, cited the following as primary index methods: 1) oxygen budgets, 2) transparency, 3) morphometric indices, 4) nutrients and associated ions, 5) biological indicators, 6) productivity, 7) diversity and stability. We find this listing useful for several reasons: it enumerates historic (pre-1969) methods, and it provides a convenient way to evaluate and compare estuarine studies with respect to their methodological choices.

Recent literature was examined to find references dealing either with index or indicator citations or to identify recent estuarine or coastal field studies; these were assigned to at least one of the categories noted earlier (32). In some cases, the effort revealed problems associated with establishment of unique delineations or clear cut separations. For instance, few workers presently use water clarity or transparency as a single eutrophication indicator to the exclusion of other methods, such as nutrient ion determinations. This difficulty was pronounced with some biological studies in which, for instance, productivity, biological indicator species, and diversity and stability index characteristics were concurrently observed and reported. In such cases, index type assignment was made based on judgment of major findings. In reporting these results, we compressed Hooper's "biological indicators" and "diversity/stability" categories to a single type.

Additionally, an attempt was made to describe each index method; major advantages and disadvantages were summarized along with relative cost and data requirements in order to estimate its potential for estuarine applications and to list citations where this method has been used in estuarine or coastal areas. Cost estimates are relative to other index methods and, along with the commentary concerning estuarine application, are our own. Other commentary is either a hybrid or is contained in Hooper's review.

These results (Tables 1-6) show that all index methods mentioned by Hooper have received some application in coastal or estuarine areas. While some methods which employ oxygen behavior as a trophic discriminator may not be transferable (Table 1), the seasonal characteristic of oxygen

TABLE 1. Index Types Based on Oxygen Budgets.

Index Types:	Oxygen budgets.
Description:	The seasonal and spatial concentration of oxygen in deep lakes is characteristic; changes relative to depth, time, and temperature can be interpretated in relationship to overenrichment.
Major Advantages/ Disadvantages:	An excellent body of scientific data exists; the technique, which Hooper suggests is phosphate-specific, integrates over an ecologically meaningful time span (a season). May be limited to specific deep lake conditions and may be insensitive to "minor" changes.
Relative Cost:	Low cost.
Data Requirements:	Few compared to some other methods, but seasonal and diurnal sampling are essential.
Estuarine Application Potential:	Probably low in the classic sense due to estuarine hydrodynamics; however, diurnal oxygen cures have been used.
Selected Estuarine References:	13, 15, 42, 54, 55.

concentration may be useful for some estuaries (personal communication, D.H. Heinle, 1979). Similarly, long term trends in water clarity (Table 2) as measured by Secchi disk may be extremely useful in interpreting trophic changes in Chesapeake Bay tributaries. The limitation here may be that of attributing change to cause, since clarity can be strongly affected by sedimentation events and other causes unrelated to overenrichment. Index types based on morphometric indices (Table 3) would seem to have little potential transfer potential. An index type that can be considered as an extension of the nutrients and associated ions index type (Table 4), which also incorporates mass balance, hydrodynamics and modeling aspects, and which is currently broadly applied in freshwater situations is

TABLE 2. Index Types Based on Transparency.

Index Type:	Transparency.
Description:	Water transparency, which integrates biotic and abiotic factors, is studied over time. Transparency is altered under "eutrophic" conditions.
Major Advantages/ Disadvantages	Simple to measure and can be highly meaningful if long term data base exists. Measurements are difficult to interpret unless causes of change can be ascribed to specific factors (runoff, sedimentation, algal blooms, etc.)
Relative Cost:	Very low.
Data Requirements:	Low compared to other methods; however, algal pigment and nutrient measures are usually necessary adjuncts.
Estuarine Application Potential:	Poor to excellent depending on the existence of data base. Turbidity not associated with overenrichment (i.e., construction or agriculture-related sediment runoff) may confound interpretation, as many location in the salinity gradient.
Selected Estuarine References:	9, 17, 19, 38, 45, 59.

the Vollenweider approach (70, 77, 78, 62). Phosphorus loading approaches and the general topic of estuarine eutrophication modeling (which are addressed elsewhere in this volume) hold great promise in the understanding of eutrophication but are considered beyond the scope of this paper.

Some index methods, either newly developed or refined from freshwater origins, have received a high level of currency of usage--sufficient to warrant specific examination as estuarine or coastal-specific index methods.

Nutrient ratios (Table 4) can be considered as explicit in Hooper's terminology (32) or as an outgrowth of nutrient and associated ions index type. Nitrogen to phosphorus ratios have been commonly cited (6, 7, 10, 20, 21, 37, 56) since their initial mention (63, 64) as useful in the interpretation of coastal nutrient processes. Silicon has also been mentioned as a potentially important nutrient in estuaries (see Ryther, this volume). Nutrient ratios have engineering appeal and are useful when an extensive time base is present. Their limitations come from imperfect knowledge concerning the role, chemical form, or identity of nutrients

TABLE 3. Index Types Based on Morphometric Indices.

Index Type:	Morphometric indices.
Description:	Physical changes in basin size, configuration, and depth occur over time; overenrichment accelerates this process.
Major Advantages/ Disadvantages:	Change rate for some lakes is predictable and can be used to estimate overenrichment processes or effects. Time scale of comparison is usually geologic.
Relative Cost:	Intermediate depending on extent of data base.
Data Requirements:	Relatively high compared to other indices. Valid characterizations may require sedimentation rate determinations, physical measures, coring, palynological determinations, etc.
Estuarine Application Potential:	Poor to possible high potential. The latter potential (coupled with sedimentary chlorophyll and other data) is in establishing effects of man-induced changes since pre-historic times. Poor because of geologic time scale, dredging of estuaries, and the high energy nature of estuarine and coastal areas.
Selected Estuarine References:	No citations which rely directly on morphometry as a measure of estuarine overenrichment could be located. Dr. Grace Brush (personal communication, 1979) is attempting to use sedimentary facies, contained chlorophyll content, lead dating, and palynology to estimate pre-historic trophic status for several Chesapeake estuaries.

and cycling phenomena in estuaries. Additionally, the choice of which chemical form or nutrient fraction to use in ratio computation has varied.

The measurement of chlorophyll and other plant pigments is emerging as an established index criterion for estuarine eutrophication (33, 34, 35, 36, 17, 76, 46, 26, 14, 45). This index measure (Table 5) is particularly useful where long term monitoring data are available and the distribution and abundance of phytoplankton species is well defined. When such an interpretational framework is lacking it may be difficult to apply, due to such factors as organism size (nannoplankton), presence of detrital chlorophyll, and the specificity of chemical tests to plant pigments.

TABLE 4. Index Types Based on Nutrients and Associated Ions.

Index Type:	Nutrients and associated ions.
Description:	The distribution, chemical form, and concentration of various nutrients within and though the various ecosystem compartments, but especially the water column and sediments, is related to the trophic condition or potential of a water body.
Major Advantages/ Disadvantages:	Mass balance approaches and phosphorus-loading models have been highly effective in many management applications. Potential and actual effect is difficult to establish due to such factors as chemical form, biological uptake, transfers, and internal cycling.
Relative Cost:	Form moderate to high dependent upon the chemistry involved.
Data Requirements:	Usually high. Often it is, not possible to obtain time series data, particularly when the measured chemical species or the measurement methodology change through time. Interpretations must sometimes be made without a comparative base.
Estuarine Application Potential:	Moderate to high depending upon specific measures and the state-of-the-art. Limiting nutrient aspect a factor. Mass balance methods and modeling may have high potential.
Selected Estuarine References:	8, 14, 15, 16, 18, 21, 24, 25, 26, 27, 29, 30, 31, 33, 34, 35, 36, 39, 45, 50, 51, 66, 67.

Nevertheless, "greenness" comparisons, seasonally and over longer time periods, may provide an excellent diagnostic of estuarine overenrichment.

Bioassays (10, 71, 5) have generally not assumed the importance, sophistication, or degree of use in estuarine and coastal areas that similar methods have in freshwater. The difficulty may be associated with selection of test species, the acceptability of test results, (Table 6), agreement concerning test conditions, or problems associated with our knowledge concerning chemical or nutrient speciation (see also Schindler, 68). Colwell (11), reviewing the present state-of-the-art of estuarine microbal indicators, paints a dismal picture---present methods are inadequate and a huge amount of basic research would be needed to develop and test new methods. Along with bioassays, some workers have

TABLE 5. Index Types Based on Productivity.

Index Type:	Productivity.
Description:	Rate changes in primary productivity can be an index of eutrophication. Chlorophyll and other indirect estimates can reflect nutrient avalability and cycling.
Major Advantages/ Disadvantages:	Productivity rate changes of algal populations may be generally applied as an index measure. Coupled with other measures and background data, these can form an established index of trophic conditions. The variability of estuarine algae populations, hydraulic phenonomena unique to tidal estuaries, and turbidity interactions may limit use or affect interpretation. Some indirect measures (chlorophyll pigments, net phytoplankton) may not account for entire population.
Relative Cost:	Moderate.
Data Requirements:	Requires long term data development and measurement as well as the establishment of species composition and seasonality.
Estuarine Application Potential	High. This has been a method of choice for estuarine and marine studies. The body of data developed and particularly the partitioning of production and consumption may approach the sophistication of oxygen budgets as an estuarine method.
Selected Estuarine References:	3, 6, 7, 13, 14, 16, 17, 19, 27, 30, 31, 37, 41, 44, 45, 46, 47, 50, 51, 64, 69, 76.

attempted to identify and measure highly specific enzymes or chemicals associated with organic discharges and nutrient mobilization (28, 69, 73, 74). Such techniques have special potential as enrichment indicators; however, many of these studies lack a historical data base, have not been widely applied, or lack quantification techniques.

Ott's recent text (57), "Environmental Indices: Theory and practice" suggests some fertile areas to pursue, given an intention of developing estuarine-specific enrichment indices. Although the point of view of the book is strongly chemical and physical rather than biological, it suggests that valid indices might be developed using common water quality determinations as a point of departure. The experiences described derive primarily from air quality and drinking water quality concerns; nevertheless, the methods described present high potential in our

TABLE 6. Index Types Based on Biological Indicators/Diversity and Stability.

Index Type:	Biological indicators/diversity and stability.
Description:	The presence or absence of a given species, an assemblage of species, or a shift in species composition and numbers within a community or system can be related to overenrichment phenomena.
Major Advantage/ Disadvantages:	Occurrence of the indicator or index phenomenom (which may be itself the complex result of previous changes) integrates these changes and may signify desirable or undesirable enrichment conditions. Many indicators are site-specific and therefore not widely applicable, or their development or establishment requires intense study.
Relative Cost:	Moderate to high dependent upon conditions; taxonomic identifications and sampling costs may be high.
Data Requirements:	May be high or even unobtainable when interpretation requires elaboration of ecological "role" or trophic placement of indicator.
Estuarine Application Potential:	Moderate to high. The potential of macrobenthos is particularly promising, as is the use of other taxonomic groups such as zooplankton, annelids, filter feeders, etc.
Selected Estuarine References:	1, 2, 3, 4, 5, 6, 7, 12, 17, 22, 28, 30, 31, 38, 40, 41, 42, 43, 44, 46, 47, 48, 53, 58, 59, 60, 64, 71, 75.

estimation. We have been unable to find many examples of this type of index development for estuarine areas, the exceptions being, Olinger et al., (56) for Escambia Bay, Florida, and the use of the National Sanitation Foundation Index for the Potomac River estuary (52). We are examining the potential for developing and applying water quality indices that are sensitive to estuarine overenrichment conditions*.

---

*A.J. McErlean and Gale J. Reed. 1979. On the application of water quality indices to the detection, measurement and assessment of nutrient enrichment in estuaries. Ref. 79-138, Horn Point Environmental Laboratory, University of Maryland, Cambridge, Md. 21613. 132 pp.

TABLE 7. Estuarine and Coastal Indicators from the Literature.

Geographic Area	Manifestation	Remarks	References
Baltic Sea (Coastal)	Macrobenthic community structure plus several indicator species. Changes in community structure. Presence or absence of specific species.	*Capitella* occurred near sewage outfall. Author feels population measures are good parameters to assess organic stress.	(1)
Puget Sound, United States	Intertidal macrofauna sampled at sewage outfalls and reference stations as well as clam condition indices, crab growth rates, etc.	Paradoxical results; substrate more a determinant than proximity to outfalls; other sublethal measures did not track well. No significant differences in polychaete distributions.	(2)
St. Johns Harbor, Newfoundland	Mussels, periwinkles, and snail growth used to track phosphorus addition. Absence of certain species, changes in population.	Investigator studied annular growth mark changes coincident with phosphorus addition. *Mytilus* had no coincident changes but some age-cohorts were not found in affected area.	(5)
Tees Estuary, Great Britain	Shifts in species distribution, species diversity and abundance compared over 38 year time period.	Pollution-tolerant species occur, and shifts of macrobenthic trophic levels. Substrate type important. Algal mat areas populated by a number of known indicator species.	(22)
Oslo Fjord, Norway	Presence or absence of marine algae.	Based on long-term surveys at a large number of sampling sites, the inner harbor area benthic algal distribution is explained. Organic enrichment a major factor.	(41)

TABLE 7, Continued.

River Mouths of Finland and Sweden	Benthic macrofauna used to measure effect of organic enrichment.	The Benthic Pollution Index is considered particularly useful for rivers which may be subjected to organic loading.	(43)
California Coast, United States	Increase in some tolerant species and increased uptake of nutrient; changes	Caloric content of invertebrates and plant species is elevated near sources of organic loading.	(44)
Mersey Estuary, Great Britain	Macrofauna seasonality, population and community structure changes.	Enrichment-associated changes and faunal gradients useful in characterizing effects.	(49)
Woods Hole, United States	Changes in population structure, animal size, abundance.	The size and abundance of macrobenthic species is proposed as a tentative indicator of environmental effect (sewage outfall).	(53)
General	Changes in macrobenthic population and community structure.	Article reviews theory and practice of community succession in macrobenthic populations as indices of organic enrichment; indicates that general theories of wide application may be forth-coming.	
Marseille, France	Changes in macrobenthic population and community structure.	Increases in sewage addition along with associated turbidity caused seagrass decline.	(59)
Lake Pontchartrain, United States	Changes in macrobenthic population and community structure.	Epifaunal invertebrates used to detect water quality changes and organic overenrichment.	(60)
Port Phillip Bay, Victoria	Changes in macrobenthic population and community structure.	Attempted use of cluster analysis to group macrobenthos populations surrounding a sewage outfall.	(61)
Panamatta Estuary, Australia	Changes in population and community structure.	Phytoplankton populations respond to increased nutrient supplies. Chlorophyll-a concentrations and primary production increase.	(64)

In estuarine and coastal science, much basic knowledge has accumulated since 1969. This is particularly true with respect to our understanding of nearshore processes and particularly with respect to population and community ecology. A number of authors have proposed elaborate and relatively sophisticated classification systems which are specific to "overenriched" or altered ecosystems. Some common characteristics of these index systems are as follows: they are usually based on a high degree of knowledge concerning the specific system with extensive, historical data; they have been enabled by the availability of taxonomic expertise, and they usually deal with organism trophic preference or energy transfer. Examples are given in Table 7.

Most of the references deal with coastal areas. Few are specific for estuarine areas where faunistic gradients and sharp physical-chemical gradients are common. None, to our knowledge, enjoys general application, although there is this potential; see especially Pearson and Rosenberg (58).

DISCUSSION AND CONCLUSIONS

Freshwater index and indicator techniques have been employed or transferred to estuarine and coastal areas with varying degrees of success. The reasons for this are manifold but probably include the following: (a) the lack of an exact and widely accepted definition of estuarine "eutrophication"; (b) a basic lack of knowledge of nutrient limitation and cycling in estuaries; and (c) possible fundamental differences between estuaries and other water bodies which invalidate transfer attempts.

Although the use and application of indices and indicators appear widespread, few authors have defined the conditions or factors which these tools are intended to respond to or measure. Indeed, many authors employ terms such as "eutrophic," "highly eutrophic," "mesotrophic," etc., without development of an evaluation or definitional basis. Thus, there is a strong presumption that limnetic terms and processes apply in estuarine areas. Often the index or indicator phenomena relate to a melange of eutrophication manifestations---periodic oxygen sags, persistent or temporary algal population blooms, high ambient nutrient levels - or to the potential for overenrichment. While problem definition is beyond the scope of this paper, it is felt to be an important limitation in the development and application of index/indicator systems.

The cycling and chemistry of both phosphorus and nitrogen in estuaries are complex and probably will not be totally elucidated in the near future. Many reports can be found wherein paradoxical results occur---ambient and presummably excessive nutrient levels occur but fail to elicit "eutrophic" responses. Yet, it is considered axiomatic that phosphorus or nitrogen are the limiting factors and special interpretations are invoked to explain why "eutrophic" conditions do not prevail. Some good examples of this are river estuaries of the highly-populated northeast where nutrient levels are extremely high, yet flushing or light limitation apparently override nutrient availability. The attempted force-fitting of liminetic methods or findings may be a source of confusion and a limitation with respect to the development of useful indicator and index systems.

Like Schindler (68), we seriously question the acceptance of processes or preconceptions which have not been validated for estuarine areas and which may have not been validated for estuarine areas and which may have significant economic penalty if misinterpreted. A high degree of specificity and rigor would to be the essential requirement of any estuarine index or indicator methodology—this goal is yet to be achieved.

ACKNOWLEDGEMENTS

The authors acknowledge the assistance and support of the following individuals: Dr. Del Barth, Steve Gage, and Pete Wagner. Jane Gilliard and Ann Meyer typed the manuscripts. Drs. Gene Cronin and Court Stevenson reviewed the manuscript and offered many helpful suggestions and criticisms. This is CEES Contribution No. 997 (HPEL) from the University of Maryland.

REFERENCES

1. Anger, K. 1975. On the influence of sewage pollution on inshore benthic communities in the south of Kiel Bay, Part 2: quantitative studies on community structure. Helgol. Wiss. Meeresunters 27:408-438.
2. Armstrong, J.W. 1977. The impact of subtical sewage outfalls of the intertidal macrofauna of several central Puget Sound beaches. Ph.D. dissertation. Washington Univ., Seattle. 233 pp.
3. Bach, S.D. and M.N. Josselyn. 1978. Mass blooms of the alga, Cladophora in Bermuda. *Mar. Pollut. Bull.* 9(2):34-37.
4. Bechtel, T.J. and B.J. Copeland. 1970. Fish species diversity Indices and indicators of pollution in Galveston Bay, Texas. Texas Inst. Mar. Sci. 15:103-132.
5. Black, R. 1973. Growth rates of intertidal molluscs as indicators of effects of unexpected incidence of pollution. *J. Fish. Res. Board Can.* 30(9):1385-1388.
6. Caperon, J., S.A. Cattell, and G. Krasnick. 1971. Phytoplankton kinetics in a subtropical estuary: eutrophication. *Limnol. Oceanogr.* 16(4):599-607.
7. ——————, W.A. Harvey, and F.A. Steinhilper. 1976. Particulate organic carbon, nitrogen, and chlorophyll as measures of phytoplankton and detritus standing crops in Kaneohe Bay, Oahu, Hawaiian Islands. *Pac. Sci.* 30(4):317-327.
8. Carpenter, J.H., D.W. Pritchard and R.C. Whaley. 1969. Observations of eutrophication and nutrient cycles in some coastal plain estuaries,

p. 210-221. *In* Proceedings of a Symposium - Eutrophication: Causes, Consequences, Correctives. National Academic Science, Washington, D.C.

9. Champ, M.A. 1975. Nutrient loading in the nation's estuaries. p. 237-255. *In* Proceedings of a Conference - Vol. I: Estuarine Pollution Control and Assessment, U.S. Environmental Protection Agency, Washington, D.C.
10. Cochrane, J.J., C.J. Gregory, and G.L. Aronson. 1970. Water resources potential of an urban estuary. NTIS PB-197 991. Boston, Mass. 110 pp.
11. Colwell, R.R. 1976. Bacteria and viruses - indicators of unnatural environmental change occurring in the nation's estuaries. p. 507-518. *In* Proceedings of a Conference - Vol. I: Estuarine Pollution Controland Assessment, U.S. Environmental Protection Agency, Washington, D.C.
12. Copeland, B.J. and D.E. Wohlschlag. 1968. Biological responses to nutrients -- eutrophication: saline water considerations, p. 65,82. *In* E.F. Gloyna and W.W. Eckenfelder, Proceedings of Water Resources Symposium No. I, Univ. of Texas, Texas. April, 1966.
13. Cory, R.L. 1974. Changes in oxygen and primary production of the Patuxent estuary, Maryland, 1963 through 1969. *Chesapeake Sci.* 15(2):78-83.
14. Duedall, I.W., H.B. O'Connors, J.H. Parker, R.E. Wilson and A.S. Robbins. 1977. The abundance, distribution and flux of nutrients and chlorophyll-a in the New York bight apex. *Estuarine Coastal Mar. Sci.* 5:81-105.
15. Duxbury, A.C. 1975. Orthophosphate and dissolved oxygen in Puget Sound. *Limnol. Oceanogr*. 20(2):270-274.
16. Eppley, R.W., C. Sapienza and E.H. Renger. 1978. Gradients in phytoplankton stocks and nutrients off southern California in 1974-76. *Estuarine Coastal Mar. Sci.* 7:291-301.
17. Flemer, D.A. and D.R. Heinle. 1974. Effects of waste water on estuarine ecosystems. CRC Publication No. 33. Chesapeake Research Consortium, Inc., 16 pp.
18. Folkard, A.R. and P.G.W. Jones. 1974. Distribution of nutrient salts in the southern North Sea during early 1974. *Mar Pollut. Bull.* 5(12):181-185.
19. Frecker, M.F. and C.C. Davis. 1975. Man-made eutrophication in a Newfoundland (Canada) harbour. *Int. Revue Ges. Hydrobiol.* 60(3):379-392.
20. Gameson, A.L.H. and I.C. Hart. 1966. A study of pollution in the Thames estuary. *Chem. Ind*. (Lond.) (Dec. 17):2117-2123.
21. Gardner, W.S. and J.A. Stephens. 1978. Stability and composition of terrestrially derived dissolved organic nitrogen in continental shelf surface waters. *Mar. Chem*. 6(4):335-342.
22. Gray, J.S. 1976. The fauna of the polluted river Tees estuary. *Estuarine Coast Mar. Sci.* 4:653-676.
23. Guide, V. and O. Villa, Jr. 1972. Chesapeake Bay Nutrient Input Study. Technical Rep. No. 47, U.S. Environmental Protection Agency, Annapolis, Md. 118 pp.

24. Haertel, L., C. Osterberg, H. Curl Jr. and P.K. Park. 1969. Nutrient and plankton ecology of the Columbia River estuary. *Ecology* 50(6):962-978.
25. Hale, S.S. 1975. The role of benthic communities in the nitrogen and phosphorus cycles of an estuary, p. 291-308. *In* Proceedings of a Symposium, Mineral Cycling in the Southeastern Ecosystems, University of Rhode Island, Kingston.
26. Harris, R.L. 1976. Processes affecting the vertical distribution of trace components in the Chesapeake Bay. Ph.D. dissertation. University of Maryland. 205 pp.
27. Harrison, W.G. and J.E. Hobbie. 1974. Nitrogen budget of a North Carolina estuary. UNC-WRRI-74-86. WRRI North Carolina 187 pp.
28. Hatcher, P.G., L.E. Keister, and P.A. McGillivary. 1977. Steroids as sewage specific indicators in New York bight sediments. *Bull. Environ. Contam. Toxicol.* 17(4):491-498.
29. Hinchcliffe, P.R. 1976. Surf-zone water quality in Liverpool Bay. *Estuarine Coast Mar. Sci.* 4:427-442.
30. Hobbie, J.E., B.J. Copeland, and W.G. Harrison. 1972. Nutrients in the Pamlico River estuary, N.C., 1969--1971. Rep. 79. WRRI North Carolina, Raleigh, N.C. 242 pp.
31. ------, and N.W. Smith. 1975. Nutrients in the Neuse River estuary, North Carolina. UNC-SG-75-21. University of North Carolina. 183 pp.
32. Hooper, F.F. 1969. Eutrophication indices and their relation to other indices of ecosystem change, p. 225-235. *In* Proceedings of a Symposium, Eutrophication: Causes, Consequences, Correctives. National Academy of Sciences, Washington, D.C.
33. Jaworski, N.A. 1969. Water quality and wastewater loadings upper Potomac estuary during 1969. Tech. Report No. 27. U.S. Environmental Protection Agency, Annapolis, Md. 104 pp.
34. ------ and L.J. Hetling. 1970. Relative contributions of nutrients to the Potomac River basin from various sources. Technical Rep. No. 31. U.S. Environmental Protection Agency, Annapolis, Md. 36 pp.
35. ------, L.J. Clark and K.D. Feigner. 1971. A water resource-water supply study of the Potomac estuary, Technical Report 35. U.S. Environmental Protection Agency, Philadelphia. 200 pp.
36. ------, D.W. Lear, Jr., and O. Willa, Jr. 1972. Nutrient management in the Potomac estuary, p. 246-273. *In* Proceedings of the Symposium on Nutrients and Eutrophication: The Limiting-Nutrient Controversy. *Am. Soc. Limnol. Oceanogr.*, Lawrence, Kansas.
37. Jeffries, H.P. 1962. Environmental characteristics of Raritan Bay, a polluted estuary. *Limnol. Oceanogr.* 7(1):21-31.
38. Jupp, B.P. 1977. The effects of organic pollution on benthic organisms near Marseille. *Int. J. Environ. Stud.* 10:119-123.
39. Kaarlgren, L. and K. Ljungsto. 1975. Nutrient budgets for the inner archipelago of Stockholm. *J. Water Pollut. Control Fed.* 47(4):823-833.
40. Kiortsis, V. and M. Moraitou-Apostolopulou. 1975. Marine cladocera (crustacea) in the eutrophicated and polluted Saronic Gulf. *Isr. J. Zool.* 24(1-2):71-74.

41. Klavestad, N. 1978. The marine algae of the polluted inner part of the Oslofjord. *Bot. Mar.* 21:71-97.
42. Knudson, K. and C.E. Belaire. 1975. Causes and probable correctives for oxygen depletion fish kills in the Dickinson Bayou estuary: a field study and simplified algal assay. *Contrib. Mar. Sci.* 19:37-48.
43. Leppaekoski, E. 1977. Monitoring the benthic environment of organically polluted river mouths, p. 125-132. *In* Biological Monitoring of Inland Fisheries.
44. Littler, M.M. and S.N. Murray. 1978. Influence of domestic wastes on energetic pathways in rocky intertidal communities. *J. Appl. Ecol.* 15:583-595.
45. Malone, T.C. 1977. Environmental regulation of phytoplankton productivity in the lower Hudson estuary. *Estuarine Coastal Mar. Sci.* 5(2):157-71.
46. McCormick, J.M. and P.T. Quinn. 1975. Phytoplankton diversity and chlorophyll-a in a polluted estuary. *Mar. Pollut. Bull.* 6(7):105-106.
47. Mihnea, P.E. 1978. Domestic wastewater effects on marine phytoplanktonic algae. *Rev. Int. Oceanogr. Med.* 49(3):89-98.
48. Miller, B.S., B.B. McCain, R.C. Wingert, S.F. Borton, and K.V. Pierce. 1977. Ecological and disease studies of demersal fishes in Puget Sound near metro-operated sewage tratement plants and in the Duwamish River. Final Rep. No. CR2231. Puget Sound Interim Studies, Municipality of Metropolitan Seattle 164 pp.
49. Moore, D.M. 1978. Seasonal changes in distribution of intertidal macrofauna in the lower Mersey estuary, U.K. *Estuarine Coastal Mar. Sci.* 7:117-125.
50. Moshiri, G.A., W.G. Crumpton, N.G. Aumen, C.T. Gaetz, and J.E. Allen. 1978. Water-column and benthic invertebrate and plant associations as affected by the physicochemical aspects in mesotrophic bayou estuary, Pensacola, Florida. WRRC-PUB-41. WRRC Gainesville, Florida. 166 pp.
51. ----- and W.G. Crumpton. 1978. Certain mechanisms affecting water column-to-sediment phosphate. *J. Water Pollut. Control Fed.* 50(2):392-394.
52. Noland, R.A. and M.A. Champ. 1979. Application of the National Sanitation Foundation Water Quality Index for the Potomac River for April to September 1977. *In* Rasin, Jr. V.J. and J.S. Lange. 1979. Potomac River Basin Water Quality: 1977. Interstate Commission on the Potomac River Basin Tech. Publication 79-2. 95 pp. (mimeo).
53. Nichols, J.A. 1977. Benthic community structure near the Woods Hole sewage outfall. *Int. Rev. Gesamten. Hydrobiol.* 62(2):235-244.
54. Odum, H.T. 1956. Primary production in flowing waters. *Limnol. Oceanogr.* 1:102-117.
55. ----- and R.F. Wilson. 1962. Further studies on reaeration and metabolism of Texas bays, 1958-1960. Pub. Inst. Mar. Sci., Univ. of Texas 8:23-55.
56. Olinger, L.W., R.G. Rogers, P.L. Fore, R.L. Todd, B.L. Mullins, F.T. Bisterfield, and L.A. Wise. 1975. Environmental and recovery studies of Escambia Bay and the Pensacola Bay system, Florida. EPA

904/9-76-016. U.S. Environmental Protection Agency, Atlanta, Georgia. 500 pp.

57. Ott, W.R. 1978. Environmental indices, theory and practice. Ann Arbor Science Publishers, Inc., Ann Arbor, Mich. 371 pp.
58. Pearson, T.H. and R. Rosenberg. 1978. Macrobenthic succession in relation to organic enrichment and pollution of the marine environment. *Oceanogr. Mar. Biol. Ann. Rev.* 16:229-311.
59. Peres, J.M. and J. Picard. 1975. Causes of decrease and disappearance of the seagrass *Posidonia oceanica* on the French Mediterranean coast. *Aquatic Bot.* 1(2):133-139.
60. Poirrier, M.A. J.S. Rogers, M.A. Mulino, and E. St. Eisenberg. 1975. Epifaunal invertebrates as indicators of water quality in southern Lake Pontchartrain. Rep. No. TR-5. WRRI, New Orleans, Louisiana. 52 pp.x
61. Poore, G.C.B. and J.D. Kudenov. 1978. Benthos around an outfall of the Werribee Sewage Treatment Farm, Port Phillip Bay, Victoria. *Aust. J. Mar. Freshwater Res.* 29:157-167.
62. Rechow, K.H. 1978. Quantitative techniques for the assessment of lake quality. Prepared for Michigan Department of Natural Resources by Michigan State University, East Lansing. 138 pp.
63. Redfield, A.C. 1958. The biological control of chemical factors in the environment. *Amer. Sci.* 46:205-221.
64. ——————, B.H. Ketchum and F.A. Richards. 1963. The Influence of organisms of the composition of sea-water, p. 27-77. *In* M.N. Hill (ed.). The Composition of Sea-Water Comparative and Descriptive Oceanography, Vol. 2. Interscience Publishers, New York.
65. Revelante, N. and M. Gilmartin. 1978. Characteristics of the microplankton and nanoplankton communities of an Australian coastal plain estuary. *Aust. J. Mar. Freshwater Res.* 29:9-18.
66. Ryther, J.H. 1954. The ecology of phytoplankton blooms in Moriches Bay and Great South Bay, Long Island, New York. *Biol. Bull.* 106:190-209.
67. —————— and W.M. Dunstan. 1971. Nitrogen, phosphorus, and eutrophication in the coastal marine environment. *Science* 171:1008-1013.
68. Schindler, D.W. 1974. Biological and chemical mechanisms in eutrophication of freshwater lakes. Annals, New York Academy of Sciences. 250:129-135.
69. Schultz, D.M. and J.G. Quinn. 1977. Suspended material in Narragansett Bay: fatty acid and hydrocarbons composition. *Org. Chem.* 1:27-36.
70. Seyb, L. and K. Randolph. 1977. Northern American Project---A study of U.S. water bodies. EPA-600/3-77-086. U.S. Environmental Protection Agency, Corvallis, Ore. 537 pp.
71. Specht, D.T. 1975. Seasonal variation of algal biomass production potential and nutrient limitation in Yaquina Bay, Oregon, P. 149-174. *In*: Proceedings of Biostimulation and Nutrient Assessment Symposium, Utah State Univ., Logan, Utah.
72. Swartz, R.C. 1972. Biological criteria of environmental change in the Chesapeake Bay. *Ches. Sci.* 13(Supplement): S17-S41.

73. Taft, J.L., M.E. Loftus, and W.R. Taylor. 1977. Phosphate uptake from phosphomonoesters by phytoplankton in the Chesapeake Bay. *Limnol. Oceanogr.* 22(6):1012-1021.
74. Taga, N. and H. Kobori. 1978. Phosphatase activity in eutrophic Tokyo Bay. *Mar. Biol.* 49:223-229.
75. Taslakian, M.J. and J.T. Hardy. 1976. Sewage nutrient enrichment and phytoplankton ecology along the central coast of Lebanon. *Mar. Biol.* 38:315-325.
76. Tilley, L.J. and W.L. Hauschild. 1975. Use of productivity of periphyton to estimate water quality. *J. Water Pollut. Control Fed.* 47(8):21572171.
77. Vollenweider, R.A. 1968. The scientific basis of lake and stream eutrophication, with particular reference to phosphorus and nitrogen as factors in eutrophication. Technical Report to O.E.C.D. Paris, DAS/CSI/68.27:1-182.
78. ------. 1976. Advances in defining critical loading levels for phosphorus in lake eutrophication. *Mem. Inc. Ital. Idrobiol.* 33:53-83.

MODELING OF EUTROPHICATION IN ESTUARIES

Donald J. O'Connor
Professor of Civil Engineering
Manhattan College
Bronx, New York

Consultant
Hydroscience, Inc.
Westwood, New Jersey

INTRODUCTION AND PURPOSE

The purpose of this paper is to present a review, summary and evaluation of nutrient-phytoplankton models which have been used in water quality management studies of estuaries. The basic concepts of eutrophication are first reviewed. These are incorporated in the conservation of mass relationship with estuarine transport equations to describe the spatial and temporal distribution of the relevant constituents. The application of these equations to eutrophication problems in various estuaries is demonstrated, particularly by studies which include both field and modeling activities. Similarly, those models whose development has been motivated by management questions are emphasized. Concluding remarks in the Applications section relate to an appraisal of the state of the art of estuarine eutrophication models and to their strengths, deficiencies, and future prospects with respect to both the scientific bases as well as engineering applications in the management context.

The term estuary, as used in this paper, refers to tidal stretches of relatively large drainage areas, which may be either saline or fresh and in which circulation and transport are affected primarily by tidal forces and runoff. These characteristics distinguish the estuary, in the sense in which it is herein defined, from tidal marshes, inlets and coves. The estuary is further characterized in terms of the impact of discharges and drainages from man's activities - municipal, industrial and agricultural. These inputs have had measurable effects on the water quality of many estuaries, the effect of nutrient discharges on eutrophication being one of the most significant.

Eutrophication, in general, may be defined as the conglomerate effect of increased nutrient discharges - including increases in abundance and changes in species composition of both microscopic and high forms of biological life in the system. Eutrophication also refers to changes in nutrient concentrations associated with the growth and decay of phytoplankton with subsequent effects on dissolved oxygen. It is the specific purpose of this paper to review the models which describe these interactions and to demonstrate their application to water quality management problems in estuaries. While quantitative mathematical relationships have been developed to define these changes, it is not yet possible to quantify such phenomena as shifts in species composition, with respect to phytoplankton, macrophytes and fish stocks. Consequently, the emphasis in this paper is directed to nutrient-phytoplankton reactions, with particular reference to those estuaries for which the modeling of the phenomena has provided a basis for water quality management.

BASIC EQUATION

The general principle of conservation of mass is used to formulate equations of the various constituents of importance in analysis of the eutrophication problem, the essential element of which is the dynamic behavior of phytoplankton. In their simplest form, these equations define the concentration of a nutrient and, either directly or indirectly, relate the growth of phytoplankton to its availability. In their most complex form they may incorporate interaction among many nutrients, specification of a number of species of phytoplankton, and predator-prey relationships between contiguous trophic levels. The nutrients, which may be undergoing additional biochemical transformations, are supplied by anthropogenic and natural sources, and by recycling from internal sources due to excretion and death of organisms and decay of the organic matter within the system. By applying the principle of conservation of mass, quantitative relationships of the progressive changes of state of the various constituents in time and space are developed. The relationships are expressed most fundamentally in terms of rates of change of the interactive substances. A series of simultaneous differential equations is thus developed, one for each constituent of the process. In this form, the greatest insight and understanding of the phenomena are provided.

The basic constituents of the analysis are nutrients, phytoplankton and zooplankton. An equation of mass balance is developed for each. Consider a segment of a natural water system of specified volume. The mass rate of change of the constituent within it is the product of the volume, V, and change of concentration, c, over the time interval, t, and is accounted for by the rates of change of three components: the physical transformations with it, R, and the inputs to or withdrawals from it, W. The mass rate of change for any segment is written:

$$V \frac{dc}{dt} = \Sigma J \pm \Sigma R \pm \Sigma W \quad (1)$$

The primary constituents to which this equation is applied are nutrients and the phytoplankton. Since the hydrodynamics of the system circulates the water as well as the substances contained within it, the transport component of the equation is similar for each of these constituents. The inputs are nutrient discharges from municipal, industrial and agricultural activities or from drainage of urban, rural and undeveloped areas. The reactions refer essentially to biochemical transformations which occur as nutrients are assimilated and released in the growth and subsequent decay of phytoplankton. The rate at which these reactions occur is generally expressed as

$$R = (G_P - D_P)P \quad (2)$$

in which

R = net rate of change of the phytoplankton, M/LT

G_p = growth coefficient - 1/T

D_p = decay coefficient - 1/T

p = concentration of phytoplankton - M/L.

REVIEW OF PREVIOUS WORK

Equation (2) is a valid representation of a natural water system, assuming spatial homogeneity of the relevant constituents. This assumption was made by those investigators who first developed and applied the kinetic approach. An excellent summary of the earlier contributions which provided the basic concepts and formulations for many current models is available (1). Fleming used a form of this equation to describe phytoplankton growth in a marine system and applied the analyses to the spring diatom bloom in the English Channel. A constant growth rate was postulated and a time variable decay rate resulting from grazing by zooplankton was used. Riley expressed the rates in a more fundamental manner. The growth was related to environmental variables of temperature, solar radiation, extinction coefficient and nutrient concentration. The decay rate included both a respiration and a grazing component. The seasonal variation of the phytoplankton was calculated using observed nutrient concentration and zooplankton levels. The validity of the approach was confirmed in three different marine environments (2, 3, 4) in the sense that the pattern and magnitude of the phytoplankton variation was reasonably reproduced. These contributions provided the basis for further development and application of the approach.

The spatial variation of phytoplankton was next introduced (5), in which mixing and settling in the vertical phase were incorporated as transport terms in equation (1). In addition, a conservation of mass equation for a nutrient (phosphate) was also introduced, as well as simplified equations for herbivorous and carnivorous zooplankton concentrations. The phytoplankton and nutrient equations were applied to a number of volume elements which extended from the surface to well below the euphotic zone. In order to simplify the calculations, a temporal steady-state was assumed to exist in each element. Thus, the equations apply to those periods of the year during which the dependent variables are not changing significantly in time.

The next significant advance (6) consisted of retaining the time derivatives of the equation and simplifying the vertical analysis by segmenting the system into two volumes to represent surface and bottom elements. Thus, both temporal and spatial variations were considered. In addition, differential equations for phytoplankton and zooplankton concentration were coupled so that interactions of the populations, as well as the nutrient-phytoplankton dependence, were explicitly included. The coefficients of the equations were not expressed as functions of time, however, so that the effects of time-varying solar radiation intensity and temperature were not included. The equations were numerically integrated and applied to the vertical distribution of chlorophyll in the Gulf of Mexico (7).

The models of both Riley and Steele have been reviewed in greater detail (1) in a discussion of their applicability and further development. Difficulties encountered in formulating simple theoretical models of phytoplankton-zooplankton population models were also discussed (8).

The principles underlying growth and decay mechanisms of phytoplankton had been qualitatively understood by limnologists and oceanographers for a number of years. The unique contribution of Riley and his collaborators and Steele lies in the quantitative description of the processes in the form of differential rate equations and in the ability of these equations to reproduce observations in natural systems. Each consider the primary dependent variables to be the phytoplankton, zooplankton, and nutrient concentration. A conservation of mass equation is written for each species, and the spatial variation is incorporated by considering finite volume elements which interact because of vertical eddy diffusion and downward advective transport of the phytoplankton. Their equations differ in some details; for example, values of growth coefficients used and the assumption of steady state. In addition, these equations were applied by the authors to actual marine situations and their solutions compared with observed data. This is a crucial part of any investigation wherein the assumptions that are made and the approximations that are used are difficult to justify a priori. The ultimate test of the validity of the equations lies in their ability to reproduce field observations.

These analyses provided the basis for development of subsequent eutrophication models. Significant advances were made over the past two decades with respect to better understanding of individual mechanisms underlying reaction rates in these models. These developments related to the optimal conditions for light, nutrient and temperature and the limiting effects at other levels of these parameters. Much of this work was directed to marine species. With respect to other levels of the food chain, developments were also made, specifically in the understanding and quantification of filtering and feeding characteristics of zooplankton.

Comparable advances in physical oceanography occurred during this period, particularly in the field of estuarine circulation. The fundamental equations of hydrodynamics were structured to define this circulation and were solved by both analytical and numerical methods. Field and laboratory studies yielded excellent data, which led to further insight and understanding of the phenomena and provided the basis for testing various hypothesis and calibrating hydrodynamic models. The development and application of physical models of estuarine systems is particularly noteworthy in this regard. By contrast to estuarine circulation, the structure of the coastal zone and the complexity of the forcing functions and boundary conditions have to date precluded comparable advances in the understanding and definition of coastal circulation. Prototype measurements and observations, however, have provided some empirical basis for assignment of the transport coefficient in these regions.

As the observations of various components increased and understanding grew, so did the development and application of eutrophication models, incorporating many of the advances in both transport and kinetics. Equations with parameters that vary as a function of temperature, sunlight, and nutrient concentrations have been presented (9) and simulated (10). A set of equations which model the population of phytoplankton, zooplankton and a species of fish has been presented (11). The application of the techniques of phytoplankton modeling to the problem of eutrophication in rivers and estuaries has been proposed (12), independently developed and applied (13) and extended through higher trophic levels (14). The interrelations between the nitrogen cycle and the phytoplankton population in the Potomac estuary have been investigated using sequential first-order kinetics, with a feedback loop to the inorganic form of the nutrients (13). A phytoplankton model was subsequently structured which employed nonlinear kinetics and applied to the Sacramento-San Joaquin delta system (14). This essential framework was applied to Chesapeake Bay systems (15). The potential, and to some degree, the actual effects of eutrophication in the Louisiana coastlands (16) and in the estuaries and tidal coves of North Carolina (17) have been reported. A summary of many of the kinetic components of nutrient-phytoplankton-zooplankton interactions has been reported (18) and a model developed for Narragansett Bay in Rhode Island. Extensive bibliographies of estuarine models, which include some eutrophication analyses, are also available (19, 20).

Considering the theoretical background and field observations of both biological and physical phenomena in estuaries, relatively few eutrophication models have been developed specifically relating to estuarine management problems. By contrast, significant developments and applications have taken place with respect to lakes, where the magnitude and impact of eutrophication have been more evident. The large estuarine systems have been, in general, more affected by bacterial contamination and by discharges of carbon and nitrogen from municipal, industrial and agricultural activities, resulting in dissolved oxygen depressions. In spite of the fact that the latter is affected by the growth and death of phytoplankton, the predominant factor is bacterial oxidation of carbon and nitrogen. The effects of eutrophication in estuaries, however, are evident further downstream after the carbon and nitrogen have been oxidized. Furthermore, recent increases in biological wastewater treatment have resulted in improvement of water quality with respect to dissolved oxygen, suspended solids and bacteria levels. Often nitrogen and phosphorus are substantially unaffected by this treatment. Consequently, phytoplankton growth continues to occur and in some cases the biophysical environment in the estuary resulting from this treatment may be more conducive to the growth of phytoplankton.

Water quality conditions in the Potomac River are characterized by the sequence of reactions where oxidation of carbon is of initial importance. In other estuaries, such as the Sacramento-San Joaquin system, where the nitrogen and phosphorus inputs are of greater magnitude than the carbon, phytoplankton activity is more pronounced than that of the heterotrophic bacteria. In each of these cases, extensive sampling and model development have proceeded simultaneously and interactively to assess water quality impacts and to provide input to management decisions. Similar activities have been initiated on Puget Sound (21). The model, based on previous work (22), is designed to simulate ecologic succession from primary production (algae) through successively higher trophic levels (zooplankton, benthic organisms and fish) in a three-dimensional, vertically stratified fjord-like estuary. Initial calibration has been performed, but application to management policies has not yet been effected. A eutrophication model of Teconic Bay in Long Island Sound has been developed (23). The relationship between nutrient inputs and chlorophyll levels and standing crop was simulated. The model has been used to examine nutrient standards and to determine treatment requirements.

Eutrophication models have been constructed for other estuarine systems both in this country; James (24), N.Y. Bight (25, 26), and in other parts of the world; North Sea (27), Thailand (28). However, these analyses have not yet provided tools for management decisions as in the cases of the Sacramento and Potomac. Consequently, in accordance with the orientation of this paper, analyses of the latter systems are emphasized, while reference to the former are made when appropriate.

As a background for the discussion, the present state of the art is reviewed. Each of the components of equation (1) is discussed separately in the subsequent sections, followed by descriptions of the specific form of the equation used in each case, and the application of the model to the particular water quality problem.

KINETIC INTERACTIONS

The assimilation of inorganic nutrients in the presence of light by phytoplankton and the respiration and predation by higher levels of the food chain underlie the eutrophication process in both estuaries and lakes. The various kinetic elements may be included in two broad categories: biological growth and death of phytoplankton, zooplankton, and upper levels of the food chain; and biochemical transformations of nutrients: nitrogen, phosphorus and silica. These regimes are linked by the growth coefficient of the phytoplankton, as shown in Figure 1. Primary production converts inorganic nutrients to phytoplankton, whose growth is a function of temperature, light and nutrient concentrations.

Secondary production of zooplankton is accomplished by their grazing on phytoplankton. Mortality and excretion pathways release organic material in particulate and soluble form. These, in turn, break down to yield inorganic nutrients which are available to the phytoplankton for growth. Ammonia is also subjected to biochemical oxidation by bacteria. The deposition pathway accounts for whatever settling of particulate organic material occurs. These kinetic interactions and pathways are presented schematically in Figure 1. A tabulation of the kinetic coefficients and parameters used in various studies is presented in Table 1.

Phytoplankton

The growth coefficient of phytoplankton is directly related to temperature in moderate climates. It is also dependent on light intensity and nutrient concentrations up to a saturating or limiting condition, greater than which it decreases with light and remains constant with nutrient concentration. Nutrient dependency is described by a Michaelis-Menton formulation whose significant parameter is that concentration at which the growth rate is equal to one-half of that at the saturated concentration. If more than one nutrient is involved, the growth rate is governed by the concentration of the limiting nutrient. A preference structure of ammonia over nitrate is often built into the analysis.

The effect of non-optimal light conditions is to reduce the growth rate. Furthermore, in a natural environment, available light decreases with depth in an exponential fashion, defined by the extinction coefficient. In addition, surface light varies throughout the day. The mean daily value for the photoperiod fraction of the day is used in these analyses. The light effect on growth is, therefore, time averaged over the day and vertically averaged over the depth.

Removal of phytoplankton is caused by endogenous respiration, settling and predation by zooplankton. The pertinent parameter in the latter is the grazing coefficient.

The kinetic equation which embodies these reactions and defines the net rate of change of the phytoplankton is.

$$R_p = (G_p - D_p)\,P \qquad (2)$$

$$G_p = \phi\,(K_T, r, n) \qquad (3)$$

$$D_p = \phi_s\,(K_I) + \phi_s\,(C_g, Z) + \phi_s\,(V_s, H) \qquad (4)$$

P = concentration of phytoplankton, M/L.

TABLE 1. Summary of Phytoplankton-Nutrient Kinetic Coefficients

Phytoplankton	N.Y. Bight	Pot.	Sac-S.J.	James	Patuxent
Saturated Growth Rate (day^{-1} @ 20°C)	2.5	2.0	2.5	2.0-3.0	2.0
Saturating Light Intensity, (Langleys/day)	300.	300.	300.	300.	350.
Half Saturation Constant for:					
Phosphorus - (μg P/l)	-	5.	5.	1.	1.
Nitrogen - (μg N/l)	25	25.	25.	5.	5.
Silica - (g Si/l)	-	-	-	-	-
Endogenous Respiration Rate (day^{-1} @ 20°C)	0.1	0.1	0.1	0.1	0.2
Settling Velocity for Chlorophyll - (m/day)	0.1			0.23	0.4
Stoichiometric Rate					
Carbon to Chlorophyll (mg C/mg chl-a)	50.	50.	50.	25.	33
Phosphorus to Chlorophyll (mg P/mg chl-a)	-	1.	0.5	1.	1
Nitrogen to Chlorophyll (mg N/mg chl-a)	7.	10.	7.	7.	7
Silica to Chlorophyll (mg Si/mg chl-a)	-	-	70.	-	-
Oxygen to Chlorophyll (mg O_2/mg chl-a)	133.	-	133.	67.	88
Organic Constituents					
Organic Carbon Mineralization Rate (day^{-1} @ 20°C)	0.1	-	-	0.2	0.2
Organic Nitrogen Mineral. Rate - (day^{-1} @ 20°C)	-	0.14	-	0.0-0.30	0.01(1)
Organic Phosphorus Mineral. Rate - (day^{-1} @ 20°C)	-	0.14	-	0.0-0.15	0.01
Settling Rate of Particulate Forms(2) - (m/day)	0.1	0.1 (decay)	0.6	0.2 (decay)	0.4
Organic Nitrogen Settling Velocity - (m/day)	-	-	0.6	0.23	0.4
Organic Phosphorus Settling Velocity - (m/day)	-	-	0.6	0.23	0.4
Inorganic Constituents					
Nitrification Rate Denitrification Rate	0.2	0.2	-	0.0-0.15	-
Denitrification Rate (day^{-1} @ 20°C)	0.1	-	-	-	-

TABLE 1. Cont'd.

Zooplankton					
Grazing Rate (1/mg C-day @ 20°C)	0.17	0.2	0.18	-	-
Half Saturation Constant for Grazing Rate Limitation (μg chl-a/l)	-	-	50.	-	-
Half Saturation Constant for Assimilation Limit (μg chl-a/l)	60.	-	-	-	-
Maximum Assimilation Efficiency	0.65	-	0.60	-	-
Respiration Rate (day^{-1} @ 20°C)	0.015	-	0.10	-	-

Range of Temperature Coefficients - θ	
Phytoplankton Saturated Growth Rate	1.04 - 1.07
Phytoplankton Endogenous Respiration Rate	1.04 - 1.08
Organic Carbon Mineralization Rate	1.045
Organic Nitrogen Mineralization Rate	1.045
Organic Phosphorus Mineralization Rate	1.045

(1) A phytoplankton cell lysing mechanism is incorporated into the kinetic structure of this model.
(2) In the Sacramento-San Joaquin a separate nonvolatile suspended solids settling velocity was specified.

The growth is made up of the saturated growth rate $K_T(T)$, which is a function of temperature, T; the reduction due to nonoptimal incident light, r, a function of the saturating light intensity, I_s, and the extinction coefficient, K_e; and the reduction factor by the nutrient, n, with Michaelis constant, K_n.

The decay is made up of the endogenous respiration rate, $K_l(T)$, the grazing rate, $C_g(T)$, the zooplankton biomass, Z, the Michaelis constant for zooplankton grazing, K_{MP}, and the apparent settling rate, V_s, and depth, H.

Although it is common to combine the various species of algae into one classification, characterized by a singular growth function, it is increasingly apparent that such a generalization is too restrictive. In the analysis of lake eutrophication, the introduction of additional species produced more realistic seasonal variations, particularly in reproducing late summer or fall blooms (29).

Zooplankton

This system is analogous to that of the phytoplankton. Zooplankton grow in accordance with the availability of their food, phytoplankton. Zooplankton, in turn, are predated upon by the upper levels of the food chain and undergo endogenous respiration and death. Their excretion

FIGURE 1. Kinetic Interrelationships.

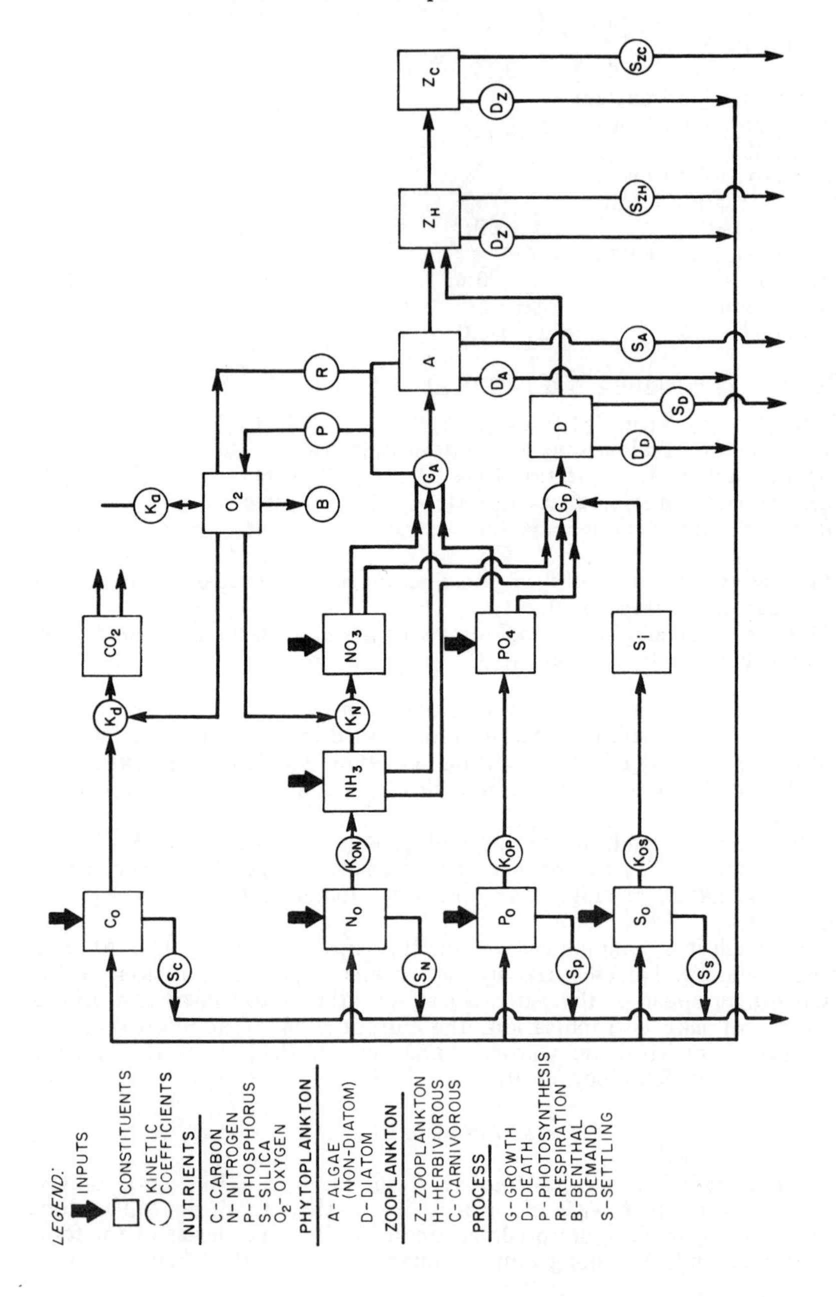

products may be significant sources of nutrients. Furthermore, to account for the fact that zooplankton graze more than they consume, a conversion efficiency coefficient is introduced into these equations. In order to simplify the analysis, the predation effect of the upper levels is often introduced as an empirical constant. The kinetic equation is identical in form to equation (2).

In the analysis of eutrophication in lakes, it sometimes is necessary to include predation by the next trophic level. Therefore, the zooplankton are divided into two classes: herbivorous, which prey on the phytoplankton, and carnivorous, which in turn prey on the herbivores (30).

Nutrients

Major components of the nitrogen system are detrital organic nitrogen, ammonia nitrogen, and nitrate nitrogen. In natural waters there is a stepwise transformation from organic nitrogen to ammonia, nitrite and nitrate, yielding nutrients for phytoplankton growth. The kinetics of these transformations are usually assumed to be first order reactions with temperature-dependent rate coefficients. The hydrolysis of organic nitrogen yields ammonia, which is oxidized through nitrite to nitrate.

Two sources of detrital organic nitrogen are considered: organic nitrogen produced by the endogenous respiration of phytoplankton and zooplankton, and the organic nitrogen equivalent of grazed but not metabolized phytoplankton excreted by zooplankton.

The phosphorus system is similar in some respects to that of nitrogen. Organic phosphorus is generated by the decay of phytoplankton. Organic phosphorus, as in the case of organic nitrogen, represents the detrital material. Phosphorus in this form is then converted to the inorganic state, where it is available to the algae. A sink of organic phosphorus, and nitrogen, is due to settling, which may be significant depending on the magnitude of vertical mixing. If settling is effective, it is assumed that only the organic form of the nutrients is susceptible to removal in this fashion. The silica cycle is comparable to that of the phosphorus.

TRANSPORT STRUCTURE

The transport structure of the analysis is determined by the hydrological, morphological and tidal characteristics of the system. Significant advances have been made over the past decade in the understanding and definition of estuarine circulation patterns. The vast majority of the hydrodynamic approaches, which involve simultaneous solution of the equations of momentum, continuity and state, define the temporal variation of the current and stage over the cycle. Since the time scale of the tidal phenomenon is much less than that of the seasonal variation of phytoplankton, approximations of the transport regime are made to make the time scales compatible. Transport components are thus usually expressed in terms of advective and dispersive coefficients, which may be fundamentally derived from the hydrodynamic solutions or empirically assigned from an analysis of the distribution of a tracer. Salinity intrusion provides an excellent basis for determination of the transport components.

The approach is identical to that which has been successfully used in the analysis of salinity and dissolved oxygen, which has provided a broad empirical basis for a realistic assignment of the transport coefficients for many estuaries. This type of analysis, which has been greatly augmented by many studies of salinity intrusion in physical models, is particularly appropriate for the one-dimensional case. The longitudinal distribution of nutrients, phytoplankton and dissolved oxygen have been realistically reproduced by one-dimensional advective-dispersive models. This essentially empirical approach, furthermore, has been substantiated by more fundamental hydrodynamic analyses, such as that presented by Harleman and Ippen (31).

The one-dimensional advective-dispersive analysis has obvious weaknesses when extended to two and three dimensional conditions. Although it is possible to assign transport coefficients to reproduce equivalent salinity distributions, these are not necessarily the best representation of the transport of other constituents. Furthermore, extrapolation to projected conditions is more tenous. For the two-dimensional analyses, the more basic approach is desirable.

One of the most significant characteristics of two-dimensional flow, which may have a marked effect on phytoplankton distributions, is attributable to classic estuarine circulation. The combined effect of less dense freshwater inflow and more dense saline intrusion produces a characteristic tidally-averaged pattern, in which the net velocity in the surface layer is seaward and that in the lower layer is landward. Continuity requires vertical upward flow to balance the difference between the flows in upper and lower layers. This circulation pattern may concentrate algae in the upstream region of the salinity intrusion in much the same way as it causes the turbidity maximum. The two-dimensional analysis, which is a more realistic representation of estuarine transport, has indicated marked differences in phytoplankton response by contrast to the one-dimensional analysis.

In Figure 2 are shown the transport components for a typical estuarine circulation and the associated phytoplankton growth characteristics.

In addition to the factors described above which are generally included in eutrophication models, there are other transport and kinetic pathways which may be of importance in defining the spatial and temporal distribution of the various constituents. Recent developments, theoretical, experimental and observational, have indicated their potential importance.

1. Adsorption - Nutrients, particularly certain forms of phosphorus are readily adsorbed to solids. Organic matter and clay species have absorptive capacities for both nutrients and other water borne constituents. Flocculation and settling of these materials in the saline region of estuaries may be an important route of removal from the water column.

2. Water-Bed Exchange - The transfer of dissolved nutrients between the water body and the bed may be of considerable importance. The decomposition processes, both chemical and biological, under anaerobic conditions, which may take place in many sediments, release dissolved

FIGURE 2. Characteristic estuarine transport regime and phytoplankton kinetic effects.

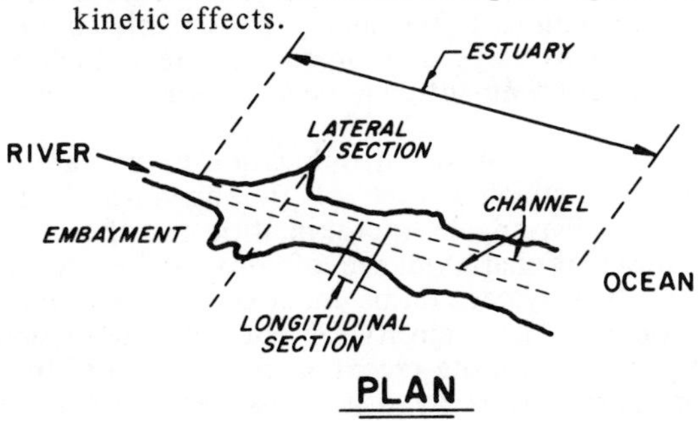

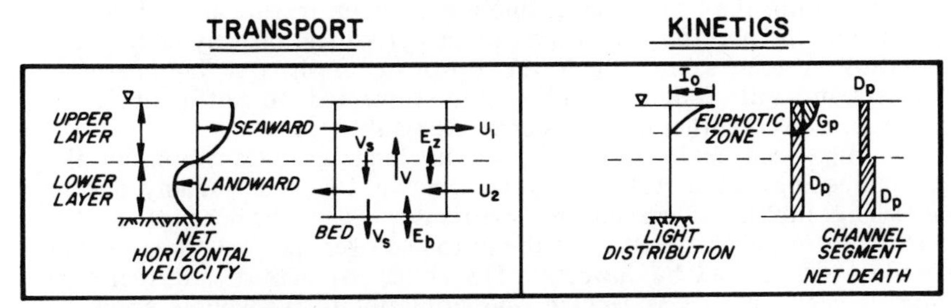

LONGITUDINAL CHANNEL SECTIONS

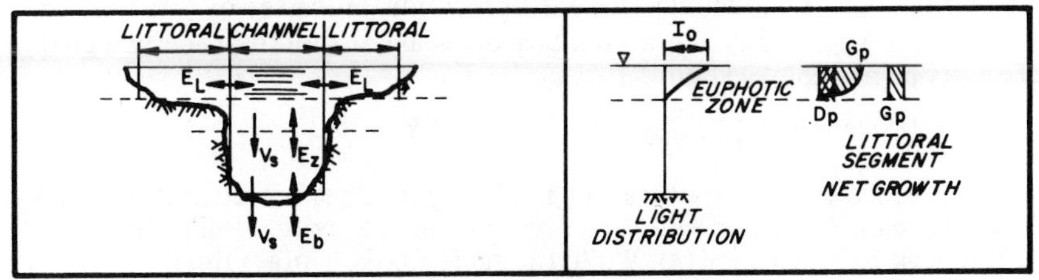

LATERAL BAY AND CHANNEL SECTIONS

forms of nutrients which diffuse through the interstitial waters and thus transfer nutrients from the bed. Conversely, under certain gradient conditions material may be transported from the water to the bed. While the potential importance of this nutrient reservoir has long been recognized in lakes and attempts made to analyze the phenomenon in a modelling form, its inclusion in estuarine models is minimal. A recent analysis by Flaherty and Harris (41) demonstrated its importance in the Potomac.

3. Sedimentation and Sediment Transports - If either or both of the above factors, adsorption or exchange with water column, are significant in a specific estuary, it follows that transport and flux conditions in the

bed must be included in the analysis in order to account for the total mass balance of nutrients. Sediment transport, particularly of the light flocculent materials such as clays and organic matter which are more susceptible to scour and entrainment, may be an important component of the balance.

4. Variable Stoichiometry - Most eutrophication models are based on the assumption that the stoichiometry of the organism is invariant. Although this may be a reasonable approximation for the many cases, there is increasing observational evidence of algal ability to store and utilize nutrients in excess of average ratios under certain conditions.

5. Seasonal variations - The majority of eutrophication models are based on a single algal species. More recent work has been addressing the question of two or more classes by distinguishing between or among diatoms greens and blue greens. In the analysis of estuarine eutrophication, the additional distinction between marine and freshwater species may be necessary. Inclusion of this classification may be required for those estuaries, in which the upstream freshwater boundary or downstream salt water boundary are subjected to different water concentrations of either nutrients or phytoplankton.

6. Long term ecological change - The long term transitions in the general ecology of a water body are known to occur - many due to increasing inputs of wastewater nutrients - e.g. the change in the eutrophication characteristics of the Potomac over the past three or four decades as described by Jaworski. The ability to analyze, and therefore project, such changes is lacking due primarily to inadequate scientific understanding of the phenomenon. It may be properly regarded a long-term goal of modelling the eutrophication processes of estuaries or any natural water body - and on a broader scale, it should properly a goal of any type of ecological model.

COMPARISON OF LAKES AND ESTUARIES

Since the art of eutrophication modeling in lakes has been developed further than in estuaries, it is instructive to note similarities and differences between the two. While the same fundamental principles apply in each case, the ecological environment and geohydrological characteristics are markedly different. Analyses of these factors result in emphasis on certain elements in the equations and elimination of others. Thus, while the principle expressed by equation (1) is generally applicable to both estuaries and lakes, the particular forms of working equations are usually quite different.

The major differences between the two relate to geomorphological structure and hydrodynamic transport. The similarity lies in the ratio of productive area to total area. Lakes are characterized by littoral zones, which feed the pelagic region. Tidal embayments contiguous to the main channel provide the same function for estuaries. This characteristic is portrayed graphically in Figure 3, which shows the productive embayment areas of various estuaries discussed in this paper. The deeper segments of each system are generally less productive due to smaller ratios of euphotic zone to total volume relative to the littoral and embayment zones.

FIGURE 3. Location maps showing channel and embayments of various U. S. estuaries.

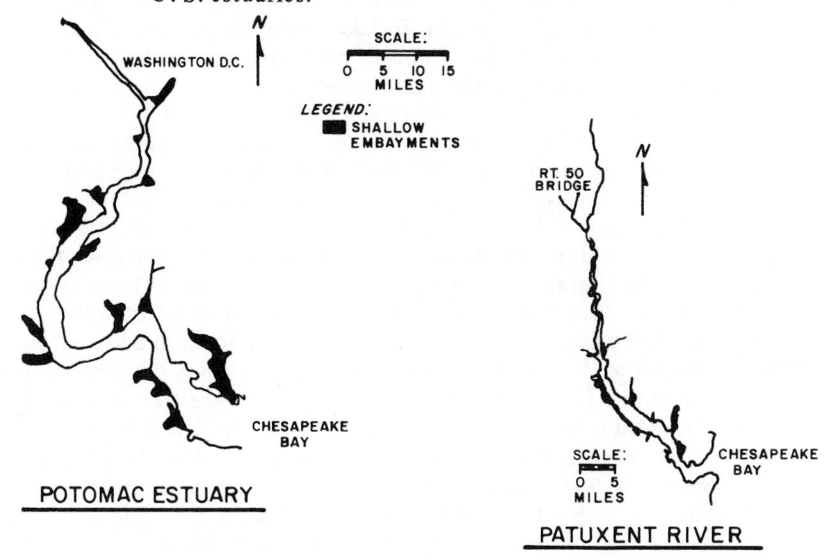

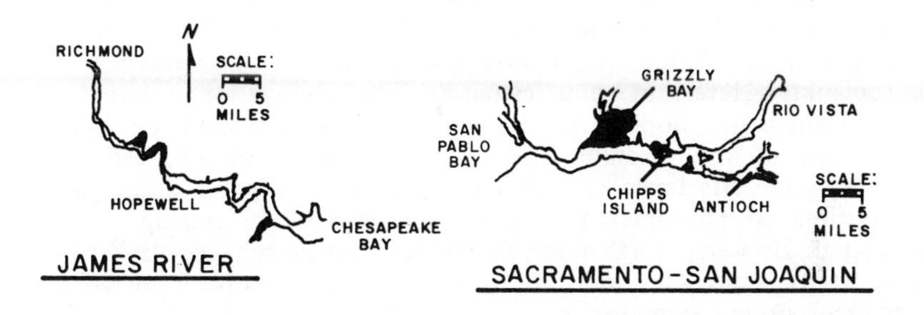

The contrast in transport associated with the hydrodynamics of each system is marked. Residence time in estuaries is often on the order of weeks to months, while that in lakes is months to years and decades in the case of the larger lakes. The direction of net transport in estuaries is longitudinal along the major axis of tidal and freshwater flow. Although there may be a primary direction in lakes due to prevailing winds, transport in two and three dimensions is more significant. During periods of stratification, mixing across the thermocline is usually less than that through the halocline of estuaries. In this regard the two-dimensional (vertical-longitudinal) circulation in the saline zone of estuaries tends to increase residence time and presumably productivity, a factor which has not yet been included in eutrophication models. A preliminary assessment of such a transport regime is discussed subsequently.

While settling of viable and detrital organic matter occurs in both systems, the effect is more pronounced in lakes. The intensity of vertical mixing is significant in this regard. In lakes much of the mixing effect is overcome by gravitational force and thermal structure, usually resulting in a greater percent of the particulate matter being deposited in the bed. This factor provides the essential element in simplified lake models and accounts for the relative appropriateness of these models. In many estuaries, tidal forces produce sufficient energy to mix upper and lower layers uniformly. In highly stratified estuaries, vertical differences exist, but not to the same degree as occurs in lakes. Because of the greater suspended solids and phytoplankton concentration, which are generally an order of magnitude greater in estuaries, the euphotic zone in these systems is shallower than that in lakes. A typical example of this effect is shown in Figure 4 for Lake Ontario and the Sacramento-San Joaquin delta. Growth rates are normalized in each case and the relative effects of temperature, surface radiation, light extinction and nutrient levels are shown. The greatest reduction in growth is evidently due to the extinction and nutrient levels are shown. The greatest reduction in growth is evidently due to the extinction coefficient in estuaries.

An additional factor is also shown in Figure 4 - the relatively insignificant effect of nutrient levels in the Sacramento estuary. This condition is typical in estuaries. Sources of nutrients from municipal and industrial inputs and upstream runoff generally produce concentrations which are greater than the Michaelis value, with the consequent effect that nutrient levels do not limit the growth coefficients. Furthermore, zooplankton predation apparently exerts a minor influence on phytoplankton levels in some estuaries. This observation is based on measurements of zooplankton biomass and substantiated by model calculations of the Sacramento-San Joaquin and Potomac estuaries. Little to no difference is calculated in the temporal and spatial distributions of phytoplankton when the zooplankton grazing term is eliminated. However, further research in phytoplankton population and transport dynamics may indicate the significance of zooplankton effects to be incorporated in models.

The variable salinity encountered in estuaries, in contrast to the freshwater environment of lakes, accounts for a marked transition in and a wide variation of phytoplankton species. These changes occur temporally and spatially with variable freshwater flow and tidal conditions. Although the distinction between fresh and marine species has long been recognized and identified by limnologists and oceanographers, this factor has not yet been included in eutrophication models of estuaries. Many of the factors described above are incorporated in the following examples and the implied or direct effects are discussed. By contrast it is usually necessary to include the predation effect in lake models to reproduce the seasonal variation. In some cases two levels of zooplankton are introduced: the herbivorous, which feed on phytoplankton, and the carnivorous which, in

FIGURE 4. Effect of light and nutrients on the phytoplankton growth-coefficient.

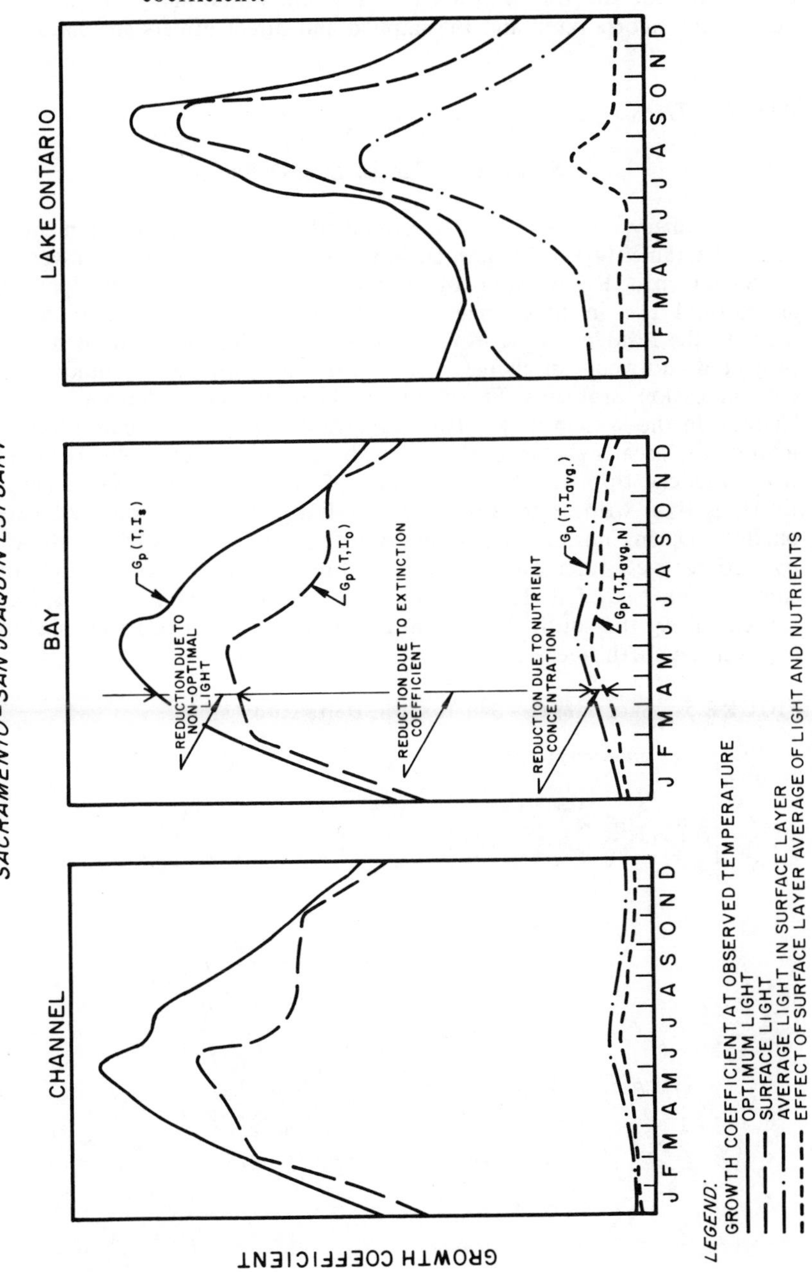

turn, prey on herbivorous zooplankton. The preceding provides a perspective for the following examples within which many of the above factors are incorporated and the implied and direct effects are discussed.

APPLICATIONS

Sacramento-San Joaquin Delta

The primary motivation for development of the eutrophication model of the Sacramento-San Joaquin Delta was to assess the effects of diversion of Sacramento River water by a peripheral canal for municipal and agricultural use in other regions of California. Such diversion would increase the residence time in the system, which, in conjunction with the projected nutrient discharges from irrigation drainage, could cause eutrophication problems. Furthermore, the proposal to deepen the ship channel in the system may further aggravate water quality conditions. A map of the area is shown in Figure 5. The area of concern, which extends from Benicia through the confluence of the Sacramento-San Joaquin rivers, is tidal with seasonally variable salinity depending on freshwater outflow. During normal high spring runoffs, the system is fresh followed by progressively increasing salinity as the flow decreases through the summer and early fall. The main channel is approximately 30 feet deep, running along the south shore, contiguous to a series of relatively shallow bays on the north side.

FIGURE 5. Sacramento - San Joaquin Delta study area.

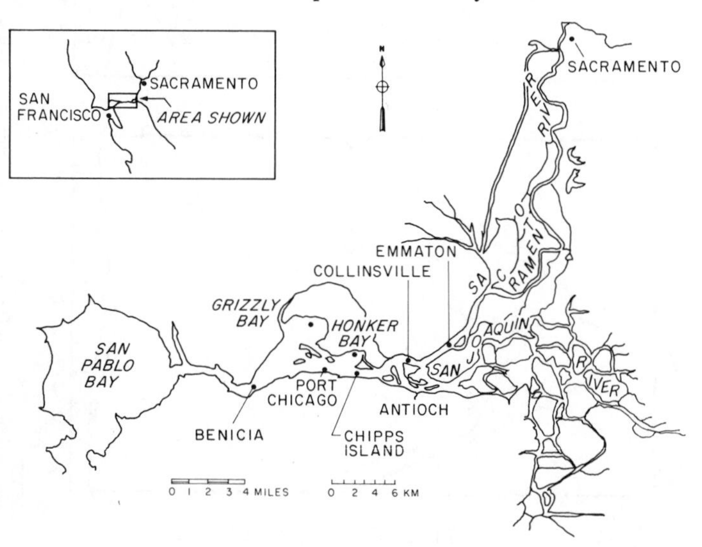

The seasonal response of phytoplankton is typical: a spring bloom due to increasing light and temperature, usually followed by a late summer-early fall growth. Phosphorus and silica are plentiful, their concentrations being far above the Michaelis value during the entire year. Nitrogen, in the form of nitrate, periodically falls in the range of the Michaelis concentration during summer after the spring bloom. The earlier version of the model was reasonably successful in reproducing the temporal distribution of chlorophyll at a number of stations (14). Since the original development, various modifications and additions have been made, relating to extinction coefficient relationships (29), suspended solids (32), growth temperature maximum, dissolved oxygen, and additional pools of nutrients: silica and phosphate. Examples of output of the time variable model and observations are shown in Figures 6 and 7. A later development of practical value is the seasonal steady-state analysis, as shown in Figure 8 by the dashed line, compared to the time-variable output, indicated by the continuous lines. The fact that the models, in general, reproduce the magnitude shape of the distributions of the various constituents is an indication that relevant phenomena are accounted for. However, occasionally secondary blooms in the order of 70 μg/l chlorophyll are not reproduced. As may be observed, concentrations of phosphate and silica, which are never limiting, are not materially affected by the successive blooms. They are included to lend additional substantiation to the analysis. In a similar vein, the predation by zooplankton has been shown to be insignificant.

During 1976 and 1977, extreme drought conditions prevailed throughout the area, resulting in increased salinity and a marked reduction in overall productivity. Figure 9 presents the flow, salinity and chlorophyll concentrations for a bay and channel location. The present version of the model cannot reproduce the levels of chlorophyll observed from the summer of 1976 through 1977. Various hypotheses are presently being investigated in order to clarify the causal mechanisms of the depressed productivity. From the transport viewpoint, relocation in the entrapment zone and reduction of vertical flow with reduced outflow are possible mechanisms. From the biological perspective, increased death rates, which have been observed in some laboratory experiments, may be associated with toxic or salinity effects. The use of an optimum temperature and a salinity effect on growth rate of freshwater phytoplankton yielded some degree of correlation, but these effects were not observed in laboratory experiments. The introduction of more than a single species appears to be a viable route. As well as being intuitively reasonable, this factor has been shown to be significant in modeling in lakes (33, 34). If a more fundamental ecological shift is occurring, the model would have to be restructured more radically. At present, there are no available models which address such a phenomenon.

That the transport associated with varying outflows may be important is demonstrated in Figure 10, which is an empirical correlation between

FIGURE 6. Temporal variations and model calculations - Antioch, 1966.

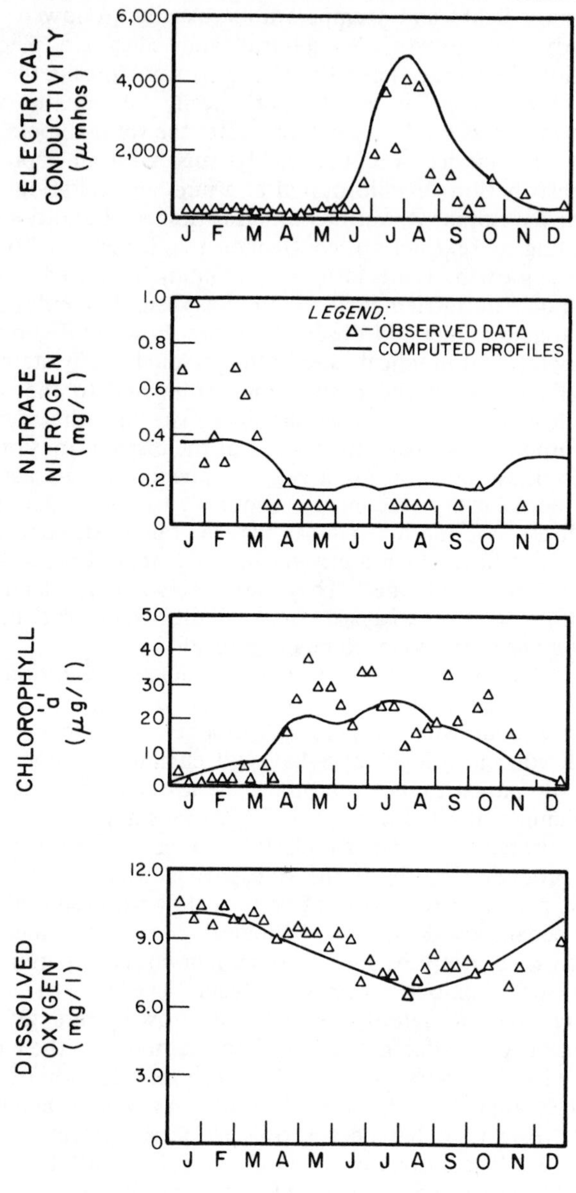

FIGURE 7. Temporal variations and model calculations - Chipps Island, 1970 and 1974.

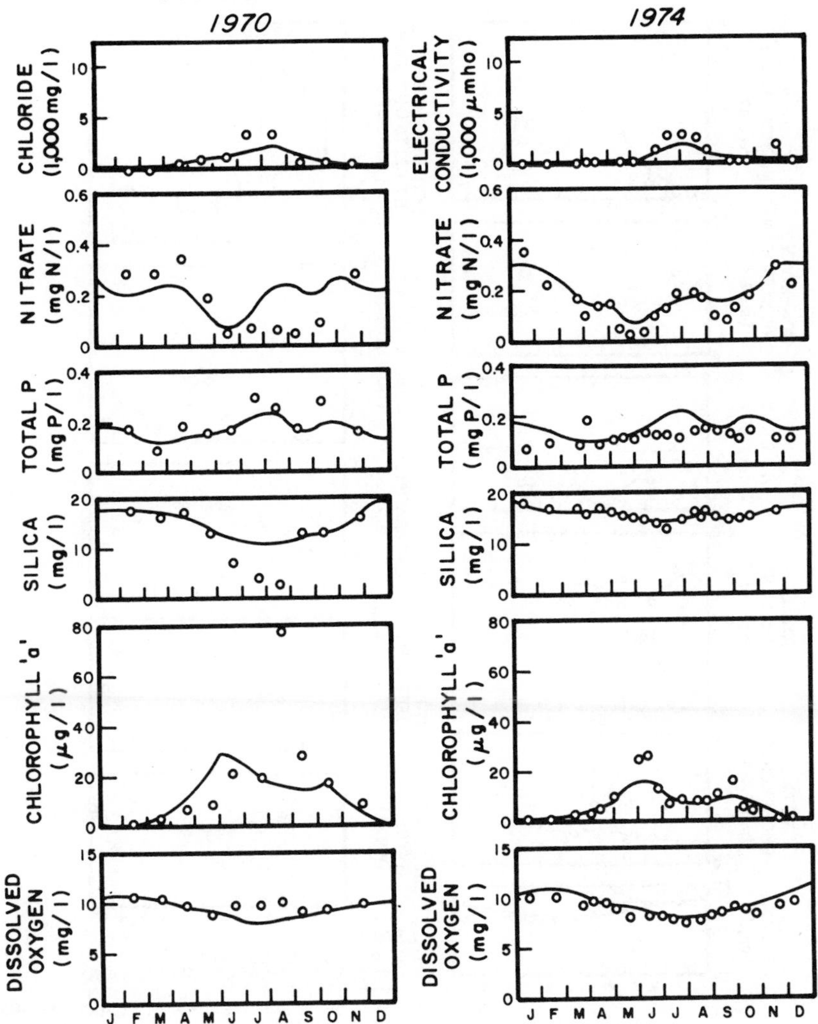

the mean values of flow and chlorophyll concentration. Data shown are averaged over the four-month period from July to October and spatially over approximately 20 miles of bays and channels (35).

FIGURE 8. Observed data time-variable calculations, steady-state calculations - Grizzly Bay, 1974.

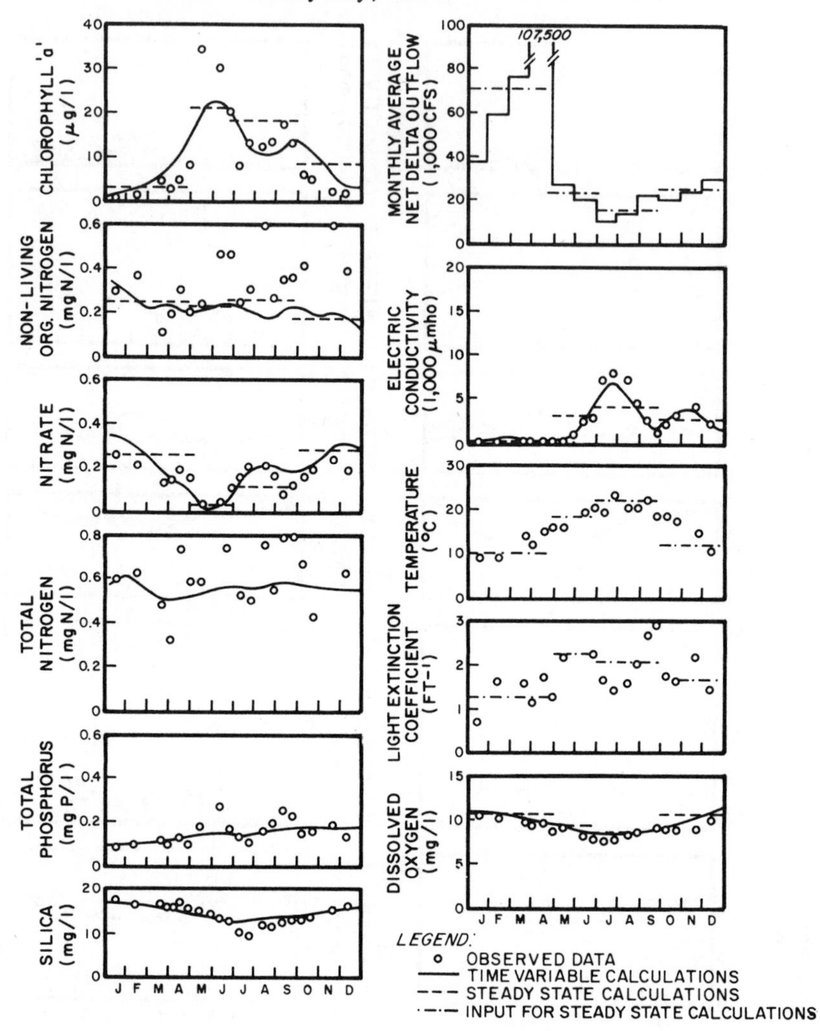

FIGURE 9. Temporal variation of flow, conductivity and chlorophyll at Grizzly Bay and Chipps Island, 1970 - 1974.

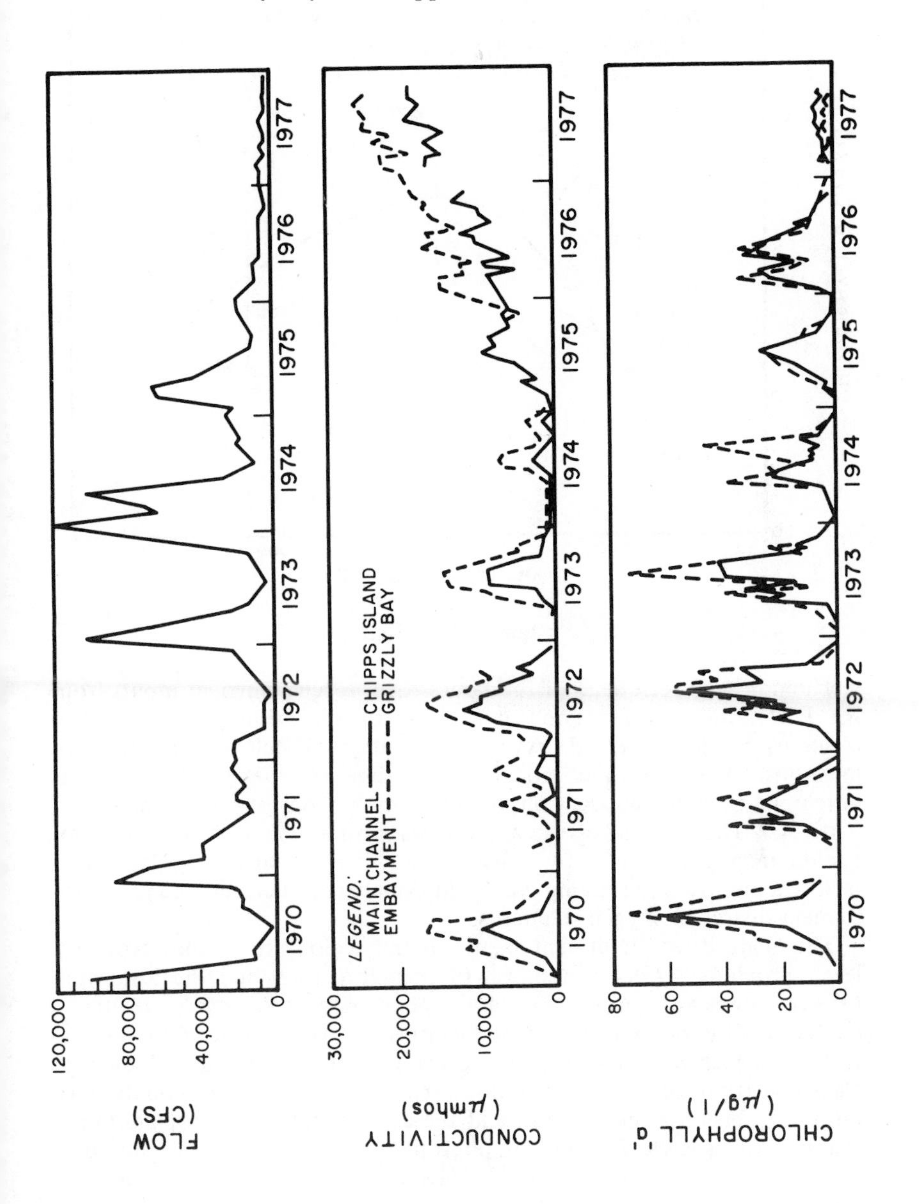

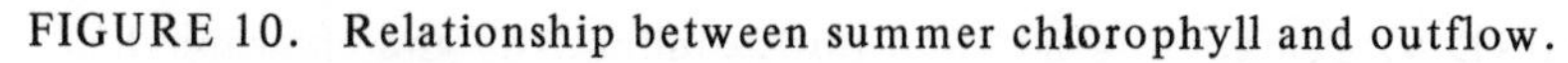

FIGURE 10. Relationship between summer chlorophyll and outflow.

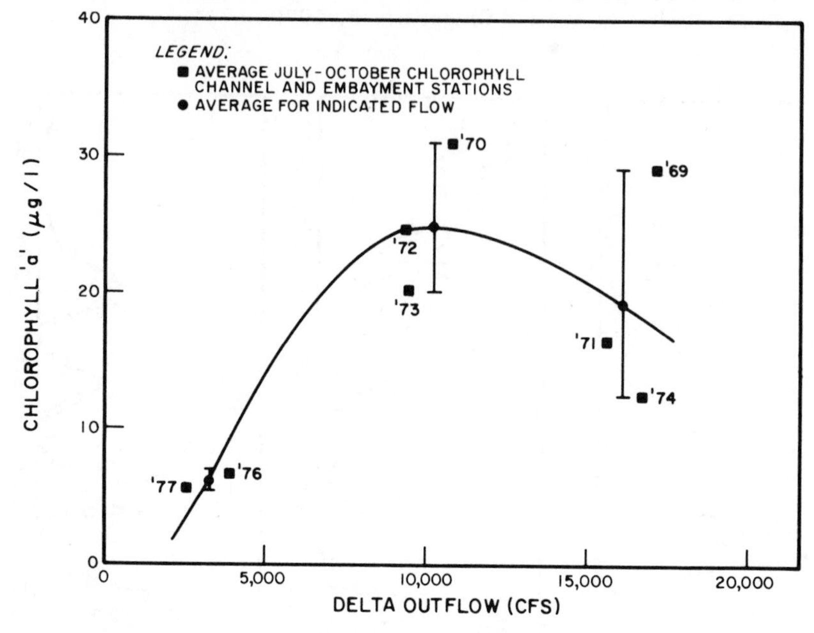

James River Estuary

The non-saline stretches of the James River receive nutrient inputs from the Richmond and Hopewell areas. One of the primary water quality concerns is the dissolved oxygen concentration below each of these locations. A steady-state model of the dissolved oxygen balance was developed in which nutrient-phytoplankton reactions played an important part (36). The purpose of the analysis was to provide a basis for a water quality management plan by assessing the adequacy of present treatment facilities and by determining what additional treatment may be necessary to meet dissolved oxygen standards.

The majority of treatment plants in the region are at the secondary level, providing effective removal of organic carbon and the associated oxygen demand. Ammonia and phosphate discharges, relatively unaffected by present treatment facilities, in conjunction with residual carbon, affect dissolved oxygen levels in the estuary. The biological transformation of ammonia follows two pathways: the oxidation to nitrate by bacterial metabolism and the assimilation by phytoplankton in the growth process. Subsequent phytoplankton decay recycles ammonia through the detrital organic pool. Phosphorus is comparably involved in these processes. Phytoplankton and organic forms of the nutrients are subjected to settling. Dissolved oxygen is utilized by the bacterial oxidation of ammonia and carbon and by the respiration and decay of the phytoplankton. In addition, benthal uptake in particular segments of the system takes place. Oxygen is replenished by transfer with the atmosphere and the photosynthetic activity of phytoplankton.

FIGURE 11. Spatial distribution of chlorophyll and nutrient systems - September, 1978.

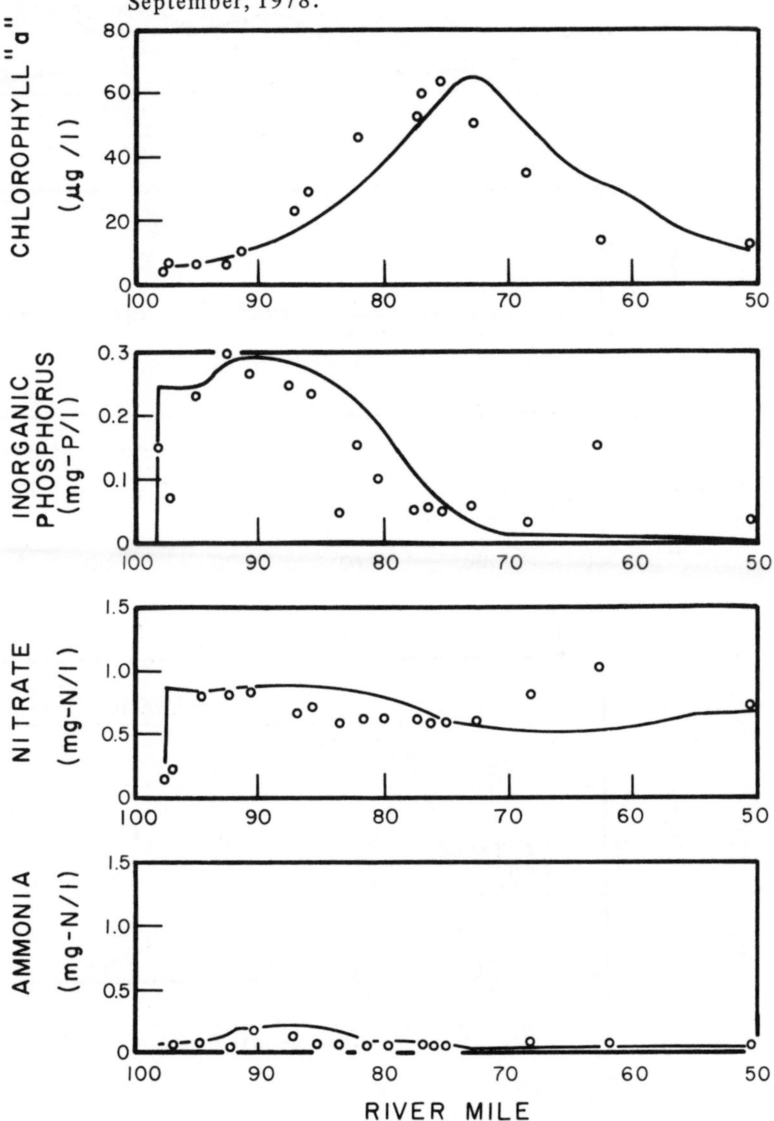

FIGURE 12. Spatial distribution of components of dissolved oxygen analysis - September, 1978.

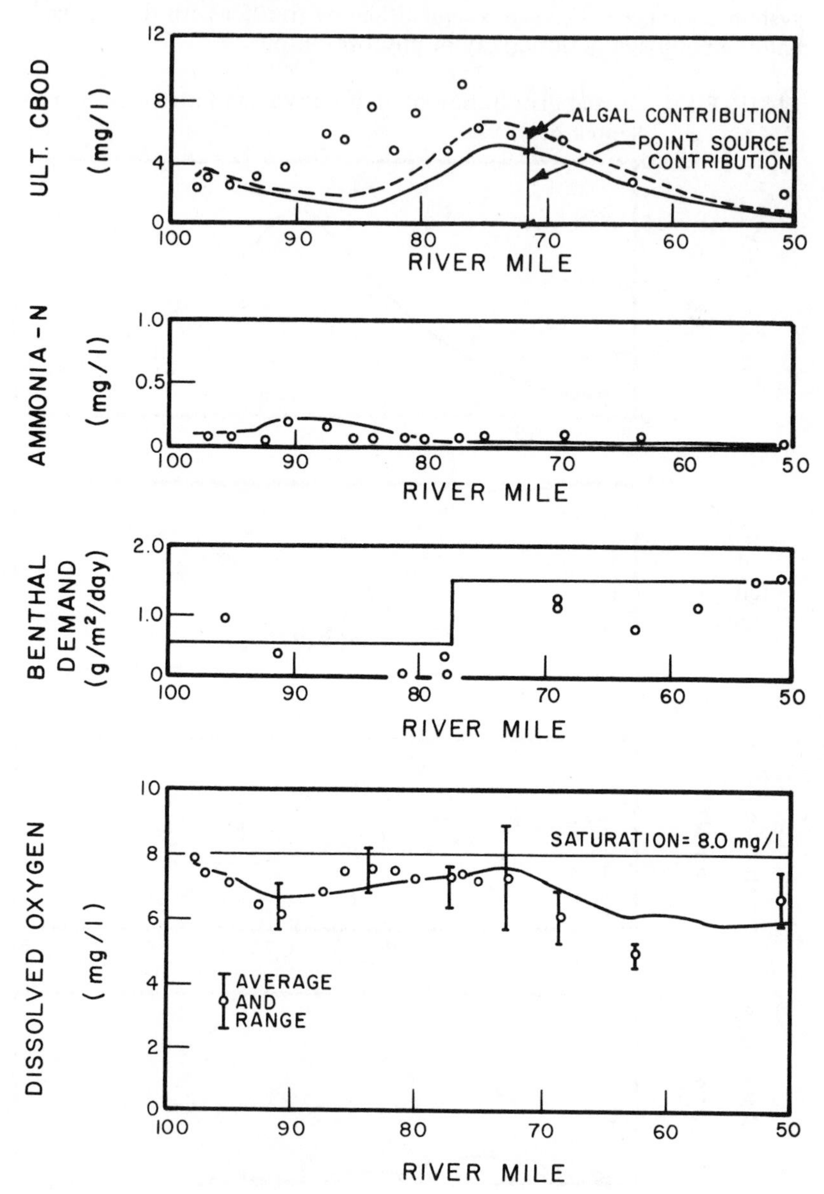

Spatial distributions of the various constituents are presented in Figures 11 and 12 for September, 1978. Chlorophyll and nutrient profiles are shown in Figure 11, and the components of the dissolved oxygen analysis in Figure 12. The kinetic interrelationships are in accordance with Figures 1 and 2. The phytoplankton increase from Richmond to Hopewell, with an associated utilization of nitrogen and phosphorus, is evident from the marked decrease of those nutrients. Settling accounts for the reduction of the total pools. Below Hopewell, the growth of phytoplankton is depressed primarily by a decrease in light and to a lesser degree by reduction of available nutrients.

Components of the dissolved oxygen analysis are shown in Figure 12. The organic carbon includes both bacterial oxidation of the input and a buildup by virtue of algal organic carbon, as measured in the BOD test. Bacterial oxidation of ammonia accounts for further utilization of oxygen. Based on a few measurements of nitrifying bacteria, it has been assumed that nitrification does not occur downstream from Richmond, but does take place below Hopewell. The photosynthesis and respiration of algae are significant, producing higher concentrations in the surface layer than in the bottom layer. There is a net contribution of oxygen from phytoplankton photosynthesis which accounts for the relatively large variation between top and bottom layers. A two-dimensional analysis would be needed to model vertical gradients. The total dissolved oxygen deficit is the algebraic sum of the individual components including that due to benthal demand.

A comparable analysis is summarized in Figure 13 for July, 1976. Chlorophyll levels are much less than in the September, 1978 survey. Productivity is much less than that observed in September, 1978, as evidenced by the concentrations of both chlorophyll and carbon. Following the previous analyses, nitrification is not effective in the upstream stretch from Richmond to Hopewell, but does occur downstream from this location.

For such analyses, the relative impact of the removal of the various nutrients - carbon, nitrogen and phosphorus - may be evaluated. The phytoplankton and benthal demand contribute most significantly to the dissolved oxygen balance. Least significant is the oxidation of ammonia and carbonaceous organic matter. Assigning removals of the various constituents, the dissolved oxygen may be readily determined by the model. The cost of treatment associated with these removals provides an essential component of a water management plan.

Potomac River

The Potomac River downstream from Washington, D.C. has a long history of water quality problems. The increasing discharge of wastewaters containing nutrients gave rise to a sequence of eutrophication problems: a proliferation of aquatic plants in the 1930s and 1940s, replaced algal

FIGURE 13. Spatial distribution of components of dissolved oxygen analysis - July, 1976.

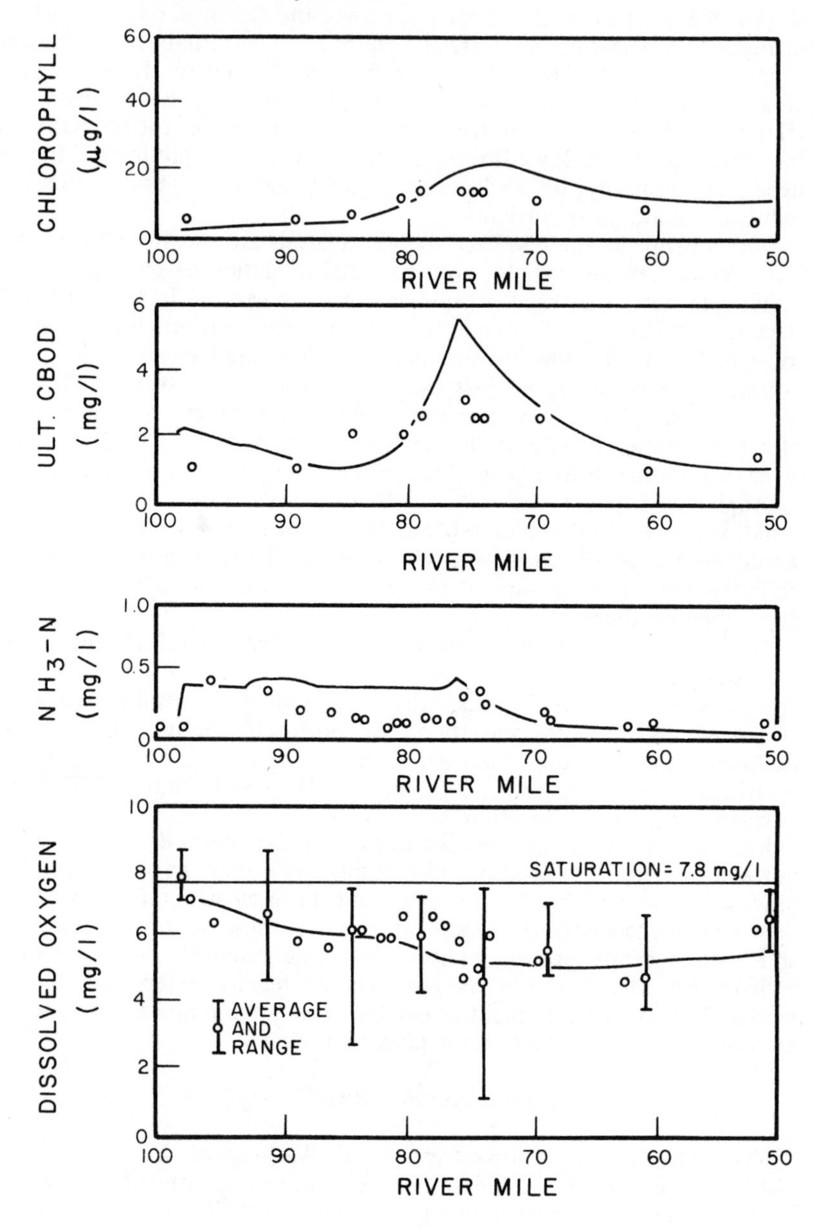

growths during the 1950s with excessive blooms in the 1960s (37). Surveys of the 1970s appear to indicate some decrease in the magnitude of the problem, probably due to reduction in nutrient inputs with additional treatment facilities. Depressed levels of dissolved oxygen and periodic high concentrations of chlorophyll persist (38).

The stretch of the Potomac in the vicinity of and directly downstream of Washington, D.C. is tidal, but non-saline. The middle segments of the system are characterized by variable salinity depending on the freshwater flow. The lower region, discharging to Chesapeake Bay, is always saline. The seasonal distribution of phytoplankton is quite variable in each of the regions. In the upper region, three seasonal blooms have been observed: certain periods in winter, late-spring early-summer, and late summer-early fall. This pattern was particularly evident during the 1960s, the period of greatest eutrophication. Preliminary evaluation of the data collected during the 1970s indicates some attenuation of these blooms, with the strongest persistence by the summer bloom. In the middle and lower saline stretches, the pattern appears to be reversed: during the 1960s, the most predominant proliferation occurred in the summer, while two or three annual blooms appear to characterize the 1970s.

The initial modeling of the process was reported by Jaworski et al. (37). The nitrification sequence, with nitrate being assimilated by algae and recycling to organic, comprises the kinetic structure of the model. It reproduced the spatial distribution of the nitrogen forms and chlorophyll for a range of flow and temperature conditions. Phosphorus was also tracked, but was not incorporated in the growth kinetics, due to the relatively high concentrations. A notable feature of the analysis was the consistent reproduction of dissolved oxygen distribution. Carbonaceous and nitrogenous oxidation and benthal demands, with the photosynthetic contribution and respiratory utilization, were the components.

A later effort (39), included the phosphorus cycle with the nitrification sequence. The predation by zooplankton was not significant, just as in the Sacramento-San Joaquin system. The calculated observed temporal distribution of chlorophyll at two Potomac estuary locations are presented in Figure 14. While the model does reproduce blooms reasonably well, it misses the atypical winter bloom. A multi-species analysis is required to address such cases. Further field and modeling work is required to fully define the nutrient phytoplankton-zooplankton relationship. As with any predictive model, however, once the validity of the calibration has been established based on available data and the model's limitations have been assessed, it can be used accordingly to provide input for management decisions.

Patuxent River

Numerous municipal treatment plant effluents are discharged into the upstream freshwater stretch of the Patuxent River. The majority of these

FIGURE 14. Temporal variations and model calculations - Potomac Estuary, 1969, phytoplankton.

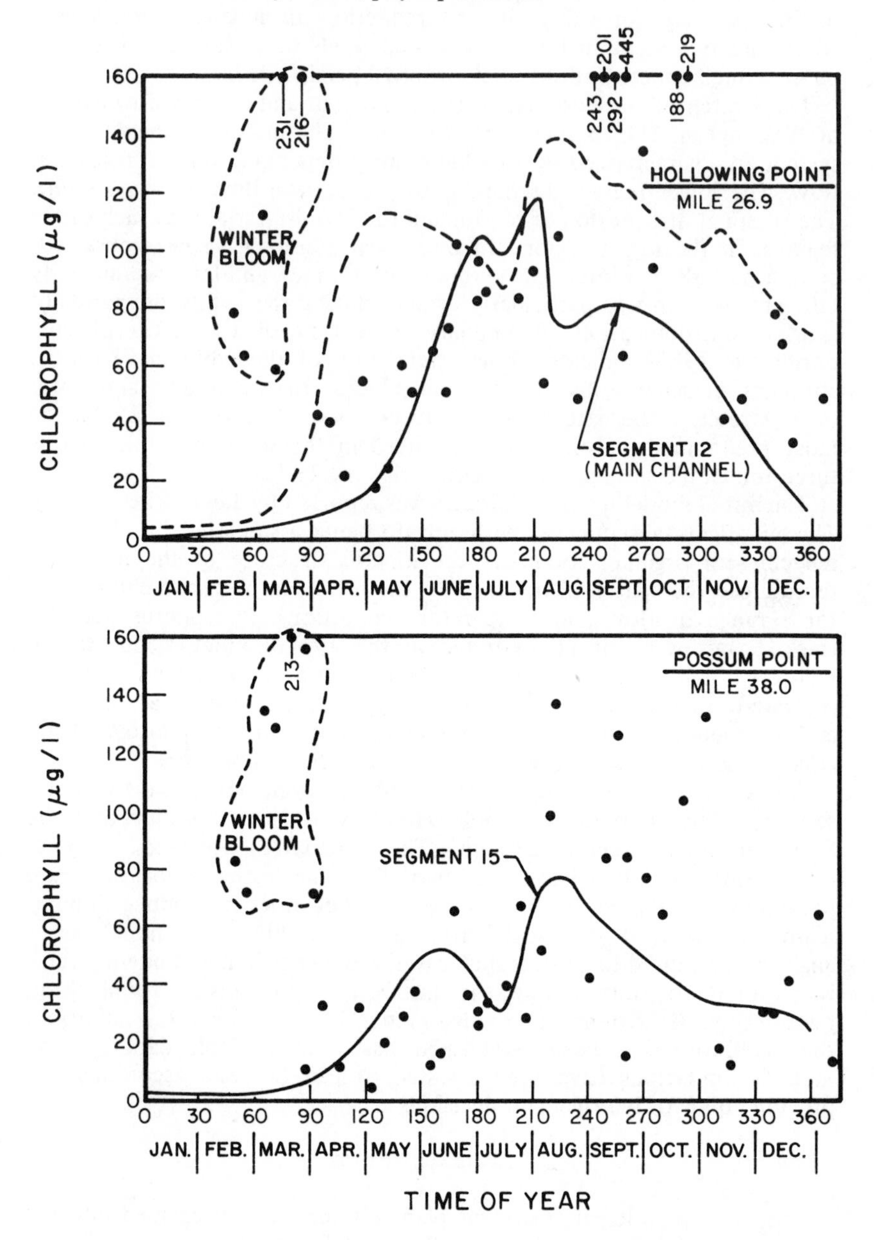

plants are of the secondary type, removing high percentages of organic carbon. The nutrients, phosphorus and nitrogen, are relatively unaffected in the treatment process. In this report, the situation is comparable to that in the James River and may be expected to occur in varying degrees in most coastal plain estuaries receiving secondary effluents. The discharge of these nutrients, in conjunction with those in the runoff from the drainage area, have produced phytoplankton blooms and associated dissolved oxygen problems.

The major purpose of this analysis is to determine the relative impact of phosphorus and nitrogen as a basis for specifying additional treatment. The model is in the first stage of development and calibration. In spite of its preliminary status, the analysis provides some insight into the effects of two-dimensional circulation, which is typical of many estuarine systems (40). It is for this reason that it is included in this report.

Water quality problems occur in both the saline and non-saline zones of the estuary. The downstream saline region is characterized by the typical two-layered estuarine transport, with a net downstream flow in the surface layer and an upstream flow in the lower layer. Vertical stratification is present, the degree depending on the magnitude of fresh-water and the horizontal salinity gradient. The benthal demand in this region causes a significant depression of dissolved oxygen, in conjunction with bacterial oxidation of carbon and ammonia and phytoplankton respiration and decay. Kinetic interactions among the various constituents are in accordance with those shown in Figure 1. These reactions are incorporated with the two-dimensional estuarine circulation.

The boundary conditions are the flux of nutrients from the upstream treatment plants and drainage and the downstream concentrations representative of Chesapeake Bay. Phytoplankton, phosphorus and nitrogen distributions are shown in Figure 15, and the BOD, benthal oxygen demand and dissolved oxygen in Figure 16. Figure 17 shows the conservative tracers used in the transport validation with the July to October, 1978, hydrograph. It is apparent that further calibration would improve the fit between observations and calculated distributions, particularly with respect to phosphorus. The present state of the model, however, does provide some insight into the significance of this type of transport with respect to relevant water quality constituents.

New York Bight

New York Bight receives treated wastewater from many industrial complexes and more than 10 million people in the New York metropolitan area, transported primarily by the Hudson River. In addition, numerous communities along the New Jersey and Long Island shoreline discharge directly to this coastal zone. The area is also used as a disposal site for sludges, construction debris and solid wastes. Over the

FIGURE 15. Spatial distribution of phosphorus, nitrogen and phytoplankton systems - August, 1978.

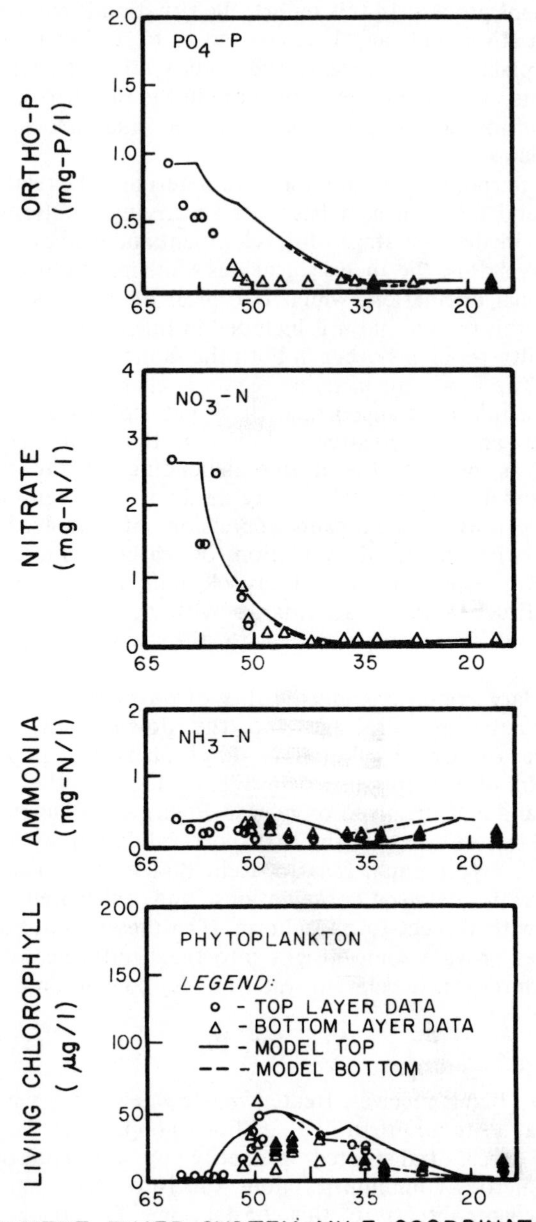

FIGURE 16. Spatial distribution of phytoplankton, BOD, benthal oxygen demand and dissolved oxygen - August, 1978.

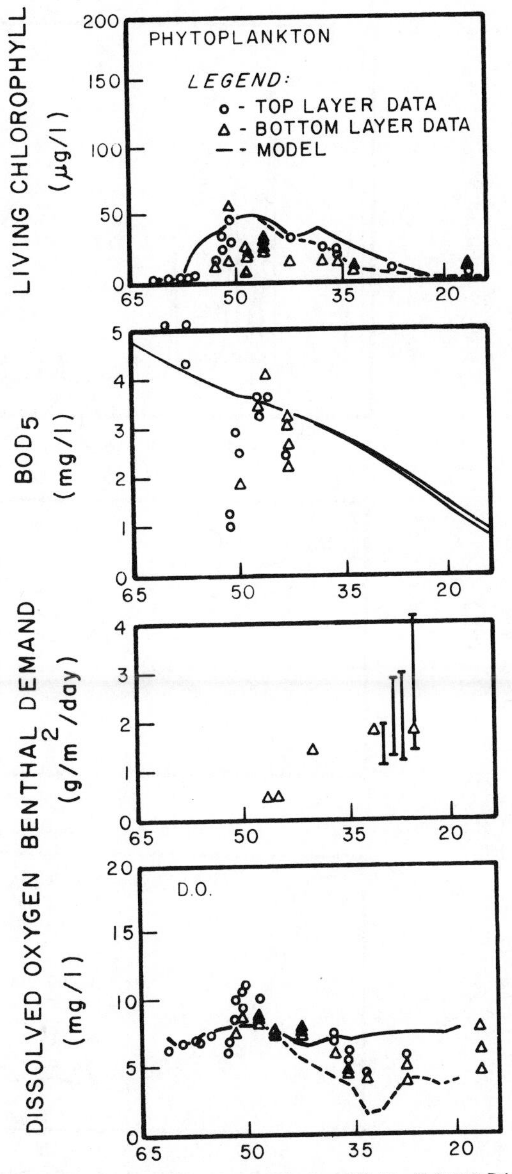

FIGURE 17. Conservative tracers used in transport validation and July-October, 1978 hydrograph.

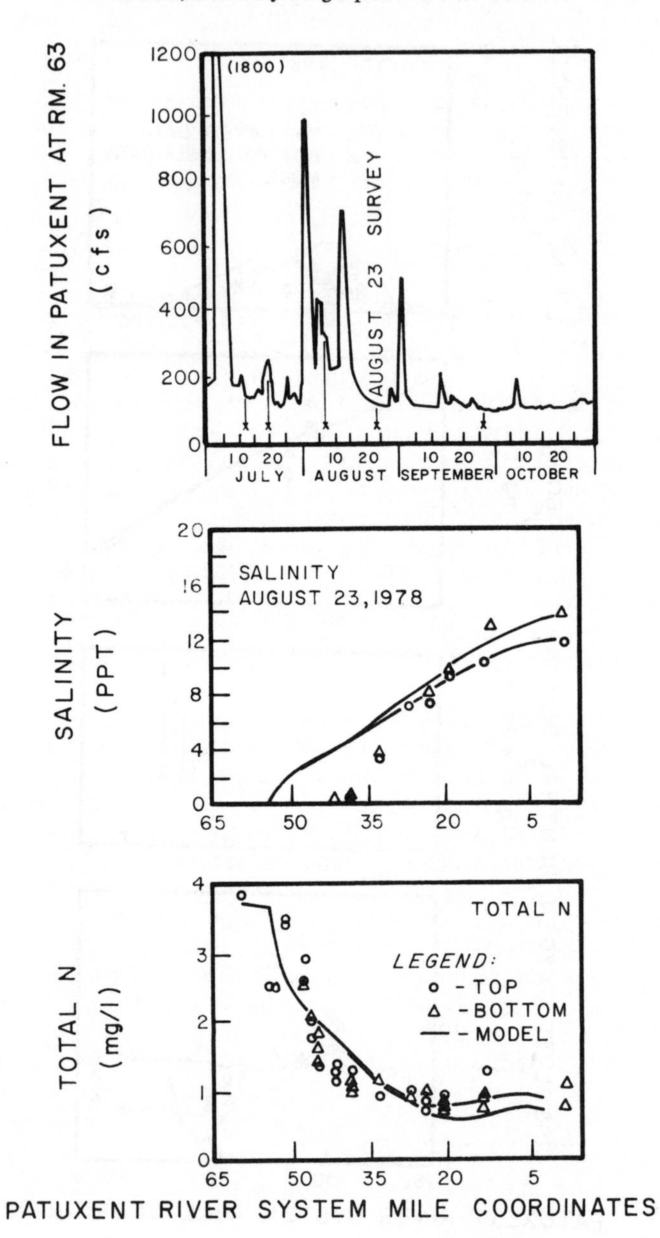

FIGURE 18. East New York Bight vertical dissolved oxygen distribution.

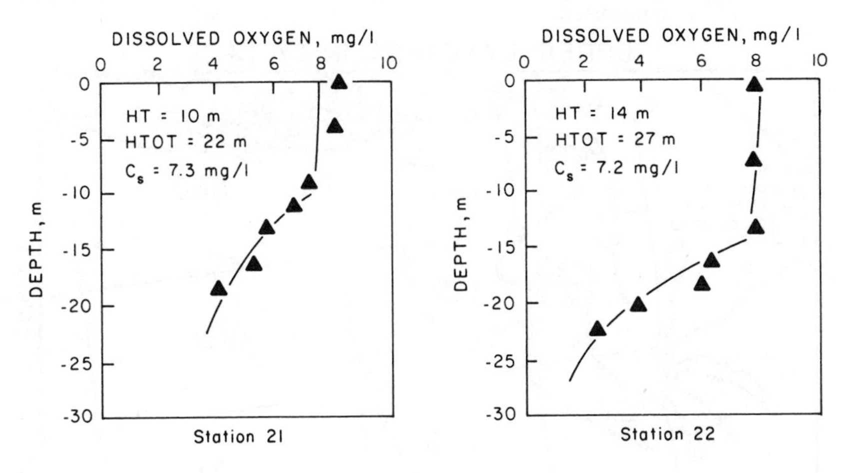

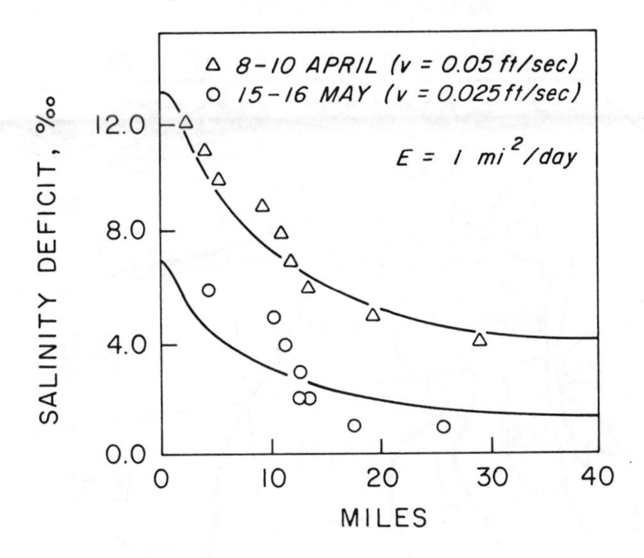

FIGURE 19. Upper and lower layer salinity distributions - observed and computed.

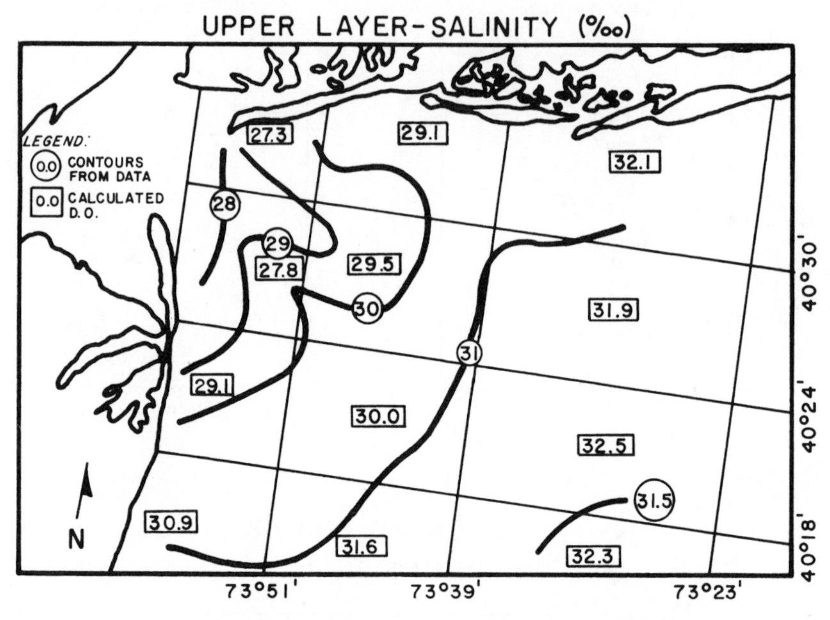

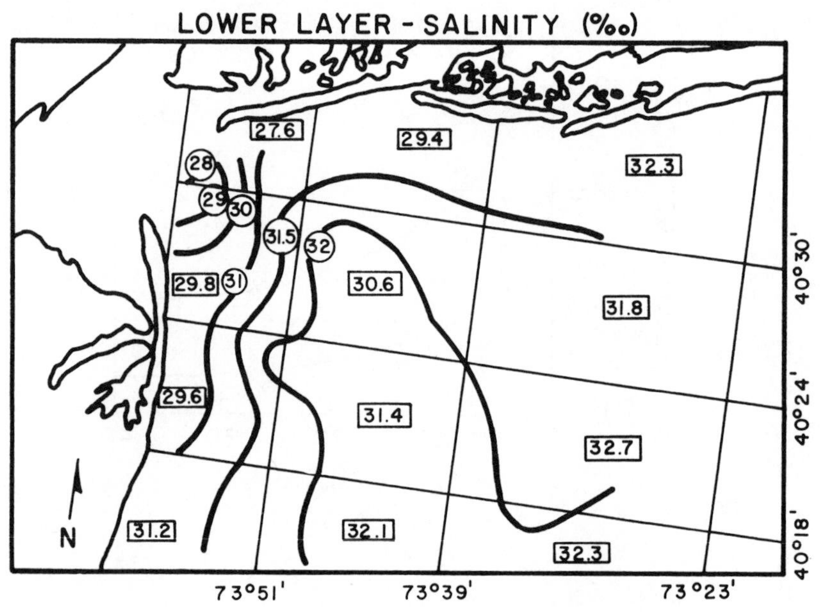

FIGURE 20. Upper and lower layer dissolved oxygen distributions - observed and computed.

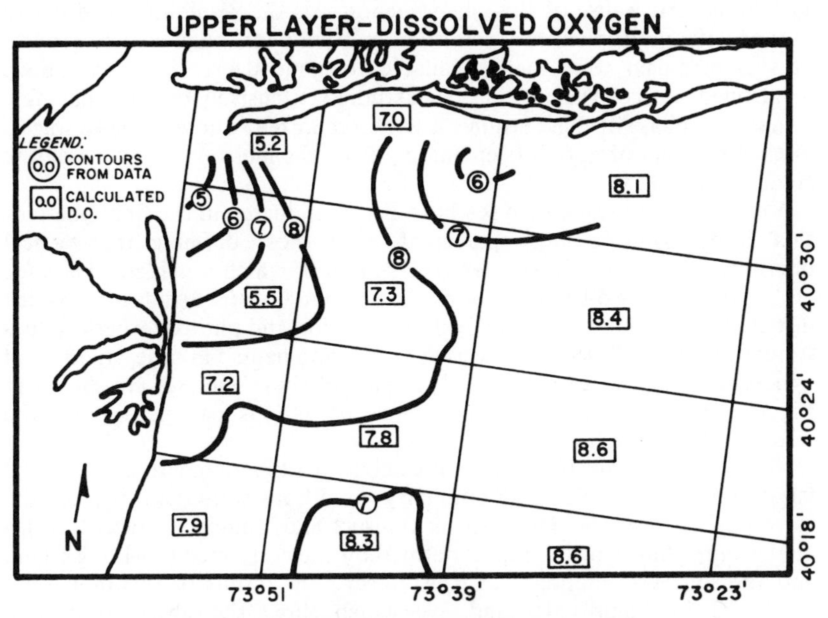

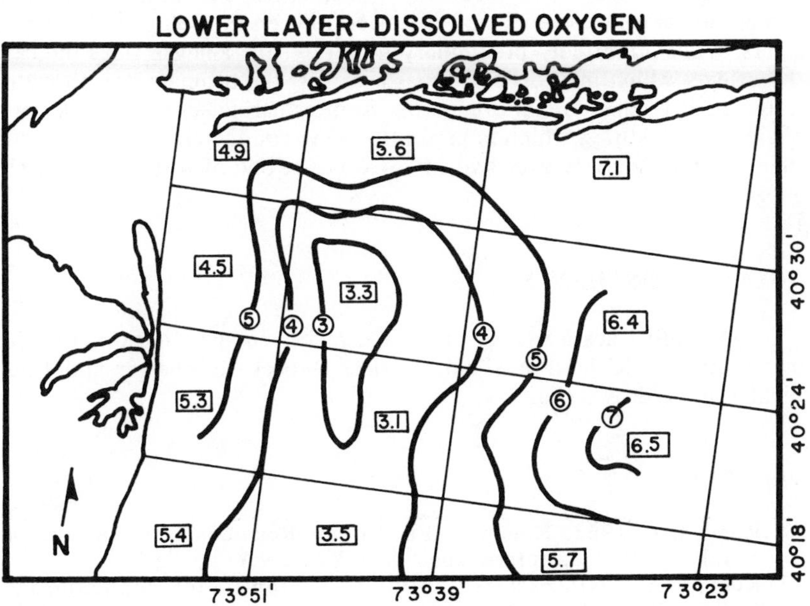

past decade, water quality deterioration has been observed, particularly during 1976 when anoxia occurred over an extensive area of the Bight. The immediate cause of this phenomenon was the respiration and decay of a specific phytoplankton, *ceratium*. Nutrients for the precedent growth of this organism were made available to some degree from the above mentioned source. The area has been studied extensively over the past few years. One phase of these studies is herein reported, whose purpose was to assess the affect of nutrients emanating from the metropolitan area via the Hudson River (25).

Various steady-state analyses have been conducted in the final phase of this study. Among most pertinent were those defining the vertical distribution of dissolved oxygen, the results of which provided a basis for an extensive two-layered horizontal model of dissolved oxygen distribution. In the vertical analysis, respiration and photosynthetic values were assigned, based on various measurements in the field and experiments in the laboratory. Examples of these analyses are shown in Figure 18, which indicates the values of various parameters and the resulting distribution of dissolved oxygen.

The vertical analysis assisted in establishing kinetic coefficients in the larger horizontal model. The boundary conditions were the flux from the New York area at the Sandy-Hook transect and boundary concentrations at the outer limits of the Bight, defined by the eastern end of Long Island and the southern boundary of New Jersey. The results of this model are presented in Figures 19 and 20, which show the observations and computed values of salinity and dissolved oxygen in the upper and lower layers. While admittedly preliminary in nature, the analysis does indicate the effects which may occur in coastal zones, the ultimate repository of water quality constituents originating from estuarine sources. The second phase of this study, which is presently being conducted, is directed to a time variable analysis over two or three years, one of which is the 1976 anoxia period.

ACKNOWLEDGEMENTS

I gratefully acknowledge the assistance in preparation of this manuscript of Richard Winfield, Patricia Ruback Kehrberger and Erich Bichler, of Hydroscience.

REFERENCES

1. Riley, G.A. 1963. Theory of Food-Chain Relations in the Ocean, *The Sea*, M.N. Hill, ed., Interscience, New York, N.Y. pp. 438-63.
2. Riley, G.A. 1946. Factors Controlling Phytoplankton Populations on Georges Bank, *Journal of Marine Research*, Vol. 6, No. 1, pp. 54-73.

3. Riley, G.A. 1947. Seasonal Fluctuations of the Phytoplankton Populations in New England Coastal Waters, *Journal of Marine Research*, Vol. 6, No. 2, pp. 114-125.
4. Riley, G.A., and Von Arx, R. 1949. Theoretical Analysis of Seasonal Changes in the Phytoplankton of Husan Harbor, Korea, *Journal of Marine Research*, Vol. 8, No. 1, pp. 60-72.
5. Riley, G.A., Stommel, II., and Bumpus, D.F. 1949. Quantitative Ecology of the Plankton of the Western North Atlantic, *Bulletin: Bingham Oceanography Collection*, Vol. 12, No. 3, pp. 1-169.
6. Steele, J.H. 1956. Plant Production on Fladen Ground, *Journal of Marine Biological Association*, Vol. 35, United Kingdom, pp. 1-33.
7. Steele, J.H. 1964. A Study of Production in the Gulf of Mexico, *Journal of Marine Research*, Vol. 22, pp. 211-222.
8. Steele, J.H. 1965. Notes on Some Theoretical Problems in Production Ecology, *Primary Production in Aquatic Environments*, C.R. Goldman, ed., 18 Supplement, Memorial Institute of Idrobiology, University of California Press, Berkeley, California, pp. 383-398.
9. Davidson, R.S., and Clymer, A.B. 1966. The Desirability and Applicability of Simulating Ecosystems, *New York Academy of Science*, Vol. 128, No. 3, pp. 790-794.
10. Cole, C.R. 1967. A Look at Simulation Through a Study on Plankton Population Dynamics, *Report BNWL-485*, Battelle Northwest Laboratory, Richland, Washington, pp. 1-19.
11. Parker, R.A. 1968. Simulation of an Aquatic Ecosystem, *Biometrics*, Vol. 24, No. 4, pp. 803-822.
12. Chen, C.W. Oct., 1970. Concepts and Utilities of an Ecological Model, *Journal of the Sanitary Engineering Divisions*, ASCE, Vol. 96, pp. 1085-97.
13. Thomann, R.V., O'Connor, D.J., and DiToro, D.M. July, 1970. Modeling of the Nitrogen and Algal Cycles in Estuaries, presented at the 5th International Water Pollution Research Conference, held at San Francisco, California.
14. DiToro, D.M., O'Connor, and Thomann, R.V. 1971. A Dynamic Model of the Phytoplankton Population in the Sacramento-San Joaquin Delta, *Advances in Chemistry*, American Chemical Society, Washington, D.C., No. 106, pp. 131-180.
15. Thomann, R.V. and Salas, H. April, 1975. The Chesapeake Bay Waste Load Allocation Study, Hydroscience, Inc. report to State of Maryland, Department of Natural Resources, Water Resource Administration.
16. Craig, N.J. and Day, J.W. June, 1977. Cumulative Impact Studies in the Louisiana Coastal Zone, Eutrophication; Land Loss, Louisiana State University, Baton Rouge, La., 166 p.
17. Boroden, W.B., and Hobbie, J.E. April, 1977. Nutrients in Albemarle Sound, North Carolina, *UNC-S6-75-25*, North Carolina State University, Raleigh, North Carolina, 202 p.
18. Kramer, J.N. and S.W. Nixon, 1978. *A Coastal Marine Ecosystem*, Springer-Verlag, New, N.Y.

19. *Estuarine Pollution, A Bibliography.* April, 1973. Office of Water Resources Research, Washington, D.C., 510 p.
20. Webb, K.L., et al. 1979. Estuarine Response to Nutrient Enrichment - An annotated Bibliography, Chesapeake Research Consortium Inc., Annapolis, Maryland.
21. Water Resources Engineers, Sept., 1975, Ecological Modeling of Puget Sound and Adjacent Waters, *WRE-11920-1*, EPA, Office of Research and Technology, Washington, D.C., 127 p.
22. Chen, W. and Orlob, G.T., Ecological Simulation of Aquatic Environments, report of Water Resources Engineers, Inc., Walnut Creek, California, 168 p.
23. Haydock, A., Johanson, P., and Lorensen, M. March, 1977. Water Qiality Model of Teconic Estuary, report of Tetra Tech, Lafayette, California.
24. Cerco, C.F., et al., August, 1978. Hydrography and Hydrodynamics of Virginia Estuaries - XVI Mathematical Model Studies of Water Quality and Ecosystems in the Upper Tidal James, *Special Report No. 155*, Applied Marine Science and Ocean Engineering, Virginia Institute of Marine Science, Gloucester Point, Va.
25. O'Connor, D.J., and Mancini, J.L. August, 1979. The Carbon-Oxygen Distribution in New York Bight; Phase I - Steady State, MESA New York Bight Project, NOAA, Marine Eco-Systems Analysis.
26. Stoddard, A., et al., October, 1979. Oxygen Depletion Within the New York Bight, An Ecosystems Model of Waste Inputs, Phytoplankton Species Succession and Climatology, presented at the October, 1979 52nd Conference of the Water Pollution Control Federation, held at Houston, Texas.
27. Nihoul, J. July, 1975. Application of Mathematical Models to the Study, Monitoring and Management of the North Sea, *Ecological Modeling in a Resources Management Framework*, C.S. Russel, ed., Johns Hopkins University Press, Baltimore, Md., pp. 125-147.
28. Liengcharernsit, W., 1979. Mathematical Models for Hydrodynamic Circulation and Dispersions of Selected Water Quality Constituents in the Upper Gulf of Thailand, Thesis presented to the Asian Institute of Technology, Bangkok, Thailand, in partial fulfillment of the requirements for the degree of Doctor of Philosphy.
29. DiToro, D.M. October, 1976. Light Propagation in Turbid Estuarine Waters: Theory and Application to San Francisco Bay Estuary, Hydroscience, Inc. report to Dept. of Water Resources, State of California, Sacramento, California.
30. Thomann, R.V., DiToro, D.M., Winfield, R.P., and O'Connor, D.J. 1975. Mathematical Modeling of Phytoplankton in Lake Ontario, 1. Model Development and Verification, *EPA-660/-375-005*, ORD, Corvallis, Oregon, 177 p.
31. Harleman, D.R.F., and Ippen, A.T. June, 1967. Two-Dimensional Aspects of Salinity Intrusion in Estuaries: Analysis of Salinity and Velocity Distributions, Technical Bulletin No. 13, U.S. Army Corps of Engineers, Committee on Tidal Hydraulics, Vicksburg, Mississippi.
32. O'Connor, D.J. and Lung, W.S. May, 1978. Assessment of the Effects of Proposed Submerged Sill on the Water Quality of Western Delta -

Suisun Bay, Hydroscience, Inc. report to U.S. Army Corps of Engineers, Sacramento, California.

33. DiToro, D.M., and Connolly, J.P. 1980. Mathematical Modeling of Water Quality in Large Lakes Part II. *Lake Erie, Ecological Research Series* (In Press), Environmental Research Laboratory, U.S. EPA, Diluth, Minnesota.
34. Bierman, V.J., and Dolan, D.M. July, 1976. Mathematical Modeling of Phytoplankton Dynamics in Saginaw Bay, Lake Huron, *EPA 600-76-016*, Proceedings of the Conference on Environmental Modeling and Simulation, Office of Research and Development, U.S. EPA, Washington, D.C.
35. California Department of Fish and Game, California Department of Water Resources, U.S. Fish and Wildlife Service and U.S. Bureau of Reclamation, March, 1978, Interagency Ecological Study Program for the Sacramento-San Joaquin Estuary, 6th Annual Report, 1976.
36. Hydroscience, Inc., Water Quality Analysis of the Upper James River Estuary, report in preparation to Commonwealth of Virginia, State Water Quality Board.
37. Jaworski, N.A., Clark, L.J., and Feigner, K.D. April, 1971. A Water Resource - Water Supply Study of the Potomac Estuary, *Technical Report 35*, Chesapeake Technical Support Laboratory, U.S. EPA, Middle Atlantic Region.
38. Villa, O., et al., December, 1977. The Potomac Estuary Current Assessment Paper No. 2, Annapolis Field Office, U.S. EPA, Region III, December, 1977.
39. Thomann, R.V., DiToro, D.M., and O'Connor, D.J., June, 1974. Preliminary Model of Potomac Estuary Phytoplankton, *Journal Environmental Engineering Division*, ASCE, Vol. 100, No. SA3, pp. 899-915.
40. Hydroscience, Inc., Water Quality Analysis of the Patuxent River Estuary, report in preparation to EPA, Region III, Philadelphia, Pa.
41. Flaherty, James P. and Robert H. Harris, December, 1979. Impact of Nutrients on the Potomac Estuary, Environmental Defense Fund, Washington, DC. 20036.

NUTRIENT ENRICHMENT AND ESTUARINE HEALTH

Rezneat M. Darnell and Thomas M. Soniat
Department of Oceanography
Texas A&M University
College Station, Texas 77843

ABSTRACT: Ecosystem health may be defined in terms of system norms or in terms of human utility. It is here defined as that state in which the components and processes remain well within specified limits of system integrity selected to assure that there is no diminution in the capacity of the system to render its basic services to society throughout the indefinite future. Knowledge of nutrient enrichment in freshwater systems is reviewed as a point of departure for understanding the response of estuaries to enrichment. Mixed and stratified estuaries respond in somewhat different fashions. Major changes associated with enrichment are changes in species succession and oxygen depletion in areas of organic accumulation, low mixing, and poor flushing. If of temporary and local occurrence, these symptoms are reversible and of minor importance. If of chronic nature or of widespread occurrence, they could lead to irreversible loss of species and genetic stocks. Measures of ecosystem health are discussed, and pertinent management recommendations are put forth. Of especial importance are the needs to establish local species reserves, to manage for total system integrity, and to develop better lines of communication between scientists and managers.

INTRODUCTION

During the past few decades, coincident with rapid increase in human population and technology, it has become apparent that the life support systems of this planet cannot survive intact without active human assistance. Broad appreciation of this fact has stimulated interest in a new endeavor, the science of ecosystem management. Ecosystems cannot be properly managed without a sophisticated input of technical information about the system being managed, the nature of its structure and processes, characteristics of the norm and symptoms of perturbation, levels of tolerance, and carrying capacities. Effective ecosystem management also depends upon a clear definition of the objectives. What do we want this ecosystem to be like, say 20 years (or approximately one human generation) from now? This coupling of management and basic science has placed upon the shoulders of the scientist a large burden. He must not only generate the facts about the nature and response capabilities of the system, but he must also aid in the interpretation of this information in terms useful to management and understandable by society in general.

ECOSYSTEM HEALTH

The Concept

"Health" is an anthropocentric term which refers to the state of well-being of the human body, its soundness and freedom from disease or injury. The term can usefully be applied to ecological systems where it would bear the same general connotations of soundness and well being. However, it also means a great deal more because ecosystem health must be judged not only in terms of the system itself, but also in terms of the utility of the system to human society.

In dealing with the state of well being of freshwater systems in relation to pollution, Cairns (4) has employed the term, "biological integrity," and this he defined as, "the maintenance of community structure and function characteristic of a particular locale or deemed satisfactory to society." This definition places a focus upon the biological system, and it leaves open the possibility of two states, the natural and that desired by society.

Somewhat the same conclusions had been reached earlier by Odum (27), who distinguished between two conceptually idealized ecological states. The first is the natural system, developed through millions of years of evolution and adaptation and essentially unmodified by man. The second is the ecological system managed for optimum human use, whether it be waste disposal, resource harvest, recreation, power generation, or a combination of these and other uses. These two states are generally hot compatible, and our concept of ecosystem health will differ depending upon which perspective we embrace. To manage for naturalness will preclude most human uses; to manage for human use will preclude naturalness. Herein lies the basic dilemma. Clearly, we cannot afford to eliminate the native species and ecological systems of the planet, yet we must also use them to the benefit of society. Therefore, we need a practical definition of ecosystem health which includes the needs of nature while recognizing legitimate, non-destructive uses by society. Our definition follows:

> *Ecosystem health is that state in which the components and processes remain well within specified limits of system integrity*

selected to assure that there is no diminution in the capacity of the system to render its basic services to society throughout the indefinite future.

Critical to this definition are three factors: 1) dedication to the proposition, espoused in the National Environmental Policy Act of 1969, that unborn generations shall not inherit a ravaged biosphere; 2) existence of a knowledge base sufficient for the determination of critical limits of tolerance; and 3) willingness of society to establish limits of human intrusion well short of the absolute limits of system tolerance, i.e., willingness to maintain a significant margin of safety. Thus, ecosystem health must be defined in terms of two limits: the one which cannot be transgressed under any circumstances without incurring long range system damage (the *categorical limit*), and the other which might be temporarily violated without major damage to the system (the *conditional limit*) (8).

Categorical limits are set by those species which inhabit the ecological systems. The loss of species and locally adapted genetic strains is irreversible. In a very practical sense, however, in order to preserve, for example, alligators and their ecological associates, we must preserve swamps. So we must be concerned with the preservation of habitats in sufficient variety, quantity and quality to insure the long range perpetuation of the native species and genetic stocks.

To some extent, the maintenance of natural genetic diversity must be carried out through special ecological reserves, set aside and maintained in relatively pristine conditions. In fact, the cornerstone of any program to maintain the health of the nation's ecosystems must be a well developed and coordinated system of national and state-wide programs for natural area preservation. We must retain the species necessary to restore damaged ecosystems. In a highly technological civilization it is certain that there will be major accidents and that some ecosystems will be damaged from time to time. However, not all species can be retained in special reserves, and this is particularly the case with estuarine species which utilize many different habitats during the course of their complicated life histories. Hence, all ecosystems heavily utilized by society must also serve a secondary function of maintaining the native genetic stocks.

Conditional limits of ecosystem health depend upon how much habitat is to be preserved in the pristine condition and what levels of habitat quality we are willing to pay for in those ecological systems heavily utilized by society. The remainder of the discussion will be addressed primarily to those estuaries which are put to heavy human use.

Factors of Ecosystem Health

The health of any ecological system may be defined in positive or negative terms, i.e., as the presence of healthy characteristics or as the absence of symptoms of trouble. Both types of traits should be considered. The healthy system could be characterized in terms of the physical/chemical environment, biological composition and structure, biological process, or a combination of these factors. A full description of health would entail a thorough analysis of the state of every structural and functional property of the healthy system. In a practical sense, one is limited to key components as indicators of health. A great many indicators of system condition have been employed. These include various measures of water quality, condition and extent of the substrate, structure or function of the biological systems, and a combination of these factors. A major drawback of most methods is that they give us only part of the required information, and what they do give is often ambiguous.

Symptoms of trouble could be represented as significant deviations from the norm if norm and significant deviations could be adequately defined. Trouble might also be represented in terms of proximity to limits of tolerance or to limits of carrying capacity, if these could be defined. Again, in a practical sense, the investigator should focus upon key components or key processes. Finally, an index of system health could be devised as a ratio between a series of healthy symptoms and a series of unhealthy ones.

THE HEALTHY ESTUARY

It is axiomatic that all biological systems of this planet tend toward equilibrium in which species composition and diversity remain relatively stable and in which nutrient fluxes remain rather constant or cyclic, but in a constant pattern. Biological stability and constancy of nutrient flux are causally related and are different faces of the same coin. However, the exact nature of the target equilibrium and the degree to which it is achieved reflect species availability and interaction, stability of the physical environment, and constancy of the nutrient supply.

System stability appears most clearly under terrestrial conditions because terrestrial vegetation itself exerts significant control over local microclimate and the cycling of critical nutrients. Local control is more complex in aquatic systems because the hydrographic regime is generally more difficult for the vegetation to modify, because of the high water solubility of many nutrients, and because much of the aquatic system is below the effective depth of light penetration and thus beyond the range of rooted macrophytes. In shallow, physically stable, tropical marine waters there has developed a biological system of great stability. This is the coral reef which exhibits structural integrity and biological complexity, but this system is made possible in large measure by a unique method of internal nutrient cycling.

Estuarine health must first be viewed in light of the prevailing physical and geological conditions. Estuaries are characterized by complex gradients of salinity, tidal action, current velocity, bottom erosion, and sediment accumulation. These factors are not constant, but shift in both short term (hours/days) and long term (seasons/years) cycles. Furthermore, they are subject to major and often unpredictable variations in response to river flow as well as wind and storm patterns. Thus, each portion of the estuary is subject to dramatic environmental variability in response to external perturbations. Both the water column and the benthic habitat of most estuaries display great temporal and spatial heterogeneity, although exceptions are known.

Against the background of a variable and heterogeneous physical environment, biological systems develop toward local biological equilibrium communities, and some of the biological solutions are relatively successful. The mangrove swamp, *Spartina* marsh, seagrass bed, and oyster reef all exhibit one or a few dominant species which provide the structural stability necessary to resist routine erosional tendencies, enhance sediment deposition, and retain soil nutrients. The mud flat, sand shoal, channel bottom, and unprotected shoreline are more exposed and subject to erosion by the force of waves and water currents. Even here, however, some biological forces are at work to stabilize bottoms and approach biological equilibria. Certain bivalves and large tube-building polychaetes aid in stabilizing the systems of the mud flats, and the accumulation of mollusk shells and other coarse biological debris retards erosion in several habitats, most notably the channel bottoms.

But as there are biological forces working toward stability and equilibrium, there are also biological forces creating local instability. Blue crabs dig holes in the bottom and disturb both the mud flats and grass beds. Cownose and other rays create large areas of disturbance in the beds and flats. Black drumfish ravage mollusk beds and have been known to destroy an oyster reef overnight. Turtles and certain other species clip and consume seagrasses. Many less spectacular species such as the croaker, spot, peneid shrimp, etc. regularly dig and root up bottoms, and, of course, numerous roving carnivores graze on the more sedentary species and keep the populations at low levels.

Although some sections of the estuary apparently proceed to equilibrium in many ways similar to the terrestrial community climax, others are kept at subclimax stages by the forces of erosion and biological activity. Local disturbance areas permit invasion by pioneer species ("opportunists" in the estuarine literature). So, the healthy estuary retains a mix of species characteristic of all stages of succession leading toward a series of different local equilibrium conditions. The healthy estuary is a very diverse place.

The unhealthy estuary would likely involve stabilization of environmental conditions and reduction of habitat diversity. Such changes would result in reduction or elimination of species which depend upon variable conditions or specific habitat types. Species diversity would be replaced by biological monotony. Unhealthiness would also result from the imposition of specific types of stress agents, and here we must consider the specific effects of nutrient enrichment.

NUTRIENT ENRICHMENT IN FRESHWATERS

A brief review of our knowledge of nutrient enrichment in freshwaters will serve as background for our discussion of nutrient enrichment in estuaries. Much of the extensive freshwater literature has been reviewed in the following books and symposium.volumes: (10), (17), (20), (22), (23), (24), (35), (40), (42), (46). The subject of nutrient enrichment and eutrophication in freshwaters is enormously complex. Since it is being dealt with in detail elsewhere in this symposium, only a few highlights will be presented here.

When freshwaters are subject to nutrient enrichment, they are generally receiving complex chemical mixtures. Sorting out a specific element or group of elements as being the key to eutrophication has been the subject of much research. Different receiving waters possess different chemical and biological properties. Hence, the basic chemical limiting factors tend to vary from one situation to another. The element phosphorus has most frequently been implicated. In other cases, nitrogen in its various forms seems to be important. In rare instances trace elements have turned out to be the key (12). Often the biological response indicates that it is really a combination of these factors that causes the greatest impact. In all cases, the real problem is the creation of a massive organic carbon load. This may result from direct organic carbon input or increased organic carbon fixation by photosynthesis. When plants die, whether by grazing or other causes, decomposition of this excess organic matter greatly increases the biological oxygen demand. To the extent that this demand exceeds the supply, anoxic conditions prevail. All aquatic systems exhibit graded response to different levels of nutrient enrichment.

The effects of mild enrichment of a lake are illustrated in Figure 1. The added nutrients stimulate growth of phytoplankton, attached algae and benthic macrophytes. This results in increased production of the animal consumers of the water column and the bottom habitats. Death occurs at all levels, and the decomposing organic matter is mineralized and oxidized, using up a fraction of the oxygen dissolved in the water. The most notable changes in the system are higher rates of plant and animal production, some shift in species composition (especially increase in blue-green algae), and increased oxygen demand by the decomposing organic matter.

The effects of severe nutrient enrichment of a lake are shown in Figure 2. Extremely heavy production of plant life outstrips the capacity of the animals to process the material. Death of the excess plants creates a large reservoir of decomposing organic matter, and this is augmented by carcasses of dead animals and any organic material associated with the incoming nutrients. This heavy organic load includes much readily-oxidizable material, resulting in an oxygen demand in excess of the

FIGURE 1. Dynamics of a freshwater system receiving low levels of nutrient. At this level of input, the system is stimulated, but it adjusts to the small nutrient increase and achieves a new equilibrium level.

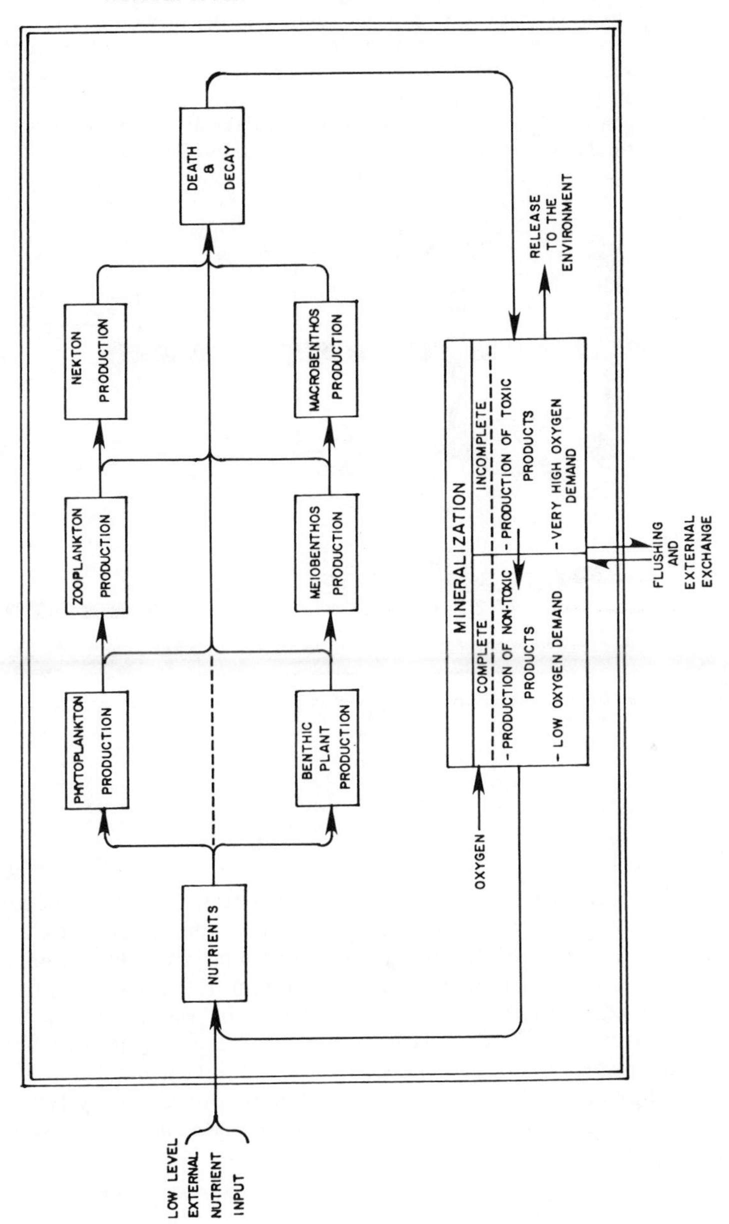

FIGURE 2. Dynamics of a freshwater system receiving high levels of nutrient enrichment. The width of the arrows (in comparison with those of Figure 1) suggests changes in the magnitude of flow through each pathway. At this level of nutrient input, the system becomes overloaded with organic matter resulting from primary production (plus added organics and increased secondary production). The increased organic load creates an extremely high oxygen demand leading ultimately to oxygen depletion. Products of incomplete mineralization rise into the water column, and the natural aerobic system collapses.

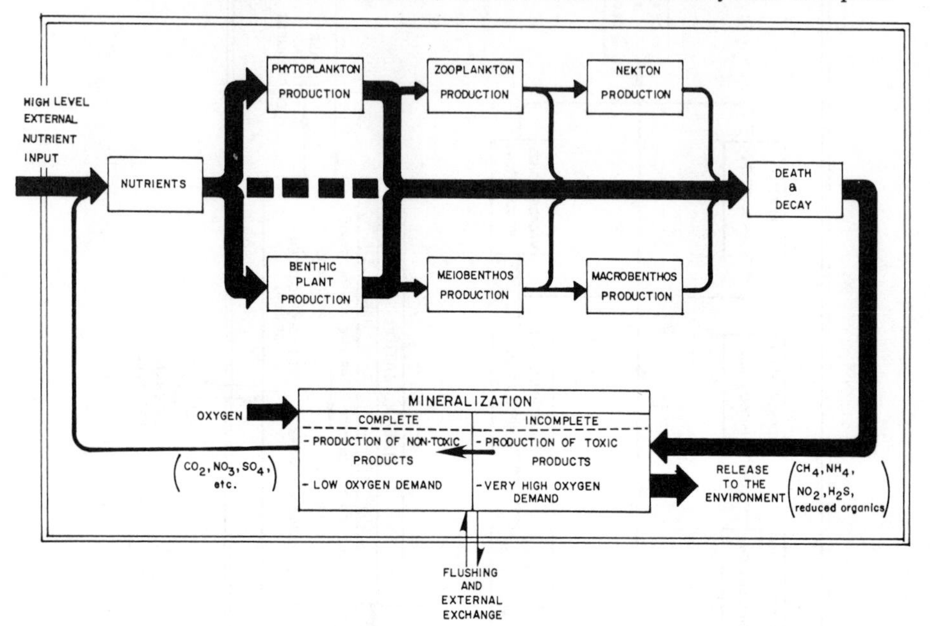

supply. The bottom becomes anoxic, and this lack of oxygen becomes evident first in the hypolimnion and then the epilimnion. Toxic and malodorous chemicals, both organic and inorganic, in various unoxidized states permeate the water column. All oxygen-requiring life disappears.

The same basic processes take place in flowing waters, but here the effects are stretched out downstream, and recovery is facilitated by the physical processes of air/water interaction and internal mixing of the flowing waters. Immediately below the nutrient outfall the system may become anoxic, but gradual recovery is observed downstream as oxygen renewal overtakes oxygen demand.

NUTRIENT ENRICHMENT IN ESTUARIES

Responses of freshwater systems to nutrient enrichment have relevance, but they are not directly translatable to the estuary for a number of reasons which are given below.

Nutrient Cycles of Unenriched Estuaries

The factors which normally determine the concentrations of biologically active elements in the estuary have been discussed by numerous authors, of which the following are especially informative; (1), (14), (25), (31), (34). In other papers in this volume, Nixon (1980) and Webb (1980) present information and conceptual models of the processes of remineralization and nutrient cycling.

Estuaries, in general, exhibit two basic physical states, mixed and stratified. Any given estuary may exist primarily in one or the other state, it may shift from one state to the other, or it may exhibit both states simultaneously in different sections of the estuary. In the mixed estuary (Figure 3), dissolved and suspended materials tend to be relatively uniform from the top to the bottom of the water column. Water moves in and out in response to tidal ebb and flow, but there may be a net movement seaward due to continued entrance of freshwater from river flow. Continual exchange takes place between the water column and the sediments.

In stratified estuaries (Figure 4), freshwater derived from river flow is carried seaward in the upper layer of the water column. A bottom counter-current of seawater moves in from the ocean to replace that which becomes entrained in the surface outflow. The dissolved and suspended materials move with each water mass, and some mixing and exchange take place between the layers. Sinking of particulate organic matter from the surface into the bottom layer creates a concentration gradient, with highest concentrations in the bottom layer up-estuary, away from the sea. Thus, as a result of the circulation pattern, the stratified estuary acts as a "nutrient trap" in which the individual nutrients reach highest concentrations toward the head of the estuary. As noted by Redfield et al. (34), the amount of accumulation varies greatly depending upon the properties of a given estuary. Accumulation increases with the rate of production of organic matter and with the length of the basin. It decreases with velocity of surface flow and with turbulence or mixing. It also varies in relation to the relative depths and velocities of the two water masses. Since these factors vary from one estuary to another and seasonally within a given estuary, the rate of upstream accumulation of biologically active nutrients also varies greatly from place to place and from time to time.

FIGURE 3. Vertical and horizontal movement of materials in the mixed estuary. The water itself moves back and forth due to tidal ebb and flow, but, as a result of river inflow, the net movement is seaward. Some of the major processes of the water column and benthos, as well as transfers between the two, are indicated. (Modified from Tenore, 1977).

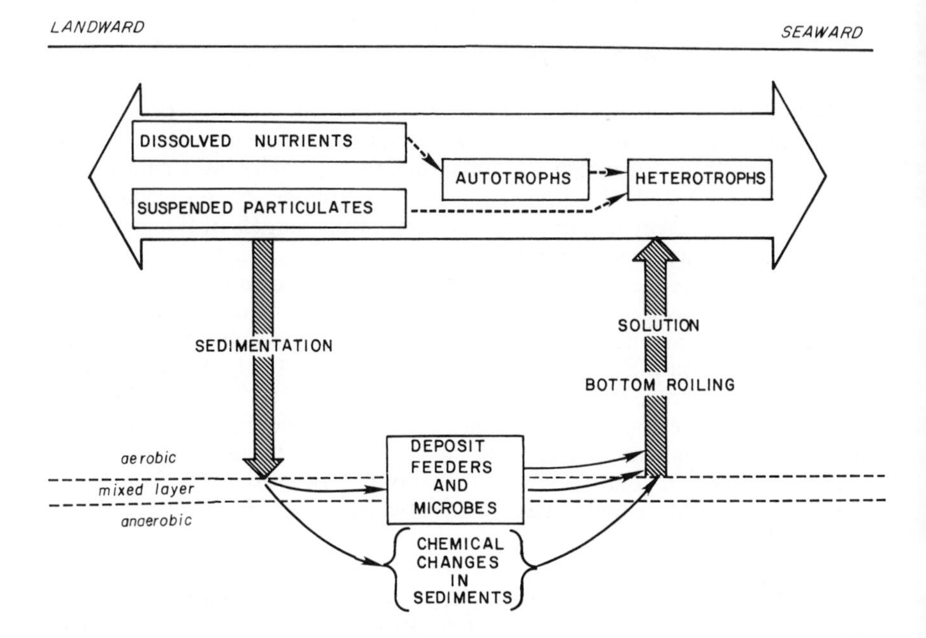

Pomeroy et al. (31) have analyzed the factors controlling phosphorus distribution in shallow, turbid, well mixed estuaries. They note that in such estuaries the concentration of dissolved phosphorus is far higher than it is in either the river or the adjacent sea. Furthermore, the sediments typically contain large reservoirs of phosphorus. Normal cycling of phosphorus involves its passage from open estuary sediments to the marsh. *Spartina* pumps it from the sediments and places it in dissolved form into the water column where it is flushed back out into the open estuary. Through adsorption onto suspended sediment particles and sinking of phytoplankton, it returns to the sediments to complete the cycle. Any concentration of phosphate in the water greater than 1 mg-at m^{-3} shifts the equilibrium toward sorption onto sediment particles. In such estuaries, physical and biological processes combine to regulate the phosphorus levels of the water column and to determine the seasonal distribution of phosphorus between the various habitat types of the estuary. In this type there is a high receiving capacity for excess phosphorus, and a high tolerance for phosphorus enrichment.

FIGURE 4. Vertical and horizontal movement of materials in the stratified estuary. Seaward flow of the surface layer and reverse flow of the lower water layer are shown. Some of the major processes of each water mass are indicated, as are transfers between water masses and between the lower water mass and the bottom.

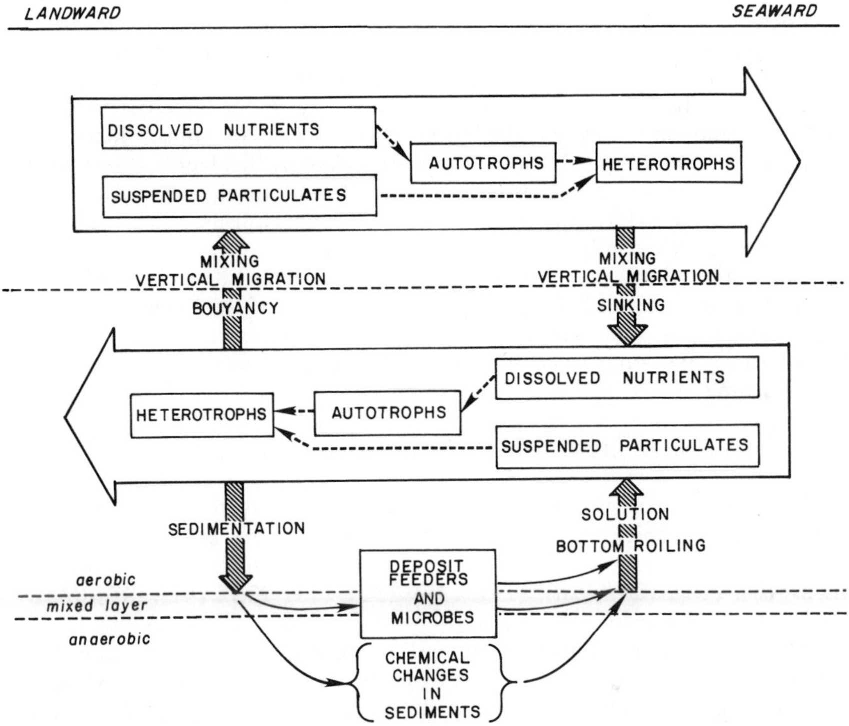

The cycling of nitrogen in estuaries is not as thoroughly understood. Apparently there is a much more direct cycling of nitrogen from sediments to the water column and back, but the role of marshes and of decomposing marsh grass, with its attendant microflora, needs further study. Nor do we known much about the role of dissolved organic nitrogenous compounds.

The cycles of carbon in estuaries are quite complex and involve interaction of organic and inorganic states. The dominant features involve production by phytoplankton, benthic algae, and rooted macrophytes. Phytoplankton sinks, and the algae and macrophytes seasonally die and decompose. Currents transport large quantities of dissolved and particulate organic matter from the marshes to the open estuary where much of the decomposition takes place. Complete oxidation places the carbon back into the water column in the inorganic state.

Organic phosphorus, nitrogen and carbon all require oxygen for complete mineralization, and all are involved in reduction of the oxygen levels of the near-bottom water. In well mixed estuaries, oxygen transport is generally adequate to handle the oxygen requirements of decomposition, but some of these come perilously close to oxygen depletion during periods of low river flow, low vertical mixing, and high temperatures, generally in late summer. In vertically stratified estuaries, the accumulation of organic matter in upstream bottom waters can readily lead to anoxic conditions if bottom water circulation is slow. If the velocity of the deeper water is rapid, the deep water may be replaced before significant oxygen depletion takes place. In those fjord-type estuaries where sills block the entrance of seawater, the deeper layers have very long residence times, providing ample time for the accumulation of organic matter. Thus, the fjord is likely to develop an oxygen deficit leading to anoxic conditions in the bottom water.

From the above considerations it is seen that the cycling of each nutrient type is dependent upon specific properties of that element as well as local physical and biological regimes. The problem ultimately comes down to relative rates of loading, physical and metabolic cycling, mixing, and flushing.

Effect of Enrichment on Estuarine Nutrient Cycles

Much work remains to be done before we are really knowledgeable about the complex effects of nutrient loading of estuaries. Salient references detailing our present state of knowledge include the following: (2), (15), (18), (26), (28), (29), (31), (32), (33), (36), (41), (44), (45). The effect of a given type of nutrient varies somewhat depending upon whether it is in the dissolved or particulate state. Particulate materials tend to sink into the bottom layer of water and be transported upstream. In some cases, smaller organic particulates coalesce and form larger flocculent particles which may cover the bottom. Dissolved nutrients tend to be more readily flushed from the estuary than particulate nutrients, but since dissolved nutrients may become incorporated in particulates (through adsorption or biological uptake), they may also be removed through sinking of particulates. While they are present in the dissolved state, the nutrients support the growth of estuarine phytoplankton populations. Garside et al. (11) noted that municipal effluents discharged into deeper waters of a stratified estuary passed upstream in the bottom waters and subsequently became entrained in the surface layer which transported them seaward.

The existing literature suggests that in organically rich estuaries, loading with additional quantities of nutrients does not necessarily result in a large increase in biological production. However, even a slight increase in production may become very important if it is enough to deplete oxygen reserves and bring about local anoxic conditions. Here it is the excess oxygen demand that is important rather than excess nitrogen or phosphorus as such. During warm summer months, poorly flushed sections of many of our estuaries stand perilously close to oxygen depletion even without the added burden of nutrient enrichment.

According to Pomeroy et al. (31), the addition of organic carbon produces eutrophic species succession in clear estuaries and anoxic zones in turbid ones. Inorganic nutrients also produce eutrophic succession in clear estuaries but may have little effect in turbid ones.

In both clear and turbid estuaries, nutrient enrichment leading to species changes may be deleterious to the system. High nutrient levels often lead to blue-green algal blooms in the fresher portions of estuaries and to dinoflagellate blooms in the more saline stretches. The ecological effects of a massive bloom of *Gymnodinium breve* on the estuarine fauna of Tampa Bay has been reported by Steidinger and Ingle (39) and Simon and Dauer (38). Both fishes and invertebrates exhibited mass mortality, and the infauna suffered a reduction of 77 percent of the species and 97 percent of the individuals.

Severe oxygen depletion leading to anoxic conditions, likewise, causes mass mortality of essentially all the normal species of the water column and of the upper sediment layer. These are replaced by species which are not oxygen dependent. Anoxia is most likely to occur in the upper reaches of an estuary where there is high oxygen demand and low flushing, but in the extreme it could render uninhabitable large areas of the estuarine bottom.

Fortunately, most of the effects of enrichment including anoxia, if of local occurrence and short duration, are not likely to engender major or permanent damage to the estuarine system because these local effects are all reversible. Severe eutrophication involving loss of species from a whole estuarine system is, of course, quite another matter.

MEASURING THE EFFECT OF ENRICHMENT ON SYSTEM HEALTH

Major elements of estuarine health are summarized in Table 1. Note that the health of the system may be described in terms of the physical (hydrological, geological and chemical), biological, or ecosystem aspects. Cairns (4) has discussed the problem of quantification of biological integrity, especially in freshwater systems subject to pollution. Pearson et al. (29) and others have addressed the problem in estuaries. Two quotes from Cairns' (ibid.) paper are especially relevant:

> "It is evident that no single method will adequately assess biological integrity nor will any fixed array of methods be equally adequate for the diverse array of water ecosystems. The quantification of biological integrity requires a mix of assessment methods suited for a specific site and problem...."

> "What is needed is a protocol indicating the way in which one should determine the mix of methods which should be used to estimate and monitor threats to biological integrity."

A series of measures which might be used to assess the integrity of estuarine systems subject to nutrient enrichment in given is Table 2. Some of the measures are already available and in use; others must be developed and tested. What we must attempt to express and quantify are the nutrient cycles themselves, oxygen availability and demand, and how these affect the various habitats. Further, we need to understand how individual species, critical communities, and the estuarine ecosystem as a whole respond to increasing levels of enrichment.

TABLE 1. Elements of estuarine health.

HYDROLOGICAL	– Adequate flushing, especially by freshwater inflow
GEOLOGICAL	– High diversity of benthic environments - an adequate mix of erosional and depositional types - an adequate mix of sediment particle-range types
CHEMICAL	– High water quality (in top and bottom layers) - nutrient levels within safe ranges - oxygen levels above safe minimum levels
	– High quality of bottom surface layer - organic content not excessive - surface layer aerobic
BIOLOGICAL	– Typical species within normal ranges of abundance (for the time of year)
	– Nuisance species within acceptable limits
	– Basic biological processes neither unduly elevated nor depressed
ECOSYSTEM	– Spatial structural diversity high
	– Nutrient cycling functioning properly
	– Temporal variability of the above factors within normal range

TABLE 2. Measures of ecosystem health.

HYDROLOGICAL	–	Flushing rate vs. oxygen demand (for the most vulnerable portion of the estuary)
GEOLOGICAL	–	Ratio of areas covered by erosional vs. depositional environments
CHEMICAL	–	Ratio of actual to critical levels of dissolved nutrients (especially P and N)
	–	Ratio of oxygen availability to oxygen demand
BIOLOGICAL	–	Chlorophyll content of water column
	–	Plankton species composition and abundance
	–	Biofouling species succession and total production
	–	Benthic species composition, diversity etc. (by habitat type)
	–	Ratio of aerobic to anaerobic species in the benthos
	–	Total respiration of unit area benthic samples
ECOSYSTEM	–	Some measure of areal structural heterogeneity of the system
	–	Some measure of areal functional heterogeneity of the system
	–	Some measure of temporal variability of the system
	–	Some index of trophic state which combines several of the above measures in a meaningful way

Of especial importance is the Trophic State Index developed by Seaton and Day (37) for the quantification of eutrophication in the Barataria Basin of the central Louisiana coast. This index was derived from related work by Brezonik and Shannon (3) for Florida lakes. Seaton and Day measured 19 physical, chemical and biological parameters a a series of stations in transects throughout the basin. Using cluster and factor analysis techniques they reduced these to the six parameters considered to be most discriminative: total organic nitrogen, total phosphorus, chlorophyll *a*, secchi disc depth, dissolved oxygen, and ammonia. Similar techniques involving the same or a different mix of parameters might be usefully applied in the study of trophic state in estuaries of a variety of types.

At the present time a great deal of attention is being given to problems of the biological structure of estuarine subsystems. A variety of techniques and measures is being used to describe species abundance, diversity, richness, evenness, equitability, etc. (6), (30). Although the ecological significance of such measures is not always clear, they do aid in describing structure and, particularly, changes in structure of subsystems with respect to space and time. However, none of these techniques can effectively address the ecosystem as such, because there is simply no effective way at this time of conceiving of the biological diversity of the total estuarine ecosystem. Biological indicator species are often used, but these must also be considered with care. For example, the polychaete *Capitella capitata* has been widely employed as an indicator of marine pollution, yet Grassle and Grassle (13) have recently demonstrated that this nominal species is really a complex of at least six different species, each exhibiting different habitat tolerances and requirements. Thus, before we can place much reliance upon indicator species, we must have considerable knowledge of who they are and what they can tell us.

Of especial importance in assessing ecosystem health are indices involving ratios between indicators of healthy and unhealthy conditions. Examples of possibly useful ratios include: nutrient levels as a function of nutrient carrying capacity, oxygen availability in relation to oxygen demand, habitat availability versus habitat need, and oxygen-requiring versus oxygen-independent species.

In the present discussion our efforts to gain a really critical focus on the complex problems of estuarine ecosystem health in response to nutrient enrichment are confounded by a series of complications:

1. We are attempting to consider all types of estuaries, even though they differ greatly from one to another;
2. Each individual estuary is highly variable in space and time;
3. When it comes to details, we are still quite short of fundamental knowledge about how general patterns of nutrient metabolism work within the local context and under the prevailing physical, chemical and biological regimes;
4. Estuaries are inhabited by enormous numbers of species, and we are still woefully ignorant of their metabolic characteristics, ranges of tolerance, the roles they play in the metabolism of the larger system, and the factors which control their life histories;
5. The systems we are dealing with thus have a great deal of variability or "background noise" which limits the precision of our focus and inhibits our ability to generalize from one system to another.

Nevertheless, we do understand basically how the systems work and we know what happens when they get badly out of kilter. Hence, we can make first order assessments of ecosystem and subsystem health, and we can put our finger on the pressure points associated with system deterioration. We know what factors to examine in assessing system response to excessive nutrient enrichment. We also know that in setting our standards for system quality we must stop well short of driving any of the species to extinction. Therefore, what we are really looking for now is precision. This can best be developed in the local site-specific context and with well funded, goal-oriented research efforts to get at the specific problems which have been outlined above.

MANAGEMENT-RELATED ISSUES

If we are to establish and maintain a high degree of health in the nation's estuaries, attention must be given to the management structure as well as to the systems being managed. The reader is reminded of the recommendations regarding coastal zone management which were elaborated by a nationwide team of interdisciplinary scientists (19). The following recommendations, more specific to estuaries, provide a guide not only for the way we should go; they are the way we must go if we are to prevent massive estuarine deterioration in the face of increasingly heavy and diverse human use (see also Cronin, 7).

Recommendations

1. Develop a nationwide program for coordinating the management of estuaries (5).
2. Develop basin-wide management plans (21). Estuarine problems begin in the land and upstream water areas, and they must be solved at this level.
3. Develop estuarine management plans with adequate care given to the protection of native biological resources. a. For each major estuarine system, local sanctuaries or species preserves must be set aside and managed against all conflicting uses (5), (7), (19), (21). b. Even in heavily utilized portions of estuaries, habitat quality must not be allowed to deteriorate beyond a level of safety dictated by local conditions.
4. Manage estuaries in terms of the whole problem, rather than in terms of piecemeal subsets. a. Control nutrient pollution at the source, on the land or in upstream areas. Once excess nutrients enter the estuary little can be done (16). b. Manage with consideration of the effects of multiple use. For example upstream water diversion reduces estuarine flushing and lowers its receiving capacity for nutrients. Dredging and spoil-placement operations should not be permitted to create areas of low flushing which could become anoxic (9).
5. Manage with consideration for local and temporal specifications of each estuarine system. Human use at one place or season could be totally inapprpriate at another place and season. Dissolved nutrients might better be placed in surface waters where they would be swept out to sea rather than in bottom waters where they would penetrate up-estuary. Receiving capacities are greater during periods of high river flow and lower water temperature than at other seasons.

6. Manage estuaries for total habitat quality, not just water quality. This requires a fix on the natural spatial and temporal variability of benthic environments as well as the water column.
7. Sponsor scientific research programs on each estuarine and drainage basin system to provide the knowledge base for effective management. Such studies should be of long term duration and should address not only in-estuary problems but also the response of estuaries to long term climatic cycles and to land and water use patterns in the drainage basin.
8. Develop more effective means of communication between manager and the scientist.

In final analysis, ecosystem health is a matter of degree of naturalness. The level of health to be maintained is a social issue to be decided on the basis of public interest and in light of existing scientific knowledge. We scientists must provide the knowledge base with as high a degree of precision as possible. We would hope our voice will be heard with increasing clarity in the back rooms of policy analysis and environmental decision making.

REFERENCES

1. Axelrad, D.M., K.A. Moore, and M.E. Bender, 1976. Nitrogen, phosphorus, and carbon flux in Chesapeake Bay marshes. Va. Polytech. Inst. Va. Water Resourc. Res. Centr. Bull. 79. Blackburg.
2. Barlow, J.P., C.J. Lorenzen, and R.T. Myren. 1963. Eutrophication of a tidal estuary. Limnol. Oceanogr. 8:251-262.
3. Brezonik, P.L., and E.E. Shannon. 1971. Trophic state of lakes in North Central Florida. Water Resour. Res. Centr., U. Fla., Publ. 13, 102 p.
4. Cairns, J., Jr. 1977. Quantification of Biological Integrity. p. 171-187. *In*: R.K. Ballentine and L.J. Guarraia (eds.), *The Integrity of Water*. U.S., E.P.A., Washington, D.C.
5. Champ, M.A. 1977. Nutrient loading in the nation's estuaries. p. 237-255. *In: Estuarine Pollution Control and Assessment*, V. 1. U.S., E.P.A., Washington, D.C.
6. Coull, B.C. 1977. *Ecology of Marine Benthos*. Univ. So. Carolina Press, Columbia. 467 p.
7. Cronin. L.E. 1977. Interactions of pollutants with the uses of estuaries. p. 739-753. *In: Estuarine Pollution Control and Assessment*, V. 2. U.S., E.P.A., Washington, D.C.
8. Darnell, R.M., and D.B. Shimkin. 1972. A systems view of coastal zone management. p. 346-364. *In*: B.H. Ketchum (ed.), *The Water's Edge, Critical Problems of the Coastal Zone*. M.I.T. Press, Cambridge, Mass. 393 p.

9. Darnell, R.M., W.E. Pequegnat, B.M. James, F.J. Bensen, and R.E. Defenbaugh. 1976. *Impacts of Construction Activities in Wetlands of the United States*. U.S. Environmental Protection Agency, Ecol. Rec. Series, 600/3-76-045. 393 p.
10. Florida Engineering and Industrial Experiment Station. 1970. Florida's environmental engineering conference on water pollution control. Fla. Eng. and Indust. Expt. Sta., U. of Fla., Gainesville.
11. Garside, C., T.C. Malone, O.A. Roels, and B.A. Sharfstein. 1976. An evaluation of sewage-derived nutrients and their influence on the Hudson Estuary and New York Bight. Est. and Coastal Mar. Sci. 4:281-289.
12. Goldman, C.R. 1972. The role of minor nutrients in limiting the productivity of aquatic ecosystems. p. 21-33. *In*: G.E. Likens (ed.), *Nutrients and Eutrophication: The Limiting-Nutrient Controversy*. Limnol. Oceanogr., Spec. Sympos., vol. 1.
13. Grassle, J.F., and J.P. Grassle. 1977. Temporal adaptations in sibling species of *Capitella*. p. 177-189. *In*: B.C. Coull (ed.), *Ecology of Marine Benthos*. Univ. So. Carolina Press, Columbia. 467 p.
14. Hale, S.S. 1976. The role of benthic communities in the nitrogen and phosphorus cycles of an estuary. Univ. Rhode Island Marine Reprint No. 57. p. 291-308.
15. Hobbie, J.E. 1974. Nutrients and eutrophication in the Pamlico River Estuary, N.C., 1971-1973. Water Resour. Res. Inst., Rept. No. 100, Univ. North Carolina.
16. Hobbie, J.E., and B.J. Copeland. 1977. Effects and control of nutrients in estuarine ecosystems. *In: Estuarine Pollution Control and Assessment*. v. 1, U.S., E.P.A., Washington, D.C.
17. Hynes, H.B.N. 1960. *The Biology of Polluted Waters*. Liverpool Univ. Press, Liverpool. 202 p.
18. Jaworski, N.A., D.W. Lear, Jr., and O. Villa, Jr. 1972. Nutrient management in the Potomac estuary. p. 246-272. *In*: G.E. Likens (ed.), *Nutrients and Eutrophication: The Limiting-Nutrient Controversy*. Limnol. Oceanogr., Special Symposium, v. 1.
19. Ketchum, B.H. 1972. *The Water's Edge: Critical Problems of the Coastal Zone*. M.I.T. Press, Cambridge. 393 p.
20. Likens, G.E. (ed.). 1972. *Nutrients and Eutrophication: The Limiting-Nutrient Controversy*. Limnol. Oceanogr., Special Symposium, v. 1.
21. Livingston, R.J. 1977. Resource management and estuarine function with application to the Apalachicola drainage system. p. 3-17. *In: Estuarine Pollution Control and Assessment*. V. 1, U.S., E.P.A., Washington, D.C.
22. Mackenthun, K.M. 1965. Nitrogen and phosphorus in water, an annotated selected bibliography of their biological effects. Robert A. Taft Sanitary Engineering Center, Cincinnati, Ohio. U.S. Public Health Ser. Pub. No. 1305. 111 p.
23. Middlebrooks, E.J., T.E. Maloney, C.F. Powers, and L.M. Kaack. 1969. *Proceedings of the Biostimulation Assessment Workshop*. Sanit. Eng. Res. Lab., U. Calif., Berkeley. 281 p.

24. Milway, C.P. 1968. Eutrophication in large lakes and impoundments. Organization for Economic Co-operation and Development, Paris. 560 p.
25. Nixon, S.W., C.A. Oviatt, and S.S. Hale. 1976. Nitrogen regeneration and the metabolism of coastal marine bottom communities. p. 269-283. *In*: J.M. Anderson and A. Macfayden (eds.), *The Role of Terrestrial and Aquatic Organisms in Decomposition Processes.* Blackwell, Oxford.
26. O'Connor, D.J. 1960. Oxygen balance of an estuary. Journal Sanitary Engineering Division, American Society of Engineers, No. 2472.
27. Odum, E.P. 1969. The strategy of ecosystem development. *Science.* 164:262-270.
28. Olson, T.A., and F.J. Burgess. 1967. *Pollution and Marine Ecology.* Interscience, N.Y. 364 p.
29. Pearson, E.A., P.N. Storrs, and R.E. Selleck. 1967. Some physical parameters and their significance in marine waste disposal. p. 297-315. *In*: T.A. Olsen and F.J. Burgess (eds.), *Pollution and Marine Ecology*. Interscience, N.Y. 364 p.
30. Pielou, E.C. 1977. *Mathematical Ecology*. Interscience, N.Y. 385 p.
31. Pomeroy, L.R., L.R. Shenton, R.D.H. Jones, and R.J. Reimold. 1972. Nutrient flux in estuaries. p. 274-291. *In*: G.E. Likens (ed.), *Nutrients and Eutrophication: The Limiting-Nutrient Controversy*. Limnol. Oceanogr., Special Symposium, v. 1.
32. Postma, H. 1967. Marine pollution and sedimentology. p. 225-234. *In*: T.A. Olson and F.J. Burgess (eds.), *Pollution and Marine Ecology*. Interscience, N.Y. 364 p.
33. Pritchard, D.W. 1969. Dispersion and flushing of pollutants in estuaries. J. Hydraul. Div., Am. Soc. Civil Engs. 95:115-124.
34. Redfield, A.C., B.H. Ketchum, and F.A. Richards. 1963. The influence of organisms on the composition of seawater. p. 26-77. *In*: M.N. Hill (ed.), *The Sea*. V. 2. Interscience, N.Y. 554 p.
35. Rohlich, G.A. 1969. *Eutrophication: Causes, Consequences, Correctives*. National Academy of Sciences, Washington. 661 p.
36. Saville, T. 1966. *A Study of Estuarine Pollution Problems on a Small Unpolluted Estuary in Florida*. Fla. Eng. Ind. Exp. Sta., Bull. 125. 202 p.
37. Seaton, A.M. and J.W. Day, Jr. 1979. The development of a trophic state index for the quantification of eutrophication in the Barataria Basin. p. 113-125. *In*: J.W. Day, Jr., D.D. Culley, Jr., R.E. Turner and A.J. Mumphrey, Jr. (eds.), *Proc. Third Coastal Marsh and Estuary Management Symposium*. L.S.U., Division of Continuing Education. Baton Rouge. 511 p.
38. Simon, J.L. and D.M. Dauer. 1972. A quantitative evaluation of red tide induced mass mortalities of benthic invertebrates in Tampa Bay, Florida. Environ. Lett. 3:229-234.
39. Steidinger, K.A. and R.M. Ingle. 1972. Observations on the 1971 summer red tide in Tampa Bay, Florida. Environ. Lett. 3:271-278.
40. Stewart, K.M., and G.A. Rohlich. 1967. *Eutrophication: A Review*. Calif. State Water Quality Control Board. Publ. No. 34. 188 p.
41. Stommel, H. 1953. Computation of pollution in a vertically mixed estuary. Sewage and Indust. Wastes. 25:1065-1071.

42. Tarzwell, C.M. 1957. Biological problems in water pollution. Public Health Service, Robert A. Taft Sanitary Engineering Center.
43. Tenore, K.R. 1977. Food chain pathways in detrital feeding benthic communities: a review, with new observations on sediment resuspension and detrital recycling. p. 37-53. *In*: B.C. Coull (ed.), *Ecology of Marine Benthos*. U. So. Carolina Press. Columbia, S.C. 467 p.
44. U.S., Environmental Protection Agency. 1977. *Estuarine Pollution Control and Assessment*. Vols. 1 and 2. U.S., E.P.A. Washington, D.C. 756 p.
45. Welch, E.B. 1968. Phytoplankton and related water quality conditions in an enriched estuary. J. Water Pollut. Contr. Fed. 40:1711-1727.
46. Yapp, W.B. 1959. *The Effects of Pollution on Living Material*. Sympos. of the Inst. Biol., No. 8. London. 154 p.

IMPACT OF NUTRIENT ENRICHMENT ON WATER USES

John H. Ryther* and Charles B. Officer**

**Contribution No. 4187 from the Woods Hole Oceanographic Institution, Woods Hole, Massachusetts 02543*

***Earth Sciences Department, Dartmouth College Hanover, New Hampshire 03755*

Nutrients do not, in themselves, influence man's use of estuaries. It is the secondary effect of the nutrients in stimulating the growth of aquatic plants, primarily the unicellular algae or phytoplankton, that make enrichment important to man. Such algal growth is usually thought of in quantitative terms: moderate amounts are good for most uses, since algae are the base of the food chains that nourish most forms of estuarine life, while too much enrichment and plant growth result in the familiar esthetic and ecological problems that are associated with the term "eutrophication." But, more and more, we have begun to view the effects of enrichment in qualitative terms, because we now recognize that, in fact, it is not the quantity but the kinds of algae that are good or bad for man's purposes.

What, then, are the characteristics of "good" and "bad" algae? Such a distinction is somewhat difficult to make without a considerable degree of subjectivity, but the latter may perhaps be excused in this case where the values to be considered are purely anthropocentric in any case. In that context, a beneficial type of alga is one that grows fast, successfully competing with undesirable species under optimal conditions for both. At the same time, it also rapidly decomposes, through autolysis or bacterial action, and its organic content is readily remineralized back to its inorganic form.

It is also typically a good food organism for filter-feeding planktonic or benthic herbivores, both with respect to size, shape and general ingestability and also in terms of digestability and nutritional value to the predator.

Because of the above characteristics, such algae do not often accumulate in the water and create unsightly or noxious "blooms." If, geographically or seasonally, they are produced out-of-phase with their filter-feeding predators, they may achieve populations dense enough to discolor the water temporarily or locally, but such manifestations are seldom sufficiently widespread to create serious esthetic problems. Similarly, if there is insufficient turbulence to maintain the occasional blooms of these organisms in the surface layers, they are either consumed or decomposed on the bottom too quickly to cause environmental stress. More characteristicly, however, these algae reach the bottom in the form of partially-digested fecal pellets of herbivorous zooplankton or fishes, where they also serve as an important food resource for the benthos. Finally, the desirable species of algae seldom if ever are toxic or otherwise harmful to aquatic animals or to man.

The typical undesirable algal species has all of the opposite characteristics. It grows more slowly than its beneficial relative, but, once produced, is often remarkably persistent, resistant to bacterial decomposition and decay, and may remain suspended in the water for weeks or even months after it has stopped growing. It is also a notoriously poor food source for herbivores, perhaps because it is too large or too small to be ingested or retained, often because it is covered with a protective sheath or membrane that resists attack by digestive enzymes. The same property of these microorganisms also explains their resistance to attack by bacteria.

Some organisms in this classification are outright toxic, sometimes lethal, to would-be predators. Others probably exercise a more subtle form of chemical mediation that may cause animals to avoid or reject them, though here the subject becomes more speculative and the evidence more empirical. A given species or group of related species of algae may, for example, have simply been found to be a poor source of food for the larvae of certain bivalves or crustacea grown in mariculture, for no obvious reason but no less conclusively for that.

Again, for the combination of reasons given above, the second group of algae tend to accumulate in the water to high population densities, often discoloring the water and giving off objectionable odors. When these organisms sink into deeper water, below photosynthetic depths, or to the bottom, they again remain for long periods uneaten by either bacteria or animals while their slow metabolism consumes oxygen and may eventually lead to anoxia and mortality of the higher life forms.

These, then, are the distinguishing characteristics of "beneficial" and "undesirable" kinds of algae. The logical question that follows is: "What are the kinds of phytoplankton that fit into these two categories?"

There are a number of publications in the contemporary literature that consider shifts in phytoplankton community structure in response to various natural and anthropogenic influences. The presumed causes of such changes in species composition will be discussed in some detail later, but the authors of these publications have usually considered changes from a community dominated by diatoms to one dominated by flagellates

(1, 2). Sometimes the distinction is based on size, a shift from net to nanoplankton (3, 4); though the succession of species may be the same, i.e., from large diatoms to small flagellates. Regulation of cell size in phytoplankton has been, in fact, the subject of a rather lively debate in recent years (see, for example, 5, 6, 7, 8).

However, such a simplistic characterization that divides phytoplankton into just two catagories does not seem to be particularly useful in the present context. Especially in estuaries, small centric diatoms are often a major constituent of the nanoplankton, certain dinoflagellates may be among the largest algae present, "flagellates" represent a broad taxonomic spectrum that includes both beneficial and undesirable species in the terms of the preceding discussion, and there may be many estuarine species, often important dominants, that are neither diatoms nor flagellates.

Furthermore, as quite correctly pointed out by Eppley and Weiler (9), small phytoplankton (i.e., those able to pass through a fine-mesh plankton net of 20-35 μm mesh diameter, or "nanoplankton") are normally the dominant kinds of algae at all times, both offshore and inshore and in areas not unduly stressed by man or nature. A similar conclusion was eariler reached by a number of other workers (10, 11, 12) and such a size distribution is a demonstrated characteristic of such inshore areas as California coastal waters (13), Long Island Sound (14), and Chesapeake Bay (15).

With respect to algae that have the "beneficial" or "undesirable" properties discussed above, it will be readily apparent to those who know them that the typical associations that inhabit most estuaries are comprised of a spectrum of species ranging more or less continuously from one extreme to the other. What determines whether the algal community is "good" or "bad" for man's purposes is the relative proportion of organisms of different values, and these, it would seem, are best categorized by conventional, at least major systematic groupings. This has been attempted in Table 1.

Although a refinement of the binary system of classification discussed above, the ranking in Table 1 is, of course, still simplistic. It does not include groups that are rare or, at least, seldom dominant in estuaries. The ranking is also, admittedly, as much intuitive as factual, for hard data concerning the role of many algal species in the estuarine ecosystem do not exist.

Information on the nutritional value of the algae was obtained from general texts on the cultivation of marine organisms (16, 17, 18) and some specific publications on feeding and nutrition (19, 20, 21). There are certainly exceptions to the ranking with respect to food value. Certain dinoflagellates (e.g., *Gymnodinium splendens*) are good food for larval anchovies (22) and copepods (23). At least one naked flagellate, the haptophyte, *Prymnesium parvum*, is toxic to marine animals (24) and probably belongs near the bottom of the list, and others in the same group may be similarly out of position in the ranking (see footnote Table 1).

TABLE 1. A ranking of major taxomic groups of phytoplankton according to their decreasing value to man, from beneficial (1) to undesirable (7).

1. Centric diatoms
2. Naked and scaled flagellates[1]
3. Green flagellates
4. Pennate diatoms
5. Dinoflagellates
6. Non-motile greens
7. Bluegreens

[1]This somewhat arbitrary grouping includes several recognized taxonomic entities which are lumped here only because too little is known of their ecological roles and significances in estuaries to treat them separately.

Improvement, perhaps including an expansion as well as some reordering of the ranking, must await more knowledge on the trophic dynamics of estuarine ecosystems. We submit that such a task represents a high priority area for future estuarine research. In the meantime, the numerical order of the ranking in Table 1 may serve as a series of tentative indices by means of which an entire phytoplankton community may be rated according to its value and uses to man. Each component may thus be rated as the product of its rank order and its numerical fraction of the total population; the community ranked as the sum of its component parts. This has been done in Table 2, the data for which (25) represent natural phytoplankton communities from Vineyard Sound, MA circulated at 10 percent exchange/day through 1000-1 fiberglass tanks. One tank was subjected to a temperature increase of 6°C, the second to 50 μg/1 of added Cu^{++}, and the third served as an unstressed control. Green and other flagellates were not differentiated between, as they are in the Table 1 ranking system, so the combined class of flagellates has been ranked 2.5. The numerical ratings of the communities, after 60 days exposure to the experimental stresses, increased in value from the control, which indicated a decrease in human use value.

It should be emphasized here that no significance should be given to the specific numbers in Tables 1 and 2, which are intended only to illustrate the very preliminary application of a concept for classifying phytoplankton communities. More and better information is needed before such a schemata may be employed with any degree of confidence, though the concept itself appears to be both valid and potentially useful. If the latter is true and it is possible to distinguish between phytoplankton communities that are beneficial and those that are detrimental to man's uses of estuaries, no matter how imperfectly we can make that distinction at present, the remaining, perhaps more important questions are: 1) What are the factors, natural and/or man-made, that determine what kind of

TABLE 2. Numerical rating of human use values[1] of phytoplankton communities from Vineyard Sound, MA calculated through 1000-liter containers with and without added stress factors.

Component	Control			Temperature (6°C)			Copper (15 μg/l)		
	%	Rank	Value	%	Rank	Value	%	Rank	Value
Centric diatoms	40	1	0.40	9	1	0.09	7	1	.07
Other flagellates	55	2.5	1.38	73	2.5	1.83	56	2.5	1.40
Pennate diatoms	2	4	0.08	5	4	0.20	19	4	0.76
Dino-flagellates	1	5	0.05	0	5	0.00	0	5	0.00
Non-motile greens	2	6	0.12	12	6	0.72	18	6	1.08
Bluegreens	0	7	0.00	0	7	0.00	0	7	0.00
Total community			2.03			2.84			3.31

1. Values are the product of the numerical percent of a given group of the total population and the rank of that group as assigned in Table 1.

algae dominate the phytoplankton community in a given estuary?, and 2) what kinds of management practices can be used to influence or control that selective process in favor of the beneficial forms? The remainder of this discussion will be devoted to those subjects.

Two shallow embayments on the south shore of Long Island, New York, Great South Bay and Moriches Bay, once supported a prosperous oyster industry. In the years immediately following World War II, two changes occurred that had a pronounced effect on the bay environment and shellfish production. First, the inlet connecting Moriches Bay to the Atlantic Ocean closed completely while that in Great South Bay became greatly restricted. Second, the Long Island duckling industry developed along the tributaries to the two bays. The waters of the bays and their tributaries thereupon became extremely brackish, warm in summer, and heavily enriched with duck wastes. Accompanying those changes, the phytoplankton grew to bloom proportions, giving the water a "pea soup" appearance, and shifted from a diverse community of diatoms and small flagellates to one strongly dominated by the small (∿2-4 μm), coccoid green alga *Nannochloris atomus*, which ranged in concentration from 10^6 to 10^7 cells/ml during most of the year. The oysters, unable to utilize the *Nannochloris*, failed to survive and the industry collapsed.

A survey of the area (26) revealed high concentrations of orthophosphate, which was clearly in excess, throughout the bays but no measurable nitrogen as ammonia, nitrate, nitrite or uric acid (the nitrogenous excretory product of ducks). However, up in the tributaries near the duck farms, where *Nannochloris* also occurred at high densities and where the salinity was ∿1°/oo and the summer water temperature was ∿30°C, both uric acid and ammonia-nitrogen could be measured. Laboratory experiments showed that *Nannochloris* grew at or near its maximum rate at the temperature and salinity that prevailed in the tributaries and as well with urea, uric acid and several amino acids as nitrogen sources as in inorganic nitrogen forms. A diatom, *Phaeodactylum tricornutum*[1], grown under the same conditions, failed to grow at temperatures above 25°C or at salinities below 10°/oo and grew poorly or not at all in the organic nitrogen sources. It was concluded that *Nannochloris* was able to out-compete diatoms and other kinds of phytoplankton in the tributaries, where it utilized all of the nitrogen as quickly as it became available. The green alga was then flushed into the bays where it grew little if at all, but persisted at high densities throughout the year, gradually being attenuated throughout the winter by slow tidal exchange and freshwater flushing action.

As in all such experiments, those reported above were biased by the choice of organisms. *Nannochloris*, isolated from the bays, was the logical and appropriate choice, but the diatom, a laboratory weed and the only diatom available at the time, is not a typical estuarine organism. It is now known that some small, centric estuarine diatoms do utilize organic nitrogen forms (27, 28) and do grow at temperatures in excess of 30°C and at salinities in the range of 1°/oo (29). One diatom in particular, *Thalassiosira pseudonana* (Strain 3H), subsequently isolated from Great South Bay, was found able to grow under all of those conditions, but apparently even it was unable to compete successfully with *Nannochloris* at the time.

Later (1958-1959), after Moriches Inlet was reopened and salinities had increased significantly, the ecological balance shifted again towards diatoms, *T. pseudonana* became strongly dominant, and a shellfish industry returned to Great South Bay (30).

It seems likely, then, that no single factor was responsible for the dominance of the undesirable green alga in Great South Bay and Moriches Bay during the post-war decade, but rather the shift was due to a combination of factors that together provided *Nannochloris* with a competive edge. Another possible factor, not considered at the time but that may also have been influential, will be discussed later. First, the effects of another class of substances will be considered.

For the past decade there has been serious concern over the consequences of discharging toxic contaminants of various kinds to the marine environment, particularly to estuaries where human activities including waste disposal tend to be concentrated. Among those concerns

[1]Incorrectly designated at the time as *Nitzchia closterium*.

have been possible effects on the phytoplankton, the major primary producers of organic matter and the base of most estuarine food chains. Although attention was initially focused on effects of toxic pollutants on the quantitative aspects of phytoplankton such as biomass and primary production, most of the ensuing environmental surveys showed little or no effect upon these parameters except in very localized situations. Concern then shifted to qualitative effects specifically on the phytoplankton community structure is the subject of the present discussion.

Toxic pollutants are, of course, not nutrients and their effect are not those of enrichment, but some consideration of them is relevant to the present discussion because they are often released in estuaries together with nutrients in wastewater discharges and their presence may thereby influence the effects of the enrichment on the resulting phytoplankton community structure.

These effects of toxic pollutants, as one would expect, result in the selection of the more tolerant and the depression of the more sensitive species. Unfortunately, the more pollution-tolerant species are usually those that are less desirable from man's point of view, though there are exceptions to that rule. Experimental studies using laboratory cultures and controlled ecosystems have shown that chlorinated hydrocarbons (DDT, PCB, endrin, deildrin, chlordane) are generally more toxic to centric diatoms than to pennate diatoms and various kinds of flagellates (31-34). Similar results were found from the effect of copper (25, 35), fuel oil (36), increased temperature (25) and chlorine (25, 37).

Occasionally, however, the selective effect of an added toxicant appears to be the opposite of that described above. Mackiernan et al. (38) found that addition of chlorine (0.05 mg/l) to plankton populations caused a sharp decrease in "nannoflagellates" and an enhancement of diatoms. Addition of mercury to natural populations in the Controlled Ecosystem Pollution Experiments (CEPEX) produced highly variable effects on phytoplankton community structure depending upon the concentration of mercury added, addition of nutrients and sediment, and time of exposure (39). Late in the experiment in which the highest concentration of mercury was used (5 μg/l), centric diatoms appeared and became strongly dominant for about 20 days.

The enclosed natural populations at the Narragansett, Rhode Island Marine Ecosystem Research Laboratory (MERL) that were exposed to oil developed a significantly higher density of phytoplankton than did the control enclosure and the enhanced algal population was also strongly dominated by diatoms (40).

The seemingly contradictory results of the CEPEX and MERL experiments, in which diatoms became the dominant component of the phytoplankton community, were in both cases attributed not to selective effects of the pollutant directly on the phytoplankton but to its lethal effect on grazing animals (zooplankton in CEPEX, that and benthos in MERL), on the assumption that removal of grazing pressure allowed the inherently faster-growing diatoms to prevail.

Selective grazing by zooplankton is regarded by many scientists to be the principal factor regulating phytoplankton community structure in the ocean, with grazing pressure selecting against the larger organisms (diatoms) in favor of the smaller nannoplankton or "flagellates" (e.g., 41-45). Since the selective process is entirely based on size, however, it seems an unlikely controlling mechanism in most estuaries, where grazing animals themselves are generally much smaller than in the open ocean and where, as mentioned earlier, phytoplankton of all kinds, including both diatoms and flagellates, come in all sizes from less than 10 to over 100 μm.

There is, however, another factor that selects for or against diatoms and that may be very important in estuaries. That is the presence or absence of silicon. As far as is known, all diatoms have a positive requirement for silicon and all other phytoplankton (except silicoflagellates, which are relatively unimportant in estuaries) do not. The amount of silicon in the tests of diatoms and the half saturation constants (k_S) for silicate uptake differ considerable for different species of diatoms (46, 47), but the relative concentration and rates of uptake and regeneration of nitrogen and silicon in the ocean suggest that, on an average, the two elements are assimilated by diatoms in approximately equal amounts, mole for mole (48-50).

The concentration of silicon, like that of nitrogen and phosphorus, shows seasonal fluctions and a vertical distribution pattern that are correlated with phytoplankton activity, often reaching undetectable levels in the surface layers of stratified waters (51-53). British limnologists long ago showed that the spring diatom bloom in English lakes is closely correlated with silicate availability (54-56), but this relationship has received surprisingly little attention by oceanographers. In his text *Plankton and Productivity in the Oceans*, Raymont (57) devotes an entire chapter to the seasonal succession of phytoplankton, describing the typical cycle from diatoms in winter-spring to dinoflagellates in summer-fall. Neither he nor the several authors he quotes (58-61) mention the possibility of silicon playing a role in this successional pattern.

In the Sargasso Sea, where silicate levels often fall below 1 μgA/l in the euphotic layer (53), diatoms are rare except during a brief period in spring when the water restratifies following late winter mixing and a small degree of surface enrichment. For a few days to a week or two at most, there is a brief diatom bloom of several species equally sharing dominance, after which the phytoplankton reverts to the typical sparse coccolithophorid-dinoflagellate community which then persists for the rest of the year (62). Incubated samples of the latter community enriched with nitrogen and phosphorus produced more of the same kind of organisms, but samples enriched with N, P and Si produced a community dominated by diatoms (63). In the only publication known to the authors that specifically addresses the role of silicon in an estuary, Pratt (64) showed that both the magnitude and cessation of the winter-spring diatom bloom in Narragansett Bay were closely correlated respectively with the quantity of silicate at its inception and the ultimate depletion of the nutrient.

From the preceding evidence, it would appear that the rapidly-growing diatoms are able to dominate the phytoplankton wherever and whenever conditions, including an adequate supply of silicon, are favorable for their growth. When the surface, euphotic layer becomes thermally stratified and nutrients are unavailable from the deeper layers, silicon becomes depleted and the relatively heavy diatoms sink, directly or in the form of fecal pellets of grazing herbivores. At that point, various kinds of flagellates and other non-diatoms that are able to subsist on recycled nitrogen and phosphorus alone become the dominant phytoplankton species.

The rate of dissolution of silicon from diatom tests, a purely chemical process in contrast to the bacteriological remineralization of other nutrients, is highly variable and dependent upon many factors (65), but the weight of evidence at present indicates that, while the time constant for nitrogen and phosphorus regeneration from diatoms is of the order of ten days, that for silicate is 50-100 days or more (66-68). Furthermore, if it is the fate of most diatoms to be eaten by zooplankton (69), the vertical partitioning of N and P from Si is more readily explained; the former nutrients being assimilated, metabolized and excreted by the animals within the euphotic layer where they are then available for recycling (70); the silicon, of no use to the herbivores and egested in their fecal pellets, sinking and becoming remineralized at greater depths.

The importance of silicon specifically with respect to the eutrophication and resulting species shift of the phytoplankton of Lake Michigan has been discussed in some detail by Schelske and Stoermer (71). Silicon, formerly supplied to the lake from unpolluted tributaries in quantities in excess of nitrogen and phosphorus, had resulted in a large surplus of the element in the lake water and, as a result, a year-round diatom-dominated phytoplankton community. In more recent times, heavy enrichment of the lake from wastewater poor in silicon stimulated additional phytoplankton growth that eventually eroded the silicon surplus to the extent that it now becomes depleted in late spring, with the resulting succession of bluegreens and other undesirable algae during the summer months.

A similar situation in estuaries and coastal marine environments affected by wastewater discharges has more recently been proposed by Officer and Ryther (72). Silicon is not often measured or reported in wastewater analyses (or, for that matter in environmental surveys), but a few available data from the northeastern United States suggest that the N:Si ratio in secondary sewage effluent is greatly in excess of the mole for mole relationship that diatoms on the average require (Table 3). The occurrences of summer blooms of non-motile green algae, bluegreens, and dinoflagellates in such places as the Potomac estuary, the Hudson estuary,

TABLE 3. Concentrations of inorganic nitrogen (NH_4^+-NO_2^-+NO_3^-), silicon ($SiO_3^=$) and phosphorus ($PO_4^{\equiv}$) and N:Si ratios in secondary sewage effluent (μgA/l). From Officer and Ryther (72).

	N	Si	P	N:Si
Tallman Island, NY	1200	80	130	15:1
Cranston, RI	2166	110	213	20:1
Warwick, RI	1342	100	295	13:1
Plymouth, MA	1176	70	243	17:1
Otis AFB, MA	1265	110	264	11:1
Wareham, MA	1205	173	142	7:1

and New York Bight were tentatively attributed to this factor (72), and it would seem probable that it may also have been another contributary cause of the *Nannochloris* bloom in Great South Bay and Moriches Bay, Long Island, described earlier (26). Although silicon was again not measured as part of the investigation, there is no known reason for it to have been a significant ingredient of the duck wastes.

This is not, however, meant to imply that the N:Si ratio in wastewater is consistently high. Weinberger et al. (73) gives values for the "average composition of effluent from secondary treatment of municipal wastewater" in which the N:Si ratio is approximately 2:1 by atoms, though they do not give the source of their data. MacIsaac et al. (74) shows ratios of 3.7 and 2.1 respectively for the White's Point and Hyperion effluents of Los Angeles, California, though the latter are the result of primary treatment in which a significant fraction of the nitrogen is organic and not included in their data. Nevertheless, silicon is undoubtedly a highly variable component of wastewater, which probably depends upon several factors, such as the source and nature of the domestic water supply and the contribution of storm run-off to the wastewater, among others. In no case, however, has an analysis of wastewater shown a N:Si atomic ratio equal to or less than one, as is typical of deep ocean water.

Except in extreme cases, it seems unlikely that any single factor normally controls species composition of phytoplankton. Several factors have been discussed above that possibly influence a shift from communities dominated by diatoms and other beneficial algae to those dominated by undesirable forms. These are high temperature, low salinity, organic forms of nitrogen, toxic contaminants for various kinds and low levels of silicon. Unfortunately, these factors frequently tend to coincide at sites of estuarine enrichment by human causes; i.e., at the heads of estuaries where centers of population are most often located and where partially treated wastewaters, rich in organic matter and toxic contaminants and poor in silicon, are discharged into brackish water areas warmed in summer by solar radiation and reduced circulation. These and probably other unmeasured as yet unsuspected factors are the conditions that together set the stage for the undesirable attributes of phytoplankton growth.

In contrast, estuaries may also be enriched from the sea by natural processes with the tidal penetration of relatively deep, nutrient-rich water up into the estuary - the so-called "salt wedge" - and its subsequent mixing into the surface, euphotic layer. Here the conditions are quite different. The introduced water is relatively cold and saline, its nutrients are well oxidized and silicon is present in concentrations roughly proportional to nitrogen. The resulting phytoplankton community that is produced from such enrichment is one dominated by diatoms, the estuary remains healthy, and its value to man is enhanced.

Perhaps there is a lesson to be learned from these contrasting conditions, and the discharge of human wastes to estuaries can be done in such a way and at such locations that it more nearly simulates natural enrichment processes and their beneficial effects.

ACKNOWLEDGEMENT

This research was partially supported by Contract No. EY-S-02-2532 with the U. S. Department of Energy.

REFERENCES

1. Parsons, T.R., P.J. Harrison, and R. Waters. 1978. An experimental simulation of changes in diatoms and flagellate blooms. J. Exp. Mar. Biol. Ecol. 32:285-294.
2. Goldberg, E.D. (ed.) 1979. Proceedings of a Workshop on Scientific Problems Relating to Ocean Pollution. Boulder, Colorado, March, 1979. NOAA Env. Res. Labs.
3. O'Connors, H.B. Jr., C.F. Wurster, C.D. Powers, D.C. Biggs, and R.G. Rowland. 1978. Polychlorinated biphenyls may alter marine trophic pathways by reducing phytoplankton size and production. Science 201:737-739.
4. Biggs, D.C., R.G. Rowland, H.B. O'Connors Jr., C.D. Powers, and C.F. Wurster. 1978. A comparison of the effects of chlordane and PCB on the growth, photosynthesis and cell size of estuarine phytoplankton. Env. Pollution 15:253-263.
5. Semina, H.J. 1972. The size of phytoplankton cells in the Pacific Ocean. Int. Rev. ges. Hydrobiol. 57:177-205.
6. Parsons, T.R., and M. Takahashi. 1973. Environmental control of phytoplankton cell size. Limnol. Oceanogr. 18:511-515.
7. Hecky, R.E., and P. Kilham. 1974. Environmental control of phytoplankton cell size (Comment). Limnol. Oceanogr. 19(2):361-365.
8. Parsons, T.R., and M. Takahashi. 1974. A rebuttal to the comment by

Hecky and Kilham. Limnol. Oceanogr. 19(2)366-368.
9. Eppley, R.W., and C.S. Weiler. 1979. The dominance of nanoplankton as an indicator of marine pollution: A critique. Oceanol. Acta 2:241-245.
10. Atkins, W.R.G. 1945. Autotrophic flagellates as the major constituent of the oceanic phytoplankton. Nature, Lond. 156:446.
11. Wood, E.J.F., and P.S. Davis. 1956. Importance of smaller phytoplankton elements. Nature, Lond. 177:438.
12. Yentsch, C.S., and J.H. Ryther. 1959. Relative significance of the net phytoplankton and nanoplankton in the waters of Vineyard Sound. J. Cons. Int. Explor. Mer:231-238.
13. Malone, T.C. 1971. The relative importance of nanoplankton and netplankton as primary producers in the California current system. Fish. Bull. 69(4):799-820.
14. Riley, G.A. 1941. Plankton studies. III. Long Island Sound. Bull. Bingham Oceanogr. Coll. 7(3):1-93.
15. McCarthy, J.J., W.R. Taylor, and M.E. Loftus. 1974. Significance of nanoplankton in the Chesapeake Bay Estuary and problems associated with the measurement of nanoplankton productivity. Mar. Bio. 24:7-16.
16. Kinne, O., and H.P. Bulnheim (eds.). 1970. Cultivation of marine organisms and its importance for marine biology. Helgol. wiss. Meeresunters 20:1-721.
17. Tamura, T. (ed.). 1966. Marine Aquaculture. Trans. from the Japanese by Mary I. Watarobe. Nat. Tech. Info. Serv., Wash. D.C. 1970.
18. Bardach, J.E., J.H. Ryther, and W.E. McLarney. 1972. Aquaculture. Wiley-Interscience, New York. 868 pp.
19. Walne, P.R. 1963. Observations on the food value of algae to the larvae of *Ostrea edulis*. J. Mar. Biol. Assoc. U.K. 43:767-784.
20. -------. 1970. Studies on the food value of nineteen genera *Ostrea, Crassostrea, Mercenaria* and *Mytilus*. Fish. Invest. Ser. II, Min. Agr., Fish. and Food, U.K. 26:1-62.
21. Ryther, J.H., and J.C. Goldman. 1975. Microbes as food in mariculture. Ann. Rev. Microbiol. 29:429-443.
22. Lasker, R. 1975. Field criteria for survival of anchovy larvae: the relation between inshore chlorophyll maximum layers and successful first feeding. Fish. Bull 73:453-462.
23. Paffenhofer, G.A. 1970. Cultivation of *Calanus helgolandicus* under controlled conditons. *In* O. Kinne (ed.), Cultivation of Marine Organisms and its Importance for Marine Biology. Helgolander wiss. Meeresunters. 20:346-359.
24. Shilo, M. 1946. Review on toxigenic algae. Verh. Internat. Verein. Limnol. 15:782-795.
25. Sanders, J.G., J.H. Ryther, and J.H. Batchelder. 1981. Effects of copper, chlorine and thermal addition on the species composition of marine phytoplankton. J. Exp. Mar. Biol. Ecol. 49:81-102.
26. Ryther, J.H. 1954. The ecology of phytoplankton blooms in Moriches Bay and Great South Bay, Long Island, New York. Biol. Bull. 106:198-209.
27. Guillard, R.R.L. 1963. Organic sources of nitrogen for marine centric diatoms, 93-104. *In* C.H. Oppenheimer (ed), Marine Microbiology.

C.C. Thomas, Springfield, Illnois.

28. Hellebust, J.A., and R.R.L. Guillard. 1967. Uptake specificity for organic substrates by the marine diatom *Melosira nummuloides*. J. Phycol. 3:132-136.
29. Hargraves, P.E., and R.R.L. Guillard. 1974. Structural and physiological observations on some small marine diatoms. Phycolog. 13:163-172.
30. Guillard, R.R.L., and J.H. Ryther. 1962. Studies of marine planktonic diatoms. I. *Cyclotella nana* Hustedt and *Detonula confervacea* (Cleve) Gran. Canad. J. Microbiol. 8:229-239.
31. Menzel, D.W., J. Anderson, and A. Randke. 1970. Marine phytoplankton vary in their response to chlorinated hydrocarbons. Science 167:1724-1726.
32. Mosser, J.L., N.S. Fisher, T. -C. Teng, and C.F. Wurster. 1972a. Polychlorinated biphenyls: Toxicity to certain phytoplankters. Science 175:191-192.
33. ———————, N.S. Fisher, and C.F. Wurster. 1972b. Polychlorinated biphenyls and DDT alter species composition in mixed cultures of algae. Ibid, 176:533-535.
34. Fisher, N.S., E.J. Carpenter, C.C. Remsen, and C.F. Wurster. 1974. Effects of PCB on interspecific competion in natural and gnotobiotic phytoplankton communities in continuous and batch cultures. Microbiol. Ecol. 1:39-50.
35. Thomas, W., and D. Seibert. 1977. Effects of copper on the dominance and diversity of algae. Bull. Mar. Sci. 27:23-33.
36. Lee, R.F., M. Takahashi, J.R. Beers, W.H. Thomas, D.L.R. Seibert, P. Koeller, and D.R. Green. 1977. Controlled ecosystems: their use in the study of the effects of petroleum hydrocarbons on plankton, 323-344. *In* F.J. Vernberg, A. Calabrese, F.P. Thurberg and W.B. Vernberg (eds.), Physicological responses of marine biota to pollutants Academic Press, New York.
37. Briand, F.J.P. 1975. Effects of power plant cooling systems on marine phytoplankton. Mar. Biol. 33:135-146.
38. Mackiernan, G.B., D.R. Heinle, and S.D. Van Valkenburg. 1978. The effects of chlorine-produced oxidants on the growth rates and survival of estuarine phytoplankton. Univ. Maryland Center for Environ. and Est. Studies Ref. 78-55CBL. 55 pp.
39. Thomas, W.H., D.L.R. Seibert, and M. Takahashi. 1977. Controlled Ecosystem Pollution Experiment: Effect of Mercury on enclosed water columns. III. Phytoplankton population dynamics and production. Mar. Sci. Commun. 3:331-354.
40. Vargo, G., G. Almquist, M. Hutchins, and D. Mongeon. 1979. Phytoplankton dynamics and production, 84-153. *In* The Use of Large Marine Microcosms to Study the Fates and Effect of Chronic Low Level Pollutants. MERL, U. Rhode Island, Kingston, Rhode Island. Report for Year 1, 1979.
41. McAllister, C.D., R.R. Parsons, and J.D.H. Strickland. 1959. Primary productivity and fertility at station "P" in the northeast Pacific Ocean. J. Cons. 25:240-259.
42. Mullin, M.M. 1963. Some factors affecting the feeding of marine copepods of the genus *Calanus*. Limnol. Oceanogr. 8:239-250.

43. Martin, J.H. 1965. Phytoplankton-zooplankton relationships in Narragansett Bay. Limnol. Oceanogr. 10:185-191.
44. Malone, T.C. 1971. The relative importance of nannoplankton and netplankton as primary producers in tropical oceanic and neretic phytoplankton communities. Limnol. Oceanogr. 16:633-639.
45. Steele, J.H., and B.W. Frost. 1977. The structure of phytoplankton communities. Phil. Trans. Roy. Soc. Lond. B. 280:485-534.
46. Kilham, P. 1971. A hypothesis concerning silica and the freshwater planktonic diatoms. Limnol. Oceanogr. 16:10-18.
47. Paasche, E. 1973. Silicon and the ecology of marine plankton diatoms. II. Silicate uptake kinetics in five diatom species. Mar. Biol. 19:262-269.
48. Richards, F.A. 1958. Dissolved silicate and related properties of some Western North Atlantic and Caribbean waters. J. Mar. Res. 17:449-465.
49. Redfield, A.C., B.H. Ketchum, and F.A. Richards. 1963. The influence of organisms on the composition of sea-water, 26-77. *In* M.N. Hill (ed.), The Sea, Vol. 1. Wiley-Interscience, NY.
50. Stephens, K. 1970. Automated measurement of dissolved nutrients. Deep-Sea Res. 17:393-396.
51. Harvey, H.W. 1928. Biological chemistry and physics of sea water. Cambridge Univ. Press.
52. Atkins, W.R.G. 1926. A quantitative consideration of some factors concerned in plant growth in water. J. Cons. Internat. Explor. Mer. 1:99-126.
53. Menzel, D.W., and J.H. Ryther. 1960. The annual cycle of primary production in the Sargasso Sea off Bermuda. Deep-Sea Res. 6:351-367.
54. Pearsall, W.H. 1932. Phytoplankton in the English Lakes. 2. The composition of the phytoplankton in relation to dissolved substances. J. Ecol. 20:241-262.
55. Lund, J.W.G. 1950. Studies on *Asterionella formosa* Haas. II. Nutrient depletion and the spring maximum. J. Ecol. 38:1-14, 15-35.
56. Lund, J.W.G., F.J.H. Mackereth, and C.H. Mortimer. 1963. Changes in depth and time of certain chemical and physical conditions and of the standing crop of *Asterionella formosa* Haas in the north basin of Windermere in 1947. Phil. Trans. Roy. Soc. Lond. B. 246:255-290.
57. Raymont, J.E.G. 1963. Plankton and productivity in the oceans. Pergamon Press, New York.
58. Braarud, T. 1961. Cultivation of marine organisms as a means of understanding environmental influences on populations 271-298. *In* M. Sears (ed.), Oceanography. Public No. 67, A.A.A.S., Washington, D.C.
59. Margalef, R. 1958. Temporal succession and spatial heterogeneity in phytoplankton, 323-347. *In* Buzzati-Traverso (ed.), Perspectives in Marine Biology. Univ. of California Press, Berkeley and Los Angeles.
60. Lillick, L.C. 1940. Phytoplankton and planktonic Protozoa of the offshore waters of the Gulf of Maine. Part II. Trans. Amer. Phil. Soc. 31:193-237.
61. Conover, S.A.M. 1956. Oceanography of Long Island Sound, 1952-1954. IV. Phytoplankton. Bull. Bingham Oceanogr. Coll.

15:62-112.
62. Hulburt, E.M., J.H. Ryther, and R.R.L. Guillard. 1959. The phytoplankton of the Sargasso Sea off Bermuda. Cons. perm. int. Explor. Mer. J. du Conseil 25:115-128.
63. Menzel, D.W., E.M. Hulburt, and J.H. Ryther. 1963. The effects of enriching Sargasso Sea water on the production and species composition of the phytoplankton. Deep-Sea Res. 10:209-219.
64. Pratt, D.M. 1965. The winter-spring diatom flowering in Narragansett Bay. Limnol. Oceanogr. 10:173-184.
65. Werner, D. 1977. The biology of diatoms. Chap. 4. Silicate metabolism, 110-149. Blackwell Scientific Publ., London.
66. Riley, G.A., H. Stommel, and D.F. Bumpus. 1949. Quantitative ecology of the Western North Atlantic. Bull. Bing. Oceanogr. Coll. 12:1-169.
67. Jorgenson, E.G. 1955. Solubility of the silica in diatoms. Physiol. Plant. 8:846-851.
68. Lewin, J.C. 1961. The dissolution of silica from diatom walls. Geochim. Cosmochim. Acta 21:182-195.
69. Harvey, H.W., L.H.N. Cooper, M.V. Lebour, and F.S. Russell. 1935. Plankton production and its control. J. Mar. Biol. Assoc. U.K. 20:407-441.
70. Johannes, R.E. 1968. Nutrient regeneration in lakes and oceans. In: M.R. Droop and E.J. Ferguson Wood, eds. Advances in Microbiology of the sea. Academic Press, N.Y. pp 203-213.
71. Schelske, C.L., and E.F. Stoermer. 1971. Eutrophication, silica depletion, and predicted changes in algal quality in Lake Michigan. Science 173:423-424.
72. Officer, C.B., and J.H. Ryther. 1980. The possible importance of silicon in marine eutrophication. Mar. Ecol. Prog. Ser. 3:83-91.
73. Weinberger, L.W., D.G. Stephens, and F.M. Middleton. 1966. Solving our water problems---water renovations and reuse. Ann. N.Y. Acad. Sci. 136-154.
74. MacIsaac, J.J., R.C. Dugdale, S.A. Huntsman, and H.L. Conway. 1978. The effect of sewage on uptake on inorganic nitrogen and carbon by natural populations of marine phytoplankton. J. Mar. Res. 37:51-66.

MANAGEMENT IMPLICATIONS OF NUTRIENT STANDARDS

Michael A. Bellanca
Deputy Executive Secretary
Virginia State Water Control Board
P.O. Box 11143
Richmond, Virginia 23230

INTRODUCTION

The promulgation of a nutrient standard has a two-fold impact and brings with it, from the regulatory point of view, a responsibility as well as an exercise of authority. On the one hand, the standard must be set correctly and based on sufficient evidence to assure the benefits will in fact derive from its imposition. On the other hand, its implementation must take cognizance of the many external forces at work over which little control can be exercised, e.g., nonpoint source runoff and the costs in energy, chemicals, and operation and maintenance to assure treatment to the level of the standard.

What follows will describe the interrelationship of these various elements and some lessons learned from past work on the Potomac River and one of its tributaries--Occoquan Creek.

BACKGROUND

Prior to the mid-1960s and early 1970s and the passage of P.L. 92-500, it would seem that compared with today all was bliss in the waste treatment arena. We were very confident in our abilities to specify a percent removal of biochemical oxygen demand (BOD), suspended solids (SS) and with them the type of treatment which would satisfy these particular needs. Our standards for the most part were those four horsemen of temperature, pH, dissolved oxygen, and coliform. Ignorance was indeed bliss.

Some researchers and engineers, however, were taking note of their environment and attempting to find the relationships between certain observable phenomena and their potential causes. Research on the Madison Lakes, Linsley Pond, Lake Washington and others were beginning to tie down nutrients, as related to sewage, as a primary if not total cause of accelerated eutrophication of these bodies of water. Attempts were made to show that diversion of waste sources from the lake environment could and would have a profound effect on the quality of these bodies. A burgeoning population in critical areas created stresses which cried, as it were, for relief as bodies of water of all sizes, from ponds to the Great Lakes, felt these stresses and responded with their tell-tale signatures. These visible incidents increased and seemingly came to a head in the period of the late 1960s and early 1970s.

The passage of 92-500 presaged an assault on water quality problems of all types -- from dissolved oxygen to nutrients to toxics. In a flurry, criteria were established with the caveat that these were not standards but the best scientific judgement then prevalent. But in the euphoric atmosphere and environmental concern which was then extant, caveats were forgotten, criteria become standards and application of these was foisted, to a large extent, on a public which did not truly understand or if they did felt that no cure was too good or too expensive for them.

The foregoing is not universally true and the historical sequence can find its exception. But in general it is correct and reflects a very human and very familiar approach to resolution of a problem -- private study, comprehension of cause and effect, slow acceptance and a rush towards implementation of a solution.

THE POTOMAC RIVER

The Potomac, because of its location, enjoys the appelation of "The Nation's River." It has a drainage area of 14,760 square miles which covers four states and the District of Columbia and, after flowing 383 miles, it empties into the Chesapeake Bay. It is a much studied river and, due to its location, a very visible one. Former President Lyndon B. Johnson once proclaimed that he would swim in it before he left office. (The Potomac is a multi-use river and swimming, oystering, fishing and boating are enjoyed and protected by the established water quality standards.)

Since the first Potomac survey, conducted in 1913, the water quality has deteriorated owing to the increasing population of the Washington, D.C. area. Dissolved oxygen levels decreased steadily but showed a slight upsurge in the early 1960s with implementation of additional treatment. Population pressures and the associated increase in waste load resulted in further downward movement of D.O. levels to the point where concentrations reached less than 1.0 mg/l during the low flow-high temperature periods.

Large populations of algae were observed in the estuary during the months of June through October and were often characterized by thick mats. These were confined principally to the upper and middle portions of the estuary and resulted in (1) large increases in ultimate oxygen demand; (2) an overall increase in D.O. due to algal respiration, and (3) creation of aesthetically objectionable nuisance conditions.

POTOMAC ENFORCEMENT CONFERENCE

In 1969, the Potomac Enforcement Conference (1) was convened for the third time in order to evaluate the Potomac's water quality problems and to recommend the steps necessary to enhance its quality and to provide for adequate treatment of area wastes.

For the Virginia side of the river (Figure 1), BOD, nitrogen and phosphorus loadings were determined for the Pentagon, Arlington, Alexandria, Fairfax-Westgate, and the then-proposed Lower Potomac Treatment Plant. Loadings were also established for the District of Columbia. These are shown in Table 1.

TABLE 1. Potomac Enforcement Conference Recommended Loadings

Facility	BOD_5	Total P	Total N
		#/day	
Pentagon	300	20	145
Arlington	1300	60	650
Alexandria	1300	60	650
Westgate	900	40	445
District of Columbia	12,700	560	6130
		% Removal	
Lower Potomac	96	96	85

One further recommendation of the Potomac Enforcement Conference was that studies be undertaken to evaluate the water quality management needs of the upper estuary, and to study its water supply potential. Both of these needs were addressed simultaneously by Chesapeake Technical Support Laboratory of the Federal Water Quality Administration and culminated in Technical Report No. 35.

POTOMAC EMBAYMENT STANDARDS

The Enforcement Conference recommendations were stiff and those of Technical Report No. 35 (2) were even more so. In part they read, "Because of the lack of assimilative capacity and transport in the upper portions of small tidal embayments and also because of ideal algae growing conditions, maximum concentrations of UOD, phosphorus and nitrogen in effluents discharged to these areas should be less than 10.0, 0.2 and 1.0 mg/l respectively. A detailed analysis for each embayment is required to determine the minimum cost of either extending the discharge outfall to the main channel of the Potomac or discharging within the embayment and providing a very high degree of wastewater treatment, approaching ultimate wastewater renovation. Unless this high degree of removal is provided, effluents from Alexandria, Arlington, Piscataway and the Lower Potomac facilities should be discharged into the main channel of the Potomac Estuary".

Once the Potomac embayment recommendations of Report No. 35 were published, Virginia's regulatory agency was faced with some difficulty; two sets of standards for BOD and nutrients had been recommended for virtually the same area of the Potomac. A comparison of these are given in Table 2. The Enforcement Conference's

TABLE 2. Comparison of Embayment Standards and Recommendations of the Potomac Enforcement Conference.

	Enforcement Conference Mg/l			Embayment Standards Mg/l		
Facility	BOD_5	Total P	Total N	UOD	Total P	Total N
Pentagon	6.4	0.32	2.65	10	0.2	1.0
Arlington	6.4	0.32	2.65	10	0.2	1.0
Alexandria	4.9	0.22	2.40	10	0.2	1.0
Westgate	4.9	0.22	2.40	10	0.2	1.0
Lower Potomac	9.6	0.40	6.75	10	0.2	1.0

recommended loadings were converted to mg/l on the basis of design flow and for the Lower Potomac Plant, influent concentrations of BOD, N and P were assumed and on the basis of the percent removal which was required the effluent concentrations were calculated.

Quite obviously the two sets of recommendations, while in the same ballpark, did present something of a dilemma in terms of regulation. The localities upon which regulation was to be imposed were not clairvoyant and could not be expected to meet their responsibilities without proper guidance.

In order to resolve the difficulty, proposed standards were submitted to the Virginia State Water Control Board and were adopted per Minute 42 (3) of the board's meeting of June 14-15, 1971. In essence, the board's standards are the embayment standards which were recommended in Technical Report No. 35 with one minor adjustment and one major variance.

BOD_5 - Not greater than 3.0 mg/l
Unoxidized N - Not greater than 1.0 mg/l from April 1 to October 31
Total P - Not greater than 0.2 mg/l

Total N - Not greater than 1.0 mg/l (when technology is available)

These standards were made applicable to all new discharges to Virginia's embayments as well as to expansions of existing plants.

In order to meet these embayment standards, it was obvious that sophisticated technology would be needed and to this end the localities moved. It should be recalled, however, that the previously quoted recommendation from Report No. 35 stated that "A detailed analysis for each embayment is required to determine the minimum cost of either extending the discharge outfall to the main channel of the Potomac or discharging within the embayment and providing a very high degree of wastewater treatment...." For the Virginia plants this approach was never utilized and the adopted embayment standards were implemented across the Board without evaluation of each individual embayment and its physical, chemical or hydraulic characteristics.

OCCOQUAN POLICY AND STANDARDS

As the embayment scenario was unfolding, a two-year study effort (4) on Occoquan Creek--a major water source for Northern Virginia--was ending. Results indicated that during the study period, 73 percent of the inorganic phosphorus and 39 percent of inorganic nitrogen were introduced into the watershed by waste discharges. Certainly phosphorus removal could have a significant impact on the amount of this nutrient which would enter the reservoir; nitrogen removal was questionable as to impact.

Two distinct options were available for resolution of the Occoquan problem: (1) high degree of treatment with discharge to the reservoir; and (2) treatment with pumpover to the Potomac Embayment of Neabsco Creek.

If the second option were chosen, then the discharge would run afoul of the Potomac Enforcement Conference recommendations and a high degree of treatment would be required in addition to the cost of conveying waste out of the Occoquan.

Discharge of highly treated waste back to the reservoir had the added effect of keeping water in the reservoir system for future use in serving the rapidly expanding water supply needs of the area.

By Minute 10 (5) of the board meeting of July 26, 1971, the Occoquan policy and associated discharge standards were adopted. The policy contained requirements for redundancy, ballast ponds, and regionalization of wastewater facilities and established effluent quality standards of:

BOD_5	1.0 mg/l
COD	10 mg/l
P	.10 mg/l
TKN	1.0 mg/l
S.S.	1.0 mg/l

In order to meet the Occoquan and Potomac embayment standards, very sophisticated treatment was needed and the affected communities responded in a variety of ways.

Treatment Plant Design

The selected designs incorporated biological treatment followed by physical chemical process trains and sludge treatment. Table 3 summarizes the treatment trains used at Arlington, Alexandria, Lower Potomac and Potomac and Occoquan.

TABLE 3. Process Trains

Type of Treatment	Arlington 30 MGD	Alexandria 54 MGD	Lower Potomac 35 MGD	Occoquan 10 MGD
Biological				
Activated Sludge	X		X	X
Rotating Bio-Discs		X		
Chemical				
Phosphorus (by lime)	X		X	X
Phosphorus (by alum)		X		
Nitrogen - breakpoint Cl_2	X		X	
- sodium hypchlorie		X		
- ion exchange				X
Physical				
Mixed media filters	X	X	X	X
Carbon adsorption	X	X	X	X
Disinfection - Cl_2	X	X	X	X
Reaeration	X		X	
Recarbonation	X		X	X
Carbon Regeneration	X	X	X	X
Ammonia Recovery				X
Sludge Treatment				
Convention	X	X	X	X
Chemical	X	X	X	X
Incineration	X	X	X	X
Landfilling		X		X
Ballast Ponds - 40 MG				X
Effluent Reservoir 160 MG				X

Operation and Maintenance

Most of the processes are either on line or under construction at the various plants. Some cost figures are now available and are instructive of the price that will have to be paid for the maintenance of the embayment and Occoquan standards. Table 4 summarizes the present and future costs.

TABLE 4. Cost of Treatment - $/1000 Gallons.

Facility	Current	Future
Arlington	1978-$0.31 1979-$0.42	$0.80 - 0.90
Alexandria	$0.155	$0.72(a)
Lower Potomac	$0.515	$0.80 - 0.90
Occoquan	$0.72(a)	$0.81 - 0.91

(a) does not include nitrogen removal costs.

The current costs reflect some of the AWT processes at all of the facilities with the exception of Alexandria whose current costs reflect operation through the biological processes.

It is expensive to construct these facilities but the costs don't end there. Operation and maintenence (O&M) charges are added to the debt service incurred by the capital outlay and form the basis for the sewer charges which are imposed by the localities. The former is fixed but the later charges move. A simple illustration based on the Occoquan experience will bear this out (Table 5).

TABLE 5. O&M Costs - UOSA

Constituent	8 MGD	10 MGD
Salary	$ 890,000	890,000
Power	530,000	630,000
Chemicals	768,000	960,000
Other	453,000	489,000
Total	$2,641,000	$2,969,000

As shown in Table 5 power and chemicals account for 20 percent and 29 percent of O&M respectively at 8 MGD and 21 percent and 32 percent respectively at 10 MGD. Without a doubt these processes are energy consumptive and costly. This high cost is demonstrated vividly in Table 6 which shows chemical costs for 1973 and 1979 and the percent increase which has occurred in the interim.

TABLE 6. Chemical Costs - $/ton.

Type	1973	1979	% Increase
Chlorine			
150 cylinder	173	342	+ 98
l-ton cylinder	125	247	+ 98
Tank Car	77	152	+ 98
Ferric Chloride	80	118	+ 48
Alum	37	87	+135
Quicklime	13	30	+130
Hydrated Lime	16	36	+125
Sodium Hypochlorite	228	806	+180

Normal inflationary pressures would surely have driven the price of chemicals upward from 1973 to 1979 and some of these costs, particularly ferric chloride, may represent such a result. However, in the case of the remaining chemicals they were affected by a major dislocation - the Arab oil embargo - and our increased dependence on and the increasingly higher costs of petroleum. It is safe to say that these costs will rise even higher and their impact will have a profound effect on the total O&M costs.

One glaring example of where we may be headed can be extracted from the experience of Arlington County whose chosen form of nitrogen removal is breakpoint chlorination and who must of necessity utilize sodium hypochlorite to accomplish it.

In 1972 when the breakpoint facilities were designed, sodium hypchlorite cost $0.18 per gallon and jumped to $0.53 per gallon in 1979. Projected O&M cost for this process rose from approximately $2.2 million to $6.4 million which equates to $0.58/1000 gallons for this process alone.

CURRENT STATUS OF AWT

Since 1970 when approximately 40,000 pounds of BOD_5 were being discharged to Virginia embayment daily, the loading has decreased to approximately 10,500 #/day. Some facilities, notably lower Potomac and UOSA, have implemented phosphorus control, and input to the system from this source has dropped significantly. The visible quality of the embayments has improved.

Since 1976 several forces have been brought to bear which have culminated in a re-evaluation of AWT fundability by EPA. In 1976 GAO issued their report (6) which questioned the benefits associated with the cost of AWT and whether grant funds could or should be put to better use elsewhere - for upgrading primary plants and providing funds for small communities which for various reasons cannot rise to the top of funding priority lists. Vertex Corporation completed a report (7) in 1977 entitled "An Analysis of Planning for AWT" which essentially confirmed the GAO report. In March, 1979 the Surveys and Investigations Staff of the House Committee on Appropriations submitted a report (8) to Congress which also took issue with the extant AWT policy and recommended:

- Preparation of an incremental cost analysis for all AWTs;

- Determination of what impact the effluent will have on the receiving waters;

- Consideration of alternatives to AWT;

- Determination that the recommended treatment process is the most economical method to meet water quality standards.

The point has been made that treatment facilities can be designed to meet very stringent standards but at a considerable capital and O&M cost. That which was designed in the early 1970s under a given set of constraints may well be inappropriate as we complete their construction and enter the 1980s.

Virginia's experience is by no means unique. The city of Las Vegas, after a six-year legal battle, signed a consent order which enabled it to convert its partially completed AWT facility into an advanced secondary plant; i.e. secondary treatment with phosphorus removal. Effluent standards will be relaxed and a two-year study undertaken to determine the relationship between the discharge and algae growth in Lake Meade.

The pendulum has begun to swing back towards the middle and EPA's response to the various reports, court cases and like criticism has been to advance a policy and a review process for AWT/AST (advanced waste treatment/advanced secondary treatment). The policy is embodied in EPA's PRM-79-7 which incorporates many of the recommendations found in the report of the congressional subcommittee.

POLICY RE-EVALUATION

The construction grants program is not funded now to a level at which Virginia and many other states can meet their needs. Without moving to AWT it is estimated that Virginia's construction needs are \$400 million. On the basis of congressional funding at \$3.8 billion nationally, the allocation formula would provide \$80 million to Virginia in 1979.

Faced with funding shortages and some improvement in the quality of

Virginia's embayments, the staff recommended and the board adopted a plan to defer funding of AWT facilities which had not been constructed yet and to delay for a three-year period enforcement of the Potomac Embayment Standards. During the interim period the staff was to conduct a detailed study of each of the embayments with a view to re-evaluating the extant standards. These are the first detailed studies of the embayments and provide an opportunity to evaluate the direction which was taken in 1971. It was estimated that the program will cost approximately $700,000 and covers physical, hydraulic, water quality and nonpoint source considerations. In the meantime, approximately $10 million of grant funds, previously designated for AWT, have been allocated to other projects on the priority list. If the studies conclude that the deferred facilities are needed and the embayment standards are confirmed, the necessary grant funds will be allocated.

In the interim and ultimately in the event that the Potomac Embayment Standards are confirmed, several alternatives are available: (1) Implmentation of nitrogen removal could be delayed. This would have the effect of postponing costly O&M, particularly at Arlington where sodium hypochlorite is to be utilized. A more definitive answer can be obtained for the question of need since the relationship of nitrogen removal to reduction of algae growth in the Potomac is much less certain than the relationship of phosphorus to algae growth. Further, the little work done in the Occoquan watershed indicates that nitrogen from nonpoint sources far exceeds that which is controllable through waste treatment. Finally, if a highly nitrified effluent is discharged, a net reduction in oxygen demand would result although the nitrogen would still be available for photosynthesis.

(2) Permits can be tiered to allow flexibility of operation of the AWT portion of the plants. A tiered permit would take cognizance of the changes in flow and climatological conditions over a year. Limits would be based on need at a particular time and would not be held constant. Stringent limits would prevail during the critical times of the year. This would allow phosphorus reduction to vary and allow for savings in chemicals and sludge handling and a concomitant reduction in O&M costs at certain times of the year when flows are high and water temperatures low. The same rationale could apply to the removal of nitrogen and BOD and would result in more savings.

IMPLICATIONS

What then does the establishment of nutrient standards imply? In the main the implications are not much different from those which would or should accrue to the establishment of any standard. It is suggested that some of these implications are: (1) Knowledge and importance of the uses to be maintained - The traditional uses are those pertinent to maintaining an environment which is suitable for primary contact recreation, water supply, propagation of fish, shellfish, and wildlife and one which is aesthetically pleasing. The establishment of nutrient standards impacts on these uses directly in some cases and indirectly in others. Are there other uses which are affected by nutrient discharges and, if so, how do they rank in the hierarchy of importance?

(2) A clear delineation and understanding of the physical, chemical and biological processes at work - Before one establishes a standard, the cause and effect relationships should be understood. There should be some assurances that the imposition of the standard will have the predicted result and will thereby maintain the uses which have been identified as being worthy of protection. A knowledge and understanding of the physical, chemical and biological processes and their interrelationships are germane. If knowledge of these processes is incomplete but it is perceived that a standard is necessary, the limitations of this standard should be acknowledged and not obscured. Too often the standard is viewed as immutable - an object of reverence and inviolability - and a definitive cure for an environmental ill. For when we establish the standard we imply that we understand the system and the cause/effect relationships.

(3) Knowledge of the benefits to be achieved - Most benefits are directly related to the use which is being maintained through the imposition of the standard and are readily discernible. Others may not be quite so apparent but deserve attention. What is said when the nutrient standard has been established is that we know what is to be protected and that worthwhile benefits will be derived from that protection.

(4) Knowledge of the costs to achieve the benefits - The nutrient standard is not implemented and the benefits are not achieved gratis. At this point a legitimate question can be asked as to who pays and how much? Assuming that the burden of the standard is placed on the waste treatment side of the equation, the costs can be obtained following design and the selection of the required processes. But the costs consist of those which derive from the capital expense (i.e., the debt service), O&M and perhaps others depending on the rate structure and overall programs and goals of the specific locality. When treatment costs are given we imply that we are aware of all of them, that some are constant or variable and we know the full range of charges to the consumer in the future.

(5) Willingness to pay - As has been seen by the previous examples the nutrient standard is achieved at considerable expense. Treatment costs which approach $1.00/1000 gallons are three-to-four times the cost for achieving secondary treatment. To a public long inured to cheap sewer rates, the levy to achieve AWT comes as quite a shock. At this point the standard is successfully implemented if the uses, benefits and costs have been properly calculated and the public is convinced that the return is worth the investment. Further, there must be a concomitant commitment to enforce the standard. Without the demonstration to support both from a financial as well as an enforcement standpoint, it will have profited little to adopt the standard.

(6) Differentiation between point and nonpoint sources - When one sets a standard, the likelihood that the standard will achieve the desired results should be high. In the case of the nutrient standard there should be the expectation that algal blooms and wide diurnal fluctuations in dissolved oxygen will be reduced and eventually disappear. Since control of nutrients is almost universally applied at the waste treatment facility, there should be evidence that control at this point will have value. A detailed evaluation of the contributions of nutrients emanating from nonpoint as well as point sources should be made to determine the percent of contribution of nutrients from both sources. The establishment of the nutrient standard implies that such an evaluation has been made and that the best evidence is that implementation of the standard - as applied to point or nonpoint sources - will achieve the desired results.

(7) Willingness to audit past decisions - Too often standards are viewed as sacrosanct and forever embalmed within the black and white of the paper upon which they are written. Since our knowledge is less than perfect and the environment changes, it is an inescapable fact that adjustments may be necessary. These adjustments may be up or down depending on several things: knowledge available when the standard was set, evaluation of environmental response to implementation of the standard, public response, re-evaluation of the uses and benefits, etc. The point is that we may strive for perfection but at the same time acknowledge that several iterations may be necessary before the precise beneficial balance we seek between the discharge and environmental response is achieved. When a standard is set, the regulatory authority implies that it has been based on the best evidence available and that its impact on the environment will be carefully reviewed.

(8) Resolution to make revisions to standards and policies when as needed -- In order to reach the goal of balance between discharge and environmental response - i.e., to obtain the best standard - regulators must be of a mind to review past decisions in light of new information and the environmental response which field audit reveals. Further, they must be prepared to make changes and to revise the standards and associated policies as dictated by this review and the associated re-evaluation. The revision of a standard, particularly one as controversial as the nutrient standard, can be a painful process, both in a personal and in a political sense. It is difficult to admit that we may have been wrong and that the standard was set too high or too low. From a political standpoint, there may be repercussions since if the standard was too stringent, the cost to meet it will have likewise been high and, if it is relaxed, the monies spent will have been considered wasted. Correspondingly, if the standard originally was lax and the thrust is to tighten it, then it might be said that the environment suffered and that we have subsidized industry and local government and perhaps caved in to political or other pressures. Regardless of the scenario, changes should be made in either direction when the evidence points toward such change. In essence this is what

those who establish standards imply to the public when the standards are set.

SUMMARY

In general, what has been said is applicable to any standard. For the present nutrient standards have generated controversy due to the costs associated with meeting them. However, decision makers and researchers should not be paralyzed by this controversy or even the prospect of future ones. It is suggested that research need not continue "ad infinitum" before a recommendation can be made to the decision makers for implementation. Likewise, the decision maker must exercise some modicum of common sense and be willing to ask for the best recommendation, and the constraints, biases, and imperfections of it. Decisions are made daily without the benefit of perfect knowledge, as we apply science, experience, and our best judgment. Where we encounter problems is when we fail to make full disclosure of the pros and cons and give the appearance that the proposed standard is "golden".

FIGURE 1. Wastewater discharge zones in the upper Potomac estuary.

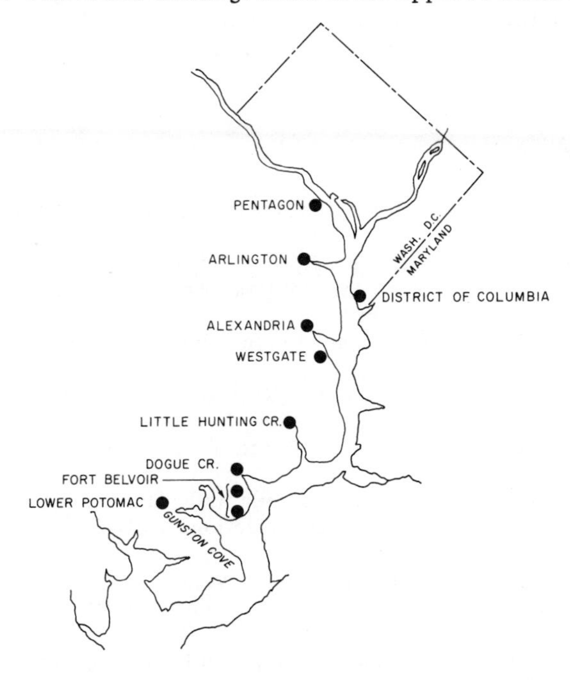

Perhaps too great a step was taken when the standards for the Occoquan and the Potomac embayments were developed and imposed. Secondary treatment was left behind and a leap in technology to chemical treatment, filtration, carbon adsorption and ion exchange was made. The tenor of the times may not have allowed any other way, but in retrospect it appears that the more sagacious approach would have been to move from one technology to another with an evaluation of each successive step before proceeding further. We may have arrived at the same point ultimately but at least we would have been in a position to say that each step was necessary and that wasted resources did not result.

The experiences of the standard setting process for northern Virginia and the Potomac River were used to illustrate an approach as well as the problems which can be encountered in the search for the best nutrient standard. These problems are surmountable and amenable to resolution provided that we are aware of the authority as well as the responsibilities which attend implemention. Time and nature do not stand still. Our environment changes and what was true yesterday may not be tomorrow. Those involved in setting standards should continue to be aware that they have a public trust to maintain and that this trust is maintained when they accept full accountability for their actions and exercise willingness to make changes when and as necessary. For it is through the process of establishing standards, implementation of controls, audit of the environmental response, and revision of the standard that we will achieve the best standard and maintain the uses, benefits and balance which is our goal.

REFERENCES

1. Recommendations of the Potomac River Enforcement Conference. May 8, 1969.
2. Jaworski, N.A., Clark, Leo J., and Feigner, K.D. A water resource - water supply study of the Potomac estuary. CTSL, MAR, USEPA Technical Report 35, April 1971.
3. Virginia State Water Control Board Minute No. 42, meeting of June 14-15, 1971.
4. Sawyer, C.N., 1969 Occoquan Reservoir Study. Metcalf and Eddy, Inc. for Commonwealth of Virginia, Water Control Board, April 1970.
5. Virginia State Water Control Board Minute No. 10, meeting of July 26, 1971.
6. GAO Report: Better data collection and planning is needed to justify advanced waste treatment construction. CED-77-12, December 12, 1976.
7. Horowitz, Jerome and Bazel, Larry. An analysis of planning for advanced wastewater treatment. The Vertex Corp., July 1977.
8. Surveys and Investigations Staff. A report to the committee on appropriations, U.S. House of Representatives on the Environmental Protection Agency Construction Grants Program. March 1979.

CASE STUDIES

THE EFFECTS OF TREATED SEWAGE DISCHARGE ON THE BIOTA OF PORT PHILLIP BAY, VICTORIA, AUSTRALIA

D.M. Axelrad[1], G.C.B. Poore[1,4], G.H. Arnott[1], J. Bauld[2,5], V. Brown[3], R.R.C. Edwards[1] and N.J. Hickman[1].

1. *Marine Studies Branch, Ministry for Conservation, P.O. Box 114, Queenscliff, Victoria 3225, Australia.*
2. *School of Microbiology, University of Melbourne, Parkville, Victoria 3052, Australia.*
3. *Department of Botany, University of Melbourne, Parkville, Victoria 3052, Australia.*
4. *Present address: National Museum of Victoria, Russell St., Melbourne, Victoria 3000, Australia.*
5. *Present address: Baas-Becking Geobiological Laboratory, CSIRO Fuel Geoscience Unit, P.O. Box 378, Canberra, A.C.T. 2601, Australia.*

ABSTRACT: The Werribee sewage-treatment farm contributes more than half of the total nitrogen and phosphorus input to Port Phillip Bay. This study attempted to determine the fate of these nutrients and their effect on the biota of the Bay. This was addressed by comparing community composition, biomass, productivity, or process rate in the Werribee area of the Bay with that in Bay areas more remote from nutrient discharge.

Rates of bacterial nitrification and denitrification were greatest in sediments closest to the sewage discharge point. Up to 15 percent of the inorganic nitrogen discharge may be lost via sequential nitrification-denitrification in a 4 km^2 area. Epibenthic microalgal biomass and productivity were found to be five times greater at Werribee than at a control station. The Werribee macrophyte community showed reduced species' diversity, dominance of fast growing opportunistic species, loss of large brown algal species, and occasional algal blooms, as compared to communities in nutrient-poor areas of the Bay. Classification analysis revealed offshore and nearshore groups of macrobenthos in a 2 km^2 area at Werribee; species diversity was greatest offshore. The Werribee offshore macrofauna was typical of that along the whole northwestern coast of the Bay. In summer, fish biomass at Werribee was equal to that found at stations remote from sewage discharge, however community composition differed. At Werribee, the nearshore fish community was dominated by juveniles and small species, while offshore older and larger fish were more abundant. Phytoplankton productivity decreased with increasing distance from sewage outfalls only during summer. Nitrogen is probably the nutrient critical to phytoplankton biomass production in the Bay but light and/or temperature may limit productivity over much of the non-summer period. Baywide phytoplankton productivity was similar to that of non-eutrophic coastal marine waters. Zooplankton standing crop was low compared to that for many estuaries and marine embayments. Densities of zooplankton in the Werribee area were highly variable, and crustacean species found at Werribee were also distributed throughout the Bay.

These findings suggest that sewage discharge has affected benthic more than planktonic communities, but that the measurable impact of the discharge is limited to a few hundred meters around the outfalls.

INTRODUCTION

Port Phillip Bay, located south of Melbourne, Victoria, Australia, (38°S, 145°E), is a 1,950 km^2, almost landlocked, body of tidal salt water (Figure 1). The Bay averages 13 m in depth with a large central section of 15-20 m. Sediments are sandy nearshore with sand-silt-clay predominating in the central section. Tides are semi-diurnal with a maximum amplitude of about 1 m. The Bay is vertically well-mixed such that prolonged salinity or temperature stratification does not occur over wide areas. Annual water temperature range is 9-22°C. Freshwater input, mostly from the Yarra River ($9 \times 10^8 m^3$/year), is approximately equivalent to net loss by evaporation so that the Bay is generally of oceanic salinity. Exchange of Bay waters with those of the Southern Ocean occurs through a 3-km-wide opening to Bass Strait; residence time of Bay water is estimated to be about 12 months (31).

The first permanent settlement in the Port Phillip Bay area was established about 1835. The Werribee sewage-treatment farm, established in 1897, is located on 11,000 hectares 35 km south-west of Melbourne, on the coast of Port Phillip Bay. Sewage from approximately 1.1 million of the Bay catchment's 2.8 million population is treated by land filtration, grass filtration and lagooning. The farm's treatment processes remove about half of the influent nitrogen and phosphorus before discharging the treated sewage. Approximately $8 \times 10^6 m^3$ of treated sewage containing 4×10^6 kg of N (75% as NH_4^+) and 1×10^6 kg of P (85% as $PO_4^{\equiv}$) is discharged annually from the farm to the Bay. This represents about 50 percent of the total N and 60 percent of the total P input to the Bay (31). The treated sewage is discharged through five main drains along the northwest coast of the Bay, an area in which the Bay's 5 m depth contour occurs between 1 and 3 km offshore.

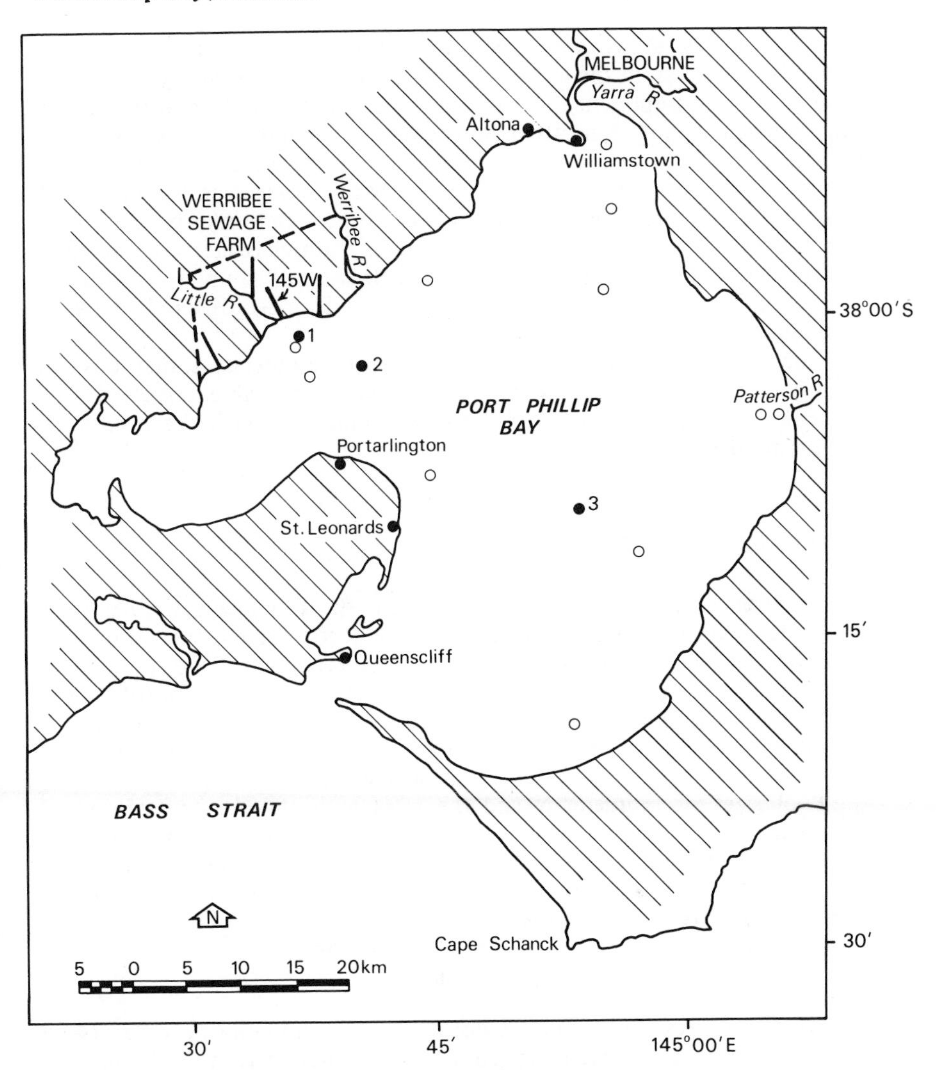

FIGURE 1. Port Phillip Bay sampling stations: closed circles (1, 2, and 3) are phytoplankton stations; open circles are zooplankton stations.

To determine the fate of nutrient discharge and its effect on the biota of Port Phillip Bay, the Victorian Ministry for Conservation, Marine Studies Branch, in 1975 initiated the Werribee Study. Data presented here are taken from this study and from previous investigations of a more general nature on Port Phillip Bay.

The Werribee Study comprised several components which addressed the general objectives in different ways. The bacterial nutrient cycling study was the only component to specifically address the question of the fate of the nutrient input. The other components attempted to determine the effect of nutrient discharge on the Bay's biota by comparison of species composition, biomass or productivity near the sewage farm with that of communities remote from the farm. The control station(s) were selected on the basis of similarity of environmental features (other than nutrient status) with those of the Werribee area. Because of sampling constraints and the scale of effect expected, control areas differed between components.

For each component of the study the specific objectives, methods and results are given briefly with a discussion of probable effects of the Werribee sewage farm discharge and a comparison with the results for other marine systems.

BACTERIAL NITROGEN CYCLING

Nitrogen entering Port Phillip Bay from the Werribee sewage outfalls may be made unavailable for primary producers through the action of a number of biological or physicochemical processes. Two potentially important bacterial processes are nitrification and denitrification. Nitrification is the oxidation of ammonium to nitrite or nitrate. During denitrification, nitrate is converted to gaseous forms of nitrogen, some of which may subsequently be lost to the atmosphere (34, 36).

Pool sizes of inorganic-N and the rates of bacterial nitrification and denitrification were determined for sediments and water from the nearshore environment adjacent to the 145W drain of the Werribee sewage-treatment farm in order to estimate the potential magnitude of nitrogen loss from this environment.

Methods

At seven sites on a transect normal to the shoreline near the 145W drain, water was collected and sediment interstitial-water was obtained by squeezing surface sediments (45). Samples were passed through membrane filters (0.45μm) prior to analysis for NH_4^+, NO_2^- and NO_3^- (48, 49).

Nitrification and denitrification rates were measured (in March and May, 1976) at a nearshore and an offshore station 50 m and 200 m distance respectively from the drain. Rates were determined by following changes in the concentrations of NH_4^+, NO_2^- and NO_3^- during incubations of samples of water, and sediment and overlying water (7). Incubations were carried out in the dark at in situ temperatures and without shaking or agitation. Incubation times were short (1 - 2 h) to prevent alterations in kinetics due to substrate limitation. N-Serve, a specific inhibitor of autotrophic ammonium oxidation (10,11), was used in parallel incubations. Denitrifying bacteria are unaffected by N-Serve at the level emplo ed in this study (1-2 ppm) (19). The method was confirmed using $^{15}NH_4$ as a substrate (7).

Results

Chemical analyses of interstitial water showed contrasting distribution of NH_4^+ and NO_3^-. NH_4^+ concentration was greatest about 125 m from the outfall while interstitial NO_3^- concentration was low close to the outfall and greatest about 200 m from the outfall (Table 1).

TABLE 1. Concentrations of NH_4^+, NO_2^-, and NO_3^- in interstitial waters from the top 2 cm of sediment along a transect originating at the Werribee 145W sewage outfall.

Distance from outfall (m)	Interstitial concentration (µg-at.N l^{-1})		
	NH_4^+	NO_2^-	NO_3^-
50[1]	506	2.7	13.1
100[1]	726	2.0	23.8
125[1]	916	5.8	35.1
150[2]	546	6.8	90.0
200[2]	422	6.8	116.0
250[2]	537	12.2	106.0
350[2]	448	8.8	56.5

[1] Silty sediment

[2] Sandy sediment

Nitrification and denitrification rates in sediments at the nearshore site were about twice those measured offshore (Table 2). The movement of ^{15}N label was consistent with loss from the $(NO_2^- + NO_3^-)$-N pool by sequential nitrification - denitrification (7). While both nitrification and denitrification proceeded at measurable rates within waters of the 145W drain proper, rates were below detection limits in the nearshore water column (7).

TABLE 2. Rates of nitrification and denitrification in surface sediments at two sites near the Werribee 145W sewage outfall.

Rates (mg-at.N m^{-2} h^{-1})	Nearshore	Offshore
Nitrification	0.25	0.12
Denitrification	0.33	0.14

Discussion

The difference in nitrification and denitrification rates between the two sites presumably refects the decreasing substrate concentrations with increasing distance from the outfall. Other experiments at sites in proximity to the 145W outfall (7) showed that the deniftrification rates (0.33-0.71 mg at.N $m^{-2}h^{-1}$) were slightly lower than those reported for sediments in a tropical mangrove estuary receiving treated sewage effluent (33).

Using these experimental data and several assumptions (7), it is possible to assess the potential nitrogen loss via denitrification in the area of the 145W drain. It is estimated that up to 15 percent of the inorganic-N discharged from the Werribee sewage-treatment farm through the 145W and Little River drains is lost via bacterial denitrification in a 4 km^2 area adjacent to the outfalls. This estimate assumes the absence of nitrogen fixation, for which there are no data presently available. The exact rates of loss will remain a matter for speculation until data are available for nitrogen fixation rates and for the seasonal and spatial variation in rates of nitrification and denitrification. However, it appears reasonable to conclude that nitrification-denitrification provides a quantitatively significant pathway for the removal of nitrogen discharged to Port Phillip Bay from the Werribee sewage-treatment farm.

EPIBENTHIC MICROALGAE

Epibenthic microalgal productivity and biomass were determined for shallow-water (2 m) stations, one offshore of the Werribee sewage discharge, and another (St. Leonards) remote from significant nutrient discharge (Figure 1).

Methods

Epibenthic microalgal productivity was determined by a ^{14}C method (6, 30). SCUBA divers using transparent plastic corers collected sediment samples. The corer bottoms were sealed with opaque caps such that the sides of the 1 cm sediment cores were just covered. After ^{14}C addition, samples in the corers were incubated for half a day at ambient seawater temperature in sunlight beneath a neutral density screen. The screen reduced light intensity by 50 percent so as to produce an intensity approximating that measured at 2 m depth at the sampling stations. Epibenthic microalgal biomass was determined by a fluorescence method (6, 48). Salinity and nutrient concentrations of waters immediately overlying the sediments were also determined (48, 49). Samplings were repeated at six-weekly intervals over one year (1977-1978).

Results

Although salinity did not differ between the two sampling sites, concentrations of nutrients, particularly of nitrogen, were much higher in water overlying sediments at Werribee than at St. Leonards (Table 3).

Mean chlorophyll a concentration in the surface centimeter of sediment showed little seasonality but was significantly greater at Werribee than at St. Leonards (Table 3). Productivity was greatest in summer at both stations (Figure 2), that at Werribee significantly greater ($P<0.01$) than that at St. Leonards for each sampling over the year. Annual productivity calculated by integrating daily productivity values over the year was five times greater at Werribee than at St. Leonards (Table 3).

Discussion

It is clear that the nutrient discharge from the Werribee farm results in elevated epibenthic microalgal biomass and productivity, increasing each about five-fold relative to the control area. The value of 236 ± 19 g C $m^{-2}yr^{-1}$ obtained for epibenthic microalgal productivity at Werribee is the highest known to be reported for shallow-water marine environments: 81 g C $m^{-2}yr^{-1}$ (29), 101 g C $m^{-2}yr^{-1}$ (9), 116 g C $m^{-2}yr^{-1}$ (18), and 143-220 g C $m^{-2}yr^{-1}$ (35). These algae may therefore have an important impact on higher trophic levels of the food chain.

MACROPHYTES

The effect of nutrient discharge on macrophyte community diversity and composition was investigated by comparing reef communities near Werribee with those in nutrient-poor areas of Port Phillip Bay.

Methods

Three basalt reef sampling sites were selected: Werribee, Williamstown, and Portarlington (Figure 1). Transects and permanent quadrats in the intertidal and subtidal zones were sampled throughout the year

TABLE 3. Annual ranges in salinity, nutrient concentrations, epibenthic microalgal biomass, and productivity at Werribee and St. Leonards stations.

	Werribee	St. Leonards
Salinity $^{o}/oo$		
minimum	34.16	34.39
maximum	35.44	35.44
$NH_4^+ + NO_2^- + NO_3^-$ (μg-at. N l^{-1})		
minimum	3.1	0.1
maximum	33.4	0.4
$PO_4^{\equiv}$ (μg-at.P l^{-1})		
minimum	3.66	0.90
maximum	7.88	2.16
Biomass (mg chl *a* m^{-2})		
minimum	60	12
maximum	130	28
Productivity (mg C m^{-2} day^{-1})		
minimum	290	50
maximum	1060	340
Annual productivity (g C m^{-2}) (95% confidence limit)	236 ± 19	46 ± 5

(1978-1979) at Werribee. Data resulting from this sampling were compared with existing information on algal communities at the other sites (22, 25).

Results

The intertidal zone at Werribee was characterized by *Centroceras clavulatum* (throughout the year) and *Ulva lactuca* and *Enteromorpha*

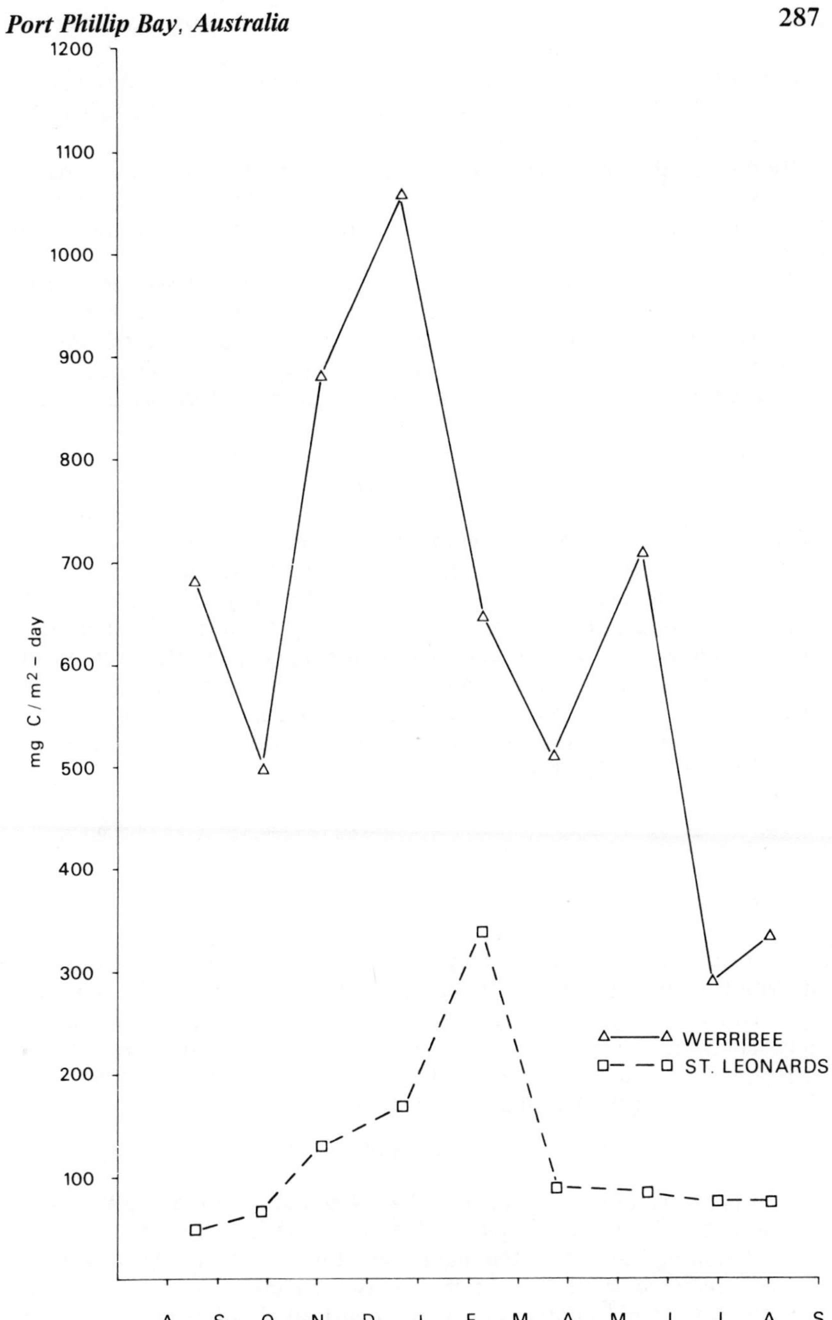

FIGURE 2. Seasonal variation in epibenthic microalgal productivity at Werribee and St. Leonards.

compressa (in summer-autumn). Subtidally, community structure varied little down to 2 m depth and was dominated by *Caulerpa remotifolia, U. lactuca* and *Caulerpa longifolia.*

Blooms of the green alga *Cladophora* were observed twice at Werribee during the study period (1976-1979) in spring, while the opportunistic species *Ulva lactuca* was consistently abundant at Werribee on a variety of substrates.

A comparison of macrophytic communities at the three basalt reefs showed 1) lower species diversity in all zones at Werribee compared to the other two sites, and 2) absence of large brown algae *Ecklonia radiata* and *Sargassum* spp., and the red alga, *Corallina officinalis* at Werribee; these algae are important zone-forming species in the sub-littoral zone at other localities.

Discussion

The Werribee macrophyte community showed many of the features of similar communities in nutrient-rich environments elsewhere in temperate regions: 1) reduction in species diversity particularly in the Phaeophyta and Rhodophyta (14, 26, 37, 52); 2) a high percentage of fast-growing species such as *Ulva lactuca* and *Cladophora* sp. (26, 37); 3) development of a structurally simple community with little vertical zonation (26); and 4) loss of large brown algae of the orders Laminariales and Fucales (14, 24, 25). It appears that lack of suitable substrate limits macrophyte distribution in the Werribee area.

MACROBENTHOS

The objective of the survey of macrofauna was to measure the distribution and abundance of species in the area of assumed influence of the treated-sewage discharge and to correlate these with selected physicochemical variables. A comparison was also made between areas offshore from the sewage drain and areas of similar sedimentary nature but remote from the Werribee discharge.

Methods

Thirty-six stations in an area of 2 km^2 adjacent to the 145W drain of the Werribee farm were sampled in November, 1975. The stations were on seven radiating transects extending up to 2 km from shore. At each station a diver-operated suction sampler was used to take a single macrofauna sample of 0.05 m^2 in area, down to a depth of about 30 cm. The sample was retained in a 1 mm mesh nylon bag from which it was washed into 10 percent buffered formalin.

Macrobenthic species were sorted from the sediment, identified to species and counted. Biomass (ash-free dry weight) was determined for the major feeding types. Samples of overlying water and of interstitial water were taken from each station and analyzed for NH_4^+, NO_2^-, NO_3^-, $PO_4^{\equiv}$ and salinity. Sediment types (mean particle size, sorting coefficient, percentages of various size fractions, percentage carbonate) were determined for each station. Methods have been described in more detail elsewhere (40).

Results

In situ examinations and sediment analysis revealed a relatively flat bottom divided into a nearshore zone (within the 1 m depth contour and less than 400 m from the shore) of well-sorted, sandy sediments and a patchy offshore zone of more poorly-sorted sediments, predominantly of sand, but containing a significant shell fraction. Chemical variables indicated high between-station variance but, in general, clear trends with increasing depth were apparent. Salinity increased with increasing depth from a minimum of 30.86 $^o/oo$ nearshore to a maximum of 35.50$^o/oo$ offshore. Concentrations of all nutrients in both interstitial and overlying water were lowest offshore, those of nitrogen by as much as an order of magnitude (40). For example, mean concentrations of nutrients measured in sediment interstitial waters within 150 m of the sewage drain and beyond 1200 m from the drain were, respectively: $PO_4^{\equiv}$, 34.0 and 5.5 μg-at.P l^{-1}; NO_3^-, 4.5 and 1.4 μg-at.N l^{-1}; NH_4^+, 128.0 and 16.4 μg-at.N l^{-1}. For nutrients in waters overlying the sediments, corresponding concentrations were: $PO_4^{\equiv}$, 7.6 and 3.3 μg-at.P l^{-1}; NO_3^-, 4.7 and 0.5 μg-at.N l^{-1}; NH_4^+, 50.9 and 4.0 μg-at.N l^{-1}.

Of the 246 species recorded, about half were deposit-feeders. However, most of the biomass was contributed by suspension-feeding molluscs. Hierarchical classification of the 36 stations using all species as attributes, Canberra metric similarity coefficient, flexible sorting (8, 40), revealed only two station-groups, the first including all but one of the stations within the 1 m depth contour (within 400 m of shore) and the second including stations at greater depths up to 2 km from shore.

Although the mean number of individuals per station and mean evenness were scarcely different between the nearshore and offshore station-groups, the mean number of species and mean diversity were higher offshore than they were nearshore (Table 4). These differences were supported by differences in faunal composition revealed by inverse classification of species and by partial correlation analysis (40). The nearshore zone was characterized by the mysid, *Gastrosaccus dakini*, the amphipods *Gammaropsis* sp., *Exoediceros* sp. and *Limnoporeia* sp., and the bivalve *Notospisula trigonella*. All of these species are known from estuarine environments elsewhere in Victoria (39).

The fauna offshore was dominated by the polychaetes *Mediomastus* sp., *Eunice antennata, Asychis glabra, Dorvillea australiensis*, the ostracods

TABLE 4. Means and standard deviations of community statistics for 17 nearshore and 19 offshore macrobenthos sampling stations (0.05 m^2 samples) at Werribee.

Statistic	Nearshore stations	Offshore stations
Number of individuals, $\log_{10}N$	2.43 ± 0.33	2.60 ± 0.15
Number of species, S	30.0 ± 12.2	50.1 ± 12.1
Diversity, H'	2.3 ± 0.5	2.7 ± 0.4
Evenness, J'	0.7 ± 0.1	0.7 ± 0.1

Euphilomedes sp. and *Rutiderma* sp., the isopod *Cirolana woodjonesi*, and the cumacean *Dimorphostylis cottoni*. The bivalve *Katelysia rhytiphora* contributed greatest biomass.

In spite of elevated nutrient concentrations in the offshore Werribee region, its fauna was typical of that found all along the northwestern coast of Port Phillip Bay. The numbers of individuals, numbers of species and dominant species in samples taken in previous surveys along 6 km of the sewage farm's coast were similar to those reported here (31).

Another survey, this time of mollusc numbers and biomass in Port Phillip Bay (41), recognized the benthos of the whole 45 km northwestern coast (between 5 and 10 meters deep) as belonging to a region of high species diversity. Unpublished data on this region show that the mean number of macrobenthic species (all taxa) per 0.1 m^2 sample range from about 40 to 100. No effect due to the farm was evident at these offshore stations although the fauna and sediments were extremely heterogenous.

The station nearest to the sewage drain differed from the remainder in possessing a particularly dense fauna dominated by deposit-feeding polychaetes *Polydora socialis, Pseudopolydora paucibranchiata* and corophiid amphipods *Corophium* sp. and *Gammaropsis* sp. (40). Other species of deposit-feeding polychaetes and oligochaetes were common at this station. Stations closest to this sample did not share its special properties and illustrated the heterogeneity and patchiness of the area.

Discussion

Though it is difficult to define the limit of the measurable effect of the sewage discharge, we tentatively conclude that the discharge modifies the nearshore benthos in a patchy fashion for about 300 m out from the drain. Subsequent studies have found that particularly high densities of spionid polychaetes prevail around the drain in the intertidal zone (J.D.

Kudenov, personal communication).

Three main components of the sewage farm's discharge are likely to affect macrofauna community structure, either directly or indirectly: dissolved nutrients, particulate organic matter, and freshwater.

Dissolved nutrients have little direct effect on macrofauna but increase its food supply by stimulating primary production. We have shown elsewhere in this paper that this is certainly true for epibenthic microalgae. Epibenthic microalgae are an important food source for many suspension-feeders and surface deposit-feeders such as spionid polychaetes and corophiid amphipods, both of which were abundant near the drain.

Particulate organic matter, the form in which about 20 percent of the farm's nitrogen and phosphorus is discharged, provides another food supply for benthos.

Fluctuating salinity resulting from the interaction of the tide and wind on direction and rate of freshwater discharge places a significant stress on the fauna around the sewage drain. That the salinity fluctuates is reflected in the occurrence of several estuarine species known from other Victorian estuaries (e.g. the Werribee River, Yarra River) (39). While the Little and Werribee Rivers must contribute a significant fraction of the fresh water to the Werribee area, much of the flow comes from the sewage farm.

Severe organic pollution of zoobenthic communities results in marked depletion of the fauna. Species populations sensitive to pollution disappear at the onset of stress while population densities of more tolerant species increase (1). Abiotic zones of different sizes (100 m to several kilometers radius) have been found around sewage outfalls in Marseilles, France (4), Kingston Harbour, Jamaica (51) and Ensenada Bay, Mexico (27). Larger areas of much modified fauna surrounded the abiotic zones in all cases. The effect of the Werribee sewage discharge on macrofauna is much less severe than that reported in these studies.

FISH COMMUNITIES

Species composition and biomass of the fish community along the Werribee coast were compared with those at similar but nutrient poor environments elsewhere in Port Phillip Bay. Because of the constraints on sampling, the Werribee nearshore and Werribee offshore fish communities were compared with communities at Altona and St. Leonards respectively (Figure 1). Estimates of food consumption and growth of some species were used to elucidate pathways of energy flow in the Werribee area.

Methods

Fish populations were sampled offshore with a 13 m Siebenhausen otter trawl and nearshore with a 180 m beach seine net. Otter trawls, 1 km from shore, covered between 3,800 and 5,200 m^2 depending on wind and tides. Trawling was undertaken in late afternoon and after dark at monthly intervals over a year (1977-1978). The beach seine net covered an area of 8,000 m^2 and was worked within 50 m of shore. Biomass estimates from netting were corrected for gear efficiency (23).

Food intake of dominant fish species was estimated by separating fish gut contents into taxonomic groups and determining wet weights. Growth was estimated from otoliths, vertebral sections and population length frequency distributions (21).

Metabolic rate of fishes was estimated through oxygen consumption measured in the laboratory. The values obtained were used to calculate food requirements (15). The accuracy of these calculations was checked through conversion efficiencies of food to growth, which ranged from 15 to 25 percent depending on average age (16).

Results

In summer the nearshore and offshore fish biomass found at Werribee (6.7 ± 2.4 and 6.5 ± 3.7 g m^{-2} live weight) did not differ significantly from the biomass found at nearshore (5.3 ± 1.9 g m^{-2}) and offshore (8.6 ± 4.0 g m^{-2}) areas at the control stations. In winter, fish biomass declined markedly at both inshore sites with 0.3 ± 0.1 and 2.1 ± 0.1 g m^{-2} live weight found at Werribee and Altona, respectively. In the Werribee offshore zone, the dominant species was the porcupine fish *Atopomycterus nichthemereus*, which represented 60 percent of total biomass. Offshore at St. Leonards, this species represented 20 percent of total biomass. In the nearshore, the euryhaline species, *Aldrichetta forsteri* (Yellow-eye mullet) and *Rhombosolea tapirina* (Greenback flounder) made up 85 percent of fish biomass at Werribee but only 56 percent at Altona.

Food intake data for dominant species from the nearshore zone at Werribee indicated that of a total 1.76 g live weight m^{-2} day^{-1} food intake, 1.32 was composed of epibenthic Crustacea. Amphipods comprised 55 percent of this crustacean total. Polychaetes comprised 0.44 g m^{-2} day^{-1} of the food intake. For Werribee offshore fish species, the dominant food resource was brachyuran and caridean crustaceans. Food requirement per unit biomass of dominant fishes in the nearshore Werribee area was six times that of the offshore Werribee area.

Discussion

It seems likely that the lower winter values of fish biomass nearshore in the Werribee region can be attributed to the high rate of freshwater input from the drains during winter. The fresh water, which is also colder than ambient sea water, makes the region less suitable even for euryhaline species. In summer the Werribee region attracts juveniles of several species, notably the flounder *Rhombosolea tapirina* feeding on amphipods and polychaetes which are abundant near the drain.

It appears from dietary analysis data that epibenthic crustaceans are the dominant pathway of energy flow between the benthos and nearshore fish populations at Werribee. This is in spite of the fact that bivalve molluscs, notably *Notospisula trigonella*, contribute most to benthic biomass (40). Molluscs were exploited only occasionally by adult fish and by the sand crab *Ovalipes australiensis*.

At Werribee the combined food requirements per unit biomass of dominant fishes in the nearshore zone were six times those in the offshore zone. This difference is due to population age structure. For example, the dominant nearshore species, *Aldrichetta forsteri*, has a similar biomass to *Atopomycterus nichthemerus* in the offshore zone but a food requirement ten times as great. This disparity results from the fact that the population of *Aldrichetta* nearshore is composed of fish less than one year old which have high metabolic and specific growth rates; *Atopomycterus* populations are largely adult fish (unpublished data). In general, nearshore populations are composed mainly of juveniles and of small species whereas offshore populations comprise older, larger fish.

PHYTOPLANKTON

The effect of the nutrient discharge on Port Phillip Bay phytoplankton was assessed by measuring biomass and productivity at three stations along a transect leading away from the area of the Werribee sewage treatment farm's discharge to the Bay.

Methods

Phytoplankton productivity was determined at three Bay stations (1, 2 and 3) which were 2, 5 and 30 km from the point of treated sewage discharge (Figure 1). Productivity was measured monthly over a year (1975-1976) at several depths at each station using 2-3 hour in situ incubations and the ^{14}C method (5, 47, 48). Sea water samples collected concurrently were analyzed for chlorophyll a, salinity, $PO_4^{\equiv}$ NO_2^-, NO_3^- and NH_4^+ (48, 49).

Results

Phytoplankton productivity ranged annually from 0.54 to 16.53, from 0.42 to 8.32, and from 0.03 to 7.38 mg C m^{-3} hr^{-1} at Stations 1, 2 and 3, respectively. As the stations differed in depth, productivity was integrated from surface to the maximum common station depth, 3 m, to facilitate comparison of productivity between stations. Productivity of this 3 m water column reached its annual maximum value in either November, December or January (Figure 3). During this three month period there was a pronounced decrease in phytoplankton productivity with increasing distance from the sewage discharge. However, for much of the remainder of the year productivity values did not differ greatly between stations. Annual phytoplankton productivity was estimated by multiplying daylight hours per month by production per hour for each month's sampling and summing over 12 months. Productivity in the 3 m water column was 73, 43 and 33 g C m^{-2} yr^{-1} at Stations 1, 2 and 3, respectively.

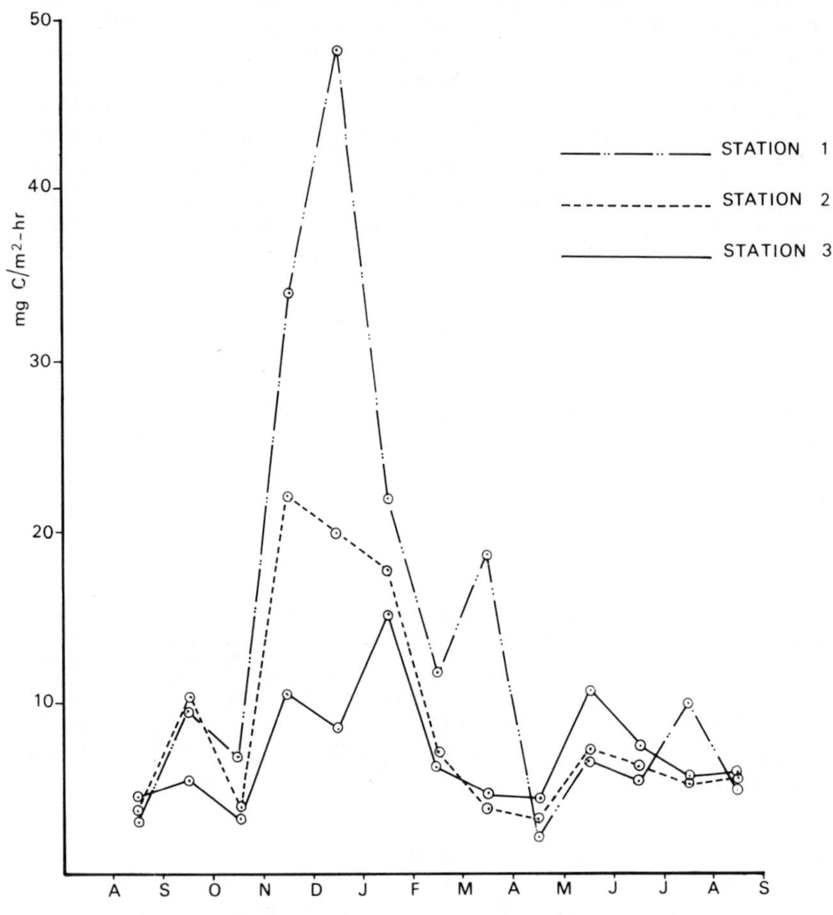

FIGURE 3. Seasonal variation in phytoplankton productivity within the surface to three meter depth water column for Stations 1, 2, and 3, Port Phillip Bay.

Chlorophyll a concentrations in the 3 m water column were similar at the three stations throughout the year, except in November when substantially greater concentrations were found at Stations 1 and 2 and values were observed to decrease with increasing distance from the sewage outfalls (Figure 4).

Nitrogen and phosphorus concentrations in the 3 m water column showed no apparent seasonal trends. Nutrient concentrations at Station 1, close to the sewage outfalls, varied erratically from month to month. Concentrations at Station 3 were relatively stable. Phosphate concentration at Station 1 was on all occasions greater than that measured at Stations 2 and 3. Nitrogen concentrations were frequently, but not always, highest at stations closer to the sewage outfalls (Table 5).

TABLE 5. Nutrient concentrations at three phytoplankton sampling stations in Port Phillip Bay.

	Stations		
	1	2	3
$PO_4^{\equiv}$ (μg-at.P l^{-1})			
minimum	2.28	1.53	1.54
maximum	5.19	3.28	2.00
NH_4^+ (μg-at.N l^{-1})			
minimum	0.3	0.2	0.2
maximum	15.8	8.0	0.6
$NO_3^- + NO_2^-$ (μg-at.N l^{-1})			
minimum	0.1	0.0	0.0
maximum	3.9	2.1	0.1

Discussion

Although annual phytoplankton production was inversely proportional to station distance from the sewage outfall, the disparity in productivity was not evident throughout the year. The observed seasonal variation in phytoplankton productivity can be better understood by examining the annual cycle of phytoplankton productivity to chlorophyll a ratio. This ratio at light saturation, termed assimilation ratio, is widely accepted as an

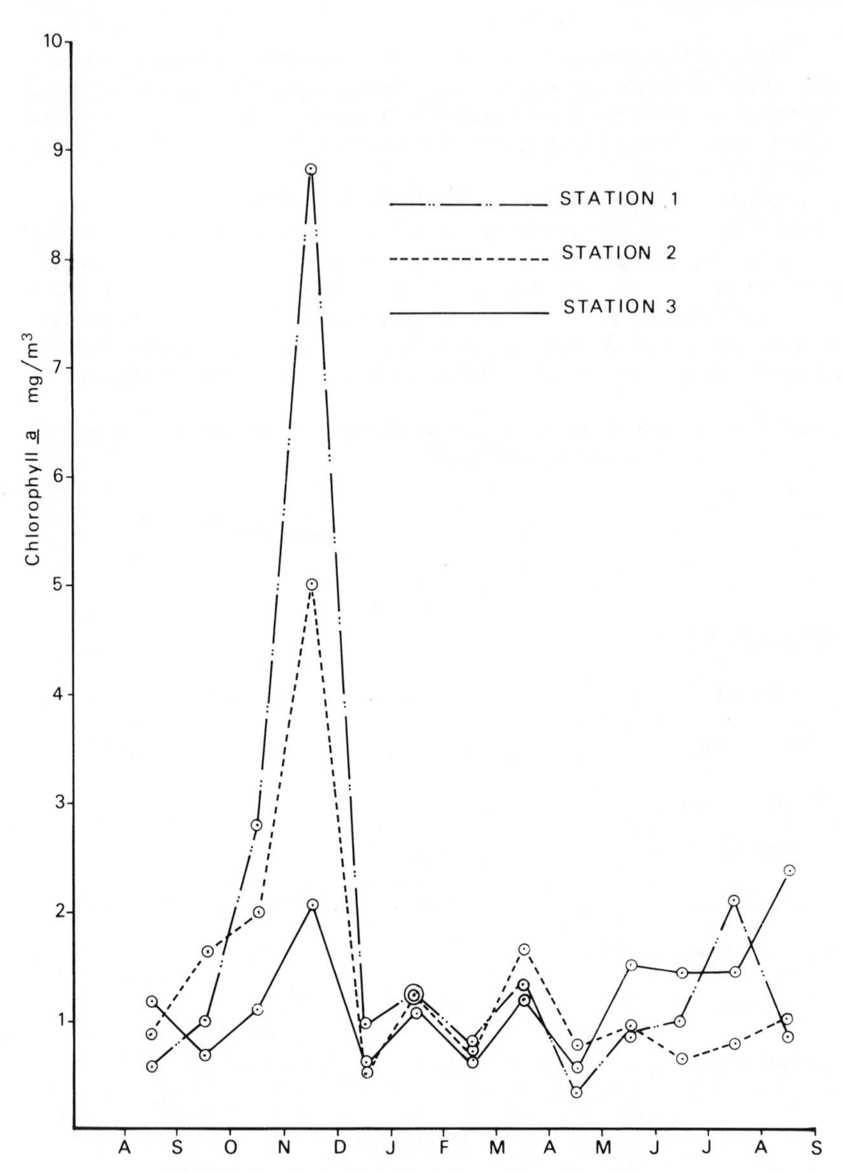

FIGURE 4. Seasonal variation in average chlorophyll a concentration in the surface to three-meter depth water column for stations 1, 2, and 3, Port Phillip Bay.

indicator of the physiological status of phytoplankton with respect to nutrients (12, 50). Assimilation ratios for Port Phillip Bay phytoplankton have been approximated by taking the maximum productivity to biomass ratios found within the water column for each sampling (Figure 5). Phytoplankton in nutrient-rich waters have been found to have assimilation ratios generally exceeding five (13). However, assimilation

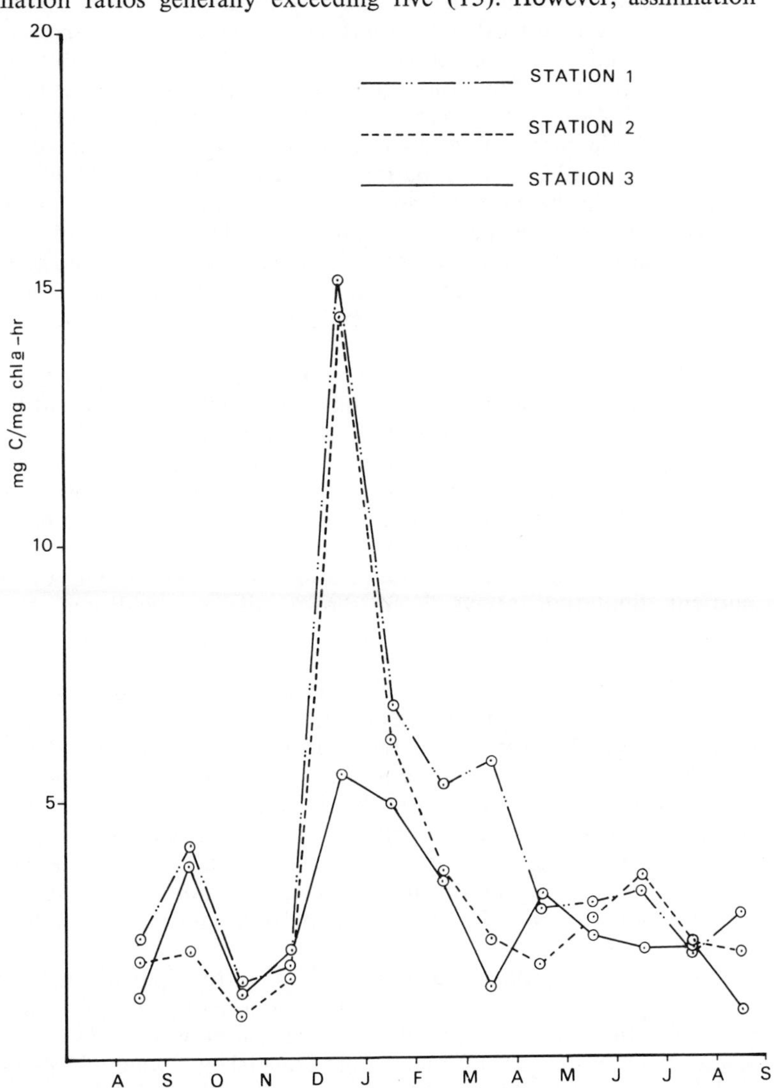

FIGURE 5. Seasonal variation in maximum phytoplankton productivity to biomass ratio at Stations 1, 2, and 3, Port Phillip Bay.

ratios greater than five occurred in Port Phillip Bay at Station 1 only from December through March despite high nutrient concentrations at this station at other times of the year. Furthermore, it was also within this December through March period that assimilation ratios at Station 1 were significantly higher than at Station 3, although Station 1 had higher nutrient concentrations than Station 3 during other months. It is thus evident that a factor other than nutrient availability was responsible for the low assimilation ratios measured at other times. One possible factor is temperature; assimilation ratios are known to vary with water temperature (17). In addition, during seasons of low light intensity, phytoplankton photosynthetic carbon fixation rates may not be light saturated even in surface waters (28). This situation is evidenced for Port Phillip Bay as maximum phytoplankton productivity-to-biomass ratio frequently occurred at 3 or 5 m depth in summer months, but most often at the water's surface in non-summer months. These facts suggest that phytoplankton productivity in the Bay is nutrient-limited between December and March but light and/or temperature-limited over the remainder of the year.

Productivity, and productivity-to-biomass ratio data suggest that for at least the summer months, phytoplankton productivity is nutrient limited. It is also apparent that with increasing distance from the point of sewage discharge, inorganic nitrogen concentration diminishes to a greater extent than does $PO_4^{\equiv}$ concentration (Table 5). This is clearly depicted in the regression of inorganic nitrogen on inorganic phosphorus (Figure 6). This regression ($r = 0.72$; $P < 0.01$) shows depletion of nitrogen at a point where phosphorus concentration is about 1.5 μg-at P l^{-1}. Indeed, phytoplankton nutrient limitation assays have shown nitrogen addition, but not phosphorus addition, results in increased phytoplankton biomass production in Bay waters (31).

The similarity in chlorophyll a concentrations at the three stations was evident despite the higher nutrient concentrations and resultant elevated summer phytoplankton productivity at the station nearest the sewage outfall. Accumulation of phytoplankton biomass in the waters near the outfall may have been prevented by several factors. However, the rapid movement of water past the point of sewage discharge to the Bay (31) appears to be the probable explanation. If the Bay mixing rate was high compared to phytoplankton division rate, the chlorophyll concentration near the outfalls would be maintained at levels similar to those for other areas of the Bay.

Phytoplankton production integrated to bottom (20 m) at Station 3 was 111 g C m^{-2} yr^{-1} with negligible productivity at 20 m depth. This value is within the range found in coastal marine waters (38), and is far lower than that measured for eutrophic Long Island Sound of 380 g C m^{-2} yr^{-1} (42, 44). Insofar as Station 3 is representative of Port Phillip Bay remote from the sewage discharge, the discharge has not resulted in high phytoplankton productivity in the Bay as a whole.

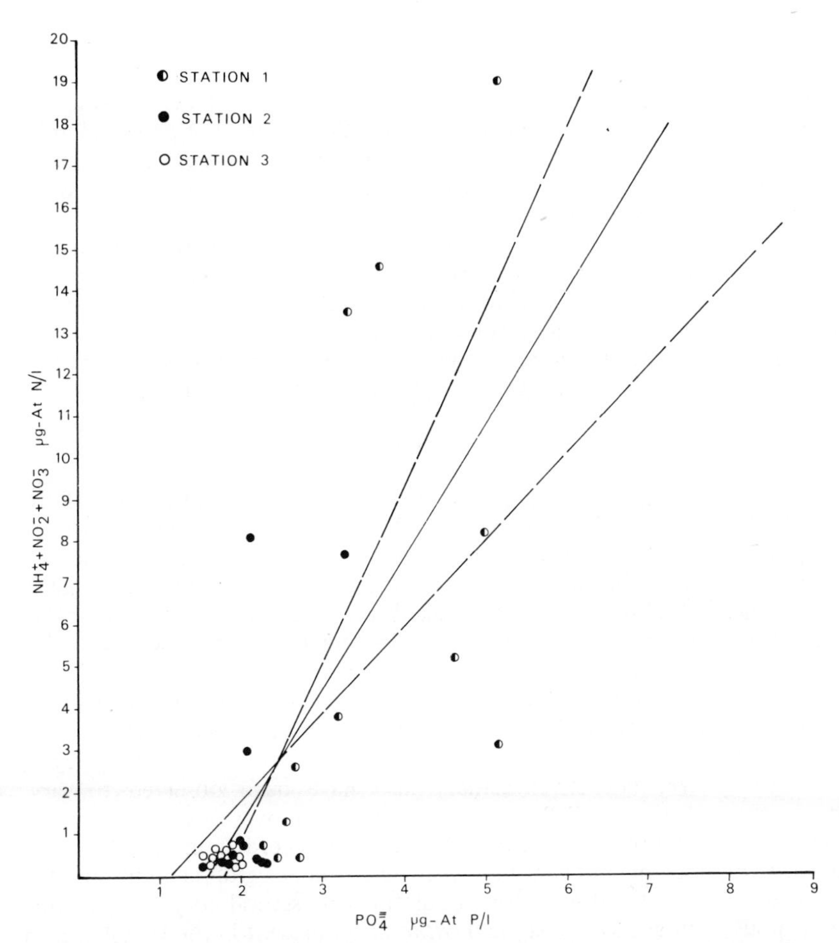

FIGURE 6. Inorganic nitrogen concentration as a function of inorganic phosphorus concentration in waters of Port Phillip Bay. Dashed lines represent 95 percent confidence limits to regression line.

ZOOPLANKTON

A specific study of the zooplankton of the Werribee region was not undertaken at the time of the Werribee Study. However, information is available from an earlier study of the fauna of Port Phillip Bay (31). These data have been used to assess the impact of the Werribee sewage discharge on Bay zooplankton communities.

Methods

Over 100 quantitative zooplankton samples were collected from 11 stations (Figure 1) throughout Port Phillip Bay during the period March, 1969 - February, 1970. Stations were sampled once during each of eight consecutive six-week sampling periods; however, not all stations were operated in each period (2).

Zooplankton was collected by oblique tows from bottom to surface using Clarke-Bumpus nets. Mesh apertures ranged from 159 to 262 μm. As a result of this variation, smaller species such as the cyclopoid *Oithona* sp. could be under-represented in some samples.

Species identifications and sample counts were restricted to the Crustacea, which was by far the most abundant planktonic group.

Results

The annual mean standing crop of the planktonic Crustacea in Port Phillip Bay was 3,690 individuals/m^3. Mean densities for individual Port Phillip Bay stations were fairly uniform, six of the eight most frequently sampled stations having densities ranging between 3,000 and 5,300 individuals/m^3. However, the mean planktonic crustacean densities for the nearshore stations off the Little and Patterson Rivers appeared to be influenced by river discharges. Four of the five samples from the Little River nearshore station had a density less than 800 individuals/m^3; however, one particularly high density (40,000/m^3) was sufficient to produce the highest station average (8,400/m^3). Zooplankton densities in the nearby Werribee River estuary also have been observed to vary markedly with time (3, 32), and are dependent on river flow. A maximum density of 300,000 individuals/m^3 has been recorded in the Werribee River (32), and this nutrient-rich river may, under certain conditions, act as a source of recruitment for populations of several euryhaline marine copepods such as *Acartia* sp. and *Oithona* sp. present in the Werribee area of Port Phillip Bay.

In agreement with many other studies of this type, the dominance of Copepoda (81.8 percent of bay-wide mean crustacean numbers) was very marked. A total of 22 species of calanoid and cyclopoid copepods have been identified from the Bay as a whole. Of these, *Acartia* sp. (42.4 percent of copepod numbers) and *Paracalanus parvus* (39.7 percent) were the two dominants. The other major species in order of abundance were *Oithona* sp. (7.9 percent), *Tortanus barbatus* (4.3 percent), *Oithona similis* (2.7 percent), *Gladioferens inermis* (1.6 percent) and *Calanus australis* (1.1 percent). All of the common copepod and cladoceran species were found to be widely distributed throughout the Bay. The only significant spatial variation in species composition resulted from the presence of four minor stenohaline species of copepods - *Acartia danae, Centropages australiensis, Ctenocalanus vanus* and *Calanus minor* - in areas closest to Bass Strait.

Discussion

Although the available Port Phillip Bay zooplankton data are somewhat limited in scope and were not specifically obtained to examine the effect of nutrient discharges on the Bay communities, several conclusions may nevertheless be drawn. First, the annual mean standing crop of zooplankton in Port Phillip Bay is low relative to that of many coastal marine systems. For example, St. Andrew Bay (40,100/m^3), Long Island Sound (62,000/m^3), Block Island Sound (16,000/m^3) and Tisbury Great Pond (52,000/m^3) all have zooplankton densities greater than those found in Port Phillip Bay. Lower values than those for Port Phillip Bay have been observed most often in more offshore waters, e.g. Tortugas (1,500/m^3), slope waters off Georgia (900/m^3) and New Jersey (310/m^3) (20). Thus, Port Phillip Bay zooplankton densities are not indicative of a highly eutrophic environment. Second, zooplankton densities in the high nutrient concentration waters of the Werribee zone off the Little River were usually lower than those elsewhere in the Bay, but were highly variable and included the highest density recorded. Third, the common crustacean species were all distributed throughout the Bay, including the Werribee area.

CONCLUSIONS

The effect of the discharge from the Werribee sewage-treatment farm on Port Phillip Bay biota differs between the different trophic levels of the ecosystem. Phytoplankton productivity has increased moderately in the vicinity of the farm's discharge but phytoplankton biomass is rarely elevated in this area. This may result from Bay water mixing rate being high relative to phytoplankton division rate. If this is the case, the impact of nutrient discharge on the phytoplankton will be Baywide, rather than confined to the Werribee area. However, despite the large discharge of nutrients from the farm to the Bay, phytoplankton productivity at stations remote from the discharge remains at a level commonly found for non-eutrophic coastal marine waters.

While the impact of the treated sewage discharge on plankton communities in the Werribee area was small, it appears that benthic communities in this area were significantly affected by the discharge. This is in part due to the treated sewage being discharged to an area of the Bay where water depth does not exceed 5 m within 1 km of shore. Epibenthic microalgal productivity and biomass in the Werribee area were increased five fold compared to an area remote from the nutrient input. This high-productivity food source enabled development of a rich zoobenthic community in the nearshore Werribee area which in turn provides a rich feeding area for juvenile fish. However, nutrient discharge is only part of the explanation for the differences between animal communities found at Werribee and at nutrient-poor areas of the Bay. Zooplankton, zoobenthos and fish communities in the Werribee area all contained several estuarine representatives. Therefore it seems probable that the freshwater discharge alone may have caused some of the modifications evident in the biota of the Werribee area. While recognizing the influence of nutrient and freshwater discharge on benthic plant and animal communities in the Werribee region, their impact seems to extend for only a few hundred meters around the discharge drains.

In marine waters proximate to urban centers discharging sewage, phytoplankton productivity is often nitrogen rather than phosphorus limited (43). This condition arises in part because the N:P ratio (by atoms) in treated sewage (about 5:1) is lower than the average N:P ratio of phytoplankton cells (about 10:1) (43). The N:P ratio of the Werribee discharge is about 9:1 and the ratio of the total nutrient discharge to Port Phillip Bay is about 11:1 (31). Owing to the similarity of N:P ratios in phytoplankton and in discharges to Port Phillip Bay, it might be expected that N and P available to phytoplankton in Bay waters would be maintained at close to a 10:1 ratio. However, the ratio of inorganic N to inorganic P a few kilometers from the area of sewage discharge is commonly less than 0.5:1. This accelerated removal from the water column of N relative to P could result from several processes, among them denitrification, high mineralization rate of detrital P relative to N, or uptake of N and P by phytoplankton in a ratio greater than 10:1. In any case, it can be strongly argued that N rather than P is the nutrient factor critical to Port Phillip Bay phytoplankton biomass production. The quantity of nutrient input and the ratio of N:P in inputs that would cause P limitation of productivity in the Bay is unknown. Freshwater ecosystems often compensate for deficiencies of N by selection for N_2-fixing blue-green algae (46). This apparently does not occur in temperate coastal marine systems such as Port Phillip Bay.

ACKNOWLEDGEMENTS

We acknowledge the considerable support given to all contributors to this study by the technical and laboratory staff of the Marine Studies Branch, Ministry for Conservation, and of the Schools of Botany and Microbiology, University of Melbourne. We thank Drs. J. Harris and C. Gibbs, Marine Studies Branch, under whose supervision the chemical analyses for this study were carried out. Dr. R. Kelly, Marine Studies Branch, co-ordinated the Werribee Study. The gift of N-Serve by the Dow Chemical Company is acknowledged.

This is paper number 243 in the Environmental Studies Series, Ministry for Conservation, Victoria. The contents do not necessarily represent the official views of the Ministry.

REFERENCES

1. Anger, K. 1975. On the influence of sewage pollution on inshore benthic communities in the south Kiel Bay. Part 1. Quantitative studies on indicator species and communities. *Merentutkimuslaitoksen Julk. Havsforskningsinst. Skr.* 239:116-122.
2. Arnott, G.H. 1974. Studies on the zooplankton of Port Phillip Bay and adjacent waters with special reference to Copepoda. Ph.D. thesis, Monash Univ., Melbourne. 268 p.
3 ——————, and S.U. Hussainy. 1972. Brackish-water plankton and their environment in the Werribee River, Victoria. *Aust. J. Mar. Freshwater Res.* 23:85-97.
4. Arnoux, A., D. Auclair and G. Bellan. 1973. Etude de la pollution chimique des sediments marins du secteur de Contion (Marseille): relations avec les peuplements macrobenthique. *Tethys* 5:115-123.
5. Axelrad, D.M. 1978. Effect of the Werribee sewage-treatment farm discharge on phytoplankton productivity, biomass and nutrients in Port Phillip Bay. Min. Conservation Victoria, Env. Stud. Ser. 170. 17 p.
6. ——————, N.J. Hickman, and C.F. Gibbs. 1979. Microalgal productivity as affected by treated sewage discharge to Port Phillip Bay. Min. Conservation Victoria. Env. Stud. Ser. 241. 14 p.
7. Bauld, J., and N.F. Millis. 1977. Role of bacteria in nitrogen cycling in the Werribee zone. Min. Conservation Victoria, Env. Stud. Ser. 166. 79 p.
8. Boesch, D.F. 1977. Application of numerical classification in ecological investigations of water pollution. U.S. Environmental Protection Agency, Ecological Research Series EPA-600/3-77-033. 115 p.
9. Cadee, G.C., and J. Hegman. 1977. Distribution of primary production of the benthic microflora and accumulation of organic matter on a tidal flat area, Balgzand, Dutch Wadden Sea. *Neth. J. Sea. Res.* 11:24-41.
10. Campbell, N.E.R., and M.I.H. Aleem. 1965. The effect of 2-chloro-6-(trichloromethyl) pyridine on the chemoautotrophic metabolism of nitrifying bacteria. I. Ammonia and hydroxylamine oxidation by *Nitrosomonas. Antonie van Leeuwenhoek J. Microbiol. Serol.* 31:124-136.
11. ——————, and ——————. 1965. The effect of 2-chloro-6-(trichloromethyl) pyridine on chemoautotrophic metabolism of nitrifying bacteria. II. Nitrite oxidation by *Nitrobacter. Antonie van Leeuwenhoek J. Microbiol. Serol.* 31:137-144.
12. Caperon, J., S.A. Cattell, and G. Krasnick. 1971. Phytoplankton kinetics in a subtropical estuary: eutrophication. *Limnol. Oceanogr.* 16:599-607.
13. Curl, H., and L.F. Small. 1965. Variations in photosynthetic assimilation ratios in natural, marine phytoplankton communities. *Limnol. Oceanogr.* 10(Suppl.):R67-R73.
14. Edwards, P. 1975. An assessment of possible pollution effects over a century on the benthic marine algae of Co. Durham, England. *Bot. J. Linn. Soc.* 70:269-305.

15. Edwards, R.R.C., D.M. Finlayson, and J.H. Steele. 1969. The ecology of O-group plaice and common dabs at Loch Eire. II. Experimental studies of metabolism. *J. Exp. Mar. Biol. Ecol.* 3:1-17.
16. ———, ———, and ———. 1972. An experimental study of oxygen consumption, growth and metabolism of the cod (*Gadus morhua* L.). *J. Exp. Mar. Biol. Ecol.* 8:299-310.
17. Eppley, R.W. 1972. Temperature and phytoplankton growth in the sea. *Fish. Bull.* 70:1063-1085.
18. Grontved, J. 1960. On the productivity of microbenthos and phytoplankton in some Danish fjords. *Medd. Dan. Fisk. - Havunders.* 3:55-92.
19. Henninger, N.M., and J.M. Bollag. 1976. Effects of chemicals chemicals used as nitrification inhibitors on the denitrification process. *Can. J. Microbiol.* 22:668-672.
20. Hopkins, T.L. 1966. The plankton of the St. Andrew Bay system, Florida. Publ. Inst. Mar. Sci. Univ. Tex. 11:12-64.
21. Jones, R. 1976. Growth of Fishes, p. 251-279. *In* D.H. Cushing and J.J. Walsh (eds.), Ecology of the Seas. Blackwell.
22. King, R.J., J.H. Black and S. Ducker. 1971. Intertidal ecology of Port Phillip Bay with systematic lists of plants and animals. Mem. Natl. Mus. Victoria 32:93-128.
23. Kjelson, M.A. and D.R. Colby. 1977. The evaluation and use of gear efficiencies in the estimation of estuarine fish abundance, p. 416-424. *In* Estuarine Processes, v. 2. Academic Press, New York.
24. Klavestad, N. 1978. The marine algae of the polluted inner part of the Oslofjord: A survey carried out 1962-1966. *Bot. Mar.* 21:71-97.
25. Lewis, J.A. 1977. The ecology of benthic marine algae at Gellibrand Light, northern Port Phillip Bay, Victoria. M.Sc. thesis, Univ. Melbourne. 234 p.
26. Littler, M.M., and S.N. Murray. 1977. Influence of domestic wastes on the structure and energetics of intertidal communities near Wilson Cove, San Clemente Island. Contrib. Wat. Res. Centre, Univ. Calif. No. 164.
27. Lizarraga-Partida, M.L. 1974. Organic pollution in Ensenada Bay, Mexico. *Mar. Pollut. Bull.* 5:109-112.
28. Malone, T.C. 1977. Environmental regulation of phytoplankton productivity in the Lower Hudson Estuary. *Estuarine Coastal Mar. Sci.* 5:157-171.
29. Marshall, N., C.A. Oviatt, and D.M. Skauen. 1971. Productivity of benthic microflora of shoal estuarine environments in south New England. *Int. Rev. Gesamten Hydrobiol.* 56:947-956.
30. ———, D.M. Skauen, H.C. Lampe, and C.A. Oviatt. 1973. Primary production of benthic microflora, p. 37-44. *In* A guide to the measurement of marine primary production under some special conditions. UNESCO Monogr. Oceanogr. Methodology 3.
31. M.M.B.W. and F.W.D. 1973. Environmental Study of Port Phillip Bay. Report on Phase I, 1968-1971. Melbourne and Metropolitan Board of Works, and Fisheries and Wildlife Department of Victoria, Melbourne. 372 p.
32. Neale, I.M., and I.A.E. Bayly. 1974. Studies on the ecology of the

zooplankton of four estuaries in Victoria. *Aus. J. Mar. Freshwater Res.* 25:337-350.

33. Nedwell, D.B. 1975. Inorganic nitrogen metabolism in a eutrophicated tropical mangrove estuary. *Water Res.* 9:221-231.
34. Painter, H.A. 1970. A review of literature on inorganic nitrogen metabolism in microorganisms. *Water Res.* 4:393-450.
35. Pamatmat, M.M. 1968. Ecology and metabolism of a benthic community of an intertidal sandflat. *Int. Rev. Gesamten Hydrobiol.* 53:211-298.
36. Payne, W.J. 1973. Reduction of nitrogenous oxides by microorganisms. *Bacteriol. Rev.* 37:409-452.
37. Pekkari, S. 1973. Effects of sewage water on benthic vegetation. The nutrients and their influence on the algae in the Stockholm Archipelago during 1970. *Oikos Suppl.* 15:185-188.
38. Platt, T. 1971. The annual production of phytoplankton in St. Margaret's Bay, Nova Scotia. *J. Cons. Cons. Int. Explor. Mer.* 33:324-333.
39. Poore, G.C.B., and J.D. Kudenov. 1978. Benthos of the Port of Melbourne: the Yarra River and Hobsons Bay, Victoria. *Aus. J. Mar. Freshwater Res.* 29:141-155.
40. ———, and ———. 1978. Benthos around an outfall of the Werribee sewage-treatment farm, Port Phillip Bay, Victoria. *Aus. J. Mar. Freshwater Res.* 29:157-167.
41. ———, and S. Rainer. 1974. Distribution and abundance of soft-bottom molluscs in Port Phillip Bay, Victoria, Australia. *Aus. J. Mar. Freshwater Res.* 25:371-411.
42. Riley, G.A. 1956. Oceanography of Long Island Sound, 1952-1954. IX. Production and utilization of organic matter. Bull. Bingham Oceanogr. Collect. Yale Univ. 15:324-334.
43. Ryther, J.H., and W.M. Dunstan. 1971. Nitrogen, phosphorus, and eutrophication in the coastal marine environment. *Science* 171:1008-1013.
44. ———, and C.S. Yentsch. 1958. Primary production of continental shelf waters off New York. *Limnol. Oceanogr.* 3:327-335.
45. Sasseville, D.R., A.P. Takacs, S.A. Norton, and R.B. Davis. 1974. A large-volume interstitial water sediment squeezer for lake sediments. *Limnol. Oceanogr.* 19:1001-1004.
46. Schindler, D.W. 1977. Evolution of phosphorus limitation in lakes. *Science* 195:260-262.
47. Steeman Nielsen, E. 1952. The use of radioactive carbon (^{14}C) for measuring organic production in the sea. *J. Cons. Cons. Int. Explor.* Mer. 18:117-140.
48. Strickland, J.D.H., and T.R. Parsons. 1968. A practical handbook of seawater analysis. *Bull. Fish. Res. Bd. Can.* 167.
49. Technicon. 1972. Operation manual for the technicon autoanalyzer (R) II system, and associated technical literature. Technicon Instruments Corporation, Tarrytown, New York.
50. Thomas, W.H. 1970. On nitrogen deficiency in tropical Pacific oceanic phytoplankton: photosynthetic parameters in poor and rich water. *Limnol. Oceanogr.* 15-380-385.

51. Wade, B.A., L. Antonia, and R. Mahon. 1972. Increasing organic pollution in Kingston Harbour, Jamaica. *Mar. Pollut. Bull.* 3:106-110.

52. Widdowson, T.B. 1971. Changes in the intertidal algal flora of the Los Angeles area since the survey by E. Yale Dawson in 1956-1959. *Bull. South Calif. Acad. Sci.* 70:2-16.

ESTUARIES AND COASTAL LAGOONS OF SOUTH WESTERN AUSTRALIA

Ernest P. Hodgkin* and R.C. Lenanton**

**Department of Conservation and Environment, 1 Mount Street, Perth, Western Australia, 6000.*

***Department of Fisheries and Wildlife, Marine Research Laboratories, P.O. Box 20, North Beach, Western Australia, 6020.*

ABSTRACT: This paper reviews briefly recent studies of estuarine ecosystems of southwestern Australia. Specifically, the paper (A) assesses the meteorologic and physiographic features which determine the unusual environmental characteristics of these estuaries, especially the extreme variation in salinity; (B) classifies the estuaries according to frequency and duration of bar closure, with consequent different hydrological characteristics; (C) discusses the more important aspects of the flora and fauna of the estuaries as related to these environmental differences; and (D) describes in general terms their nutrient status, as an introduction to the paper by McComb et al. (10) in this Symposium on the Peel-Harvey estuarine system.

The estuaries and coastal lagoon systems of southwestern Australia are of bar-built type (12). Of some 80 such systems located along 1,200 km of coastline of temperate Western Australia between the Murchison River (28°S) and Israelite Bay (124°E), only nine remain open continuously. Another 64 are only open to the sea seasonally or less frequently, and seven systems are now permanently closed coastal lagoons with their tributary rivers. The systems discussed in this paper are shown in Figure 1.

All systems are subject to extreme changes in salinity. Even permanently open estuaries may be fresh throughout for several months in winter and brackish to hypersaline in summer, while barred estuaries may stay fresh or become hypersaline. In these circumstances, salinity is the limiting ecological factor for what are "physically controlled communities" (13) of plants and animals.

FIGURE 1. Estuaries and coastal lagoons of southwestern Australia. Isohyets - mm.

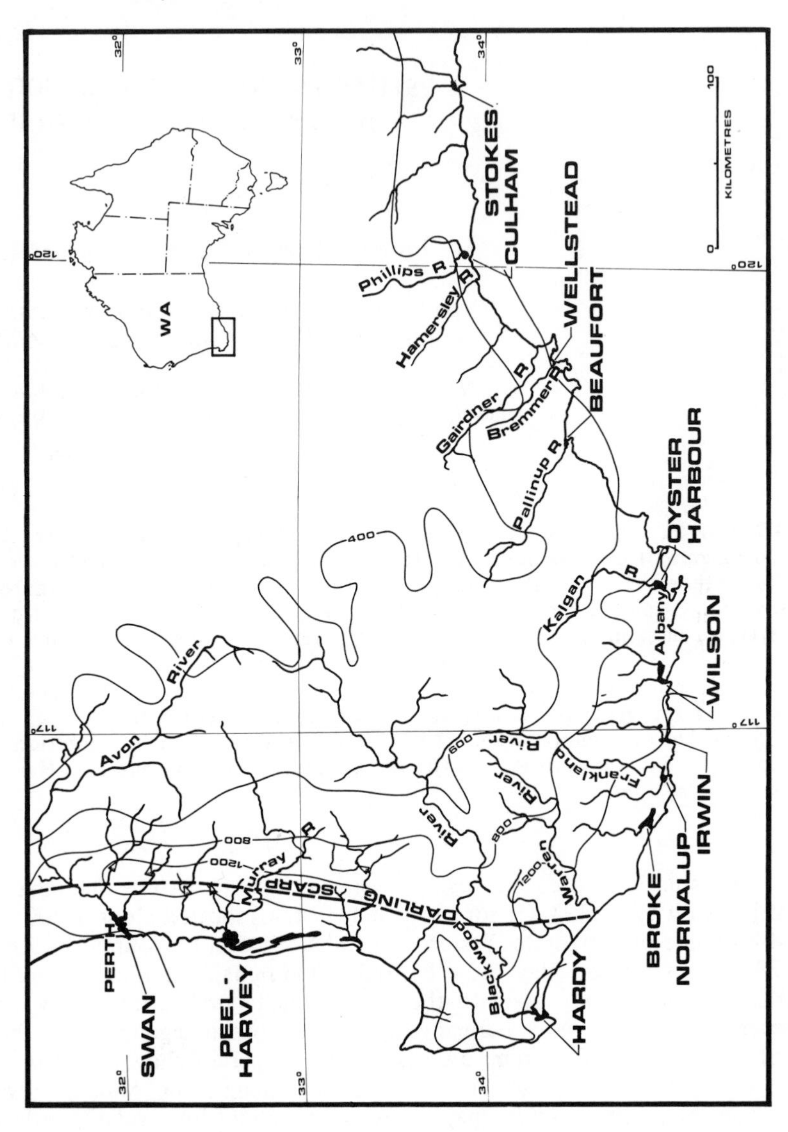

These extreme hydrological conditions result from: (a) the seasonality of rainfall and river flow; the climate is of Mediterranean-type (Figure 2); (b) seasonality of evaporation; (c) great variation of annual volume of river flow, and the impact of flash floods; (d) salinity of the river water (up to 10°/oo or more), and seepage of seawater through closed sea bars; (e) small tidal range; the astronomic tide never exceeds 90 cm; (f) the characteristic geomorphology of: a riverine stretch (tidal river); an estuarine basin which is generally shallow; a narrow inlet channel which restricts tidal exchange with the sea (Figure 3); (g) sea bars which are liable to close when river flow is low, and may stay closed for extended periods.

ESTUARINE TYPES

We recognize four major categories of estuarine systems: those *permanently open* to the sea; *seasonally closed*, open one or more times each winter; *normally closed*, only open following unusually heavy rains; *permanently closed* coastal lagoons.

The *permanently closed* systems (e.g. Culham Inlet) have probably been estuarine during the Holocene. These lagoons become grossly hypersaline, or dry out completely in summer, and have a very restricted invertebrate fauna similar to that of inland saline waters. When they fill, the fauna is recruited from the saline streams which enter them.

The *normally closed* estuaries are in the area of low rainfall, east of Albany. Heavy cyclonic rains and river flow are required to breach the bars. This may happen in winter or summer and the composition of the fauna recruited to them, from both sea and rivers, depends very much on the season and duration of opening.

Two examples illustrate this point. The bar of Beaufort Inlet, had been closed for seven years before it opened in July, 1978. It remained open for six weeks. Before the rains, salinity was about 55°/oo and the estuary had an impoverished fauna with some 12 macro-invertebrate species (three molluscs) and eight species of fish. In October salinity was 20°/oo. There had been no change in the invetebrate fauna. However, there were 21 species of fish. As discussed later, some of the more marine species will disappear rapidly but others will survive and grow.

At Wellstead Estuary nearby, the bar broke at the same time and remained open through the following spring and summer. Salinity had been about 50 °/oo in the previous summer and in October, 1978, it was 25 °/oo - beyond the influence of tidal exchange. Before the opening the benthic invertebrate fauna was even more impoverished than that of Beaufort Inlet (eight species, one mollusc). The prolonged opening through spring and summer allowed establishment of 11 mollusc species at the marine end of the estuary. Some were casual marine invaders, but others were species characteristic of permanently open estuaries.

FIGURE 2. Peel-Harvey estuarine system, 1977-78. Weekly salinity, evaporation (mean daily pan), rainfall, river flow.

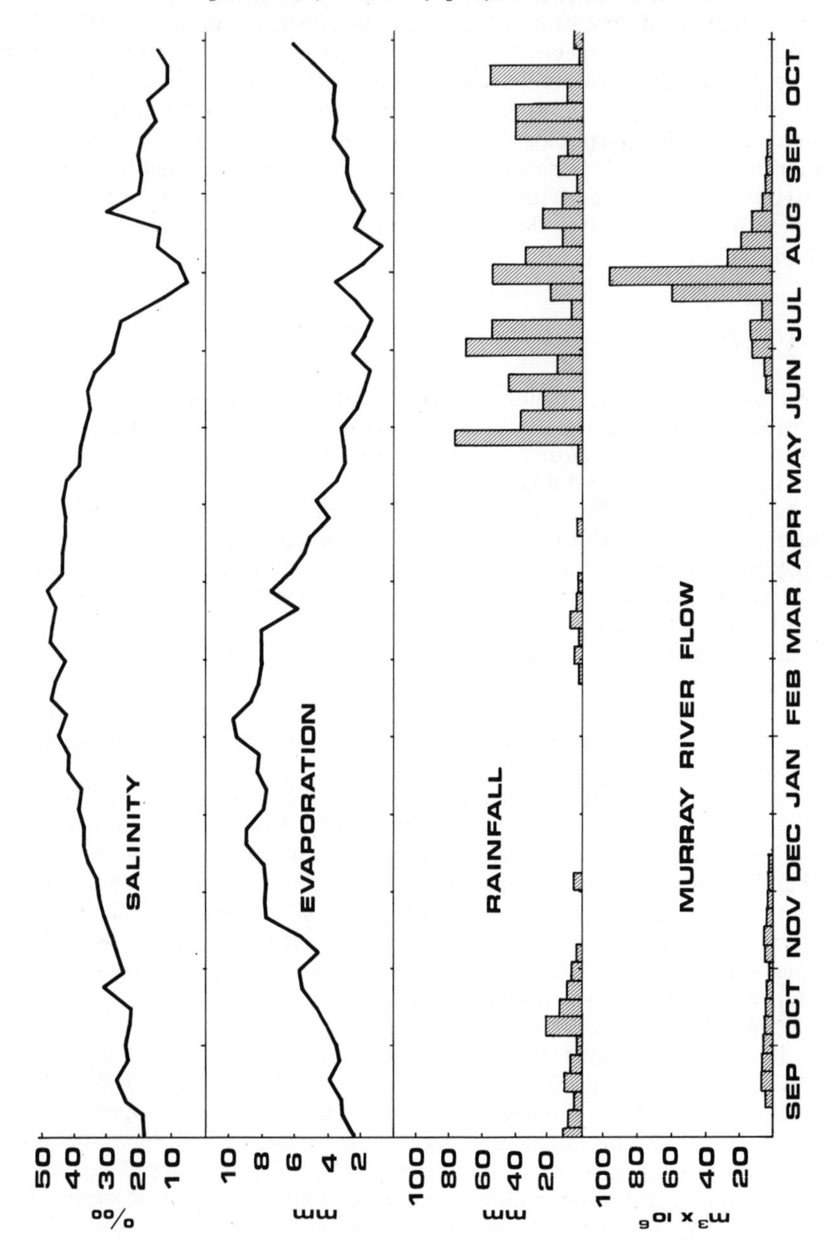

FIGURE 3. Four estuaries of southwestern Australia: Peel-Harvey, Blackwood, Wilson Inlet, Oyster Harbour. The Swan estuary is shown in Figure 4.

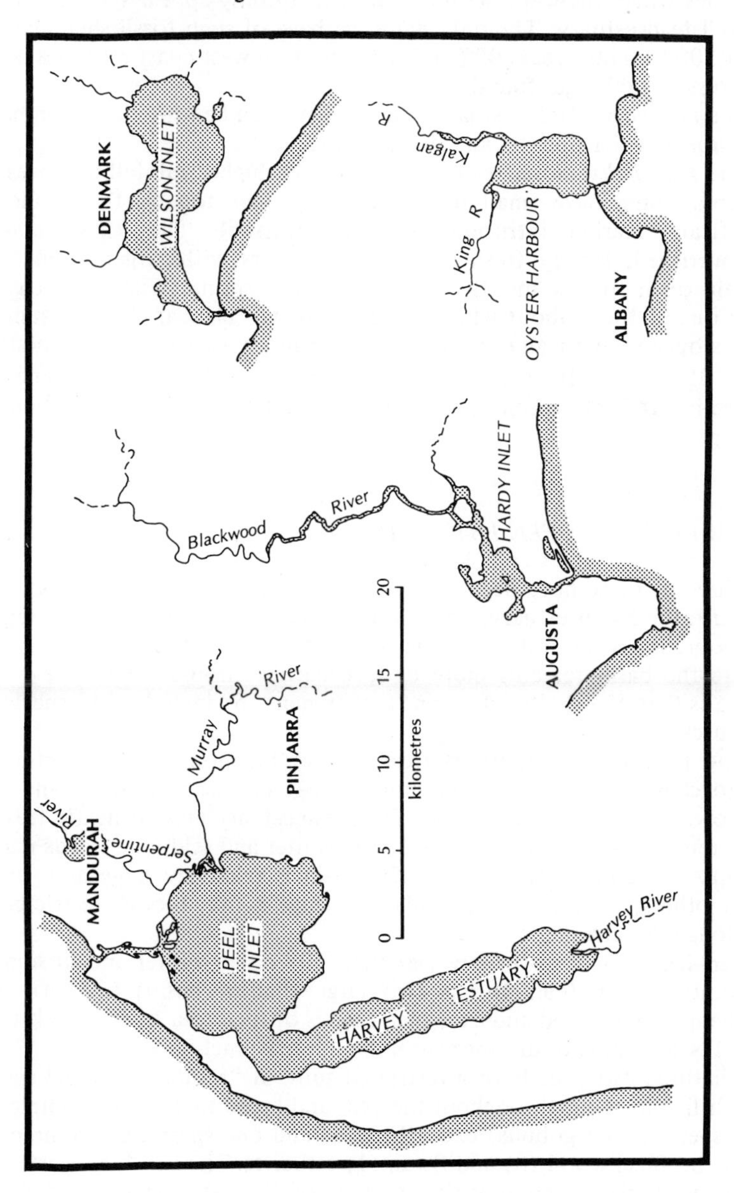

Two species of *Katelysia*, a cockle (Veneridae) are of particular interest because there are enormous deposits of shells of this bivalve in many estuaries where these species no longer live, or only appear sporadically, as in Wellstead estuary. The only dates we have of such fossil shell material are 4,500 to 6,600 years BP (5). These are from west coast estuaries where *Katelysia* is no longer found.

The *normally closed* estuaries have a restricted benthic flora dominated by *Ruppia*, a charophyte, and *Acetabularia*.

The *seasonally closed* estuaries are in the higher rainfall area west of Albany. Some have small basins and stay open too briefly to allow a significant marine intrusion (e.g. the Warren). They have a very impoverished, largely freshwater fauna. Others with larger basins may remain open for six months over winter and spring. Wilson Inlet is the only one of these about which much is known. The bar closes in summer and is opened artifically in June or July. Salinity varies between about 10 and 35 $^{0}/oo$. It has a moderately rich fauna with some 32 macro-invertebrate species (17 molluscs including *Katelysia*) and about 40 fish species.

PERMANENTLY OPEN ESTUARIES

Three of these have been studied in some detail - the Blackwood River estuary, the Swan estuary, and the Peel-Harvey system. A fourth, Oyster Harbour, is an interesting reference point because it is the only system where the basin remains more or less marine throughout the year. The fauna is that of a sheltered marine embayment and includes 112 species of molluscs.

The Blackwood estuary is fresh for several months each winter and approaches marine salinity in summer (3). The basin is small and very shallow. Tidal exchange is good. The Swan estuary may be nearly fresh or may remain brackish (15-20$^{0}/oo$) over winter and seldom exceeds marine salinity in summer (Figure 4). The basin is relatively large and, unlike most others, is deep (20 m). Tidal exchange is fair, though restricted by the long inlet channel.

Peel-harvey is only briefly less than 10$^{0}/oo$ in winter and in summer becomes hypersaline. It has two large shallow basins (<2 m). Tidal exchange is restricted and exchange time is of the order of one week to six months according to distance from the inlet channel.

All three estuaries have a restricted suite of "resident" animal species which live in them throughout the year and breed there. For example five fish species, two gobiids, two atherinids and one sparid, are common to both the Blackwood (8) and Swan (1) estuaries. The benthic invertebrate fauna is similar in composition to that of the seasonally closed Wilson Inlet, and in all three a few species constitute the bulk of the biomass; for

FIGURE 4. The estuary of the Swan River. Surface Salinity, $^{0}/oo$, 1965 and 1969-1970.

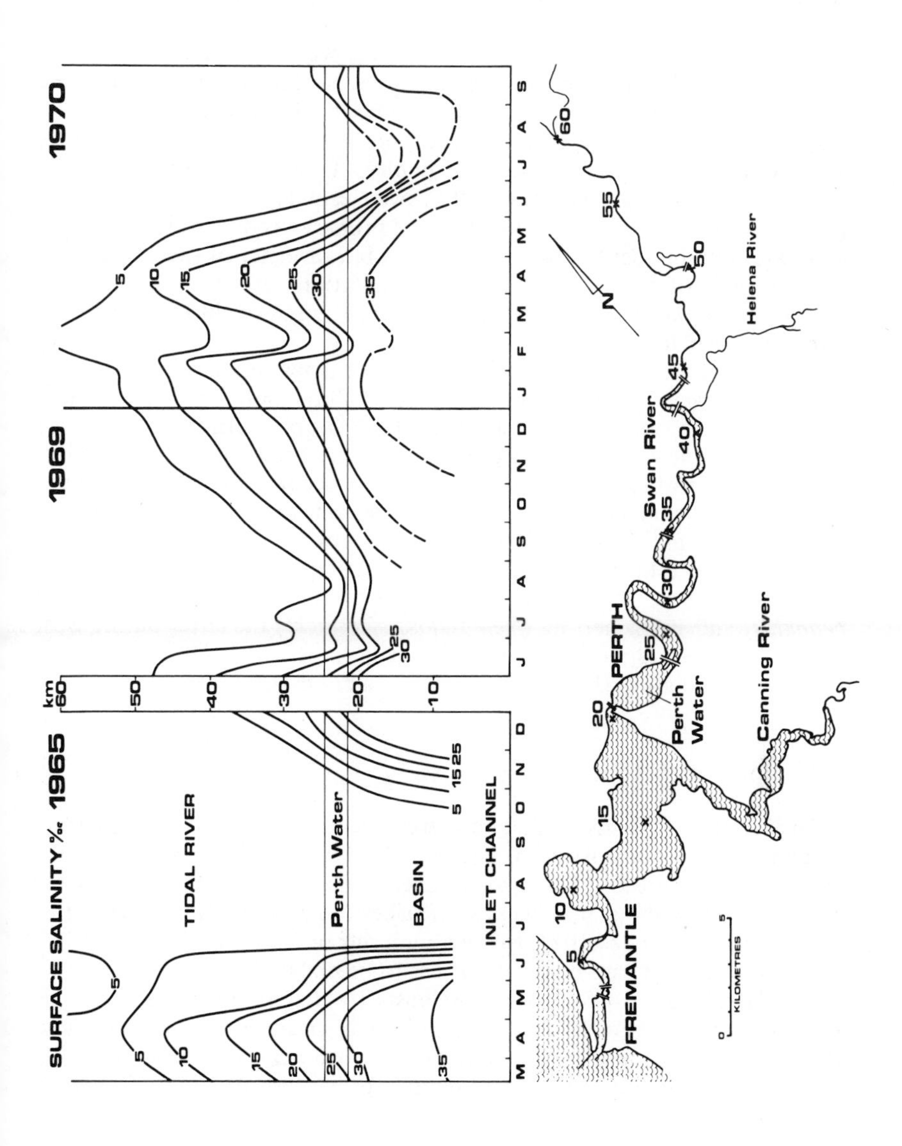

example in the Blackwood three polychaetes, four molluscs and three amphipods (3). In contrast to this poverty of resident species, these estuaries have a rich and abundant non-resident fauna of crabs, penaeid prawns and fish (Blackwood 57, Swan 112 species of fish) which have varying degrees of euryhalinity. Their use of the estuaries is discussed later.

Figure 5 is a simplified diagram of what we believe to be the main features of the food chains of these estuaries. Nothing is known about the contribution made by benthic diatoms. In the Blackwood estuary, the major photosynthetic input is from *Ruppia, Potamogeton*, a charophyte, and swamp vegetation (rushes, samphire, trees); phytoplankton and macroscopic algae are sparse. In the Swan estuary with its deep basin, phytoplankton are probably the principal primary producers. Macroscopic algae are not abundant, although there is a fair diversity of species. During the first half of the 20th century, accumulations of green algae were a problem and the disappearance of this algae nuisance during the 1960s remains unexplained.

Phytoplankton and benthic algae are the principal primary producers in the Peel-Harvey estuary with its large shallow basin and poor tidal exchange. The macroscopic algae, mainly a ball-forming species of *Cladophora*, and other green algae have become a nuisance during the last decade and great quantities accumulate and decompose in the shallows. There is some evidence that aquatic angiosperms (*Ruppia* and *Halophila*) were formerly more abundant than they are now.

In all three estuaries, however, the most important item in the food chain is detritus. Two very abundant fish species, sea mullet (*Mugil cephalus*) and Perth herring (*Nematalosa vlaminghi*) the latter on the west coast only, feed directly on detritus, while most other species feed on benthic invertebrates that are themselves mainly detrital feeders.

THE FISH FAUNA

During the 1970s fish collections were made in many of the estuarine systems of temperate Western Australia. The systems which have received most attention are the Blackwood River estuary (8) and the Swan River system (1). A preliminary checklist of fishes of many south coast estuaries was prepared as a result of surveys undertaken between 1971-1972 (6), and recent surveys have enabled this list to be updated.

To date, some 183 species of teleosts (representing 79 families), 13 species of elasmobranchs (ten families), one species of cyclostome, and three species (two families) of commercial crustaceans have been recorded from the 80 estuarine systems of temperate Western Australia (Table 1).

FIGURE 5. Food web of the Peel-Harvey estuarine system. Simplified scheme of probable main pathways.

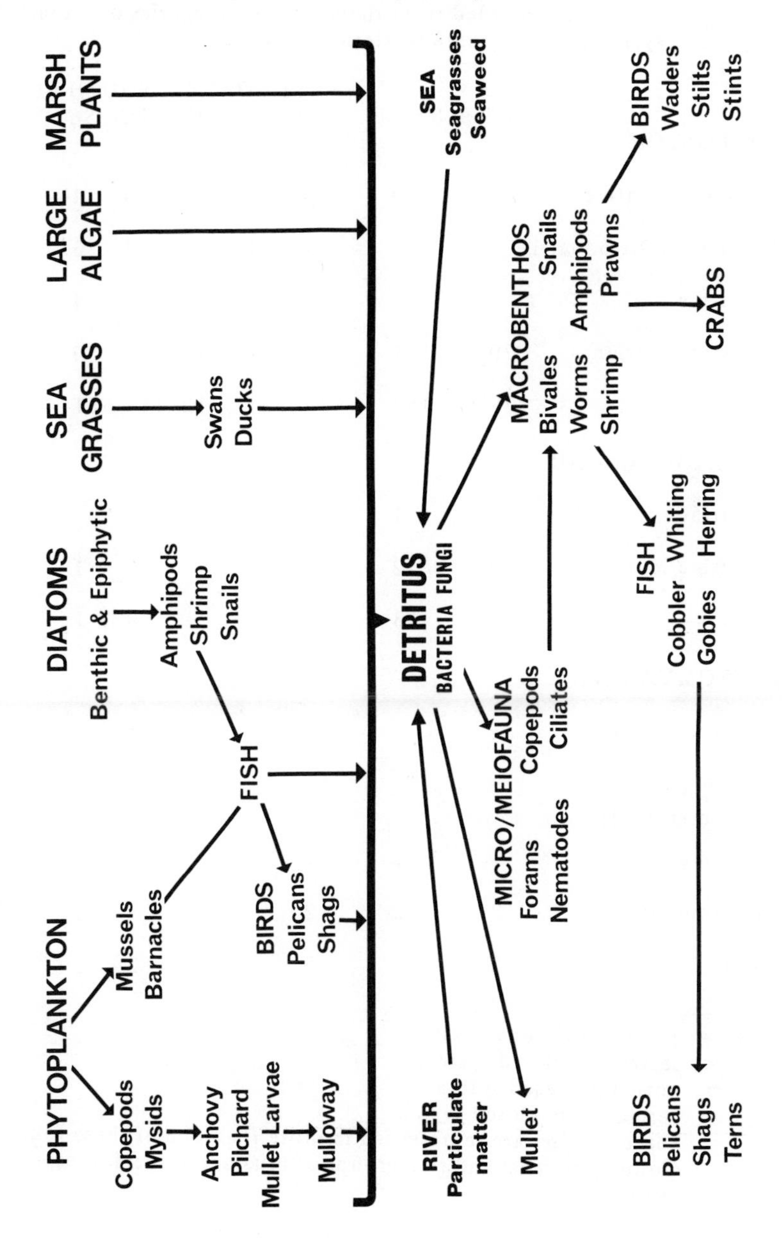

TABLE 1. The number of families and species of the classes Teleostomi, Elasmobranchii, Petromyzones, and of commercially important crustacea, recorded from the different categories of estuarine systems of temperate Western Australia.

			WEST COAST			SOUTH COAST			
	Families	Species	*PO	SC	NC	PO	SC	NC	PC
TELEOSTOM									
Freshwater native	4	9	4	4		4	4	3	1
Freshwater introduced	6	8	7	1		2	3		
Estuarine	3	4	4	4	2	4	4	3	3
Estuarine/Marine	70	161	97	12	10	103	50	35	1
Total	†79	183	112	21	12	113	61	41	5
ELASMOBRANCHII									
Estuarine	1	1	1						
Estuarine/Marine	9	12	3			9	1		
Total	10	13	4			9	1		
PETROMYZONES									
Estuarine/Marine	1	1	1			1	1		
CRUSTACEANS (commercial)									
Estuarine	1	1	1			1			
Estuarine/Marine	2	2	2		1	1	2	1	
Total	†2	3	3		1	2	2	1	

*PO = permanently open system
SC = seasonally closed system
NC = normally closed system
PC = permanently closed system
† = Some families are represented in more than one of these four groups of teleost and two groups of crustacean species respectively.

The Estuarine Environment as a Fish Habitat

Fishes of the estuaries adopt a number of different life history strategies which enable them to utilize the estuarine habitat (7, 9). It should be emphasized that grouping species into these strategies is simply a convenient way of dealing with a wide range of different species and that, under the extreme environmental conditions experienced, departures from the three "typical" situations described below are to be expected.

A. *A permanent breeding and nursery habitat* - Two groups of species typify this life history strategy. One group includes species such as black bream (Sparidae) and some atherinid and gobiid species which normally complete their life cycle within an estuarine system. Some of these have specialized breeding habits which make them well adapted to life in estuaries; gobiids build nests under stones and other objects, while the atherinids attach their eggs to seagrass fronds.

The other group comprises species such as cobbler (Plotosidae), gobbleguts (Apogonidae), and yellowtail trumpeter (Teraponidae) which can pass through their entire life cycle either in estuaries or marine embayments. Perth herring (Clupeidae) is also included in this group because it is believed to reproduce within estuaries and is known to spend its maturing years, and probably longer, in them (1). The only exploited crustacean in this group is the greasy-back or school prawn (Penaeidae).

The native freshwater fish fauna of this part of Australia is extremely impoverished (1). Only a few small perch and galaxid species and a single plotosid have been recorded from the estuaries, together with a small number of introduced freshwater species (Table 1).

B. *A nursery habitat for juveniles* - A considerable number of species fall into this category, including many that are commercially important. Such species include sea and yelloweye mullet (Mugilidae), King George, western sand and trumpeter whiting (Sillaginidae), silver bream (Sparidae), tailor (Pomatomidae), mulloway (Sciaenidae), some flathead (Platycephalidae) and flounder (Bothidae, Pleuronectidae), sea garfish (Hemiramphidae), striped perch (Teraponidae), whitebait (Clupeidae), as well as many less important commercial and non-commercial species.

Abundant food supplies and shelter from predators are important factors which encourage the yound stages of many of these species to use the barred coastal lagoon type of estuaries (8, 15). Current research has shown that the shallow banks, particularly those covered with *Ruppia* and *Halophila*, are important habitats for these fish. This mostly because of the abundant invertebrate fauna which inhabits these areas. There is evidence that shallow areas of Western Australian estuaries are more productive than deeper parts.

Exploited crustaceans included in this group are the king prawn (Penaeidae) and the blue swimming crab (Portunidae).

C. *Occasional feeding area for maturing and mature individuals* - Many species adopt this strategy. However, although their presence in the estuaries is reflected by a high diversity, the annual biomass of individual species is extremely low. These basically marine species are confined to the seaward end of estuaries during the summer, saline phase. For example, in the Swan (1) and Blackwood (8) they include almost all the elasmobranchs and most members of such families as Monocanthidae, Ophichthidae, Odacidae, Cheilodactylidae, Triglidae, Syngnathidae, and Scorpaenidae.

Factors which influence fish utilization of the estuaries

Recruitment to estuarine populations presents no problem for species which breed in the estuaries (Strategy A). However, for the majority of species (Strategies B and C) it involves migration into an estuary from the adjacent ocean. Both season and duration of bar opening influence species composition of the fish fauna which enter seasonally and normally closed estuaries.

Once fish have entered such estuaries, survival of the species trapped in them is determined by duration of closure and changes in salinity. Stenohaline-marine species of Strategy C disappear rapidly as estuarine salinity diverges from seawater salinity. Those which adopt Strategy B survive and grow; including commercially and recreationally important species which may ultimately reach an exploitable size. With prolonged bar closure, populations decline and ultimately disappear in the absence of recruitment from the sea; or, as is often the case, salinity becomes too extreme for all but the most euryhaline to survive.

Thus, one can expect that permanently open systems will have the greatest number of species and permanently closed systems the least (Table 1). Biomass of individual species may nevertheless be high in any estuary, provided hydrological conditions are favorable.

In order to cope with the enormous salinity extremes experienced in our estuaries, from the prolonged winter freshwater flushes of permanently open systems to the extreme hypersalinity of closed systems, fishes of Strategy A obviously need to be extremely good osmoregulators. This is certainly true of black bream, the atherinids, and gobies (8). The ability of other species to survive and grow in the extreme salinities encountered in our estuaries is rather better for fishes of Strategy B than it is for those of Strategy C.

Other factors such as temperature, dissolved oxygen, turbidity (including suspended solids), available food and shelter are discussed by Lenanton (8, 9).

NUTRIENT ENRICHMENT

There are only limited data on the nutrient status of estuaries other than the Peel-Harvey system and the Swan and Blackwood River estuaries. Most seasonally closed estuaries appear to be oligotrophic, though periodic blooms of green algae in Wilson Inlet suggest eutrophy there. Some normally closed estuaries have an abundance of phytoplankton and relatively high nutrient levels have been observed; e.g. Beaufort Inlet 3500 $\mu g\ l^{-1}$ total N, 120 $\mu g\ l^{-1}$ total P, 200 $\mu g\ l^{-1}$ chlorophyll.

Effective river flow and the major external input of nutrients is confined to not more than four months each year. In 1978, 96 percent of nitrogen and 94 percent of phosphorus (total N and P) input to the Peel-Harvey system came in a nine week period in July and August, bringing 1,632 tonnes N and 119 tonnes P to an estuarine water body of $100 \times 10^6 m^3$. Concentrations of both nutrients in estuary water decreased rapidly after the rivers ceased to flow (10).

The Swan River estuary experiences the same massive input of nitrate nitrogen and the same almost total disappearance from basin water when river flow ceases (14, 4). However, there was not the same periodicity of phosphate levels in estuary water; concentrations varied erratically. Spencer (14) suggests that bottom muds contribute much phosphate to the overlying water.

Although more than half the catchment of the Blackwood River is cleared for agriculture, 80 percent of the river flow to the estuary is from forest land in the lower part of the catchment (3). Nutrient levels are low and estuary water is oligotrophic; however, nutrient concentrations at least treble during river flow, with nitrates increasing to 25 times the normal (2).

River flow varies greatly from year to year and nutrient input presumably varies proportionately, with an unknown amount lost to the sea. Both 1976 and 1977 were dry years and river flow to Peel-Harvey was only 1.5 times the volume of the estuary. In 1978, a year of near average flow, it was little more than four times, but in flood years (1945, 1964) flow was 20 times the volume.

Size of the catchments and their nature, whether forested or cleared for agriculture, also greatly influence nutrient input. In 1978, half the flow to the Peel-Harvey estuary was from the Murray River, the catchment of which is mainly on the Pre-Cambrian shield east of the Darling Escarpment and is drainage from forest or passes through a wide forest belt. While 73 percent of nitrogen input was from this source, it contributed only 21 percent of the phosphorus. The remaining 79 percent came via rivers and agricultural drains with catchments on the coastal plain. This is grazing land; it is partly irrigated and has received generous applications of phosphate fertilizer over the last 30 years.

The massive plankton blooms observed in certain estuaries are not known to cause severe problems for fish, by world standards. They encourage the predominance of planktivorous fishes, perhaps at the expense of benthic feeding species, and the recent increased abundance of pilchard (Clupeidae) in Harvey Estuary may be attributable to an increase in phytoplankton.

The green alga *Chaetomorpha* spp. sometimes forms dense beds in shallows of the Nornalup estuarine system, and daytime oxygen concentrations of 8.5 mg/l have been found to fall to 2.0 mg/l before dawn; this is below the level reported to cause fish deaths (11). Dead and dying algae accumulate on the bottom and create anoxic conditions in areas which were previously relatively free from algal growth. Current investigation of the fish fauna of the Peel-Harvey system may show what effect accumulations of living and decaying *Cladophora* have on fish populations there.

SUMMARY

Although study of the estuaries of southwestern Australia is still at an early stage, it is evident that the extreme seasonality of river flow and nutrient input, together with the geomorphic conditions associated with bar formation and closure, create potentially eutrophic conditions. That more estuaries have not shown obvious signs of nutrient enrichment is probably due mainly to the absence of large urban complexes (other than Perth) or intensive agriculture in the effective catchment of tributaries.

There is no evidence yet that the large planktonic or benthic blooms experienced in some estuaries are associated with any marked change in composition or abundance of the fauna. The small number of invertebrate species present is attributable mainly to the extreme hydrological conditions; it contrasts with the relatively large number of fish species which use the estuaries when salinity is favorable.

ACKNOWLEDGEMENTS

It is a pleasure to acknowledge the help given us in preparation of this paper by colleagues involved in the Peel-Harvey estuarine system study: Peter Birch, Ron Black and Arthur McComb; also the help we have had from Marion Cambridge and John Wallace in our field studies. We are grateful, too, to Mrs. Ann Ross for her elegant draftsmanship.

REFERENCES

1. Chubb, C.F., J.B. Hutchins, R.C.J. Lenanton and I.C. Potter. 1979. An annotated checklist of the fishes of the Swan-Avon River system, Western Australia. *Rec. West. Aust. Mus.* 8(1):1-55.

2. Congdon, R.A., and A.J. McComb. 1979. Nutrient pools of an estuarine ecosystem - the Blackwood River estuary in south-western Australia. *J. Ecol.* 68:287-313.
3. Hodgkin, E.P. 1978. An environmental study of the Blackwood River estuary. West. Aust. Dept. Conserv. Envir. Rept. 1.
4. Jack, P.N. 1977. Seasonal variations in the water of the Swan River. West. Aust. Govt. Chem. Labs. Rept. 14.
5. Kendrick, G.W. 1977. Middle Holocene marine molluscs from near Guildford, Western Australia, and evidence for climatic change. *J. Roy. Soc. West. Aust.* 59:97-104.
6. Lenanton, R.C.J. 1974. Fish and crustacea of the Western Australian south coast rivers and estuaries. *Fish. Res. Bull. West. Aust.* 13:1-17.
7. Lenanton, R.C.J. 1974. Biological aspects of coastal zone development in Western Australia. II. Fish Crustacea and birds. *In* The impact of human activities on coastal zones. Proceedings of Australian UNESCO Committee for man and the biosphere. National Symposium Sydney May 1973.
8. Lenanton, R.C.J. 1977. Aspects of the ecology of fish and commercial crustaceans of the Blackwood River estuary, Western Australia. *Fish. Res. Bull. West. Aust.* 19:1-72.
9. Lenanton, R.C.J. 1978. Fish and exploited crustaceans of the Swan-Canning estuary. *Fish. Rept. West. Aust.* 35:1-36.
10. McComb, A.J., P.R. Atkins, P.B. Birch, D.M. Gordon, and R.J. Lukatelich. Eutrophication in the Peel-Harvey estuarine system, Western Australia. *In* B.J. Neilson and L.E. Cronin (eds.) Nutrient Enrichment in Estuaries. Humana Press, New Jersey.
11. Perkins, E.J. 1974. The biology of estuaries and coastal waters. Academic Press, London and New York.
12. Pritchard, E.J. 1967. What is an estuary: physical viewpoint. *In* G.H. Lauff, *Estuaries*, Amer. Assoc. Adv. Sci. 83:3-5.
13. Sanders, H.L. 1968. Marine benthic diversity. *Amer. Naturalist* 102(925):243-282.
14. Spencer, R.S. 1956. Studies in Australian estuarine Hydrology. II The Swan River. *Aust. J. Mar. Freshwater Res.* 7:193-253.
15. Wallace, J.H., and R.P. Van der Elst, 1975. The estuarine fishes of the east coast of South Africa. IV. Occurrence of juveniles in estuaries. V. Ecology, estuarine dependence and status. *Invest. Rept. Oceanogr. Res. Inst.* 42:1-63.

EUTROPHICATION IN THE PEEL-HARVEY ESTUARINE SYSTEM, WESTERN AUSTRALIA.

A.J. McComb, R.P. Atkins, P.B. Birch, D.M. Gordon and R.J. Lukatelich

Department of Botany
University of Western Australia
Nedlands, Western Australia, 6009

ABSTRACT: The most obvious symptom of eutrophication in this estuarine system is a green alga, *Cladophora*, which was sparse in 1966 but now accumulates and rots on the shores. The work is part of a continuing study designed to assess the relationships between nutrient input and the growth of *Cladophora* and phytoplankton.

The system consists of two shallow basins, interconnected and linked to the ocean by a narrow channel. It is fed by three rivers; 90 percent of river flow occurs during four winter months.

Phytoplankton and water nutrient levels are low in summer, but high during and after an input of river nutrients in winter. Nitrogen:phosphorus ratios, regression analyses, and nutrient limitation assays suggest that nitrogen is potentially limiting in summer and autumn, phosphorus in winter and spring. The Harvey typically supports higher phytoplankton levels than the Peel. Diatom populations may be replaced by blue-greens in summer in the Harvey.

Cladophora forms detached spheres of branched filaments and is only prominent in the Peel. Changes in biomass and growth of confined populations in the field, together with laboratory experiments, show that growth occurs when temperatures and light intensities are high, not in winter when water column nutrient levels are high. Water from between the algal spheres has increased levels of phosphorus compared with the water column above, emphasising the possible importance of nutrient release from decaying material below. It is suggested that phytoplankton are important in trapping water-column nutrients during and after river nutrient input, and that subsequent *Cladophora* and phytoplankton growth depends on nutrient recycling.

INTRODUCTION

The Peel-Harvey Estuarine System consists of two shallow basins (at most 2.5 m deep) which are linked together and which communicate to the ocean through a narrow channel (Figure 1). The most obvious eutrophication problem occurs in the Peel, where there is excessive growth of the green alga, *Cladophora*. This forms small spheres of radiating, branched filaments (Figure 2a) which rise to the surface of the water as they photosynthesize and which are driven by wind-induced currents towards the shore where they may foul what were once sandy beaches (Figure 2b), or accumulate in large banks offshore. For example, the algal bed shown in Figure 2c was 140 m long, estimated to have a dry weight of about 75 metric tons and to contain 230 kg of phosphorus and 2,000 kg of nitrogen. *Cladophora* was virtually absent from the estuary in 1966, but now it is bulldozed from certain beaches and carted away.

This paper summarizes part of a continuing study of the relationship between nutrient levels and the growth of phytoplankton and *Cladophora*. The system is of particular interest because, as shown below, the major nutrient input occurs during a restricted period of the year, and because it undergoes very marked seasonal changes in salinity.

The system is fed by three rivers, the flow of which is highly seasonal because of the mediterranean climate; 90 percent of river flow reaches the estuary in the four winter months, June to September. The basins are of similar volume: Peel 61 x $10^6 m^3$, and Harvey 56 x $10^6 m^3$. In an average year the Murray River contributes a volume of water equivalent to about four times that of the Peel. The Harvey River has been dammed, and the lower reaches drain irrigated pasture. The Serpentine River has also been dammed. The relative contributions of volume flow during 1978 were Murray 54 percent, Harvey 30 percent, and Serpentine 16 percent.* The morphology, hydrology, and ecology of the system are discussed by Hodgkin and Lenanton (3); in its general features of seasonal salinity changes, shallow basin, and restricted ocean exchange it is typical of a series of estuaries in southwestern Australia.

PHYTOPLANKTON/NUTRIENT RELATIONSHIPS

In order to assess nutrient input and its effect on phytoplankton growth, water samples have been taken each week since August 31, 1977, at six sites in the Peel and one in the Harvey (Figure 1). Figure 3a shows the mean concentration of chlorophyll at these sites, and it is clear that a

*Data on river flow computed by R.E. Black, Dept. of Physics, Western Australian Institute of Technology.

FIGURE 1. The Peel-Harvey estuarine system, Western Australia. The town of Mandurah (lat. 115°43' E, long. 32°32' S, population 12,000) lies to the east of the inlet channel. The 0.5 m contour is included. The arrow shows the site of *Cladophora* work illustrated in Figure 8.

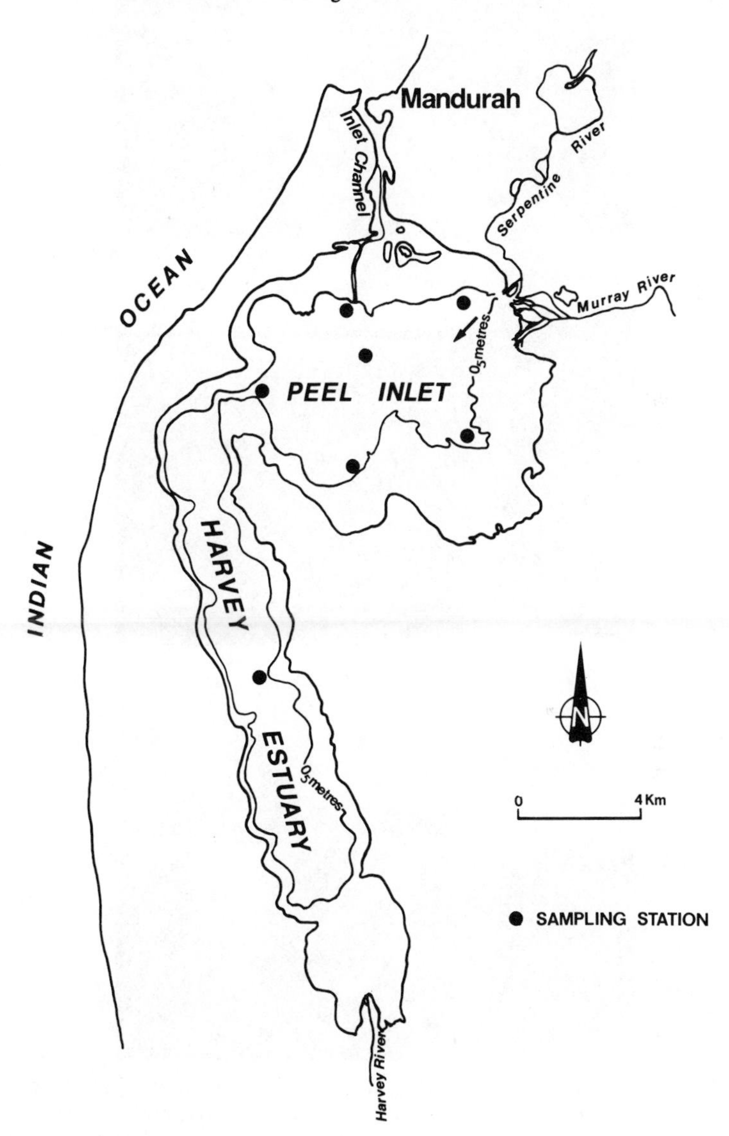

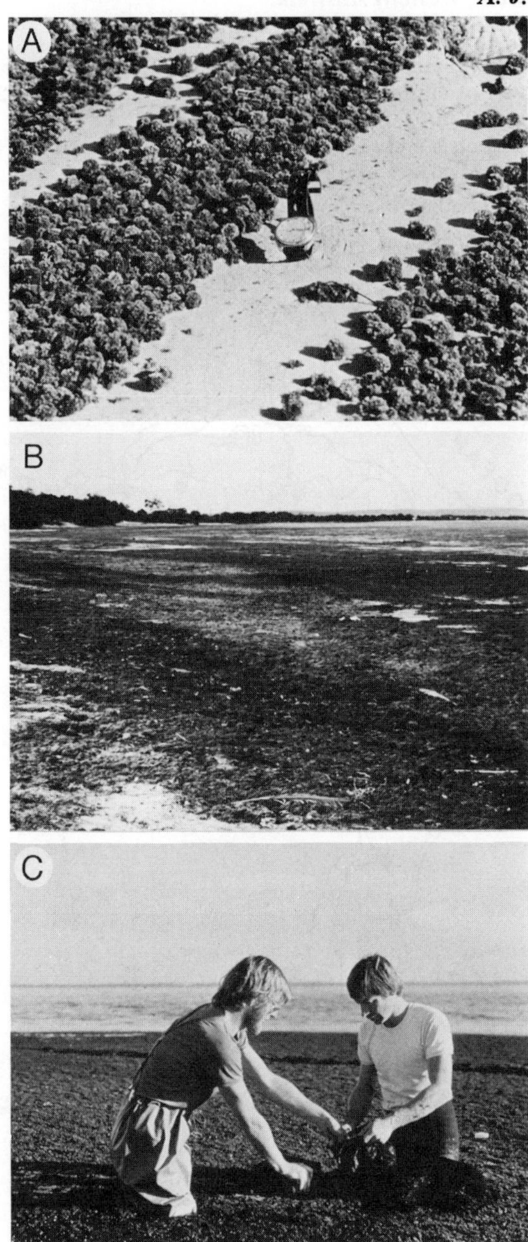

FIGURE 2. *Cladophora* on the shores of Peel Inlet. (A) Algal balls washed onto beach sand; (B) Algal drift, mainly *Cladophora*, on a sandy beach, north-east Peel; (C) An off-shore accumulation of *Cladophora*, near the north-east shore.

series of algal blooms was initiated in mid-winter of 1978; in contrast, no such blooms were initiated during the previous winter. The increase in chlorophyll was quite marked; the mean level rose from 1-2 $\mu g.l^{-1}$, which we find in ocean water, to almost 60 $\mu g.l^{-1}$, levels which are aesthetically unsatisfactory. The species were almost entirely diatoms, except for the early summer peak, which was due to the blue-green *Nodularia* (see below). Common diatom genera were *Pleurosigma, Synedra, Navicula, Rhizosolenia* and *Chaetoceros*, with a high species diversity in summer and autumn, and a low species diversity in winter.

Figure 3b shows mean water temperature for the same sites. In both winters the temperature fell to about 12°, rising to a maximum of 27° in summer. The phytoplankton blooms (Figure 3a) were initiated at the time of lowest temperatures in 1978, but similar low temperatures in 1977 were without effect; clearly temperature alone does not account for these changes in phytoplankton.

Figure 3b also includes salinity. As it happened 1977 was a particularly dry winter, with a low Murray River flow; the total discharge was 80 x 10^6 m^3. In 1978, however, the flow was three times greater (250 x $10^6 m^3$), which is approximately the yearly average. There was nevertheless a marked decrease in the salinity of the estuary water in 1977, but that was largely due to the direct fall of rain (339 mm, June-September) on the shallow estuary (mean depth about 1 m). After fresh water input ceases, salinity rises because of evaporation and tidal exchange with the ocean; evaporation is so high in summer that the estuary becomes hypersaline with respect to seawater, reaching about 45-50 $^o/oo$. Salinity then falls in autumn because evaporation falls with temperature, and so tidal exchange begins to reduce salinity towards $35^o/oo$ well before there is any significant rainfall input. Finally, salinity falls below that of seawater as rain falls on the estuary, and later as rivers flow. As with temperature, salinity change was not in itself responsible for triggering algal blooms. Light penetration was poor at the time of the phytoplankton bloom, mainly because of the presence of stained river water, and in part because of the bloom itself. Attenuation coefficients (calculated from vertical profiles through the water of photosynthetically-active radiation, PAR (4)) in the Peel in summer are $0.46 \pm 0.03 m^{-1}$, but in winter are $0.74 \pm 0.04 m^{-1}$; the Secchi depths are more than 2 m in summer, but as low as 0.3 m in winter.

In contrast, the algal blooms are related to nutrient levels in the water, which rise markedly with river flow. For example, nitrate + nitrite-nitrogen levels are given in Figure 3b. During much of the year these are less than 2 $\mu g.l^{-1}$, but when the rivers flowed in winter 1978, the level increased to 2,000 $\mu g.l^{-1}$; and reference to Figure 3a shows that the increase in phytoplankton in winter immediately followed this increase. In synchrony with nitrate, levels of ammonium-nitrogen (NH_4-N) increased some 15-fold, from 20 to 300 $\mu g.l^{-1}$; the two inorganic forms dominated the total nitrogen levels in the estuary, which rose from 400 to almost

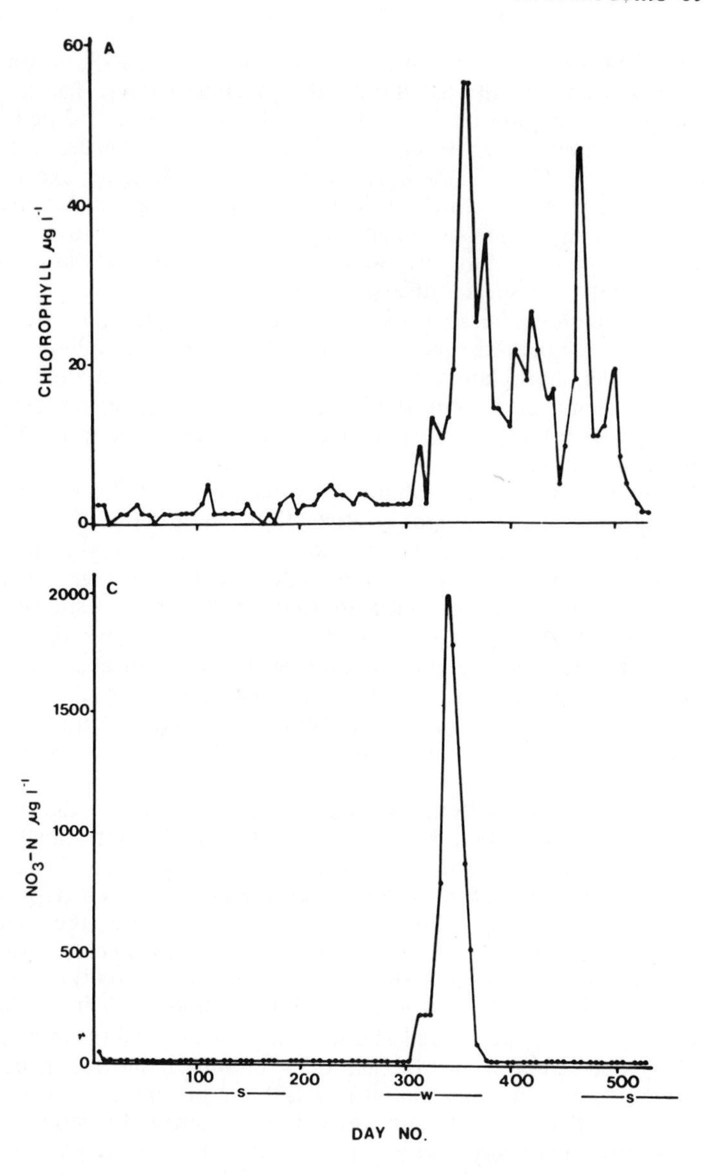

FIGURE 3. Chlorophyll and environmental variables. Each point is the mean for surface water collected at the sites shown in Figure 1. (A) Chlorophyll. Water samples were filtered through glass-fibre papers, and extracted chlorophyll measured spectrophotometrically (9); (B) Temperature and salinity, mea sured in the field (Auto-lab, model 602); (C) Nitrate+nitrite-

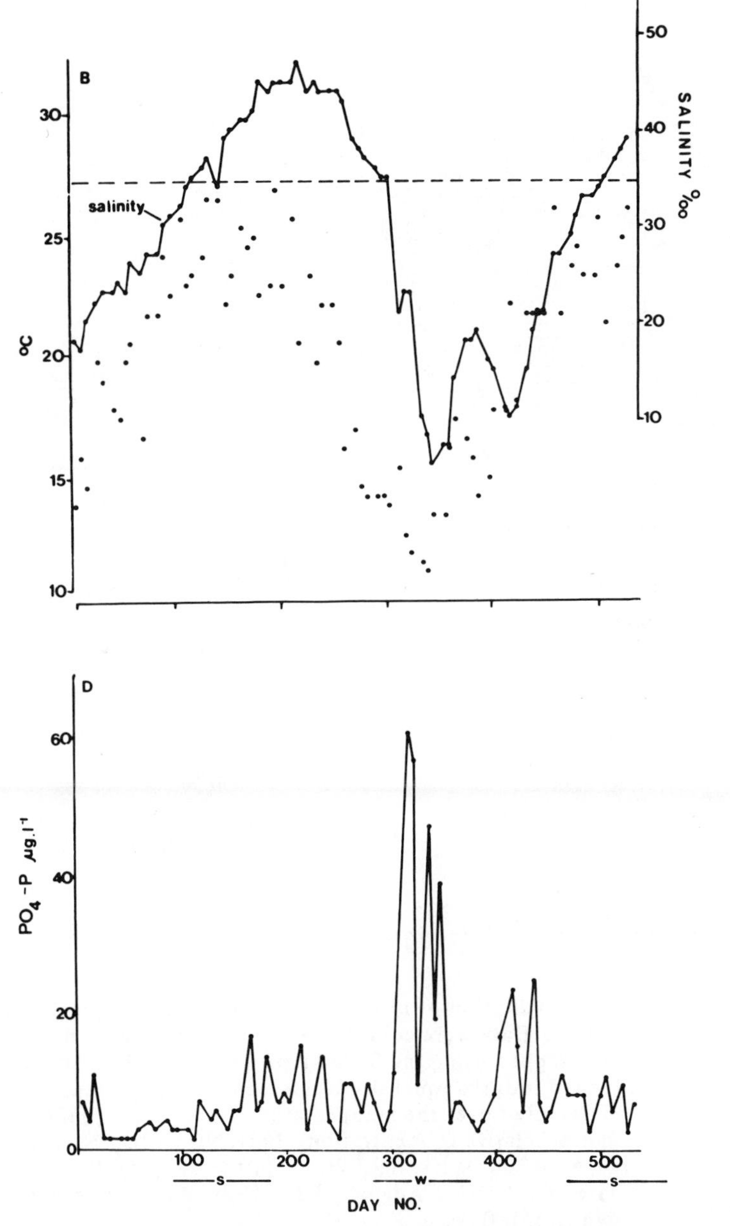

FIGURE 3 (cont'd). nitrogen determined after copper-cadmium reduction with a Technicon Autoanalyzer II; (D) Phosphate-phosphorus, by the single solution method (5). S = summer; W = winter; Day 0 is August 31, 1977.

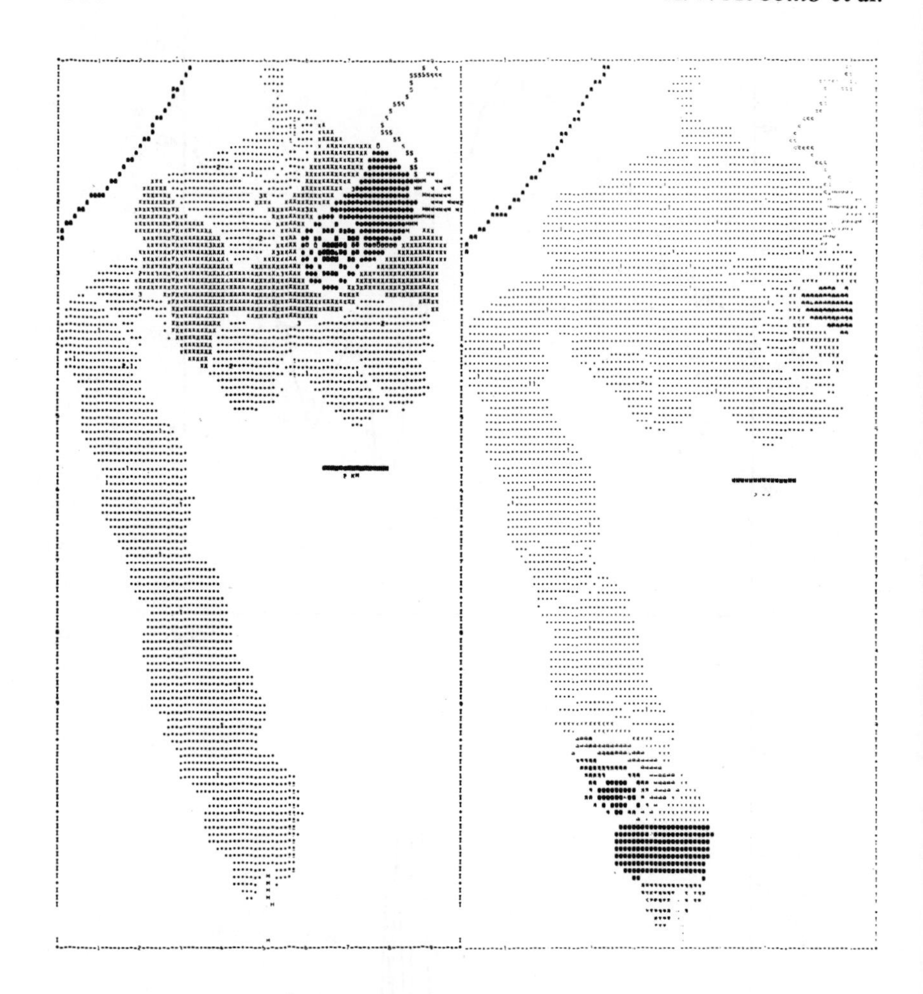

FIGURE 4. The distribution of nitrate and phosphate in the Peel-Harvey system. Data were collected at 36 sites (indicated by units and mapped using the SYMAP program (1). The data are divided into six equal size-classes, and "running means" constructed over the mapped area. Dark shading indicates highest relative concentration. Left, NO_3+NO_2-N in the range 0-1700 $\mu g.l^{-1}$; right PO_4-P in the range 0-50 $\mu g.l^{-1}$. Date of collection August 14-15, 1978, during a period of high river inflow.

4,000 $\mu g.l^{-1}$ at the peak of river flow. Figure 3d shows phosphate-phosphorus (PO_4-P), and total phosphorus rose from 25 to 300 $\mu g.l^{-1}$. It therefore seems reasonable to conclude that the phytoplankton blooms are triggered by river nutrient inputs.

Not surprisingly, the nutrients are not uniformly distributed over the estuary. Figure 4 consists of two computer-drawn maps based on 36 sampling sites in the system at a time of river flow. The NO_3+NO_2-N level was particularly high off the Murray delta, while PO_4-P was relatively high near the Harvey River. This suggests that the Harvey may contribute relatively more P than the Murray, and that the Murray contributes relatively more N. Preliminary estimates of river loading of N and P, based on river nutrient concentrations and flow rates, support this suggestion (3).

LIMITING NUTRIENTS

One may ask whether nitrogen or phosphorus is the primary limiting nutrient which could potentially control phytoplankton levels. N to P ratios were calculated by atom using the inorganic species ((NO_3+NO_2+NH_4)/PO_4). In the summer/autumn period the ratios were relatively low, compared with winter, with a mean of 11:1 for the Peel and 10:1 for the Harvey; it is unlikely that phosphorus would be limiting under these conditions, but it is possible that nitrogen levels may be (7). In contrast, in the period beginning with river flow, "winter and spring", N to P ratios are very high, 240:1 for the Peel and 50:1 for the Harvey, suggesting that at those times nitrogen availability would not be limiting, but that phosphorus may be.

These suggestions are borne out by multiple regressions of results taken during those two periods (Table 1). Regressions for summer/autumn suggest that nitrate/nitrite level is the most important nutrient factor in accounting for chlorophyll variance in both the Peel and the Harvey. In winter/spring it is less important and, if anything, somewhat less so than phosphate.

Algal nutrient limitation assays are also consistent, in general, with these interpretations. For example, the assay summarized in Figure 5a was carried out when the N to P ratio was about 4:1. Addition of phosphorus had no significant effect on phytoplankton level; the addition of nitrogen gave a spectacular increase in phytoplankton; the addition of both, a synergistic effect; and the addition of a complete medium, containing nitrogen and phosphorus plus other macro- and micro-supplements, gave no further increase. Figure 5b is the result of an assay where the N to P ratio was 30:1. The addition of nitrogen initially brought about no effect, a later rise being attributed to the slow release of phosphorus from organic forms in the estuary water. The addition of phosphorus alone gave a small

TABLE 1. Multiple regression analyses of factors accounting for chlorophyll a variance in the Peel-Harvey system.

Winter and Spring[1]

Peel factor[3]	Peel variance[4]	Harvey factor	Harvey variance
Salinity	28.7	Salinity	42.2
PO_4-P	42.3	Oxygen	48.2
NO_3+NO_2-N	44.3	PO_4-P	49.5
NH_4-N	44.6	NH_4-N	50.6
Temperature	45.1	Temperature	51.1
Oxygen	45.1	NO_3+NO_2-N	52.0
n = 21		n = 21	

Summer and Autumn[2]

Peel factor	Peel variance	Harvey factor	Harvey variance
NO_3+NO_2-N	41.3	NO_3+NO_2-N	83.1
Temperature	51.8	Salinity	87.2
Salinity	56.1	Temperature	89.6
Oxygen	57.0	Oxygen	90.1
NH_4-N	58.0	PO_4-P	90.4
PO_4-P	58.2	NH_4-N	90.5
n=28		n=28	

[1] Day numbers 1-70, 301-364 in Fig. 3.
[2] Day numbers 77-266 in Fig. 3.
[3] Relation between chlorophyll a and environmental factors analyses by multiple regression (6).
[4] Percentage of chlorophyll a variance (R^2 value), cumulative, accounted for by listed environmental factors.

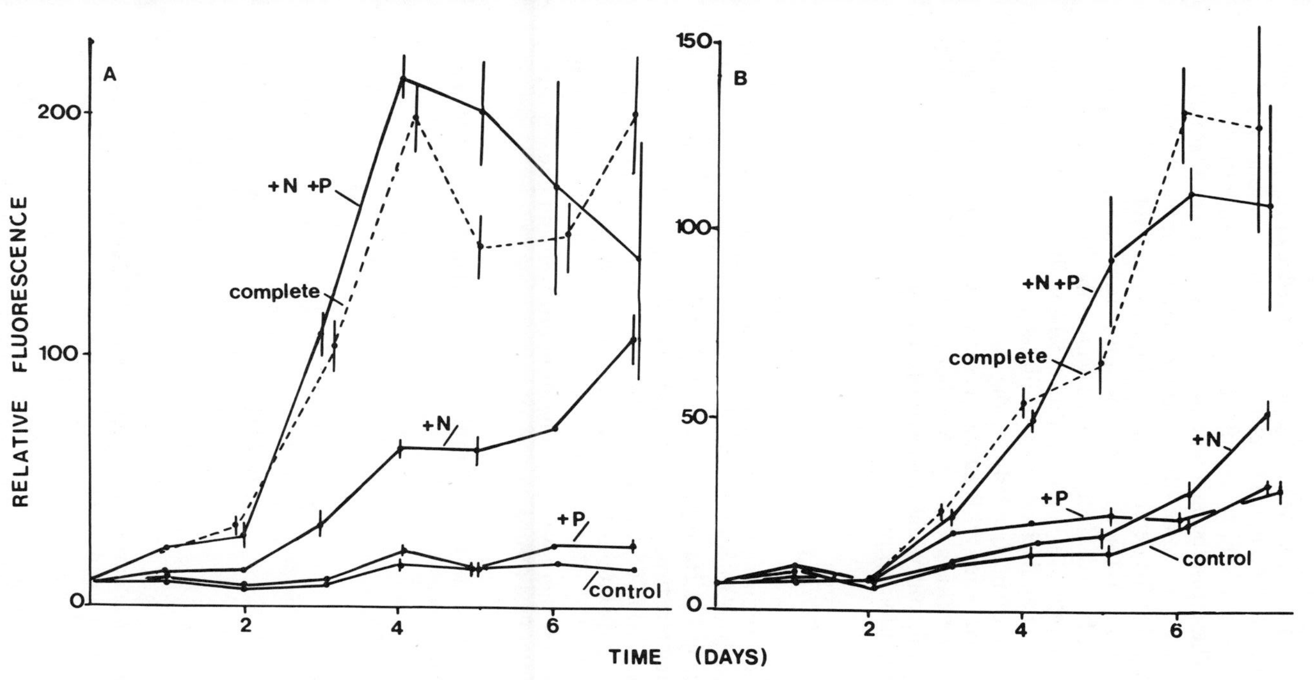

FIGURE 5. Limiting nutrient assays. 500 ml samples of estuary water were placed in a 2l flask, held in a water bath at the temperature of the estuary and under natural illumination in a glasshouse. Nutrient additions increased the natural levels of the water by P, 155 $\mu g.l^{-1}$ as KH_2PO_4; N, 310 $\mu g.l^{-1}$ as NH_4NO_3; complete medium, P and N at stated levels plus trace elements and vitamins (10). Phytoplankton was read as "relative fluorescence" using a fluorometer (Turner Designs Model 10) after addition of 3-(3, 4 dichlorophenyl)-1, 1-dimethyl urea, DCMU (8); (A) Assay for water with low N:P ratio; (B) Assay for water with high N:P ratio.

effect; while both nutrients together gave a marked increase, comparable with that of complete medium. The effect of phosphorus alone was small, presumably because both nitrogen and phosphorus were initially at low concentrations. These results, incidentally, provide evidence that nitrogen and phosphorus, and not some other element, are the nutrients which are potentially responsible for determining phytoplankton levels.

Changes in phytoplankton species sometimes occurred late during these assays and often reflected changes which were subsequently observed in the field. For example, in some cultures when nitrogen had become exhausted, diatom populations were replaced by blue-greens. Such a replacement occurred spectacularly in the summer of 1978-1979 in the Harvey (Figure 3a), when N:P ratios were initially low. There was a massive bloom of the blue-green alga *Nodularia*, which reached almost 300 $\mu g.l^{-1}$ chlorophyll a, and which was actively fixing nitrogen as shown by the acetylene reduction assay*.

COMPARISON OF PEEL AND HARVEY

The *Nodularia* bloom mentioned above was largely confined to the Harvey, but blooms of diatom-dominated phytoplankton populations are also often higher in the Harvey than the Peel (Table 2). The water of the Harvey was characteristically more turbid than the Peel; for example, in summer the attenuation coefficients for the Peel were $0.45 \pm 0.03 m^{-1}$, and for the Harvey, $0.61 \pm 0.09 m^{-1}$. This was due not only to the higher levels of phytoplankton, but also to other suspended material. During the summer and autumn there was typically more nitrogen and phosphorus in the Harvey than the Peel (Table 2).

TABLE 2. Concentrations of chlorophyll and nutrients in the Peel and Harvey basins[1].

	Peel	Harvey
Chlorophyll	2.42 (0.23)	6.22 (1.18)
Total Nitrogen	488 (41)	770 (59)
Total Phosphorus	35.8 (4.4)	93.7 (15.8)

[1] Data for 28 weeks in spring and autumn, days 77-266 in Fig. 3. All data in $\mu g.l^{-1}$, with standard errors.

*Acetylene assays by A. Huber and D. Kidby, Department of Soil Science and Plant Nutrition, University of Western Australia.

Figure 6 shows the levels of chlorophyll a in September, 1978, accompanied by the results of a productivity assay in which ^{14}C fixed is expressed per unit chlorophyll a. The dominant planktonic organism was the diatom, *Rhizosolenia*. The phytoplankton from the Peel were behaving less efficiently than those from the Harvey under the same light and temperature conditions. The observations are consistent with the suggestion that phytoplankton standing crops are generally lower in the Peel, not because of relatively higher grazing or flushing rates there, but because of some property such as lower nutrient levels there.

CLADOPHORA LIFE HISTORY AND DISTRIBUTION

Most of the *Cladophora* in the estuary is due to vegetative propagation, even though we have on rare occasions observed sporing of *Cladophora* in the field. There are beds of detached balls and other fragments of the alga more or less permanently in some regions of the estuary, and these populations are not lost and recruited from spores each year. The alga can be shaken or teased apart, and the fragments will readily form new balls when cultured in the laboratory.

Figure 7 shows the distribution of the alga in autumn; accumulations were present in three shallow regions, and there is an area of the alga in somewhat deeper water (>1 m). The map does not include banks of alga on the shores. Using maps of biomass generated in this way, the total amount of *Cladophora* in the estuary was estimated as very approximately 20,000 metric tons dry weight.

Because the alga is a perennial which may drift about because of the interplay between density change and water movement, it is difficult to characterize the features which control its growth. One approach has been to chose an area where a bed of alga in water about 1.5 m deep appears more or less permanent, and simply measure the biomass at monthly intervals (Figure 8). There was a marked increase in the alga in the early summer of 1976, and we may hypothesize that this represented a natural period of growth. However, biomass remained fairly stable until the winter of 1978, when the water of the estuary became very turbid; then there was a marked depletion of alga, representing transport out of the site.

To help interpret these observations, we placed samples (about 2 g dry wt) of the alga in perforated flasks on the algal bed, and harvested and weighed them after they had grown a month. Growth of these imprisoned populations of the alga are also given in Figure 8; in two successive years growth occurred in the warmer period of the year, providing evidence that the organism is essentially a summer-growing one, and that the increase in biomass observed in spring and summer of 1976 was due to growth at the site. Thus, in marked contrast to phytoplankton, *Cladophora* grows in spring and summer when nutrient levels are relatively low in the water column, but when temperatures and light intensity are high. Broadly

FIGURE 6. Distribution of chlorophyll (left) and assimilation number (right) at 20 sites in the Peel-Harvey system in September, 1978. Mapping technique as described under Figure 4. Water samples were collected and analyzed for chlorophyll (as in Figure 3a) (the incorporation was measured (2)) with 100 ml samples incubated for 3-4.5 hr at 23°C and 200 $\mu E.m^{-2}.sec^{-1}$, after addition of 2 $\mu Ci.ml^{-1} NaH^{14}CO_3$. The range of chlorophyll is 4 - 50 $\mu g.l^{-1}$, and assimilation numbers (carbon fixed per unit chlorophyll) 1.1 - 8.8 mg $C.mg\ chl^{-1}.hr^{-1}$.

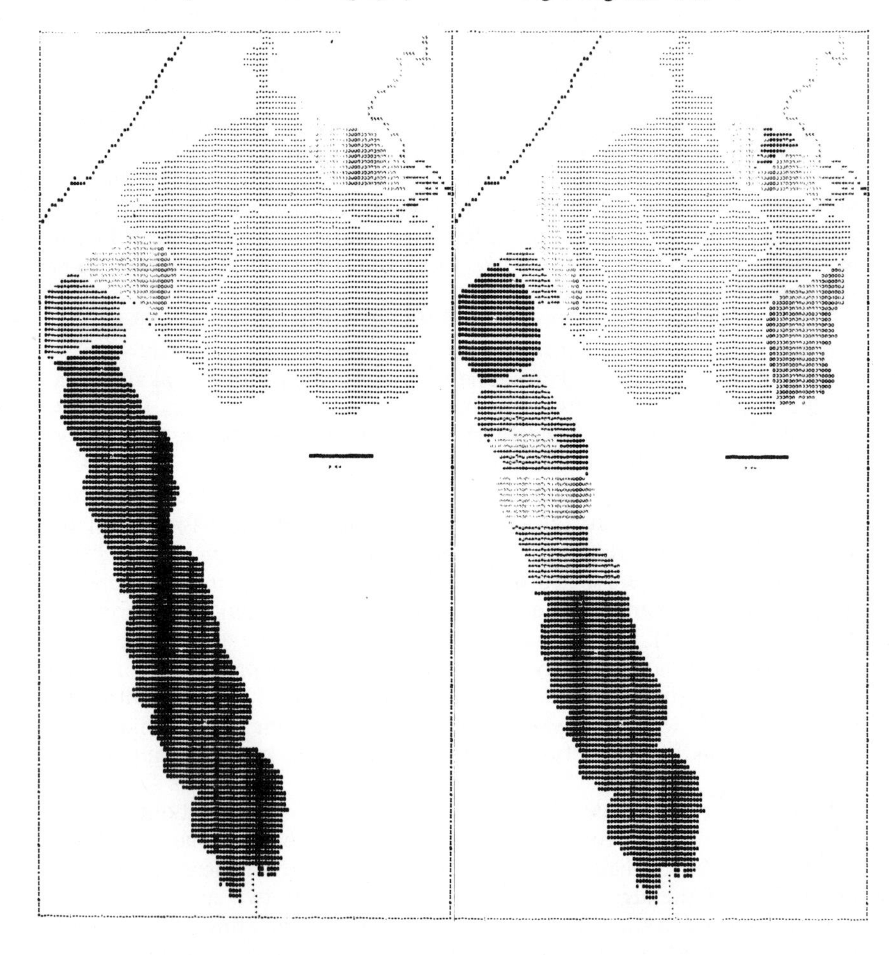

FIGURE 7. The distribution of *Cladophora* in the Peel-Harvey system, March, 1978. Data for percentage cover were collected from the 36 sites represented by numerals, and graphed by the method explained under Figure 4. The range is from 0-100 percent, divided into five equal divisions.

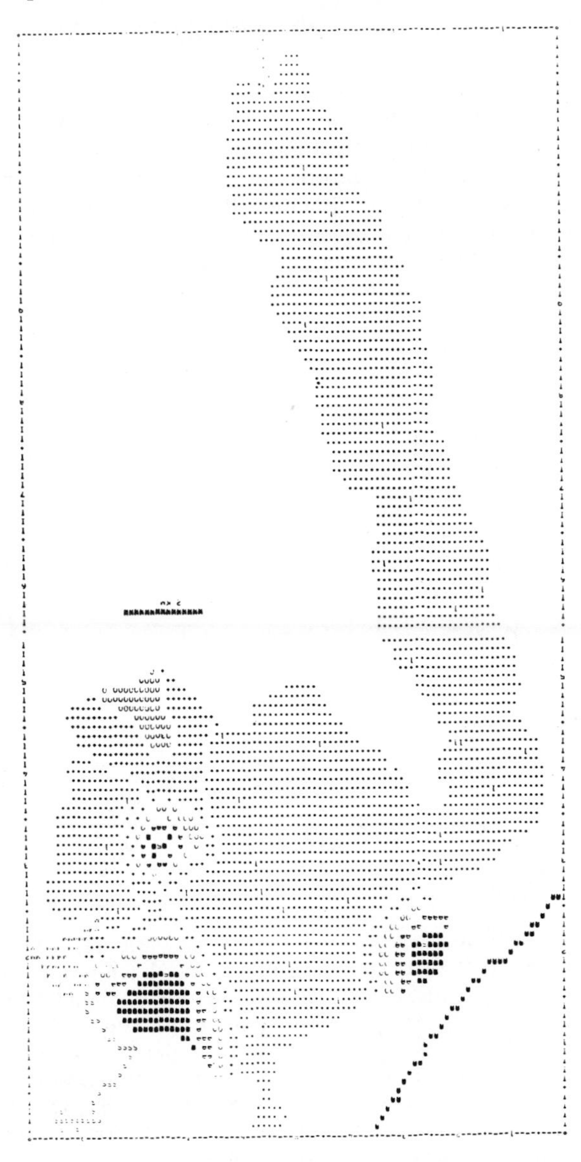

speaking, the absence of *Cladophora* from the Harvey appears to be related, at least in part, to the higher water turbidity there, resulting from phytoplankton and other suspended material.

ENVIRONMENTAL FACTORS AND CLADOPHORA *GROWTH*

Further evidence on control of growth comes from laboratory studies. Salinity in the range 2.5-60°/oo, exceeding that encountered in the estuary, did not significantly affect growth of the alga over a period of 14 days. The effect of temperature and light intensity on rate of photosynthesis is shown in Figure 9; high rates occur at high temperatures, while low temperatures (at the levels encountered in the estuary in winter) reduce photosynthesis markedly.

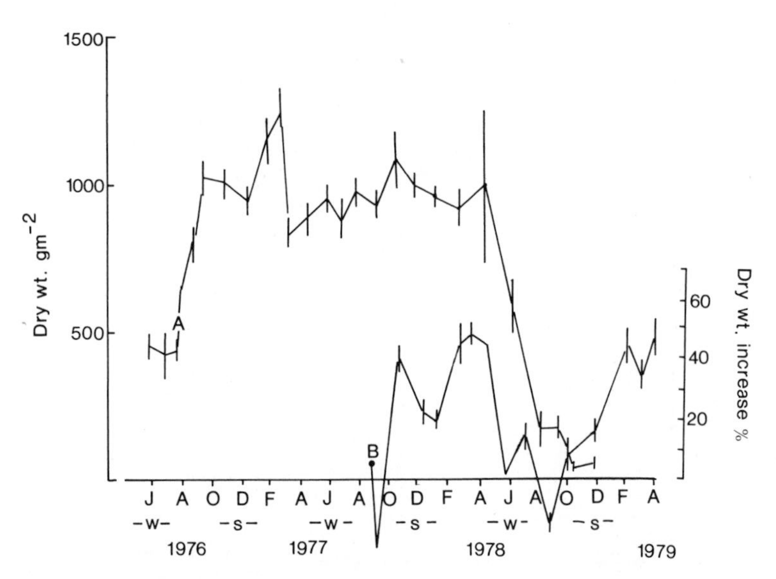

FIGURE 8. Changes in *Cladophora* biomass. (A) Biomass of a field population at the site shown in Figure 1a. Each point is the mean dry weight of 20 samples taken with a 625 cm^2 sampler. (B) Change in biomass in imprisoned populations. Samples of the alga, 1-2 dry weight, were placed in perforated culture flasks; 20 initial samples were harvested, and 20 placed in the algal bed and harvested some 30 days later. Dry weights were expressed at percent of mean initial dry weight. Vertical bars show standard errors. S = summer; W = winter.

In such experiments, which use small pieces of alga, the compensation point is 15-20 μE.m^{-2}.sec^{-1} (about 1 percent of full sunlight PAR) while photosynthesis is saturated at 250-300 μE.m^{-2}.sec^{-1} (14-17 percent of full sunlight PAR). In the field these "saturating" levels are often reached at the floor of the estuary in summer, but we should not be misled into thinking that at those times, therefore, light is not limiting *Cladophora* productivity. On the contrary, even if full sunlight reached the surface of an algal bed, the algal cells 1 cm into the bed would be below the compensation point (Figure 10), and beds are often more than 10 cm deep. There is little doubt that light must be the primary limiting factor in the estuary, even in shallow water.

FIGURE 9. Effect of light and temperature on *Cladophora* photosynthesis. A sample of alga 5.8 mg fresh weight was placed in estuary water and initially bubbled with .03 percent CO_2 in nitrogen in the cell of a Clark-type oxygen electrode (Rank Bros. Bottisham, U.K.). Illumination was provided with a fibre-optics light source (KL150B, Schott, Mainz, West Germany) and measured with a quantum sensor (Lambda Instruments, Nebraska, USA). Each point is from the slope of a 5-10 min chart record.

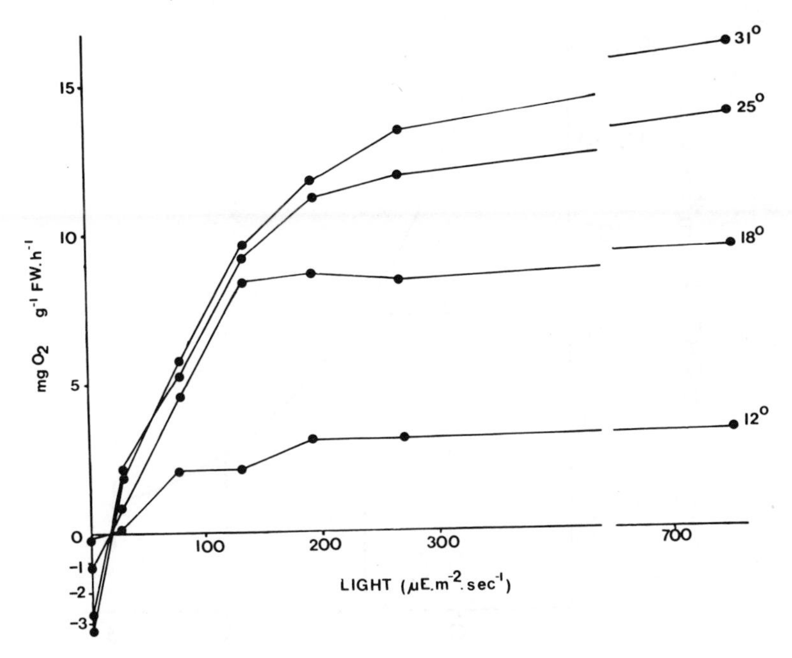

Nevertheless, the alga requires nutrients to grow and to maintain the observed populations, and the amounts of nutrients incoporated are quite large. The increase in biomass in 1976, shown in Figure 8, represents the utilization by the alga, in 12 weeks, of 30 g N and 3.5 g P per square metre, in the summer period when concentrations of nutrients in the water above the alga were very low. On the other hand, levels of total nitrogen and phosphorus in the rotting organic layer below the living alga were at levels quite comparable with those in the algal bed.

This suggests that the alga obtains most of its nutrients not from the water column above, but from decomposing material below. For this reason measurements have been made of the concentrations of nutrients in the water surrounding the algal balls ("interalgal" water; Table 3). Higher concentrations occur in the algal bed than in the water column, consistent with the suggestion that there is indeed a diffusion of nutrients away from the rotting material beneath the algal bed, through the alga, to the water column above. The very occurrence of these relatively higher concentrations supports the conclusion that the level of nutrients may not in fact be limiting algal growth in the beds under field conditions.

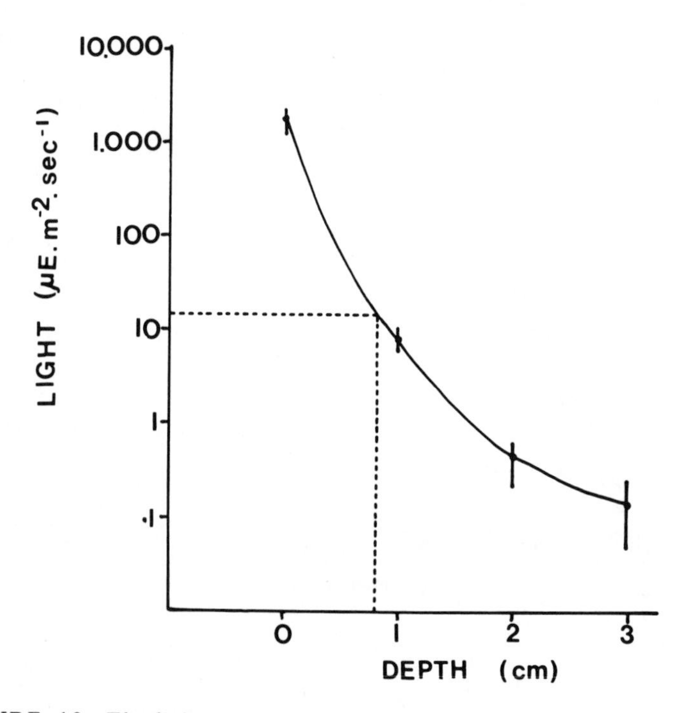

FIGURE 10. The light intensity (PAR) reaching different depths in a bed of *Cladophora*.

TABLE 3. Concentrations of nutrients in the water between *Cladophora* balls, compared with the water above tha algal bed[1].

	Water µg.l-1	Interalgal µg.l-1	Ratio. Interalgal water:water column
NH_4-N	43 (16)	186 (52)	4.3
NO_3+NO_2-N	3 (0.6)	7 (1.3)	2.3
Organic N	828 (260)	1845 (352)	2.2
Total N	874 (253)	2038 (352)	2.3
PO_4-P	6 (1)	93 (25)	15.5
'Organic'-P	27 (11)	145 (67)	5.4
Total P	33 (10)	238 (69)	7.2

1 Each datum is the mean of 5 determinations, (with standard error) taken at monthly intervals in summer and autumn of 1978.

CONCLUSIONS

Our studies so far suggest that *Cladophora* grows best in summer when light and temperature are high, and that growth in the algal beds relies upon recycled nutrients from below rather than on nutrients from the water column above. On the other hand, the major nutrient input to the system occurs in winter, during river flow, and so a mechanism must be sought for the long-term accumulation of the bank of nutrients in the *Cladophora* and associated sediments. One possibility under investigation is that *Cladophora* takes up N and P during winter for later redeployment in growth. Another is that there is an input of particulate nutrients during winter. However, there is very little particulate nitrogen and phosphorus in the river water, and we suggest, therefore, that the phytoplankton blooms which follow river nutrient input offer a mechanism by which nutrients become trapped, sediment out, and are later recycled, becoming available for the growth of *Cladophora* under conditions of high light and temperature.

ACKNOWLEDGEMENTS

The Department of Conservation and Environment of Western Australia provided financial assistance. We are indebted particularly to Dr. E.P. Hodgkin of that department, and to other members of the study team, for their cooperation and advice; and to Dr. R.B. Humphries, Australian National University, for introducing us to the SYMAP technique.

REFERENCES

1. Dougenik, J.A. and Sheehan, P.E. 1977. Symap User's Reference Manual. Fifth edition. Laboratory for Computer Graphics and Spatial Analysis, Harvard University, Cambridge, Mass.
2. Hall, C.A.S., and Moll, R.A. 1975. The measurement of primary productivity in water. *In* H. Leith and R. Whittaker (eds.), Primary Productivity of the Biosphere. Springer-Verlag, Berlin, Heidelberg and New York.
3. Hodgkin, E.P., and Lenanton, R.C. 1979. Estuaries and coastal lagoons of south western Australia. *In* B. Neilson and A. Cronin (eds.), Nutrient Enrichment in Estuaries. Humana Press, New Jersey.
4. Kirk, J.T.O. 1977. Use of a quanta meter to measure attenuation and underwater reflectance of photosynthetically active radiation in some inland and coastal south-eastern Australian waters. *Aust. J. Mar. Freshwater Res.* 28(1):9-21.
5. Major G.A., Dal Pont, G.K., Klye, J., and Newell, B. 1972. Laboratory Techniques in Marine Chemistry - A Manual. Report No. 51, Division of Fisheries and Oceanography, Commonwealth Scientific and Industrial Research Organization, Cronulla, N.S.W., Australia.
6. Nie, N.H., Jenkins, C.H., Steinbrenner, K., and Bent, D.H. 1975. Statistical Package for the Social Sciences, second edition. McGraw-Hill, New York.
7. Ryther, J.H., and Dunstan, W.M. 1971. Nitrogen, phosphorus and eutrophication in the coastal marine environment. *Science* 171(3975):1008-1013.
8. Slovacek, R.E., and Hannan, P.J. 1977. In vivo fluorescence determinations of phytoplankton chlorophyll a. *Limnol. Oceanogr.* 22(5):919-925.
9. Strickland, J.D.H., and Parsons, T.R. 1972. A Practical Handbook of Seawater Analysis, second edition. Bulletin No. 167. Fisheries Research Board of Canada, Ottawa.
10. Thayer, G.W. 1973. Identity and regulation of nutrients limiting phytoplankton production in the shallow estuaries near Beaufort, N.C. *Oecologia* 14(1):75:92.

THE USE OF NUTRIENTS, SALINITY AND WATER CIRCULATION DATA AS A TOOL FOR COASTAL ZONE PLANNING

Y. Monbet*, F. Manaud*, P. Gentien, M. Pommepuy*, G.P. Allen*, J.C. Salomon**, J. L'Yavanc***

**Centre Oceanologique de Bretagne, Departement Environnement Littoral et Gestion du Milieu Marin, B.P. 337, 29273 Brest Cedex, France*

***Universite de Bretagne Occidentale, 20 avenue Le Gorgeu, 29200 Brest, France*

ABSTRACT: An oceanographic study of the Bay of Brest (France) was conducted in 1977, together with a hydrological study of the rivers entering the Bay.

Samples were collected from rivers and analyzed for dissolved nutrients (nitrate, phosphate, ammonium, and reactive silicate), in order: 1) to assess the extent of the area influenced by terrigeneous inputs within the Bay; 2) to elaborate management zoning recommendations.

The relationship between river discharge and variations of nutrient concentration was examined by regression analysis. Tentative solute budgets (observed and calculated) for nitrogen and phosphorus in the drainage basins show, for the two main rivers discharging into the Bay, a specific contribution of both nutrients: the southern river (Aulne) discharges 66 percent of the total input of N, and the northern river (Elorn) contributes 70 percent of the total input of P.

With the aid of land use analysis and watershed runoff studies, loading rates have been calculated from land use categories (agriculture, industries, and domestic sewage). The relationships between N and P from agriculture, industry, and urban activity show a difference between the southern and the northern part of the Bay. The predominating type of activity in the drainage basin controls the N/P ratio in the areas of the Bay affected by the runoff from each watershed. These values are lower in the northern part of the Bay (intensive urbanization and industry) than in its southern part (intensive agriculture).

At the same time, samples were collected to investigate the processes controlling the distribution of nutrients in the Bay. Detailed measurements of nutrients, salinity, and currents were conducted during high river flow. Observations of the surface distribution of the silicate to phosphate ratio enabled the identification of distinct water masses originating from the different fluvial drainage basins; it was also possible to materialize the movement and dispersion patterns of the different water masses.

Finally, a comprehensive zoning plan for the development of the activities around the Bay was established. This zoning was based on water-quality criteria, existing and future pollution hazards, and the oceanographic mechanisms affecting the Bay.

INTRODUCTION

The Bay of Brest is located on the northwest coast of France (Figure 1). The Bay is approximately 11 km wide and 27 km long from the extreme eastern end to the extreme western end.

Three main tributaries, the Aulne River, Elorn River and Penfeld River enter the Bay respectively in southern, northeastern, and northern part. The total area at high tide is approximately 180 km^2 and the total drainage basin has an extent of 2,655 km^2. Perhaps the most striking feature of the Bay is the tidal influence, resulting in strong tidal currents up to 1 $m.s^{-1}$ and large water movements. Average depth of the Bay is 10 m, with a maximum depth increasing to 40 m at the mouth.

In recent years, there has been an increase in livestock-raising, use of fertilizers, population, sewage networks, and industry. By contrast, fishing activities, mainly represented by oyster production and scallop fishing, have been declining.

In order to avoid mismanagement of the local resources, a program of studies called *Schema d'Aptitudes et d'Utilisation de la Mer* (S.A.U.M.) has been established to develop an understanding of the oceanographic environment of the Bay and its relations with the activities of the region.

The National Center for the Exploitation of Oceans (CNEXO) has been designated to conduct the oceanographic study. Although some previous studies have been made on the physical oceanography of the Bay of Brest, there have been few chemical investigations until recent years. Le Corre and Treguer (1), who studied the chemical characteristics of the Bay of Biscay, commented upon the paucity of chemical data on the Bay. In 1966-1967, an extensive study of water circulation, salinity, and suspended sediments was conducted by Auffret and Berthois (2), who pointed out for the first time that the Bay has estuarine characteristics. Since 1974 a monitoring program including six sampling stations has recorded major oceanographic and pollution parameters twice a month (3). Finally, a recent pollution study of the Bay has given some insight on the behavior of dissolved nutrients (4).

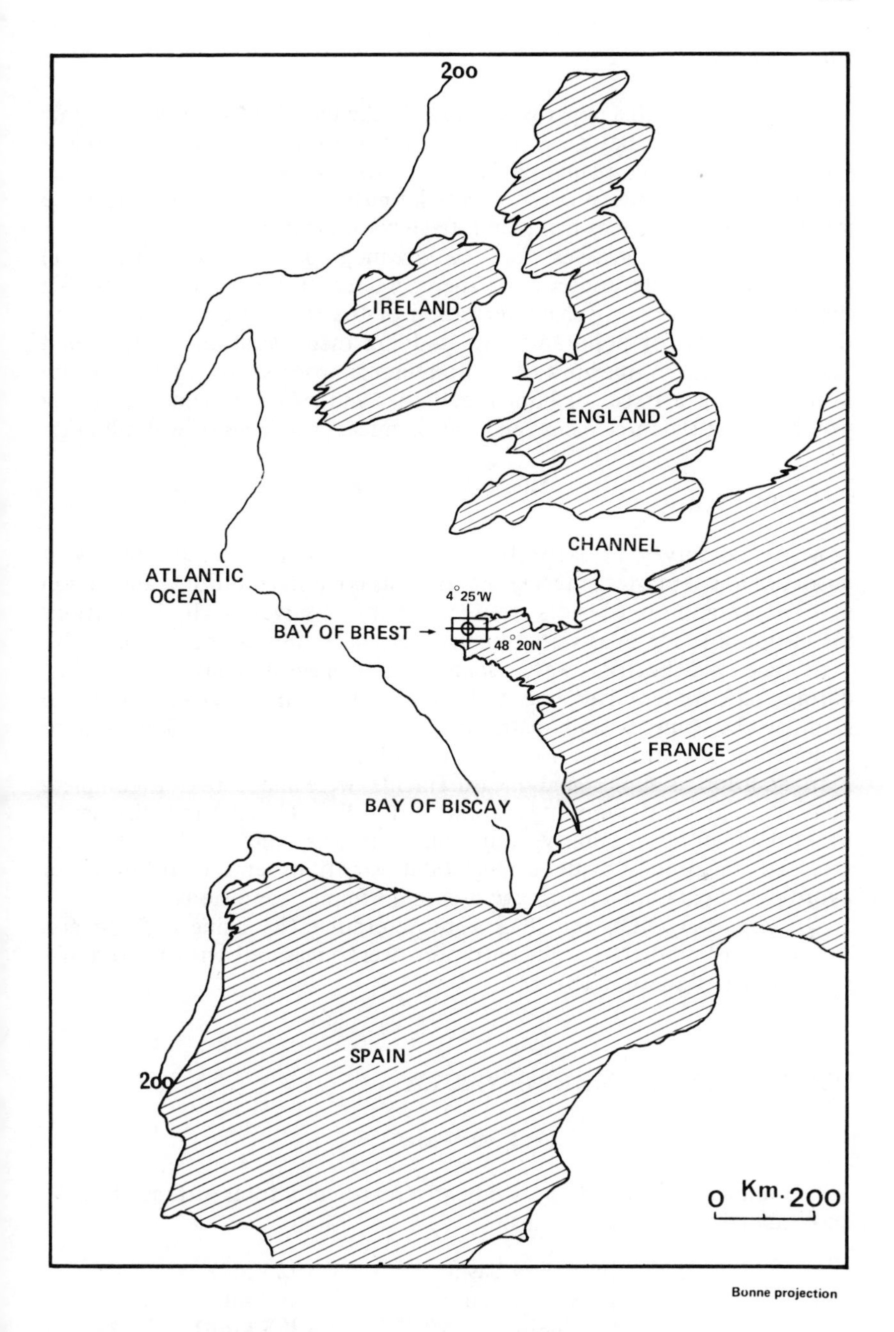

FIGURE 1. Geographical setting of the Bay of Brest.

OBJECTIVES

The purpose of this paper is to illustrate the use of major oceanographic characteristics of a Bay (i.e., nutrients, circulation, and salinity data) as a tool for coastal zone planning. The basic idea is that, like other estuarine systems, the dynamic processes and the nutrient distribution reflect the interaction of influences from land, fresh water, and the sea.

The specific objectives for the oceanographic program were: 1) identification of the sources and the characteristics of nutrients entering the Bay (N and P); 2) characterization of the transport mechanisms for these nutrients (or pollutants) by studying their variations in time and space within the framework of hydrodynamic processes of the Bay; 3) use of nutrients as a tracer of human activities and establishment of a zoning plan based on the movements and the chemical properties of water bodies.

Methods and Materials

A general survey to assess the distribution, abundance, and inflow of nutrients in tributaries entering the Bay was conducted during each season of 1977. In addition, results from systematic observations by institutions (5) responsible for river water quality were taken into account (Figure 2).

Surface and bottom water samples and measurements were taken during slack water after ebb and flood. These measurements were to identify the sources and distribution of nutrients in the Bay and to evaluate the seaward dispersion rate of fresh water inflow.

In addition, samples and measurements were made over a complete tidal cycle at eight sampling stations in the Bay. Current measurements, water temperature, salinity, nutrients, and suspended sediments were analyzed in order to determine tidal variations and distribution of hydrological parameters during winter and summer conditions.

$Si/(OH)_4$-NO_2^--NO_3^- and PO_4^{3-} were determined by using a Technicon Autoanalyzer while NH_4 concentrations were measured with the aid of a non-automated method (8).

INPUTS TO THE ESTUARINE SYSTEM

River Flow

The drainage basins of the study area represent a total surface of 2,655 km^2 subdivided into five basins:

-northern Bay basin	(93 km^2)
-Elorn River basin	(402 km^2)
-middle basin	(288 km^2)

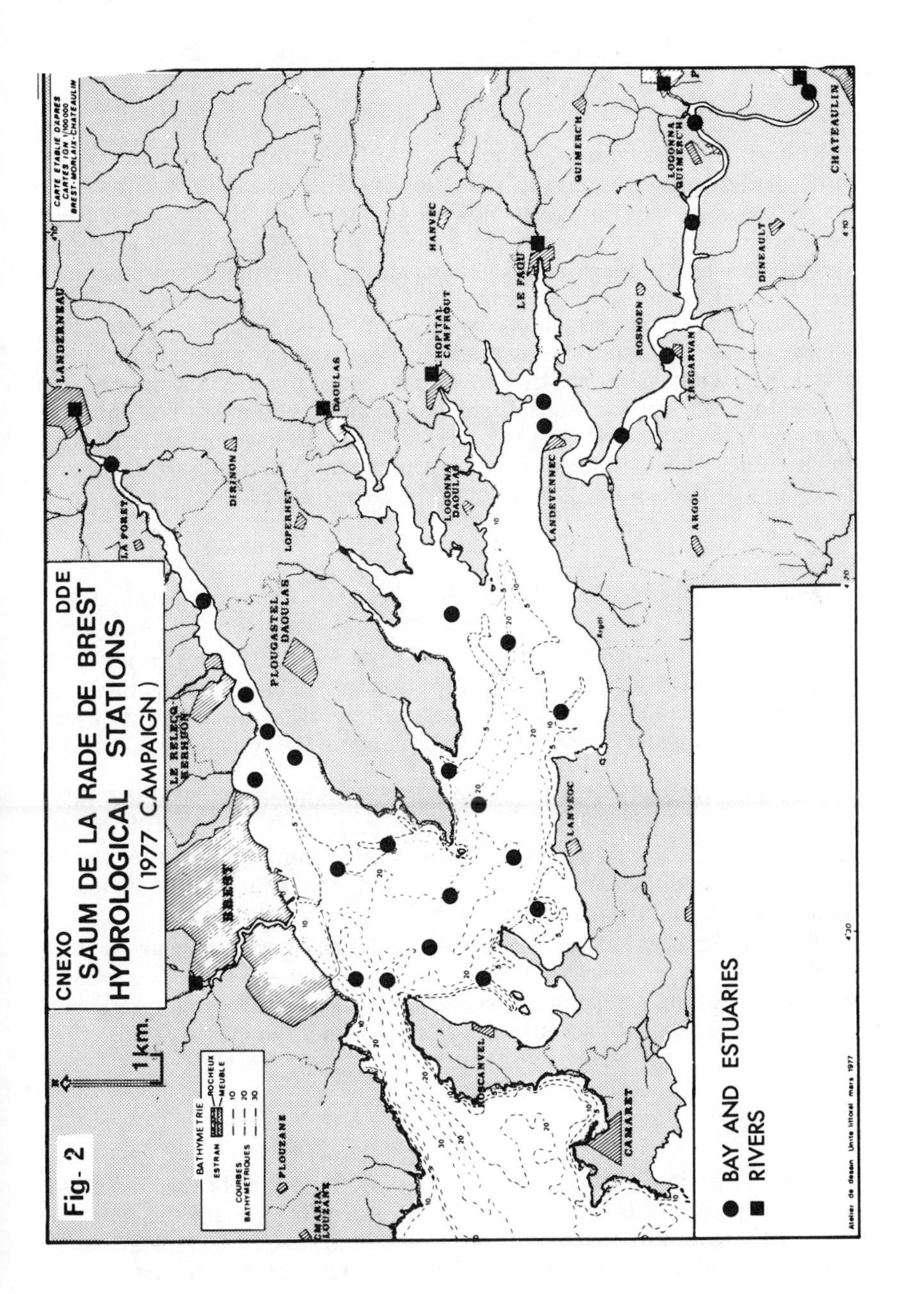

FIGURE 2. Hydrological and nutrient sampling stations in the Bay of Brest-winter and summer, 1977.

-Aulne River basin (1842 km^2)
-southern Bay basin (30 km^2)

Runoff and heavy seasonal rainfall, both significant inputs of major nutrients, enhanced an erratic river flow regime. The ratio between the highest average monthly river flow (February) and the lowest average monthly river flow is five for the Penfeld River. It increases southward and reaches 18 for the Aulne River. As shown in Figure 3, seasonal river flow fluctuations are also very large.

During river floods, which occur between November and February, the average combined monthly discharge of the rivers is approximately 90 m^3s^{-1}. The Aulne River contributes 66 percent of the total inflow. During heavy flood the instantaneous discharge can attain 400 m^3s^{-1} but only for short periods of time. During summer early fall, average monthly flow is on the order of 20 to 30 m^3s^{-1} with a minimum value of 5 to 10 m^3s^{-1}. Total annual fresh water inputs for 1977 are estimated at 1.27 x 10^9 m^3.

TIDES

The tides in the Bay are semi-diurnal with a 12 h 25 mn period. Tidal amplitudes vary from 3 m during neap tides to 7 m during spring tides. As shown by Bassoullet (6), a marked asymmetry of the time height curves develops in the Aulne upper estuary, prolonging the ebb and reducing the flood.

The tidal prism, i.e., volume of water introduced at the flood tide, varies from 0.5 x 10^9 m^3 during neap tides to more than 1.0 x 10^9 during spring tide (2). Figure 4 shows the ratio of tide volume to river flow in the Bay of Brest and the Aulne estuary. It appears that stratification increases with river flow and decreases with increasing tide range.

These differences between neap-spring tides and low and high river flow indicate markedly different hydrological behavior of the estuaries and Bay, with varying fluvial and tidal regimes. Any oceanographic sampling program must take into account these seasonal and tidal fluctuations.

THE NUTRIENT LOAD

Both inorganic nitrogen (i.e., N. NO_2 - N.NO_3 - N.NH_4) and dissolved inorganic phosphorus were measured in order to estimate the nutrient load discharged into the Bay. Two methods were used to evaluate total nutrient loading: 1) a calculated budget, and 2) a measured budget.

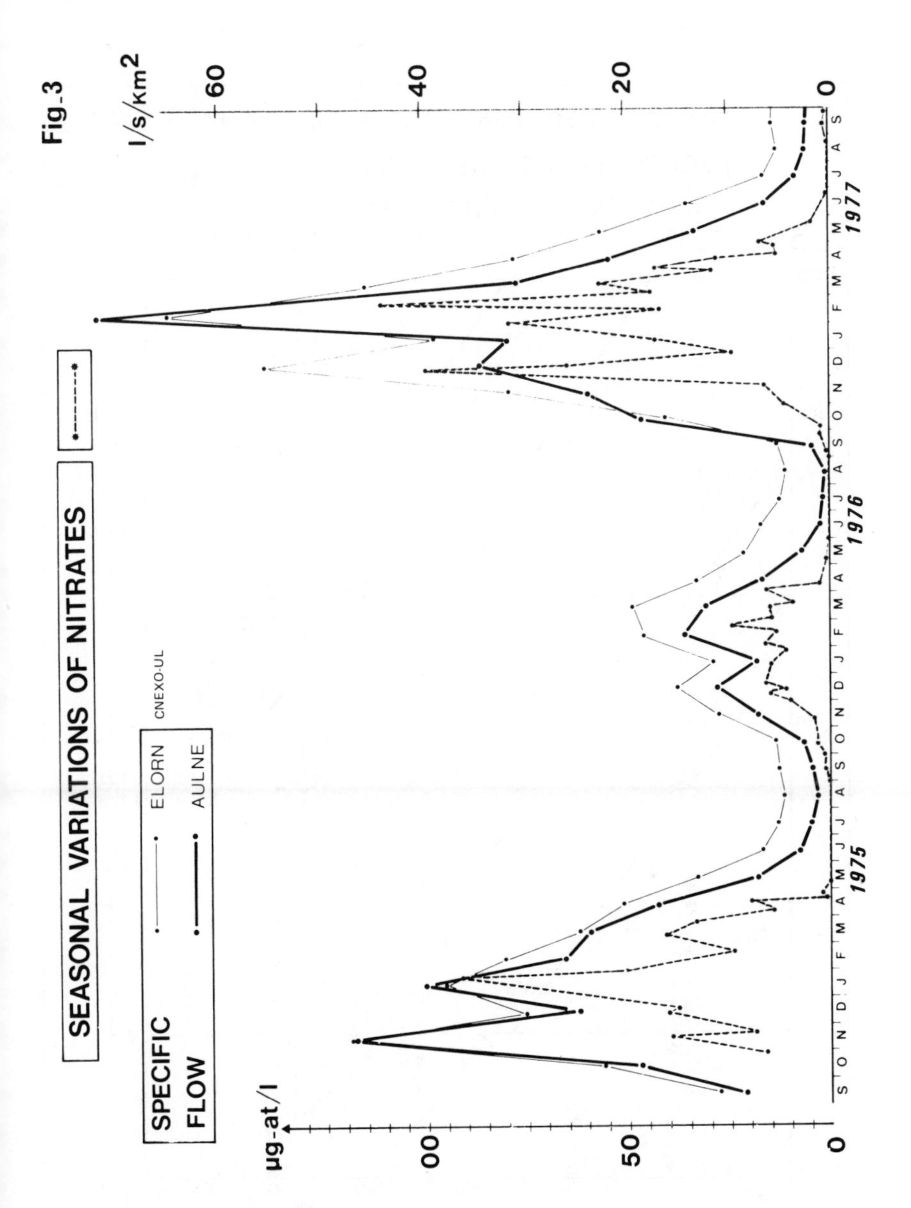

FIGURE 3. Seasonal variation of river flow and nitrate concentration in the bay.

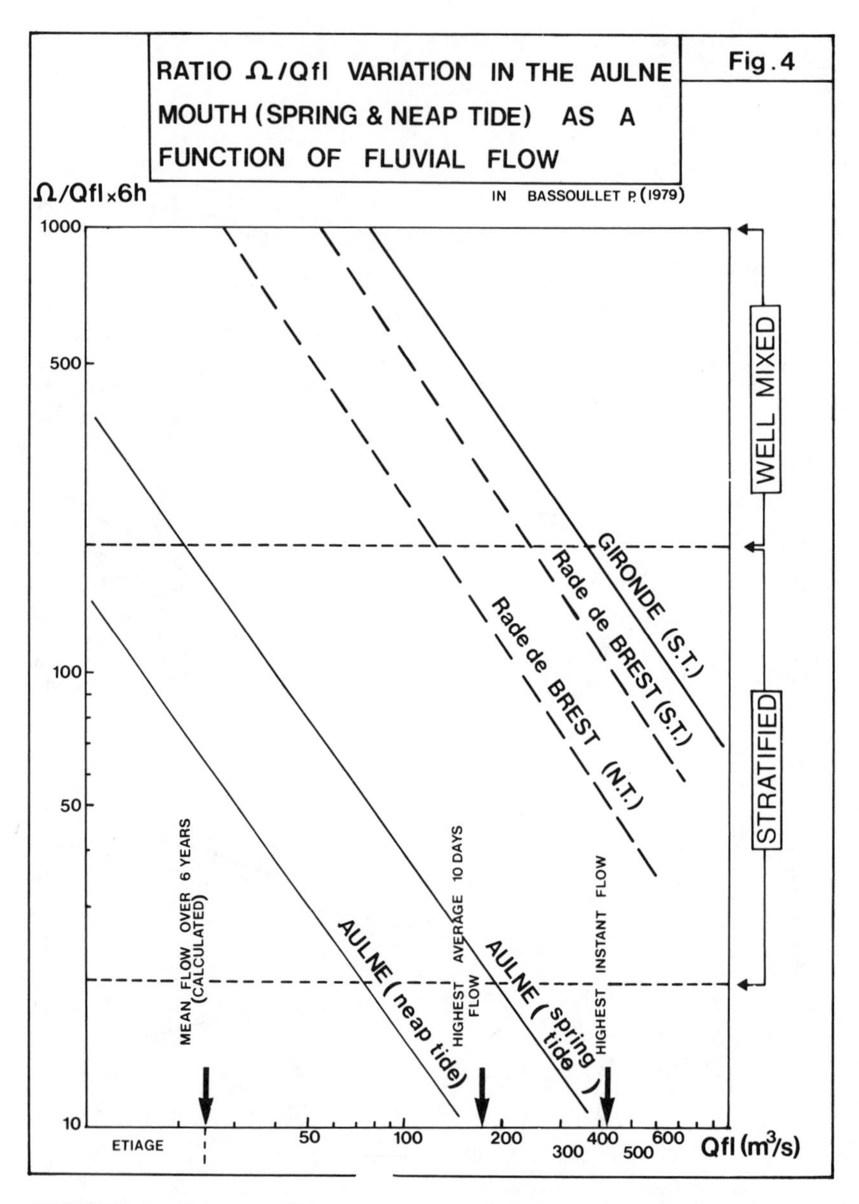

FIGURE 4. Ratio of tide volume to river flow in the Aulne estuary and Bay of Brest as function of river flow, for spring and neap tides. After Bassoullet (1979).

Calculated Budget

Using land analysis and watershed runoff studies, loading rates have been calculated from each land use category (agriculture, industry, domestic sewage, miscellaneous). This budget was based on statistical data and estimates of losses from data provided by O.E.C.D. (7).

Average losses from agricultural areas were estimated at 24 kg N/ha/year and 0.8 kg P/ha/year and 10 percent of total excreta (from livestock).

So, for agriculture and livestock, average combined losses and discharges vary between 39.6 kg N/ha/year and 2.4 kg P/ha/year for the Aulne River and 52 kg N/ha/year and 6 kg P/ha/year for the Elorn River.

Inputs originating in urban activities contribute 12 g N per inhabitant per day and 2.25 g P per inhabitant per day. Results are summarized in Table 1.

According to data from the Loire-Brittany Regional River Basin Commision, results of industrial inputs are presented in Table 2. These data show that inputs of N and P originating from industries are fairly equal for the northern and southern parts of the Bay.

In order to make this evaluation more accurate, inputs from losses supplied by non-harvested soils and from direct rainfall onto the Bay were added, although they accounted for a small percentage of the total inputs.

Results of this analysis indicate that: 1) livestock and agriculture supply the Bay with 70 percent of the total input of nitrogen; 2) urban and industrial activities supply 74 percent of the phosphorus; 3) natural sources (proceeding from non-harvested soils and direct rainfall) account for only 3 percent of N and 2 percent of P; 4) the average yearly value of the N/P ratio is 16.5 in rivers.

Distribution of inputs in each drainage basin is as follows: The Aulne River basin contributes 60 percent of the total input of N but supplies relatively small amounts of P. This is due to the predominance of agricultural activities in this watershed.

The northern Bay and Elorn River basins supply more phosphorus (42 percent of total P inputs) than nitrogen (31 percent of total N inputs).

As a result, inputs from the northern part of the Bay are characterized by a greater proportion of phosphorus and ammonium compounds than those supplied by the southern part. In addition, northern inputs, dominated by urban and industrial discharges, are less seasonally erratic than southern inputs.

Observed Budget

Total load of nitrogen and phosphorus has been evaluated by multiplying the river flow and the nutrient concentration during one year, from October 1976 to October 1977, at sampling points located in the lowest part of the rivers' stream (seaward).

TABLE 1. Calculated budget (N & P) from the fluvial drainage basins into the Bay of Brest. (Tons/year)

		Ag	L	Urb	Ind	Soil	Rain	Total by Basin
North Bay	N	128	68	850	269	3.8		1319
	P	4.2	3	165	53			226
Elorn	N	577	666	287	301	32		1863
	P	19.8	32	93	97	0.8		243
Middle basins	N	465	247	89	14	18		833
	P	14	12	36	8	0.4		71
Aulne	N	2934	1940	353	616	123		5966
	P	97	94	106	288	3		588
South Bay	N	27	20	12	7	3		69
	P	0.9	0.6	8	3	0.07		13
Total	N	4131	2941	1591	1207	180		10050
	P	136	142	408	449	4		1140
Total + rainfall	N						55	10105
	P						7	1147
		AGRICULTURE		URGAN + INDUSTRIES		NATURAL LOSSES		
Total by activities	N	7072	70%	2798	27%	235	3%	
	P	278	24%	857	74%	11	2%	

Ag = Agriculture, L = Livestock, Urb = Urban Activities, Ind = Industries, Soil = Non-Harvested Soils, Rain = Direct Rainfall.

TABLE 2. Industrial Inputs (Converted to Grams per Inhabitant Equivalent)

	Oxidizable kg/day	Population equivalents	Total N Tons/year	Inorganic N Tons/year	Total P Tons/year	Inorganic P Tons/year
Northern Bay	3,636	60,600	265	225	49	28
Elorn River	3,843	64,050	280	238	52	30
Middle Basins	134	2,233	9.7	8.2	1.8	1
Aulne River	7,778	129,633	567	481	106	61
Southern Bay	135	2,250	9.8	8.3	1.8	1
Total				960		121

From Loire-Brittany Regional River Basin Commission

Population Equivalents based on 60 g oxidizable matter, 12 g total nitrogen and 2.25 g total phosphorus per person per day, and the assumptions that Inorganic N = 0.85 Total N and Inorganic P = 0.58 Total P

The concentrations of nitrate show a fairly good positive correlation with the discharge of the Aulne River, but the correlation is less clear for the Elorn River (Figure 5). This effect can be explained by the different proportion of agricultural inputs evacuated by runoff (50 percent of N from the Aulne River and only 30 percent from the Elorn River).

Results are summarized in Table 3.

TABLE 3.

	Tons/Year (1977)			
	N	%	P	%
Northern Bay basin	526	6	47	42
Elorn River basin	1684	20	34	31
Middle basin	578	7	1	1
Aulne River basin	5493	66	27	24
Southern Bay Basin	18	0	2	2
Total	8299	99	111	100

Comparison Between Calculated and Observed Budget

Figure 6 shows a comparison between the calculated and measured budgets. The deviation of the predicted load from the sample estimates is less than 15 percent for nitrogen and 80 percent for phosphorus. When inputs from point source discharges and nonpoint source discharges are integrated in one fluvial inlet, and when the measured nutrient is conservative with respect to the river flow, budgets are in good agreement.

The lack of agreement between calculated and measured P values probably reflects the existence of numerous unmeasured point source discharges. The phosphorus output is therefore underestimated. The deficit of the observed budget increases with increasing agricultural activities. Removal of phosphorus may also be explained by the fact that phosphorus concentrations in many fresh water environments are significantly affected by mineral water reactions (9).

"The estimation of solute budgets, though possibly imprecise, is useful in interpretation of the factors governing river chemistry and assessing the effect of a change in one variable on the whole system" (10); likewise in our case, the relationship between dissolved N and river discharge permitted the prediction of nutrient output from a river basin, if river flow is sufficiently gauged.

Results obtained from this analysis show that the relationship between N and P produced from agriculture, industry, or urban activity causes a difference between the southern and the northern parts of the Bay. This influence is noticeable when considering the distribution of reactive phosphate (Figure 14) and the N/P ratio in the surface water bodies. The predominating type of activities in the drainage basin determines the N/P

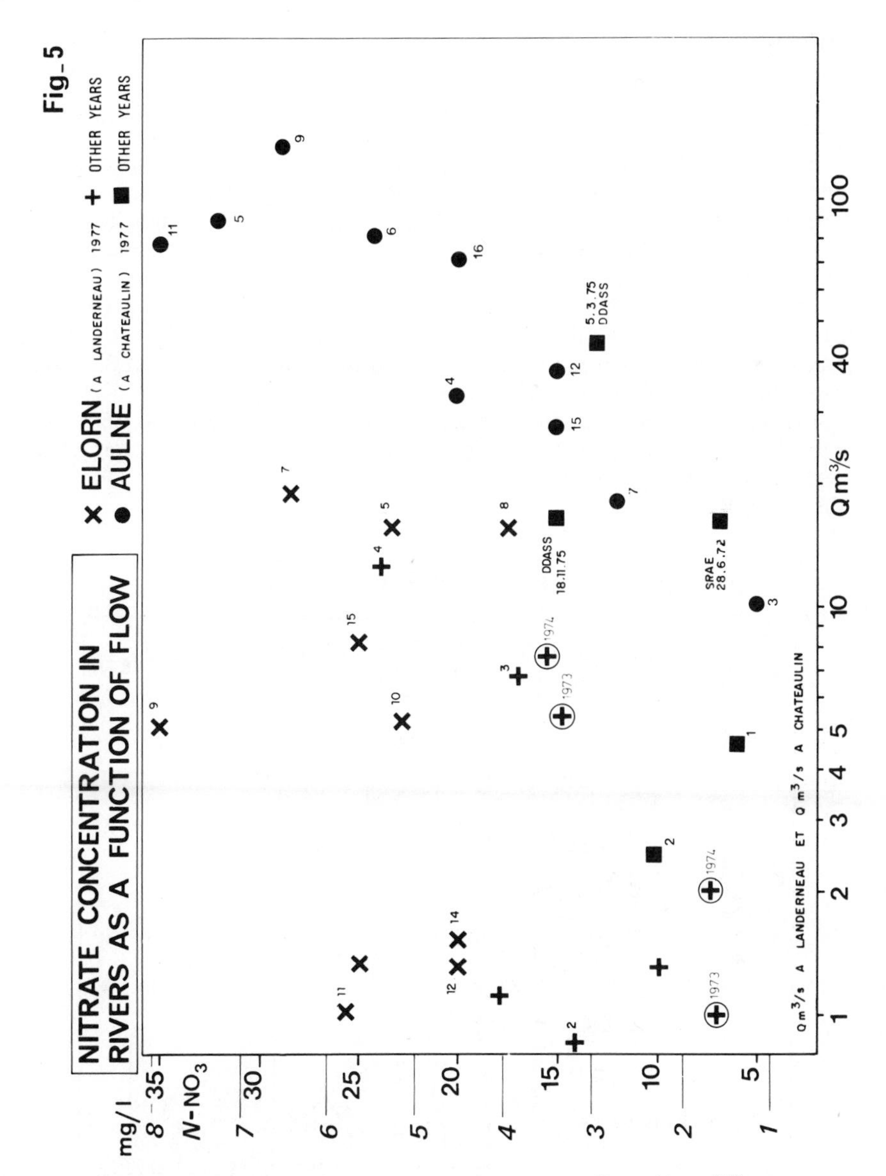

FIGURE 5. Nitrate concentration in the river as function of flow.

ratio in the areas of the Bay affected by the runoff of each watershed. In the northern part, where urban and industrial activities are more developed, N/P values are lower than in the southern part of the Bay.

HYDROLOGY

Salinity Intrusion

Seasonal variations or river inflow cause marked changes in the salinity distribution and water stratification. Fresh water mainly comes in the eastern part of the Bay; surface salinity is generally less than 20 $^{o}/oo$ (Figure 7). Near the inlet the influence of oceanic seawater increases the salinity to 33.8 $^{o}/oo$.

Pronounced vertical stratification occurs during high river flow (Figure 8) with a vertical salinity gradient greater than 26 $^{o}/oo$ in the upper reaches of the Bay. During neap tides, bottom salinity constant at 33 $^{o}/oo$ varies little throughout the entire Bay.

During low river flow, the surface salinity values range up to 35 $^{o}/oo$, and the Bay undergoes the major influence of oceanic seawater. The vertical salinity gradient is very low, but some stratification may occur in the upper Bay.

DRAINAGE BASINS OF THE BAY OF BREST

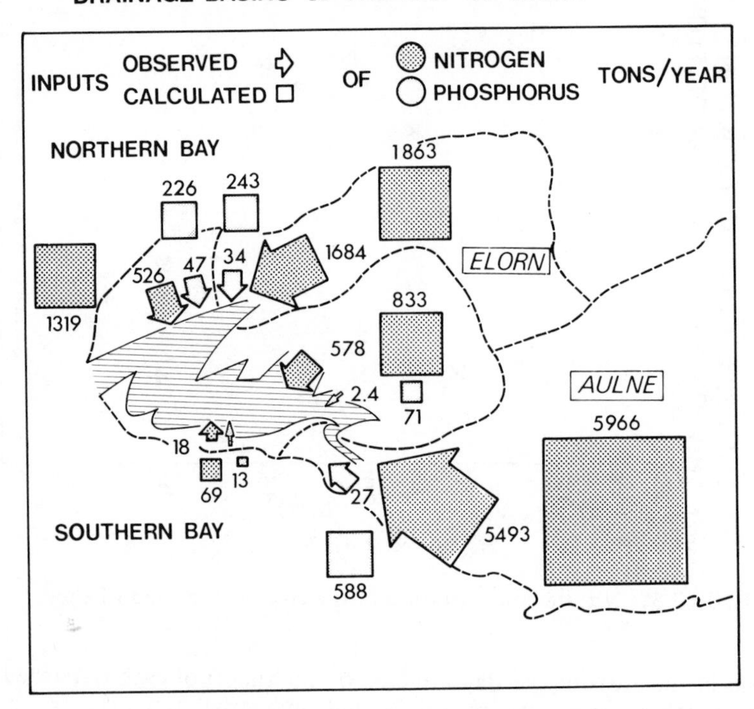

FIGURE 6. Total yearly nutrient input in the Bay from the river basins.

In addition to the general decrease in salinity upstream from the Bay inlet, there are slight horizontal salinity gradients. The southern part of the Aulne inlet shows systematically higher salinities than the northern shore (Figure 7).

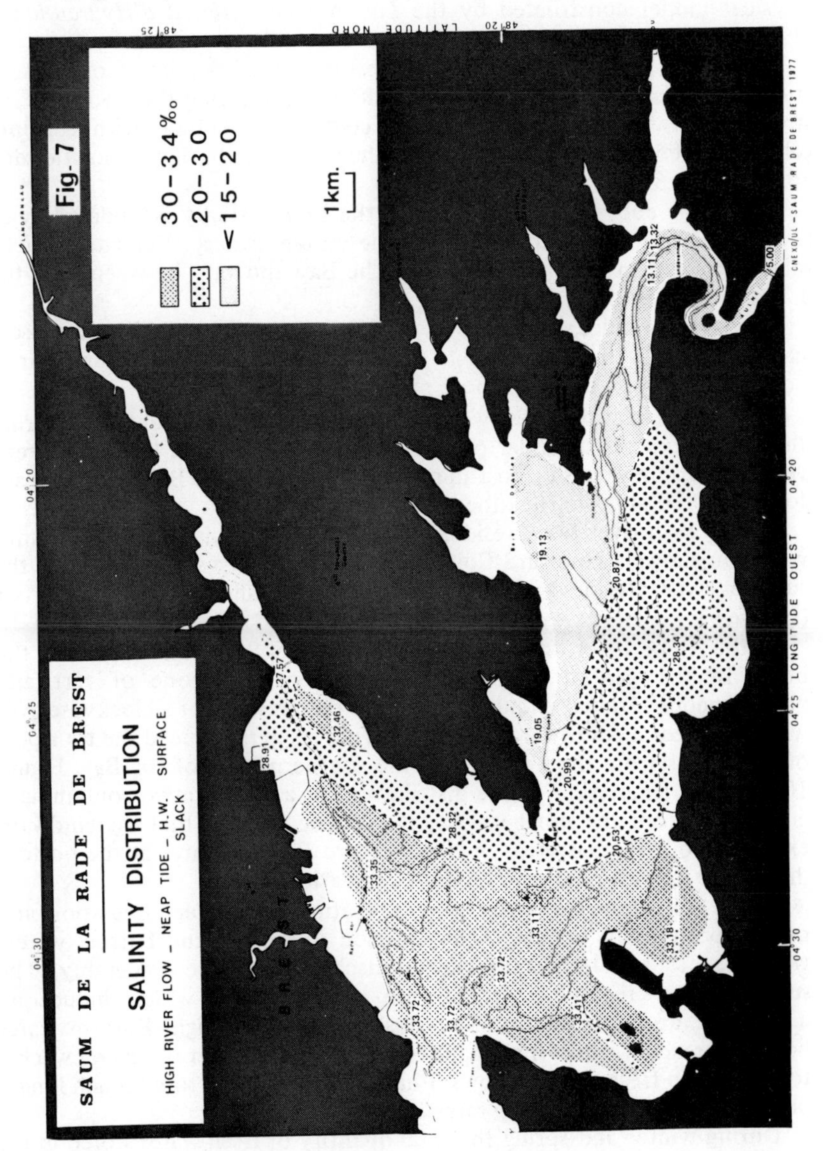

FIGURE 7. Surface salinity distribution in the Bay of Brest during high river inflow, neap tide, high water slack.

Water Circulation

Surface water circulation in the Bay is well known, due to numerous data from the *Service Hydrographique de la Marine* and from results of a physical model constructed by the *Laboratoire National d'Hydraulique* (12).

Using these data, Salomon (11) made a numerical simulation of surface currents for a spring tide. This model shows that during the flood tide, a strong clockwise gyre occurs in the center of the Bay, with current velocities attaining 1.5 m s^{-1}. In the other areas of the Bay, velocities do not exceed 0.50 m s^{-1}.

During the ebb, seaward flushing of the Elorn estuary is hindered by a large water movement coming from the Aulne estuary. Velocities reach values up to 1 m s^{-1} in the center of the Bay and vary between 0.40 to 0.20 m s^{-1} in the other areas.

During neap tides, current observations show that velocity decreases markedly (0.20 to 0.50 m s^{-1}), the central gyre is weaker, and wind induced circulation becomes more pronounced.

Bottom circulation is less well known, but some results obtained during this study indicate a similar pattern of current directions. Bottom current velocities attain values up to 1 m s^{-1} in the center of the Bay, and decrease to 0.40 - 0.20 m s^{-1} in the other areas.

During high river flow, residual, or non tidal, circulation in the salinity intrusion shows a landward flow at the bottom and a seaward flow in the surface layer. Residual velocities are important and average 0.15 m s^{-1} (Figure 9).

In the Aulne River, the nodal point is located 40 km upstream from the inlet. In the central part of the Bay there is a zone of vertically homogeneous residual circulation, due to the presence of a clockwise gyre which induces a southward drift of water. This gyre induced the transport of pollutants from the northern to the southern parts of the Bay. Figure 10 shows a simulation of the distribution of a conservative contaminant emanating from the northern Bay. It can be noticed that the landward extension of affected areas in the southern Bay is greater at the bottom due to the upstream bottom residual circulation.

Figure 11 summarizes the general features of water circulation and mixing patterns in the Bay. The reaction of surface and bottom waters with respect to changes in fluvial discharge varies considerably. The surface water reacts in synchronism with floods, which induce an immediate drop in salinity. The flushing rates are high. Bottom water however, reacts with an important time lag (on the order of several weeks) to changes in fresh water inflows. Fluvial water is then stored much longer on the bottom when flushing rates are very low.

During winter and spring the total quantity of fresh water stored in the Bay is approximately 100 to 150 x 10^6 m^3 and during summer this volume is reduced to 1- to 200 x 10^6 m^3.

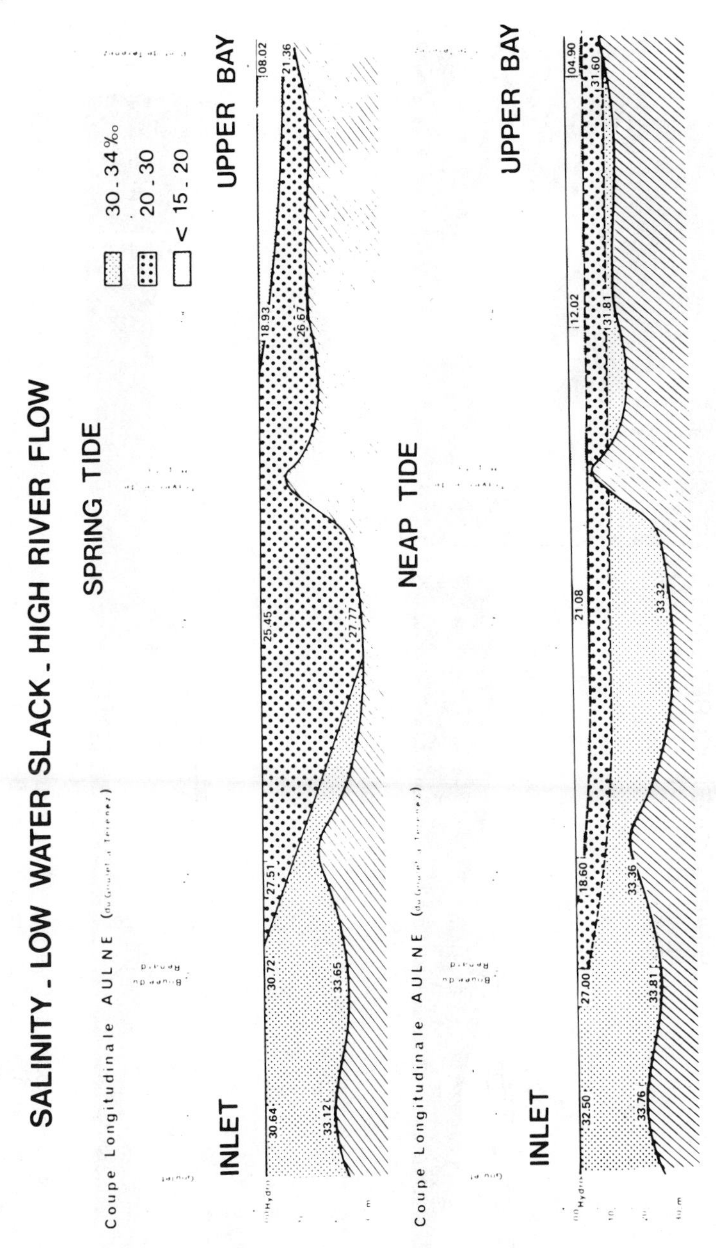

FIGURE 8. Longitudinal salinity profiles in the Bay of Brest, extending from the Aulne estuary (right) to the Bay Inlet (left). Measurements carried out during high river flow, low water slack; spring tide (top), neap tide (bottom).

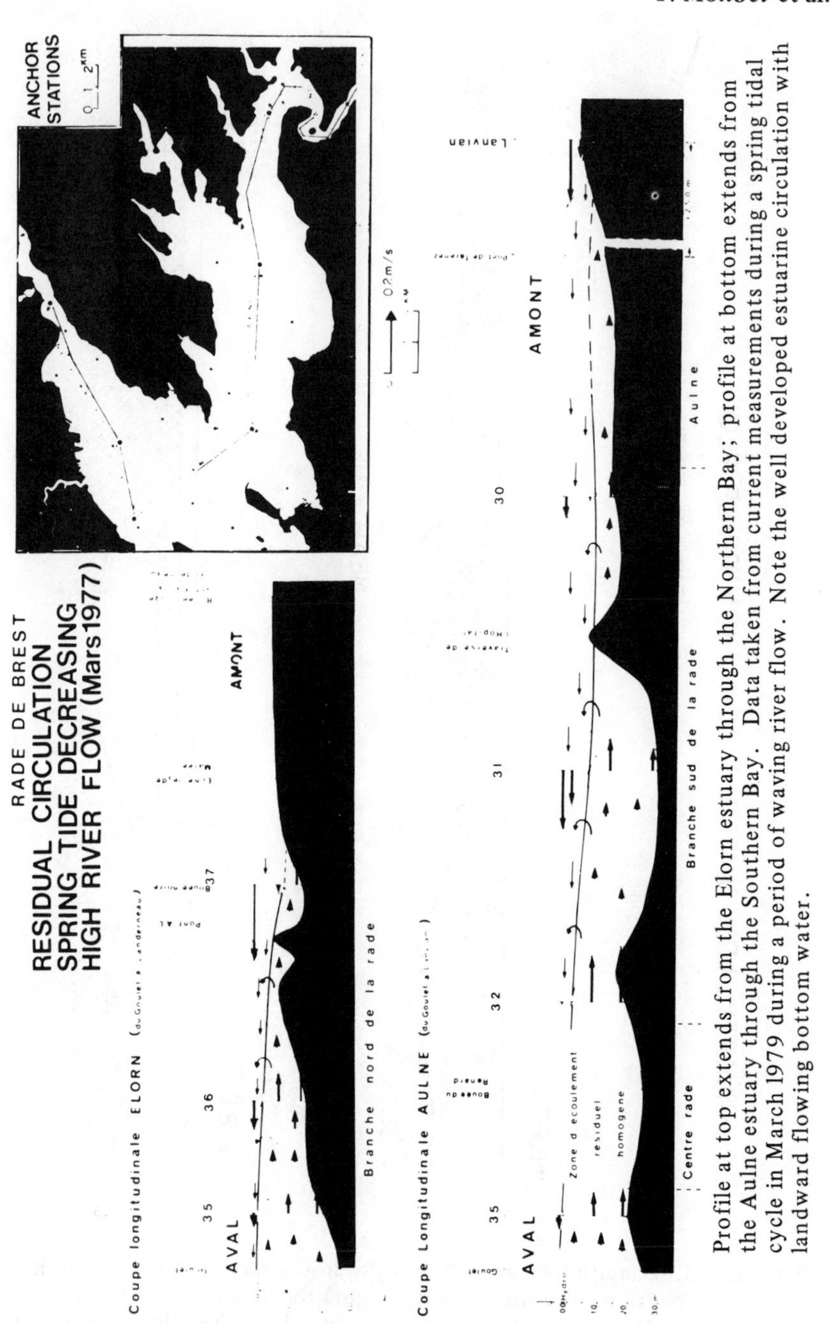

Profile at top extends from the Elorn estuary through the Northern Bay; profile at bottom extends from the Aulne estuary through the Southern Bay. Data taken from current measurements during a spring tidal cycle in March 1979 during a period of waving river flow. Note the well developed estuarine circulation with landward flowing bottom water.

FIGURE 9. Longitudinal residual current profiles in the Bay of Brest.

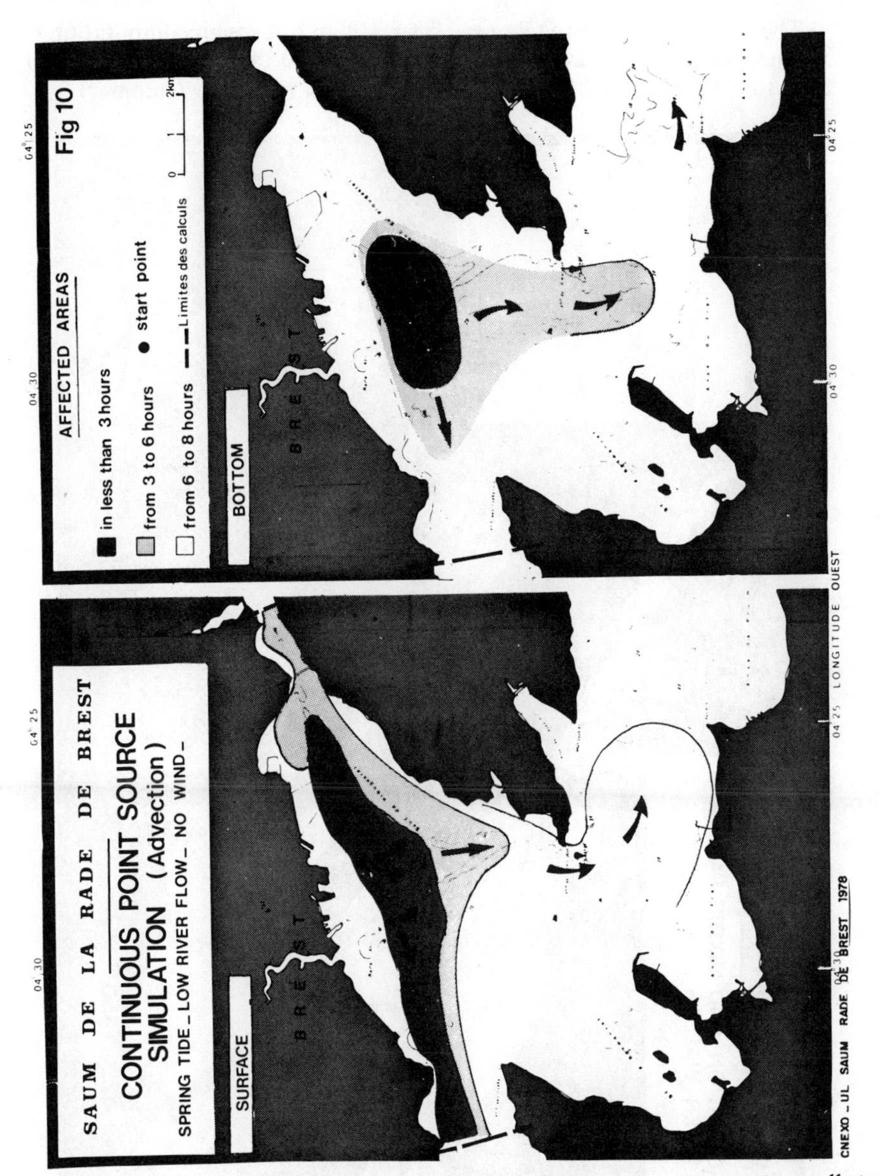

FIGURE 10. Simulation of the geographical dispersion of a conservative pollutant emanating from a point source adjacent to the Brest Harbor; conditions of spring tide and low river flow. The source is assumed to be vertically distributed over the water column. Affected zones are indicated in terms of the minimum time required for pollutant to attain them. Note the difference in surface and bottom dispersion due to the differing surface and bottom residual circulation.

The total average of the Bay is 1.58×10^9 m^3. A rough computation of flushing rate gives an average flushing time of 10 to 15 days during high river flow (winter) and 20 to 35 days during low river flow (summer).

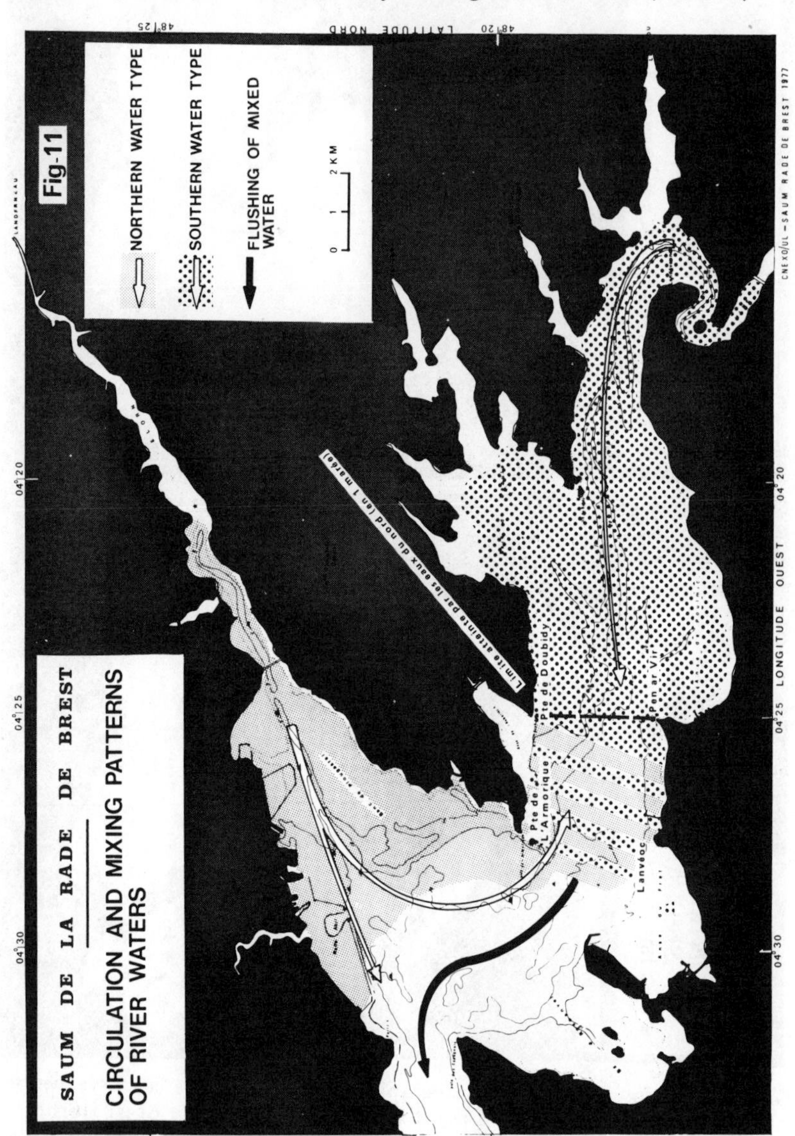

FIGURE 11. Schematic surface water circulation in the Bay of Brest. Main mixing zone of northern and southern water is in the south central bay; and main evacuation occurs in the SW branch of the Bay.

THE DISTRIBUTION OF NUTRIENTS IN THE BAY

In order to assess the fate of nutrients discharged into the Bay, a study of dissolved inorganic nutrients was conducted during winter and summer conditions. The most significant results were obtained during the high river flow period, before the uptake by the phytoplanktonic bloom appeared. During low river flow, nutrient concentrations remain very low. The restoration of the nutrient reserve begins in early fall for NO_3 and PO_4 and in summer for silicates.

Nitrate Distribution

The spatial distribution of nitrate in the Bay is simple and indicates the effective mixing of fresh and salt water. Nitrate concentrations decrease from upstream to downstream (Figure 12). In the northern bay, nitrate values range from 79.3 μg at.l^{-1} to 17.8 μg at.l^{-1}. In the southern area, values are higher (100 to 200 μg at.l^{-1}) and associated with lower salinities Near the inlet, concentrations are on the order 20 μg at.l^{-1}. (During summer, nitrate and phosphate concentrations drop to average values of, respectively, 0.24 μg at.l^{-1} and 0.07 μg at.l^{-1}.) Plotting nitrate as a function of salinity shows (Figure 13) a high correlation. Extrapolation to the zero value of salinity indicates that the nitrate concentration in the rivers are similar and equal to 500 μg at.l. This is in good agreement with observed value in rivers.

Phosphate Distribution

During low tide and winter conditions, concentrations of reactive phosphate range between 0.5 and 1.15 μg at.l^{-1} and do not follow the same seaward gradient as the nitrate distribution pattern.

The northern Bay concentrations as shown in Figure 14, are generally higher than those of the southern Bay. Contrary to the previous scatter diagram, the plotting of phosphate as a function of salinity shows a different pattern, and does not behave conservatively. The trend of P-PO_4^{-3} data with salinity is likely due to adsorption of phosphate from solution onto solid phases, which are abundant in the water during periods of high river flow (9).

Bassoullet (6) has shown that sediments accumulate in the form of a "turbidity maximum" at the upper limit of salinity intrusion in the Aulne estuary. Maximum concentration of suspended sediments reaches 1 g/l. The author suggests that the asymmetry of the tidal wave creates a trap for sediments and that tidal cycles control the sedimentation and resuspension of particulates in the estuary.

Chemical Identification of Water Masses

Within the Bay, the relationship between reactive phosphate and reactive silicate can be used to identify surface water bodies. The scatter

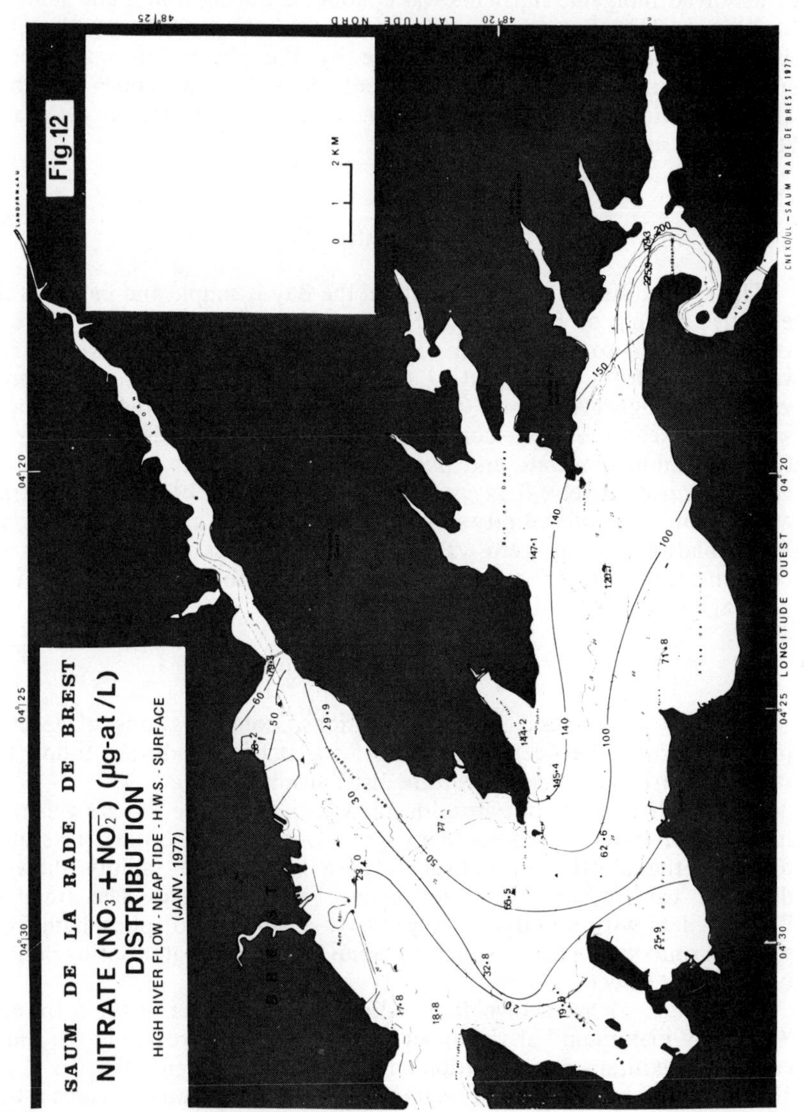

FIGURE 12. Surface dissolved nitrate ($NO_3 + NO_2$) distribution, in μg at.l^{-1} during high river flow, neap tide, high water slack (January, 1977).

diagram of these parameters distinguishes two linear trends, a high SiO_2/PO_4 trend and a low SiO_2/PO_4 trend (Figure 15). The high silicate-phosphate ratio trend characterizes water from the southern Bay region, while the low ratio trend is indicative of water originating from the urbanized and industrial basins of the northern Bay region.

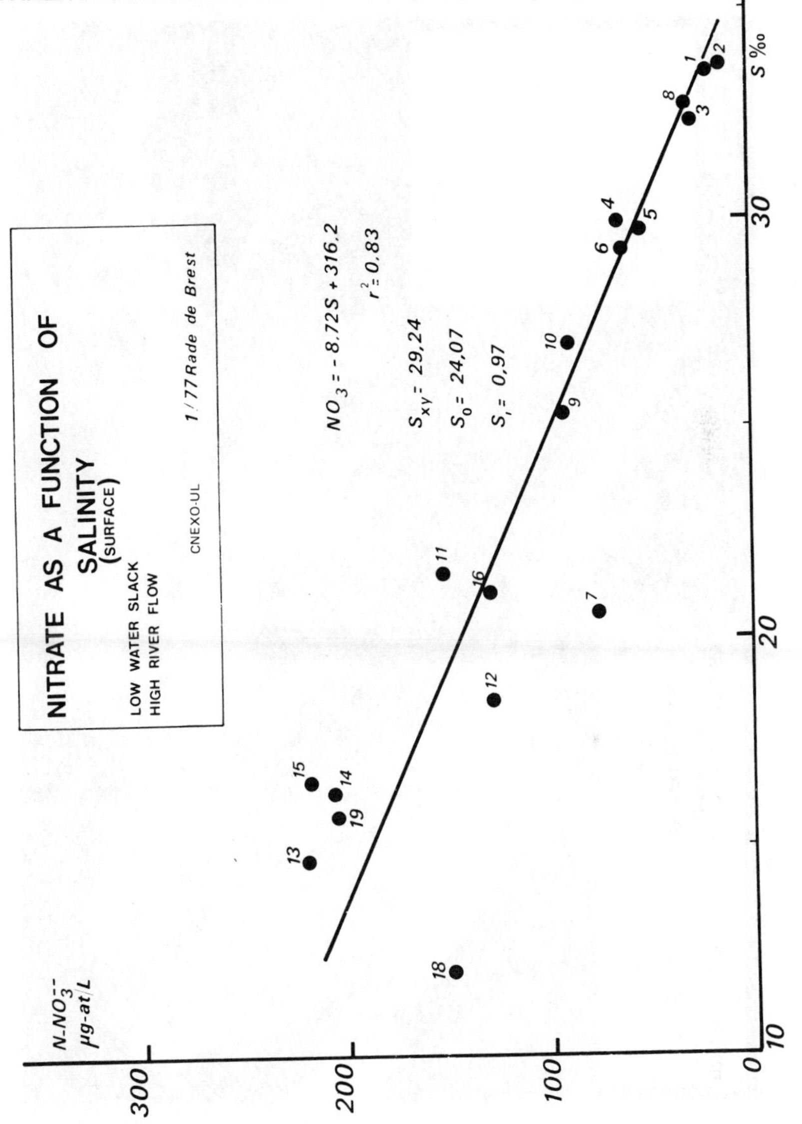

FIGURE 13. Plot of surface nitrate concentration as a function of surface salinity (January 1977).

Analysis of these relationships permits the computation of the percentage of fresh water coming from either the northern or southern basins (Table 4 and Figure 15).

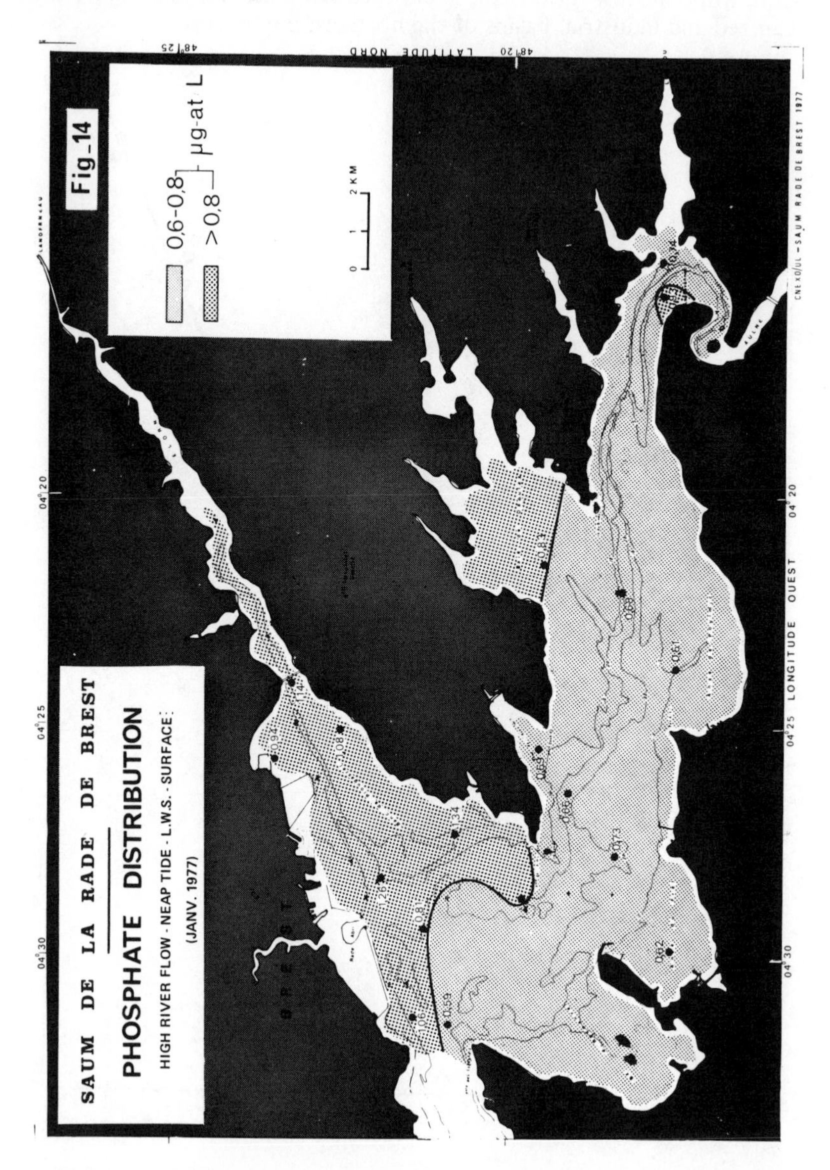

FIGURE 14. Surface phosphate distribution in the Bay of Brest; high river flow, neap tide, low water slack (January 1977).

TABLE 4.

B.M.M.E.	S^o/oo	% E.d.m.	% E.d.	X a	X b	% E.S.	% E.N.
Station 1	33.506	99.4	0.6	-	-	0	0.6
Station 2	33.677	99.9	0.1	-	-	-	0
Station 3	32.455	96.6	4.4	0	-	0	4.4
Station 4	29.895	88.7	11.3	-	-	0	11.3
Station 5	29.770	88.7	11.3	-	-	0	11.3
Station 6	29.375	87.2	12.8	38	6.5	1.9	10.9
Station 7	20.676	61.3	38.7	-	-	0	38.7
Station 8	32.794	97.3	2.7	-	-	0	2.7
Station 9	25.415	75.4	24.6	3	35.0	22.7	1.9
Station 10	27.009	80.1	19.9	18	31.0	12.6	7.3
Station 11	21.560	64.0	36.0	-	-	36.0	0
Station 12	18.600	55.2	44.8	8	41.0	37.5	7.3
Station 13	14.770	43.8	56.8	-	-	56.2	0
Station 14	16.350	48.5	51.5	-	-	51.5	0
Station 15	16.583	49.2	50.8	-	-	50.8	0
Station 16	21.083	62.6	37.4	-	-	37.4	0
Station 17	23.418	69.5	30.5	14.0	44.0	23.1	7.3
Station 18	12.071	35.8	64.2	2.0	56.0	62.0	2.2
Station 19	15.178	45.0	55.0	12.0	6.9	46.9	8.1

B.M.M.E. = Low water level - Neap tide
% E.d. = % fresh water
S^o/oo = % Salinity
% E.S. = % South water type
% E.d.m. = % Seawater
% E.N. = North water type

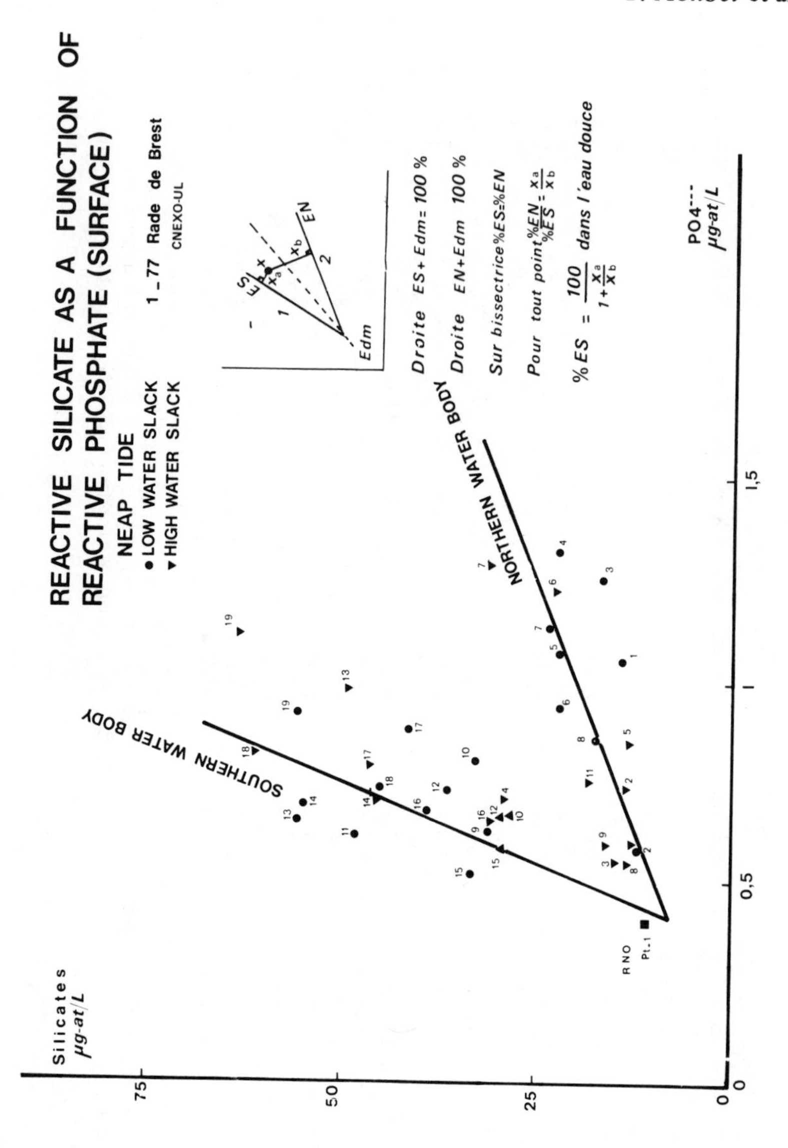

FIGURE 15. Scatter diagram of reactive silicate as function of reactive phosphates in the surface water of the Bay, during a neap tide (January, 1977). The two linear segments indicate the southern and northern Bay waters, characterized respectively by high and low silicate to phosphate ratios. This difference is due to the more developed industrial and urban activities of the northern Bay, with a higher phosphate load.

Plotting the percentage of sea water in each sample as a function of the ratio % north water/% south water enables the distinction of four water types: northern water, southern water, coastal water, and mixed waters.

Figure 16 shows the spatial distribution of each water type within the Bay. This chemical zoning is in good agreement with salinity and current data, but its use must be restricted to the winter period, as a consequence of non-conservative behavior of nutrients in the estuaries during times of sustained primary production.

Bay Responses to Nutrient Additions

The Bay of Brest is a fairly healthy body of water, and field measurements show that essential processes remain intact. No eutrophication problem has been identified during recent years in the Bay itself, but we lack knowledge on detailed chemical and biological data in the upper reaches of small estuaries which are tributary to the Bay. Field observation has shown that high values of nitrites and ammonium may occur in the Elorn estuary as well as in the Aulne estuary. Substantial problems have been seen in the depletion of dissolved oxygen < 3 mg.l^{-1} observed during August, 1978 in the Aulne estuary. This summer oxygen depletion is associated with high levels of suspended sediments (ca. 300 mg.l^{-1}) and can be an indicator of overloading and a cause of damage to biota and to aesthetic quality of the estuary. There is an exceptional need for definition of problems occuring in the upper estuaries. Fortunately no reports of mortalities were associated with these conditions and no serious concern has been expressed for the Bay itself. Winter data suggest that nutrients are diluted and washed out without being utilized. In early spring, when high concentrations of nitrate reach the middle part of the Bay, a phytoplankton bloom occurs and chlorophyll content increases to 10 μg Chla.l^{-1}. By the time phytoplankton populations build up (60 days), the nitrate concentration is reduced and decreases down to almost zero values at the beginning of June. Low levels of chlorophyll content, associated with complete depletion of nitrate, suggest that nitrate might be a limiting factor and lead to the hypothesis that the Bay, with its rapid flushing rate, can handle large amounts of added nutrients as they are transported away and rapidly diluted in ocean water.

CONCLUSION

Agreement between the observed and calculated nutrient solute budgets verifies the estimated percentage and impact of each activity's contribution to the total input.

In addition, the correlation between N concentrations and fresh water flow gives precise information about the environmental role of agriculture in these inputs, a reality often difficult to ascertain.

Specifications of nutrient compounds (NH_4-NO_2-PO_4) found at the same time in river and the adjacent Bay waters illustrate the close

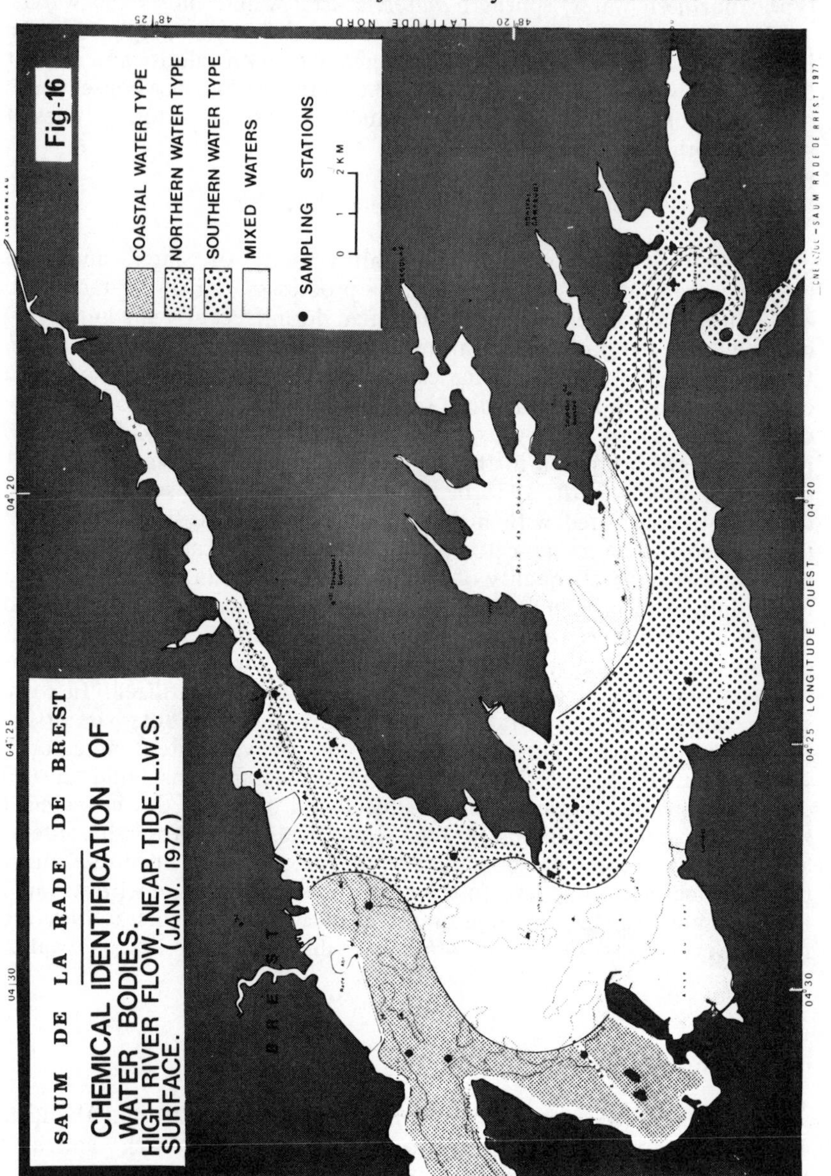

FIGURE 16. Classification of surface Bay waters according to silicate-phosphate ratios; January 1977, neap tide, low water slack.

relationship with the particular activities carried on in the river drainage basins.

The morphology of the central Bay creates a current gyre which diverts waters from the northeastern Bay to the southern Bay.

Strong seasonal variations in fresh water flow from rivers affecting mixing processes and regulate nutrient concentrations in the main water masses. In winter, during periods of high river flow, the Bay acts as a stratified estuary and large amounts of nitrates are evacuated out to sea by the strong net seaward flow of surface water. During summer, salinity stratification breaks down and the low nutrient inputs are well mixed throughout the entire Bay. Salinity and nutrient concentration in bottom waters react with one or two months delay to these terrigenous impulses.

Hydrochemical zonation of the bay is in good agreement with salinity and current data.

Conclusions based on the oceanographic characteristics of the Bay lead to the following zoning recommendations. In order to avoid contamination of the southern Bay (designed for aquaculture projects), the development of industrial activities should be restricted to the northern Bay and a buffer zone should be established between the north and south parts. This buffer zone would act to dilute effluents coming from the northern industrial zone. Future development in the buffer zone should be limited, with no major aquaculture projects. Systematic water quality monitoring should be established in these areas in order to control the efficiency of the buffer (Figure 17).

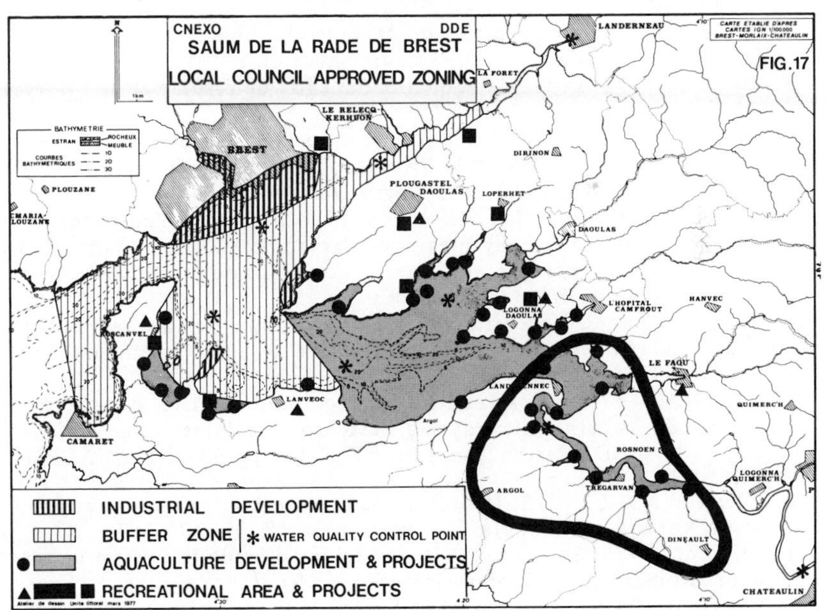

FIGURE 17. Proposed development zoning plan for the Bay of Brest. (Schema d'Aptitude et d'Utilisation de la Mer - S. A. U. M.)

Current patterns and non-tidal circulation suggest that sewage waste input should be discharged near the surface and in the lower part of the Bay to provide a better seaward flushing.

High nutrient values in the upper estuary can lead to eutrophication problems during periods of low river flow, and to contamination of drinking water. It appears necessary to improve fresh and estuarine water quality by controlling nonpoint sources of nutrients and constructing new sewage networks.

For the well being of the population, we finally recommend that municipal wastewater treatment facilities be improved or created and that major efforts be focused on the elaboration of water quality standards.

REFERENCES

1. Le Corre P. et P. Treguer. 1976: Contribution a l'etude de la matiere organique et des sels nutritifs dans l'eau de mer. Caracteristiques chimiques du Golfe de Gascogne et des upwellings de l'Afrique du Nord Ouest. These doct. etat. Fac. sc. Universite de Bretagne Occidentale Brest. 490 p.
2. Berthois L. et G.A. Auffret. 1968. Contribution a l'etude des conditions de sedimentation dans la rade de Brest. Cahiers Oceanographiques n°9 pp 893-920, n° 5 pp 469-485, n° 7 pp 701-725, n° 10 pp 981-1010.
3. Reseau National d'Observations de la qualite du Milieu Marin. C.N.E.X.O.
4. S.E.P.N.B. 1976. Pollution de la rade de Brest. Rapport pour la Direction Departementale de l'Equipement du Finistere. 187 p.
5. S.R.A.E. 1976. Etude de la qualite des eaux de l'Aulne et de l'Hyeres. 66 p.
6. Bassoulet P. 1979. Etude de la dynamique des sediments en suspension dans l'estuaire de l'Aulne (Rade de Brest). These de 3eme cycle. Fac. sc. Universite de Bretagne Occidentale. Brest. 136 p.
7. Vollenweider R.A. 1971. Les bases scientifiques de l'eutrophisation des lacs et des eaux courantes pour l'aspect particulier du phosphore et de l'azote, comme facteurs d'eutrophisation. O.C.D.E. Paris. Rapport DAS 45 L.
8. Treguer P. et P. Le Corre. 1975. Manuel d'analyses des sels nutritifs dans l'eau de mer. Utilisation de l'autoanalyser II. Technicon. 2eme edition 1975. 110 p.
9. Liss P.S. 1976. Conservative and non conservative behaviour of dissolved constituents during estuarine mixing. In Estuarine Chemistry. J.D. Burton and Liss P.S. Edit. pp 93-130. ACADEMIC PRESS N.Y. 229 pp.
10. Edwards A.M.C. 1973. Dissolved load and tentative solute budgets of some Norfolk catchments. J. of hydrobiol. 18. pp. 201-217.
11. Salomon. 1978. Sur un procede numerique d'exploitation des donnees courantometriques (Rapport interne CNRS).
12. Electricite De France. Direction des Etudes et Recherches 1968-1969. Etude de la diffusion en Rade de Brest d'un rejet pollue.

REVERSAL OF THE EUTROPHICATION PROCESS: A CASE STUDY

Gerald A. Moshiri, Nicholas G. Aumen* and William G. Crumpton**

**Department of Biology*
University of West Florida
Pensacola, Florida 32504

***Kellogg Biological Station*
Michigan State University
Hickory Corners, Michigan 49060

ABSTRACT: Bayou Texar, Pensacola, Florida, is a bayou estuary which was advancing toward eutrophy, due to nutrient loadings from various sources. The occurrence of characteristic symptoms of eutrophication, coupled with the closing of the bayou for water contact recreation, led to the initiation of an intensive seven-year study. Initial results indicated that Carpenter's Creek is a prime source of most nitrogen species and that an exchange exists between dissolved phosphates and those adsorbed onto sediments. Carbon fixation rates varied, with the stations closest to Carpenters Creek exhibiting the most productivity. It also appeared that nitrates may be more important in controlling rates of carbon fixation than phosphates. An ensuing study directed at phosphate exchange suggested that reducing conditions in the muds cause substantial release of iron-bound PO_4 from the sediment-water interface followed by adsorption onto particles in the aerobic flocculent layer above the interface. This mechanism could allow sufficiently rapid exchange of PO_4 between sediment and water resulting in a low, stable concentration in the water as observed. Other investigations involving bacterioplankton, dissolved glucose, and heterotrophic production indicated that algal primary productivity is the major source of dissolved glucose. A final study demonstrated a paucity of benthic macroinvertebrates, probably due to a graded suspension of particles at the sediment-water interface and the absence of a sharply defined bottom. The zooplankton community exhibited low diversity while numbers of individual species were high. The phytoplankton community was composed primarily of dinoflagellates, with diatoms, cryptophytes, chlorophytes and microflagellates occurring in lesser quantities. It was also suggested that the importance of toxins from algal blooms to fish kills may be greater than previously indicated. Specific recommendations made after the fourth year of the study were implemented in most part and led to a substantial improvement of water quality and the subsequent opening of the bayou to the public for recreational use.

INTRODUCTION

In northwest Florida a number of extensive estuaries and bayous interlace urban areas. The larger ones provide spawning waters for fish and shellfish and are, therefore, of major significance to the economy of the region.

Years of abuse, however, have severely degraded and damaged northwest Florida's vast estuarine waters, resulting in the depression of water quality and the occurrence of extensive fish kills. As a consequence, many such areas have been closed to commercial fishing and public recreation. Factors that have contributed to this lowering of water quality have been numerous. The leading causes, however, have been:

1. Construction of roads and bridges which have interfered with the normal circulation and tidal patterns, and have thus augmented the detrimental effects of siltation and nutrification from various industrial and real estate developments.

2. Rapid increases in real estate developments, including residential, commercial, and industrial; and consequential overloadings of wastewater lines and sewage treatment facilities in the vicinity, with the invariable results of ruptures and spills.

3. Alterations of watersheds and increases in the extent of stormwater runoff and consequential increases in inorganic nutrients, organic substances, and coliform inputs.

4. Direct inputs of untreated industrial and domestic wastewater.

Results have been the typical signs of water quality degradation such as algal blooms, shallow depths, increased concentrations of sediment organics, highly variable oxygen profiles, and elevated community respiration. Consequences have been frequent in the form of extensive fish kills, particularly during periods of summer stratification and stagnation following extensive periods of summer stratification and stagnation following extensive periods of warm, dry weather. In the case of the subject of this study, Bayou Texar in Pensacola, Florida, for example, a kill which lasted for five weeks during August and September, 1972 resulted in the harvesting of 2-1/4 tons of dead fish per day. Under these circumstance, factors contributing to such lowering of water quality have so altered the natural energy flow of these systems that established procedures for recovery, which usually attack the superficial symptoms, now often prove ineffective. Recovery of these systems, therefore, depends upon developing an understanding of their altered energy flow. Such as understanding requires first a knowledge of pertinent physico-chemical relationships of the system in question followed by an investigation of its trophic dynamics. From this base, an elucidation of the biotic-abiotic interactions can be made, an energy flow pattern thereby determined, and the pattern of community development and evolution established.

Over the past seven years, the author and his associates have studied three northwest Florida estuaries with major emphasis on mesotrophic Bayou Texar, a shallow inlet from Pensacola Bay, Escambia County (Figure 1). Here studies have steadily progressed from preliminary investigations of inorganic nutrient inputs, assessments of inorganic and organic nutrient flow, producer trophic dynamics, heterotrophic uptake of dissolved carbohydrates by bacteria and algae, and benthic invertebrate and plant associations toward the formulation of a comprehensive profile of the trophic status of the system (6, 7, 9, 10, 11, 12, 13, 14, 15). The

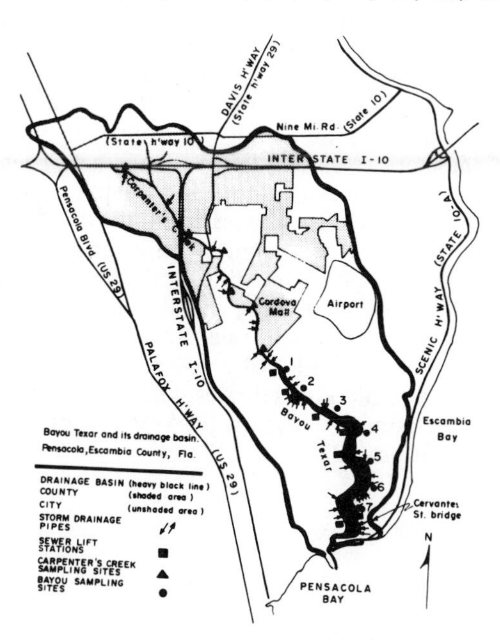

FIGURE 1. Bayou Texar and its watershed (6).

purpose of this paper is not to cite detailed specifics of the seven years of investigation of the eutrophication trend and processes in Bayou Texar, but to present an overview of some of the more salient points of the complete study in order to show how information thus acquired was used to make recommendations for the improvement of water quality and the retardation of the eutrophication process in Bayou Texar.

METHODOLOGY

A number of parameters were routinely monitored throughout the complete duration of the study. Among these were:

Temperature	Light penetration
pH	Organic and inorganic carbon
Salinity	
Dissolved oxygen	Five-day BOD

Appropriate analytical instruments were used for the measurement of these factors as detailed elsewhere (7, 15).

Other factors measured in both water and sediment included organic and inorganic carbon; Kjeldahl, ammonia, nitrite and nitrate nitrogens; organic and inorganic phosphorus; sulfides; iron; and glucose. Methods for the analysis of these are essentially standard and have been referred to by Moshiri, et al. (15); Hannah, Simmons, and Moshiri (6); and Moshiri, et al. (7). In some instances, such as the measurement of sediment Eh, original modifications of existing techniques were employed (8). Primary productivity and phytoplankton qualification and quantification involved the use of standard ^{14}C and settling chamber methods, respectively (6). The ^{14}C technique was also used to estimate bacterioplankton heterotrophic productivity (11, 14). A special sediment sampler was designed to collect benthic invertebrates, and a modification of the high volume pumping techniques described by Aron (2) was used to determine zooplankton distribution (12).

RESULTS AND DISCUSSION

Examination of the extent of the Bayou Texar watershed and its orientation in relationship to the location of Carpenter's Creek, the Bayou's principal source of fresh water, reveals that this stream is the primary source of silt and sand from the various real estate developments in the watershed. At the uppermost regions of the Bayou, sedimentation has been primarily in the form of coarser sand and silts, down to a depth of 25-30 cm, covering a layer of partially decomposed organic matter mixed with silt and loam. The large volume and rapid flow of Carpenter's Creek water resulting from rechannelization for mosquito control purposes, as well as from increased runoff from extensive real estate developments in the watershed, had caused the transport of finer and lighter organic particles further down the bayou. These lighter particles have created a loose, flocculent layer of suspended organic particles 5-10 cm above the hard sediment-water interface. This contrasts with relatively undisturbed systems where silt and organic loading would be expected to be substantially less.

Nutrients

Periodic analysis showed nitrite concentrations to be undetectable. Therefore, assays for this parameter were discontinued. Assays for organic nitrogen and phosphorus were also discontinued because their values did not exhibit definitive trends.

Surface nitrates and Kjeldahl nitrogens showed a pattern of decreasing concentrations from upper to lower stations throughout the duration of the study. The same pattern, though less pronounced, was evident in bottom waters. Surface nitrate concentrations exceeded those of the bottom waters during the spring months. Highest and lowest surface concentrations occurred during the summer and winter months, respectively (Figure 2).

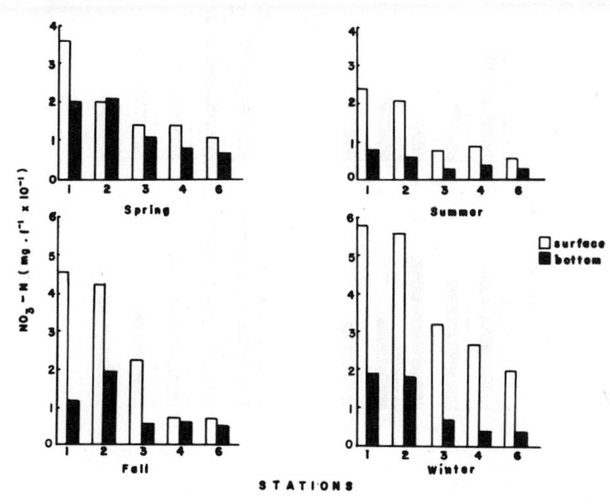

FIGURE 2. Spatial and temporal variations of nitrate nitrogen in Bayou Texar (6).

Similar spatial and temporal patterns were also observed for surface ammonia (Figure 3). Bottom water samples, however, did not show a clear spatial decrease during spring, summer, and fall, as was the case for nitrates. There was, however, a noticeable decrease from upper to lower stations during the winter months. Bottom ammonia concentrations were generally found to be higher than those of surface waters-a phenomenon that is believed to be caused by the decomposition of organic detritus in the sediments.

Seasonal and spatial variations and trends cited above for nitrogen species were not evident in the case of orthophosphates. All evidence indicated that while Carpenter's Creek is the primary source of nitrate and ammonia nitrogens (primarily from overloaded sewage and storm water lines), it was not the principal source of inorganic phosphates into Bayou Texar (Figure 4).

Results of studies conducted to delineate the relationships between water column and sediment phosphorus indicated the existence of an equilibrium between dissolved and adsorbed phosphates. Silt-size or smaller particles and their associated phosphate ions also possessed a substantial buffering capacity for dissolved orthophosphates (6) (Figure 5). Dissolved oxygen data suggested that during all seasons sufficient concentrations of oxygen were present in the water column to prevent reduction of ferric to ferrous iron. In spite of this oxygen availability, analysis of sediment cores showed reducing conditions in this region, a circumstance which repeatedly resulted in the release of substantial amounts of sediment-adsorbed phosphates into the water column.

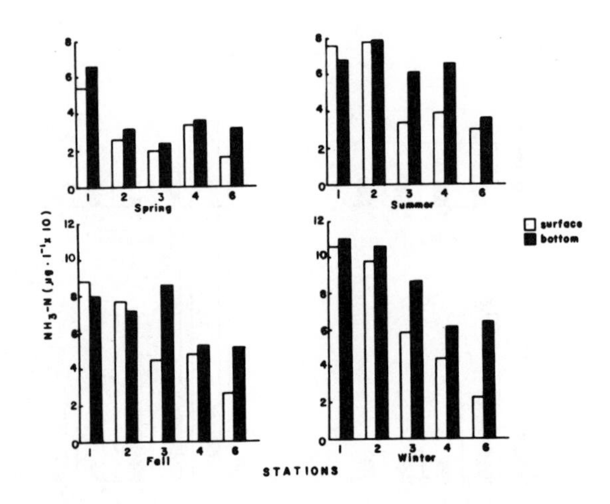

FIGURE 3. Spatial and temporal variations of ammonia nitrogen in Bayou Texar (6).

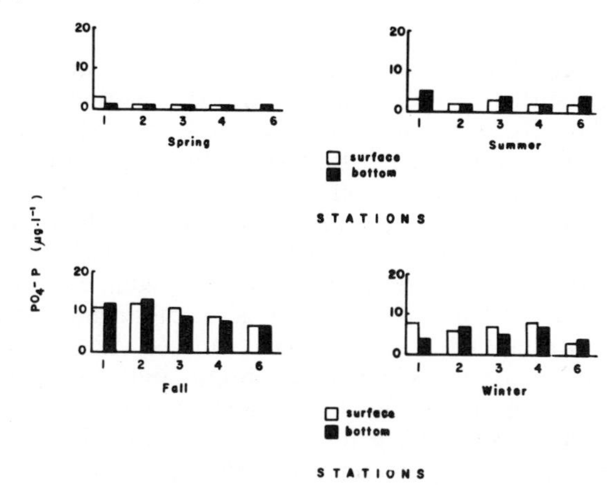

FIGURE 4. Spatial and temporal variations of orthophosphate phosphorus in Bayou Texar (6).

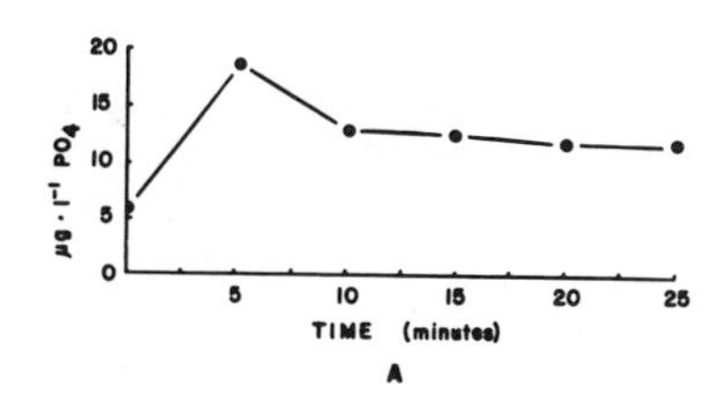

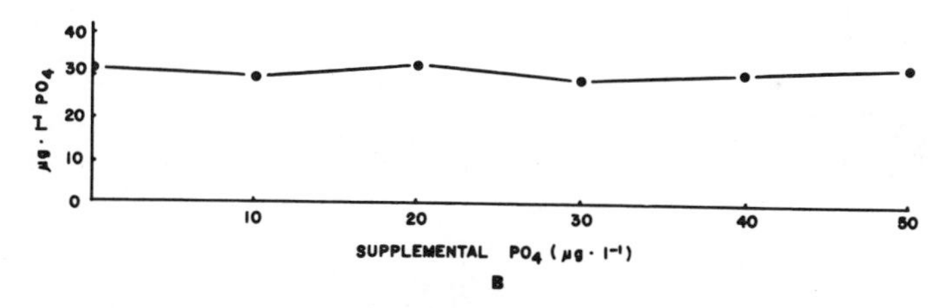

FIGURE 5. Buffering capacity of Bayou Texar sediment particles and their release of inorganic phosphates into the water column.

A. Augmentation of the initial bayou water orthophosphate concentration of 6 $\mu g.l^{-1}$ to 20 $\mu g.l^{-1}$ after agitation for five minutes with bayou sediment. Subsequent agitation lowered the concentration to 13 $\mu g.l^{-1}$ and no more.

B. Stability of water column orthophosphate over a range of supplemented orthophosphate concentrations and 25 minutes of agitation with one gram of bayou sediment (6).

Therefore, even though oxygen concentrations at the sediment-water interface were sufficient enough to prevent the reduction of iron, reducing conditions did exist in the muds for this reduction and the consequential release of ironbound phosphates from the sediment (Figure 6).

The fate of this released phosphate is not clear, however. Evidence from our studies suggests that these phosphates are rapidly readsorbed onto particles of organics in the previously-cited loose, flocculent layer suspended into oxygenated water above the sediment-water interface. Therefore, it is possible that such a layer of organic particles could increase substantially the total surface area involved in adsorption-desorption processes, although our studies did not address this phenomenon. Such a potential increase in exchange surfaces could conceivably allow for a sufficiently rapid phosphate exchange, even under aerobic conditions, to maintain the rather low but stable phosphate concentrations in the Bayou Texar water column (Figure 4) even before or after extensive algal blooms (6, 10).

Redox

As already stated, the Eh profiles of sediment cores from Bayou Texar showed some interesting relationships with those of dissolved oxygen and

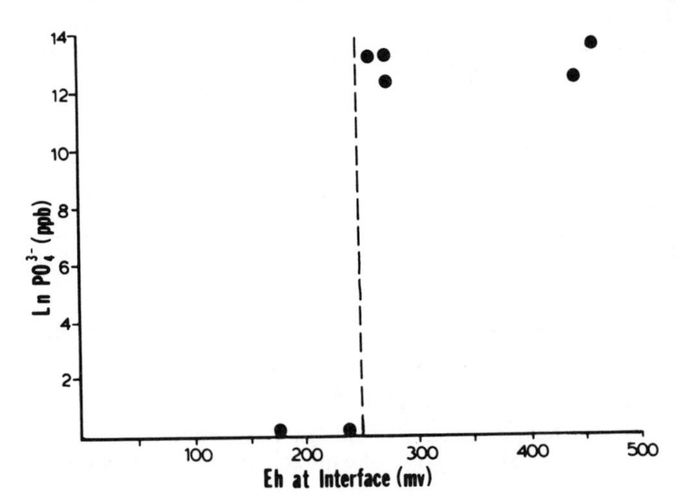

FIGURE 6. Variations of orthophosphate in the sediment-water interface as related to Eh in Bayou Texar (10).

the distribution of sediment bacteria and organics (Figure 7). Bacterial numbers decreased from the sediment-water interface down to a depth of approximately 35 cm. This was accompanied by a corresponding decrease in the fraction of sediment organics, trends which, when considered together, suggested a decrease in bacterial activity with sediment depth. Considering these phenomena, Eh would have been expected to have a minimum value near the interface, where bacterial activity was the highest, and to have increased as decreasing organics became limiting to bacterial activity in the deeper sediments (Figure 8A, also Bella's (3) type 2 system).

As previously stated, however, in Bayou Texar there was always a high oxygen concentration (7-9 $mg.l^{-1}$) in the water overlying the sediment-water interface. The effect of this high oxygen content was to increase the Eh near the interface, regardless of bacterial activity, thereby creating the parabolic Eh profile exemplified in Figure 8B. Therefore, it was postulated that during the warm months (up to December), high bacterial activity near the sediment-water interface prevented oxygenation of the deeper sediments by increased microbial respiration and thus contributed to the lowerings of the redox potentials of these sediments. Lower bacterial activity during colder months, however, permitted

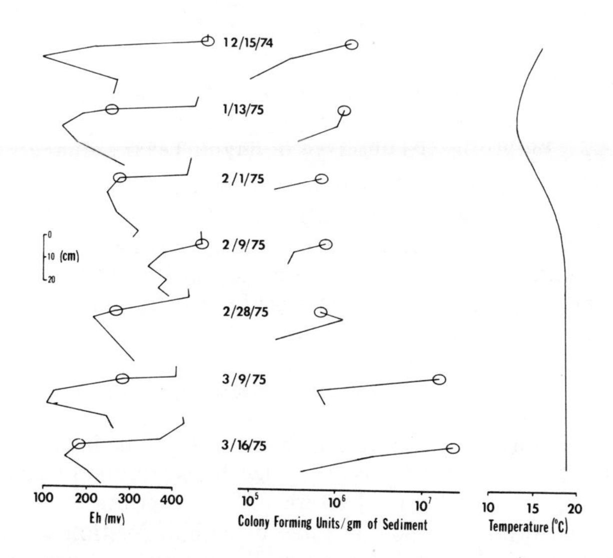

FIGURE 7. Temperature trends in Bayou Texar bottom water and accompanying profiles for Eh and bacterial colony-forming units at the sediment-water interface (indicated by circles) and in underlying sediments. Data from cores taken on the dates indicated (9).

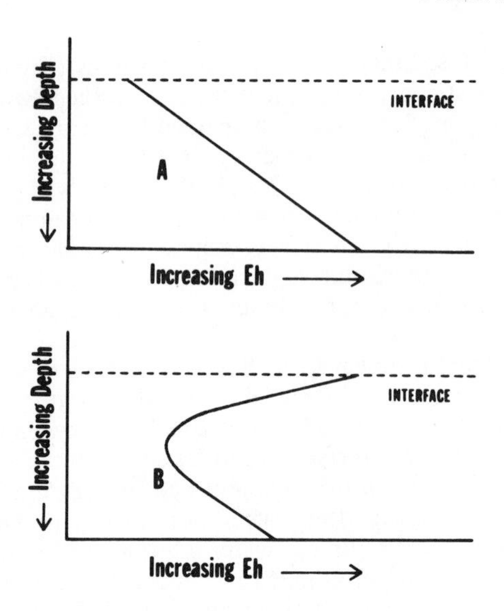

FIGURE 8. A comparison of the expected Eh profile (A) based on declining bacterial activity with sediment depth and the actual Eh profile (B) observed in Bayou Texar sediments (9).

sufficient oxygenation of these sediments to yield the observed high redox potentials. Change of seasons, with associated warming of the water and increasing bacterial activity, repeated the cycle, resulting in the lowering of redox potentials (9). The significance of these redox patterns has already been discussed in relationship to the release of adsorbed phosphates.

Dissolved Glucose and Heterotrophic Productivity

Dissolved glucose concentrations were found to be high and displayed a seasonal pattern which was accompanied by corresponding changes in the rates of glucose uptake by bacteria. A relationship between dissolved glucose concentrations and the rates of carbon fixation was apparent at the upper, more restricted reaches of the bayou, but was less evident at the more pelagic stations (Figure 9). This phenomenon may have been due to the increased effect of mixing of the water columns by wind action at the more pelagic locations. These stations are characterized by greater surface area as compared to the upper stations, which are restricted in morphology and less exposed to wind action. Relationships between

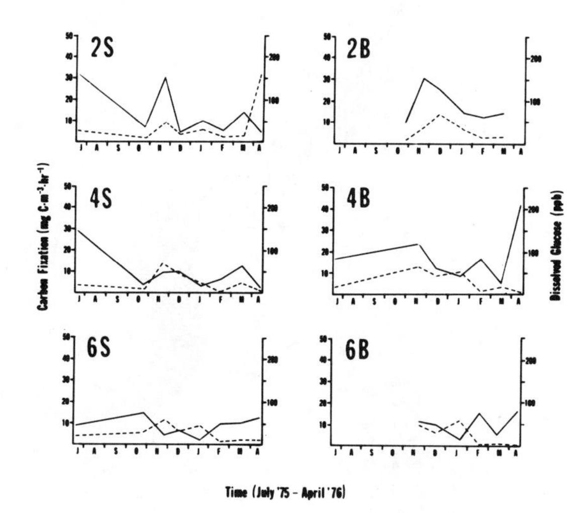

FIGURE 9. Temporal trends in dissolved glucose (dashed line), and carbon fixation (solid line) in surface (S) and bottom (B) waters at Stations 2, 4, and 6 (Figure 1). Only those dates having data for both parameters were included (11).

bacterial numbers or biomass and glucose uptake cited by Wright and Hobbie (16), and Allen (1) were generally confirmed by our studies. Primary productivity was shown to be the major source of dissolved glucose in the Bayou Texar waters (11).

Carbon Fixation

Results of in situ carbon fixation studies for six stations are shown in Figure 10. Concerted efforts to measure variations in carbon fixation, however, were directed only at Stations 2, 4, and 6. Generally, the upper bayou stations were found to be more productive on a per volume basis (Figure 10). This was attributed to the input of significant amounts of major nutrients into upper Bayou Texar waters by Carpenter's Creek. The nutrient composition of surface bayou waters, as compared with creek waters, was found to be highest at the creek end and to decrease southward. This pattern was also found to correlate with the high nitrogen concentrations in the upper bayou, and a gradual decrease toward Station 6 (6) (Figure 11). It is likely that this nutrient distribution gradient was caused by a combination of factors such as utilization of the nutrients by algae and dilution by tidal and wind actions. At the lower ends of the bayou, nitrate and ammonia concentrations were low during the summer

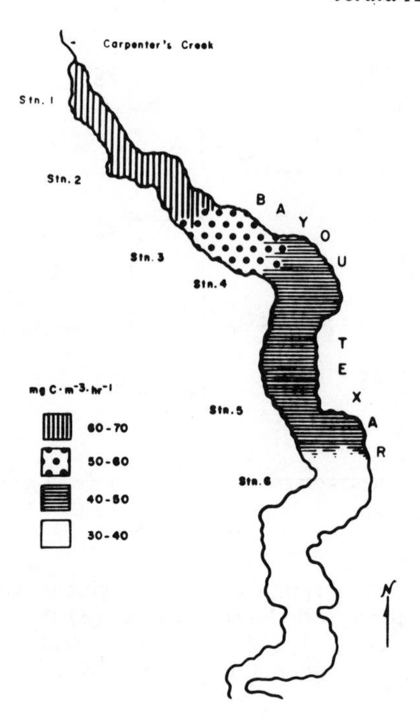

FIGURE 10. Spatial variations in carbon fixation in Bayou Texar between 1971 and 1973.

and fall and high during the winter months. This suggested a direct correlation between surface carbon fixation rates and utilization of nitrate and ammonia by phytoplankton in Bayou Texar (Figure 11).

It was therefore concluded that Carpenter's Creek water, through its nutrient inputs, affected the rate of carbon fixation by the bayou phytoplankton. Confirmation of this conclusion was to some extent aided by a series of culture experiments in which the stimulating effects of Carpenter's Creek water on the primary productivity of the bayou waters were determined (6).

Phytoplankton

In terms of biomass, the phytoplankton populations of Bayou Texar were represented by four major groups: dinoflagellates, chrysophytes, diatoms, and cryptophytes. While no clear seasonal patterns were evident, a notable long-term shift was observed. This shift was from a diverse phytoplankton community to almost unispecific blooms of the first three of the above-mentioned groups. The fish kill in August, 1972 was accompanied by an extensive bloom of dinoflagellates composed primarily of the genera *Ceratium* and *Gymnodinium*. During the fall of 1973, the

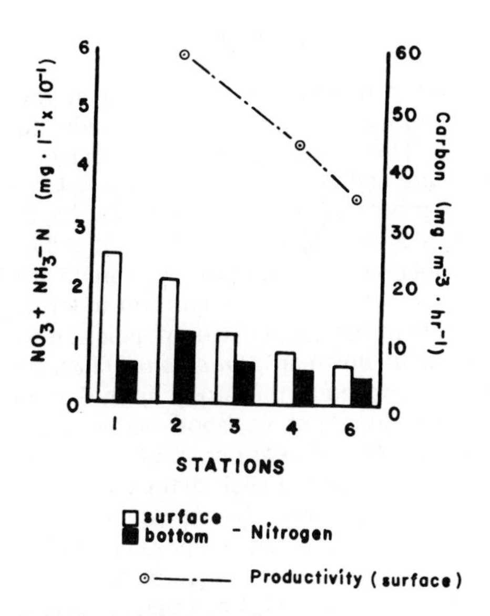

FIGURE 11. Correlations between inorganic nitrogen and carbon fixation in Bayou Texar (6).

dinoflagellate populations fluctuated and eventually gave way to a bloom of chrysophytes including the genera *Chrysochromulina* and *Chromulina*. These organisms were replaced in winter of 1974 by a moderate bloom of the genera *Navicula* and *Cyclotella*. This was finally succeded by a second, smaller dinoflagellate bloom in the summer of 1975.

Throughout the sampling period, the sizes of the blooms declined as did the total phytoplankton biomass. These trends were accompanied by a decrease in dissolved organics and BOD maxima as well as by an increase in DO minima. These patterns might be expected if, as stated later in this presentation, efforts at reducing organic loading to the system were successful, and might superficially suggest a general trend toward the bayou's recovery. This has obvious significance in relation to the 1972 fish kill, which was generally attributed by state and local sources to DO stress.

However, it is necessary that other possible, more important factors also be considered. As stated previously, the 1972 dinoflagellate bloom consisted almost entirely of the genera *Ceratium* and *Gymnodinium*, the latter of which is well known for the toxicity of some of its metabolites to fish. Although the DO concentrations did, at later dates, reach levels as low or lower than those observed during the 1972 fish kill, at no other time did *Gymnodinium* occur in such large numbers, nor did comparable fish mortalities occur. Although the evidence is circumstantial, we suggest that the importance of toxins from algal blooms to the Bayou Texar fish kills may be greater than first suspected.

Zooplankton

Although zooplankton diversity was found to be low throughout the sampling period, numbers and biomass of individual species were seasonally very high. This low diversity/high biomass characteristic has been reported elsewhere and is a feature of many estuaries (5). The most frequently occurring groups of zooplankton in the bayou were copepods (adults and nauplii), rotifers, polychaetes (larvae), and a composite group of miscellaneous zooplankton (tintinnids, infrequent rotifer species, and barnacle nauplii) which, at times, were important in terms of numbers.

The principal grazers in the zooplankton population were *Acartia tonsa* (Dana), *Oithona colcarva* (Bowman), *Brachionus plicatilus*, and *Synchaeta* sp., which occurred throughout the study in rather high numbers. *A. tonsa*, the consistently dominant copepod, showed peaks in the early spring and late summer. *O. colcarva* occurred somewhat less frequently, but in the late summer did show a peak that exceeded that of *A. tonsa* at some of the stations. The rotifers, *Synchaeta* sp. and *B. plicatilus*, were found to be almost totally asympatric temporally. *Synchaeta*, which feeds largely on living algae, was present throughout the winter and spring and declined at the onset of summer. This abundance early in the year could have corresponded to the rising numbers of algae in the spring. By early summer, *Synchaeta* had totally disappeared from all stations and *B. plicatilus*, a detritivore, was present in increasing numbers. *B. plicatilus* peaked in the later summer when organics, in the form of detritus, were high, and was itself beginning to decline by late August.

Although no adult polychaetes were found during the regular sampling regimen, polychaete larvae were found to be consistently present in the bayou throughout the year. These larvae were present at all stations in low but consistent numbers.

The final category of zooplankton found in Bayou Texar was the miscellaneous group. This category was clearly dominated by tintinnids, which showed an uncharacteristically high degree of diversity (ten species encountered) and low species numbers for Bayou Texar. The two dominant species, *Tintinnopsis turbo* and *Favella* sp., were the major constituents of an extreme peak at all bottom stations in the late summer, presumable due to their affinity for high organics.

Benthic Invertebrates

In the first survey, a set of five cores was taken from each of Stations 2, 4, and 6 (Figure 1). Subsequent examination of these samples showed few macroinvertebrates at these stations. In order to determine if a spatial distribution of such organisms exists according to depth from shore to the center of the bayou, a transect consisting of 60 cores was established at Station 4 (Figure 1). Again, examination of cores indicated a paucity of macroinvertebrates in the substrate, and further benthic sampling was discontinued.

It should be noted that the design of the coring device employed in this study did not permit collection of cores from sandy substrates as exist in the nearshore areas within an approximate distance of 10 meters from the shore line in Bayou Texar. It is possible that colonization by macroinvertebrates occurs only in this sandy habitat and not in the organic-rich ooze of the three mid-bayou sites where our original sampling efforts were concentrated.

While there are no known direct sources of chemical pollutants into Bayou Texar, there exists a small pleasure boat marina just north of the Cervantes Street bridge (Figure 1). Therefore, the use of Bayou Texar waters for recreational purposes could have added small amounts of petrochemicals to the water and sediment. It is also possible, however, that tides could have brought other chemical pollutants into the bayou from Pensacola Bay. Although the extent of such pollutants were not determined in the present study, it is possible that they may have adversely affected the benthic invertebrate community.

The consistent presence of polychaete larvae in the water column suggested that these larvae are transported by tidal inflow from Escambia Bay, or are the result of colonization within the bayou. A possible explanation for the observed paucity of adult polychaetes, and benthic macroinvertebrates in general, lies in the inherent nature of the benthic substrates of Bayou Texar. As stated earlier, these substrates were characterized by a graded suspension of particles and a subsequent absence of a sharply delineated bottom. This extended water column-substrate interface could present an effective barrier to inhabitation by macroinvertebrates. It has also been suggested that in estuaries, the alternating exposure of the bottom to fresh and salt water is a major factor responsible for the apparent absence of benthic flora and fauna (4).

CONCLUSIONS AND RECOMMENDATIONS

The extremely poor water quality Bayou Texar, along with periodic and extensive fish kills, were indicative of long-term and permanent cultural eutrophication. Careful examination of our information, however, suggested that such symptoms may indeed be transitory and may dissappear if certain contributing factors were to be removed. For instance, occurrences of fish kills were accompanied by dinoflagellate blooms and preceeded by a number of weeks of very warm, dry, and stagnant weather. These factors suggested that these kills were not necessarily caused by anoxic conditions in the bayou. A prolonged natural mixing and turnover of the stratified and stgnant water column by a change in meterological conditions, however, always abated such fish kills. Therefore, it was deemed logical that implementation of conditions that would improve circulation and tidal flows, and would reduce long-term stagnation would reduce the effects of nutrient inputs.

Presentation of the results of our investigations after the first year resulted in the establishment of a moratorium on building activities in the watershed in December, 1972. This action was aimed at the slow-down of circumstances that were accelerating the eutrophication process in Bayou Texar. In 1974, we suggested four major steps toward the recovery of the bayou. These were:

1. a. Clearing of the entrance to the bayou from Pensacola Bay.

   b. Dredging and straightening of the present channel.

   c. Opening of a new channel east of the present one currently in operation in order to improve flushing.

   d. Removal of excess and abandoned tressel pilings, also to improve flushing.

2. Alleviation of the runoff problem, and associated siltation and nutrification, by the establishment of a holding reservoir to serve as a retention system to enable eventual release of storm water back into the bayou. Repair and augmentation of the inadequate sewage and other waste water systems in order to reduce or stop overflows into the bayou.

3. Retardation of the high speed of water flow from Carpenter's Creek by restoration of meanders, or construction of retardation dams.

4. Selected dredging at sites of excess siltation from storm water drains to restore the bayou to a uniform depth.

The implementation of a number of these suggestions since 1975 has already resulted in a significant improvement in the water quality of Bayou Texar. Only one fish kill has occurred since 1975 and the extent of carbon fixation has been reduced by nearly 90 percent as compared with data collected earlier. Algal blooms have become almost non-existent and algal cell concentrations have been reduced substantially. The bayou has since been reopened for public use and the building moratorium lifted.

ACKNOWLEDGEMENT

This research was supported in part by grants No. B-016-FLA, No. B-021-FLA, and No. B-024-FLA provided by the U.S. Department of the Interior, Office of Water Resources Research Act of 1964. The authors gratefully acknowledge the participation and support provided by numerous faculty colleagues and graduate students throughout the seven-year duration of the study.

REFERENCES

1. Allen, H.L. 1969. Chemo-organotrophic utilization of dissolved organic compounds by planktic algae and bacteria in a pond. *Internat. Rev. ges Hydrobiol.* 54(1):1-33.
2. Aron, W. 1958. The use of a large capacity portable pump for plankton samplings with notes on plankton patchiness. *Sears Found. Journ. Mar. Res.* 16(2):158-173.
3. Bella, D.A. 1972. Environmental considerations for estuarine benthal systems. *Water Research.* 6:1409-1418.
4. Campbell, P.H. 1973. Studies on brackish water phytoplankton. Sea Grant Publication UNC-S6-73-07.
5. Darnell, R.M. 1961. Trophic spectrum of an estuarine community, based on studies of Lake Ponchartrain, Louisiana. *Ecology* 42:553-568.
6. Hannah, R.P., A.T. Simmons, and G.A. Moshiri, 1973. Some aspects of nutrient-primary productivity relationships in a bayou estuary. *Journal of Water Pollution Control Federation.* 45(12):2508-2520.
7. Moshiri, G., D. Brown, P. Conklin, D. Gilbert, M. Hughes, M. Moore, D. Ray, and L. Robinson, 1974. Determination of a nitrogen phosphorus budget for Bayou Texar, Pensacola, Florida, No. 29, Florida Water Resources Research Center, Univ. of Florida, Gainsville.
8. Moshiri, G.A., D.P. Brown, and W.G. Crumpton. 1977. An inexpensive and easily fabricated sampler for collecting sediment cores to measure Eh potential. *Journ. Florida Acad. Sci.* 40(2):203-205.
9. Moshiri, G.A., and W.G. Crumpton. 1978b. Some aspects of redox trends in the bottom muds of a mesotrophic bayou estuary. *Hydrobiologia* 57(2):155-158.
10. Moshiri, G.A., and W.G. Crumpton. 1978a. Certain mechanisms affecting water column-to-sediment phosphate exchange in a bayou estuary. *Jour. Water Poll. Control Fed.* 50:392-394.
11. Moshiri, G.A., W.G. Crumpton, and N.G. Aumen. 1979. Dissolved glucose in a bayou estuary; possible sources and utilization by bacteria. *Hydrobiologia* 62:71-74.
12. Moshiri, G.A., W.G. Crumpton, and N.G. Aumen, C.T. Gaetz, J.E. Allen, and D.A. Blaylock. 1978. Water column and benthic invertebrate and plant associations as affected by the physico-chemical aspects in a mesotrophic bayou estuary, Pensacola, Florida, Publ. No. 41, Florida Water Resources Research Center, Univ. of Florida, Gainesville.
13. Moshiri, G.A., W.G. Crumpton, and D.A. Blaylock. 1978. Algal metabolites and fish kills in a bayou estuary: an alternative

explanation to the low dissolved oxygen controversy. *Jour. Water Poll. Control Fed.* 50:2043-2046.

14. Moshiri, G.A., W.G. Crumpton, D.P. Brown, P.R. Barrington, and N.G. Aumen. 1976. Interrelationships between certain micro-organisims and some aspects of sediment-water nutrient exchange in two bayou estuaries - phases I & II. Publication No. 37, Florida Water Resources Research Center. Univ. of Florida. Gainesville.
15. Moshiri, G.A., R.P. Hannah, A.T. Simmons, G.C. Landry, and Whiting. 1972. Determination of a nitrogen-phosphorus budget for Bayou Texar, Pensacola, Florida. Publ. No. 17, Water Resources Res. Center, Univ. of Florida, Gainsville.
16. Wright, R.T., and J.T. Hobbie. 1966. Use of glucose and acetate by bacteria and algae in aquatic ecosystems. *Ecology* 47:447-464.

RESPONSES OF KANEOHE BAY, HAWAII, TO RELAXATION OF SEWAGE STRESS

S.V. Smith

Hawaii Institute of Marine Biology
P.O. Box 1346
Kaneohe, Hawaii 96744

ABSTRACT: Kaneohe Bay is a subtropical coral reef/estuary complex which was subjected to increasing sewage loading; that sewage was diverted in 1977 and 1978. We have treated the loading and diversion as a controlled, total-ecosystem experiment to evaluate the chemical and biological responses to external subsidy of nutrients. We here consider the Bay's response to sewage loading and to its diversion largely in the context of a nitrogen budget.

Even the most heavily impacted portion of the Bay showed only modest increases in dissolved nitrogen levels within the water column. Particulate materials including plankton biomass, as well as dissolved inorganic phosphorus, were elevated substantially. The benthos showed increased biomass and metabolic rates, especially of heterotrophs responding to fallout of organic particulate materials. Nutrient recycling within the Bay was the major immediate source for the nutrients for the observed rapid metabolic activity.

Sewage accounted for approximately 80 percent of the inorganic nitrogen and 90 percent of the inorganic phosphorus delivery to Kaneohe Bay. Diversion lowered the land-derived inorganic nitrogen and phosphorus input to the Bay by 70-80 percent. Virtually all components of the system have responded to this diminished nutrient subsidy, but the water column nutrient washout and biological recovery are predictably occurring more rapidly than the benthos responses.

Kaneohe Bay, Hawaii (Figure 1), is an estuary and coral-reef complex which, until recently, has been subjected to sewage loading and other human perturbations (1, 4, 11, 12). The sewage has been largely diverted from the Bay to a deep-ocean outfall, and we are treating that diversion as a controlled, total-ecosystem experiment. The experiment is based on the premise that sewage input is a nutrient source whose influence can be partially explained by a chemical mass balance for the Bay. The sewage largely enters the system as dissolved inorganic nutrients; accounting for its conversion to particulate organic material, dispersion and recycling within the system, and washout from the system provides insight into the environmental influence of this sewage and of its diversion.

The study has been underway since early 1976, with the diversion occurring in two major steps in December 1977 and May 1978. At the time of this presentation, field work on the present program is essentially completed; our post-diversion data analysis is only barely underway. This paper provides a conceptual and largely qualitative overview based on this as yet incomplete data analysis.

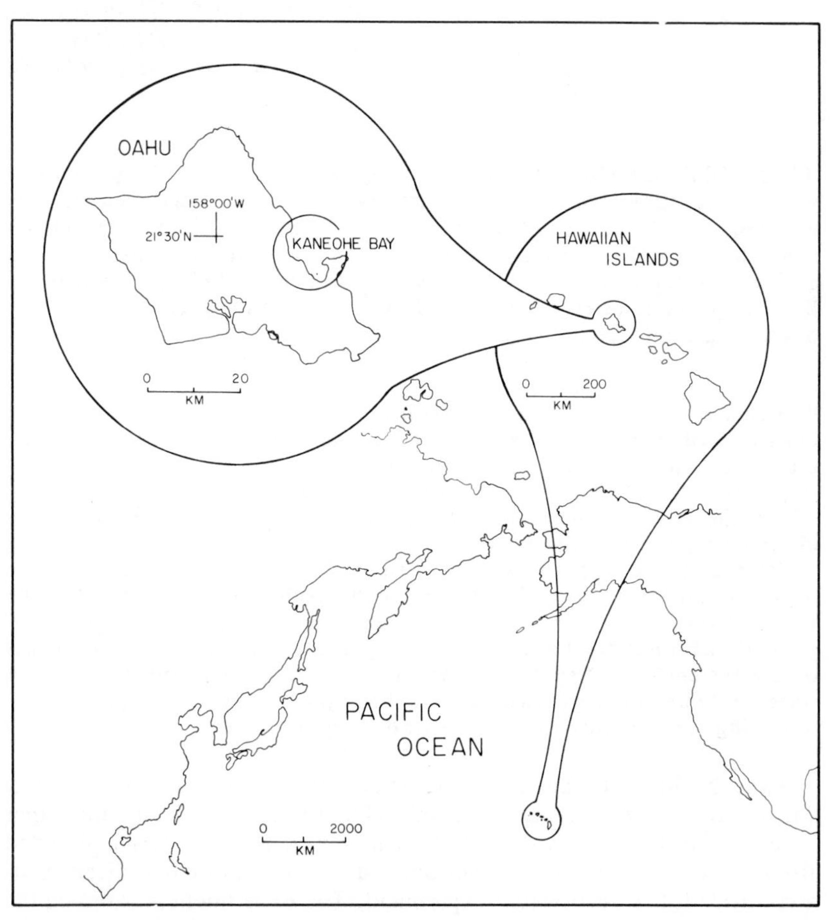

FIGURE 1. Kaneohe Bay, Hawaii, in global perspective.

Before discussing Kaneohe Bay further, I will review briefly why we consider the Kaneohe Bay study to be a controlled, total-ecosystem experiment. It is an "experiment" rather than a routine monitoring exercise because the stations were chosen to test expectations of ecosystem response rather than to document general ecosystem status. The experiment is "controlled" in the sense that the major nutrient inputs (sewage and runoff) for the Bay are well known, and the advanced scheduling of the sewage diversion allowed us to design a sampling program around that predictable and quantifiable event. The "control" we exercised was the manipulation (conceptually, at least) of the sewage nutrient input function. The experiment involves the total ecosystem in being a relatively comprehensive analysis, from the perspective of nutrient dynamics, of a physiographically well-defined and distinct ecosystem. Observations at individual sites are scaled up to a view of the Bay in its totality. A number of other "before versus after" monitoring studies have been or are being performed (e.g., 5). However, we have not seen this kind of study recognized to be a powerful general class of environmental experiments, nor are the results from individual stations usually scaled up to an understanding of the total ecosystem.

Kaneohe Bay is particularly appropriate for such a total-ecosystem experiment (Figure 2). The Bay is rather small, only about 30 km^2 in area, and thus is amenable to monitoring. Mean residence time of water in the Bay is only about two weeks (12), so responses to perturbations are rapid. The physiographic boundaries of the Bay and of its watershed are easily defined; watershed inputs to the Bay are well-known, relatively homogenous, and (except for the sewage which has been the focal point of this study) ordinarily relatively small (12).

The Bay is only weakly estuarine; short periods of high freshwater input cause distinctive low-salinity spikes in a pattern of otherwise relatively constant, near-oceanic salinity (Figure 3).

Because Hawaii is semitropical, the biological dynamics of the Bay are not entrained in as strongly oscillating a seasonal oceanographic or meterological cycle as temperate systems. There is some seasonality, however. Figure 4 illustrates variation in water temperature, ambient solar radiation, wind speed, and wind direction. Note particularly that there is a substantial variation in wind direction; this variability has direct bearing on water flow, and particularly flow of the major sewage plume, in the Bay (8).

It is possible to identify diverse biological communities in Kaneohe Bay and to recognize specific, significantly different external controlling variables acting on otherwise very similar communities. These communities largely coincide with physiographic boundaries within the Bay. Figure 5 illustrates the bimodal distribution of water depths. The shallow water is dominated by reef flat biota living in or on both limestone and loose sediments, and the deep water has a prominent

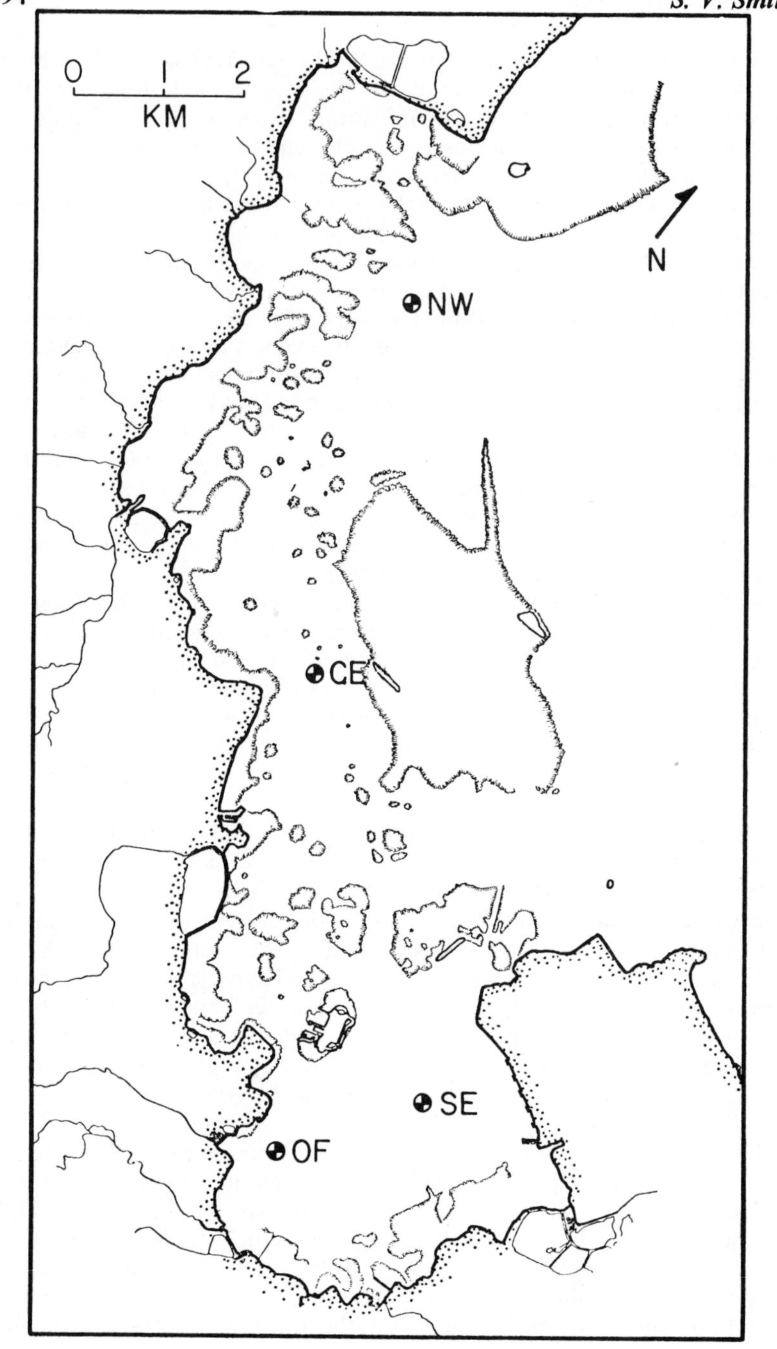

FIGURE 2. Sample locations and distribution of reefs in Kaneohe Bay.

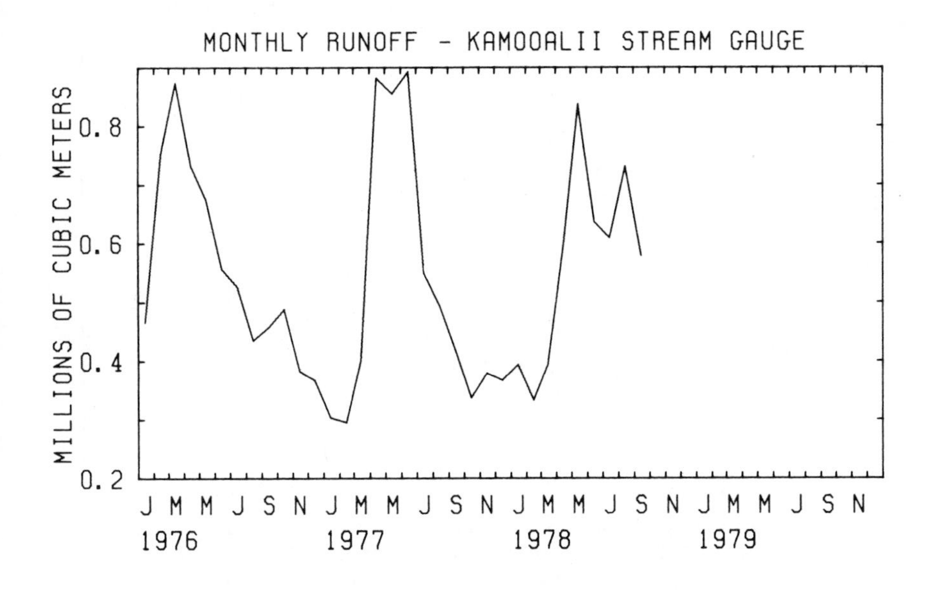

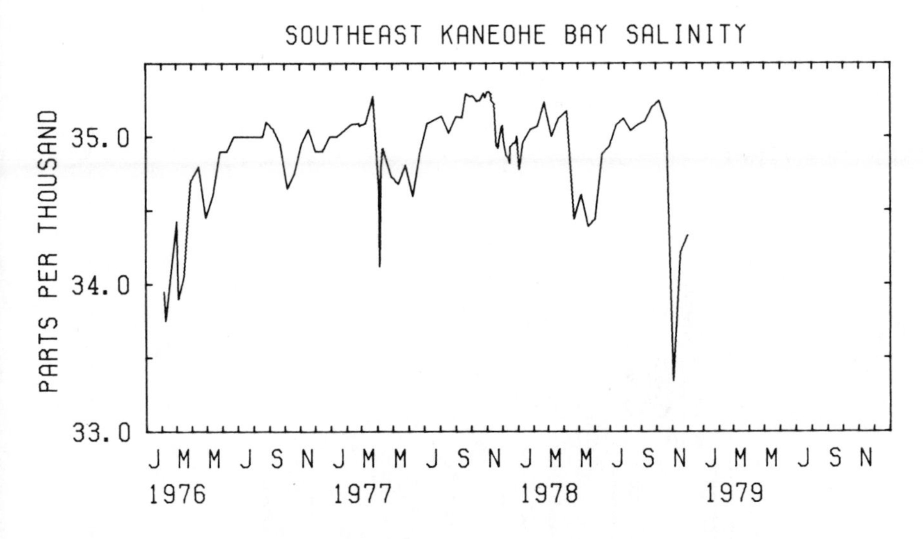

FIGURE 3. Temporal variation in stream runoff at a representative stream discharging into the SE sector and depth-averaged salinity in that sector.

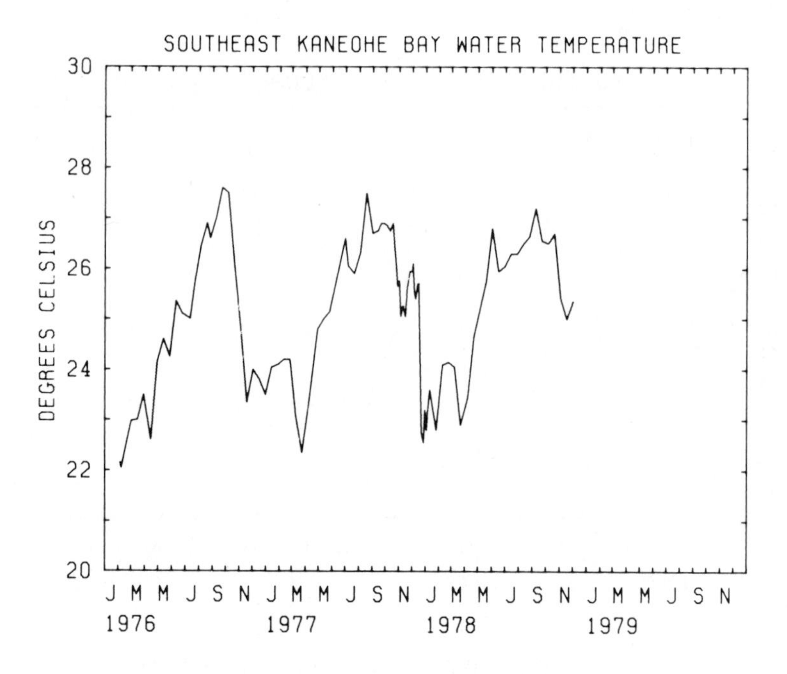

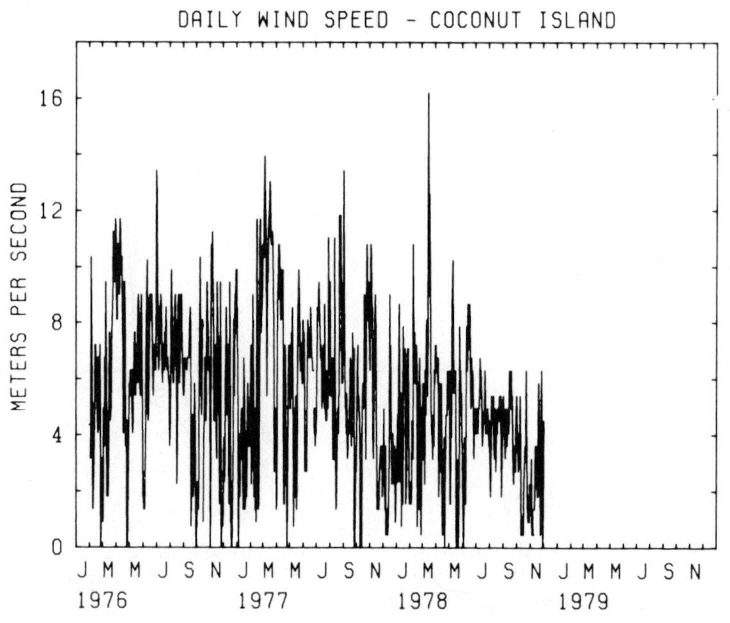

FIGURE 4. Temporal variation in water temperature, solar radiation,

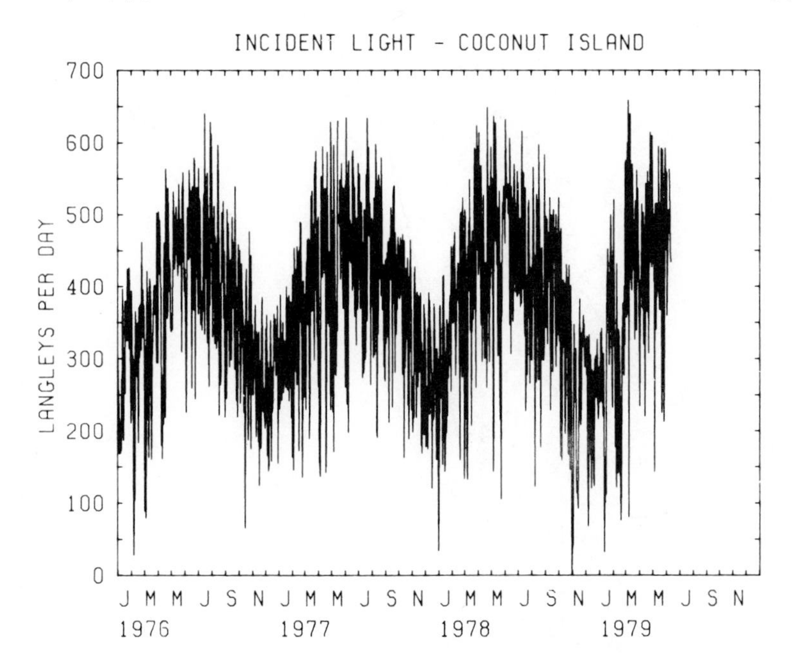

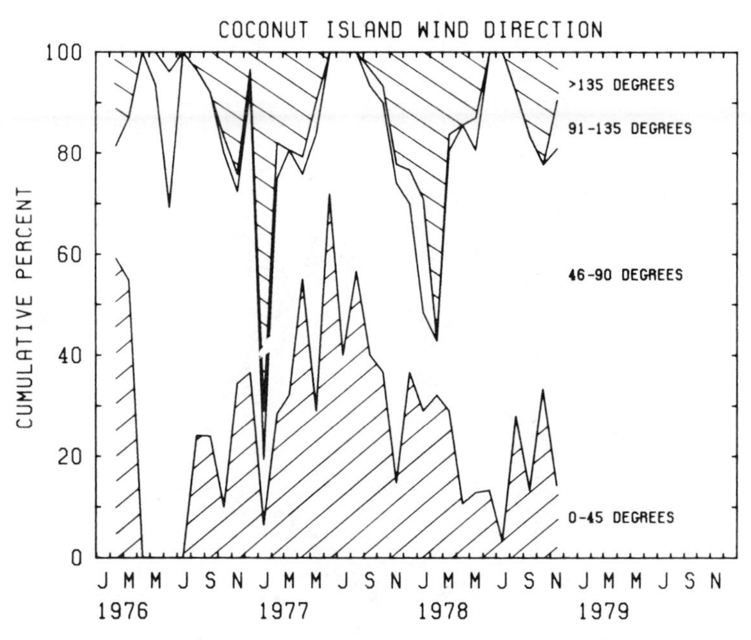

FIGURE 4. (Continued) wind speed, and wind direction.

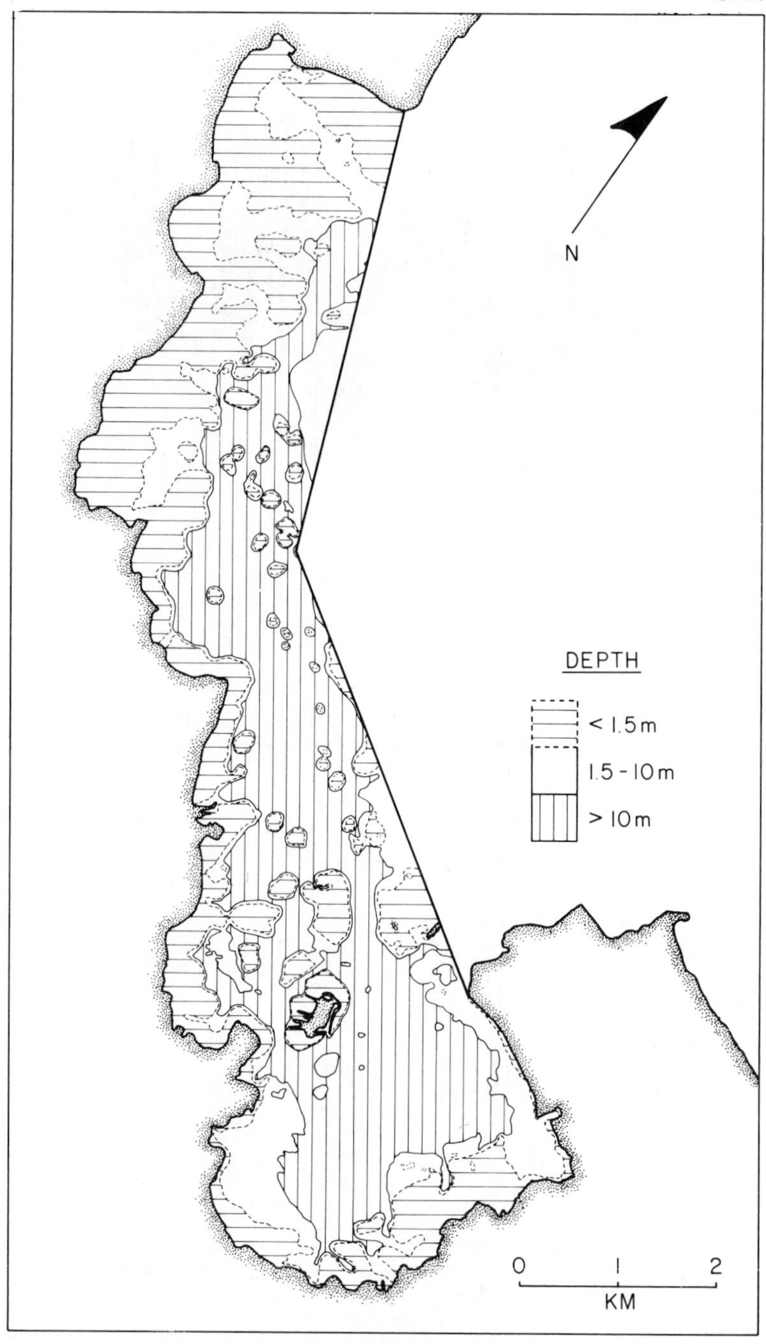

FIGURE 5. Simplified map of depth distribution in Kaneohe bay.

plankton community as well as an infaunal community within the lagoon-floor mud.

There is no evidence that significant components of the biotic communities have been totally eliminated from any major portion of the Bay by the stresses which have been imposed, although the communities have been severely modified (2, 7, 13). Therefore, our analysis deals with the varying structure, biomass, and metabolism of initially very similar biotic communities modified by sewage and other perturbations.

Much of the remainder of this presentation will be an analysis, largely from the vantage of nitrogen mass balance, of point-source nutrient input to Kaneohe Bay, cycling within the Bay, and output from the Bay. This analysis is virtually identical with an analysis of sewage loading: (a) most of the sewage entered the Bay at one of two point sources; (b) the sewage was the major exogenous nutrient source for the Bay; (c) the sewage lacked significant toxic effects from low dissolved oxygen, high biological demand, heavy metals, chlorine, or anthropogenic biocides; (d) the sewage was sufficiently concentrated to be an essentially pure nutrient input with little effect on salinity.

Figure 6 summarizes calculated sewage production in the watershed, based on human population data (12) and point-source sewage discharge.

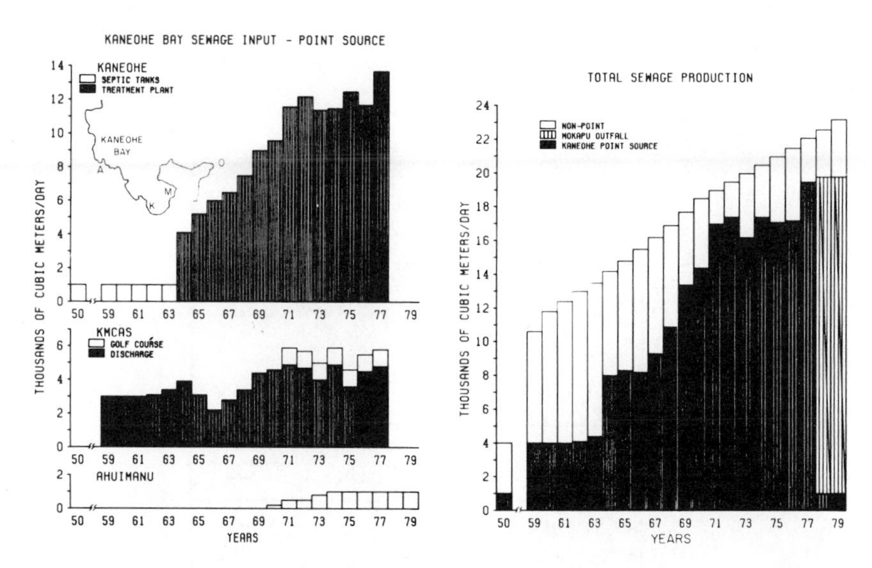

FIGURE 6. Production of sewage in the Kaneohe watershed, as calculated from population data, and discharge of that sewage into the Bay. The inset map shows the locations of the sewer outfalls; A = Ahuimanu; K = Kaneohe municipal; M = Marine Corps: and O = Open ocean (Mokapu).

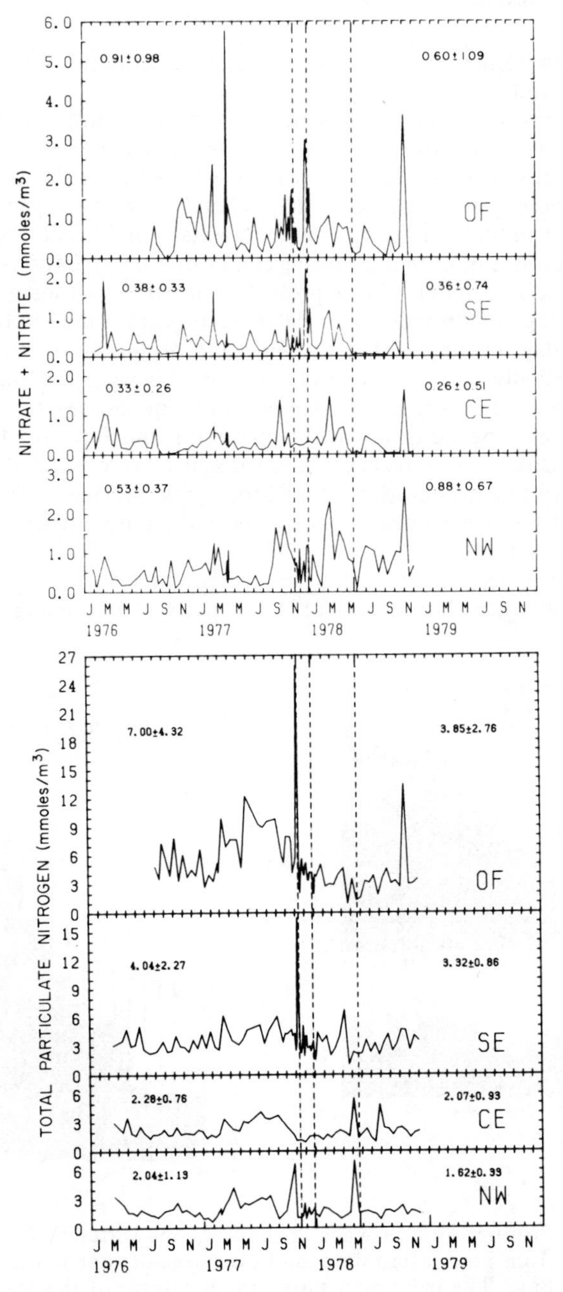

FIGURE 7. Temporal variation in depth-averaged nitrogen concentration of Bay waters. The three vertical lines in 1977-78 indicate onset of an initial period of temporary diversion of the

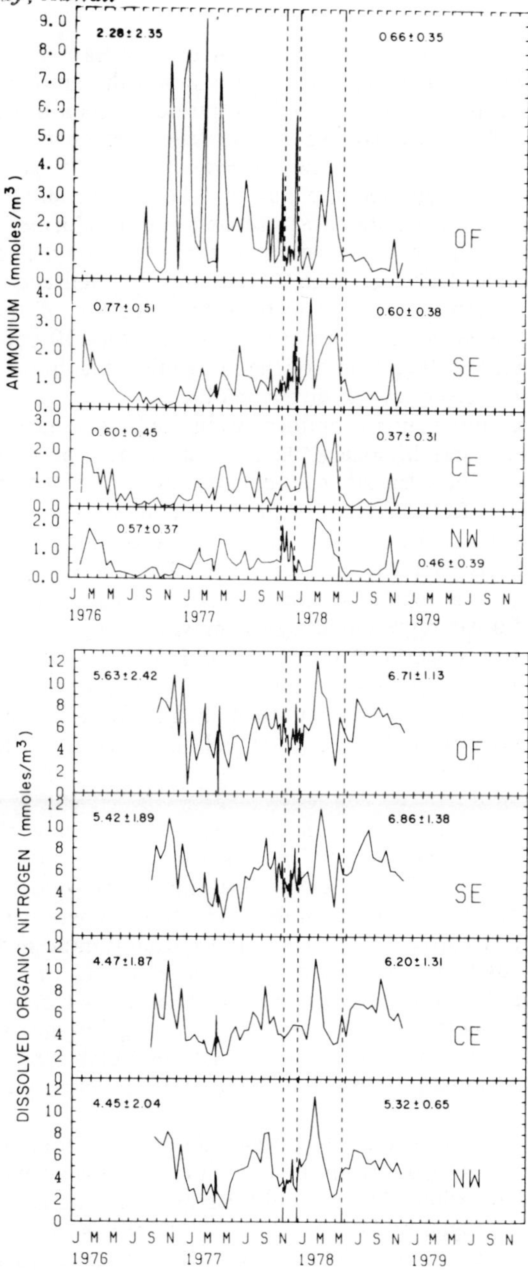

FIGURE 7 (cont'd).
Kaneohe Municipal sewage, permanent diversion of that sewage, and permanent diversion of the Marine Corps sewage. The pre-diversion and post-diversion mean concentration ±1 standard deviation unit are also given.

The nonpoint production probably does reach the Bay via groundwater seepage along the shoreline. The figure also shows the amount of sewage diverted by the construction of the Mokapu deep-ocean outfall. Present discharge to the Bay is approximately equivalent to that of 1950, well below the discharge of the past two decades. About 70 percent of the point-source discharge entered the Bay at a single outfall (the Kaneohe STP); another 25 percent entered at a second outfall (Marine Corps STP). Both of these were in the southeastern portion of the Bay. The remaining point-source sewage discharge, which continues, is small and is delivered to the Bay via a stream in the northwestern portion of the Bay (the Ahuimanu STP). Sewage accounted for approximately 80 percent of the inorganic nitrogen and 90 percent of the inorganic phosphorus delivery to the Bay from land before sewage diversion (12).

We established four major sampling sites in the Bay (Figure 2) and have collected chemical and biological data at those sites periodically since 1976. Station OF is near the site of the major sewer outfall and was under its direct influence. Station SE is in the center of the relatively enclosed southeastern basin and was under the influence of diffuse effects from both major outfalls. Station CE is removed from direct effects of the outfalls, although this region of the Bay does exchange water with the sewage-impacted areas represented by the previous stations. Station NW is in the northwestern portion of the Bay, removed from significant sewage effects but receiving substantial freshwater input from runoff. Stations CE and NW are very similar to one another chemically, although there are substantial biological differences. Water column data are collected at all sites, and benthos data are collected at sites immediately landward of all but Station SE.

From the data, we are generating time-trend plots such at those shown in Figure 7. Between-station differences can be seen for particulate nitrogen; before-versus-after diversion differences are most marked at Station OF. Note the lack of seasonal oscillations. There is a somewhat weaker before-versus-after NH_4 pattern and virtually none in the NO_3 or DON. The main nitrogen input from the sewage is NH_4; the NO_3 largely enters via streams. The phosphorus data (Figure 8) show a between-station trend as well as a before-versus-after trend in PO_4 but not in dissolved organic P. Most of the within-treatment variations which can be seen (e.g., the nutrient peaks most prominent at Station OF) are closely tied to wind patterns (8).

We are also using these data to consider the balance of material masses in the Bay and fluxes to and from the Bay. Such mass-balance budgets can be informative even when they are only partly complete, as illustrated in Figure 9 with a nitrogen budget. Some of the numbers in this budget are only very roughly approximated. However, the patterns are sufficiently clear that the roughness of the approximation is not a particular problem to general interpretation of the data as presented here.

This analysis reduces the nitrogen compartments of the system to the

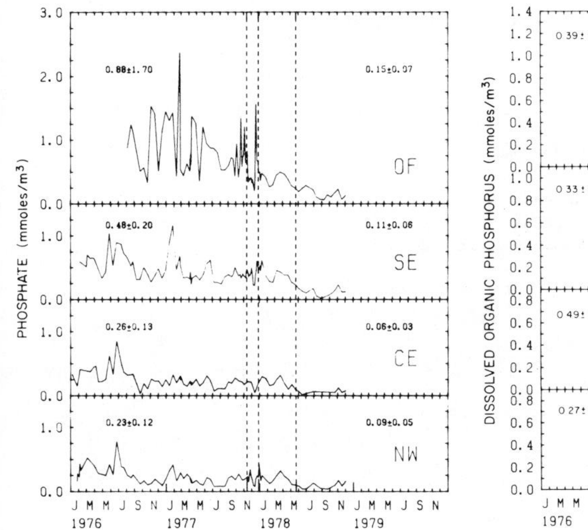

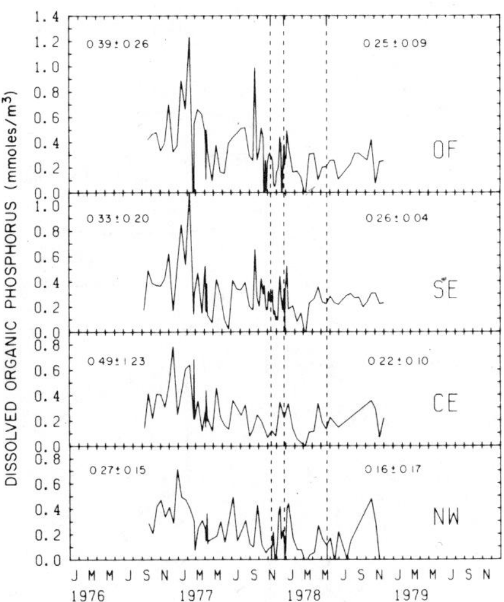

FIGURE 8. Temporal variation in depth-averaged phosphorus concentration of Bay waters. The three vertical lines in 1977-78 indicate onset of an initial period of temporary diversion of the Kaneohe municipal sewage, permanent diversion of that sewage, and permanent diversion of the Marine Corps sewage. The pre-diversion and post-diversion mean concentrations ± 1 standard deviation unit are also given.

following: Terrigenous nitrogen is largely sewage; water column dissolved nitrogen includes NH_4, NO_3, and NO_2; dissolved organic N is excluded from this analysis on the assumption that an insignificant proportion of it (relative to other nitrogen forms) is metabolized. It is thus assumed in our analysis to be acting as an essentially conservative constituent of seawater; this is undoubtedly an oversimplification of a poorly understood nutrient reservoir. Water column particulate nitrogen includes both living and non-living material; benthic dissolved nitrogen is the inorganic interstitial nitrogen in the upper 30 centimeters of the sediment column; benthic particulate nitrogen includes the biota plus particulate nitrogen in this portion of the sediment column. This 30 cm depth is our estimate of the depth of active biological activity and homogenization and oxidation of the lagoonal sediments (6). Nitrogenous material in the Bay is exchanged with an oceanic nitrogen reservoir.

First consider compartment sizes. Virtually all nitrogen in the Bay is in the benthic particulate nitrogen compartment. About 96 percent of that

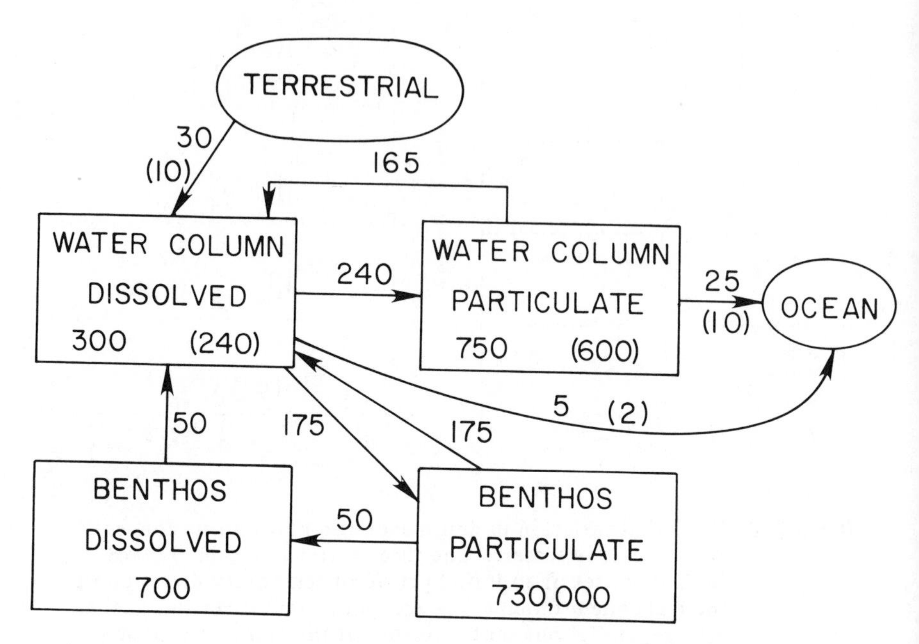

FIGURE 9. Simplified compartment diagram indicating the nitrogen budget of the Bay. Unparenthesized numbers on the diagram represent pre-diversion figures; parenthesized numbers are post-diversion. Compartment sizes are in kilomoles; fluxes are kilomoles/day.

compartment is non-living detritus (Table 1). Even the remaining 4 percent, which is living, is a much larger subcompartment than the sum of the water column compartments. A compartment size divided by the present terrigenous input rate gives some feeling for a shorter limit on how fast that compartment might be "filled up" or "emptied". For the benthic detritus subcompartment, the result is about 60 years. This figure may be severely in error if the average "active sediment" thickness is less than 30 cm. However, the term is so much larger than the other compartments that such an error (even by a factor of two to three on reservoir size) would not significantly alter our interpretations. Clearly this subcompartment can be no more than slightly under the influence of sewage. That benthic detritus must represent long-term nitrogen accumulation, and most of the material is probably refractory. By a similar argument, the benthic organism subcompartment is equivalent to about three years of terrigenous input at the input rate immediately

TABLE 1. Partitioning, by percentage, of pre-diversion particulate nitrogen in ecosystem. Derived from Ref. 10 plus unpublished data.

	Water Column	Benthos
Detritus	64	96
Plants	25	2
Animals	11	2

before sewage diversion. It is therefore probable that the benthic organism subcompartment is strongly under the long-term influence of the sewage input rates. The water column compartments are small, equivalent to only a few days' terrigenous input, and obviously under short-term potential influence of this sewage input rate coupled with oceanic exchange.

The interpretation that the water column compartments are strongly influenced by the sewage is borne out by the 20 percent drop in the size of those compartments after sewage diversion (Figure 9). That change is not particularly large summed over the entire Bay, but concentrations of nitrogenous materials within sectors demonstrate that much of the change is reflected by a large concentration change in the OF sector (Figure 7).

Next consider fluxes between compartments. Land-derived nitrogen input, mostly from the sewage (Table 2), is small in comparison with net nitrogen uptake by water column particulate materials (as approximated from primary productivity data) (10) or by benthic particulate materials (approximated from ΔO_2 and ΔCO_2 data for reef-flat communities) (9). The only way these fluxes can be forced to balance is if most nitrogen uptake in the system is sustained by internal cycling rather than by the input of new nitrogen. Caperon (3) arrived at similar conclusions with respect to the importance of cycling in the planktonic part of the system alone.

If the exogenous nitrogen, including sewage, does not particularly support higher productivity, what role does that input play in the nutrient economy of the system? As a first answer to this question, the standing crop of particulate materials is enhanced by nutrient loading. The washout of particulate matter from the Bay to the ocean is proportional to the standing crop of particles in the water column, so the Bay is effectively a chemostat producing and exporting particulate materials at the rate of dissolved nitrogen addition. The primary productivity in the chemostat is apparently limited by light and nutrient turnover rate, whereas the biomass is limited by nutrient input (10).

TABLE 2. Land-derived inorganic nitrogen and phosphorus delivery to Kaneohe Bay, 1976-1977. Derived primarily from Ref. 12.

	moles/day			% of total delivery Inorganic	
	NO_3+NO_2	NH_4	PO_4	N	P
sewage	0	18000	3000	67	79
streams	4000	1000	200	18	5
ground-water*	1000	3000	600	15	16
total	5000	22000	3800	100	100

*(estimated to include diffuse sewage seepage)

The point-source injection of nutrients causes development of a concentrated plume of particulate organic material. The benthos living downstream from the outfall and within the plume respond to this particulate material in two ways. The proportion of ambient light which reaches the bottom is lowered (Figure 10), depressing biomass and productivity of benthic algae. Sedimentation of the particulate organic materials provides nutrition to enhance heterotrophic activity (6, 11). This enhancement is particularly evident on hard-bottom areas, where animal biomass is greatly elevated (Table 3). The animal biomass buildup is not as readily apparent in the soft sediments which blanket most of the Bay floor, because these sediments do not provide a suitable habitat for most of the larger organisms which account for the bulk of the biomass within the ecosystem. The larger organisms there are probably limited by available habitat rather than by food. However, the effect is felt in the soft-bottom areas if, instead of large-organism biomass, nutrient release rate is used as the index of choice. There is about a three-fold increase in nitrogen release from the sediments from the northwest end of the Bay to the southeast (Figure 10) (6, 11). This variation in nutrient release probably reflects variation in meiofauna and bacteria whose biomass is not directly sampled by our analysis. Both the hard-bottom animal biomass and the nitrogen release rates from the sediments have apparently decreased in the year since sewage has been diverted from the Bay, although data processing has proceeded insufficiently for us to report that decrease as hard numbers at this writing.

Benthic plant biomass was low near the outfall before sewage diversion; the biomass increased at intermediate distance, and decreased with further distance from the outfall (Figure 10). We interpret this pattern largely to

FIGURE 10. Mean hard-bottom algal biomass on the reef flat, lagoon-floor inorganic nitrogen flux and light at 1.5 m (i.e., the maximum reef-flat depth) before sewage diversion.

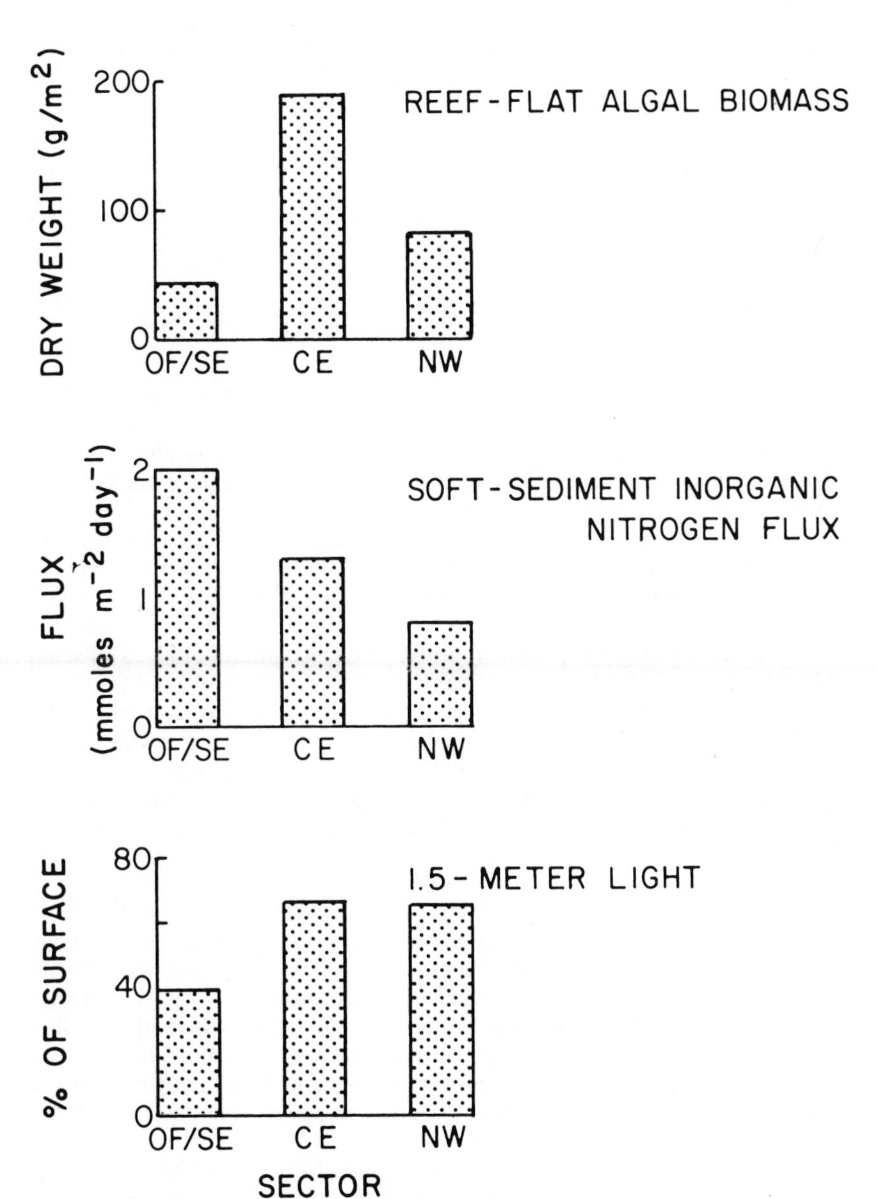

TABLE 3. Benthic infaunal biomass, by sector (grams dry weight/m^2). From Ref. 12.

Physiographic Zone	Substratum Type	Site Designation OF/SE	CE	NW
Reef flat	hard	330	73	70
Reef flat	soft	17	19	no data
Reef crest	hard	410	28	58
Reef slope	hard	220	44	46
Lagoon floor	soft	7	3	3

represent light limitation of algal productivity near the outfall, light-enhanced productivity and sufficient nutrient flux supporting elevated biomass at intermediate distances, and eventually nutrient limitation with increasing distance from the outfall. We have not yet seen a significant response to sewage diversion.

Thus, the immediate effect of sewage injection into Kaneohe Bay was enhanced plankton biomass near the sewer outfall, with little inorganic nutrient enhancement. The intensity of the plankton bloom which develops in a sewage plume will be a function of the temporal coherence of the plume relative to plankton growth rate. Kimmerer et al. (8) discuss the coherence of the Kaneohe sewage plume in terms of a wind-mediated front of low-density water trapping sewage along the shoreline. The location and intensity of the plume restricted the impact of the turbid water to a relatively small part of the Bay during prevailing wind conditions. The physical front which trapped the sewage remains in the absence of the sewage, but the nutrient delivery to that plume is now absent.

Overall, the major biological effect of the sewage was enhanced biomass; productivity is mostly dependent upon nutrient cycling within the system rather than on exogenous nutrient delivery. The biomass buildup sustained by any particular level of nutrient loading into an ecosystem is a function of the nutrient input and washout characteristics of that system. Thus, the plankton biomass, which constantly washes from the Bay by water exchange, adjusts rapidly to a changing regime of nutrient loading. The benthos biomass adjustment to a changing nutrient loading regime is slow, because material slowly accumulates or breaks down with little ongoing washout.

The decisions to treat sewage diversion from Kaneohe Bay as an experiment, to sample within the region of significant ecological impact, and to establish an extended timespan of data collection, are allowing us to fine-tune our qualitative understanding of how the system works until we can quantify major known processes and perhaps unearth major unknown ones as well.

ACKNOWLEDGEMENTS

This paper is presented as a sole authorship paper but is the product of a team investigation. I am grateful to my many colleagues who have contributed to the study. The program has been funded by U.S. Environmental Protection Agency grant R803983 and by funds from the Hawaii Marine Affairs Coordinator. We have also cooperated extensively with personnel at the Naval Ocean Systems Center, Kaneohe, Hawaii. Personnel in various local and state agencies have also been most helpful. Hawaii Institute of Marine Biology Contribution No. 576.

REFERENCES

1. Banner, A.H. 1974. Kaneohe Bay, Hawaii: Urban pollution and a coral reef ecosystem. Proc. 2nd Internat. Symp. Coral Reefs, Brisbane. 2:685-702.
2. Brock, R.E., and J.H. Brock. 1977. A method for quantitatively assessing the infaunal community in coral rock. *Limnol. Oceanogr.* 22:948-951.
3. Caperon, J. 1975. A trophic level ecosystem model analysis of the plankton community in a shallow-water subtropical estuarine embayment. *Estuarine Research*, Academic Press, New York. 1:691-709.
4. Devaney, D.M., M. Kelly, P.J. Lee, and L.S. Motteler. 1976. Kaneohe: A history of change (1778-1950). B.P. Bishop Museum, Honolulu, 271 pp.
5. Edmondson, W.T. 1977. Recoveiy of Lake Washington from eutrophication, 102-109. *In* J. Cairns, Jr., K.L. Dickson, and E.E. Herricks (eds.), Recovery and Restoration of Damaged ecosystems. U. Press of Virginia, Charlottesville.
6. Harrison, J.T. 1979. University of Hawaii Ph.D. dissertation. In preparation.
7. Hirota, J., and J.P. Szyper. 1976. Standing stocks of zooplankton size-classes and trophic levels in Kaneohe Bay, Oahu, Hawaiian Islands. *Pac. Sci.* 30:341-361.
8. Kimmerer, W.J., T.W. Walsh, and J. Hirota. 1980. The effects of sewage discharged into a wind-induced plume front.
9. Kinsey, D.W. 1979. Carbon turnover and accumulation by coral reefs. U. Hawaii Ph.D. dissertation, 248 pp.
10. Laws, E.A., and D.G. Redalje. 1979. Effect of sewage enrichment on the phytoplankton population of subtropical estuary. *Est. Coast. Mar.*

Sci. (in press).

11. Smith, S.V. 1977. Kaneohe Bay: A preliminary report on the responses of a coral reef/estuary ecosystem to relaxation of sewage stress. Proc. 3rd Internat. Coral Reef Symp. 2:577-583.
12. Smith, S.V., R.E. Brock, E.A. Laws. 1980. Kaneohe Bay: A coral reef ecosystem subjected to stresses of urbanization. Manuscript to be published in D.R. Stoddart and R.W. Grigg (eds.). Coral reef ecosystems under stress. Academic Press, London.
13. Smith, S.V., K.E. Chave, and D.T.O. Kam. 1973. Atlas of Kaneohe Bay: A reef ecosystem under stress. U. Hawaii Sea Grant Pub. TR 72-01. 128 pp.

CONTRIBUTED PAPERS

NITRIFICATION IN THE UPPER TIDAL JAMES RIVER

Carl F. Cerco
Assistant Marine Scientist
Virginia Institute of Marine Science
Gloucester Point, Virginia 23062

ABSTRACT: A field and model study of the nitrification process in the upper tidal portion of the James River, Virginia, has been completed. Attention was devoted to the enumeration of nitrifying bacteria and to the determination of the fate of constituents involved in the nitrification process. Ammonia and nitrite oxidizers are present in the James River water column in concentrations of 10^{-1} to 10 mpn/ml and in the bottom sediments in concentrations of 10^2 - 10^5 mpn/ml. Elevated populations are observed in the vicinity of waste discharges suggesting that nitrification of these wastes begins immediately upon discharge. Examination of three sets of field data supports the hypotheses that nitrification occurs. Of particular interest is the observation, during two instances, of nitrate removal concurrent with nitrification.

The detrimental effect of the nitrification of ammonia on stream dissolved oxygen has been recognized since the early studies of Courchaine (3) on the Grand River. Contemporaneous work by Stratton and McCarty (9) and Wezernak and Gannon (13) proposed kinetic models for the nitrification process and applied these models to laboratory and streamflow conditions. Since the time, a number of investigations have been conducted into the occurrence of the nitrification process in free-flowing rivers and streams (e.g. Curtis et al., (4); Finstein and Matulewich, (5). Attention has also been devoted to nitrification in tidal and estuarine waters. Berdahl (1) conducted laboratory investigations into the effect of salinity and residence-time on nitrification while Tuffey et al. (11) hypothesized about the importance of nitrification in estuarine waters. With the exception of Tiedmann (10), however, little in-situ investigation of nitrification in tidal waters has taken place.

This paper details the results of a study to investigate the occurrence of nitrification in the upper, tidal portion of the James River, Virginia. Particular attention is devoted to enumerating the nitrifying bacteria and to examining the fate of nitrogen components within the system.

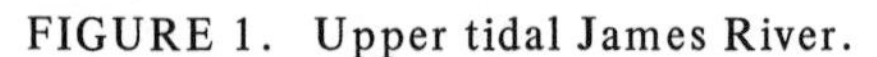

FIGURE 1. Upper tidal James River.

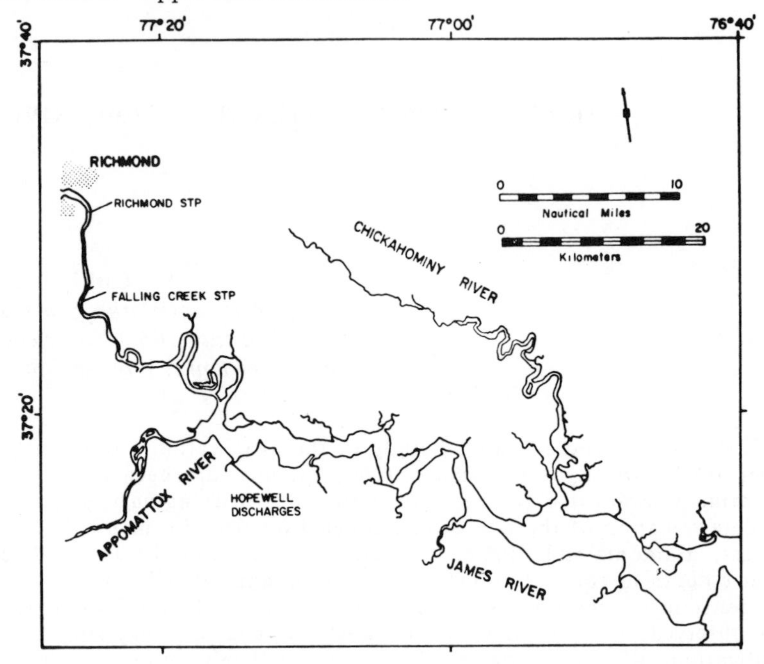

The study section of the James River (Figure 1) extends from the fall-line at Richmond (km 160) to the confluence with the Appomattox River (km 124) just above the city of Hopewell. The cross-sectional area of the river in this region varies from 560 m^2 to 2,800 m^2 with average depths of from 4.0 to 8.5 m. Annual average freshwater flow at the fall line is 211 m^3/sec and the tidal range varies from 80 cm at Hopewell to 98 cm at Richmond.

The upper tidal James River receives discharges from a number of point sources including sewerage treatment plants at Richmond (km 157) and Falling Creek (km 148). Several smaller sources exist as do major municipal and industrial wastefalls at Hopewell. Although these latter wastefalls are out of the study area, tidal action carries a portion of these pollutants upstream into the region of interest. Typical values of point source inputs to the James are presented in Table 1.

Bacterial Surveys

The first portion of the study was devoted to detecting and enumerating the nitrifying bacteria within the study area. From the relative abundance and distribution of bacteria, the location and extent of the nitrification process may be determined. If the nitrifiers are found far downstream of the major point sources, nitrification of wastes will be delayed until the wastes flow downstream to the nitrifying region and/or until nitrifiers found within the wastes multiply to effective levels. Conversely, if elevated concentrations of nitrifiers exist in the vicinity of the wastefalls, the onset of nitrification will be immediate.

On August 17, 1978 bacterial samples were taken at ten stations along the upper tidal James and cultured for ammonia oxidizers (principally *Nitrosomonas*) and nitrite oxidizers (principally *Nitrobacter*) according to the method of Matulewich et al. (7). Samples were withdrawn from the surface and mid-depth of the water column and from the upper 5 cm of the bottom sediments. The results of the survey are presented in Figure 2-4.

TABLE 1. Point Source Inputs.

	Org N	NH_4-N	NO_3-N
Richmond STP	180	1400	1300 kg/day
Falling Creek STP	30	220	60 kg/day
Other Sources	120	120	200 kg/day
Hopewell	540	6000	1940 kg/day

FIGURE 2. Ammonium oxidizers in the James River water column.

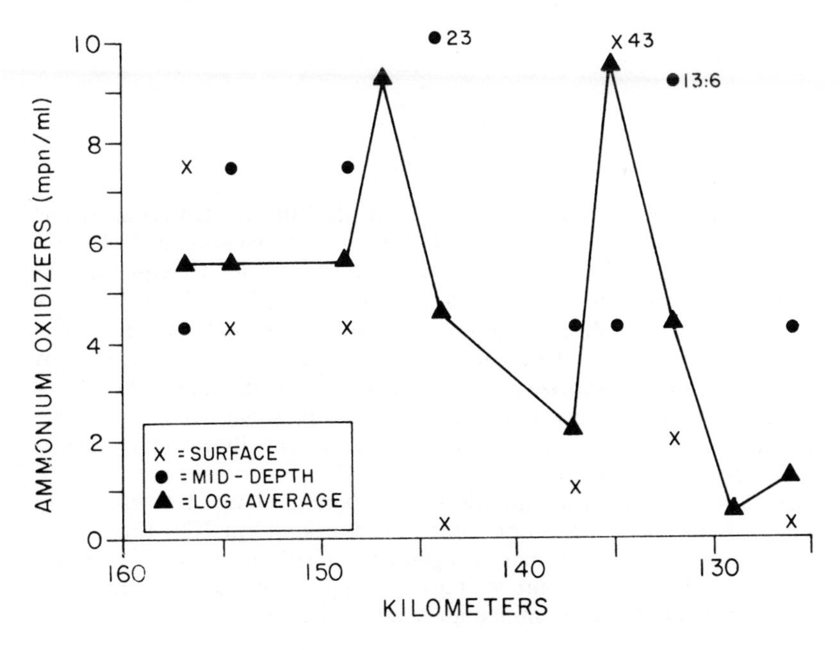

FIGURE 3. Nitrite oxidizers in the James River water column.

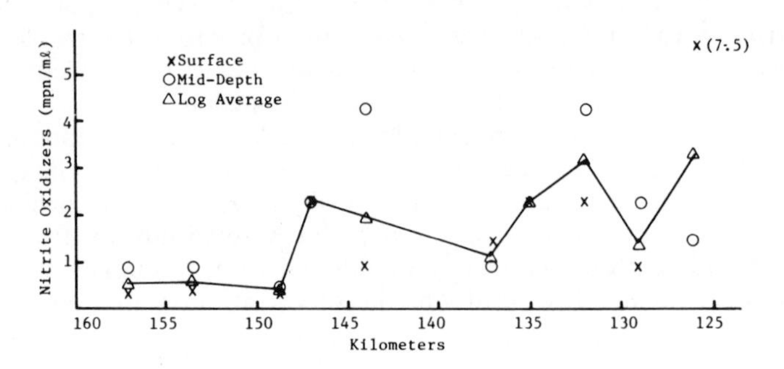

FIGURE 4. Ammonium and nitrite oxidizers in James River sediments.

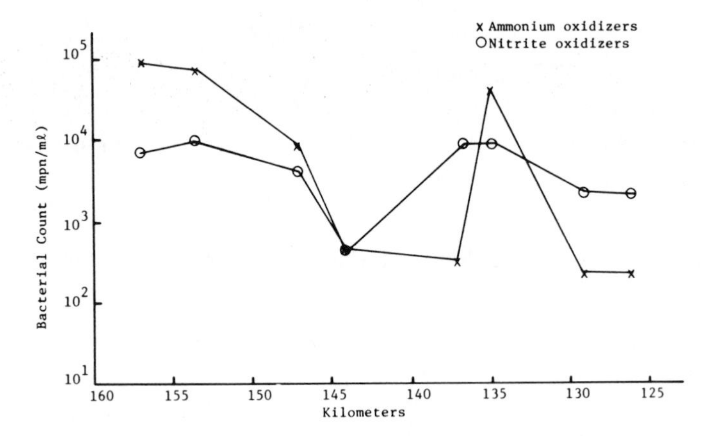

Examination of the population levels of planktonic nitrifiers shows no uniform longitudinal trend. Ammonia oxidizers are well-established in the vicinity of the point sources (km 157-145) and decrease in the downstream direction except for a peak at km 135. A similar peak of planktonic nitrite oxidizers is noted in the same vicinity. No point sources exist in this region, but the adjacent drainage area is largely occupied by dairy farms and unauthorized discharges from these operations have been reported (Virginia State Water Control Board, personal communication). Thus this population peak is likely to represent localized runoff of nitrifiers from dairy operations.

Populations of benthic nitrifiers also exhibit elevated levels in the vicinity of the upstream point sources and around km 135. From these data it may be deduced that a significant nitrifying population exists adjacent to the major point sources and that if nitrification of wastes occurs, it is likely to begin immediately upon discharge.

TABLE 2. Average Bacterial Population Levels.

Planktonic Ammonia Oxidizers	Benthic Ammonia Oxidizers	$\frac{\text{Water Column}}{\text{Sediment Column}}$
3.2 mpn/ml	3990 mpn/ml	$\simeq \frac{1}{10}$
Planktonic Nitrite Oxidizers	Benthic Nitrite Oxidizers	
1.3 mpn/ml	3328 mpn/ml	$\simeq \frac{1}{20}$

Equally important as the longitudinal distribution of nitrifiers is their vertical distribution within the water column and sediments. Table 2 presents the log-average population levels of planktonic and benthic nitrifiers and the ratio of these averages. It can be seen that the benthic nitrifiers are approximately 1,000 times more abundant, by volume, than the planktonic nitrifiers. Even when the entire water column is considered, benthic nitrifiers still comprise 90-95 percent of the nitrifying population, suggesting that nitrification in the upper tidal James is predominantly a benthic phenomenon. Similar results have been noted in the Trent River by Curtis et al. (4) and in the Passaic River by Mutulewich and Finstein (1978).

The Occurrence of Nitrification

Bacterial surveys are costly and difficult to conduct and they provide information only about the relative occurrence of nitrification rather than the overall significance of this phenomenon on nutrient and dissolved oxygen concentrations within the water column. Rather than bacterial surveys, the classic evidence of nitrification is the decline in ammonia concentration downstream of a point source concurrent with an increase in the concentration of nitrate, the end product of the reaction. This trend is difficult to observe in a tidal river, however, and may be disguised by distributed nitrogen sources, tidal dilution, varying geometry, and algal uptake of ammonia and nitrate.

Figure 5 shows the concentration of ammonia and nitrate sampled in an intensive survey conducted during July, 1976. There is an observable decline in the ammonia concentration downstream of the principal point source at Richmond, suggestive of nitrification, but no concurrent increase in nitrate, the end product of nitrification, is noted. Thus the visual evidence for nitrification is ambiguous and additional investigation is necessary. This investigation is conducted through the use of a predictive mathematical model which permits isolation of the various physical and biochemical processes influencing the concentrations of ammonia and nitrate.

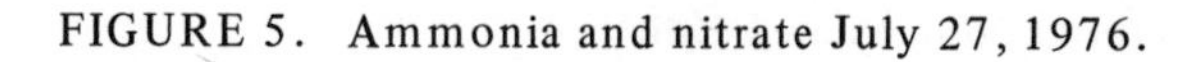
FIGURE 5. Ammonia and nitrate July 27, 1976.

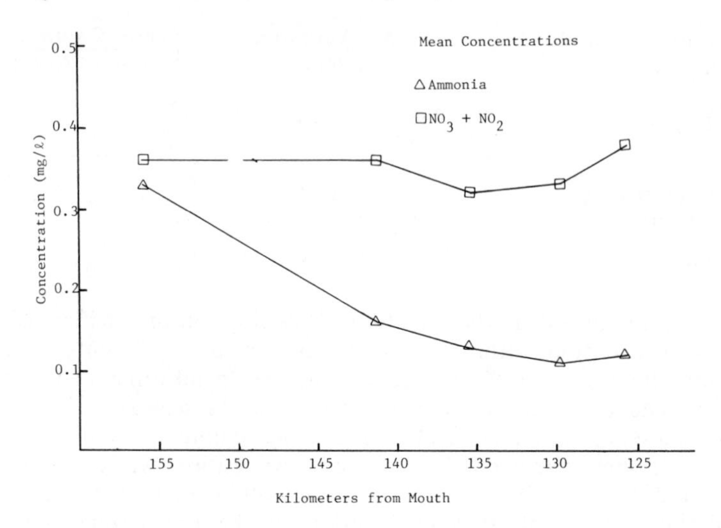

To be useful in this fashion, a model must account for all the phenomena which interact to determine the ammonia and nitrate concentrations in the tidal James. These include freshwater and tidal flow, estuarine geometry, external inputs, hydrolysis of organic nitrogen to ammonia, nitrification of ammonia to nitrate, algal uptake of ammonia and nitrate, and production of organic nitrogen by algal excretion and death.

Such a model has been calibrated and applied to the upper tidal James River. The model is real-time and simulates mass transport and biogeochemical substance transformations through a finite difference solution to the one-dimensional conservation of mass equation. Details of the model formulation and application to the James River may be found in Cerco et al. (2).

The model has been calibrated and applied to three sets of data collected during August 1975, July 1976, and July 1977. All data were collected during dry, steady conditions. The 1975 and 1976 data sets represent eight-to-ten samples collected over a 24-hour period from the surface, mid-depth, and bottom of the water column. The 1977 data set is comprised of the results of four "slackwater" surveys conducted over a three-week period. The data points are depth-averaged concentrations.

Results of the model application are presented in Figures 6-8 for the 1975, 1976, and 1977 data, respectively, Field measurements and predictions of organic nitrogen, chlorophyll, ammonia, and nitrate are shown. The kinetics parameters utilized in the simulations are presented in Table 3.

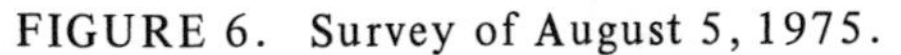
FIGURE 6. Survey of August 5, 1975.

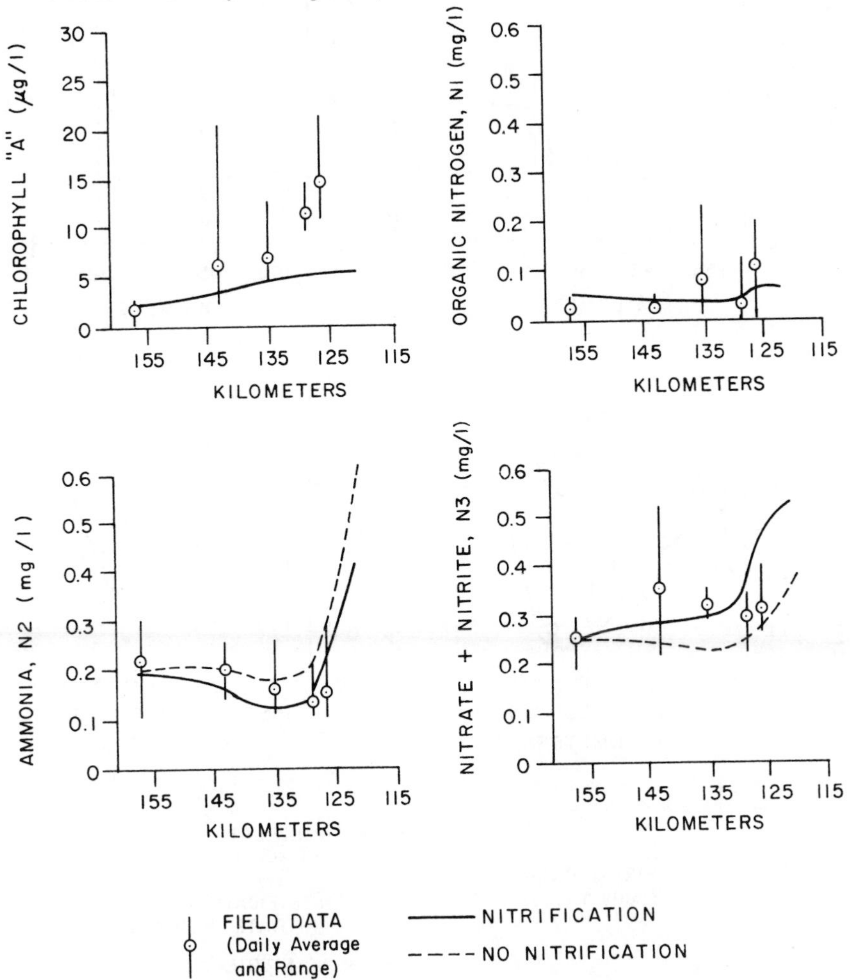

The 1975 ammonia concentrations can be fitted about equally well, given the range of the field data, under conditions of nitrification or no nitrification. There was, however, more nitrate observed in the river than would be expected if nitrification was not taking place. Thus, on the basis of an excess of end-product, the occurrence of nitrification during the August, 1975 field survey is suggested.

FIGURE 7. Survey of July 27, 1976.

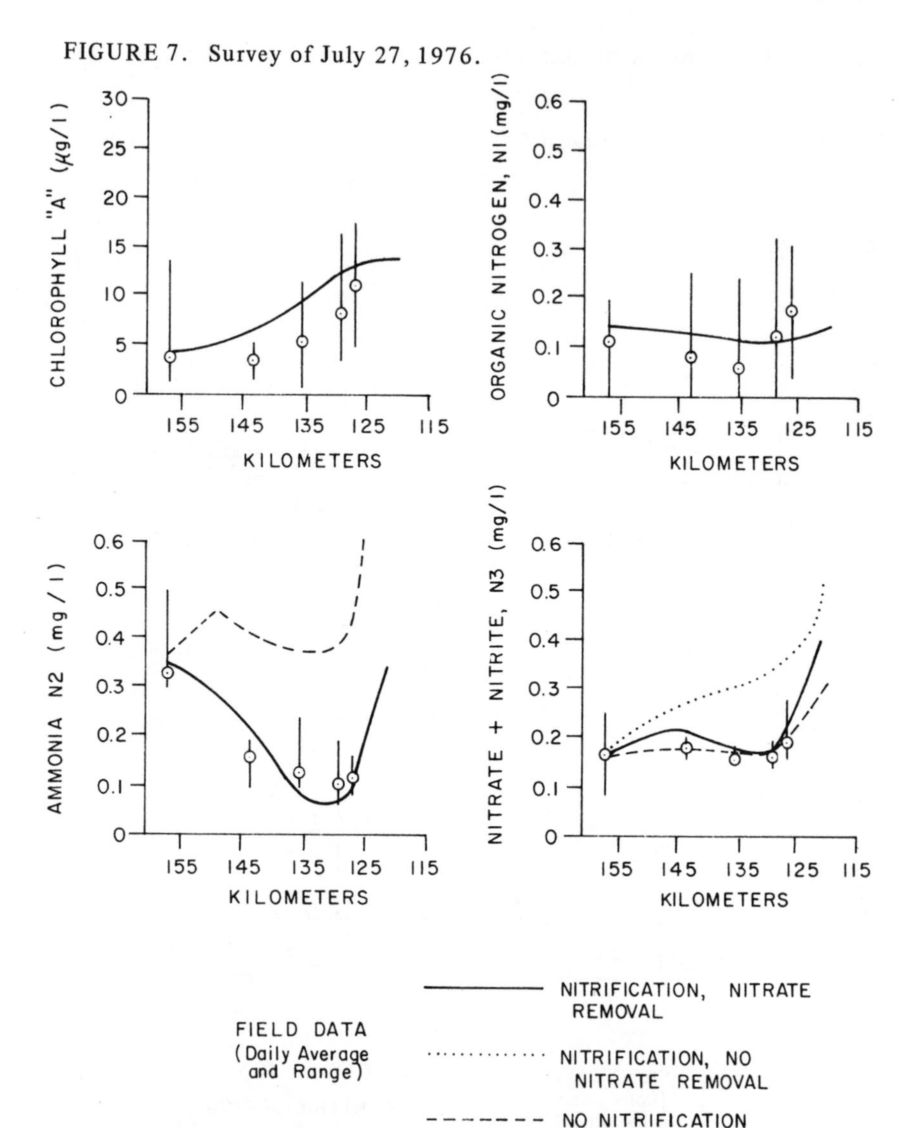

Data observed during the 1976 survey exhibit different characteristics than that observed during 1975. During July, 1976, in the absence of nitrification, much higher ammonia concentrations should have been observed than were present. Predictions based on the presence of nitrification show good agreement with the ammonia field data. An excess of nitrate would be expected, however, unless nitrate removal occurs simultaneously with nitrification.

FIGURE 8. Survey of July, 1977.

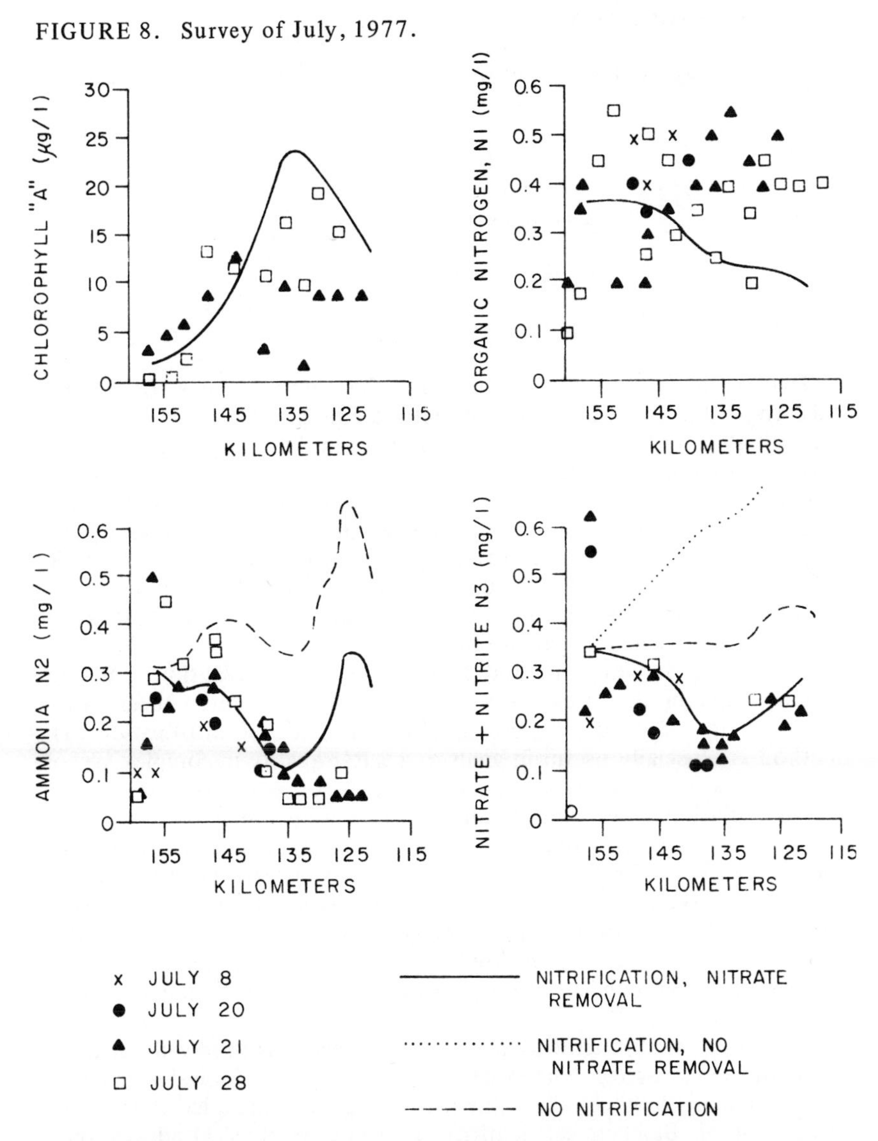

The 1977 data support the hypotheses of both nitrification and nitrate removal. As in 1976, in the absence of nitrification, an excess of ammonia would be present in the river. An excess of nitrate would also be expected, even without nitrification, unless nitrate removal is assumed. A similar loss of nitrate concurrent with nitrification has been observed by Finstein and Matulewich (5) in a freshwater stream.

TABLE 3. Simulation Kinetics Parameters.

	Temperature (C^o)	Hydrolysis (day^{-1})	Nitrification (day^{-1})	Nitrate Removal (day^{-1})
1975	31.	0.04	0.12	0.0
1976	28.	0.03	0.22	0.10
1977	31.	0.04	0.12	0.15

DISCUSSION

Two methods have been utilized for the detection of nitrification in tidal rivers and estuaries. The first method consists of direct counts of the nitrifying bacteria and is a useful technique for detecting longitudinal and vertical zonation of the process. Bacterial sampling detected elevated population levels in the vicinity of the upper James River point sources, suggesting that nitrification of wastes in this water body begins immediately upon discharge. The preponderance of bacteria in this region was noted in the bottom sediments rather than the water column, confirming that nitrification is a benthic rather than planktonic phenomenon in the upper tidal James.

The second detection method relies on the use of a real-time, predictive water quality model to examine the ammonia and nitrate concentrations expected to occur in the presence and absence of nitrification. This method is especially useful in analyzing water quality in complex systems such as tidal rivers where trends in parameter concentrations are obscured by tidal dilution, flow reversals, and multiple point source inputs.

The model has been applied to three sets of data collected during 1975, 1976 and 1977. The occurrence of nitrification is evidenced in the 1975 data by an excess of nitrate over that which would be expected based on point source and upstream inputs. From the 1976 and 1977 data, nitrification is inferred by the low concentrations of ammonia observed in the river; higher concentrations would be expected in the absence of nitrification.

Concurrent with nitrification, nitrate removal was observed during the 1976 and 1977 surveys. The exact mechanism of this removal is unknown. Algal uptake is unlikely since this loss is explicitly included in the model mass balance. Biochemical denitrification and/or physical adsorption to bottom sediments are two alternative pathways.

Denitrification is usually considered to be negligible in the water column except under conditions of elevated organic carbon concentrations and/or extreme oxygen depletion (Nelson, et al., (8)). Denitrification has been observed in bottom sediments, however, independent of the dissolved oxygen concentration of the overlying waters (Van Kessel, (12), suggesting that denitrification may occur as a benthic process in the James.

Harms et al. (8) have observed the adsorption of mineral phosphorous to stream bed sediments downstream of a wasteflow to a South Dakota river. A similar adsorption of nitrate to the sediments of the upper tidal James can be hypothesized.

It is clear both from observations of the data and from interpretation of the model results that the nitrification process in a tidal river may not be evidenced by the classic decline of ammonia and increase of nitrate. Additional study is necessary before the occurrence of nitrification can be confirmed or excluded.

ACKNOWLEDGEMENTS

This study was sponsored by the Virginia State Water Control Board through the Cooperative State Agencies Program.

Bacterial analyses were performed by the VIMS Department of Microbiology and Pathology under the direction of Dr. Howard Kator.

REFERENCES

1. Berdahl, B.J. 1972. Estuarine nitrification. Doctoral dissertation submitted to The Graduate School of Rutgers University. New Brunswick, N.J.
2. Cerco, C.F., A.Y. Kuo, C.S. Fang, and A. Rosenbaum. 1978. Mathematical model studies of water quality and ecosystems in the upper tidal James. Special Report No. 155. Va. Inst. of Mar. Sci., Gloucester Point, Va.
3. Courchaine, R.J. 1968. Significance of nitrification in stream analysis - effects on the oxygen balance. J. of Wat. Poll. Cont. Fed. 40:835-847.
4. Curtis, E.J.C., K. Durrent, and M.M.I. Harman. 1975. Nitrification in rivers in the Trent Basin. Wat. Res. 9:255-268.
5. Finstein, M.S., and V.A. Matulewich. 1977. Nitrification potential of river environments. Dept. of Envr. Sci., Cook College, Rutgers University, New Brunswick, N.J.
6. Harms, L.L., P.H. Vidal, and T.E. McDermott. 1977. Release and sorption of phosphorous by sediments in a moving watercourse, OWRT State University, Brookings, S.D. 57706.
7. Matulewich, V.A., P.F. Strom, and M.S. Finstein. 1975. Length of incubation for enumerating nitrifying bacteria present in various environments. App. Microbiology. 29:265-268.
8. Nelson, D.W., L.B. Owens, and R.E. Terry. 1973. Denitrification as a pathway for nitrate removal in aquatic systems. Tech. Report No. 42. Purdue University Water Resources Research Center, West Layfayette, Ind.

9. Stratton, F.E., and P.L. McCarty. 1967. Prediction of Nitrification effects on the dissolved oxygen balance of streams. Envir. Sci. and Tech. 1:405-410.
10. Tiedemann, R.B. 1977. A study of nitrification in the Delaware River estuary. Nat. Tech. Info. Ser. No. PB290-488. U.S. Dept. of Commerce, Springfield, VA.
11. Tuffey, T.J., J.V. Hunter and V.A. Matulewich. 1974. Zones of nitrification. Wat. Res. Bull. 10:555-164.
12. Van Kessel, J.F. 1977. Factors affecting the denitrification rate in two water sediment systems. Wat. Res. 11:259-267.
13. Wezernak, C.T. and J.J. Gannon. 1968. Evaluation of nitrification in streams. J. of San. Eng. Div., A.S.C.E. SA5:883-895.

EUTROPHICATION TRENDS IN THE WATER QUALITY OF THE RHODE RIVER (1971-1978)

David L. Correll
Chesapeake Bay Center for Environmental Studies
P.O. Box 28
Edgewater, Maryland 21401

ABSTRACT: Five to eight-year data sets on turbidity and a series of nutrient parameters have been taken in the Rhode River, a small tidal river tributary to Chesapeake Bay. Trends of change at headwater stations indicate changes due primarily to local watershed runoff, while changes at stations near the river's mouth are indicative of changes in Chesapeake Bay.

The data have been analyzed first by summarizing as monthly and seasonal means and variances, then by looking for year-to-year trends. Thus, for example, linear regressions of summer and fall turbidity versus time for 1971-1978 had very low slopes and low coefficients of determination. An interesting finding was the pattern for total phosphorus in surface waters. At the mouth of Rhode River, seasonal mean concentrations increased steadily for each season each year from the fall of 1971 to the fall of 1976. The most dramatic increases (four-fold) were observed in the summer and fall. A much smaller increase occurred in winter and spring values. In contrast, year-to-year concentrations of total phosphorus in surface waters in upstream stations over the same time period showed less clear-cut trends. Also, phosphorus loadings from local watershed runoff fluctuated widely, but had no steady rapid rise with time. Although total phosphorus increased dramatically at the mouth of Rhode River, dissolved orthophosphate, nitrate, and dissolved ammonia remained essentially constant, especially in the summer and fall. These data could indicate an increasing impact of summertime anoxic bottom waters on the phosphorus dynamics of the upper western shore of the Bay.

INTRODUCTION

The Rhode River is a small subestuary, tributary to Chesapeake Bay on its western shoreline in Maryland (Figure 1). The Smithsonian Institution

FIGURE 1. Map of the Chesapeake Bay Region showing the location of the Rhode River subestuary (arrow).

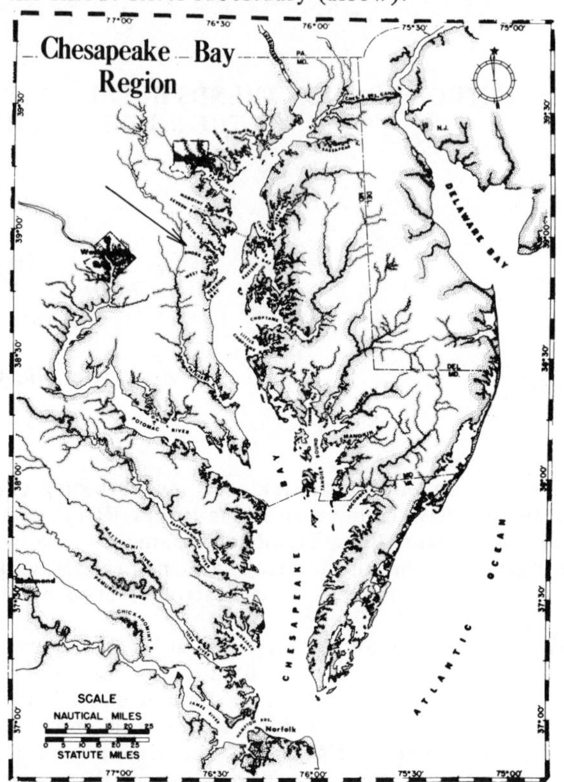

has been conducting long-term environmental studies at this site for over a decade. Relatively long, intensive water quality data sets have been gathered at a series of stations throughout the estuary and on land runoff from a series of watersheds that discharge into Rhode River. No point sources of pollutants were present on Rhode River. Parameters measured included total-P, total dissolved-P, orthophosphate, dissolved orthophosphate, nitrate, dissolved ammonia, organic nitrogen, and turbidity. For the purposes of this paper, estuarine data are presented from only two stations (Figure 2). Station 5 is near the head of a small tidal creek channel through which land runoff is discharged and is representative of Rhode River watershed effects. Station 13 is at the mouth of Rhode River; it is strongly influenced by surface waters of the western shoreline of upper Chesapeake Bay that move south down the shoreline and exchange rapidly with the outer parts of Rhode River (9). All data were taken on surface waters, since Rhode River is not deep enough to communicate hydrodynamically directly with Chesapeake Bay bottom waters.

FIGURE 2. Map of the Rhode River showing Stations 5 and 13.

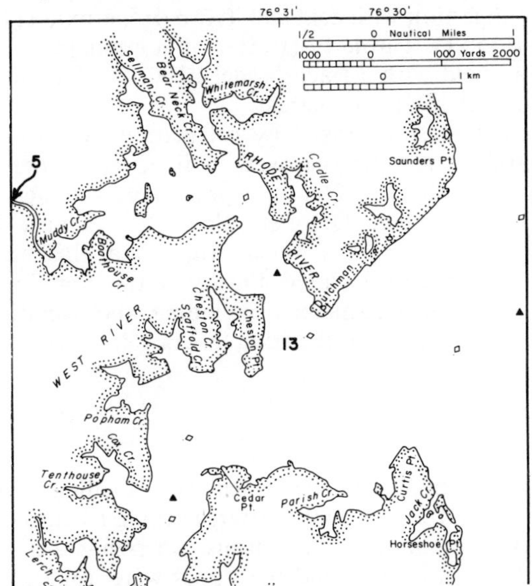

The intent of these data analyses was to look for trends over a period of five to eight years in the 1970s. The general approach was to summarize data by season and to then carry out linear regression analysis of concentrations versus time. Some of the more interesting results are presented.

METHODS

Estuarine surface water was sampled from a depth of about 10 cm and a portion of each sample was immediately filtered through Millipore HA membrane filters which had been thoroughly prewashed with distilled water. Filtrate concentrations are referred to as dissolved. Stream water samples were collected as grab samples (5). Water sampling frequencies varied from several times a week to monthly, but averaged about once a week.

Orthophosphate was analyzed by the stannous chloride method (1) with a turbidity correction on whole samples. Total-P was digested with perchloric acid (10), then analyzed by the same method as orthophosphate. Nitrate plus nitrite were analyzed colorimetrically by reduction of the nitrate on a cadmium amalgam column, then coupling to sulfanilamide (1, 14). Nitrite was determined separately without the reduction step and was substracted from the nitrate plus nitrite values to give nitrate. Dissolved ammonia was determined by oxidation to nitrite with hypochlorite (13) and then as for nitrite. Total Kjeldahl nitrogen was determined by digestion with acid and hydrogen peroxide (11) distillation and Nesslerization (1). Turbidity (scatter of columnated white light) was measured in the field with a Hach Model 2100A turbidimeter, calibrated before each use with sealed standards.

All data were first summarized by calculation of seasonal means and standard deviations. The seasons were defined as: winter - December, January, February; spring - March, April, May; summer - June, July, August; fall - September, October, November. Linear least squares regressions of seasonal means versus time and their coefficients of determination were then calculated. Finally, in the more interesting cases, a series of statistical parameters, including the significance of the correlations, was calculated on the complete raw data sets (12).

RESULTS

Since changes in suspended particle loading and plankton concentrations would affect light scatter, Station 13 turbidity data were analyzed for the summer and fall seasons. No trends were evident. Both regressions had low R^2 values and the slopes were low (0.027 and 0.063 unit/year, respectively).

FIGURE 3. Total Phosphorus data for surface water at Station 13. Each point is a seasonal mean.

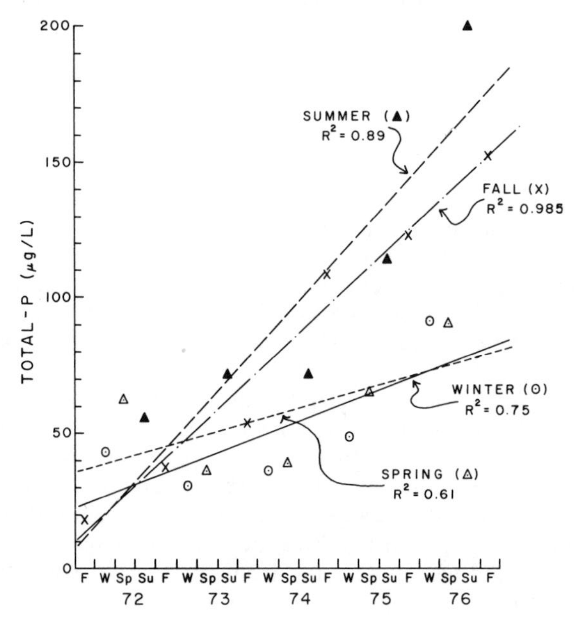

In contrast, total-P at Station 13 (Figure 3) quite clearly increased. Regression slopes were highest for summer, followed by fall, winter, and spring. These regressions were significant at probabilities of 0.001, 0.00001, 0.01 and 0.10, respectively. Total-P concentrations at Station 5 (tidal headwaters) also increased (Figure 4), but the correlations and

FIGURE 4. Total Phosphorus data for surface water at Station 5. Each point is a seasonal mean.

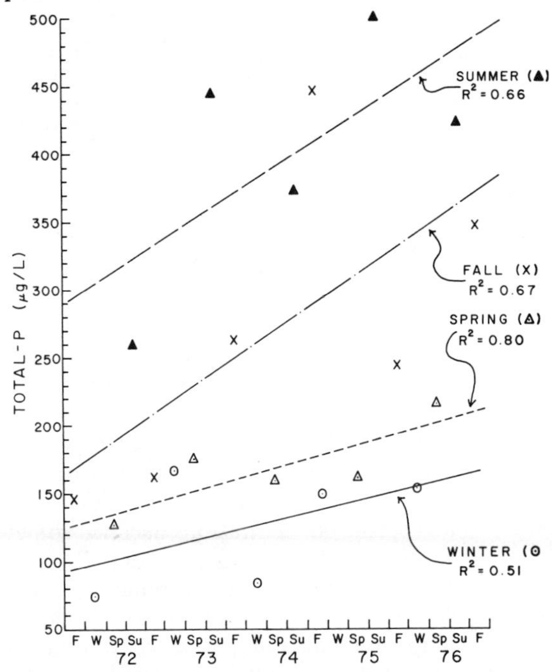

relative slopes of the regressions were much lower in general. Slopes were highest in the fall, followed by summer, spring, and winter. These regressions were significant at probablities of 0.0002, 0.09, 0.05, and 0.25, respectively. Total-P concentrations in streams discharging from the watershed to the estuary (Figure 5) had even lower R^2 values and slopes, with the possible exception of the summer season. Increases in summer and fall total-P concentrations at Station 13 (Figure 3) were four-fold or more, while those at Station 5 (Figure 4) were about two-fold; streams (Figure 5) showed about a two-fold increase in the summer and a small decrease in the fall..

While it is clear that total-P at Station 13 increased dramatically in the summer and fall, dissolved orthophosphate at this station did not (Figure 6). Although some increase may have occurred, it was 50 percent or less. In the spring, the regression slope was -1.1 μgP ℓ^{-1}yr^{-1} (R^2=.0.50). Nitrate

FIGURE 5. Mean total phosphorus concentrations in grab samples of Rhode River watershed streams. Each point is a seasonal mean for all data from all streams sampled.

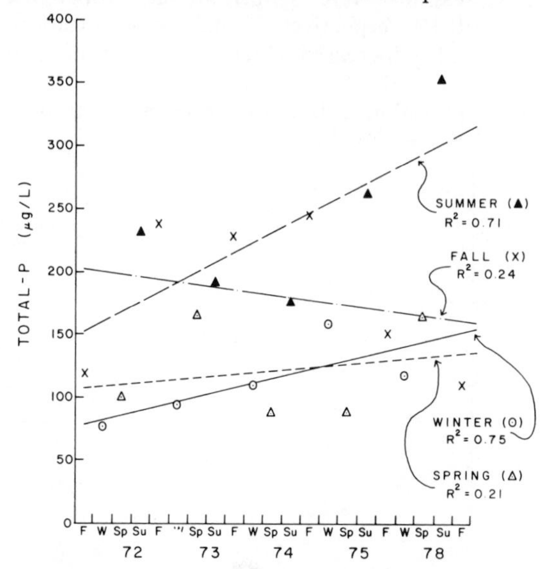

concentrations at Station 13 (Figure 7) were essentially constant at about 20 μgN ℓ^{-1} except in 1972 when the effects of tropical storm Agnes resulted in a high seasonal mean (6). Fall nitrate concentrations showed no significant trend and spring concentrations declined somewhat.

If benthic regeneration of phosphorus is potentially implicated in the rapid increase in total-P at Station 13, then trends in dissolved ammonia concentrations should also be of interest. In fact, a significant trend of increase (P=0.12) was apparent for summer concentrations (Figure 8). The regression slope was about 4μg ammonia N $\ell^{-1}yr^{-1}$ and the R^2 was 0.91. However, this amounted to only about a 50 percent increase. Also, fall concentrations had almost no correlation with time; spring concentrations had a low negative correlation (P=0.27); winter concentrations declined significantly (P=0.006) with a regression slope of -31 μg ammonia N ℓ^{-1} yr^{-1} (R^2=0.80).

DISCUSSION

The large, rapid increase in total-P at the mouth of Rhode River is an interesting and important finding with respect to the eutrophication of Chesapeake Bay.

If one extrapolates the regressions for summer and fall total-P concentrations backwards in time (Figure 3), they would have been zero

FIGURE 6. Dissolved orthophosphate concentrations in surface waters at Station 13 in the summer and fall. Each point is a seasonal mean.

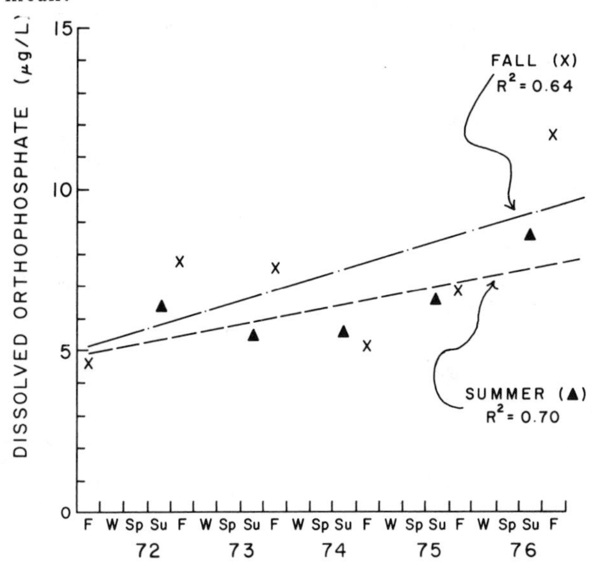

in 1970 or 1971. This obviously was not the case, so one is forced to assume that the measured rapid increase began in about 1971 or 1972. Only two prior relatively long-term intensive studies are available for comparison (3, 4). In the mid-1960s, summer total-P values in the surface waters of the upper Chesapeake Bay area were reported to be in the range of 31 to 62 μgP ℓ^{-1} (3), which corresponds with the 1971-1972 values at all seasons and most of the winter and spring values found in this study. In the period from 1968 to 1971, total-P concentrations in the upper bay were tracked (4). No seasonal patterns were apparent in 1968 but, beginning in 1969, a distinct summer-fall peak was found which agrees with the seasonal pattern reported for this study. The conclusion was also reached (4) that total-P concentrations had increased from 1968 to 1971. In the mid-1960s (3), dissolved orthophosphate rarely exceeded 6 μgPℓ^{-1} and averaged about 3 μg Pℓ^{-1} in the summer with no seasonal or spatial patterns. No seasonal patterns of dissolved orthophosphate were found in this study, and no strong temporal trends were apparent (Figure 6). A low positive slope for summer and fall regressions (0.56 and 0.86 μgP ℓ^{-1} yr^{-1}, respectively) was found, and if these regressions are extrapolated backwards in time to the mid-1960s, mean seasonal concentrations would have been about 2-4 μgP ℓ^{-1}. The actual data seasonal means for the five year period were 6.5 and 7.3 μgP ℓ^{-1} for summer and fall, respectively. Whether or not this magnitude of increase in immediate by biologically

FIGURE 7. Nitrate concentrations in surface waters at Station 13 in spring, summer, and fall. Each point is a seasonal mean.

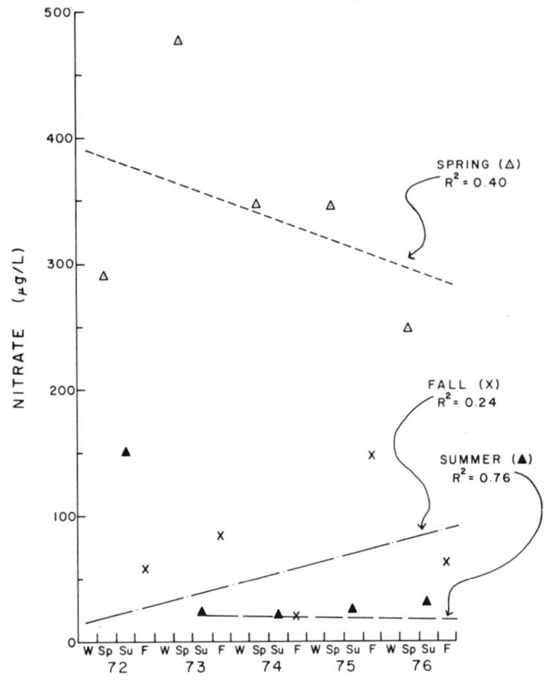

available phosphorus is significant in terms of Chesapeake Bay plankton responses should be considered carefully. Nitrate concentrations during the growing season do not seem to have changed to a noticeable extent since 1965 (3).

It should be noted that the observed increase in total-P is probably not primarily due to increased loading in runoff from the Rhode River watershed. Neither the tidal headwaters nor the land runoff total-P concentrations increased as much (Figures 4 and 5) as total-P concentrations at the mouth of Rhode River. In fact, to explain such a large increase at the mouth of the river, a much larger increase in the concentrations from watershed sources would be required, since a rapid exchange of Rhode River and Chesapeake Bay surface waters prevails at this location. Thus, one is forced to consider alternative sources, such as increased loading from the Baltimore metropolitan area (Figure 1) or increased rates of regeneration from benthic sediments.

The probability of a large increase in phosphorus loading from the Baltimore metropolitan area seems low. Total sewage treatment plant discharge (volume) has increased, but less than two-fold between 1965 and 1973 (2, 3). No sudden population increases have occurred and levels of treatment have improved over this time interval. The effects of these discharges were not observable below Baltimore in 1965, probably due to interaction with iron sulfate effluents from a steel plant (3). Unless some major change has occurred in the Baltimore area of the bay to cause the release of large amounts of phosphorus in effluents or from sediments, this explanation of the observed data is unlikely.

FIGURE 8. Dissolved ammonia concentrations in surface waters at Station 13 in spring, summer, and fall. Each point is a seasonal mean.

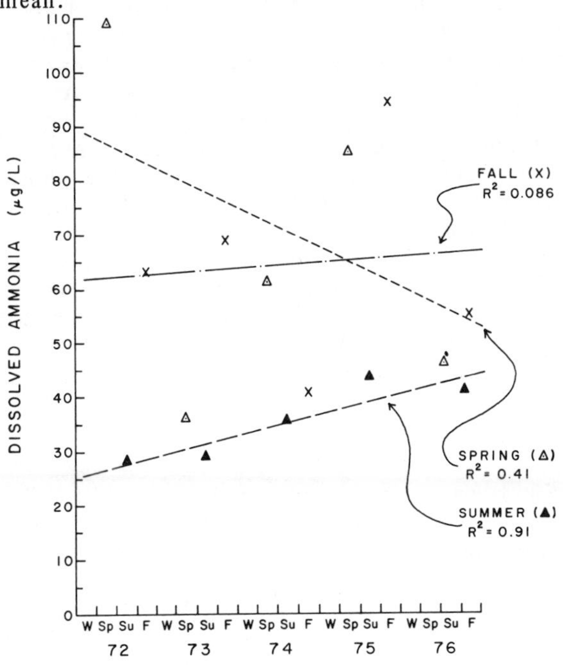

An increase in the extent and duration of anoxic bottom water conditions in the bay, which resulted in increased mixing of phosphorus into the surface waters, is another possible mechanism to explain the observed changes. This is an attractive possibility but is hard to document. The relatively small increases in dissolved ammonia over time at the mouth of Rhode River could be due to increased biological assimilation. If both phosphorus and available nitrogen concentrations were increased, plankton would remove a portion of both. The ratio of assimilation of nitrogen to phosphorus would approximate the Redfield ratio of about 10. Mean total-N to total-P ratios in the summer and fall at Station 13 were about 9, but from 80-95 percent of the total-N was present as organic N and probably did not recycle very rapidly, while total-P is rapidly recycled by these populations (7, 8). Both prior studies (3, 4) concluded that available nitrogen was probably limiting algal production in the upper bay in summer. Thus, it seems reasonable that nitrate and ammonia were not observed to increase rapidly in this study. This benthic regeneration mechanism of phosphorus increase is also supported by the nature of the seasonal patterns observed. The concentration of total-P did not increase much in winter or spring, although point sources would also be high at those times. Instead, the increases were greatest in summer and fall, when anoxic bottom water conditions are most likely to occur.

In summary, the most likely hypothesis to explain the observed rapid increase in total-P is the following series of events: 1) A steady but slow increase in nutrient loading from both point and nonpoint sources has been occurring over the last ten to twenty years. 2) As a result, plankton populations have increased (3). These plankton have been able to assimilate the increased load, convert much of the nitrogen to relatively refractory compounds, and deposit much of the phosphorus in the bottom sediments. 3) About 1970, the increased loading of organic matter from increased plankton productivity began to produce temporally and spatially more extended anoxic bottom water conditions. 4) This resulted in increased release of phosphorus from bottom sediments and more rapid transport of phosphorus into surface waters. It should be stressed that the efficiency of exchange from bottom sediments to surface waters need not be high; indeed, even an extremely small change could be significant.

ACKNOWLEDGEMENTS

This research was supported in part by the Smithsonian Institution's Environmental Sciences Program, and by a series of grants from the NSF-RANN program to the Chesapeake Research Consortium, Inc.

REFERENCES

1. American Public Health Association. 1976. Standard Methods for the Examination of Water and Waste Water, 14 Ed. APHA, N.Y.
2. Brush, L.M., Jr. 1974. Inventory of Sewage Treatment Plants for Chesapeake Bay. Chesapeake Research Consortium Publ. No. 28, 62 pp.
3. Carpenter, J.H., Pritchard, D.W., Whaley, R.C. 1967. Observations on Eutrophication and Nutrient Cycles in Some Coastal Plain Estuaries. *In* Lauff, G.H. (ed.) *Estuaries*. Publ. No. 83, Amer. Assoc. Adv. Sci., Washington, D.C. pp. 210-221.
4. Clark, L.J., Donnelly, D.K., Villa, O., Jr. 1973. Nutrient Enrichment and Control Requirements in the Upper Chesapeake Bay. EPA Report No. 903/9-73-002-a.
5. Correll, D.L. 1977. An Overview of the Rhode River Watershed Program. *In* Correll, D.L. (ed.), Watershed Research in Eastern North

America. Smithsonian Institution Chesapeake Bay Center for Environmental Studies, Edgewater, Md. pp. 105-120.

6. Correll, D.L. 1977. Indirect Effects of tropical storm Agnes upon the Rhode River. *In* The Effects of Tropical Storm Agnes on the Chesapeake Bay Estuarine System, Chesapeake Research Consortium, Publ. No. 54, pp. 288-298, Johns Hopkins Univ. Press, Baltimore, Md.
7. Correll, D.L., Faust, M.A., Severn, D.J. 1975. Phosphorus flux and cycling in estuaries. *In* Cronin, L.E. (ed.), *Estuarine Research* 1:108-136. Academic Press, N.Y.
8. Faust, M.A., Correll, D.L. 1976. Comparison of bacterial and algal utilization of orthophosphate in an estuarine environment. *Marine Biology* 34:151-162.
9. Han, G. 1975. Observations on the hydrodynamics of the Rhode River and a Kinematic salt balance model. Chesapeake Bay Institute Tech. Report 89 (Ref. 75-1).
10. King, E.V. 1932. The colorimetric determination of phosphorus. *Biochem. J.* 26:292-297.
11. Martin, D.F. 1972. *Marine Chemistry*, Marcel Dekker, Inc., N.Y. pp. 174-176.
12. Nie, N.H., Hull, C.J. Jenkins, J.G. Steinbrenner, K., Bent, D.H. (eds.). 1975. Statistical Package for the Social Sciences, 2nd Ed. McGraw-Hill, N.Y. (KU version 6.02B-1) 675 pp.
13. Richards, F.A., Kletsch, R.A. 1964. The spectrophotometric determination of ammonia and labile amino compounds in fresh and seawater by oxidation to nitrite. Sugawara Festival Volume, Maruza Co., Ltd. Tokyo. pp. 65-81.
14. Strickland, J.D., Parsons, T.R. 1965. A manual of seawater analysis. *Bull. Fish. Res. Bd. Canada* 125, 2nd Ed.

ABOVEGROUND NET PRIMARY PRODUCTIVITY OF THREE GULF COAST MARSH MACROPHYTES IN ARTIFICIALLY FERTILIZED PLOTS

Armando A. de la Cruz*, Courtney T. Hackney, and Judy P. Stout*****

**Department of Biological Sciences*
Mississippi State University
P.O. Drawer GY, Mississippi State, MS 39762

***Department of Biology*
University of North Carolina at Wilmington
Wilmington, NC 28406

****Dauphin Island Sea Lab*
P.O. Box 386
Dauphin Island, AL 36358

ABSTRACT: Plots (100 m^2) of four tidal marsh communities (*Juncus roemerianus* and *Spartina alterniflora* in Alabama, *J. roemerianus* and *Spartina cynosuroides* in Mississippi) common in the Gulf Coast were enriched with commercial NH_4NO_3 (34 percent N). The fertilizer was applied once at the beginning of the 1978 growing season to simulate a farm-plantation operation at a dosage (136 g/m^2) estimated to return to the soil approximately the same amount of nitrogen contained in the plants. Six 0.25 m^2 quadrats were harvested monthly from each community from April through November. The annual net productivity was estimated with a maximum minus minimum standing crop technique based on a predictive periodic model (PPM). A correction for plant mortality during the sampling period is provided in the PPM technique. Annual aboveground net primary productivity increased by 59 percent in the Alabama *J. roemerianus*, 84 percent in the Mississippi *J. roemerianus*, 82 percent in the *S. alterniflora* and 26 percent in the *S. cynosuroides*. It appears that short form or high marsh macrophytes responded more to nitrogen enrichment than tall form or low marsh plants.

INTRODUCTION

Experimental enrichment of marshes with nitrogenous fertilizer has resulted in a dramatic increment in biomass production (1), (2), (4), (10), (12), (14), indicating the role of nitrogen as a limiting factor in the growth of marsh plants. Repeated fertilization of a *S. alterniflora* (short form) marsh in Delaware with commercial grade ammonium nitrate at 20 g/m^2 increased biomass production by at least 100 percent (14). A study done on a *Juncus gerardi* marsh in Sweden (15) showed 30 percent biomass increase following a single application of nitrogen as ammonium chloride. Fertilization studies conducted on *S. alterniflora* and *S. patens* marshes in Massachusetts using urea and commercially available sludge also resulted in increased productivity (16). Similarly, studies on both natural and artificially propagated *S. alterniflora* in North Carolina showed that the addition of nitrogen as ammonium sulfate doubled the yield of aboveground shoots (1). But, an enrichment study on marsh communities similar to the ones fertilized in the present study treated with 13-13-13 (N-P-potash) fertilizer at the recommended single dosage of 71 g/m^2 (9.2 g N/m^2) did not change the annual biomass yield of the plants (13). Nitrogen in the ammonium form promotes better growth than nitrate nitrogen (10).

Our current investigations of management procedures applicable to coastal wetlands for purposes of cultural activities prompted our interest in artificial enrichment of marshes. Development of certain marshlands into farm-plantations for the cropping of marsh grass is likely, should our attempts to recover chemicals of potential pharmacological value and by-products relative to extraction of pulp from marsh vegetations prove feasible. In either case, harvesting marshland meadows for products of direct value to mankind implies a cropping system that will produce maximum biomass yield. Obviously, any cultural use of a marsh will likely be done on the high marsh through fertilization, especially of high elevation areas which are the marsh type most likely to be altered.

The objective of this effort was to determine the effect of a commercial fertilizer (NH_4NO_3) on the annual biomass yield of marsh plants common to the north central Gulf Coast when applied once at the beginning of the growing season.

MATERIALS AND METHODS

Four marsh communities were studied: a *Juncus roemerianus* and a *Spartina cynosuroides* marshes in Mississippi and a *J. roemerianus* and *S. alterniflora* marshes in Alabama. The *J. roemerianus* communities were both short-medium forms (about 1.0 m in height), although the *J. roemerianus* in Mississippi is more of a high marsh community than the one in Alabama. The *S. alterniflora* community was an inland high marsh with short (0.5 m) plants while the *S. cynosuroides* was the typical giant cordgrass community growing to about 2-3 m in height. The substrates of the two Alabama marshes are of higher salinity, higher sand content, and lower organic matter concentration. Tide ranges in the Alabama and Mississippi marshes are similar. However, the degree and frequency of inundation are different due to topographic differences. The *S. alterniflora* marsh is regularly and completely inundated. The *J. roemerianus* marsh in Alabama has a lower profile than the one in Mississippi and thus is more frequently inudated.

Two 100 m^2 plots were established in each of the marsh communities. One of the two plots in each marsh type was used as a control plot. The other plot was enriched with commercial ammonium nitrate (34 percent N) fertilizer at the rate of 136 g/m^2, which is about five times the recommended per area dosage of 25 g/m^2 for normal lawn-farm treatment in Mississippi and Alabama. The fertilizer was applied in mid-February, 1978, when the marsh was beginning its annual spring regrowth. A single application of fertilizer was adhered to in compliance with common practice and for economy.

To make our results comparable with results of another study whereby enrichment was caused by ashes remaining after the marsh is burned, 136 g/m^2 of NH_4NO_3 (34 percent N) was applied. This dosage was estimated to return to the soil the same amount of nitrogen contained in the plant shoots above a square meter area, and was intended to simulate a post fire condition. The fertilizer was applied by hand broadcast during low tide as soon as the marshes became exposed. Tides along the north central Gulf Coast region are mostly diurnal so that there was at least 24 hours before the next high tide. During the month of February, northerly winds predominate and the marshes in Mississippi were seldom flooded. The fertilizer had adequate time to become incorporated into the moist mud.

Six 0.25 m^2 quadrats were harvested monthly from the control and from the fertilized plots of each marsh community from April through November. The samples were sorted into dead or living plants, dried at 100°C to constant weight, and weighed. Dead and decaying litter on the ground was also collected in the Mississippi marshes but not in Alabama, where the marsh floor was almost always virtually free of litter. The marshes in Alabama were monospecific while a few associated minor species were found in the Mississippi marshes.

The annual net aboveground primary productivity (NAPP) was estimated from changes in standing biomass using a maximum minus minum technique based on a predictive periodic model (PPM) devised by Hackney and Hackney (5). The PPM fits a periodic regression curve of:

$$y = c_0 + c_1 \sin(cti) + c_2 \cos(cti)$$
$$+ c_3 \sin(2\,cti) + c_4 \cos(2\,cti),$$

where $c = 2\pi/n$,

t_i = 1 to 9,

c_0 = overall mean,

c_1 and c_2 are coefficients of the first harmonic term, and

c_3 and c_4 are coefficients of the second harmonic term,

to the total biomass. Productivity was calculated by substracting the minimum expected value of standing biomass in summer. The NAPP was corrected for die-back or loss due to plants dying during the intervals between sampling periods according to the PPM method (5). This involved analysis of the monthly standing dead biomass from another plot previously cleared of all plant material, and the periodic max-min value obtained was added to the control and fertilized data for each marsh community.

RESULTS AND DISCUSSION

The annual aboveground net primary productivity of the control (i.e., natural) and experimental (i.e., fertilized) plots is summarized in Table 1. Fertilization with NH_4NO_3 significantly increased primary productivity by 84 percent for Mississippi *J. roemerianus*, 59 percent for Alabama *J. roemerianus*, 82 percent for *S. alterniflora* and 26 percent for *S. cynosuroides*. With the exception of *S. cynosuroides*, the increase in

TABLE 1. Summary of annual net aboveground primary productivity of four marsh communities which received a single application (136 g/m^2) of commercial ammonium nitrate (34 percent N) fertilizer.

	Net Aboveground Primary Productivity ($g/m^2/yr$)					
	Control			Fertilized		
Marsh Community	Live[1]	Dead[2]	Total	Live[1]	Dead[2]	Total
Juncus roemerianus (Mississippi)	549	214	763	1191	214	1405
J. roemerianus (Alabama)	253	80	333	449	80	529
Spartina alterniflora	441	180	621	948	180	1128
S. cynosuroides	1766	558	2324	2369	558	2927

[1]Periodic max-min values derived from Figures 1 and 2.

[2]Values based on PPM analysis of monthly standing dead biomass collected fr a clipped plot previously cleared of all plant material.

primary productivity in the fertilized plots was statistically significant from the control in the means and/or harmonic components of the periodic model (Table 2). Table 2 shows the summary of the PPM statistics. With the exception of the two *Juncus* control plots, all r^2 values were significant. Due to the growth pattern of *J. roemerianus* in Mississippi (Figure 1), a four-harmonic equation was used for this marsh community.

The net aboveground primary productivity of 333-763 $g/m^2/yr$ for the control *J. roemerianus* in Alabama and Mississippi is comparable to the values reported for a high marsh *J. gerardi* (616 g/m^2) in Maine (9) and a high marsh *J. roemerianus* (595 $g/m^2/yr$) in north Florida (8). The 621 $g/m^2/yr$ NAPP for the control *S. alterniflora* is comparable to the data obtained for inland marsh (581 $g/m^2/yr$) in Barataria Bay, Louisiana (7). The 2324 g/m^2 annual NAPP of *S. cynosuroides* was close to the 2190 g/m^2 previously determined for a similar marsh (3). The productivity values obtained for the fertilized plots, while they were significantly higher than that of the control (Table 2), nevertheless fall within the maximum ranges of values reported earlier (6), (9), (17) for natural marshes from different geographic regions, although different methods of determining NAPP were used.

In this study, a comparison was made between a natural marsh and a marsh which received N enrichment, in the form of NH_4NO_3. Comparing our observations with those made earlier by other investigators (1), (14), (15), it appears that N enrichment in the form of ammonium is better utilized by marsh plants than if the N is in the NO_3 form (10), (11), (13).

As can be seen in Figures 1 and 2, the standing crop of live plants in the fertilized plots were higher than the control from May to the end of the growing season (October-November) in all four marsh communities. The growth patterns of *J. roemerianus* in Mississippi during the 1978 growing season showed two growth peaks, in June and October, as reflected by the two-harmonic curves in the periodic model (Figure 1). It appears that there is a high *Juncus* die-back during the months of July and August, as indicated by the high biomass of standing dead plants during these months. Unlike the Mississippi marsh, the Alabama *J. roemerianus* marsh did not show a bimodal growth pattern (Figure 1). Standing crop peaked only in September. NAPP of the Mississippi *Juncus* community is more than twice that of Alabama for both control and fertilized plots. The growth curves (Figure 2) for the two *Spartina* species generally followed the same pattern with peak biomass in August. The monthly standing crop of *S. cynosuroides* was twice that of *S. alterniflora* in both control and fertilized plots. Productivity values of *S. cynosuroides* was 120 percent (in the control) and 118 percent (in the fertilized plot) more than that of *S. alterniflora*.

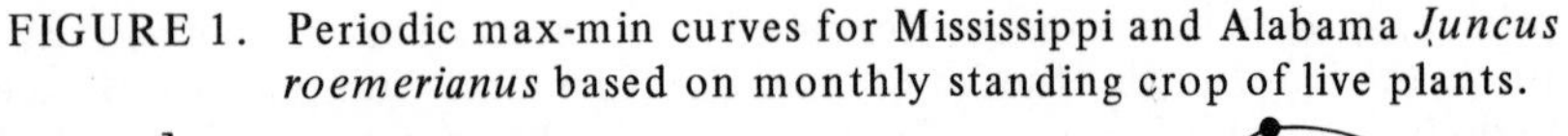

FIGURE 1. Periodic max-min curves for Mississippi and Alabama *Juncus roemerianus* based on monthly standing crop of live plants.

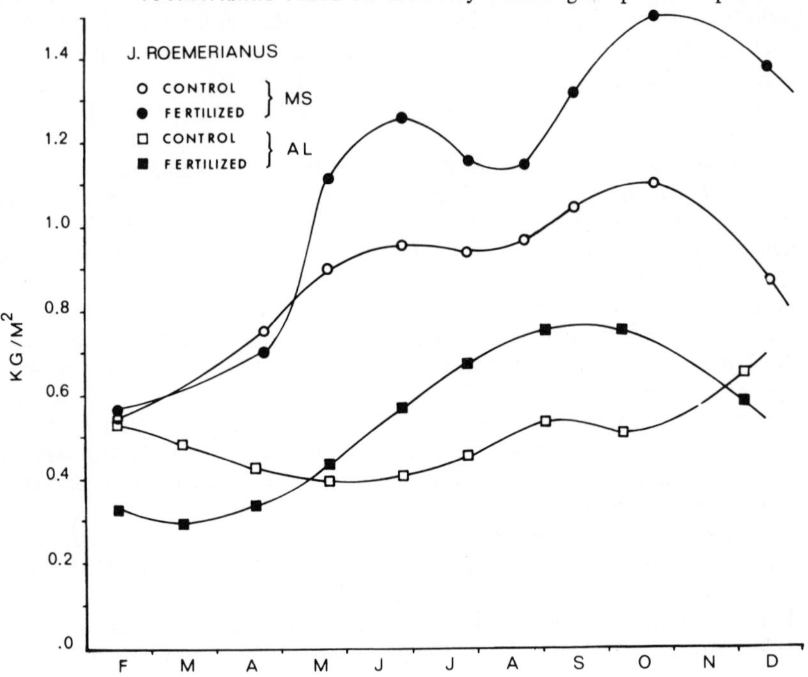

The data presented in Table 1 and Figures 1 and 2 do not indicate any variability because they are derived from the predictive model. There was variability among the six replicates collected monthly from each marsh community. Standard deviations of ± 200 g/m^2 were observed in the maximum raw values during peak biomass production of *Spartina spp.* and ± 50 g/m^2 in the minimum raw data at the beginning of the growing season of *Juncus*.

The higher response of the high marsh *J. roemerianus* (84 percent) in Mississippi and of the shore form *S. alterniflora* (82 percent) to fertilization compared to the low marsh *J. roemerianus* (59 percent) in Alabama and the giant cordgrass *S. cynosuroides* (26 percent) indicated that there may be a differential reaction by different growth forms of marsh plants to nitrogen enrichment. A study in North Carolina (10, 11) showed that nitrogen fertilization increased the aerial standing crop of the short form of *S. alterniflora* as much as 172 percent, but had no significant effect on that of the tall form. These observations suggest that marsh management by fertilization is most applicable to high marsh areas because of suitable hydrology and the potentially higher biomass production increment of the plant types that grow there.

TABLE 2. Periodic models and summary statistics for the four marsh communities.

Juncus (Miss) $y = c_0 + c_1 \sin(cti) + c_2 \cos(cti) + c_1 \sin(2cti) + c_2 \cos(2cti)$
Others $y = c_0 + c_1 \sin(cti) + c_2 \cos(cti)$

Marsh Type	c_0	c_1	c_2	c_3	c_4	r^2
J. roemerianus control (MS)	858.3	-209.0	-54.7	-106.6	33.8	.373
J. roemerianus fertilized (MS)	1092.8	-400.1	35.3	-217.6	199.0	.548*
J. roemerianus control (AL)	129.8	-17.6	26.4	--	--	.208
J. roemerianus fertilized (AL)	133.8	-55.5	-13.6	--	--	.515*
S. alterniflora control	107.2	-52.6	19.7	--	--	.720*
S. alterniflora fertilized	136.5	-118.6	-19.2	--	--	.780*
S. cynosuroides control	847.1	-764.6	-441.5	--	--	.750*
S. cynosuroides fertilized	957.5	-1018.5	-605.2	--	--	.690*

Comparisons between control and fertilized	c_0	c_1	c_2	c_3	c_4
J. roemerianus control vs. fertilized (MS)	*	--	--	N.S.	N.S.
J. roemerianus control vs. fertilized (AL)	N.S.	*	*	--	--
S. alterniflora control vs. fertilized	*	*	N.S.	--	--
S. cynosuroides control vs. fertilized	N.S.	N.S.	N.S.	--	--

* = Significant at 0.05.

N.S. = Not significant.

Significance of c_0 indicates a significant difference in the means, c_1, the first harmonic component, c_2, the second harmonic component, c_3, the 3rd harmonic component, and c_4, the 4th harmonic component.

FIGURE 2. Periodic max-min curves for *Spartina alterniflora* and *S. cynosuroides* based on monthly standing crop of live plants.

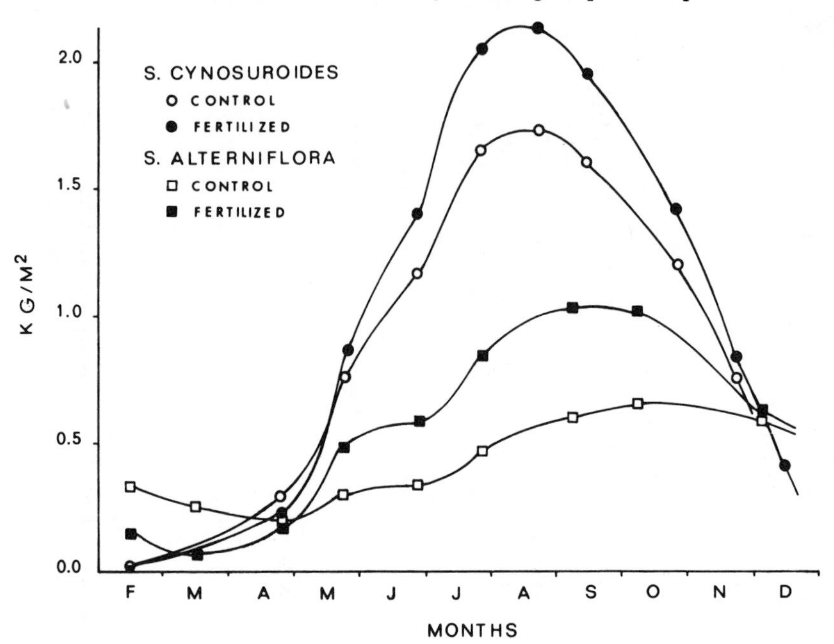

ACKNOWLEDGEMENTS

We are grateful to Eileen Duobinis, Susan Invester, and Keith Parsons for their invaluable assistance in the fieldwork. We thank Olga Pendleton for providing statistical and computer counseling.

This research was supported by a grant from the NOAA Office of Sea Grant, U.S. Department of Commerce, under Grant No. 04-8-MOI-92. Mississippi-Albama Sea Grant Consortium M-ASG-78-047.

REFERENCES

1. Broome, S.W., W.W. Woodhouse, Jr., and E.D. Seneca. 1975. The relationship of mineral nutrients to growth of *Spartina alterniflora* in North Carolina: II. The effects of N,P, and Fe fertilizers. *Soil. Sci. Soc. Amer. Proc*. 39:301-307.
2. Buresh, R.J., R.D. DeLaune, and W.H. Patrick, Jr. 1980. Nitrogen and phosphorus distribution and utilization by *Spartina alterniflora* in a Louisiana Gulf Coast marsh. *Estuaries*, 3:111-121.

3. Cruz, A.A. de la. 1975. Primary productivity of coastal marshes in Mississippi. *Gulf Res. Rpts.* 4:351-356.
4. Gallagher, J.L. 1975. Effect of ammonium pulse on the growth and elemental composition of natural stands of *Spartina alterniflora* and *Juncus roemerianus*. *Amer. J. Bot.* 62:644-648.
5. Hackney, C.T. and O.P. Hackney. 1978. An improved, conceptually simple technique for estimating the productivity of marsh vascular flora. *Gulf Res. Rpts.* 6:125-129.
6. Hopkinson, C.S., J.G. Gosselink, and R.T. Parrondo. 1978. Aboveground production of seven marsh plants species in coastal Louisiana. *Ecology*. 59:760-769.
7. Kirby, C.J. and J.G. Gosselink. 1976. Primary production in a Louisiana Gulf Coast *Spartina alterniflora* marsh. *Ecology* 57:1052-1059.
8. Kruczynski, W.L., C.B. Subrahmanyam, and S.H. Drake. 1978. Studies on the plant community of a north Florida salt marsh. Part I. Primary production. *Bull. Mar. Sci.* 28:316-334.
9. Linthurst, R.A. and R.J. Reimold. 1978. Estimated net aerial primary productivity for selected estuarine angiosperms in Maine, Delaware, and Georgia. *Ecology* 59:945-955.
10. Mendelssohn, I.A. 1979. The influence of nitrogen level, form, and application method on the growth response of *Spartina alterniflora* in North Carolina. *Estuaries* 2:106-112.
11. Mendelssohn, I.A. 1979. Nitrogen metabolism in the height forms of *Spartina alterniflora* in North Carolina. *Ecology* 60:574-584.
12. Patrick, Jr. W.H. and R.D. Delaune. 1976. Nitrogen and phosphorus utilization by *Spartina alterniflora* in a salt marsh in Barataria Bay, Louisiana. *Estuar. and Coastl. Mar. Sci.* 4:59-64.
13. Stout, J.P., C.T. Hackney, and A.A. de la Cruz. 1979. Vascular plant productivity, decomposition and tissue composition. *In* Evaluation of the Ecological Role and Techniques for the Management of Tidal Marshes on the Mississippi-Alabama Gulf Coast. Interim Report to the Mississippi-Alabama Sea Grant Consortium on Project No. 40(3).
14. Sullivan, M.J. and F.C. Daiber. 1974. Response in production of cord grass, *Spartina alterniflora*, to inorganic nitrogen and phosphorous fertilizer. *Ches. Sci.* 15:121-123.
15. Tyler, G. 1967. On the effect of phosphorous and nitrogen, supplied to Baltic Shore - meadow vegetation. *Bot. Notices* 120:433-447.
16. Valiela, I., J.M. Teal and W.J. Sass. 1975. Production and dynamics of salt marsh vegetation and the effects of experimental treatment with sewage sludge. *J. Appl. Ecol.* 12:973-981.
17. White, D.A., T.E. Weiss, J.M. Trapani, and L.B. Thein. 1978. Productivity and decomposition of the dominant salt marsh plants in Louisiana. *Ecology* 59:751-759.

[illegible] 1979. [illegible] climate [illegible] affects the [illegible] Marsh [illegible] [illegible] 353–359.

5. [illegible] 1978. [illegible] the growth and [illegible] composition [illegible] and [illegible].

6. [illegible] 1974. [illegible] on the productivity [illegible] salt [illegible]. Vol. 11.

7. [illegible] and [illegible] 19[illegible]. [illegible] vegetation [illegible] *Spartina alterniflora* [illegible].

8. [illegible], [illegible] and [illegible] 19[illegible]. [illegible] of [illegible] fertilization of a [illegible] marsh. [illegible] production. [illegible]

9. [illegible] 197[illegible]. [illegible] nutrients as a primary [illegible] production in [illegible] in [illegible] marsh [illegible] Maine. [illegible] 255.

10. [illegible] 197[illegible]. The influence of [illegible] on [illegible] *Spartina alterniflora* in North Carolina. [illegible] 211–[illegible].

11. [illegible] 197[illegible]. [illegible] in the [illegible] of [illegible] North Carolina salt marsh [illegible].

12. [illegible] and [illegible] 1976. [illegible] [illegible] [illegible].

13. [illegible] 197[illegible]. [illegible] composition of [illegible] and [illegible] Coast [illegible] Report [illegible] 197[illegible].

14. Sullivan, M.J. and [illegible] 1974. [illegible] and [illegible] to [illegible] and [illegible] [illegible] 1[illegible]–1[illegible].

15. [illegible] of [illegible] vegetation [illegible] 455–4[illegible].

16. [illegible] and [illegible] 197[illegible]. [illegible] [illegible] [illegible].

17. [illegible], [illegible] 1[illegible]. [illegible] and [illegible] of the [illegible] [illegible].

NITRIFICATION AND PRODUCTION OF N_2O IN THE POTOMAC: EVIDENCE FOR VARIABILITY

James W. Elkins , Steven C. Wofsy, Michael B. McElroy, Warren A. Kaplan
Center for Earth and Planetary Physics
Harvard University
Cambridge, Massachusetts 02138

ABSTRACT: Extensive measurements were carried out during the summers of 1977 and 1978 to define concentrations of inorganic nitrogen, O_2 and N_2O in the Potomac River. The chemistry of the river varied significantly between 1977 and 1978, with nitrification rates slower near the city of Washington D.C. by more than a factor of 10 in 1978. The nitrification rate was inversely correlated with the rate of fresh water flow into the estuary. It appears that production of N_2O in 1978 occurred mainly as a by-product of nitrification. The quantity of N_2O released to the atmosphere represented approximately 0.3 percent of sewage nitrogen. Conversion was more efficient in the summer of 1977, about 1-5 percent, reflecting either additional mechanisms for production of N_2O or larger yields for gas production in nitrification.

INTRODUCTION

The Potomac estuary is under intense stress due to inputs of sewage effluent from treatment plants near Washington, D.C. The river receives approximately 90 percent of its sewage load ($\sim$3.5 x 10^4 kg NH_4^+-N day^{-1}) over a short distance, 17-23 km below the head of the estuary at Chain Bridge. Concentrations of NH_4^+ and NO_3^- often exceed 1 mg N $liter^{-1}$ in the river, with concentrations of PO_4^{3-} as high as 0.8 mg P $liter^{-1}$ (2). The Blue Plains treatment plant, located 18.3 km below Chain Bridge, provides the largest input of sewage, processing waste from a population of about 2.3 million people ($\sim$70 percent of the population in the drainage basin). The principal form of nitrogen released from this plant is NH_4^+, on average about 1.8 x 10^4kg N day^{-1}. Phosphorus is removed with relatively high efficiency (70 percent) during treatment and release rates of P have decreased from about 1.1 x 10^4 kg P day^{-1} in 1970 to about 2 x 10^3 kg P day^{-1} in 1979 (15, 16).

FIGURE 1. The upper Potomac estuary with shallow areas (≤ 1m) stippled and river distances below the head-of-tide (Chain Bridge) indicated.

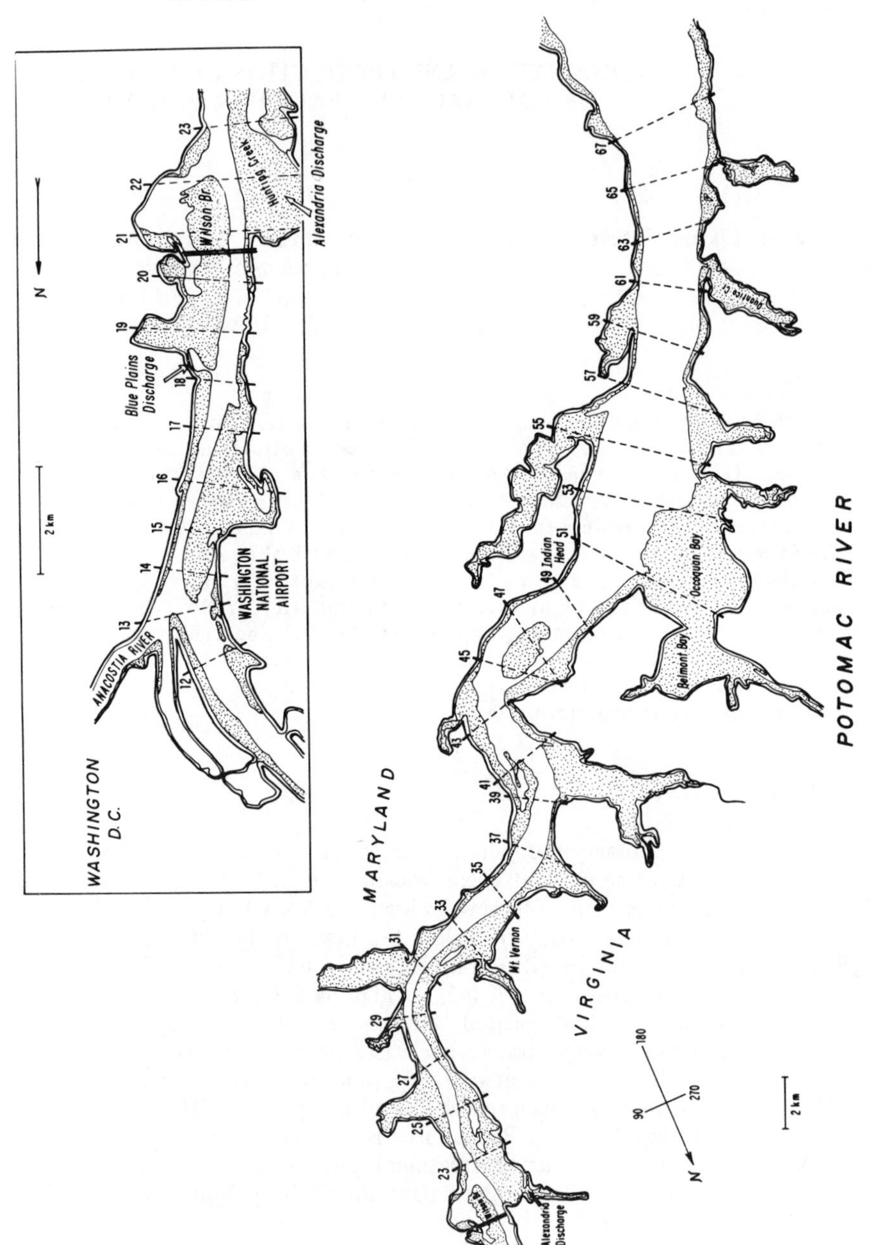

Sewage nitrogen undergoes a series of complex transformations in the Potomac. Ammonium is oxidized (nitrified) by bacteria, first to nitrite, and then to nitrate. Both ammonium and nitrate may be taken up by phytoplankton, with ammonium released to the water as the plankton decompose. In addition to loss due to algal uptake, nitrate is transported out of the estuary and may also be removed by denitrification. Rates for these various processes may change markedly from year to year and even from week to week.

The present work was motivated initially by concern that human perturbations to the nitrogen cycle might lead to an enhanced abundance of atmospheric N_2O (7, 8, 18, 20, 21) with consequent effects on the chemistry of the stratosphere. Change in the concentration of atmospheric N_2O would also significantly alter transmission of infrared radiation through the troposphere (25). Assessment of these problems has been hampered by major gaps in our understanding of processes which regulate fixed nitrogen and N_2O in natural systems and in environments affected by human activity. The experimental studies reported below shed some light on the manner through which disposal of human wastes may affect the evolution of N_2O to the atmosphere. Information is obtained also on several aspects of the estuarine nitrogen cycle, including response to inputs of NH_4^+ and to variations of river flow rate.

EXPERIMENTAL METHODS AND RESULTS

The upper Potomac estuary was sampled at frequent intervals during time periods shown in Table I. Sampling runs totaled 16 in 1977, 27 in 1978, and 4 in 1979. Selection of sampling dates emphasized summertime, steady-flow conditions. Additional data are available from extensive EPA surveys (4) conducted during 1977 and 1978.

The measurements discussed here cover a section of the estuary between National Airport (11 km below Chain Bridge) and Quantico, Virginia (61 km). A map of the upper estuary is given in Figure 1. When the tide is coming in, effluent from the Blue Plains Sewage plant mixes with water in the main channel at a point about 18 km below Chain Bridge. On the outgoing tide or during high-flow conditions, most of the effluent flows downstream into an embayment before mixing with river water at about 23.5 km. Additional inputs of sewage occur at Alexandria (22 km) and at Arlington (Four Mile Run, 15.5 km).

TABLE I. Sampling Periods 1977-1979.

1977	1978	1979
17-23 July	1 June	27 March
16-25 September	20 June - 3 Sept.	2 June
13-14 December	14 October	15 July
		4 August

Depth casts were taken along the length of the estuary during the initial phases of the investigation. Vertical profiles of dissolved species were nearly constant except for salinity and, occasionally, oxygen. Subsequent studies were restricted to samples of surface water.

Samples for determination of nutrients and total N were placed in 60 ml, acid-washed linear polyethylene bottles and were then quick-frozen on dry ice. They were analyzed within three weeks using standard procedures (23). Persulfate digestion was used for determination of total N. Tests on frozen and unfrozen samples showed that artifacts introduced by the freezing procedure were negligible. Samples for determination of dissolved gases were placed in 60 or 300 ml BOD bottles. Oxygen was measured by polarographic electrode and/or the Winkler procedure, with good agreement between the two methods. Samples taken for analysis of nitrous oxide were preserved by addition of 1 ml saturated solution of $HgCl_2$. Gas concentrations were determined by gas chromatography using procedures described in detail elsewhere (9, 17, 22).

The data presented here were selected to illustrate typical patterns observed under a variety of conditions in the estuary. The period May-October, 1977, was characterized by relatively steady freshwater flows in the estuary. Discharge rates were significantly lower than normal. Higher flows interspersed by major transients occurred during May-August, 1978, but river flows returned to low values in the fall. Figure 2 summarizes daily mean freshwater inputs for the two years.

Figure 3 shows the amount of dissolved N_2O present in September, 1977, in excess of that expected if the river were in equilibrium with the atmosphere. It shows also simultaneous measurements of nitrogenous nutrients and dissolved oxygen in surface waters between Quantico and National Airport. It seems likely that the narrow peak in N_2O observed near Blue Plains is due in part to rapid biological production in the embayment near the plant, and in part to generation of N_2O during chlorination of sewage at Blue Plains (3, 9, 22). The large peak in N_2O near 24 km, in excess of 14 μg l^{-1}, appears to result mainly from in situ biological production. The influence of NH_4^+ from sewage effluent is evident in Figure 3, but NH_4^+ was removed within a short distance downstream. Ammonium was oxidized efficiently, first to NO_2^-, and then to NO_3^-. More than 50 percent of the ammonium consumed appeared in the water eventually as nitrate. High concentrations of dissolved N_2O were correlated with low concentrations of oxygen.

Observations such as those given in Figure 3 may be used to obtain an estimate for the yield of N_2O associated with disposal of human waste. The input rate of NH_4^+ is known. We need an estimate for the rate at which N_2O is transferred from water to atmosphere. A number of different approaches indicate transfer times in the range of 0.5-3 days. Residence times of this magnitude imply that 1-5 percent of the total nitrogen supplied to the river by sewage treatment plants should be released to the atmosphere as N_2O (22).

FIGURE 2. The daily average discharge of freshwater into the upper Potomac estuary for 1977 and 1978. Units are cubic feet per second (cfs, 1 cfs = 0.0283 $m^3 sec^{-1}$).

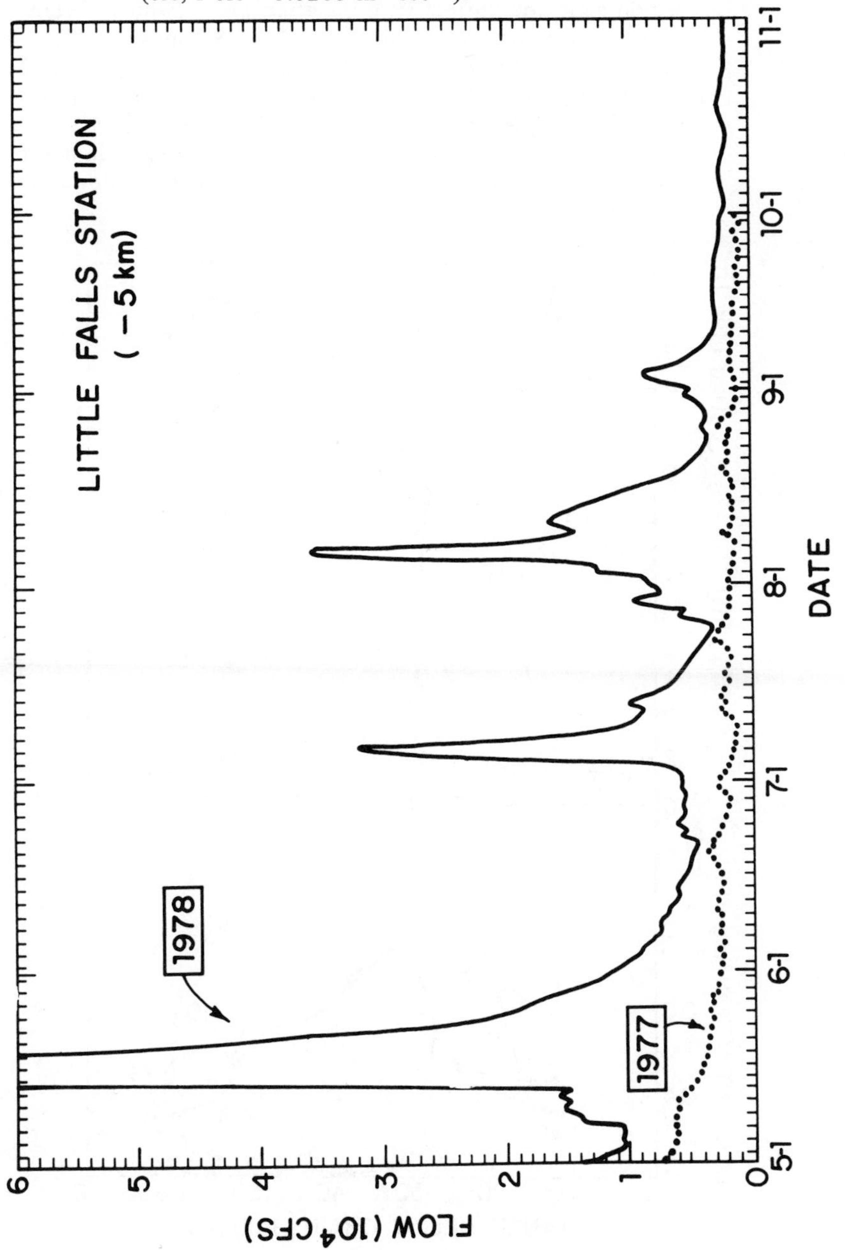

FIGURE 3. A. Measured concentrations of dissolved O_2 (mg liter^{-1}) and N_2O (μg liter^{-1}) in surface waters on September 20, 1977, plotted against river distance. The amount of N_2O which would be in equilibrium with tha atmosphere (which contained 290 ppb) is indicated by the dashed line. Solubility of N_2O in fresh and salt water was calculated from Weiss and Price (27). Water temperatures were relatively uniform, 22-25°C. The river flow at Little Falls dam (∿5 km above Chain Bridge) was 40 m^{3}sec^{-1}. Lower panel shows inorganic nutrients (in mg N liter^{-1}) along the length of river. The Blue Plains wastewater treatment plant is located at 18.3 km.

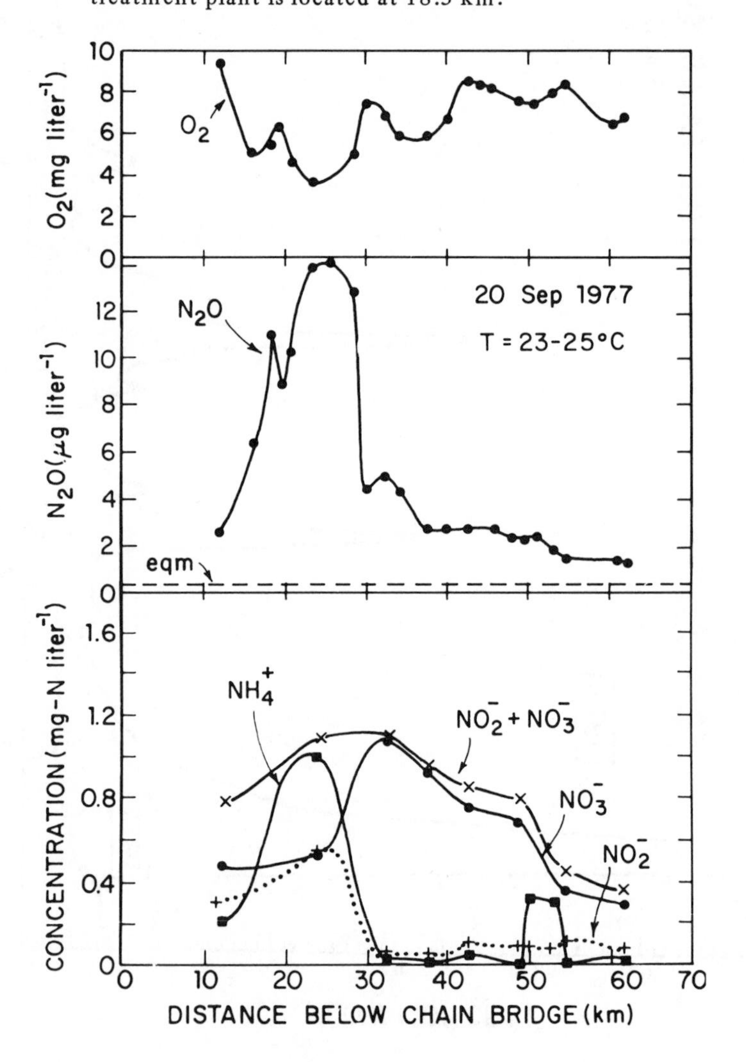

It seems probable that the dominant production of N_2O in the Potomac during the summer of 1977 occurred near the water-sediment interface. Concentrations of N_2O were measured in replicate samples from surface waters, incubated for periods between one and three days in 60 ml BOD bottles. Rates for production of N_2O were observed to be substantially slower than rates inferred from the considerations outlined above. Incubation studies implied production rates between 0.3 and 0.5 μg liter^{-1} day^{-1} in the water column, as compared to rates of 5-30 μg liter^{-1} day^{-1} required to account for the river observations. The apparent conflict could be resolved if one were to invoke an important role for the near-sediment region. Indeed, highest concentrations of N_2O were observed in this environment (22).

Measurements during 1978 present a different picture. River flow rates were high due to heavy rains which fell during 1978. Residence times for water in the primary study region were much shorter (eight days) in 1978 than in 1977 (40 days). The concentration of NH_4^+ remained high over a more extensive region of the river in 1978 (Figure 4). The nitrite maximum occurred downstream of the peak ammonia concentration, nearly 25 km below the location of the nitrite peak in 1977. The appearance of the concentration maxima for NH_4^+ and NO_2^- suggests that oxidation of NH_4^+ occurred mainly between 40 and 60 km in 1978. The concentration of downstream NO_3^- increased only slightly in response to input of NH_4^+, in contrast to the situation observed in 1977 (Figure 3). The patterns observed for nutrients on July 4, 1978 are consistent with data acquired June 20, 1978 and August 4, 1979, when river flow rates and water temperatures were similar.

The rate of oxidation of NH_4^+ was slow between 15 and 40 km during June and July, 1978. If we assume first-order kinetics for nitrification, we infer an oxidation rate for NH_4^+ slower than 0.05 day^{-1} on July 4, 1978, in contrast to the rapid oxidation (0.5 day^{-1}) inferred from analysis of 1977 data. Nitrification rates have been reported (15) to lie between 0.5 and 0.15 day^{-1} for the Potomac under summer conditions, when water temperatures are between 23° and 30°C. It appears that the rate of ammonia oxidation may be influenced by factors in addition to water temperature, in particular the river discharge rate. To explore this matter further, we examine ammonia distributions observed for a range of flow conditions in the estuary.

Figure 5 shows the distribution of ammonia in the estuary for ten days in summer, five in 1977, four in 1978, and in 1979. In 1977, EPA measurements (4) of chlorophyll showed a massive bloom of algae extending from 24 to 80 km. The bloom arose during August and died off abruptly about September 6. The presence of the bloom is reflected by very low levels of NH_4^+ in the data from August 22, 1977 (panel 3 of Figure 5a), and its subsequent decomposition is signaled by the appearance of anomalous concentrations of NH_4^+ in the downstream region (panels 4 and 5, September 8 and 20, 1977). A similar

FIGURE 4. The distributions of dissolved N_2O, O_2, and nitrogen nutrients in the Potomac on July 4, 1978. Note the association of peaks for NO_2^- and N_2O, which occur in the region of rapid disappearance of NH_4^+. Water temperatures were uniform, 26-28°C. The river flow at Little Falls dam was about 200 $m^3 sec^{-1}$.

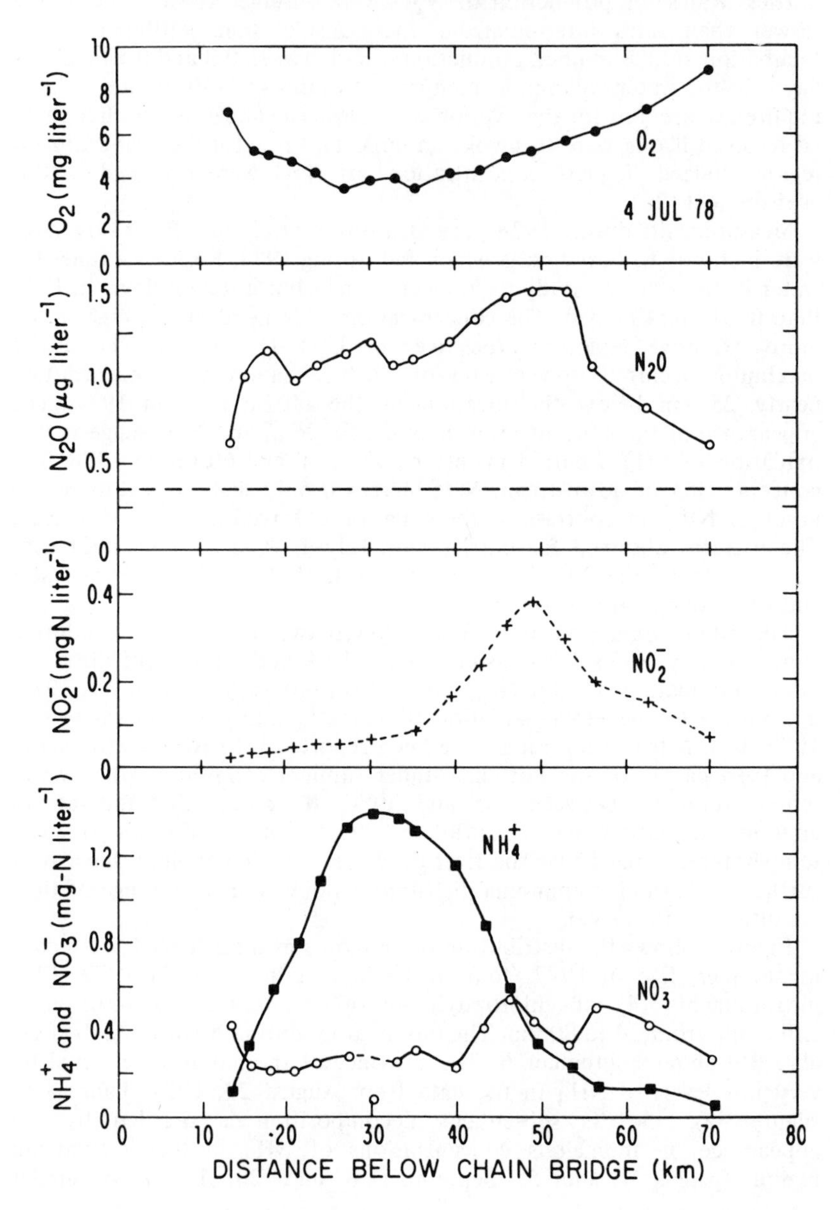

phenomenon occurred in mid-September, 1978 where chlorophyll data indicate senescence of an algal bloom in the downstream region, accompanied by elevated concentrations of ammonia (panel 9).

The most striking feature of Figure 5a is the enhancement of the ammonia content of the estuary observed with increasing river flow rate. As fresh-water discharges approach intermediate values (4,000-6,000 cfs = 120-300 m^3sec^{-1}), the plume of sewage ammonia becomes extended and the peak of the mass distribution rises to more than double that observed at lower flow rates. Dilution of wastes by river water would be expected to produce the opposite trend. It appears that the residence time for ammonia in the estuary is markedly increased as the discharge rate increases (Figure 5c), implying strong inhibition of the first stage nitrification under these flow conditions.*

At flow rates above 3,000 cfs (100 m^3sec^{-1}), consumption of ammonia and production of nitrite are not associated with a corresponding rise in the nitrate concentration (for example, compare Figures 3, 4, and 7). Evidently, rates for the second stage of nitrification are also suppressed at river discharge rates in the ragne 4,000-6,000 cfs.

Oxidation of nitrite in the Potomac often lags behind oxidation of ammonium, resulting in considerable accumulation of nitrite. Peak concentrations of nitrite may exceed 0.5 mg N $liter^{-1}$ and may be larger than ambient concentrations of NO_3^-. We have observed these large concentrations of NO_2^- both at low and at intermediate flow rates. High concentrations of a species as labile as nitrite could have potentially significant chemical and biochemical effects. For example, denitrification in sediments is more rapid if NO_2^- rather than NO_3^- is supplied to the overlying water (24).

The data from 1978 and 1979 suggest removal of NO_3^- from the water column, especially near 19 and 50 km (for example, Figure 4). Ambient concentrations of NH_4^+ were such that uptake of NO_3^- by phytoplankton should have been slow (19). The concentration of oxygen was too high to permit denitrification in the water column. Removal of NO_3^- by denitrification in the sediment appears to be indicated, with as much as 20 percent of the total nitrogen input lost by this mechanism (Figure 6). It should be cautioned, however, the quantitative conclusions concerning denitrification cannot be drawn at this time due to complications introduced by mixing between various water masses in the estuary. Computer modelling studies are currently underway to resolve this problem.

* The larger ammonia content at intermediate flow rates cannot be attributed to increased inputs from sewage plants, non-point sources or bottom scour. Measured sewage inputs varied by less than 30 percent over the period discussed here. Urban runoff from the Washington area enters the estuary mainly above National Airport, but no enhancement of NH_4^+ concentrations were noted at upstream stations. Velocities associated with tidal currents are more than ten times as large as the mean flow velocity in the study region for discharges ≤ 300 m^3sec^{-1}.

FIGURE 5a. The content of NH_4^+ (metric tons N) is shown for 8 km segments of the Potomac estuary. Ten summer days in 1977, 1978 and 1979 are represented. Data for September 25, 1978 and from July 20, 1977 to September 8, 1977 are taken from reference 4; the remainder were determined in the present work. The freshwater discharge in cubic feet per second is given for each date, averaged over the preceding seven days (1 ft^3 = 0.0283 m^3).

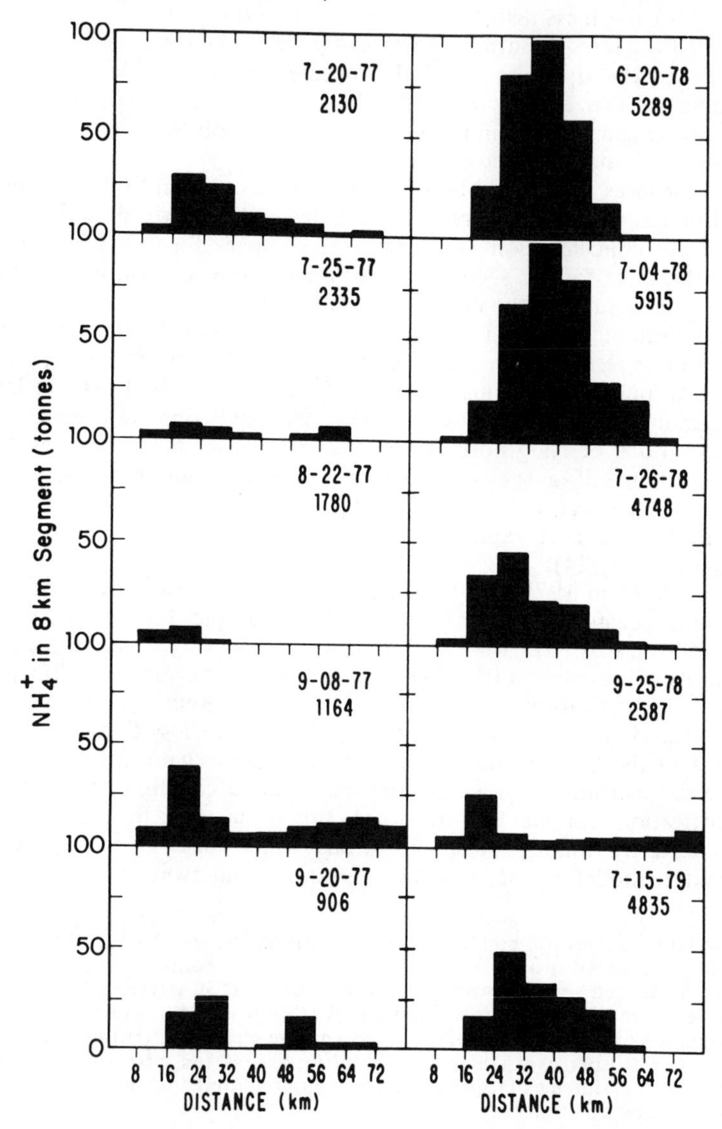

FIGURE 5b. Water volumes are given for the 8 km segments shown in Figure 5a.

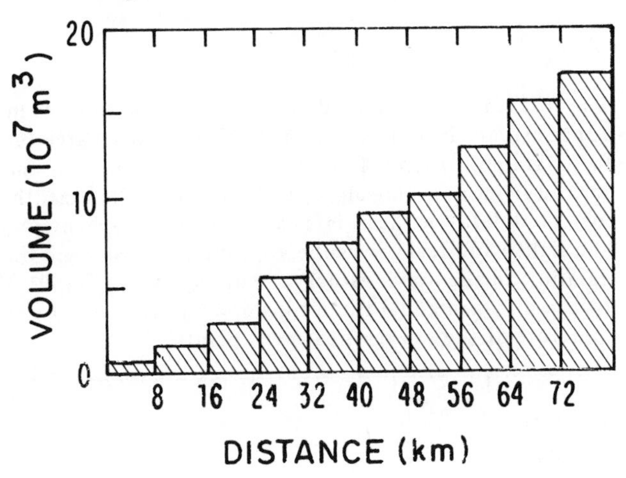

Production of atmospheric N_2O was smaller in 1978 than in 1977 by a factor between three and five (compare Figures 3 and 4). Note that the river widens below 40 km, so that the broad peak of N_2O seen in 1978 represents a disproportionately large quantity of N_2O. Shapes of the NO_2^- and N_2O distributions suggest a correlation between these two constituents in 1978. This matter is discussed below.

Figure 7 shows measurements for dissolved N_2O, O_2 and inorganic nitrogen on October 14, 1978. The river flow rate on this occasion (1,780 cfs = 50 m^3sec^{-1}) was intermediate between flows observed on September 20, 1977 and July 4, 1978, and the water temperatures ($\sim$17°C) were somewhat lower than summer values. The data in Figure 7 illustrate what one might regard as a classic example of estuarine nitrification. We shall argue that the observations imply a yield for N_2O, i.e., the fraction of sewage nitrogen converted to this gas, mid-way between yields inferred for the summers of 1977 and 1978.

DISCUSSION OF N_2O YIELDS

The coincidence of the peaks for N_2O and NO_2 in Figures 4 and 7 indicates that N_2O may have formed primarily as a by-product of the oxidation of NH_4^+. The data suggest that nitrification may have represented the dominant path for production of N_2O over the period June, 1978 to October, 1978.

FIGURE 5c. The total mass of dissolved ammonia (metric tons N) in the upper estuary is plotted against freshwater discharge (cfs). Two data points from 1977 are shown corrected for ammonia derived from decaying algae, as discussed in the text; the uncorrected data for these dates are shown by dashed circles. The right-hand scale gives the approximate residence time of ammonia in the upper estuary, assuming average input of sewage nitrogen equal to 26 tons day^{-1} as reported by the operators of the sewage treatment plants.

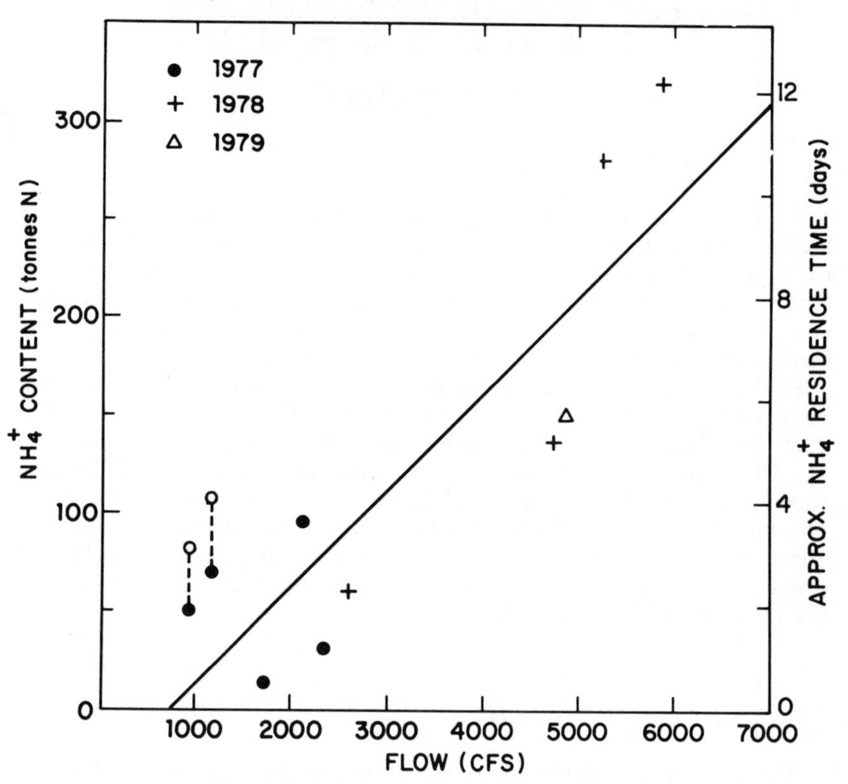

FIGURE 6. The profiles of inorganic nitrogen ($NH_4^+ + NO_2^- + NO_3^-$) and total nitrogen (persulfate digestion) are plotted against river distance. The curve labelled "organic N" is calculated by subtracting measured inorganic nitrogen from total nitrogen.

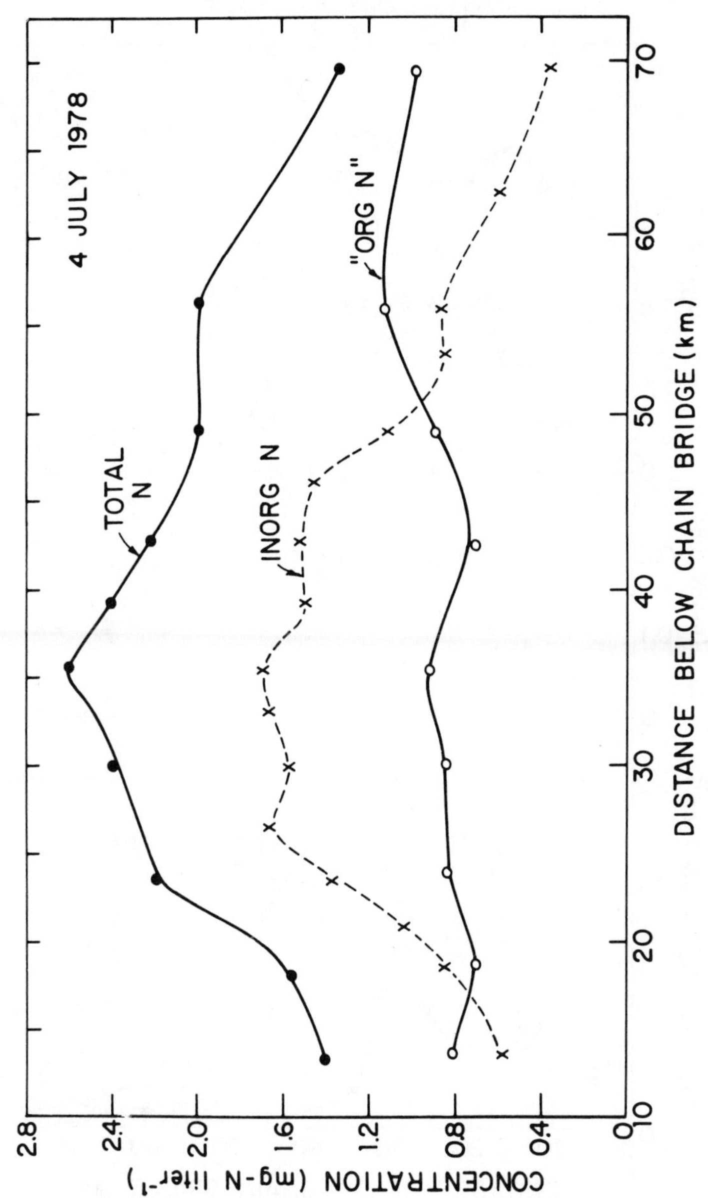

FIGURE 7. The distribution of dissolved species in the Potomac on October 14, 1978. Note the efficient oxidation of NH_4^+ to NO_2^- and NO_3^-, and the association of maxima for N_2O and NO_2^-. Water temperatures were 18°C. The river flow at Little Falls dam was approximately 50 $m^3 sec^{-1}$.

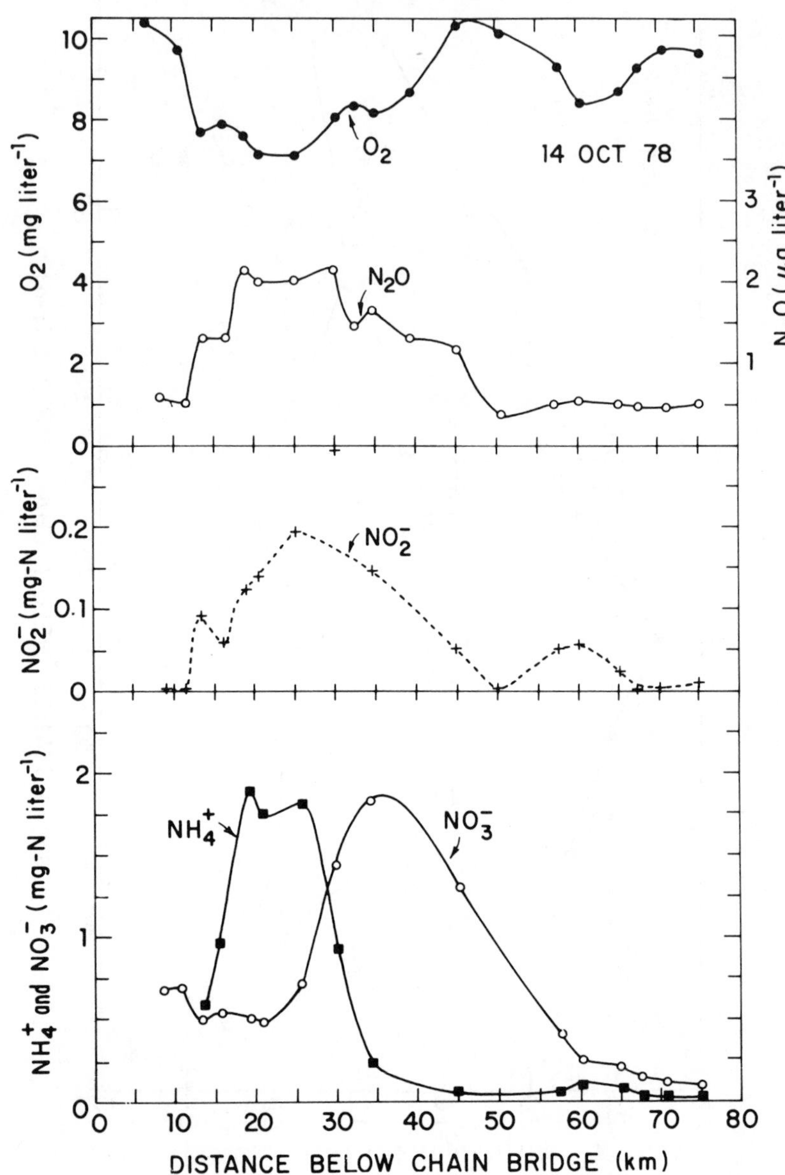

If production of N_2O is associated mainly with nitrification, it would be reasonable to assume that N_2O should be released to the atmosphere at a rate proportional to the rate for nitrification. An estimate for the yield of atmospheric N_2O, Y, would be given in this case by

$$Y = \frac{2\,(N_2O)}{(NO_2^-)}\;\frac{\tau_{NO_2^-}}{\tau_{N_2O}}$$

where (X) is the peak molar concentration of species X in the river, τ_X is the lifetime of species X, and Y is the number of moles N incorporated into N_2O for each mole of NO_2^- produced. We assume here that N_2O and NO_2^- are products of microbial oxidation of NH_4^+, and that N_2O is removed mainly by evasion to the atmosphere. The coincident location of the peaks for N_2O and NO_2^- implies that $\tau_{NO_2^-} \simeq \tau_{N_2O}$. Yields estimated in this manner are 0.2 percent for July, 1978, and 0.4 percent for October, 1978. These estimates are in remarkable agreement with yields inferred from oceanic data (5, 6, 9, 10, 27) and from laboratory studies of nitrifying bacteria (12). Measurements by Bremner and Blackmer (1) suggest a similar yield for nitrification in soils.

It appears that the yield for N_2O was much higher (1 to 5 percent) during the low-flow conditions typical of 1977. Higher yields may reflect the contributions of additional sources for N_2O, assoicated, for example, with denitrification near the water-sediment interface. Alternately, they may indicate that the yield for N_2O associated with nitrification may be enhanced under conditions of low oxygen. Data from the Central Pacific Ocean (9, 10) provide some evidence to support the notion that the yield of N_2O should respond in a non-linear fashion to reduction in the concentration of oxygen, and indeed Bremner and Blackmer (1) and Freney et al. (11) observed that the yield of N_2O in fertilized soils appeared to rise in response to increased rates of fertilization with reduced forms of fixed N, such as NH_4^+, urea, or organic N.

CONCLUDING REMARKS

For conditions that prevailed in the Potomac during the summer and fall seasons of 1977 and 1978, we have shown that:

1. Oxidation of NH_4^+ to NO_2^- was strongly inhibited when river discharge rates exceeded $\sim$3,000 cfs (100 $m^3 sec^{-1}$);
2. Oxidation of NO_2^- to NO_3^- also became inefficient at river discharge rates in excess of 3,000 cfs. Thus both stages of nitrification were sensitive to river flow rates in the range 1,000-6,000 cfs;
3. Rather large accumulations of NO_2^- ($\sim$0.5 mg N $liter^{-1}$) were observed at various times under all flow regimes;
4. Sewage nitrogen was converted to N_2O with a yield in the range

0.2-5 percent (as moles N in N_2O, per mole of N in sewage);

5. The observed concentrations of N_2O were strongly enhanced under low-flow conditions. Conversion of sewage N to N_2O occurred with a significantly higher yield under these conditions.

Our data suggest that oxidation of NH_4^+ may be the major source of N_2O in the estuary when the water column is well-oxygenated.

There is some indication of NO_3^- loss by benthic denitrification. The role which denitrification might play in production of N_2O is unclear. Particular interest attaches to the large yields observed for N_2O in the Potomac under low flow conditions.

Decomposition of human and animal waste could have a significant impact on atmospheric N_2O. Current production of N_2O associated with these processes may be estimated in the range 0.1-1 x 10^6 tons N yr^{-1} using the lower yield fractions implied by the Potomac measurements of 1978. On the other hand, the source could be higher by a factor between five and ten if we were to use yields inferred from data of 1977. To the extent that higher yields may be regarded as a consequence of waste concentration, and to the extent that concentration of waste may become more prevalent as population increases, future production of atmospheric N_2 might be expected to grow at a rate faster than the growth of population.

ACKNOWLEDGEMENTS

We wish to acknowledge help offered by R.A. Rasmussen, C.E. Kolb, J. Waterbury, E. Jones, J. Ledwell, A. Duran, and L. Hashimoto. This work was supported by the National Science Foundation and the National Aeronautics and Space Administration under contracts NSF-ENV76-24239 and NAS-NASW-2952, respectively, to Harvard University.

REFERENCES

1. Bremner, J.M. and A.M. Blackmer. 1978. Nitrous oxide: emission from soils during nitrification of fertilizer nitrogen. *Science* 199:295-296.
2. Carpenter, J.H., D.W. Pritchard, and R.C. Whaley. 1969. Observations of eutrophication and nutrient cycles in some coastal plain estuaries, 210-221. Eutrophication: Causes, consequences, correctives. National Academy of Sciences, Washington, D.C.
3. Cicerone, R.J., J.D. Shetter and S.C. Liu, 1978. Nitrous oxide in Michigan waters in U.S. municipal waters. *Geophy. Res. Lett.* 5:173-176.

4. Clark, L.J. and S.E. Roesch. 1978. Assessment of 1977 water quality conditions in the upper Potomac estuary. EPA 903/9-78-008: 74 pp.
5. Cohen, Y. and L.I. Gordon. 1978. Nitrous oxide in the oxygen minimum of the Eastern Tropical North Pacific. *Deep-Sea Research* 25:509-524.
6. ———. 1979. Nitrous oxide production in the ocean. *Journal of Geophysical Research* 84:347-354.
7. Crutzen, P.J. 1974. Estimates of possible variations in total ozone due to natural causes and human activities. *Ambio* 3:201-210.
8. ———. 1976. Upper limits on atmospheric ozone reduction following increased application of fixed nitrogen to the soil. *Geophys. Res. Lett.* 3:169-170.
9. Elkins, J.W. 1978. Aquatic sources and sinks for nitrous oxide. Ph.D. thesis, Harvard University.
10. ———, S.C. Wofsy, M.B. McElroy, C.E. Kolb and W.A. Kaplan. 1978. Aquatic sources and sinks for nitrous oxide. *Nature* 275:602-606.
11. Freney, T.R., O.T. Denmead and J.R. Simpson. 1979. Nitrous oxide emission from soils at low moisture contents. *Soil Biol. Biochem.* 11:167-173.
12. Goreau, T.J., W.A. Kaplan, S.C. Wofsy, M.B. McElroy, F.W. Valois and S.W. Watson. 1979. Laboratory studies on the nitrifying bacterium *Nitrosococcus oceanus*: Implications for marine and atmospheric N_2O. (unpublished).
13. Hahn, J. 1974. The North Atlantic Ocean as a source of atmospheric N_2O. *Tellus* 26:160-168.
14. Hesslein, R.H. 1976. An *in situ* sampler for close interval pore studies. *Limnology and Oceanography* 21:912-914.
15. Jaworski, N.A., D.W. Lear, Jr. and O. Villa, Jr. 1972. Nutrient management in the Potomac estuary. American Society of Limnology and Oceanography Special Symposium 1:246-273.
16. Jones, E.R. 1979. Wastewater Division monthly performance report--June 1979. Government of District of Columbia.
17. Kaplan, W.A., J.W. Elkins, C.E. Kolb, M.B. McElroy, S.C. Wofsy and A.P. Duran. 1978. Nitrous oxide in fresh water systems: An estimate for the yield of atmospheric N_2O associated with disposal of human waste. *Pure and Applied Geophysics* 116:423-438.
18. Liu, S.C., R.J. Cicerone, T.M. Donahue and W.L. Chameides. 1976. Limitation of fertilizer induced ozone reduction by long lifetime of the reservoir of fixed nitrogen. *Geophys. Res. Lett.* 3:157-160.
19. McCarthy, J.J., W.R. Taylor, and J.L. Taft. 1975. Dynamics of nitrogen and phosphorous cycling in the open waters of Chesapeake Bay. ACS Symposium Series #18, Marine Chemistry in the Coastal Environment (ed. T.M. Church): 664-681.
20. McElroy, M.B. 1976. Chemical processes in the solar system: A kinetic perspective, 127-211. *In* D.R. Herschbach (ed), *Chemical Kinetics*, Butterworths, London.

21. ———, S.C. Wofsy and Y.L. Yung. 1977. The nitrogen cycle: Perturbations due to man and their impact on atmospheric N_2O and O_3. Philosophical Transactions of the Royal Society of London. 277(B): 159-181.
22. ———, J.W. Elkins, S.C. Wofsy, C.E. Kolb, A.P. Duran and W.A. Kaplan. 1978. Production and release of N_2O from the Potomac estuary. *Limnology and Oceanography* 23:1168-1182.
23. Strickland, J.D.H. and T.R. Parsons. 1972. A Practical handbook of seawater analysis. Fisheries Board of Canada Bull. 167:310 pp.
24. van Kessel, J.F. 1977. Factors affecting the denitrification rate in two water-sediment systems. *Water Research* 11:259-267.
25. Wang, W.C., Y.L. Yung, A.A. Lacis, T. Mo and J.E. Hanson. 1976. Greenhouse effects due to man-made perturbations of trace gases. *Science* 194:685-690.
26. Weiss, R.F. and B.A. Price. 1979. Nitrous Oxide Solubility in Water and Seawater. Submitted to *Marine Chemistry*.
27. Yoshinari, T. 1976. Nitrous oxide in the sea. *Marine Chemistry* 9:189-202.

PHOTOSYNTHESIS, EXTRACELLULAR RELEASE, AND HETEROTROPHY OF DISSOLVED ORGANIC MATTER IN RHODE RIVER ESTUARINE PLANKTON

Maria A. Faust and Ryszard J. Chrost
Chesapeake Bay Center for Environmental Studies
Smithsonian Institution
P.O. Box 28
Edgewater, Maryland 21037

ABSTRACT: Rates of phytoplankton photosynthesis, extracellular release of dissolved organic matter (DOM), and the assimilation of DOM by bacterioplankton during ***in situ*** incubation were measured in the Rhode River estuary. Phytoplankton photosynthesis was in the range of 3 to 147 μg C l^{-1} h^{-1}. The release of DOM appeared to be inversely related to the photosynthetic activity of phytoplankton. A large percentage of the total carbon (C) fixed was released as ^{14}C-DOM, ranging from 4 to 66 percent of the total. Ultrafiltration was used to fractionate the ^{14}C-DOM into size classes from less than 500 to 300,000 molecular weight (MW). Bacterioplankton readily assimilated a wide range of ^{14}C-DOM MW size fractions. Both photosynthesis and phosphorus uptake were higher in nanoplankton. A positive linear relationship existed between log surface to volume ratio and log phosphorus uptake per biomass of phytoplankton. Bacterioplankton assimilated phosphorus more rapidly than phytoplankton. The species composition of plankton populations also constantly changed. The ^{14}C-DOM represented only 0.4 to 2.3 percent of the total DOM in the Rhode River. Its quantity and quality appear to be determined by the changing phytoplankton population which produces it, and the changing bacterioplankton population which uses it. Preliminary data such as these may yield important insight into changes occurring in metabolic patterns of brackish ecosystems due to biological enrichment with nutrients.

INTRODUCTION

Enrichment of coastal aquatic systems occurs in various ways. In addition to input of organic and inorganic nutrients from the land, from sediments, and from the ocean, there is also some evidence that dissolved organic matter is released extracellularly by phytoplankton during photosynthetic activity and is readily available to at least part of the heterotrophic plankton population. In marine and fresh waters, the amount of DOM released extracellularly is variable, but can be as high as 50 percent of the total photo-assimilated carbon (1, 2, 4, 11). This biological nutrient enrichment could be significant in controlling productivity because it represents a source of readily available organic compounds to organisms which are adapted to utilize them. At present, our knowledge is far from complete on the production, transformation, and decomposition of DOM in coastal waters.

The relationship between photosynthesis and extracellular release of DOM by phytoplankton is not simple. The environmental and biological parameters constantly change. The light spectra and intensity, inorganic carbon and oxgyen concentrations, temperature, and salinity all affect the rate of DOM release. The species composition of plankton communities and their metabolic state also change constantly (8, 9, 19). These conditions result in the release of a spectrum of organic compounds (16, 20, 23) ranging from simple organic molecules to larger polymers and biologically active substances (13, 17). In turn, these biologically active molecules are very readily available to the heterotrophic population of plankton, mainly bacteria (1, 4, 6, 11, 17, 19).

Research presented in this paper consists of preliminary evaluations of a comparative study in which the interdependence of algae and bacteria was examined. The processes studied simultaneously were primary production, excretion, and heterotrophic uptake of photosynthetic products. An attempt was made to determine the molecular size distribution of DOM which regulates activity and productivity of a brackish coastal area.

MATERIALS AND METHODS

Description of the Study Area

The Rhode River is a tidal subestuary on the western shore of the Chesapeake Bay, south of Annapolis, Maryland. The Rhode River receives runoff from its watershed primarily through Muddy Creek, the largest fresh water source. The brackish water enters the system from the Chesapeake Bay. The Rhode River has a length of 6.7 km, a mean depth of 1.5 m, and a volume of 349 x 10^4 m^3 at mean low tide. The salinity varies from about 5 ppt in the spring to 16 ppt in the fall. Water temperatures reach a low of 1.3^oC in January, and a high of 30-33^oC in summer. Experiments were carried out at the Smithsonian dock during the past several years.

Enumeration of Algae and Bacteria

The initial sample of plankton was taken from a depth of 1 m with a peristaltic pump (8). Cell numbers of bacteria were estimated by direct count procedures as described by Rodina (18) and algae by the method of Campbell (3) on samples fixed in gluteraldehyde by the methods of Faust and Correll (8). Volume and surface area of phytoplankton cells were calculated from length and width measurements using standard formulas for either a rod, a sphere, or oblate spheroid. Most of the dinoflagellates were assumed to be oblate spheriods, with an eccentricity of 0.9 (10).

Measurement of Phosphorus Assimilation by Plankton

Phosphorus assimilation by the plankton was measured according to the method of Faust and Correll (8). Particulates were separated according to size into larger fractions (algal) greater than 5 μm using a Nitex screen and into smaller fractions (bacterial) less than 5 μm in size containing a 0.45 μm Millipore membrane filter. The rate of utilization of orthophosphate by individual phytoplankton species has been estimated using liquid emulsion autoradiography as described in detail by Faust and Correll (9).

Measurement of Primary Production and Extracellular Release

Primary production was estimated by adding 100 μCi ^{14}C-bicarbonate (specific activity sodium ^{14}C-bicarbonate 48.0 mCi/m mole) into 2 liter light and dark bottles containing water samples. Bottles were incubated at 10 cm below water surface for four hours. Immediately after the addition of radioactivity (0 time) and after the incubation period, particules were separated for the determination of radioactivity using 0.45 μm pore size Amicon filters with gentle suction. Filters were washed with 10 ml of 0.1 N HCl acid to remove unbound ^{14}C-bicarbonate. The filters were placed directly into scintillation vials (8). The initial specific activity of ^{14}C was calculated from the total inorganic C concentrations of the water after filtration and the cpm/ml of whole water at 0 time. The photosynthetic rate (μg C h^{-1} l^{-1}) was then estimated according to Vollenweider (22).

The filtrate was then acidified to pH 3.0 with 0.1 HCl and was exposed to bubbling nitrogen gas for 30 minutes to remove $^{14}CO_2$, and the remaining radioactivity in the filtrate was determined (^{14}C-DOM). The percentage of total photosynthetate released into the filtrate as extracellular ^{14}C-DOM after *in situ* incubation with ^{14}C-bicarbonate for four hours was calculated (15).

Molecular Weight Fraction

An Amicon Diaflo ultrafiltration cell was used to estimate the molecular weight of ^{14}C-organic compound fractions in filtrates. Filtrate containing $DO^{14}C$ was filtered through an Amicon model 202 (200 ml capacity) ultrafiltration cell containing a 76 mm diameter Diaflo membrane. The membranes used were (with their nominal molecular weight (MW) cutoffs in parentheses) as follows: UM-05 (500 MW), UM-2 (1,000 MW), UM-10 (10,000 MW), PM-30 (30,000 MW), XM-50 (50,000

MW), XM-100A (100,000 MW), and XM-300 (300,000 MW). The membranes are organic polymers (polyelectrolyte complexes) supported on porous, spongy polyethylene bases. Nitrogen gas was used to pressurize the cell (2 kg/cm^2) and constant stirring was maintained during fractionation.

The fractionation procedure for $DO^{14}C$ was performed by employing the membrane discriminating the high MW fractions followed by membranes in decreasing order of MW. The first and last 70 ml ultrafiltrate was discarded (23) and the remaining 60 ml was collected for analysis. One ml of this middle fraction was used to estimate $DO^{14}C$ radioactivity in each fraction.

Measurements of Heterotrophic Activity

The heterotrophic activity of plankton was estimated using sodium ^{14}C-acetate. Water samples were prefiltered through a 3 μm Nitex plankton net to remove phytoplankton and zooplankton. Filtered water (100 ml in each of three replicates) was placed in 250 ml pyrex bottles and to each 2 μCi sodium ^{14}C-acetate (specific activity 54.15 mCi/m mole) was added. The bottles were then incubated *in situ* at 10 cm below the water's surface in light and dark bottles. A subsample of each was immediately withdrawn after the addition of radioactivity (0 time), filtered through a 0.45 μm Amicon filter which was then washed with 10 ml 0.05 N HCl acid and placed into scintillation liquid. Heterotrophic uptake of acetate was terminated after four hours *in situ* incubation of water. The radioactivity of the filters was determined using methods outlined above. The amount of the substrate taken up by bacterioplankton was determined. The remainder of the sample was acidified and returned to the laboratory. The $^{14}CO_2$ formed during mineralization of acetate was estimated by the difference between isotope added and remained in the sample after 30 minutes exposure to bubbling gaseous nitrogen. This enabled the amount of acetate respired to be determined. Details of these procedures are described in detail by Chrost and Siuda (5). The ability of bacterioplankton to utilize the extracellularly released $DO^{14}C$ molecular weight fractions was also estimated following above procedures. Radioactivity was determined in a Packard model 3320 liquid scintillation spectrophotometer (8).

Water Chemistry

Chlorophyll concentrations were estimated by the method of Jeffrey and Humphrey (14). Total inorganic carbon determination was performed according to Strickland and Parsons (21). Phosphorus analyses were done on whole and filtered water samples as described by Faust and Correll (8). Physical parameters were monitored during the experiments with recording instruments at the Smithsonian dock. Total dissolved organic carbon was estimated by wet dichromate digestion in sealed ampules (12).

RESULTS

Association of Microorganisms of Plankton

Populations of phytoplankton and bacteria are constantly changing in estuarine water (8, 9, 10). Usually the most dominant algal species in the Rhode River are the dinoflagellates, comprising 18-99 percent of total cell numbers and often forming blooms (Table 1). In winter, when the water is much cooler, the phytoplankton community has the most diverse population, composed of small flagellates, cryptophytes, chrysophytes, and nanoplankton less than 10 μm in size. High populations of algae were associated with high bacterial numbers. The highest number of bacteria was found in May when the dominant species was *Prorocentrum mariae-lebouriae*, and in July when it was *Gymnodinium nelsonii* (Table 2). Bacterial populations were predominantly cocci (less than 1 μm in size) during spring and summer months, while rods (0.5 x 1.5 to 8.5 μm in size) predominated during fall and winter. Although the species composition of bacteria and phytoplankton changed constantly, a relationship was found between these groups of microorganisms. A linear least-squares regression calculation gave the result: number of bacterial cells x 10^6 ml = 23.6 + 0.69 (algal cells x 10^3/ml) ($r = 0.95$), indicating that the conditions that favor bacterial growth also favor algal growth.

Role of Bacteria in P-Assimilation

Bacterial utilization of orthosphate in estuarine water has been differentiated from algal utilization by using a differential filtration procedure (Table 3). Phosphorus assimilation by phytoplankton was highest in July, representing 42 percent of P-assimilation, whereas bacterial assimilation of phosphorus was high throughout the study, in the presence and absence of light, ranging from 58 to 99 percent of total P-assimilation. This variation in P-assimilation was related to the species composition and metabolic activity of the microorganisms in the water (8, 9).

Surface:Volume Ratio

The relationship between phytoplankton size and P-assimilation in a natural algal community (Figure 1) was examined by the incorporation of ^{33}P-orthophosphate into phytoplankton and autoradiography grains per cell were estimated for metabolically active species (10). Phosphorus assimilation was estimated per unit of cell volume from the increase in grain counts per cell over time as described in detail by Friebele et al. (10). The results indicate that phosphorus uptake per μm^3 cell volume is the highest for smaller cells. This relationship is illustrated more clearly in a plot of log P uptake per cell mass against log surface/volume ratio for various phytoplankton species.

TABLE 1. Total cell numbers (cells x 10^3/ml) of various algal classes and their seasonal occurrence in the Rhode River.

	Date (1973-1974)										
	Mar 28	May 16	June 14	July 9	Aug 7	Sept 6	Oct 3	Nov 26	Dec 18	Jan 29	Feb 28
Algal Classes	% of Total Numbers										
Dinophyceae	0*	99	47	80	36	74	18	72	18	19	1
Chrysophceae	0	0	26	0	6	4	16	2	2	25	21
Prasinophyceae	0	0	0	0	0	0	6	5	17	4	5
Chlorophyceae	17	0	5	0	3	7	32	0	5	0	5
Bacillariophyceae	7	0	19	0	3	3	12	8	5	8	21
Cryptophyceae	4	1	0	2	4	2	6	3	6	37	16
Euglenophyceae	7	0	0	12	36	2	0	0	16	0	0
Haptophyceae	0	0	0	0	4	0	0	7	21	0	9
Flagellates	10	1	0	2	2	1	6	0	0	2	12
Nanoplankton	55	1	3	4	6	7	4	3	10	5	10
Total cells/ml	1.4	200.4	1.5	10.1	2.6	0.8	1.5	0.8	3.7	0.3	1.1

*None present

TABLE 2. Total cell numbers (cells x 10^6/ml) of various bacterial size classes and their occurrence in the Rhode River.

Bacterial Size Classes µm	Mar 28	May 16	June 14	July 9	Aug 7	Sept 6	Oct 3	Nov 26	Dec 18	Jan 29	Feb 28
					% of total numbers						
0.6 x 0.6	18	0*	0	12	0	44	42	0	0	2	3
1.0 x 1.0	15	56	13	86	52	45	15	17	7	2	1
1.7 x 1.7	1	0	40	1	0	0	9	0	0	2	7
0.5 x 1.5	16	6	0	0	32	0	20	32	31	82	82
0.5 x 5.0	50	0	0	0	0	1	5	33	50	7	5
0.5 x 8.5	0	38	0	1	13	6	4	15	12	1	1
1.0 x 5.0	0	0	47	1	3	3	5	3	0	4	1
1.0 x 100	0	0	0	0	0	1	0	0	0	0	0
2.0 x 4.0	0	0	0	0	1	0	0	0	0	0	0
Total cells/ml	0.5	17.6	1.0	8.3	7.8	2.5	1.4	3.1	1.6	2.3	2.6

*0 = none present

TABLE 3. Phosphorus assimilation of plankton at a selected time of the year in the Rhode River.

Date 1974	Treatment	P-Uptake µg P $l^{-1}h^{-1}$ Algae	%	Bacteria	%
June 14	Light	0.73	9	7.6	91
	Dark	0.46	15	2.6	85
July 9	Light	1.50	42	2.1	58
	Dark	0.28	7	3.7	93
Sept 6	Light	0.06	3	1.8	97
	Dark	0.01	1	1.5	99
Nov 26	Light	0.46	3	15.0	97
	Dark	0.27	2	18.0	98
Feb 28	Light	0.01	1	0.8	99
	Dark	0.01	1	0.7	99

Role of Nanoplankton in C-Assimilation

We have estimated primary production per volume of water and per algal biomass (Table 4) using the procedure of Faust and Correll (8). The first two sampling dates, primary production was high; 222 and 269 µg C l^{-1} h^{-1}; while in December it was only 20 µg C l^{-1} h^{-1}. The first two dates,

FIGURE 1. Relationship between phosphorus uptake per μm3 cell volume per hour for various phytoplankton species and their surface/volume ratios.

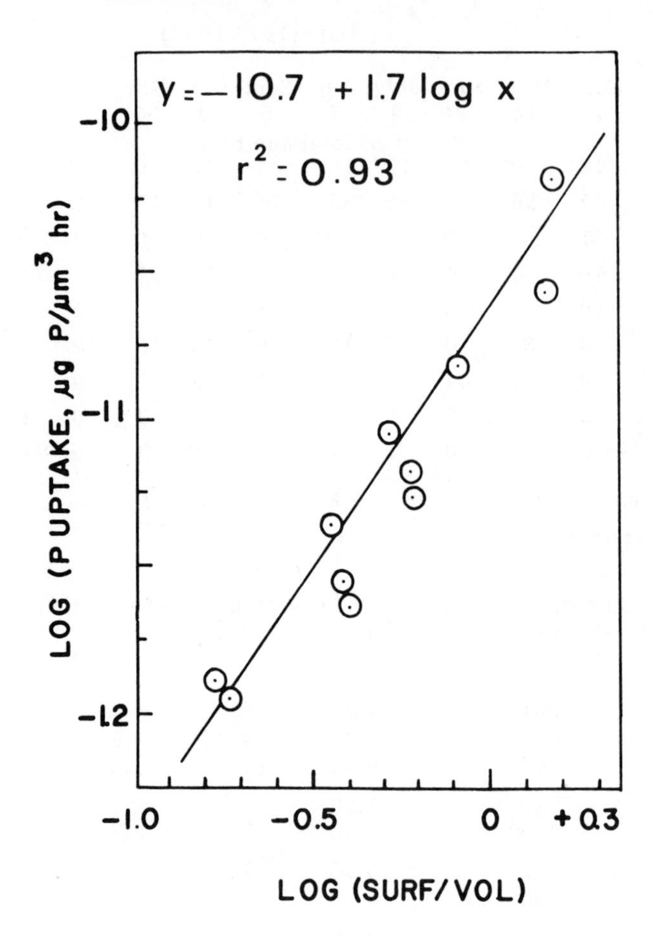

the dominant algal species was *P. mariae lebouriae* (20 μm size), present in relatively large numbers and biomass, whereas on the latter day, most of the algal population was nanoplankton (less than 10 μm sizes), which had lower numbers and biomass. We have estimated photo-assimilation of carbon per biomass of algae; the resulting data indicate that the nanplankton algal community has an advantage in fixing more carbon per biomass (3.5 μg C mm^3 l^{-1}) than larger cell populations of algae (2.6 to 3.1 μg C mm^3 l^{-1}).

TABLE 4. Comparison of primary production estimated per volume of water and per biomass.*

Date	Phytoplankton Number Cells	Total Biomass	Biomass per cell	C-uptake	
1977	10^3 x ml^{-1}	mm^3 l^{-1}	x 10^{-3}	µg C l^{-1} h^{-1}	µg C mm^3 l^{-1}
Nov. 10	23**	104	4.5	269	2.6
Nov. 16	35**	70	2.0	222	3.1
Dec. 1	4***	2	0.5	20	3.5

*Determined by one hour incubation
***Prorocentrum mariae-lebouriae* dominant species
***Nanoplankton community

Primary Production and Extracellular Release

Primary production ranged from 3.1 to 147.1 µg C l^{-1} h^{-1} and the percent of the total production released as dissolved organic matter ranged from 4 to 66 percent (Table 5). The highest primary production occurred on March 14, and the lowest percent of $DO^{14}C$ was released into the water at the same time. On this date, the dominant alga was *Chlamydomonas vernalis*. In contrast, a low primary production (8.8 µg C l^{-1} h^{-1}) occurred on December 28, when the extracellularly released $DO^{14}C$ was 66 percent of total carbon fixed. The phytoplankton community on this date included *Skeletonema costatum, Asterionella japonica, Synedra* sp., *Calycomonas ovalis*, and *Prymnesium parvum* species. The chlorophyll concentrations were the lowest (11.4 µg Chl_a l^{-1}) in January and the highest (41.8 µg Chl_a l^{-1}) in March.

TABLE 5. Primary production and extracellular release of labelled organic compounds.

Date 1978-79	Primary production (µg C l^{-1} h^{-1}) Retained in Particulates	Released Extracellularly	Total	PER*	Chl_a µg l^{-1}
Dec. 28	8.8	16.3	25.1	66	25.2
Jan. 18	3.1	1.3	3.4	29	11.4
Mar. 14	147.1	6.0	153.1	4	41.8
Apr. 19	16.2	2.7	18.9	14	38.9

*PER = Percentage of the total production released as dissolved organic matter.

Molecular Weight Distribution of ^{14}C-DOM

The distribution of molecular weights of ^{14}C-DOM was determined (Figure 2). The largest amount of radioactivity was estimated in the 10 to 30,000 MW size fractions of ^{14}C-DOM in three of the four experiments, but in January no radioactivity was found in that fraction. A relatively large percentage of radioactivity was estimated in fractions less than 500 MW. Radioactivity declined proportionally with the increase of MW fractions of ^{14}C-DOM. Despite large variations in these experiments, the 500 to 30,000 MW fractions of ^{14}C-DOM were responsible for at least 70 percent of the total amount of radioactivity detected.

Utilization of Acetate and ^{14}C-DOM

The heterotrophic activity of bacterioplantkon was also measured by the uptake and mineralization of sodium ^{14}C-acetate (Table 6). On March 14, when the primary production was high (Table 5), acetate uptake and mineralization was also the highest, indicating a metabolically active bacterial population in the water. However, a relatively high acetate uptake on April 19 by bacteria resulted in a four-fold lower mineralization rate. It appears that heterotrophic and autotrophic activity of bacteria and phytoplankton are closely related, since the percentage of total carbon production released as dissolved organic matter was higher (14 percent) on April 19 than on March 14 (4 percent). The temperature of the salinity was low (2°C) in January and warm (16°C) in April (Table 6). The salinity of the water ranged from a high of 14.5 ppt to a low of 4.5 ppt, and dissolved oxygen from 12.8 to 13.6 mg l^{-1}.

TABLE 6. Sodium ^{14}C-Acetate uptake and utilization by bacterial populations in Rhode River water.

Date 1978-79	Sodium ^{14}C-acetate utilization, Uptake		Mineralization		Temp. ^{o}C	Salinity ppt	D.O.* mg l^{-1}
	Light	Dark	Light	Dark			
	µg C x 10^{-2}l^{-1}h^{-1}						
Dec. 28	0.74	0.91	---	---**	4	14.5	12.8
Jan. 18	2.77	4.30	18.05	17.15	2	12.5	13.0
Mar. 14	4.01	4.15	25.65	24.93	8	7.6	12.8
Apr. 19	3.95	0.93	5.39	6.45	16	4.5	13.6

* D.O. = Dissolved Oxygen
**Not tested

The production of ^{14}C-DOM and its utilization by bacterioplankton was also estimated (Figure 3). ^{14}C-DOM less than 1,000 MW was preferentially utilized by bacterioplankton on both days, but as the MW size fractions increased, ^{14}C-DOM utilization also changed. Additionally, the amount of extracellularly released ^{14}C-DOM in the various MW size fractions varied from one experiment to the other.

The total amount of dissolved organic matter in the water ranged from a low of 3.1 mg C l^{-1} h^{-1} to a high of 8.8 mg C l^{-1} h^{-1} (Table 7). The extracellularly released ^{14}C-DOM represented only a small fraction of total DOM in the water, amounting to 6 to 50 μg C l^{-1}. The total DOM and ^{14}C-DOM was also estimated in the whole Rhode River. The total DOM pool size was 10,993 to 30,781 kg and ^{14}C-DOM production rate varied from 42 to 270 kg C day^{-1} respectively. The extracellularly released ^{14}C-DOM represented a relatively small fraction of the total pool ranging from a low of 0.34 percent to a high of 2.35 percent of total DOM estimated for the whole river volume.

FIGURE 2. Distribution of extracellularly released ^{14}C-DOM into various molecular weight size fractions.

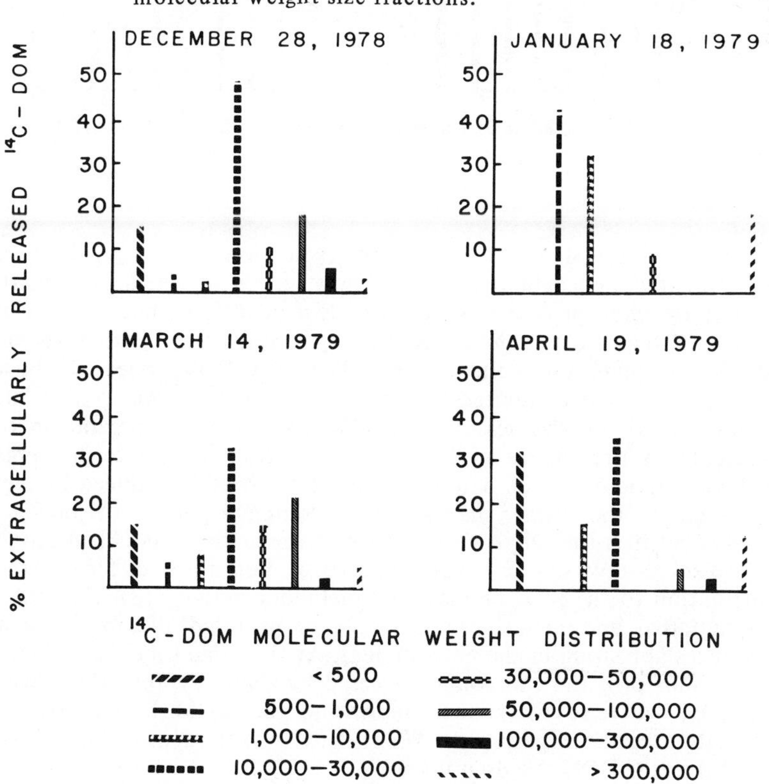

FIGURE 3. Production of ^{14}C-DOM and its utilization by bacterioplankton.

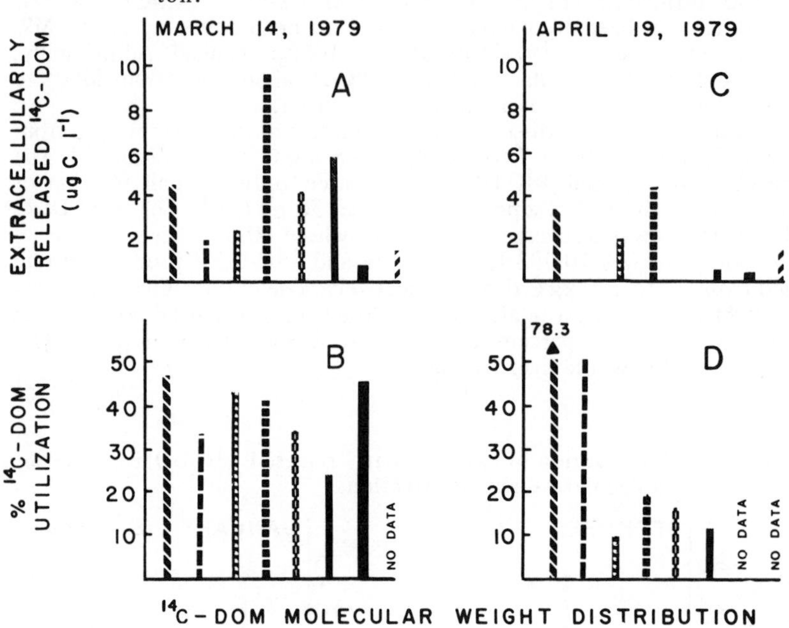

DISCUSSION

Our study consisted of a preliminary evaluation of biologically driven nutrient enrichment, about which we know very little. The organic matter released by phytoplankton occurred in low quantities, but may have a significant regulatory role in controlling productivity, structure, and growth of plankton communities. The ^{14}C-DOM released during photosynthesis was composed of a mosaic of organic compounds, a large portion of which was less than 1,000 MW. Dissolved organic matter represents a source of energy for bacterioplankton adapted to the uptake of these compounds. The determination of ^{14}C-DOM produced by algae and found in the filtrate may give an incomplete picture of metabolic activity, yet it is the only one we can measure at this time. Plankton not only display diversity in shape and size, but they differ in their contribution to the total biomass and metabolic activity (8). This results in qualitative and quantitative changes in excreted dissolved organic substances appearing in the environment. At the same time, bacteria may use or transform the compounds released by algae (7), thus the released ^{14}C-DOM we have determined represents the DOM found after the termination of the experiments. Whether bacteria transformed or released any of the ^{14}C-DOM, we do not know.

TABLE 7. Total amount of Dissolved Organic Matter (DOM) and extracellularly released ^{14}C-DOM in Rhode River water.

Date	DOM Concentrations		DOM in Rhode River		
	Total mg C l^{-1}	Extracellularly Released µg C l^{-1}	Total kg C	Extracellularly Released kg C day^{-1}	% of Total
Dec. 28	4.2	50.0	14,800	348	2.35
Jan. 18	3.1	6.0	10,993	42	0.38
Mar. 14	5.9	31.0	10,730	270	1.30
Apr. 19	8.8	12.0	30,781	105	0.34

We have continued our investigation to separate the role of algae from that of bacteria in the assimilation of phosphorus. It appeared that P-uptake by algae was insignificant compared with uptake by bacteria, indicating the important role of bacterioplankton in this process. We have evidence (8) that bacteria release orthophosphate within a short time after uptake. There is a good possibility that the bacteria are making phosphorus readily available to the algae.

Among the phytoplankton, the smaller species which dominated the plankton community at certain times were more active in assimilating phosphorus. This was explained by the larger surface/volume ratios of the smaller cells (10). The small algal cells were also more productive in assimilating CO_2. Since 80 percent of the total phytoplankton population in the Rhode River belong to the less than 20 µm size range, high productivity is probably due to the high metabolic rate of these microorganisms (8, 9).

We attempted to estimate the photosynthetic and heterotrophic processes in the same sample, involving the release and subsequent uptake of ^{14}C-DOM. We measured a sequence of events: primary production, extracellular release of ^{14}C-DOM, and its subsequent utilization by bacterioplankton. Bacterioplankton readily utilized ^{14}C-DOM released extracellularly by algae. The bacterial utilization of ^{14}C-DOM was equal between 500 and 300,000 MW fractions taken up by bacteria on March 14, in contrast to April 19 when less than 1,000 MW ^{14}C-DOM was preferably consumed. On both dates, acetate uptake, which measures the heterotrophic activity of bacteria, was equally high, indicating active bacterial populations. Mineralization of acetate, however, was four-fold higher on March 14 than on April 19, when a larger range of ^{14}C-DOM MW fractions were utilized by bacteria. This perhaps indicates that when the metabolic activity of bacteria is higher, they can assimilate a wider range of MW size fractions of ^{14}C-DOM.

Preliminary analysis of the data appears to reveal important differences in "DOM pool" size and MW fractions. While it remains important to evaluate DOM dynamics and roles of specific classes of DOM in estuaries, it is recommended that field studies be combined with measurements of changes in total dissolved organic compounds, plankton population dynamics, and heterotrophic mineralization studies before the roles of DOM can be adequately evaluated as a source of nutrient enrichment.

ACKNOWLEDGEMENT

This research has been supported partially by the Smithsonian Institution's Post Doctorate Fellowship and Environmental Science Programs.

REFERENCES

1. Allen, H.L. 1978. Low molecular weight dissolved organic matter in five soft water ecosystems: A preliminary study and ecological implication. *Verh. Internat. Verein. Limnol.* 20:514-524.
2. Antia, N.J., C.D. McAllister, T.R. Parsons, K. Stephens, and J.D.H. Strickland. 1963. Further measurements of primary production using a large-volume plastic sphere. *Limnol. Oceanogr.* 8:166-183.
3. Campbell, P.H. 1973. Studies on brackish water phytoplankton. Sea Grant Publication UNC-SG 73-07. University of N. Carolina, Chapel Hill, NC. 403 pp.
4. Chrost, R.J. 1978. The estimation of extracellular release by phytoplankton and heterotrophic activity of aquatic bacteria. *Acta Microbiol. Polonica* 27:139-146.
5. ---------, and W. Siuda. 1978. Some factors affecting the heterotrophic activity of bacteria in lake. *Acta Microbiol. Palonica* 27:129-138.
6. Derenbach, J.B. and P.J. Le B. Williams. 1974. Autotrophic and bacterial production: fractionation of plankton populations by differential filtration of samples from the English Channel. *Marine Biol.* 25:263-269.
7. Dunstall, T.G. and C. Nalewajko. 1975. Extracellular release in planktonic bacteria. *Verh. Internat. Verein. Limnol.* 19:2643-2649.
8. Faust, M.A. and D.L. Correll. 1976. Comparison of bacterial and algal utilization of orthophosphate in an estuarine environment. *Mar. Biol.* 34:151-162.
9. ---------, ---------. 1977. Autoradiographic study to detect metabolically active phytoplankton and bacteria in the Rhode River Estuary. *Mar. Biol.* 41:293-305.
10. Friebele, E.S., D.L. Correll, and M.A. Faust. 1978. Relationship between phytoplankton cell size and the rate of orthophosphate uptake: in situ observations of an estuarine population. *Mar. Biol.* 45:39-52.

11. Fogg, G.E., C. Nalewajko, and W.D. Watt. 1965. Extracellular products of phytoplankton photosynthesis. *Proc. R. Soc. (Ser. B)* 162:517-534.
12. Gage, S.J. 1978. The ampule C.O.D. digestion and analysis system under EPA's alternative method regulations. *Federal Register 43*: No. 45.
13. Hellebust, J.A. 1974. Extracellular products, 838-63. W.D.P. Stewart (ed.) *Algal Physiology and Biochemistry. Botanical Monogr.* 10:838-863.
14. Jeffrey, S.W. and G.F. Humphrey. 1975. New spectrophotometric equations for determining chlorophylls a, b, c, and c_2 in higher plants, algae, and natural phytoplankton. *Biochem. Physiol. Pflanzen (BPP) Bd. 167*, S:191-194.
15. Nalewajko, C. and D.W. Schindler. 1976. Primary production, extracellular release, and heterotrophy in two lakes in the ELA, Northwestern Ontario. *J. Fish. Res. Bd. Canada.* 33:219-226.
16. Ogura, N. 1974. Molecular weight fractionation of dissolved organic matter in coastal seawater by ultrafiltration. *Mar. Biol.* 24:305-312.
17. ————————. 1977. High molecular weight organic matter in seawater. *Mar. Chemistry* 5:535-549.
18. Rodina, A.G. (ed.) 1972. *Methods of Aquatic Microbiology*, 461 pp. (Translated, edited, and revised by R.R. Colwell and M.S. Zamburski). University Park Press, Baltimore, Maryland.
19. Sapers, A.B.J. 1977. The utilization of dissolved organic compounds in aquatic environments. *Hydrobiol.* 52:39-54.
20. Saunders, G.W., Jr. 1972. The kinetics of extracellular release of soluble organic matter by plankton. *Internatl. Assoc. of Theoretical and Applied Limnol.* 18:140-146.
21. Strickland, J.D.H. and T.R. Parsons. 1972. *A Practical Handbook of Seawater Analysis.* Bulletin 167 (2nd ed.) 310 pp. Fisheries Research Board of Canada, Ottawa, Canada.
22. Vollenweider, R.A., J.F. Talling, and D.F. Westlake. 1969. *A Manual on Methods for Measuring Primary Production in Aquatic Environments.* IBP Handbook No. 12. 213 pp. Blackwell Science Publications, Oxford.
23. Wheeler, J.R. 1976. Fractionation by molecular weight of organic substances in Georgia coastal water. *Limnol. Oceanogr.* 21:846-852.

ENRICHMENT OF A SUBTROPICAL ESTUARY WITH NITROGEN, PHOSPHORUS AND SILICA

Thomas H. Fraser and William H. Wilcox
Environmental Quality Laboratory, Inc.
Port Charlotte, Florida

ABSTRACT: Seasonal pulses of nutrients are delivered to Charlotte Harbor from the Peace River and other tributaries. Greatest loads occur during the summer wet season. Phosphate and nitrate values are much higher in the Peace River than in adjacent tributaries while ammonium and silica values are similar among the streams.

Phosphate dilution curves in the estuary suggest that changing concentrations along the chloride gradient are the result of mixing processes. Inorganic nitrogen and silica dilution curves suggest non-conservative processes are occurring, since their concentrations decrease faster along the chloride gradient than can be explained by dilution alone. Thus, both inorganic nitrogen and silica could become limiting factors in different parts of Charlotte Harbor during portions of the annual cycle, while phosphorus is always abundant.

Phytoplankton populations respond positively to seasonal pulses of nutrients with higher productivity occurring during or just after high river flow. Productivity increases from the lower harbor to near the river mouth in all seasons. General population level and productivity are similar to other Florida estuaries, with diatoms dominating the phytoplankton (blue-green algae less than 1 percent) community even with the increased nutrient levels.

INTRODUCTION

Charlotte Harbor is the second largest estuarine-bay complex in Florida (excluding Florida Bay) and is located on the southwestern coast. Two rivers feed the estuary with a dominating freshwater discharge from the Peace River. The major opening to the Gulf of Mexico, Boca Grande, is at the southwestern end of the harbor (Figure 1). General characteristics of this harbor have been summarized by Taylor (24).

FIGURE 1. Station locations are shown for nutrient sampling and primary production studies in the Charlotte Harbor area, Florida. Transect data are from Stations 2, 4, 7, 9, 10, 12, 14, and 18 as well as intermediate stations for data in Figure 8.

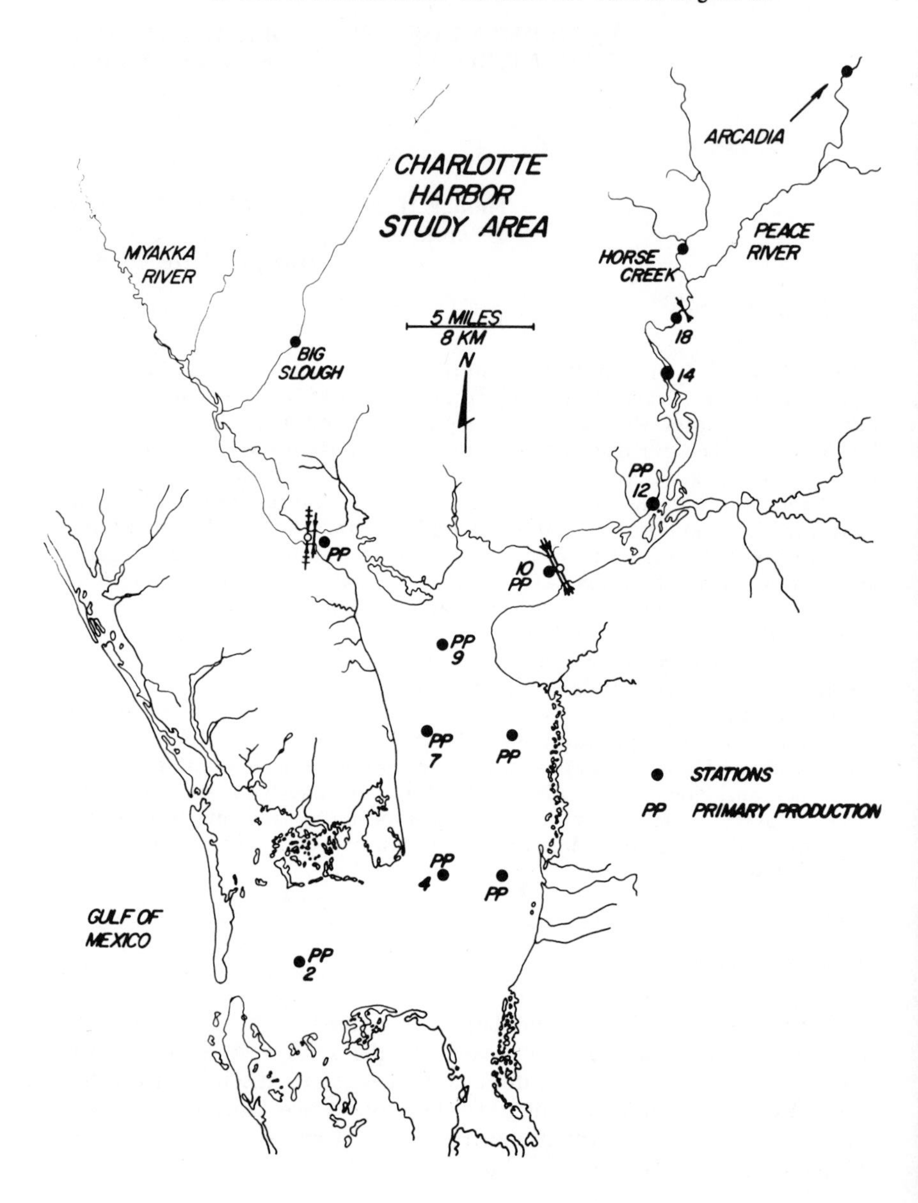

The Peace River has a superabundance of phosphate (1, 2, 3, 5, 6, 7, 8, 9, 11, 12, 13, 14, 20, 24). The Peace River may also be enriched with nitrogen (19) and silica (12). Several workers believe that the high levels of phosphate compared with nitrogen could lead to nutrient limitation caused by the depletion of nitrogen (2, 3, 7, 9). None of these authors provides much data on nitrogen and their statements are viewed as predictive. Browder (3) cites Dragovich et al. (6) for several comments about nitrogen but they provide no nitrogen data or discussion of nitrogen. Alberts et al. (2) predicted that in the absence of nitrogen limitation, dissolved silica might become a limiting nutrient for diatoms.

This paper reports three years of data on phosphorus, nitrogen silica, and color values in the Peace River and Charlotte Harbor together with a one-year study of phytoplankton productivity in Charlotte Harbor. These data demonstrate how pulses of nutrients delivered to the harbor by seasonal river flow produce obvious changes in nutrient concentrations and phytoplankton responses. No previous work has combined chemical and biological aspects to determine how river-borne nutrients change harbor characteristics.

METHODS

Samples of near surface water were taken once a month from each station (Figure 1) and analyzed the same day for ammonia, nitrate-nitrite, ortho-phosphate, silica, color and chlorophyll-a. Sampling in the Peace River, Horse Creek, Big Slough, and Charlotte Harbor was not always done on the same day. During the phytoplankton study, nutrient analyses were performed on samples at the same stations. All water samples for laboratory chemical analyses were collected in the morning with a Kemmerer water sampler and placed in acid-washed polypropylene or linear polyethylene bottles. The samples were immediately iced and brought back to the laboratory within four hours of collection. No chemical preservative was added.

The following list briefly outlines each specific method used.

Chlorophyll-a - the Strickland and Parsons (23) method of extraction, Chlorophyll-a concentration was then determined with a Turner Fluorometer Model 111 high sensitivity door, F4T4-blue lamp. General volume extracted was 50 mℓ. *Ammonia* - automated phenate method of Rand et al. (16). *Modifications*: Sodium citrate rather than EDTA used in buffer (both are per Technicon recommendations.) *Nitrate and nitrate* - automated cadmium reduction method (16). *Ortho-Phosphate* - automated molybdophosphate-ascorbic acid method, Technicon Industrial Method 329-74W. *Silica* - automated molybdosilicate method (16). *Color* - spectrophotometric at 455 nm (16). *Phytoplankton* - primary production experiments were performed with C-14 inoculations of two light bottles and one dark bottle at the surface and bottom at 8 to 12 stations in

Charlotte Harbor (Figure 1) and incubated in situ for three hours. Primary production and chlorophyll values were determined by methods given in Strickland and Parsons (23). For estimates of cell counts and volume, 500 mℓ of sample water was fixed with Lugol's solution and stored in the dark. Fifty mℓ of well-mixed sample was poured into a settling chamber, allowed to stand for 24 hours, and counted on an inverted microscope. *Loads* - nutrient load exported from the Peace River was estimated by multiplying concentrations by mean discharge recorded at the USGS Arcadia station for each month.

RESULTS AND DISCUSSION

USGS records (25) show that the Peace River at Arcadia delivered 45-93 percent of the total gaged fresh water to Charlotte Harbor each month and averaged about 70 percent each year. Below Arcadia, three tributaries increase the contribution of the Peace River to about 88 percent of all gaged fresh water delivered to the harbor. River discharge at Arcadia varied from less than 5 $m^3.s^{-1}$ to about 170 $m^3.s^{-1}$ during the study and showed a consistent wet season high flow pattern in the summer (Figure 2). Particulate and organic phosphorus, ortho-phosphate, nitrate plus nitrite, and silica concentrations during the period of study were consistently higher in the Peace River than two other streams (Table

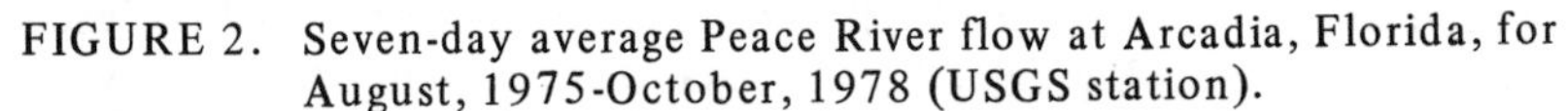

FIGURE 2. Seven-day average Peace River flow at Arcadia, Florida, for August, 1975-October, 1978 (USGS station).

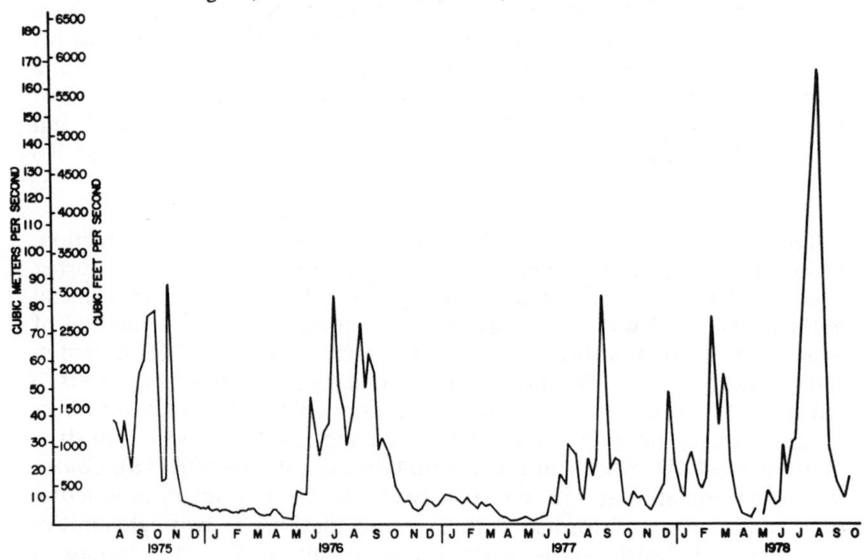

1). Data presented by Donnelly et al. (5), Dragovich et al. (6), Kaufman (11), and Slack and Goolsby (19) indicate that higher levels of nutrients occur in the Peace River than in the Myakka River and smaller streams entering Charlotte Harbor every year. Bulk precipitation rates reported by Coleman (4) for Port Charlotte indicate that this source makes a minor contribution of nitrogen and phosphorus when compared with Peace River input.

Inorganic Phosphorus

There are distinct seasonal variations in the concentration and load of ortho-phosphate in the Peace River (Figure 3). Variations in concentration and load are inverse to each other because of the much greater quantity of water conveyed in the wet season. Dry season concentrations (November-May) are apparently diluted each year during higher wet season flows (June-October), indicating a constant, non-flow related source of ortho-phosphate in the Peace River. Although river concentration decreases during the wet season, it is still an order of magnitude higher than marine values. This fact, along with much greater wet season flows, produces a consistent wet season increase in harbor concentrations.

Dilution along the chloride gradient is a major factor controlling ortho-phosphate concentrations in the harbor (Figure 8); also see Alberts et al. (2). Our data show no deviations from dilution control for any month of the year. Thus, inorganic phosphate does not appear to be a limiting factor for harbor phytoplankton.

FIGURE 3. Ortho-phosphate concentrations (as mg/ℓ P) and load in the Peace River at Arcadia and concentrations in Charlotte Harbor at Station 9.

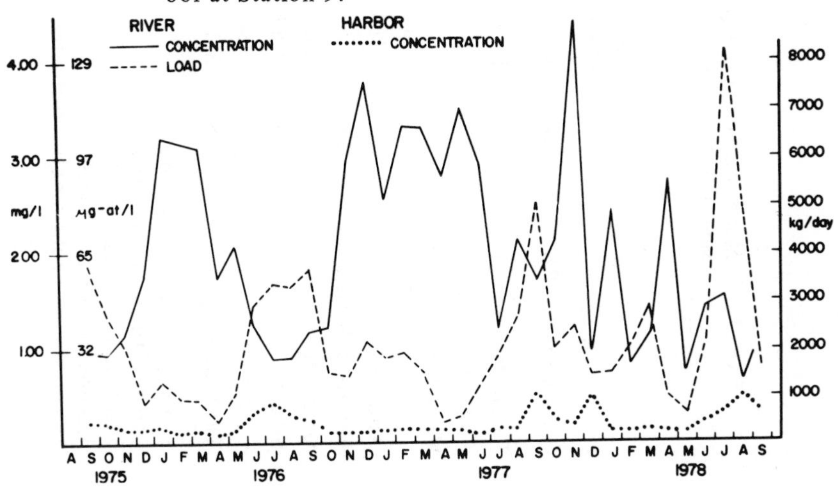

TABLE 1. Mean concentrations (mg/l as C, N, P, Si) of selected parameters in flowing freshwater streams sampled monthly.

PARAMETERS	PEACE RIVER[1] at Arcadia			HORSE CREEK[2] at Fla. 761 Bridge			BIG SLOUGH[3] at McCarthy Blvd. Bridge		
	N	Mean	SD	N	Mean	SD	N	Mean	SD
Total Organic Carbon	37	25.9	9.9	27	33.1	10.9	13	37.6	13.8
Organic Nitrogen	38	1.28	1.10	28	1.03	.63	13	1.40	.70
Particulate & Organic Phosphorus	27	.49	.28	23	.19	.12	11	.15	.16
Ammonia	38	.13	.13	28	.15	.19	13	.14	.09
Nitrate & Nitrite	38	.63	.41	28	.07	.10	13	.04	.05
Ortho-Phosphate	38	2.01	1.01	28	.49	.20	12	.24	.11
Silicate	38	2.57	1.18	28	1.94	1.19	13	2.47	.80
Color[4]	38	150	106	28	227	132	13	177	102
Turbidity[5]	38	3.2	2.1	28	1.3	.3	12	3.5	1.5

[1]12 Sep 1976 - 13 Sep 1978
[2]30 Jun 1976 - 13 Sep 1978
[3]10 Nov 1977 - 12 Sep 1978
[4]Expressed as Co-Pt units
[5]Expressed as nephelometric turbidity units.

Inorganic Nitrogen

There is a poorly defined seasonal variation in ammonium concentration in the Peace River but a distinct seasonal pattern for load (Figure 4). Greatest loads occur with high river flow during the summer. A

FIGURE 4. Ammonium concentration (as mg/ℓ N) and load in the Peace River at Arcadia and concentrations in Charlotte Harbor at Station 9.

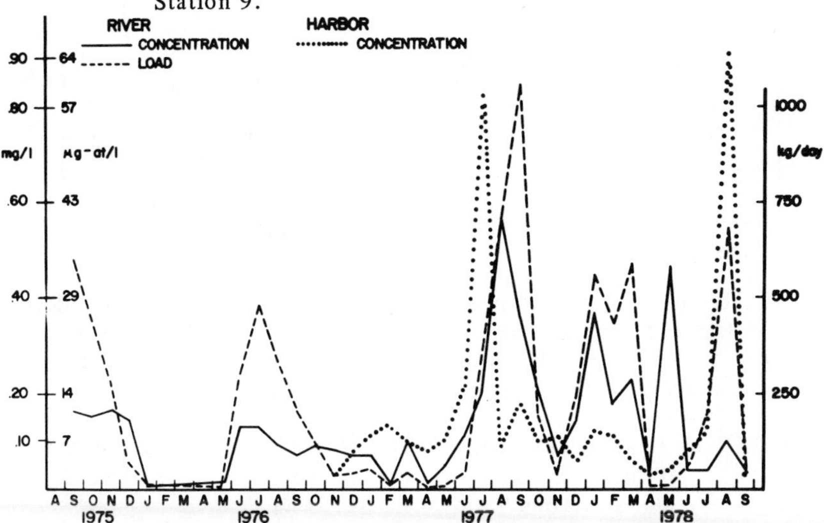

distinct peak was apparent during the unusually wet winter of 1978 when river flow was high. Ammonium concentrations in the harbor are strongly seasonal and tend to be higher than river concentrations. Variation of harbor concentration does not consistently follow either the river load or concentration. Other sources, such as regeneration from sediments, must be sought to explain seasonal variation of ammonium in Charlotte Harbor.

Highest concentrations of nitrate plus nitrite in the river usually occur in the dry season, but with a strong seasonal decrease near the end of the dry season in April and May (Figure 5). Peak river flows show a variable dilution effect. Loading is least during the end of the dry season and greatest some time during the wet season. Because of the effect of dilution during peak high flows, wet season loads can be variable. Harbor values of nitrate plus nitrite are very low, generally less than .02 $mg \cdot l^{-1}$ (1.4 µg $at \cdot l^{-1}$) and show virtually no seasonal variations.

The elevated nitrate plus nitrite concentration in the Peace River (Table 1), but not other area streams, may be primarily the result of discharges from phosphate chemical processing plants and municipal wastes. Our data, while lower in average concentration, are consistent with those presented by Slack and Goolsby (19) for the Peace River.

FIGURE 5. Nitrate plus nitrite concentration (as mg/ℓ N) and load in the Peace River at Arcadia and concentration in Charlotte Harbor at Station 9.

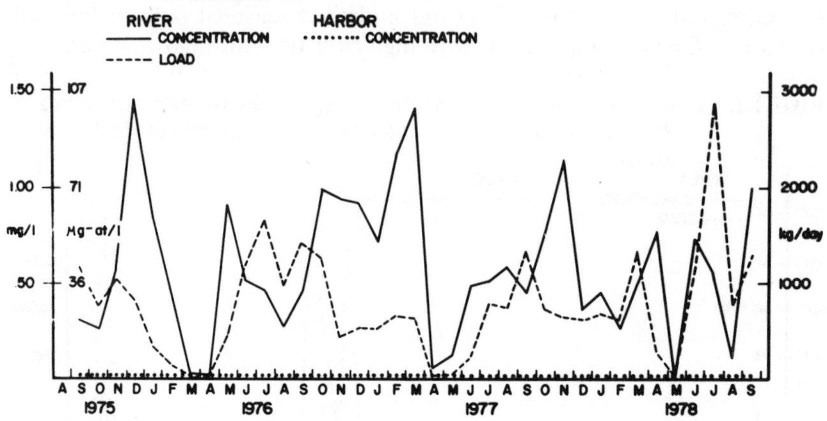

The behavior of nitrate plus nitrite concentrations along the chloride gradient shows that factors other than dilution rapidly decrease levels (Figure 8). We believe this is the result of biological activity in the fresh and low brackish water, principally from phytoplankton.

FIGURE 6. Reactive silica concentration (as mg/ℓ Si) and load in the Peace River at Arcadia and concentration in Charlotte Harbor at Station 9.

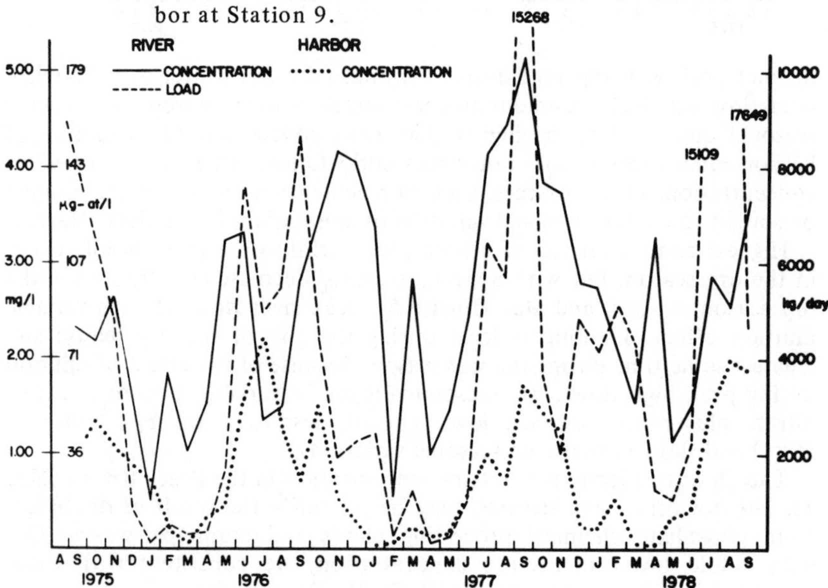

Reactive Silica

Silica concentration is variable with highest and lowest values occurring during each wet and dry season respectively. River loads are strongly seasonal with greatest loads occurring during the wet season (Figure 6). The harbor concentration curve is strongly seasonal and is somewhat similar to the load curve. However, even though river load increased sharply in 1977 and 1978, the harbor silica concentrations apparently did not respond by increasing to greater values. This observation is similar to that for phosphate (Figure 3). Some factor may be preventing the concentration from rising further, possibly increased estuarine circulation during peak flows with greater export of nutrients into the Gulf of Mexico.

Dilution curves along the chloride gradient show that although some concentration change is due to dilution, sharp decreases can occur with increasing chloride values (Figure 8). We believe these sharp decreases are the result of growing diatom populations that remove soluble silica in the presence of other adequate nutrients.

Color

Water color is strongly seasonal in the Peace River and Charlotte Harbor (Figure 7). Water color increases as river flow increases. This naturally-occurring high color (10) may reduce the potential primary productivity in the water column and on the bottom. Our evidence suggests that highly colored water results in lower productivity rates at or below 1-2 meters by reason of diminished light levels. Dilution along the chloride gradient appears to control the level of color in the harbor (Figure 8).

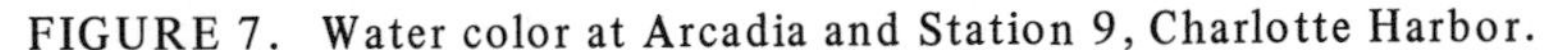

FIGURE 7. Water color at Arcadia and Station 9, Charlotte Harbor.

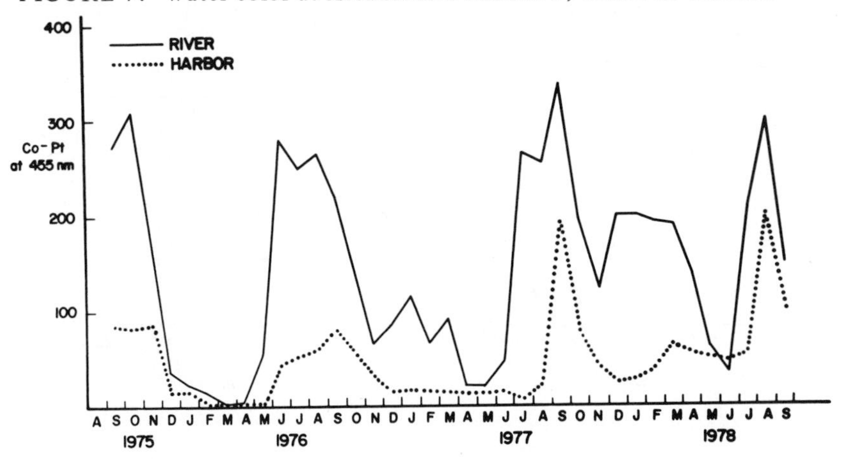

FIGURE 8. Dilution curves (17 stations between stations 2-14) for ortho-phosphate, ammonium ion, nitrate plus nitrite, silica (all as elemental mg/ℓ) and atomic ratios taken on April 11, 1979, near the end of the dry season and distribution of chlorophyll-a value. Chloride values are the same for each x-axis.

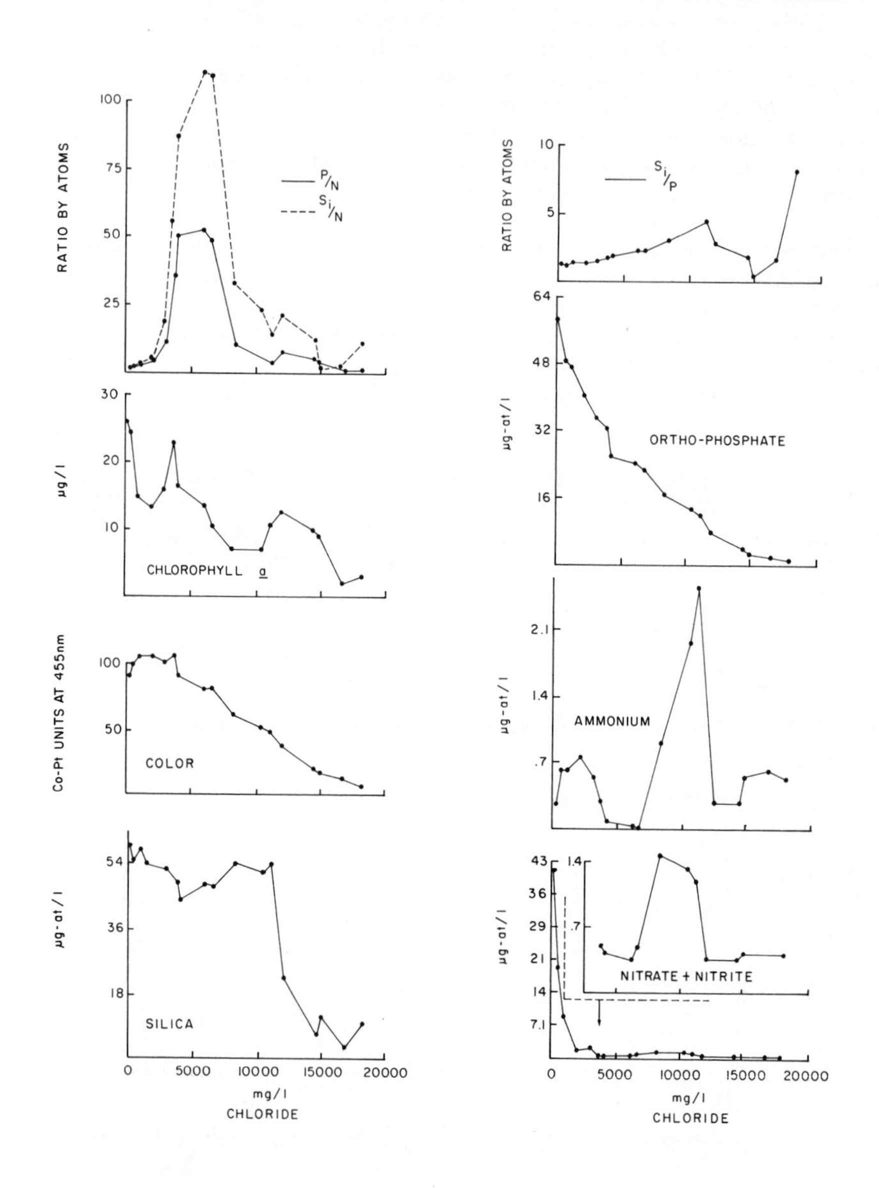

Phytoplankton

Primary production in Charlotte Harbor is seasonal but the patterns of seasonality varies from the river mouth the lower part of the harbor. Near the river mouth at Station 10, the estimated carbon fixation was high and variable throughout the year, with a major peak in November (Figure 9). Station 9, farther from the river mouth, showed a summer increase with a carbon peak in May and a major peak in October. In the lower part of the harbor, near the harbor mouth, Station 2 consistently had lower carbon values and had a major peak in April and a minor one in September. Productivity decreases from the river mouth toward the Gulf of Mexico, and the productivity curve for the harbor mouth is more like open coastal water.

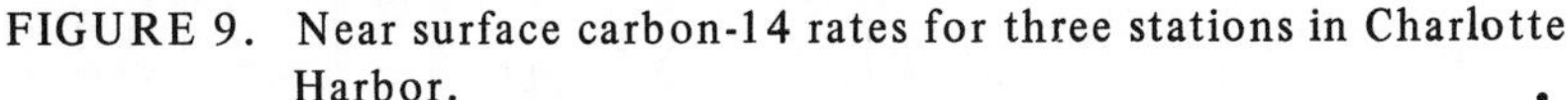

FIGURE 9. Near surface carbon-14 rates for three stations in Charlotte Harbor.

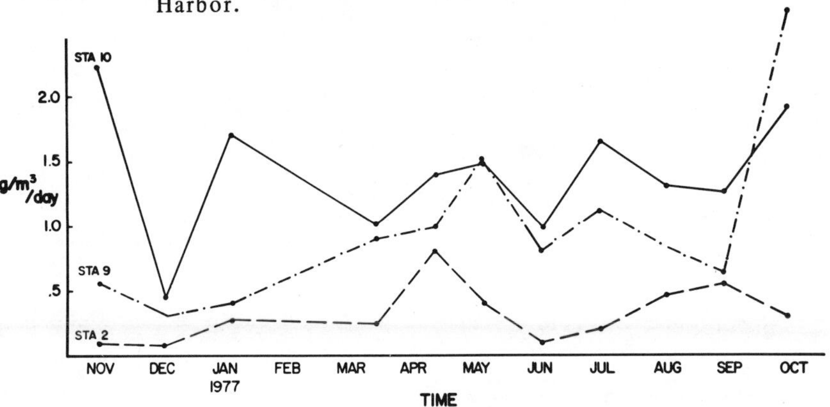

Table 2 presents a summary of the minimum and maximum values for selected parameters. These ranges are similar to those summarized by Steidinger (22) for other Florida areas. The carbon-14 values for near-bottom productivity generally were 10-35 percent of the near-surface rates during most of the year. When water color was least (May-June) in the harbor, near-bottom productivity was 6-124 percent of the near-surface rates, and when color was the greatest (September-October), near-bottom rates were 1-15 percent of the near-surface rates.

Peak cell numbers, volumes, and chlorophyll-a content did not coincide well with one another (Table 3) or with carbon-14 fixation rates (Figure 9).

The major phytoplankton group trhoughout the year were diatoms. Dinoflagellates and golden-brown algae occasionally became dominant at a few stations. Genera that accounted for 25 percent or more of the individuals at one or more stations were *Asterionella, Chaetoceros,*

TABLE 2. Near surface minimum and maximum phytoplankton values in Charlotte Harbor during 1976-1977.

	$\mu g\ C\ l^{-1}\ hr^{-1}\ ly^{-1}$	Total Cell Numbers $x\ 10^6\ l^{-1}$	Total Cell Volume μl^{-1}	Chlorophyll ($\mu g\ l^{-1}$) a	b	c
Nov	.0446 - 1.105	.95 - 4.5	.02 - 3.5	.99 - 8.29	.04 - 1.65	.14 - 4.14
Dec	.0710 - .370	.75 - 3.0	.20 -34.3	1.50 - 6.43	.38 - 1.54	.95 - 4.46
Jan	.0560 - .535	1.3 - 4.6	.55 -75.5	1.11 -23.7	.36 - 1.12	.95 -12.4
Mar	.0300 - .178	.74 - 2.1	.79 -14.7	1.02 - 6.14	.21 - .58	.61 - 2.53
Apr	.0541 - .296	1.4 - 4.0	.24 - 3.3	2.11 -15.4	.72 - 3.14	2.65 -11.4
May	.0188 - .234	.69 - 3.3	.06 -15.2	1.14 -12.5	.41 - 5.02	.75 -13.3
Jun	.0157 - .163	.88 - 2.4	.07 - 9.5	1.28 -10.1	.32 - 1.34	.97 - 4.92
Jul	.0278 - .293	1.1 -10.6	.05 -32.1	2.78 -24.9	.41 - 3.80	1.25 - 6.11
Aug	.0273 - .203	1.4 -23.2	.32 -53.3	2.13 -25.8	.45 - 5.43	1.30 -16.6
Sep	.0206 - .365	1.4 -34.3	.22 -18.2	3.43 -30.7	.29 - 1.20	1.61 -11.2
Oct	.0614 - .525	2.2 -14.7	.31 -14.9	5.16 -30.4	.13 - 1.05	1.88 -13.9

	Diatoms Number %	Volume %	Dinoflagellates Number %	Volume %	Other[1] Number %	Volume %	Number of Stations
Nov	6-28	6-93	<1-3	<1-52	72-94	1-94	8
Dec	11-73	49-99	<1-9	<1-8	27-83	<1-43	8
Jan	35-84	91-99	1-4	<1-5	16-64	<1-17	8
Mar	61-98	74-99	<1-3	<1	1-43	<1-25	8
Apr	65-86	70-98	<1-1	<1-5	13-34	2-27	8
May	55-84	63-99	<1-3	<1-7	16-44	<1-37	9
Jun	52-61	72-99	<1-1	1-7	32-48	<1-35	9
Jul	61-94	47-99	<1-3	<1-40	6-44	<1-39	9
Aug	56-98	90-99	<1-3	<1-4	2-41	<1-7	9
Sep	57-94	53-99	<1	2-35	7-43	1-42	10
Oct	57-93	17-99	<1-9	3-82	4-43	<1-30	10

[1] Blue-green algae less than 1%

Pleurosigma, Skeletonema, Nitzschia, and *Rhizosolenia.* Species that accounted for more than 50 percent of the total cell volume at one or more stations were *Biddulphia mobiliensis, B. regia, Rhizosolenia fragilisima, Skeletonema costatum, Coscinodiscus centralis, Guinardia flaccida, Leptocylindrus danicus, Pleurosigma strigosum, Gymnodinium breve,* and *Dinobryon* sp. These results suggest that diatoms still make up the bulk of the phytoplankton community even with the increased nutrient levels.

Spence (21) reported that Charlotte Harbor had lower diatoms counts than other Florida estuaries, with an average of about 0.534 x 10^6 cells·l^{-1}. She sampled four periods during 1970 (April-November). Our diatom counts ranged from 0.12 - 31.9 x 10^6 cells·l^{-1} and averaged 3.5 x 10^6 cells l^{-1} for the period November, 1976 to October, 1977. Our values were similar to the other estuaries that Spence cited as were values reported by Saunders et al. (18) for Charlotte Harbor from January to October, 1964.

The highest diatom counts for the harbor as a whole were measured in September after the highest river flow for the wet season in 1977 and averaged 18.3±10.2 x 16^6 cells·l^{-1} for Stations 2, 4, 5, 6, 7, and 9. Figure 2 shows that the September flow peak was the only significant pulse of fresh water and nutrients into the harbor during the 1977 wet season. Although this pulse of nutrients produced the highest diatom counts measured during 1977, it did not alter the phytoplankton community composition as illustrated in Table 2. Minimum counts for the same stations occurred in December and averaged 0.26±.09 x 10^6 cells·l^{-1}.

TABLE 3. Near surface values for three stations in Charlotte Harbor during 1976-1977.

	Sta. 2			Sta. 9			Sta. 10		
	No.	Vol.	Chl a	No.	Vol.	Chl a	No.	Vol.	Chl a
Nov	1.4	.15	1.0	2.3	3.5	3.9	3.6	.26	8.3
Dec	.75	.22	1.5	1.5	10.3	3.1	3.0	13.8	6.4
Jan	1.8	2.2	3.0	2.0	35.8	3.5	4.6	75.5	23.7
Mar	1.7	1.2	1.3	1.9	7.4	4.5	2.1	2.1	6.1
Apr	3.7	.24	4.7	2.4	.61	3.7	1.7	3.3	15.4
May	1.9	1.4	2.0	1.4	.06	3.5	3.3	.19	4.4
Jun	1.2	.61	1.7	2.4	.17	3.8	2.3	.18	5.0
Jul	10.6	2.9	6.1	2.4	.16	5.8	2.2	.38	12.7
Aug	1.4	.31	2.1	9.7	17.6	12.2	15.2	5.4	25.8
Sep	8.0	3.7	4.6	21.6	10.0	14.9	4.5	1.1	23.0
Oct	5.5	8.2	7.1	14.7	6.6	30.4	8.8	5.2	13.0

No. = Number (x 10^6 l^{-1})
Vol. Cell = Volume (μl^{-1})
Chl a = Chlorophyll a (mg m^{-3})

Peak chlorophyll-a values moved along a transect from fresh water toward the Gulf of Mexico in response to changes in river flow (Table 4). The peaks remained in the riverine portion (Stations 12, 14, 18) of the estuary during low flow periods and moved into the harbor (Stations 4, 7, 9, 10) during high flow. Since nutrient inputs to the harbor of ammonia silica and phosphate increase in the wet season, the increase in chlorophyll-a content may be interpreted as partially a nutrient response.

Inorganic Nutrient Ratios

The average nutrient values and atomic ratios demonstrate that high levels of phosphate occur in the Peace River and Charlotte Harbor (Table 5). When we follow the assumption of 10:1 N:P atomic ratio for phytoplankton assimilation (17), then both harbor Stations 2 and 9 would show an excess of phosphorus present after all nitrogen is assimilated. At Station 9, 1.44 $\mu g\text{-at}\cdot l^{-1}$ of phosphorus and at Station 2, 0.68 $\mu g\text{-at}\cdot l^{-1}$ could be taken up at the 10:1 ratio, leaving 5.69 and 2.29 $\mu g\text{-at}\cdot l^{-1}$ respectively of available phosphorus unused.

In Charlotte Harbor, inorganic nitrogen may be approximately .2 to 14.7 times as abundant (by atoms) as phosphorus (Table 5). However, if the generalization about the ratios of nitrogen to phosphorus in marine algae (17) holds for estuaries, then nigrogen limitations may be expected to occur in parts of the harbor during portions of an annual cycle. Ratios (by atoms) taken near the end of the dry season along the chloride gradient show a decrease for inorganic nitrogen atoms that apparently do not occur for silica or phosphorus (Figure 8) in the region between 2,000 - 10,000 $mg\cdot l^{-1}$.

Inorganic nitrogen is frequently more abundant (1.2 to 9.7 times by atoms) than silica during the dry season in Charlotte Harbor. At these times silica limitations could occur for diatoms. The dilution curves (Figure 8) show that available silica atoms drop sharply and become scarce with respect to nitrogen between 11,000 - 16,000 mg/ℓ chloride.

The distinctions that Pomeroy et al. (15) made between clear and turbid estuaries are important. Charlotte Harbor is a relatively clear body of water and neither the Peace nor Myakka rivers typically carries much suspended material. The fresh water, however, is highly colored as a result the harbor has characteristics more like a turbid estuary, i.e., light limited, sufficient nutrients, and tendency toward low dissolved oxygen. In addition, Charlotte Harbor becomes stratified each wet season (June-October) layering highly colored water at the surface, both helping to depress bottom dissolved oxygen (unpublished data).

Our data, together with the other data cited, demonstrate that inorganic phosphate and nitrate plus nitrite concentrations are higher in the Peace River than in adjacent tributaries and streams. Reactive silica concentrations do not appear very different in the Peace River than in adjacent tributaries and streams. Seasonal changes in harbor concentrations that are related to high river flow occur for phosphate and silica, but not for nitrate plus nitrite. Ammonium concentrations show changes in the harbor, but these changes apparently are not related to river-derived ammonium.

Ortho-phosphate is in excess of biological requirements in the harbor as a result of the rich natural and anthropogenic sources (particularly phosphate strip mining and chemical fertilizer plants) in the Peace River. This means that among the three macro-nutrients, only inorganic nitrogen and silica could become limiting factors. The analysis of atomic ratios indicate that both inorganic nitrogen and silica can become scarce in Charlotte Harbor.

TABLE 4. Chlorophyll-a values (mg/m^3) along A transect from freshwater in the Peace River (Station 18) to lower Charlotte Harbor (Station 2).

Cumulative Distance km	Station	1977 Feb	Apr	May	June	July	Aug	Sept*	Oct	Nov	Dec*
0	18	1.0	5.8	8.5	13.1	4.8	.9	.6	1.6	6.0	1.3
6.8	14	3.6	20.9	21.3	24.1	12.0	1.1	1.2	2.3	11.0	3.6
14.4	12	16.7	21.0	23.1	19.1	11.8	15.6	12.3	7.9	199	5.1
22.6	10	10.3	12.3	8.4	9.8	11.9	34.9	53.3	16.4	61.3	25.0
30.2	9	2.1	2.7	11.5	7.6	7.0	8.7	23.5	47.1	13.5	14.6
36.7	7	2.0	2.2	1.7	4.3	5.2	16.0	27.3	15.0	6.1	23.6
46.7	4	2.2	2.8	2.6	1.9	3.5	2.1	10.6	7.1	8.8	13.9
54.8	2	1.9	1.3	3.2	1.6	6.0	.7	5.3	7.3	2.7	9.4

km	Station	1978[1] Jan	Feb	Mar*	Apr	May	June	Aug*	Sept*
0	18	2.9	.8	3.2	37.6	18.5	4.9	-	.1
6.8	14	2.6	1.5	3.0	47.3	80.4	6.3	1.9	.1
14.4	12	18.0	9.4	3.4	16.9	29.4	27.1	5.3	-
22.6	10	17.5	1.9	15.5	10.5	16.8	15.3	12.2	8.6
30.2	9	8.5	7.0	22.9	8.4	9.7	12.0	18.2	22.1
36.7	7	7.8	8.4	21.3	6.9	10.0	13.9	40.1	11.4
46.7	4	9.1	6.3	12.1	5.0	5.0	6.8	7.8	1.5
54.8	2	4.7	7.0	7.1	3.1	3.1	1.6	4.3	9.0

* High flow periods.

1 No data for July.

TABLE 5. Mean inorganic nutrient values (μg-at· ℓ$^{-1}$) and ratios by atoms to phosphorus for November, 1976 to September, 1978.

Peace River at Arcadia

	P	N	Si
n	23	23	23
$\bar{x}$	62.26	53.77	92.81
SD	35.21	27.60	54.32
Mean Ratio[1]	1	.78	1.34
Range of Ratios	-	.06-1.81	.20-4.03

Harbor Station 9

	P	N	Si
n	23	23	23
$\bar{x}$	7.13	14.36	23.90
SD	4.22	18.24	22.82
Mean Ratio[1]	1	2.51	3.27
Range of Ratios	-	.22-14.7	.24-6.88

Harbor Station 2

	P	N	Si
n	23	19	23
$\bar{x}$	2.97	6.82	9.90
SD	2.67	3.32	7.17
Mean Ratio[1]	1	2.14	4.44
Range of Ratios	-	.31-6.89	.24-13.9

[1]Ratios for each date were determined and then averaged.

ACKNOWLEDGEMENTS

Richard L. Iverson and David W. Conally of Florida State University performed the carbon-14 and other laboratory analyses of the phytoplankton study. Larry G. Maurer and Alexander Padva of our laboratory supervised all chemical analyses and Steven W. Osborne was in charge of the field work. This work was funded by General Development Corp., Miami, Florida as part of it on-going studies of water and water quality issues in Charlotte Harbor.

REFERENCES

1. Alberts, J.J. 1970. Inorganic controls of dissolved phosphorus in the Gulf of Mexico. Ph.D. dissertation. Florida State University, Tallahassee, 1-88.
2. Alberts, J., H. Mattraw, R. Hariss, and A. Hanke. 1970. Studies on the geochemistry and hydrography of the Charlotte Harbor estuary, Florida. Charlotte Harbor Estuaries Studies, Mote Marine Lab., Sarasota, 1-34.
3. Browder, J. 1977. The estuary viewed as a dynamic system, 144-170. *In*: W. Seaman, Jr. and R. McLean (eds.), Freshwater and the Florida Coast: Southwest Florida. Sea Grant Report 22.
4. Coleman, J.M. 1979. Present and past nutrient dynamics of a small pond in southwest Florida. Ph.D. dissertation, University of Florida, Gainesville, 157 pp.
5. Donnelly, P.V., R.A. Overstreet, M.A. Burklew, and J.H. Viulle. 1967. A chemical study of southwest Florida river water, 1965-66. 98-141. *In* Red Tide Studies, Pinellas to Collier Counties 1963-66. Fl. Bd. Conserv. Prof. Pap., Series 9.
6. Dragovich, A., J.A. Kelly, Jr., and H.G. Goodell. 1968. Hydrological and biological characteristics of Florida's west coast tributaries. Fish. Bull. 66(3):463-477.
7. Gilliland, M.W. 1973. Man's impact on the phosphorus cycle in Florida. Ph.D. dissertation. University of Florida, Gainesville, 268 pp.
8. Graham, H.W., J.M. Amison, and K.T. Marvin. 1954. Phosphorus content of waters along the west coast of Florida. U.S. Fish Wildlife Serv., Special Scient. Rept. Fish. 122, 1-43.
9. Hobbie, J.E. 1974. Ecosystems receiving phosphate wastes, 252-270. *In* H.T. Odum, B.J. Copeland and E.A. McMahan (eds.), Coastal Ecological Systems of the United States. Vol. III, Conservation Foundation, Washington, D.C.
10. Kaufman, M.I. 1969. Color of water in Florida streams and canals. U.S. Geol. Surv. Map Series 35.
11. ————————. 1975. Generalized distribution and concentration of orthophosphate in Florida streams. U.S. Geol. Surv. Map Series 33 (revised).

12. LaRock, P. and H.L. Bittaker. 1973. Chemical data of the estuarine and nearshore environments in the eastern Gulf of Mexico, IIC, 8-86. *In* J.I. Jones, R.E. Ring, M.D. Rinkel, and R.E. Smith (eds.), A Summary of Knowledge of the Eastern Gulf of Mexico. State Univ. System Florida Institute of Oceanography, St. Petersburg.
13. Martin, D.F. and Y.S. Kim. 1977. Long term Peace River characteristics as a measure of a phosphate slime spill impact. *Water Resource* 11:963-970.
14. Odum, H.T. 1953. Dissolved phosphorus in Florida waters. Fla. Geol. Surv. Rept. Invert. 9(1):1-40.
15. Pomeroy, L.R., L.R. Shenton, R.D.H. Jones, and R.J. Remold. 1972. Nutrient flux in estuaries, 274-291. *In* G.E. Likens (ed.), Nutrients and Eutrpohication: The Limiting-Nutrient Controversy. Special Symposia. Limnol. Oceanogr. 1.
16. Rand, M.C., A.E. Greenberg, and M.J. Taras (eds.). 1976. Standard Methods for the Examination of Water and Wastewater, 14th Ed., American Public Health Association, Washington, D.C.
17. Ryther, J.H. and W.M. Dunstan. 1971. Nitrogen, phosphorus, and eutrophication in the coastal marine environment. *Science* 171:1008-1013.
18. Saunders, R.P., B.I. Birnhak, J.T. Davis, and C.L. Wahlquist. 1976. Seasonal distribution of diatoms in Florida in shore waters from Tampa Bay to Caxambas Pass, 1963-1964. 48-78. *In* Red Tide Studies, Pinellas to Collier Counties, 1963-66. Florida Board Conservation Prof. Pap. Series, No. 9.
19. Slack, L.J. and D.A. Goolsby. 1976. Nitrogen loads and concentrations in Florida streams. U.S. Geol. Surv. Map Series 75.
20. Specht, R.C. 1950. Phosphate waste studies, Fla. Industr. Engrg. Expt. Sta. Bull 32, Engr. Prog. Univ. Fla. 4(2):1-28.
21. Spence L.L. 1971. *Cyclotella* Kutzing Taxonomic and ecological relations of the diatom genus in south Florida waters. Florida State University, M.S. thesis. 1-80.
22. Steidinger, K.A. 1973. Phytoplankton, III-E, 1-17. *In*: J.I. Jones, R.E. Ring, M.O. Rinkel, and R.E. Smith (eds.), A Summary of Knowledge of the Eastern Gulf of Mexico. State University System Florida Institute of Oceanography, St. Petersburg.
23. Strickland, J.D.H. and T.R. Parsons. 1972. A practical handbook of seawater analysis, 2nd Ed. Fish Res. Bd. Canada, Bull. 167, 310 pp.
24. Taylor, J.C. 1975. The Charlotte Harbor estuarine system. Florida Scientist 37(4):205-216.
25. U.S. Dept. Interior. 1976-1979. Water resources data for Florida U.S. Geol. Surv., Tallahassee.

A SUGGESTED APPROACH FOR DEVELOPING ESTUARINE WATER QUALITY CRITERIA FOR MANAGEMENT OF EUTROPHICATION

Norbert A. Jaworski* and Orterio Villa, Jr.**
U.S. Environmental Protection Agency
**Environmental Research Laboratory*
Diluth, MN 55804

***Region III, Annapolis Field Office*
Annapolis, Md. 21401

ABSTRACT: A conceptual approach for developing water quality criteria for eutrophication management is suggested. The three basic components of the framework include source ambient relationships, effects, and impact analyses. The approach focuses on a conceptual method for developing decision-making criteria as opposed to the classical water quality criteria of a single value of limitation. The approach to developing water quality criteria for eutrophication management provides an analysis framework of response relationships which can be readily incorporated into water quality standard-setting processes that include environmental considerations and technological and economic factors.

INTRODUCTION

The objective of the U.S. Federal Water Pollution Control Act, as amended, is to "Restore and maintain the chemical, physical, and biological integrity of the Nation's waters." An integral part of the Act is the promulgation of water quality standards and the issuing of water quality criteria accurately reflecting the latest scientific knowledge on the kind and extent of all identifiable effects on health and welfare which may be expected from the presence of pollutants in any water body.

The three major scientific considerations in developing water quality criteria for a pollutant in a given water body and water use are:

1. An analysis of the sources, concentration, dispersal, transformation, and availability through physical, biological, and chemical processes;

2. The kind and extent of all identifiable effects on the health and welfare of humans and their environment;
3. The impact of these effects on humans and on the diversity, productivity, stability, and utility of the biological community.

The above three considerations are generally applicable to the developing of water quality criteria for any pollutant.

APPLICATION TO EUTROPHICATION

Since the 1960s, considerable research effort has been conducted to aid in developing sound approaches to the control of eutrophication in fresh water systems, with considerably less effort on estuarine systems. With prudent use much of the information and analysis techniques developed for fresh water systems can be applied to estuarine systems.

The overall goal of eutrophication control can be generally stated as "to maintain a balance between the various nutrients and a balance of the biota conducive to the production and/or protection of a healthy ecosystem." Some general indicators of excessive eutrophication in marine systems and fresh water systems are:

1. An increase in the dissolved oxygen deficit in the hypolimnion in lakes, or in the lower water column depths in estuaries;
2. An increase in dissolved solids (not for marine systems), especially nutrients, such as nitrogen and phosphorus;
3. An increase in suspended material, especially organic;
4. A shift from diatom-dominated plankton populations to those dominated by blue-green or green algae;
5. A decrease in light penetration;
6. An increase in the concentration of organic compounds and nutrients in the benthic deposits;
7. A potential production of toxins mainly from massive blue-green algal blooms in fresh water systems and from "red tides" in marine systems;
8. An increase in biomass and primary production;
9. The presence of nuisance seasonal algal blooms.

In the management of eutrophication, the development of specific water quality criteria is much more difficult to formulate than the management of most toxic pollutants. This difficulty arises in that overenrichment may have both beneficial and detrimental impacts on an ecosystem. In addition, no definitive indicators have been developed which can be used as specific "end points" to assess a given criteria. Some indicators often used in eutrophication management are the maintenance of a given chlorophyll concentration, productivity level, or transparency level. In developing specific requirements of eutrophication management for fresh water systems, limits on phosphorus and nitrogen are often suggested.

Many approaches in developing nutrient criteria in fresh water are applicable to estuarine systems. However, there are considerable differences between fresh water and estuarine systems including nutrient cycling, transport and transformation, physical hydrodynamics, indigenous species, and degree of salinity intrusion.

Although considerable research has been conducted in determining how to remove nutrients from wastewater discharges and how to better understand the ecosystem, little effort has gone into determining to what degree the ecosystem should be managed--that is, to quantify water quality criteria for eutrophication control, including control cost and ecosystem response and benefits. This paper discusses an approach to aid in quantifying water quality criteria for eutrophication management requirements. The approach focuses on a conceptual method for developing decision-making criteria as opposed to the classical water quality criteria of single value limitation.

VARIOUS APPROACHES TO DEVELOPING WATER QUALITY MANAGEMENT CRITERIA

Review and Extension of Historical Data

Various investigators studying nuisance conditions in fresh and estuarine water systems have suggested criteria for nitrogen and phosphorus needed to control algal blooms. In a study of the Occoquan Reservoir on a tributary of the Potomac, Sawyer (21) recommends limits of inorganic nitrogen and inorganic phosphorus of 0.35 and 0.02 mg/l, respectively. Mackenthun (13) cites data indicating the upper limits of inorganic nitrogen at 0.3 mg/l and inorganic phosphorus at 0.01 mg/l at the start of the growing season to prevent algal blooms. Pritchard (17), studying the Chesapeake Bay and its tributaries, suggests that if the total phosphorus concentration in estuaries is less than 0.03 mg/l, biologically "healthy" conditions will be maintained. Jaworski, et al. (10), reviewing the historical data for the upper Potomac estuary, suggests that if concentrations of inorganic phosphorus ranging from 0.03 to 0.1 mg/l and inorganic nitrogen from 0.3 to 0.5 mg/l were maintained in the upper estuary, a chlorophyll level of about 25 μg/l could be maintained under normal conditions during the summer growing season. There are no current federal water quality criteria for estuarine waters (18).

Use of Indices

Another approach to developing water quality management criteria is the integration of numerous water quality and ecosystem parameters into an index. The use of indices to set eutrophication control requirements for studies has been mainly limited to fresh water ecosystems (19). Included in the relationships are nitrogen and phosphorus concentration, transparency, turbidity, chlorophyll, dissolved oxygen, and productivity. Frank F. Hooper (8) suggests several criteria for a good index. It should differentiate between changes in enrichment levels and seasonal and climatic cycles. It should be sensitive to enrichment levels and indicate the presence or absence of organisms over a period of time to reduce the extent of monitoring. It should indicate values that are widespread geographically and sensitive to cultural changes, and it should be practical to use in terms of simplicity, interpretation, and collection of data. McErlean (12) expands further on this concept for marine systems.

Use of Ecosystem Mathematical Models

In both freshwater and marine systems, mathematical models of ecosystems have been used to help delineate eutrophication control management requirements. Approaches ranging from simple to very complicated models have been applied to the Potomac estuary (10), Chesapeake Bay (20), and Narragansett Bay (14) in marine systems and to fresh water systems such as the Great Lakes (5). Further elaboration of ecosystem estuarine modeling is presented by O'Connor (15).

Subjective Analysis

To aid in developing water quality criteria in freshwater systems, three terms have been suggested: oligotrophic, mesotrophic, and eutrophic. These subjective terms were used by Vollenweider (23) in suggesting total phosphorus loading in grams per square meter of surface area per year that would be "permissible" or "excessive" for a lake ecosystem. Included in the approach were physical parameters such as mean depth of the lake and hydraulic residence time. This approach was applied to the Potomac estuary by Jaworski (9) as part of the North American project of OECD. The effort was further extended by Lee and Jones (11).

Use of Benefit/Cost Analysis

One of the first attempts at using this method, employing both cost of nutrient control and benefit of nutrient control in developing water quality criteria, was for the San Joaquin drainage basin by Bain, et al., in 1968 (1). In this approach, rather than trying to set a nutrient level based solely on historical data, the use of indices, or subjective analyses, an attempt was made to relate the nutrient level in the system to a given biomass and then relate the biomass to a dollar value of benefits.

Analysis of Various Approaches

In each of the above approaches, except for the benefit/cost analysis, a "single value" parameter usually evolves; e.g., a suggested maximum concentration level for phosphorus or nitrogen. As suggested in a recent report by the Council on Wage and Price Stability (4), water quality criteria should reflect "estimates of the effects on human and aquatic life of different concentrations of various substances instead of setting maximum suggested concentration levels." The different levels (dose-response) can be more rapidly utilized in developing water quality standards which involve determining the use attainable for specific bodies when environmental, technological, and economic factors are taken into consideration. While all of the above approaches can aid in developing water quality criteria for estuarine systems, they all are somewhat lacking (in the opinion of the authors) in addressing the three major scientific considerations presented earlier in this paper.

The establishment of the U.S. Regulatory Council also reflects the current concern in this country to more rigorously determine environmental benefits from environmental control systems. Therefore, it is now more prudent than ever to emphasize the third component of developing water quality criteria: the impact on the ecosystem. The next sections present a suggested procedure for developing water quality for eutrophication management for estuarine systems.

BASIC CONSIDERATIONS FOR DEVELOPING WATER QUALITY CRITERIA FOR EUTROPHICATION MANAGEMENT

Although the approach presented in this paper is mainly conceptual, the various relationships suggested are those often utilized in ecosystem studies. This approach links the key relationships into a hierarchical framework allowing for successive decision making.

Listed below are some of the basic considerations that were used in developing the approach:

1. Inclusion of key physical/chemical/biological processes of estuarine systems;
2. Relationships to key existing water quality parameters (e.g., dissolved oxygen);
3. Provision of trade-off options for management consideration;
4. Indication of the sensitivity of the various relationships;
5. Translation into simple and inexpensive monitoring parameters upon implementation;

6. Readily presentable to and understandable by decision makers.

Considerable effort has been made to keep the framework relatively simple. If future studies indicate that other relationships are needed, they may be added to the conceptual framework.

The proposed framework includes the following basic elements:

1. Physical attributes of the estuary;
2. Designated water uses of the estuary;
3. Nutrient source/ambient relationship;
4. Nutrient/aquatic biomass relationship;
5. Biomass/oxygen resource relationship;
6. Biomass/estuarine productivity relationship;
7. Biomass/aesthetics relationship.

In the approach that follows, the three general scientific considerations required in developing water quality are linked in a framework through the above elements.

SUGGESTED FRAMEWORK FOR DEVELOPING WATER QUALITY CRITERIA

Segmentation of the Estuarine System

The first step in this approach to developing water quality criteria is to segment the estuary according to its physical, chemical, and biological properties including designated water uses. For example, Lake Erie was divided into three segments based on depth as part of the International Joint Committee eutrophication study. Factors which should be taken into consideration are the hydrodynamics, salinity, and the various uses of the water such as recreation, oystering, and fishing. A segment surface area/depth relationship should be developed as shown in Figure 1. This relationship has utility in the analysis of oxygen resources and in the delineation of those areas which have important aquatic considerations, including commercial and recreational potential. For example, those segments which mainly attract recreational use can be examined critically for aesthetic aspects.

Nutrient Source/Ambient Relationship

For a given nutrient and segment, the development of a source/ambient water column concentration relationship (as shown in Figure 2) indicates the chemical, biological, and physical response of an ecosystem to external sources. Such responses can be developed by examining existing data and by using simple to very complex models. Because of uncertainty,

FIGURE 1. Typical Segment Surface Area/Depth Relationship.

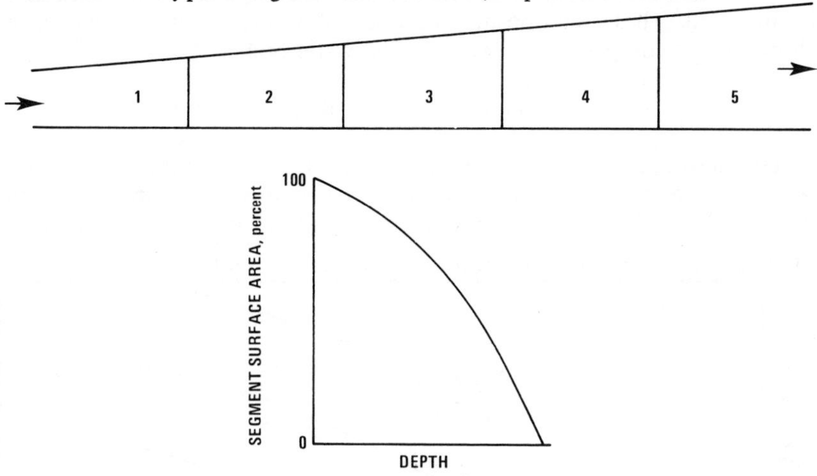

high, most probable, and low estimates should be developed. These types of relationships have been developed for the Great Lakes (22).

FIGURE 2. Source/ambient water column concentration relationship.

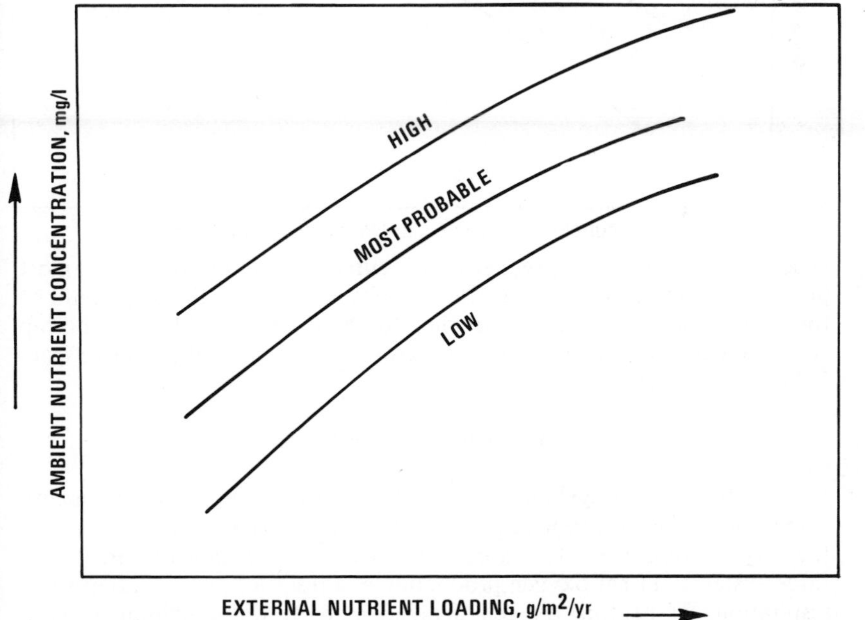

This relationship can be for annual loading in grams per square meter of surface area per year or on a more short term basis such as seasonally or monthly. If atmospheric or benthic sources are important, they can also be included. These relationships have an additional utility in that they allow an easy comparison between similar water bodies.

Estimate of Nutrient/Biomass Relationship

After having developed the source/ambient relationship, the next step is to develop a nutrient concentration versus biomass potential for each segment in a given time frame (Figure 3). For this relationship, it is

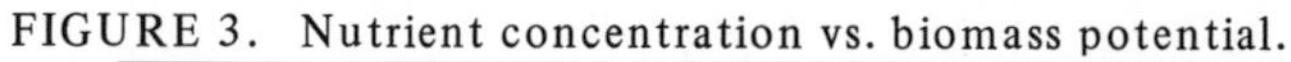
FIGURE 3. Nutrient concentration vs. biomass potential.

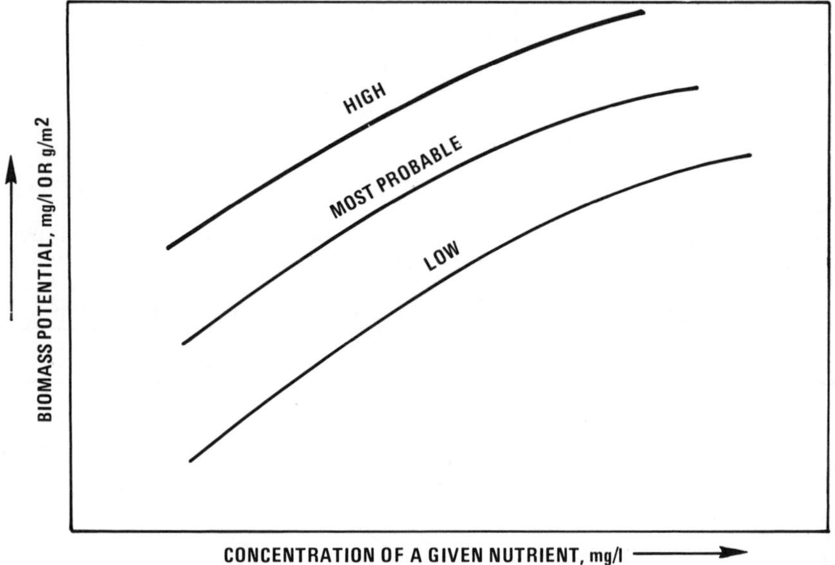

necessary to develop a sequence of curves including the high, most probable, and low estimates for various seasons. This relationship could be for rooted aquatic and/or for planktonic organisms. The relationship could be developed either from laboratory chemostat studies and/or from field observations.

Oxygen Resources Considerations

As in fresh water systems, one of the key parameters in water quality management for eutrophication control is the impact of the biomass on the oxygen resources. This approach divides the oxygen resources into three systems: 1) net oxygen production as a result of photosynthesis and respiration, 2) benthic oxygen demand, and 3) water column oxygen demand. Treating the three systems separately is facilitated by incorporating the surface area/depth relationship shown in Figure 1.

FIGURE 4. General net oxygen production/depth relationship.

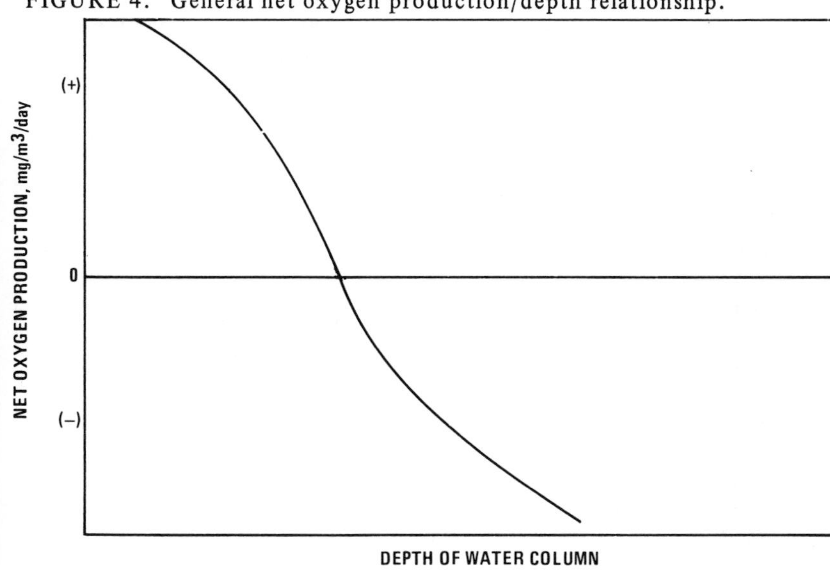

Figure 4 presents a general net oxygen production/depth relationship due to either planktonic or rooted aquatic growth. Examples of this were used in the San Joaquin study (1) and in the Potomac estuary studies (3).

The second component of the oxygen resources considerations is a general benthic oxygen demand/time relationship for given biomass concentrations as shown in Figure 5. Studies of the Narragansett Bay (16), suggest that relationship as shown in Figure 5 can be developed for each segment of the estuary. Additional verification of this can be developed from benthic respirometer studies such as those conducted in the Potomac estuary (10) and Narragansett Bay (16).

The third component is the estimate of water column oxygen demand attributed to the eutrophication process. Analyses of data in field studies of the Potomac estuary (3), Narragansett Bay (16), Albemarle Sound (2) and New York Bight (7) have shown that a large percentage of inorganic carbon is converted to the organic form during eutrophication. Eutrophication not only significantly increases the amount of organic carbon in these estuarine ecosystems, but also can significantly add to the water column dissolved oxygen demand. Relationships, as shown in Figure 6, can be developed for each segment of the estuarine system.

By using the theory of superposition, a general total oxygen deficit/biomass relationship for a given season can be evolved as shown in Figure 7. This total relationship includes the photosynthesis/respiration, water column oxygen demand, and benthic oxygen demand terms. It may also be possible to develop this relationship directly as suggested by the efforts developed by Dobson (6) for Lake Erie.

FIGURE 5. General benthic oxygen demand/time relationship.

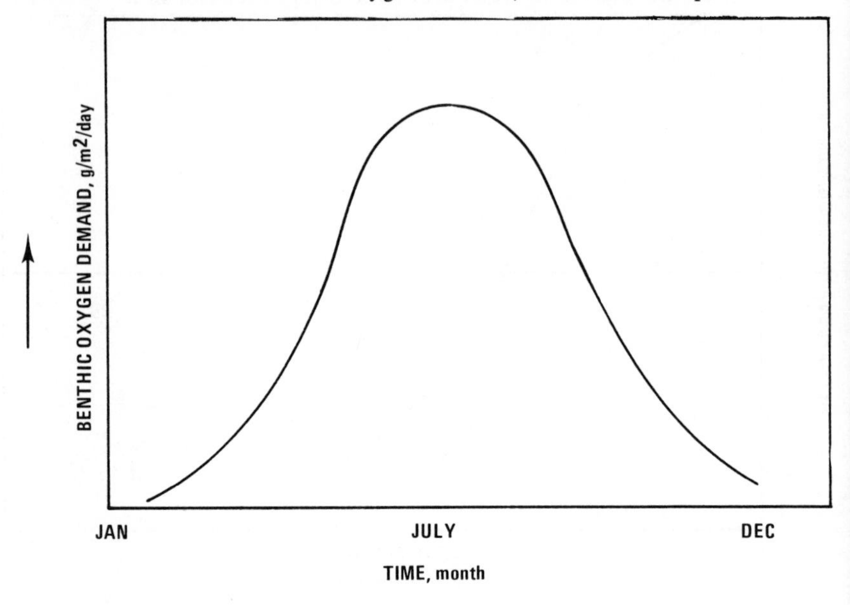

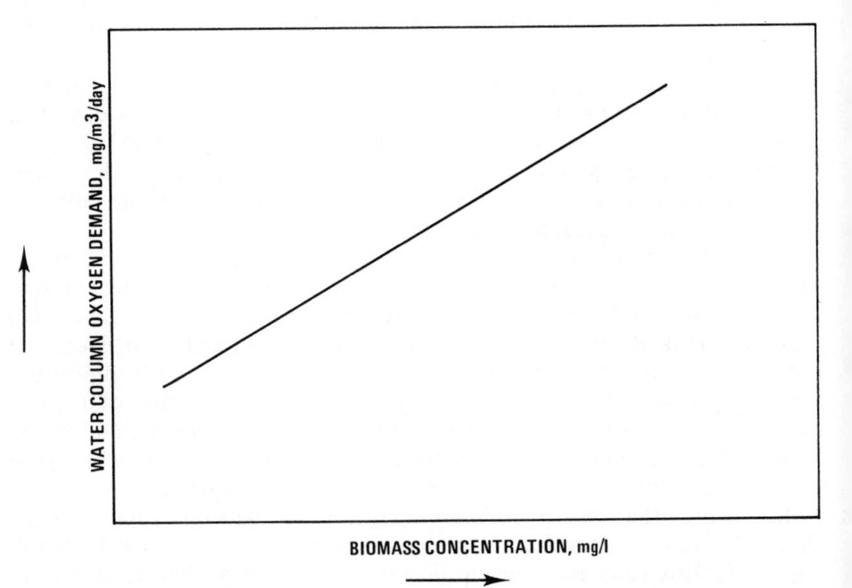

FIGURE 6. Water column/dissolved oxygen demand relationship.

FIGURE 7. General total oxygen deficit vs. biomass relationship (by superposition).

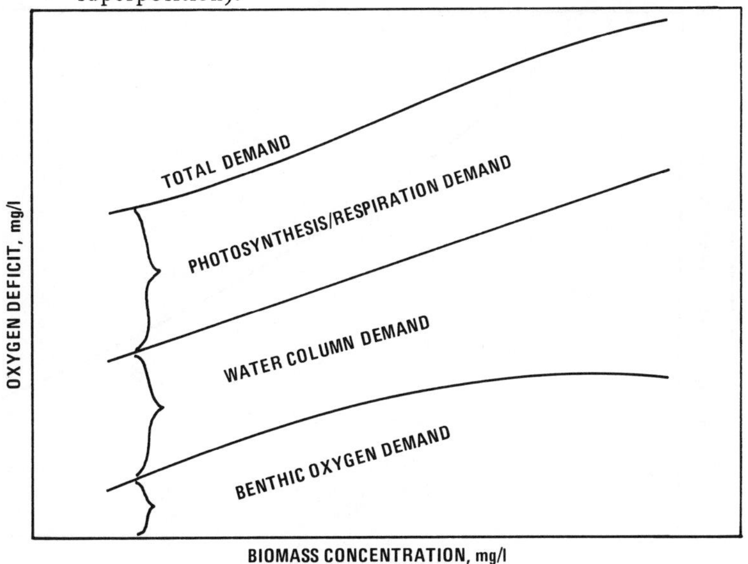

Ecosystem Impact

This approach incorporates three ecosystem impact considerations. Figure 8 shows a general relationship between biomass concentration, dissolved oxygen concentration, and surface area. The dissolved oxygen concentration/biomass relationship is a further incorporation of the surface area/depth relationship shown in Figure 1, and the relationships developed in Figures 4-7. These relationships permit quantification on an area basis of the impact of biomass on the dissolved oxygen resources of the segment. Further expansion of this concept is required to fully implement this impact.

A second impact, indicated in Figure 9, is the transparency/biomass component, a major aesthetic and ecosystem stress consideration when considering such activities as boating and swimming. Further expansion of this impact could include the water transparency versus nutrient relationship developed by Lee and Jones (11).

A major indicator of the eutrophication process is the increase in the biomass of the ecosystem and its biological value. This increase can have a beneficial or detrimental impact on the ecosystem food chain depending on the "desirability" of the increase in the food chain. If the ecosystem is at a desired balance, it can be argued that there will be a minimal buildup in unusable cellular material. It can further be argued that if the cellular material increase is not readily assimilated in the food chain, the increase

FIGURE 8. General biomass concentration vs. dissolved oxygen concentration vs. surface area concentration.

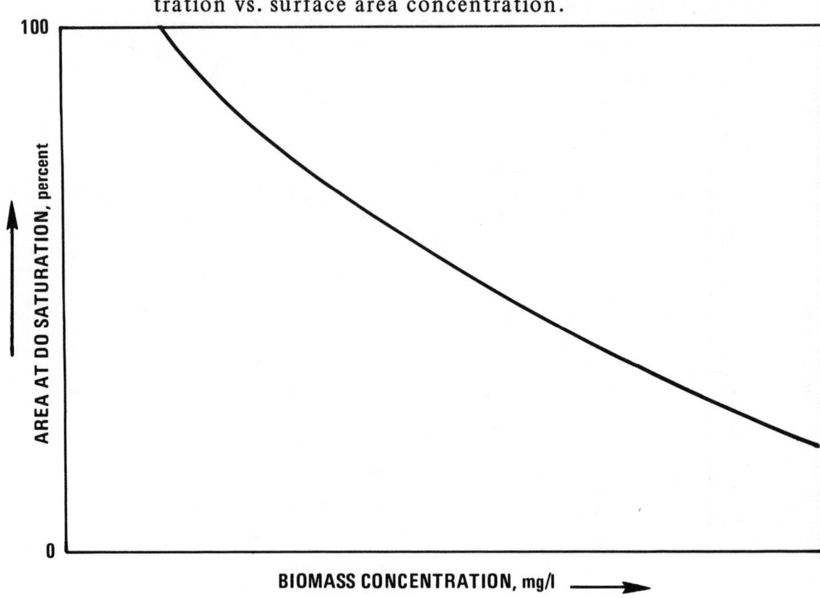

FIGURE 9. Transparency vs. biomass relationship.

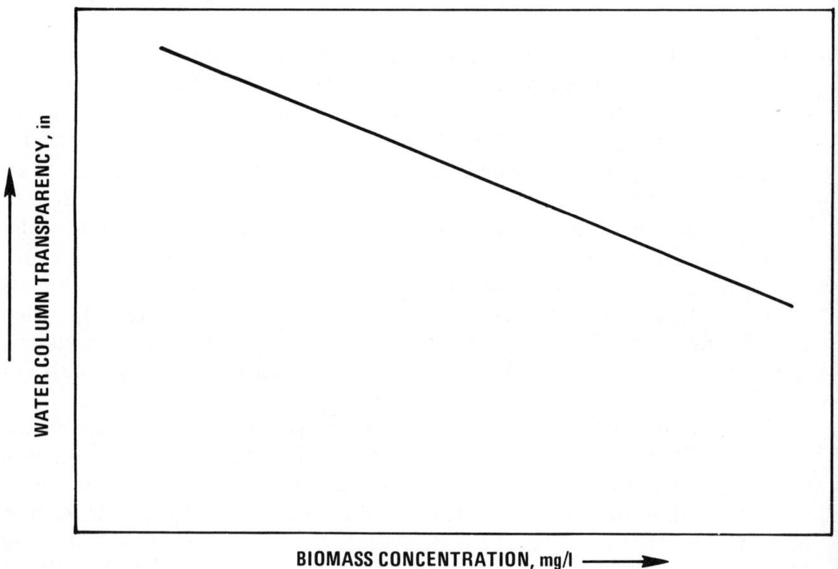

will be particulate cellular material, as indicated in the particulate total organic carbon measurement. Figure 10 suggests a relationship between

FIGURE 10. Biomass vs. particulate total organic carbon relationship.

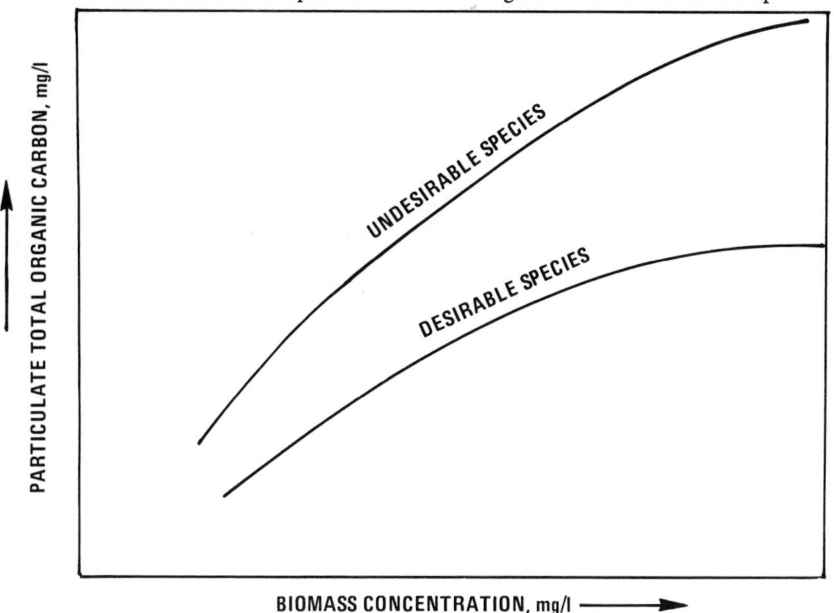

biomass and particulate total organic carbon. Further expansion of this concept is required to fully implement the incorporation of this impact.

INTEGRATION OF VARIOUS RELATIONSHIPS

The integration of the general relationships developed, as presented in Figures 2-10, is incorporated into an overall analysis framework as shown in Figure 11. This integration includes the three general major scientific considerations employed in developing water quality for any pollutant.

The framework can be considered a hierarchical analysis flowing from Level I (source/ambient relationships) to Level II (ecosystem effects) to Level III (ecosystem impacts). In this hierarchical concept, the complexity and interdependence increase as the level of analysis increases. However, it allows decisions to be made at each level, depending on the degree of analysis required.

To illustrate the utility of the approach, three hypothetical cases are presented in Tables I-III which could represent typical conditions in east coast estuaries. Hypothetical Case I represents the response of three

FIGURE 11. Overall Analysis Framework.

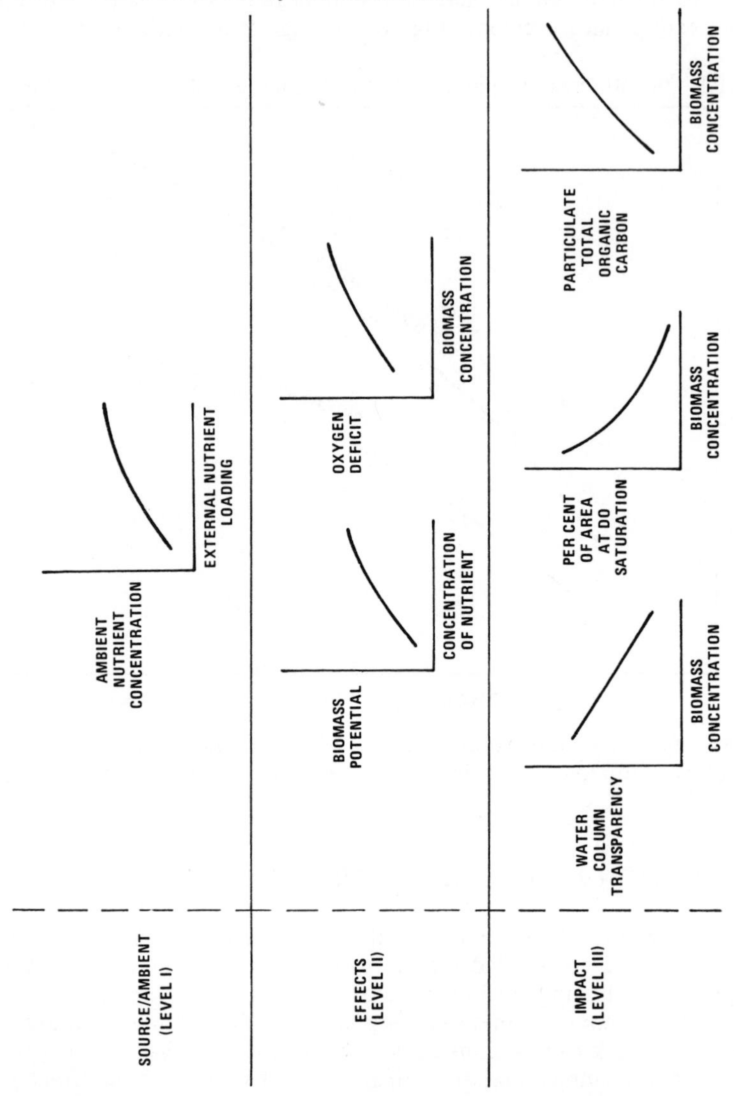

phosphorus loadings on the three ecosystem impact considerations. In this case, a desired aquatic community is assumed. Case II is similar to Case I except the aquatic community is assumed to be "undesirable." Case III is the same as Case I except the average depth is 15.0 m instead of 5.0 m. (A further assumption is that the nutrient and biomass concentrations are the same).

TABLE 1. Hypothetical Case I. - (Average Depth = 5.0 m) (Major Use: Recreation and Commercial Fishing)

Ecosystem Impact	Phosphorus Loading, $g/m^2/yr$ 1.0	2.0	5.0
Transparency, in.	40	30	10
Percent of Area at DO Saturation	90	90	90
Particulate Organic (mg/l) Carbon ("desirable" community)	2	3	5

Cases I and II represent the trade-off between recreation and desirable productivity (commercial fishing). Case 1 would favor recreation while Case II does not. Case II would probably not have any impact on commercial fishing. The trade-off is between an increase in particulate organic carbon and the decrease in transparency. Cases I and III represent the trade-off between desirable productivity and oxygen resources. The trade-off is between an increase in food chain organic carbon in Cases I and III and a decrease in oxygen resources in Case III. The decrease in oxygen resources could significantly impact commercial fishing. These three hypothetical cases vividly represent the need to include ecosystem impact in the decision framework.

This analysis framework (Figure 11) overcomes one of the historically most difficult aspects of developing water quality criteria for eutrophication management - the inability to develop well defined parameters indicative of the impact of overenrichment. Instead of the maximum concentration concept, this approach provides an analysis framework of response relationships which can be readily incorporated into the water quality standard-setting process to include not only environmental considerations but also technological and economic factors.

TABLE 2. Hypothetical Case II. - (Average Depth = 5.0 m) (Major Use: Recreation and Commercial Fishing)

Ecosystem Impact	Phosphorus Loading, $g/m^2/yr$ 1.0	2.0	5.0
Transparency, in.	50	40	30
Percent of Area at DO Saturation	90	90	90
Particulate Organic (mg/l) Carbon ("undesirable" community)	5	7	12

While the approach does not yield specific values for "excessive" or "permissible" concentrations for eutrophication parameters, it does provide a framework for analysis on a multi-level response basis. It also

TABLE 3. Hypothetical Case III. - (Average Depth 15.0 m) (Major Use: Recreation and Commercial Fishing)

Ecosystem Impact	Phosphorus Loading, $g/m^2/yr$ 1.0	2.0	5.0
Transparency, in.	40	30	10
Percent of Area at DO Saturation	80	60	30
Particulate Organic (mg/l) Carbon	2	3	5

provides for a trade-off between the various impacts of overenrichment which may be both beneficial and detrimental to a given ecosystem.

SUMMARY

A conceptual approach for developing water quality criteria for eutrophication management is suggested. This approach incorporates:

1. A multi-level hierarchical framework for developing water quality criteria for eutrophication management on a multi-response basis.
2. Key nutrient source/ambient, ecosystem effects, and ecosystem impact relationships.
3. Linkages which can be used in a simplified form or expanded into ecosystem models.
4. Uncertainties of the various relationships.

To aid in further development of water quality criteria for eutrophication management on a scientific basis, the following are suggested:

1. Greater development of ecosystem impact relationships due to overenrichment. Three possible relationships presented in this approach require further development and field testing.
2. Greater development and application of simple and inexpensive monitoring parameters, including such possible measurements as various nutrient forms, chlorophyll, and various forms of total organic carbon to aid in ecosystem impact analysis.

The approach has reduced many complex ecosystem relationships into simple forms. The intent is to put development of water quality criteria aspects in a usable form for decision-makers. It is also intended to provide an opportunity for the biologist, chemist, engineer, and hydrodynamist to understand the linkages between the various scientific fields required in developing water quality criteria for eutrophication management.

It is not the intent of this approach to minimize the complexities of the physical, chemical and biological processes of eutrophication; rather the intent is to establish a simplified and orderly approach so that future research can be organized into a framework that is logical, uncomplex, and understandable. The goal of this approach is to aid in developing a scientific basis for the developing of water quality criteria for the management of overenrichment of estuarine systems.

REFERENCES

1. Bain, Richard C., Jr., et al. 1968. Effects of the San Joaquin master drain on water quality of the San Francisco Bay and Delta. Appendix, Part C: Nutrients and biological response. U.S. Department of Interior. San Francisco, California.
2. Bowden, William B. and John E. Hobbie. 1977. Nutrients in Albemarle Sound, North Carolina. University of North Carolina Sea Grant College publication UNC-SG-75-25.
3. Clark, Leo J. and Stephen E. Roesch. 1978. Assessment of 1977 water quality conditions in the Upper Potomac Estuary. U.S. Environmental Protection Agency, Annapolis Field Office.
4. Council on Wage and Price Stability. Executive Office of the President News. 1979. Comments of the Council on Wage and Price Stability before the Environmental Protection Agency. Water Quality Criteria. 23 pp.
5. DiToro, M.C., et al. 1978. Report on Lake Erie mathematical model. U.S. Environmental Protection Agency, Ecological Research Series.
6. Dobson, H.F.H. 1976. Eutrophication status of the Great Lakes. Canada Centre for Inland Waters. Burlington, Ontario.
7. Garside, C., et al. 1976. Estuarine and coastal marine science, 4(3):281-289.
8. Hooper, Frank F. 1969. Eutrophication indices and relation to other indices of ecosystem change. *In* Eutrophication: Causes, Consequences, Correctives. Proceedings of a Symposium, National Academy of Sciences, Washington, D.C.
9. Jaworski, N.A. 1977. Limnological characteristics of the Potomac Estuary. *In* Les Seyb and Karen Randolph (comps). North American project--a study of U.S. water bodies. A report for the Organization for Economic Cooperation and Development. Office of Research and Development, Environmental Protection Agency, Corvallis, Ore.
10. Jaworski, Norbert A., et al. 1971. Water resource-water supply study of the Potomac Estuary. U.S. Environmental Protection Agency Water Quality Office, Technical Report No. 35.
11. Lee, G. Fred and R. Anne Jones. 1979. Application of the OECD eutrophication modeling approach to estuaries. Paper presented at the International Symposium on the Effects of Nutrient Enrichment in Estuaries. Williamsburg, Va.

12. McErlean, Andrew J. and Gale J. Reed. 1979. On the application of water quality indices to the detection, measurement and assessment of nutrient enrichment in estuaries. Paper presented at the International Symposium on the Effects of Nutrient Enrichment in Estuaries. Williamsburg, Va.
13. Mackenthun, K.M. 1965. Nitrogen and phosphorus in water. U.S. Public Health Service. Department of Health, Education and Welfare.
14. Nixon, Scott W. 1979. Remineralization and estuarine nutrient dynamics. Paper presented at the International Symposium on the Effects of Nutrient Enrichment in Estuaries. Williamsburg, Va.
15. O'Connor, Donald J. 1979. Nutrient and phytoplankton analysis in estuaries. Paper presented at the International Symposium on the Effects of Nutrient Enrichment in Estuaries. Williamsburg, Va.
16. Olsen, Stephen and Virginia Lee. 1979. A summary and preliminary evaluation of data pertaining to the water quality of Upper Narragansett Bay. Coastal Resources Center, University of Rhode Island.
17. Pritchard, Donald W. 1969. Dispersion and flushing of pollutants in estuaries. *Journal of Hydraulics Division, ASCE* 95(HY1).
18. Quality criteria for water. 1976. U.S. Environmental Protection Agency Report No. EPA-440/9-76-023 (NTIS No. PB 263943). Washington, D.C.
19. Rast, Walter and G. Fred Lee. 1978. Summary analysis of the North American (U.S. portion) OECD eutrophication project: Nutrient loading - lake response relationships and trophic state indices. University of Texas at Dallas Center for Environmental Studies. Richardson, Texas.
20. Salas, Henry J. and Robert V. Thomann. 1978. A steady-state phytoplankton model of Chesapeake Bay. *Journal of Water Pollution Control Federation* 50(12):2752-2770.
21. Sawyer, C.N. April 1970. 1969 Occoquan Reservoir study. Metcalf and Eddy, Inc. for Commonwealth of Virginia Water Control Board.
22. Vallentyne, J.R. and N.A. Thomas, Canadian and U.S. Co-Chairmen. 1978. Report of Task Group III, a Technical Group to review phosphorus loadings. Fifth year review of Canada-United States Great Lakes Water Quality Agreement.
23. Vollenweider, R.A. 1968. Scientific fundamentals of the eutrophication of lakes and flowing waters, with particular reference to nitrogen and phosphorus as factors in eutrophication. OECD Technical Report DAS/CSI68(27). Paris, France.

THE SIGNIFICANCE OF DREDGING AND DREDGED MATERIAL DISPOSAL AS A SOURCE OF NITROGEN AND PHOSPHORUS FOR ESTUARINE WATERS

R. Anne Jones and G. Fred Lee
Department of Civil Engineering,
Environmental Engineering Program
Colorado State University
Fort Collins, Col. 80523

ABSTRACT: Several hundred million cubic meters of waterway sediment are dredged each year in the United States to maintain adequate navigation depth. A significant part of this dredging and dredged sediment disposal takes place in freshwater tidal, estuarine, and marine waters. U.S. waterway sediments contain sufficient quantities of nitrogen and phosphorus compounds which, if released in available forms during dredged material disposal, could stimulate the growth of excessive amounts of algae and other aquatic plants in estuaries. This paper discusses the results of that portion of a comprehensive study conducted by the authors which assessed the significance of dredged sediments as a source of nutrients for U.S. estuarine waters.

Laboratory studies were conducted on waterway sediments from over 20 marine, estuarine, and freshwater tidal areas to evaluate the suitability of the elutriate test for estimating contaminant release from dredged sediments during open water disposal. Disposal operations involving sediments from nine of these areas were monitored for nutrient release to the water column.

Sediment kjeldahl nitrogen concentration generally ranged from $\sim$ 200 to 4,000 mgN/kg with a mean concentration of $\sim$ 1,550 mgN/kg. Elutriate tests generally showed a release of soluble nitrogen (ammonia plus nitrate) to 0.5 to 35 mgN/l. Total phosphorus concentrations of the sediments studied averaged $\sim$ 950 mgP/kg and ranged from $\sim$ 100 to 3,750 mgP/kg. Response of sediment-associated and soluble phosphorus to elutriation ranged from a decrease in soluble ortho P concentration by 0.6 mgP/l to an increase by 1.6 mgP/l. Generally, relatively small amounts of nitrogen and phosphorus are released during elutriation and would be expected to be released during open water disposal of hydraulically dredged sediments; this was substantiated by intensive field monitoring of disposal operations. While there is no relationship between the bulk sediment content of N or P and the amounts of available forms released in elutriate tests, the maximum sediment nitrogen released in available forms was about 1 percent; maximum soluble orthophosphate released was less than 0.1 percent of the total P content of the sediment. Even less release would be expected during barge-scow disposal of mechanically dredged sediment.

This study demonstrated that modified elutriate tests provide an indication of sediment-associated contaminant behavior during open water disposal of hydraulically dredged sediment. The key to the proper use of these procedures is having a knowledge of the hydrodynamic characteristics of the disposal area, the limiting nutrient in the area of concern, and, for phosphorus, a knowledge of the iron system in the sediments being dumped.

INTRODUCTION

One of the potential sources of nutrients which has been of concern in the eutrophication of surface waters is dredged sediment. On the order of 400 million cubic yards of sediments are dredged each year in the development and maintenance of United States waterways. A substantial part of this material is dredged from and disposed of in freshwater tidal and estuarine areas. U.S. waterway sediments contain highly variable levels of nutrients, typically on the order of several hundred to several thousand mg/kg N and P. These quantities are sufficient to stimulate the growth of excessive amounts of algae and other aquatic plants if they were released in available forms during dredging and disposal.

Because of the concern about algal available nutrient release associated with dredged sediment disposal, this topic was included as one aspect of a five-year study undertaken by the authors under sponsorship of the Corps of Engineers Dredged Material Research Program (DMRP). This study consisted of two parts. One part was devoted to evaluating the release of a wide variety of chemical contaminants including available forms of nutrients (ammonia, nitrate, and soluble orthophosphate) in the laboratory using bulk sediment analysis and the elutriate test. The second part involved determination of the amounts of these contaminants released to the water column during a number of dredged material disposal operations. The second part was also designed to assess the ability of the laboratory tests to predict contaminat release. The detailed results of these studies and discussion of their significance to dredged material disposal have been presented in a number of reports and papers (1, 2, 3, 5, 6, 7, 8). This paper discusses the results of that portion of the study conducted by the authors designed to determine the significance of dredged sediments as a source of nutrients for U.S. estuarine waters. Data obtained from parts of the James River and New York Bight studies are discussed herein to illustrate the types of information collected.

FIGURE 1. Sampling locations.

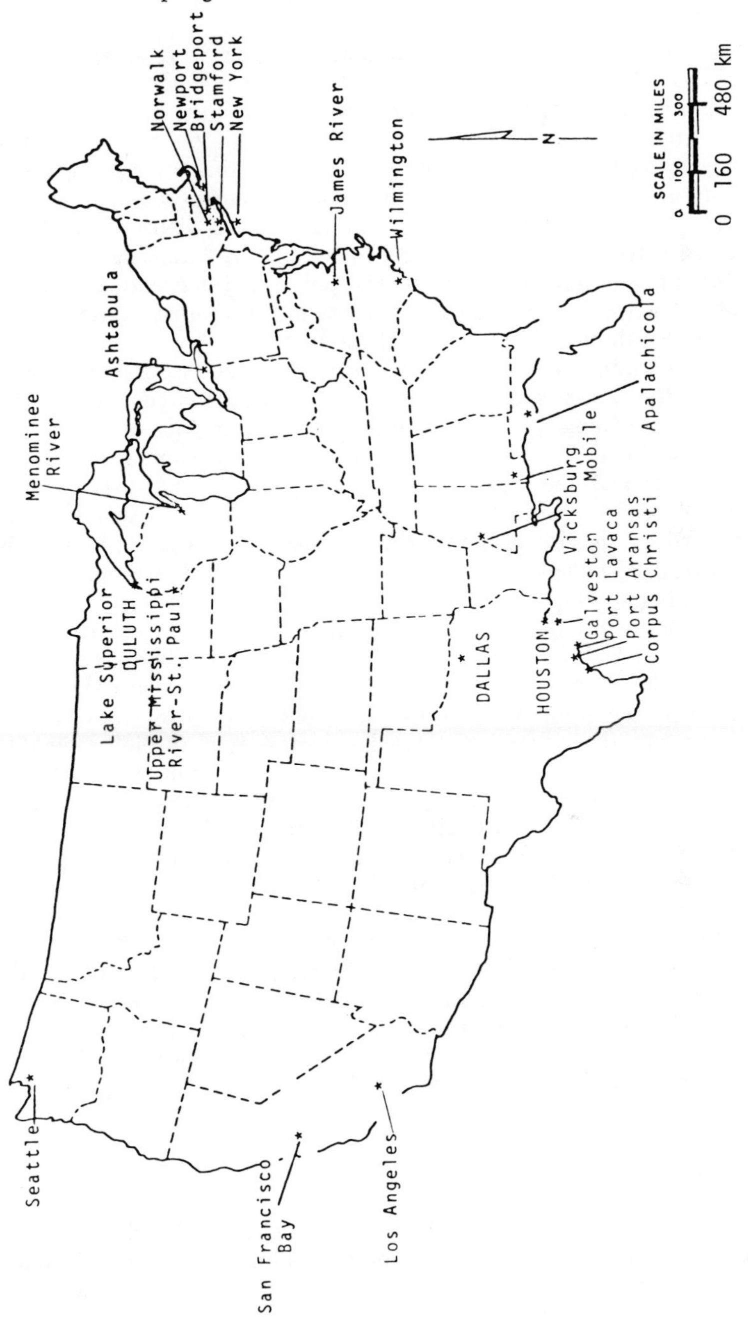

LABORATORY AND FIELD STUDIES

The laboratory studies were conducted on over 50 sediments from the approximately 20 areas across the U.S. shown in Figure 1. Freshwater tidal and estuarine systems investigated included areas on the Gulf coast, North Atlantic coast, and the western U.S. coast. More than 325 elutriate tests were conducted during this study in evaluating and refining this test procedure, which consists of combining one volume of sediment with four volumes of water, mixing for 30 minutes using vigorous aeration or bubbling with nitrogen gas, settling for one hour, decanting and filtering the supernatant through a half micron pore size filter, and analyzing the filtrate. Each sediment processed through the elutriate test was also analyzed for about 30 bulk chemical parameters.

Field studies were conducted during about 20 disposal operations at nine of the areas shown in Figure 1. Hopper dredge dumping of hydraulically dredged sediment and/or barge dumping of mechanically dredged sediment were studied in Elliott-Bay Puget Sound, Seattle; New York Bight; and the Galveston Bay Entrance Channel disposal area in Texas. Pipeline disposals of hydraulically dredged sediment were monitored in the James River (near Hopewell), Virginia; Mobile Bay, Alabama; Apalachicola Bay, Florida; and the upper Mississippi River near St. Paul., Minnesota.

In order to monitor a hopper dredge or barge dump event, sampling vessel(s) were positioned down current from a marker buoy such that the turbidity plume created by the disposal would pass beneath the sampling vessel. Near the dump-disposal site the turbidity is a conservative tracer of contaminants that may be released, so maximum turbidity would be expected to coincide with maximum released contaminant concentration. For hopper dredge or barge disposal operations, water samples were collected shortly before (as early as 30 minutes before) disposal, in rapid succession (as frequently as every 15 to 20 seconds) during passage of the turbid plume created by the disposal, and for as much as an hour to 1.5 hours following the dump, depending on the persistence of the increased turbidity. For pipeline disposal operations, water samples were collected directly down current from the discharge, perpendicular to the direct downstream path, and upstream from the discharge in a nearby area not affected by the dredging or disposal operations. Presented below is a discussion of data collected in connection with two site studies to demonstrate the types of results found.

NEW YORK BIGHT

The Mud Dump site in the New York Bight off the New York-New Jersey coast, receives on the order of 10 million cubic meters of dredged material each year from New York-New Jersey waterways, in addition to chemical and other waste materials. As discussed by Lee and Jones (4), this region had experienced a massive fish die-off due to low dissolved oxygen water resulting from algal growth and death.

Monitoring vessels were positioned downcurrent from the dump site as shown in Figure 2, such that the turbid plume of dredged sediment would pass beneath them for sampling. Figure 3 shows changes in turbidity in the bottom waters at the Mud Dump site at the position of the sampling vessel Hatton during disposals of sediment from Perth Amboy channel (No. 1 - hopper disposal), and Perth Amboy Anchorage (No. 2 - barge disposal); the dump events are obvious. Turbidity in the bottom waters decreased to ambient levels within about one hour after the dump. At this location the ammonia concentrations (which include ammonium and unionized ammonia) were found to increase coincident with the increase

FIGURE 2. Positions of sampling vessels at Mud Dump Site during monitoring of New York Bight Dump No. 1.

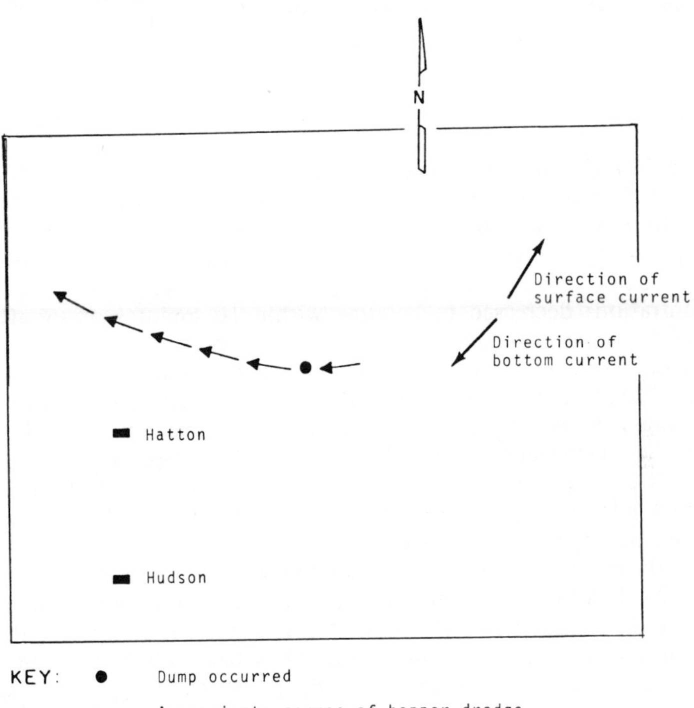

FIGURE 3. Turbidity in near bottom water during New York Bight Dump Nos. 1 and 2.

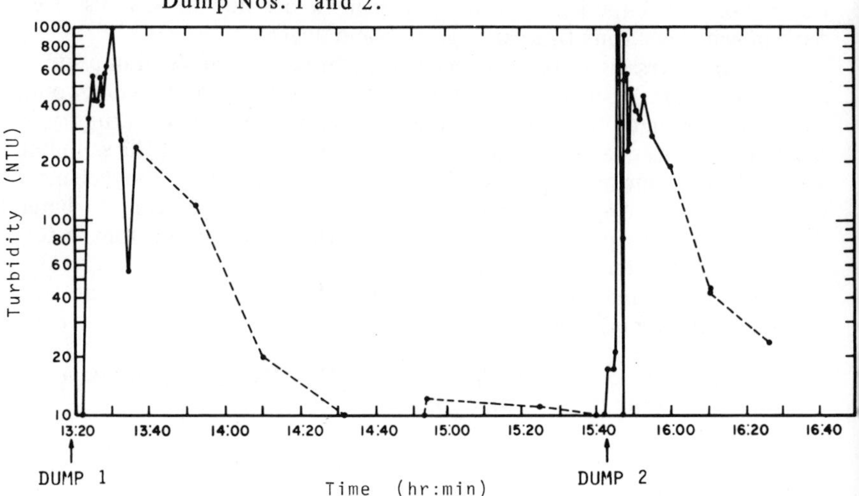

in turbidity (Figure 4). Concentrations reached about 1.5 mg/l N and ∿ 2.5 mg/l N in association with the two dumps and returned to ambient at the dump site within 10 to 20 minutes of the dump. Soluble orthophosphate concentrations (Figure 5) increased to about 200 to 250 μg/l P from a baseline of about 50 μg/l P. With the first dump, concentrations decreased to baseline within 10 minutes or so; after the second dump they remained elevated for the duration of sampling.

Elutriate tests run on the sediments being dumped in the New York Bight during the monitored dumps (Perth Amboy Channel and Anchorage) showed releases of ammonia on the order of 15 to 30 mgN/l. While the magnitude of release in the elutriate test was considerably greater than what was found during the passage of the turbid plume, the fact that release occurred was predicted. Further, the elutriate test was designed to approximate concentrations found in the hopper bins rather than in the open water where large amounts of dilution can occur.

The Perth Amboy sediment anoxic elutriate tests showed soluble ortho P release to about the level found at the disposal site, whereas the oxic ones showed uptake. This is in accord with what could be expected when anoxic sediments are dumped into oxic waters. Phosphate would be released from the sediment during the anoxic mixing period during dredging and in the hopper in transit to the disposal site. When it comes in contact with oxic waters, the reduced iron would be oxidized and sorb the released phosphate, causing a decrease in phosphate concentration in the water. This seemed to be the case with the New York Bight dumps, as evidenced by the rapid decrease in concentration of soluble ortho P between two sampling vessels about 100 m apart, both downcurrent from another New York Bight dump.

FIGURE 4. Ammonia concentrations in near bottom water during New York Bight Dump Nos. 1 and 2.

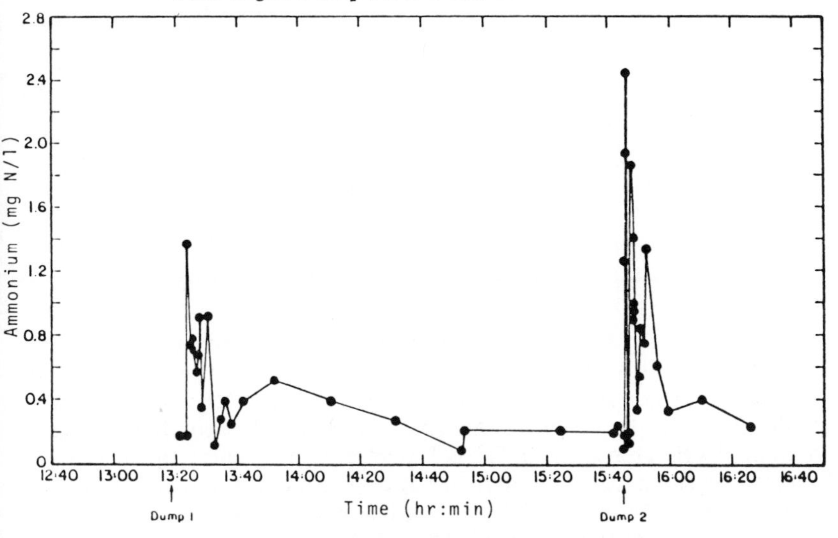

Similar nutrient release patterns were found at other hopper dredge disposal sites; there was typically an increase in concentration followed by a rapid decrease to ambient levels. At the Galveston Bay Entrance Channel disposal site (7) a number of water samples were collected over a several year period in order to detect long term contaminant release from the deposited sediments. There was no indication that there was sufficient release of contaminants from the deposited dredged sediments to affect water quality in the disposal area.

JAMES RIVER

The primary concern at the James River dredging-disposal area near Hopewell, Virginia, was the release of sediment-associated kepone during dredging. Monitoring of this pipeline disposal of hydraulically dredged sediments, however, included measurement of aquatic plant nutrients as well. Figure 6 shows the dredging-disposal area and sampling locations. A drogue designed to move with the near bottom current was released at the discharge point. Samples of water were collected both in the path of the drogue to measure variations in concentrations with time-distance, and also in a number of other areas perpendicular to the path of the drogue and upstream from the discharge to determine the extent of the impact of the disposal.

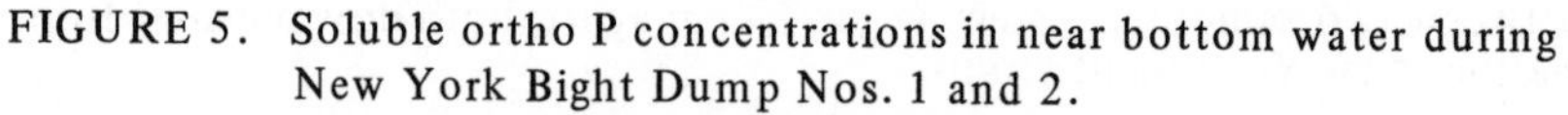

FIGURE 5. Soluble ortho P concentrations in near bottom water during New York Bight Dump Nos. 1 and 2.

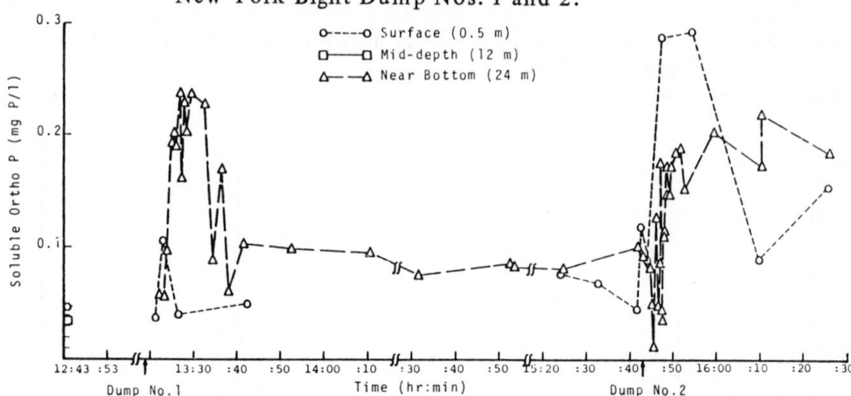

The turbidity with depth and distance from the discharge (Figure 7) shows that the discharged sediment sank to the bottom and flowed downcurrent as a density current, decreasing in intensity from over 60,000 NTU to the background level by 350 m or so downcurrent. Ammonia levels presented in Table 1 were also elevated at the discharge, Station 8 reaching 20 to 25 mg/l N, but decreasing to ambient within 200 or so meters (Station 9) and remaining below detection. Soluble ortho P concentrations (Table 2) were also elevated in the area of the discharge, reaching levels of 35 μg/l P in the bottom waters and dispersing downcurrent.

FIGURE 6. James River study area.

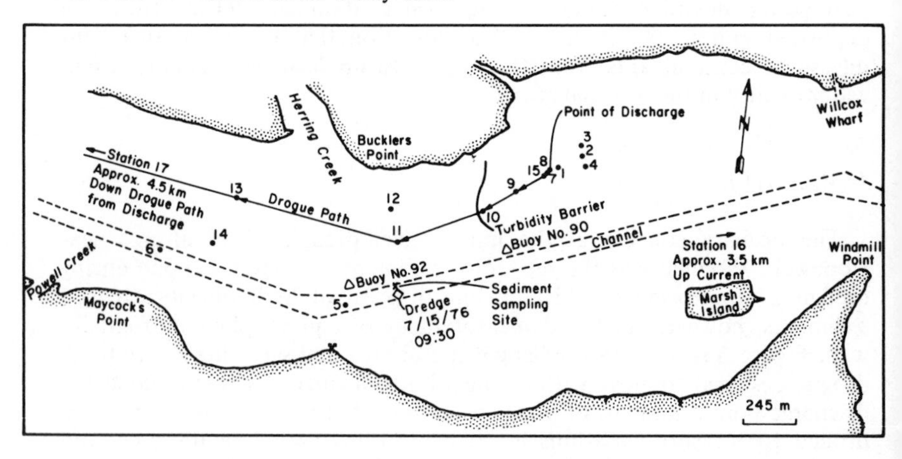

FIGURE 7. Turbidity near James River disposal area.

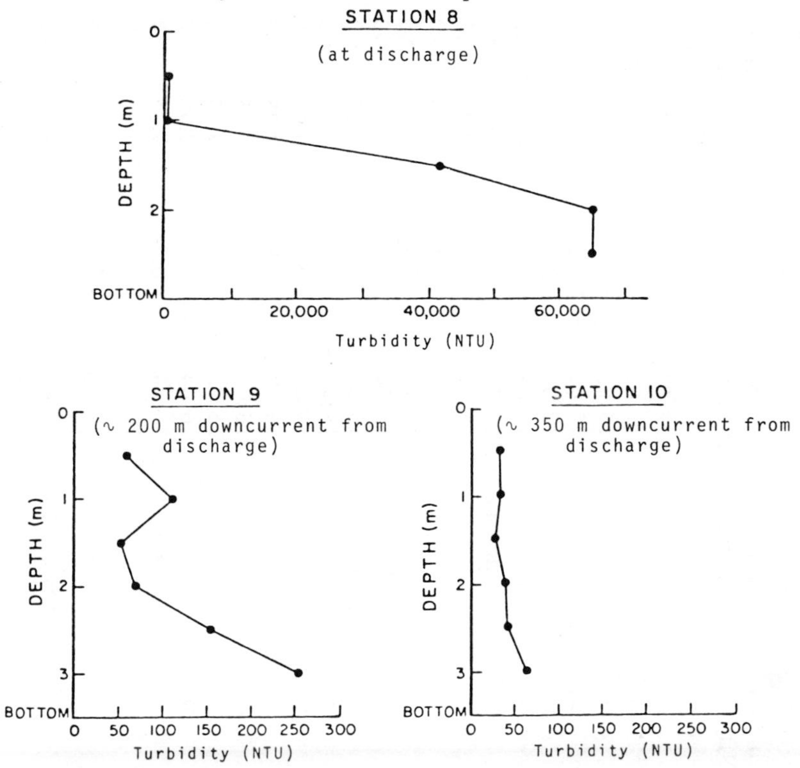

Elutriate tests run on the James River sediments showed ammonia release on the order of 5 to 6 mgN/l. This level is considerably below that found in the disposal site water column. Elutriation also resulted in a small uptake of soluble ortho P whereas some release was found near the discharge point. These seemingly deviant results point to the importance of proper interpretation and use of the data in terms of the disposal operation characteristics. In the case of the James River disposal operation, these results may have been due in part to the fact that the elutriate test which was designed for hopper dredge operations had a considerably longer sediment-water contact time than was found in the field. Also, there were large amounts of finely divided (<0.45 μ diameter) particulate matter in the discharge. The additional sediment-water contact time in the elutriate test may have allowed a greater amount of soluble ortho phosphate to sorb onto the fine materials, thereby showing a lower soluble ortho P concentration than actually found in the field. The factors to be considered in interpretation of elutriate test results are discussed below.

TABLE 1. Concentrations of ammonia near James River disposal area. (After Lee *et al.*[8])

Sampling Location	Time (hr:min)	Depth (m)	Ammonium (mg N/l) $\overline{x}$	SD	Sampling Location	Time (hr:min)	Depth (m)	Ammonium (mg N/l) $\overline{x}$	SD
Station 5	12:19	0.5	<0.05	∿0	Station 10	15:01	0.5	<0.05	∿0
	12:40	3.0	<0.05	∿0		15:02	1.0	<0.05	∿0
	12:41	6.0	<0.05	∿0		15:03	1.5	<0.05	∿0
	12:43	8.5	<0.05	∿0		15:03	2.0	<0.05	∿0
Station 6	13:09	2.0	<0.05	∿0		15:04	2.5	<0.05	∿0
	13:10	0.5	<0.05	∿0		15:05	3.0	<0.05	∿0
Station 7	13:54	0.5	<0.05	∿0	Station 11	15:27	0.5	<0.05	∿0
	13:56	1.0	<0.05	∿0		15:27	1.5	<0.05	∿0
	13:57	1.5	<0.05	∿0		15:28	2.5	<0.05	∿0
	14:08	0.5	<0.05	∿0		15:29	3.5	<0.05	∿0
	14:10	1.0	<0.05	∿0		15:31	4.0	<0.05	∿0
	14:11	1.5	0.43	0	Station 12	15:39	0.5	<0.05	∿0
	14:34	Surface	<0.05	∿0		15:40	1.5	<0.05	∿0
	14:43	0.5	<0.05	∿0		15:41	2.5	<0.05	∿0
	14:44	1.0	<0.05	∿0		15:42	3.5	<0.05	∿0
	14:45	1.5	1.66	0.05	Station 13	16:02	0.5	<0.05	∿0
Station 8	13:56	0.5	<0.05	∿0		16:03	1.5	<0.05	∿0
	13:58	1.0	<0.05	∿0		16:04	2.0	<0.05	∿0
	14:00	1.5	16.45	0.64	Station 14	16:14	0.5	<0.05	∿0
	14:02	2.0	23.40	0.42		16:16	2.5	<0.05	∿0
	14:04	2.5	24.15	0.21		16:18	4.5	<0.05	∿0
Station 9	14:53	0.5	<0.05	∿0		16:19	6.0	<0.05	∿0
	14:54	1.0	<0.05	∿0	Station 15	16:42	0.5	<0.05	∿0
	14:54	1.5	<0.05	∿0		16:43	1.0	<0.05	∿0
	14:55	2.0	<0.05	∿0		16:44	1.5	0.29	0.01
	14:55	2.5	<0.05	∿0		16:46	1.75	2.93	0.06

TABLE 2. Soluble ortho P concentrations at James River disposal site.

Sampling Location	Time (hr:min)	Depth (m)	Soluble Ortho P $\overline{X}$	SD
Station 7	13:54	0.5	0.024	0
	13:56	1	<0.01	0
	13:57	1.5	<0.01	0
	14:08	0.5	0.12	0
	14:10	1	<0.01	0
	14:11	1.5	<0.01	0
	14:34	0.5	<0.01	0
	14:43	0.5	<0.01	0
	14:44	1	<0.01	0
	14:45	1.5	0.040[1]	0
Station 8	13:56	0.5	0.020	0.001
	13:58	1	0.021	0
	14:00	1.5	0.032[1]	0.001
	14:02	2	0.13[1]	-
	14:04	2.5	0.35[1]	0.001
Station 9	14:53	0.5	<0.01	0
	14:54	1	0.029	0
	14:54	1.5	0.033	0
	14:55	2	<0.01	0
	14:55	2.5	<0.01	0
	14:56	3	0.025	0
Station 10	15:01	0.5	<0.01	0
	15:02	1	0.017	0
	15:03	1.5	<0.01	0
	15:03	2	<0.01	0
	15:04	2.5	<0.01	0
	15:05	3	<0.01	0
Station 13	16:02	0.5	<0.01	0
	16:03	1.5	<0.01	0
	16:04	2	<0.01	0

[1]Sample contained fine particulate matter after 0.45 μ pore size filtration. Samples were refiltered through a 0.2 μ pore size membrane filter prior to analysis. After Lee et al.[8]

INTERPRETATION OF ELUTRIATE TEST RESULTS

In order to effectively use elutriate test results, they must be interpreted in terms of the characteristics of the disposal operation and the disposal site. Without such interpretation, elutriate test concentrations are of very limited value in making a decision about the potential impact on water quality associated with disposal of dredged sediment in a particular location.

The elutriate test was designed to predict contaminant release associated with hopper-dredge disposal of hydraulically dredged sediment. When it is used in conjunction with other types of dredging-disposal operations, such as barge disposal of mechanically dredged sediment, and even with hopper dredge disposal, the sediment-to-water ratios used in the elutriate test should be compared to and adjusted for what is found under actual dredging conditions. Likewise, the sediment-water contact time should be representative of what takes place in the field. It has been found that these two factors can have a significant impact on elutriate test results for some chemicals, which would not be predictable by means other than altering the elutriate test conditions.

A number of characteristics of the disposal area must also be considered. If a thermocline or chemocline is present at the disposal site, then the dredged material "turbid plume" and associated released chemical contaminants are likely to be trapped in the hypolimnetic waters. This would affect the impact on water quality of released nutrients, since there could be considerable time during which available nutrients released could be made unavailable for use by aquatic plants and could be diluted before being mixed into an area where algal uptake occurs. The redox conditions of disposal area waters will influence the chemistry of iron which can affect the uptake and release of some chemical contaminants such as phosphorus. Waterway sediments are typically anoxic, so when they are disposed of in oxic open waters, reduced iron can be oxidized forming a ferric hydroxide floc, an effective scavenger of a number of chemicals. The running of both oxic and anoxic elutriate tests may better define this situation for a system.

One of the most important factors to consider in defining potential impact of dredged sediment disposal on eutrophication-related water quality is the dilution character of the disposal area, including that associated with tides, wind, and current mixing. An evaluation must be made of the expected duration of elevated contaminant concentrations and the critical concentration of available forms-duration of exposure relationship for the contaminant of concern. In the case of predicting the impact of released phosphorus on water quality, as discussed by Jones and Lee (2), the dilution as expressed as the hydraulic residence time is a crucial component. It must be noted that as currently designed, the elutriate test does not give consideration to dilution and therefore typically represents a worst case with respect to contaminant release.

When release of nutrients is of concern with respect to dredged material disposal, the nutrient limiting algal growth in the disposal area during the time of year of concern must be determined. If, as is the case with many coastal marine waters, the disposal area waters are nitrogen-limited, then the release of phosphorus from dredged sediment may be of little consequence. Further, highly turbid water, such as is typical of many estuarine and freshwater tidal waters, may preclude growth limitation by nutrients. If a nutrient of potential concern is expected to be released with dredged material, other sources of this nutrient should be evaluated for their relative significance in stimulating algal growth.

Finally, it may be determined that the water quality significance of other contaminants released during disposal, such as toxic materials, may dwarf the overall water quality significance of increased aquatic plant growth.

CONCLUSIONS

The results presented here are typical of sediment-associated nutrient behavior at the seven other disposal sites studied. Sediment Kjeldahl nitrogen concentration generally ranged from ∿ 200 to 4000 mgN/kg with a mean concentration of ∿ 1,550 mgN/kg. Elutriate tests generally showed a release of soluble nitrogen (ammonia plus nitrate) to 0.5 to 35 mgN/l. Total phosphorus concentrations of the sediments studied averaged ∿ 950 mgP/kg and ranged from ∿ 100 to 3750 mgP/kg. Response of sediment-associated and soluble phosphorus to elutriation ranged from a decrease in soluble ortho P concentration by 0.6 mgP/l to an increase by 1.6 mgP/l. Detailed data presentation and discussion of the study results are presented by Lee et al. (8) and Jones and Lee (2). Conclusions that can be drawn based on these studies include the following:

Based on total amounts of N and P, dredged sediments appear to be a potentially significant source of nutrients if the nutrients are released in available forms during disposal.

No general relationship can be developed between bulk composition of a sediment and the amounts of contaminant release either in the laboratory elutriate test or in the field. This was based on intensive study of contaminant release from over 50 sediments from across the country.

The elutriate test can provide a prediction of the behavior of available nitrogen and phosphorus during open water dredged material disposal and will also provide an estimate of the concentrations if interpreted properly in light of disposal operation characteristics.

Based on the results of this study as well as a review of the literature, it appears that in general, dredged sediment-associated nutrients will rarely have an adverse effect on eutrophication-related water quality at the disposal site mainly because the events are short-lived, there is typically fairly rapid dilution of the disposed of sediment, and, relative to the dilution, nutrient release is small. The potential impact must be evaluated on a site-by-site basis.

For several contaminants including phosphorus, elutriate test results must be interpreted in terms of the aqueous environmental chemistry of iron. Iron present in the anoxic dredged sediment is oxidized when it comes in contact with oxic disposal area water, forming a ferric hydroxide floc which is an effective scavenger for phosphate.

ACKNOWLEDGEMENTS

We wish to acknowledge the Corps of Engineers Dredged Material Research Program, Vicksburg, Miss. for support of this study. Many individuals contributed to the success of the overall study. Of particular impoıtance to the work on nitrogen compounds was the assistance of Pinaki Bandyopadhyay. The preparation of this paper was supported by the Department of Civil Engineering, Colorado State University, Fort Collins, and EnviroQual Consultants and Laboratories, Fort Collins, Col.

REFERENCES

1. Jones, R.A. 1978. *Release of Phosphate from Dredged Sediment*, Ph.D. dissertation, University of Texas at Dallas.
2. Jones, R.A., and G.F. Lee. 1978. Evaluation of the elutriate test as a method of predicting contaminant release during open water disposal of dredged sediment and environmental impact of open water dredged material disposal, Vol. I: Discussion, Technical Report D-78-45, U.S. Army Engineer Waterways Experiment Station, Vicksburg, Miss.
3. Lee, G.F., and R.A. Jones. 1977. An assessment of the environmental significance of chemical contaminants present in dredged sediments demped in the New York Bight, Environmental Engineering, Colorado State University, Fort Collins, Occasional Paper No. 28.
4. Lee, G.F., and R.A. Jones. 1979. Application of the OECD eutrophication modeling approach to estuaries. Presented at International Symposium in Nutrient Enrichment in Estuaries, Williamsburg, Virginia.
5. Lee, G.F., M. Piwoni, J. Lopez, G. Mariani, J. Richardson, D. Homer, and F. Saleh. 1975. Research study for the development of dredged material disposal criteria. Technical Report D-75-4, U.S. Army Corps of Engineers, WES, Vicksburg, Miss.
6. Lee, G.F., J.M. Lopez, and M.D. Piwoni. 1976. An evaluation of the factors influencing the results of the elutriate test for dredged material disposal criteria. Proc. ASCE Specialty Conference on Dredging and Its Environmental Effects, Amer. Soc. Civil Engr. pp. 253-288.
7. Lee, G.F., P. Bandyopadhyay, J. Butler, D.H. Homer, R.A. Jones, J.M. Lopez, G.M. Mariani, C. McDonald, M.J. Nicar, M.D. Piwoni, and F.Y. Saleh. 1977. Investigation of water quality parameters at the offshore disposal site, Galveston, Texas, Technical Report No. D-77-20, U.S. Army Corps of Engineers, WES, Vicksburg, Miss.
8. Lee, G.F., R.A. Jones, F.Y. Saleh, G.M. Mariani, D.H. Homer, J.S. Butler, and P. Bandyopadhyay. 1978. Evaluation of the elutriate test as a method of predicting contaminant release during open water disposal of dredged sediment and environmental impact of open water dredged material disposal. Vol. II: Data Report, Technical Report D-78-45, U.S. Army Engineer Waterways Experiment Station, Vicksburg, Miss.

THE EFFECTS OF SEWAGE DISCHARGE INTO A WIND-INDUCED PLUME FRONT

W.J. Kimmerer, T.W. Walsh, and J. Hirota
Hawaii Institute of Marine Biology
P.O. Box 1346
Kaneohe, Hawaii 96744

ABSTRACT: Enhanced concentrations of particulate matter and nutrients are often associated with fronts. In Kaneohe Bay, Hawaii, runoff, sewage discharge, and persistent trade winds produced a plume front which was readily visible because of high chlorophyll concentrations. About 50 percent of the time the plume advected approximately all of its nutrient supply out of the south sector of the bay into the apparently more thoroughly flushed central sector. Thus about half of the nitrogen from secondary treated sewage, discharged until 1978 into the low-density side of the front, was lost to the south sector almost immediately. Several models of circulation and material flux in the bay have assumed that the south sector is well mixed and that all of the nutrient input from sewage and streams remains there for several weeks. These models should be revised to reflect the loss of about half of the nutrient input from the south sector. These results also emphasize the importance of wind in controlling physical events in estuaries.

INTRODUCTION

Fronts in oceanic and coastal waters are often sites for enhanced primary production and phytoplankton biomass (2,7). In particular, plume fronts, which result from the discharge of low density water into the ocean, can trap terrigenous nutrients, producing bloom conditions on the shoreward side (3). The seaward flux of terrigenous materials can be substantially altered by plume fronts, particularly if they are persistent.

The shape of a plume front depends on the flow rate of the discharge and on the actions of wind and tidal flow. In the absence of wind, the front can be maintained by friction between the spreading surface layer and the ambient fluid (8). Onshore wind stress can maintain the front by preventing the seaward spread of the lower-density layer. Where winds blow persistently onshore, a plume front can be maintained even when discharge rate is low. Such a condition occurs in Kaneohe Bay, Hawaii, where onshore winds blow about 85 percent of the time.

FIGURE 1. Kaneohe Bay, showing approximate division into north, central, and south sectors, and sampling stations NW, CE, SC, and OF, and area shown in Figure 2.

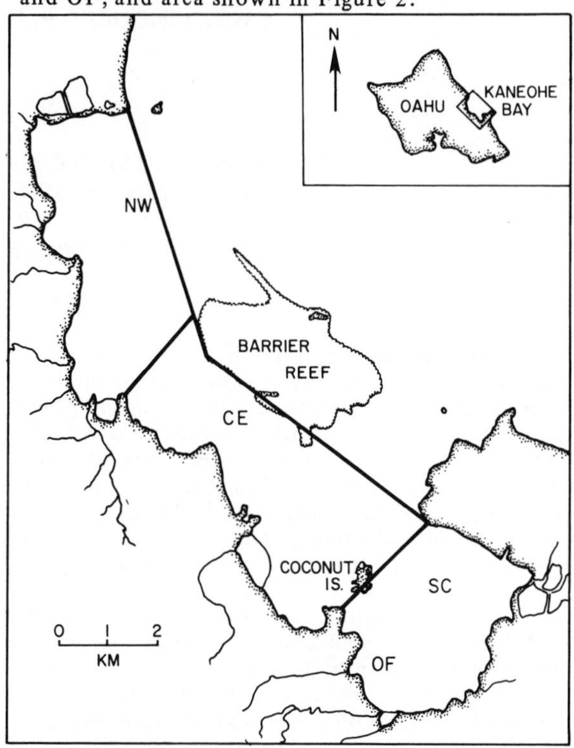

Kaneohe Bay (Figure 1) is a shallow (8m) subtropical estuary on the northeast or windward coast of Oahu (5). Until the end of 1977 about 12,000 m^3/day of sewage was discharged into the southwest corner from a municipal secondary treatment plant. An annual average of 35,000 m^3/day fresh water enters the bay from three stream mouths near this outfall (19). The flow is seasonal, with generally higher flows during the rainy winter season. Tides in Kaneohe Bay are mixed semidiurnal, with spring tidal amplitudes up to 90 cm.

Kaneohe Bay can be conveniently divided into three sectors on the basis of physiography and probable circulation (Figure 1) (5). Water turbidity and chlorophyll levels have been highest in the south sector, which has the most restricted circulation and received most of the sewage. Models of circulation, plankton dynamics, and material flux in Kaneohe Bay (4, 6, 12) have been developed with the assumption that the south sector was internally well mixed. The presence of a plume of discolored water in the vicinity of the outfall belied the assumption of thorough mixing in south Kaneohe Bay. This plume was observed only when northeast trade winds were blowing. It could be tracked northward along the shore, beginning at the outfall and often passing out of the south sector through Lilipuna Channel. Caperon et al. (5) had previously reported steep gradients in chlorophyll concentrations between a station near the outfall and stations further from shore. The purpose of this research was to further elucidate the physical and chemical nature of the plume, to determine whether its effect on material flux in south Kaneohe Bay was significant, and to observe changes in the plume resulting from the termination of sewage discharge.

METHODS

Figure 2 shows the locations of sampling stations and transects. Stations SC and OF were selected for a long-term study of Kaneohe Bay (in preparation). Positions were determined with a hand bearing compass or by using a sextant to measure angles between landmarks. A Hydro Products Model 612S transmissometer with a 1-meter path length, calibrated at 92 percent in air, was used to measure percent light transmission, which was converted to volume attenuation coefficient α (10). Surface plankton tows were taken with a 1/2 m 333 μm mesh net equipped with a flow meter. Salinity and temperature were measured in situ with a Beckman RS5-3 portable salinometer. In vivo chlorophyll fluorescence was measured with a Turner Model 111 fluorometer with a flow-through door, and expressed as relative units. On several occasions water samples for fluorescence were first passed through a 35 μm Nitex mesh for crude size fractionation. Water samples were taken with a battery-powered pump, screened through a 100 μm Nitex gauze, and stored in dark polypropylene bottles at surface temperature until they were processed, usually within four hours. Particulate matter 200 mℓ aliquots was collected on precombusted GF/C filters which were then dried at 60°C and stored in a dessicator. Filters were analyzed for total carbon and nitrogen and inorganic carbon using Hewlett-Packard Model 185 and 185B CHN Analyzers (9). The filtrate was frozen immediately for nutrient analysis, which was done within one to two weeks. Nitrate plus nitrite, ammonium, phosphate, and silicate were determined on a Technicon Autoanalyzer II system (20, 21). Total dissolved nitrogen (TDN) was determined by UV-peroxide oxidation followed by analysis for nitrate plus nitrite (20).

Salinity of filtered samples was determined using a Plessey Model 6230N laboratory salinometer, calibrated with standard Copenhagen water. These salinity values were used to correct the field salinities, which were then combined with the temperature values to calculate density (σ_t).

FIGURE 2. South corner of Kaneohe Bay showing location of Coconut Island and the outfall pipe, Stations SC and OF, and sampling transects A, B and C.

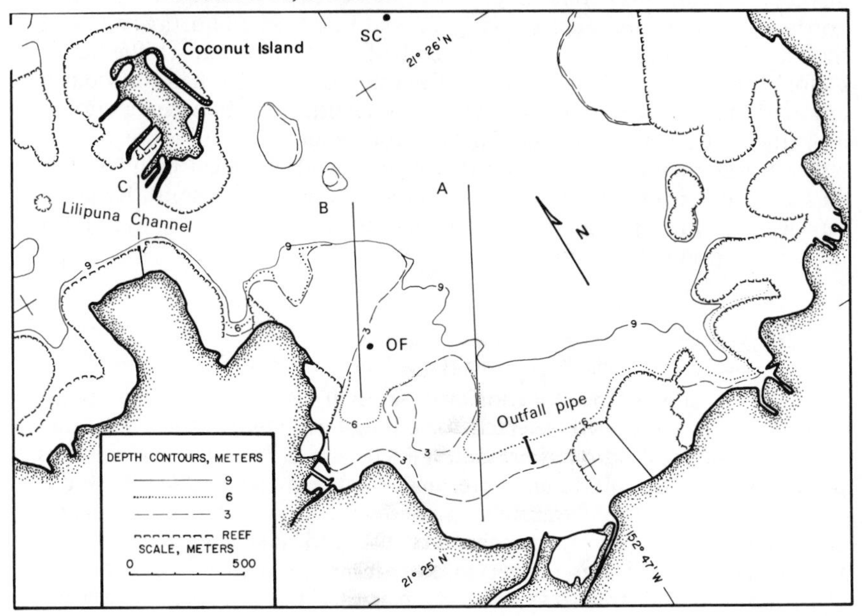

Counts of suspended particles in unscreened samples were made on a Coulter Counter Model TA-II, using 70 and 400 micron orifice tubes to count and size particles in the 2-128 μm size range.

Wind speed and direction were obtained at Coconut Island (Figure 1) using a Weather Measure Corp. Model W121-SD wind recorder, and from the records of the Kaneohe Marine Corps Air Station, which is located on the peninsula northeast of Coconut Island. Incident light was measured at Coconut Island with an Eppley Precision Radiometer. Runoff from the major stream near the outfall contributes about 50 percent of the fresh water flow to the area; data from Geological Survey stream gauge records were extrapolated to estimate total input by all streams in the area. Current estimates were made using 80 cm square cruciform drogues with small styrofoam floats.

RESULTS

Table 1 shows median wind speed and direction, duration of direction, incident light energy, and stream runoff for the sampling dates. On most of those dates vertical profiles of salinity, temperature, and light transmission were taken at six to eight stations along transect A (Figure

TABLE 1. Summary of environmental data for sampling dates. Asterisks indicate dates from which data appear in Figures 3, 4, 6, and 7.

Date	Wind			Light	
	Median speed m/sec	Direction	Duration of direction	cal $cm^{-2} day^{-1}$	Stream flow $10^3 m^3/hr$
*Sept. 30, 1976	2	variable	0	459	.69
Oct. 13, 1976	6	NE	6 days	344	.78
Oct. 20, 1976	6	E	13 days	348	.64
Mar. 31, 1977	1	variable	0	261	1.84
*May 17, 1977	2	E	8 hours	499	1.12
*May 24, 1977	9	ENE	6 days	571	.95
*July 6, 1977	9	ENE	14 days	510	.67
Aug. 9, 1977	4	NNE-ENE	10 hours	360	.64
Aug. 23, 1977	7	NE-ENE	11 days	579	.67
Aug. 30, 1977	6	NE-ENE	18 days	272	.67
Sept. 26, 1977	2	ESE	4 hours	333	.56
*Mar. 16, 1979	6	NE	8 days	570	.63

2). Horizontal gradients in salinity, σ_t (Table 2) and α were always present when trade winds were blowing. Figure 3 shows vertical profiles of σ_t on two trade wind days, one before and one after termination of sewage discharge, and on one calm day. Profiles for both trade wind days show tilting of isopycnals, although this was more pronounced before sewage diversion. Under calm conditions, density stratification spread across the bay, with only slight tilting of the isopycnals very near the shore (Figure 3b). Profiles of α (Figure 4) on the same dates showed a steep horizontal gradient on both trade wind days, with much higher values while sewage was being discharged. α remained uniformly low on calm days even before sewage diversion (Figure 4b). In general, values taken more than 1 meter from the bottom were higher on windy days than on calm days irrespective of location, because of resuspension of sediments.

TABLE 2. Maximum and median values of salinity and density gradients along transect A for trade wind days.

Date	ΔS ($^o/oo$ $m^{-1} \times 10^4$)		$\Delta \sigma t$ ($m^{-1} \times 10^4$)	
	max.	med.	max.	med.
Oct. 13, 1976	85	23	72	27
Oct. 20, 1976	36	11	36	11
May 17, 1977	22	0	15	3.9
May 24, 1977	18	6.7	14	5.9
Aug. 9, 1977	14	5.3	11	5.3
Aug. 23, 1977	25	32	15	9.8
Aug. 30, 1977	23	4.4	18	5.9
Mar. 16, 1979	7	1.2	5.6	2.0

FIGURE 3. Vertical profiles along transect A for σ_t on: **a**) a trade wind day;

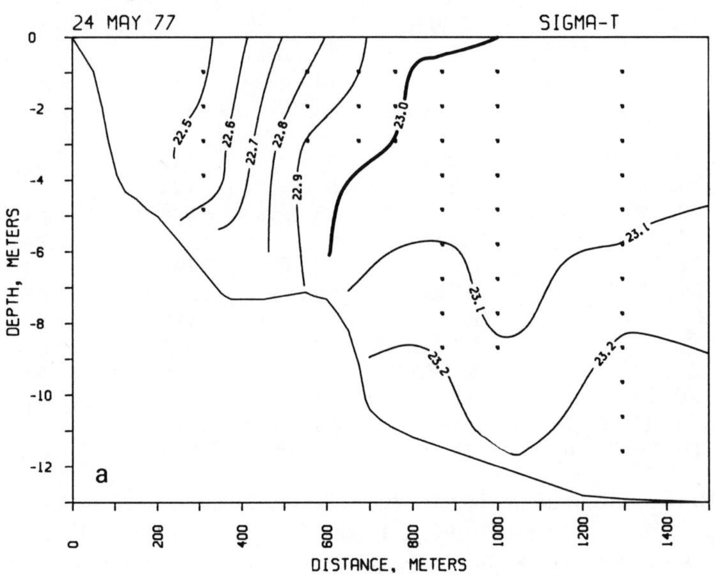

b) a calm day;

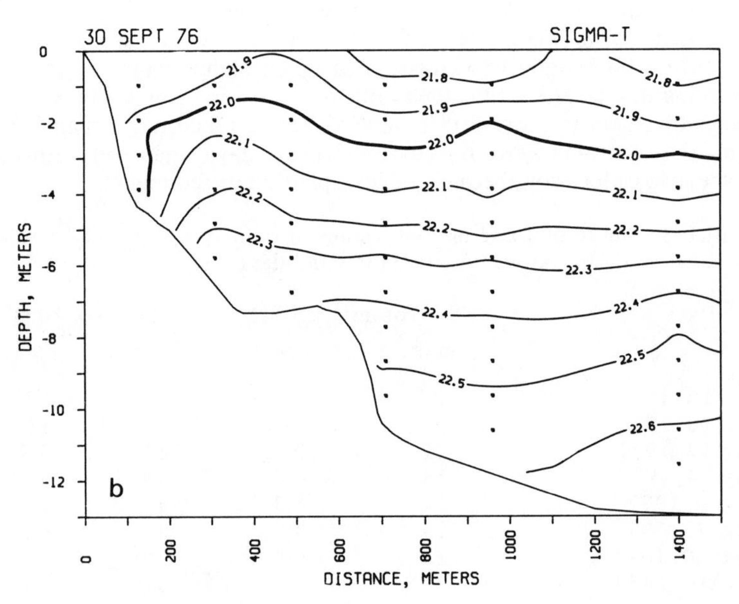

c) a trade wind day after termination of sewage discharge.

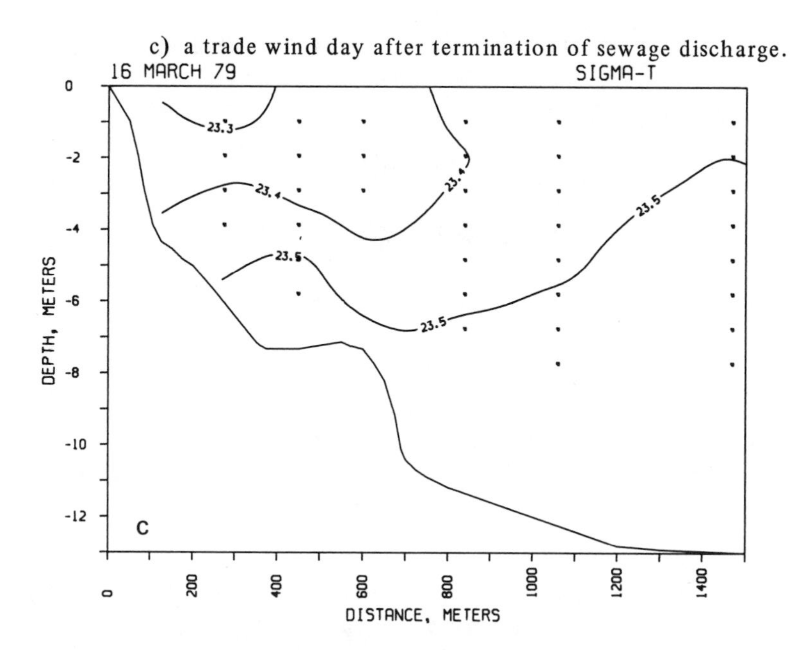

Samples for temperature and salinity taken between August, 1976 and March, 1979 from 1 meter depth at stations OF and SC were used to determine the effect of wind speed and direction on differences in σ_t (OF - SC). This σ_t gradient had a significant negative correlation with the component of daily mean wind speed normal to the shoreline near the outfall, taken to be 45° (Figure 5). The regression equation was $\Delta \sigma_t = .41 - .81$ W, ($r = -.53$ $p < .001$) before sewage diversion; after sewage diversion, the range of wind speeds was small and the slope was not significantly different from 0 or from the pre-diversion slope. There was no evidence of any effect of tidal amplitude or tidal height at the time of sampling on the difference in σ_t between SC and OF. Before sewage diversion, differences in depth-averaged values of α between these stations also showed a significant positive correlation with the onshore component of wind speed ($r = .52$, $p = .0004$).

Wind speeds and directions and onshore wind components for 1976 and 1977 are summarized in Table 3. Approximately 85 percent of the time northerly to easterly winds blow persistently above 1 m/sec.

On several of the dates listed in Table 1 we took samples along transect A from 1 meter for nutrients, total dissolved and particulate nitrogen, and in vivo fluorescence. When the front was present before sewage termination there were steep gradients in total nitrogen coincident with the σ_t gradients (Figure 6). Table 4 shows medians and ranges for total nitrogen and its components in and out of the plume. The greatest fractional change on entering the plume was in ammonium. On calm days

FIGURE 4. Vertical profiles along transect A for beam attenuation coefficient on the same dates as in Figure 3.

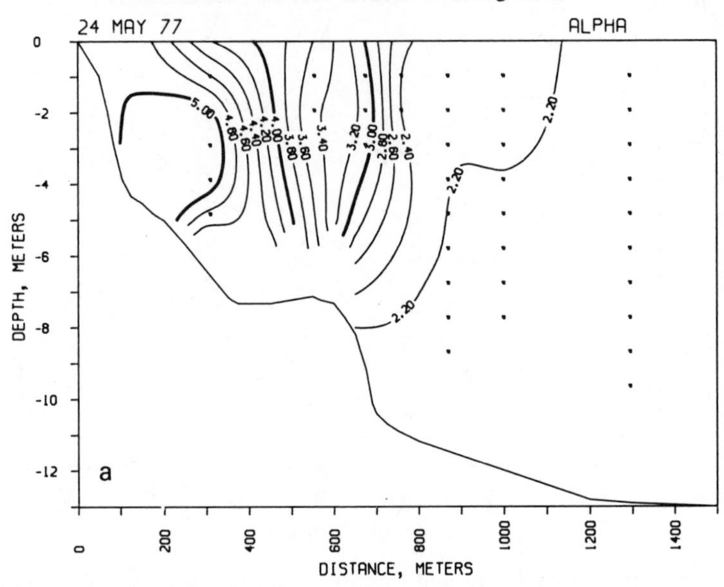

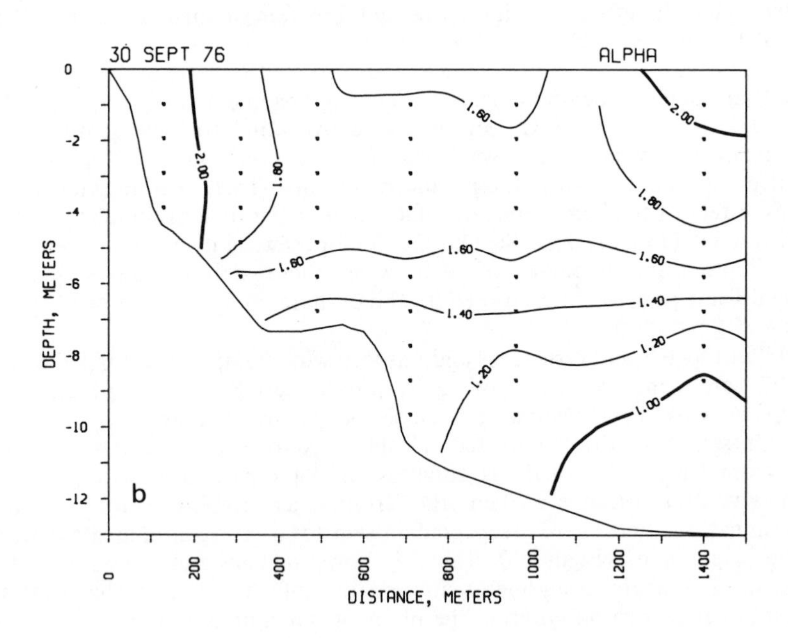

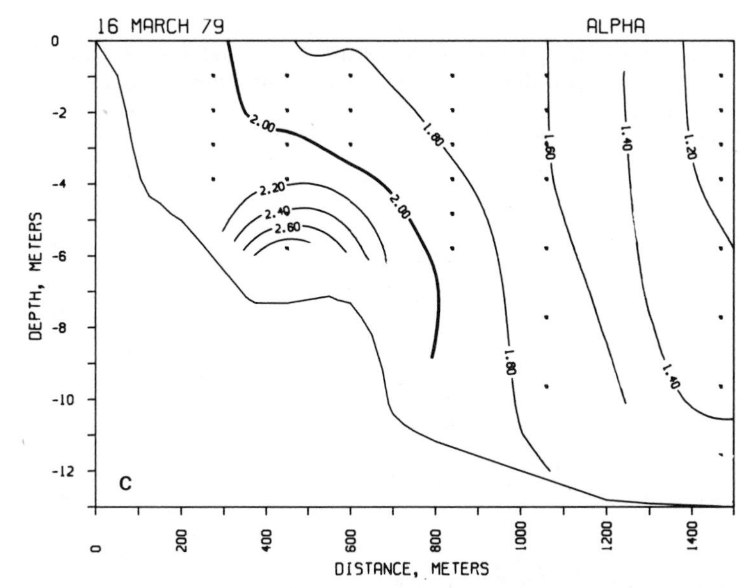

there was still a slight gradient in dissolved organic nitrogen (DON) but ammonium and nitrate did not decrease along the transect. Phosphate behaved similarly to ammonium, with sharp gradients on trade wind days and little change on calm days. In the trade wind data taken after sewage termination, the total nitrogen gradient was slight, with all of the decrease along the transect attributable to DON. Dissolved inorganic nitrogen accounted for only 4 percent of the total along the transect.

Chlorophyll concentrations as measured by in vivo fluorescence decreased substantially from landward to seaward along transect A on trade wind days, with a mean decrement of 60 percent (standard error = 8) of the inshore reading for seven trade wind days. On the two calm days the mean decrease was only 18 percent. Fluorescences of the >35 μm size fraction was greater than in the smaller size fraction on several sampling dates, comprising 10-87 percent of the total. On trade wind days the proportion of pigments in the larger fraction was higher on the nearshore side of the front. This is further borne out by particle counts which show that the greatest change in particulate volume along transect A usually occurred in volume peaks of 10-65 μm diameter; such peaks were produced by blooms of large chain diatoms including *Skeletonema costatum, Chaetoceros socialis* and *C. curvisetus*. Two plankton samples taken 20m apart, one in and one out of the plume, showed little difference in most of the common taxa, except that decapod (crab and shrimp) zoeas were 75 times more abundant inside the plume than out.

A plume which developed during a period of strong trade winds following heavy rainfall (four-day mean stream flow 2.1 m^3/hr) had a readily visible edge and a dark olive-brown color. The plume could be traced visually along the shore to the north almost to station CE. Table 5

FIGURE 5. Relationship between the gradient in σ_t between Stations OF and SC and the onshore component of wind speed at the outfall.

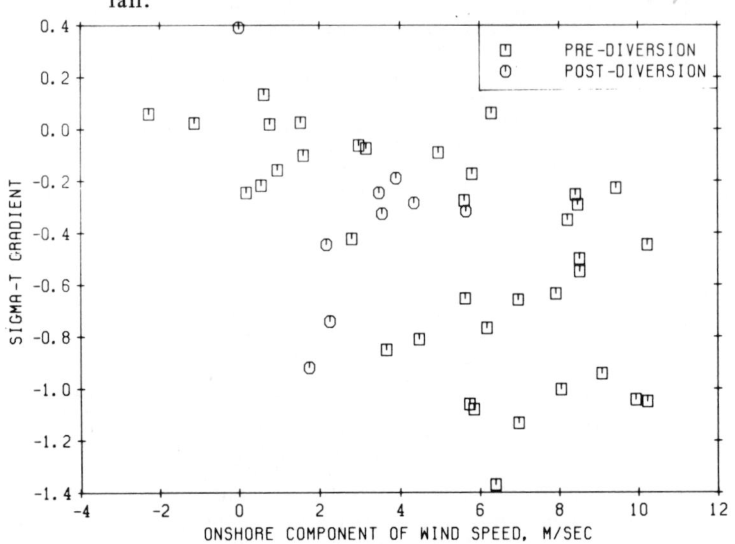

shows comparisons of salinity, total fluorescence, total dissolved nitrogen, and nutrients for stations CE, SC and OF. The extremely high levels of nitrate and silicate at OF are characteristic of stream nutrient input. Elevated fluorescence and depressed salinity could be detected as far away as station CE; these conditions must have originated in the south sector plume, as there are no major streams on the central bay, and the salinity at station SC was normal. Most of the excess fluorescence in the plume was in a broad peak in particle concentration centered around 10 μm.

On several occasions we sampled along transects across the plume to determine its extent and shape. On July 6, 1977 we measured light transmission and relative fluorescence continuously and temperature, salinity and ammonium at approximately 50m intervals at 1 meter depth along five transects. We used aerial photographs taken concurrently with the sampling to supplement the transect data for α and fluorescence. The plume was well developed, passing through Lilipuna Channel and around Coconut Island (Figure 7). Elevated values of fluorescence and α were detected on every transect; ammonium decreased very rapidly away from the outfall pipe, but values above 1 μg-at/ℓ could still be detected in Lilipuna Channel. The narrowing of the plume south of Coconut Island was observed on many occasions, and may have resulted from wind-driven currents over the reefs in that area. The edges of the plume were usually distinct except in the area just east of that point. The boundary there was usually ragged, often with long intrusions of cleaner southeast bay water, or long fingers of plume water extending to the northeast.

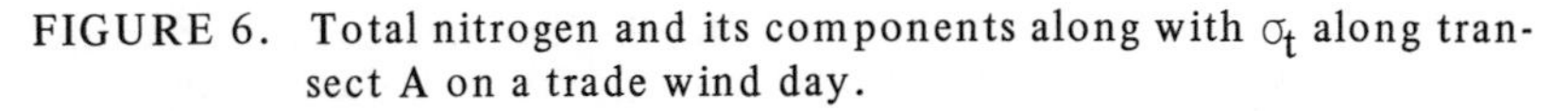

FIGURE 6. Total nitrogen and its components along with σ_t along transect A on a trade wind day.

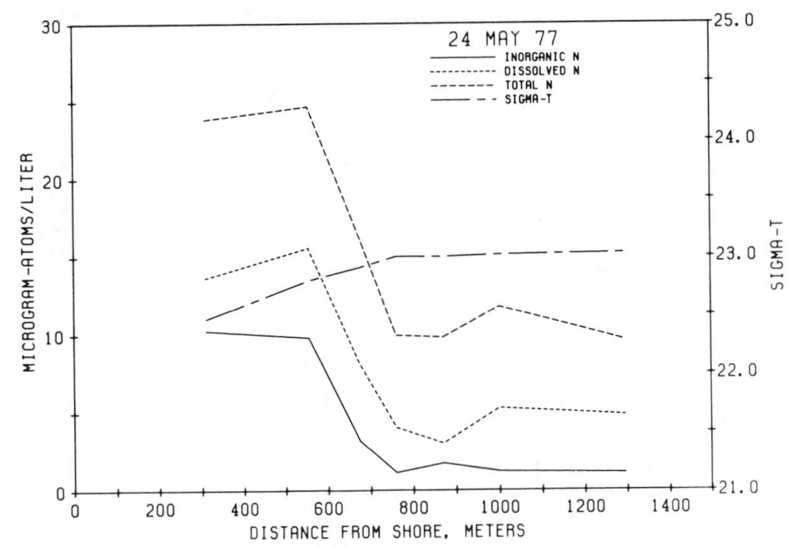

TABLE 3. Percent frequencies at various wind speeds for trades ($45^{\circ} \leq D < 135^{\circ}$), north winds ($335^{\circ} \leq D$ or $D < 45^{\circ}$), and south to west winds ($135^{\circ} \leq D < 335^{\circ}$), and for the component of wind speed normal to the shore at the outfall. Based on daily mean wind speeds for 1976 and 1977.

Direction	Wind Speed <1	1-3	3-6	6-9	>9	All Speeds
Trade	1.6	6.3	25.1	34.1	9.4	76.5
North	.9	2.9	3.8	2.9	.7	11.2
South to West	1.5	6.0	4.4	.3	.1	12.3

On August 9, 23, and 30, we measured salinity, temperature and light transmission at stations along transects A, B and C (Figure 2), and tracked drogues near each transect to estimate the fluxes of materials in the plume. The was present on all dates (Figure 8) but varied in shape and in steepness of the gradients. On August 9, the plume was least developed, ending near transect B. Drogue tracks in Lilipuna Channel showed a slow southerly tidal flow with some southwestward surface drift. Apparently the northerly (20-45°) wind direction prevented flow through the channel. On August 23 and 30, gradients in α and σ_t were steeper and the

FIGURE 7. Variables measured at 1 m depth along five transects on July 6, 1977, a trade wind day: a) σ_t; b) α; c) in vivo fluorescence; d) ammonium. The α and fluorescence data were

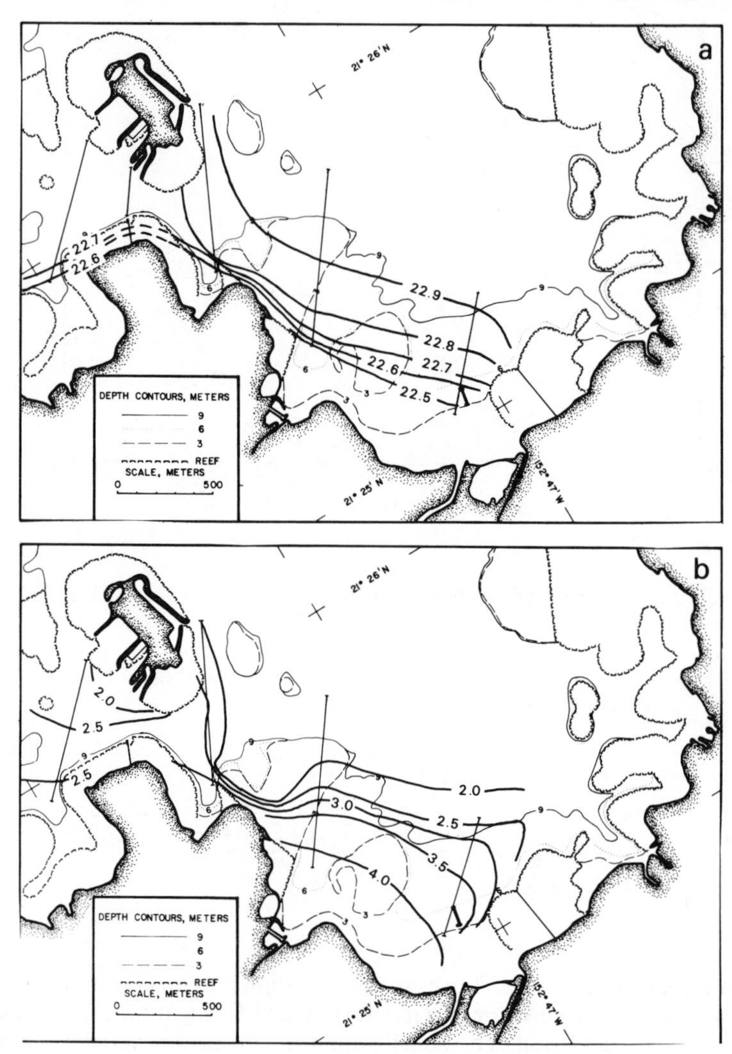

plume could be detected in Lilipuna Channel. Drogue data from the channel showed that a northward surface current of 9-15 cm/sec was flowing in the channel, although on August 30 a northward wind shift in the afternoon stopped the current. The surface flows in Lilipuna Channel were not affected by the stage of the tides.

FIGURE 7. supplemented with aerial photography by tracing the visible edge of the plume and drawing contours with reference to that line.

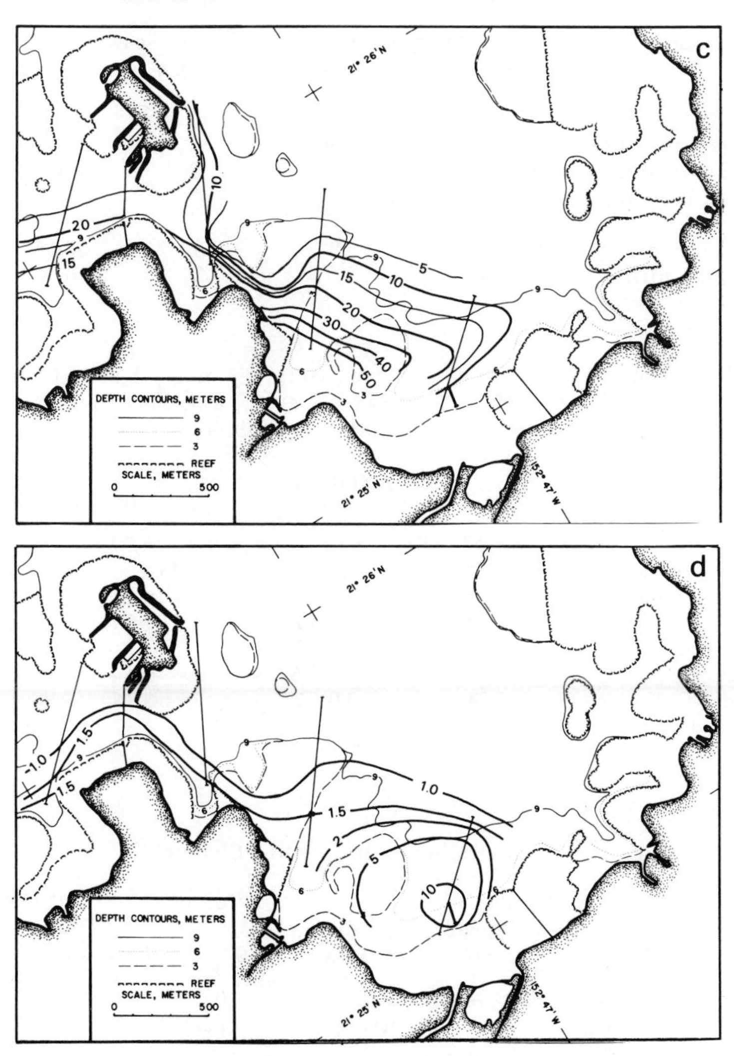

Nitrogen fluxes in Lilipuna Channel were estimated from the August, 1977 data and from data taken on September 26, 1977 and on March 16, 1979, after termination of sewage discharge. Total nitrogen was available from 1 meter samples at two stations in August and March and from five stations in September. Since total nitrogen and α were correlated, linear

TABLE 4. Total nitrogen and components (μg-at/ℓ) at shoreward (in plume) and seaward (out of plume) stations on transect A. Medians and ranges for 6 trade wind days.

Component	Shoreward		Seaward	
	Median	Range	Median	Range
Total nitrogen	26	22-34	13	9-19
Particulate nitrogen	11	5-19	5	3-7
Dissolved organic nitrogen	11	3-15	8	4-12
Dissolved inorganic nitrogen:				
Ammonium	5	1-10	0.7	.05-.9
Nitrate + nitrite	0.5	.4-1.1	0.2	.06-.4

TABLE 5. Comparison of values between samples 1 m depth at stations CE, SC, and OF on April 8, 1977 during a period of intense plume development following heavy rainfall.

Variable	Station		
	CE	SC	OF
Salinity	33.27	34.75	26.21
Total fluorescence	44.7	15.3	176
Total dissolved N	4.78	6.15	14.25
DON	4.19	5.47	5.77
Ammonium	.30	.42	1.94
Nitrate	.29	.26	6.54
Silicate	5.54	10.03	64.35
Phosphate	.22	.30	.37

TABLE 6. Summary of flux estimates in Lilipuna Channel.

Date	Wind Direction	Wind Speed m/s	Predicted tides	Volume flux 10^5 cu m/hr	Nitrogen flux g-at/hr
Aug. 9, 1977	NNE	4	Rising	0.0	0
Aug. 23, 1977	NE	7	Rising	1.3	760
Aug. 30, 1977 a.m.	ENE	5	Low Slack	1.5	1100
Aug. 30, 1977 p.m.	NE	9	High Slack	0.0	0
Sept. 26, 1977	ENE	3	High Slack	1.1	800
March 16, 1979	NE	6	Rising	1.6	100

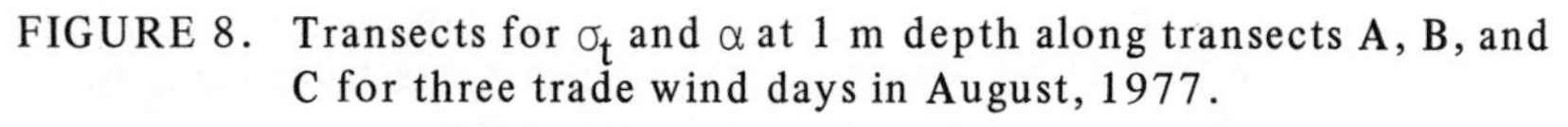

FIGURE 8. Transects for σ_t and α at 1 m depth along transects A, B, and C for three trade wind days in August, 1977.

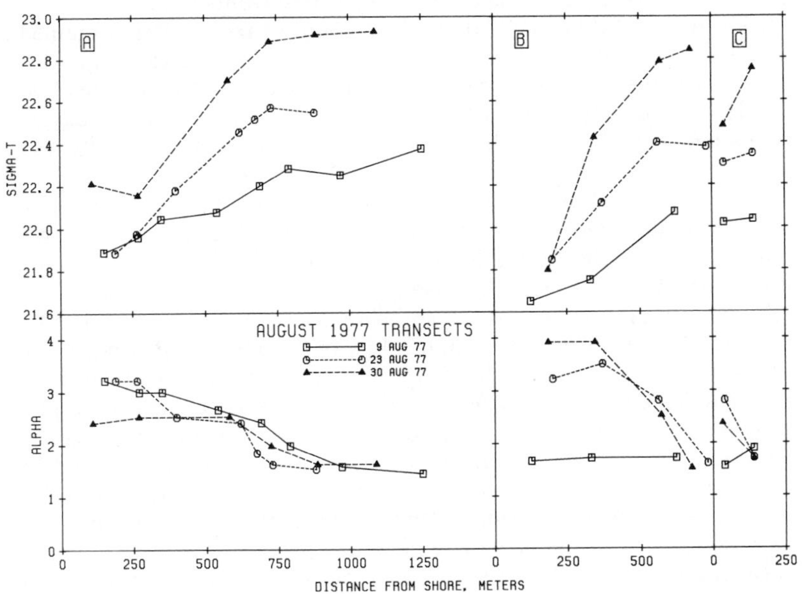

relationships between them were used to estimate total nitrogen from α values from vertical profiles at each station. Volume flow rate was calculated assuming a 200m wide stream, and using the drogue vectors at each depth to get volume fluxes for 1-meter layers. Velocity was small below 3 meters and was assumed to be 0. Return flow to balance the outward volume flux was assumed to have a total nitrogen concentration equal to the mean value for the water column at station SC (10.6μg-at/ℓ before diversion, 9.8 after) from three years of biweekly sampling at that station. This was substracted from the channel flux to get a net outward flux. Table 6 shows the calculated volume and material fluxes through the Lilipuna Channel.

DISCUSSION

Concentrations of phytoplankton caused by wind have been reported before, but the mechanism has usually involved either buoyancy (1) or phototaxis (16) of the organisms. In frontal areas such as that in Kaneohe Bay, the same effect is produced by buoyancy supplied by reduced density of the water mass. Much of the phytoplankton biomass in the plume was in the larger chain diatoms, as has been observed in eutrophic coastal waters (13) and in some river plumes (15).

The density difference across the front is small but occurs over a short distance; hence the gradient is larger than typical values for oceanic fronts (2, 7, 17, 18) but smaller than in some estuarine fronts (11).

Based on the July 6, 1977 data, the amount of fresh water contained in the plume is about 3×10^4 m^3 or about one day's stream input. This suggests that the time for complete frontal development should be about a day. This agrees with visual observations of the plume within a day after trade winds have begun. Dissipation can occur much more rapidly, as the plume is usually not visible after only a few hours of calm or southerly winds. When the plume is flowing through Lilipuna Channel, the volume flux is around 1.3×10^5 m^3/hr, about 100 times the rate of fresh water input. Evidently the plume is being entrained in a larger surface water mass being advected out of the channel.

Sensitivity of a plume front to wind speed and direction has been observed in the Hudson River plume (13). In the absence of wind, a plume front develops in Kaneohe Bay only during storms, when stream flow is very high. This front is similar in cause and dynamics to the typical plume front (8): it propagates rapidly across the bay surface and persists only as long as stream flow remains high. When trade winds blow onshore persistently, a plume front will develop regardless of stream flow. Thus the wind stress in effect causes frontal development, and maintains the frontal slope against the pressure gradient.

During 1976 and 1977 the rate of input of sewage nitrogen was about 22 kg-at/day or 900 g-at/hr (19). When the plume was flowing through Lilipuna Channel the flux of nitrogen was close to that value (Table 6). Thus when the plume was flowing it carried most of the sewage-derived nutrients into the central sector, where more thorough circulation apparently dissipated the plume. Mean input of nitrogen from the streams is about 110 g-at/hr (19) which is close to the value (100 g-at/hr) estimated for the plume after sewage diversion.

The calculations of flux out of the Lilipuna Channel were deliberately made to include only the flux directly attributable to the plume; that is, the contribution of the ambient south sector concentration of nitrogen to the outward flux was subtracted out. Thus we have not considered the indirect fluxes of material from sewage to south sector to central sector.

Wind directions between 45° and 135°, necessary for outward flux through the channel, occur about 50 percent of the time, if wind directions reported as equal to 45° are excluded. Therefore about half of the time nearly all of the terrigenous nitrogen supply to south Kaneohe Bay is advected within a few hours into the central bay. The impact on the bay is tremendous. The high concentration of particulate matter in the plume resulted in elevated respiration: production ratios on coral reefs underlying the plume (D.W. Kinsey, pers. comm.). We observed heavy growth of the benthic alga *Ulva* on the reef flats west of Lilipuna Channel, apparently caused by the high nutrient concentration in the plume. The most readily apparent effect of the plume, though, is on the nutrient supply to south Kaneohe Bay. Instead of a steady input of nutrients, there is a pulsed input with on the average about half of the nutrients reaching the south sector. Models of nutrient flux in south Kaneohe Bay have contained the assumption that all nutrients entering the south sector remain there for several weeks (4, 6). Such models should be revised either by reducing the average input of nutrients by half (e.g., 19) or by explicitly considering the flux of materials under various wind regimes.

Until recently (14), the importance of wind to estuarine circulation has received little attention. In Kaneohe Bay both wind speed and direction influence circulation and material fluxes. This finding has implications for the siting of waste disposal facilities, which should be placed to take advantage of mixing phenomena such as fronts.

ACKNOWLEDGEMENTS

Funds for this study were provided by the U.S. Environmental Protection Agency Contract R803983 and by the State of Hawaii Marine Affairs Coordinator. We thank Dr. Stephen V. Smith, Principal Investigator, for his valuable assistance in this project and for reviewing the manuscript, and Dr. Chris Welch, Dr. Wayne Esaias, and an anonymous reviewer for their constructive comments. HIMB Contribution No. 582.

REFERENCES

1. Baker, A.L., and K.K. Baker. 1976. Estimation of plankton wind drift by transmissometry. *Limnol. Oceanogr.* 21:447-452.
2. Bowman, M.J. and W.E. Esaias. 1978. Oceanic fronts in coastal processes. Springer-Verlag, New York.
3. Bowman, M.J. and R.L. Iverson. 1978. Estuarine and plume fronts. *In* Bowman and Esaias 1978.
4. Caperon, J. 1975. A trophic level ecosystem model analysis of the plankton community in a shallow-water subtropical estuarine embayment. *Estuarine Research* 1:691-709.
5. Caperon, J., S.A. Cattell and G. Krasnick. 1971. Phytoplankton kinetics in a subtropical estuary: eutrophication. *Limnol. Oceanogr.* 16:599-607.
6. Dames and Moore. 1977. Kaneohe Bay urban water resources study. Kaneohe Bay Computer Modeling, Kaneohe, Hawaii, for U.S. Army Engineer District, Honolulu, Oahu. Prepared by Dames and Moore, 2875 S. King St., Honolulu. 133 pp plus supplements.
7. Fournier, R.D., J. Marra, R. Bohrer and M. Van Det. 1977. Plankton dynamics and nutrient enrichment off the Scotian Shelf. *J. Fish. Res. Bd. Can.* 34:1004-1018.

8. Garvine, R.W. 1974. Dynamics of small-scale oceanic fronts. J. Phys. Oceanogr. 4:557-569.
9. Hirota, J., and J. Szyper. 1975. Separation of total particulate carbon into inorganic and organic components. *Limnol. Oceanogr.* 20:896-900.
10. Kiefer, D.A. and R.W. Austin. 1974. The effect of varying phytoplankton concentration on submarine light transmission in the Gulf of California. *Limnol. Oceanogr.* 19:55-64.
11. Klemas, V. and D.F. Polis. 1977. Remote sensing of estuarine fronts and their effects on pollutants. *Photogramm. Eng. Rem. Sens.* 43:599-612.
12. Niemeyer, G. 1978. Numerical methods for the simulation of hydrodynamic and ecological processes. Univ. of Hawaii, Hawaii Inst. of Geophys. Rept. 78-1, 127 pp.
13. Parsons, T.R. and M. Takahashi. 1973. Environmental control of phytoplankton cell size. *Limnol. Oceanogr.* 18:511-515.
14. Pritchard, D.W. and J.R. Schubel. 1980. Physical and geological processes controlling nutrient levels in estuaries. This volume.
15. Revelante, N. and M. Gilmartin. 1976. The effects of Po River discharge on phytoplankton dynamics in the northern Adriatic Sea. *Mar. Biol.* 34:259-271.
16. Seliger, H.H., J.H. Carpenter, M. Loftus and W.D. McElroy. 1970. Mechanisms for the accumulation of high concentrations of dinoflagellates in a bioluminescent bay. *Limnol. Oceanogr.* 15:234-245.
17. Simpson, J.H. and J.R. Hunter. 1974. Fronts in the Irish Sea. *Nature* 250:404-406.
18. Simpson, J.H. 1976. A boundary front in the summer regime of the Celtic Sea. *Est. Coastal Mar. Sci.* 4:71-81.
19. Smith, S.V., R.E. Brock and E.A. Laws. In press. Kaneohe Bay: a coral reef ecosystem subjected to stresses of urbanization. *In* D.R. Stoddart and R. Grigg, *eds.* Coral reef ecosystems under stress. Academic Press, London.
20. Strickland, J.D.H. and T.R. Parsons. 1968. A practical handbook of seawater analysis. Fisheries Research Board of Canada, Ottawa.
21. Technicon Industrial Systems. 1973. Autoanalyzer II Industrial Methods. Tarrytown, N.Y.

APPLICATION OF THE OECD EUTROPHICATION MODELING APPROACH TO ESTUARIES

G. Fred Lee and R. Anne Jones
Department of Civil Engineering
Environmental Engineering Program
Colorado State University
Fort Collins, Colorado 80523

ABSTRACT: Approximately five years ago, the Organization for Economic Cooperation and Development (OECD) initiated a 22 country, 200 lake and impoundment study of nutrient load-eutrophication response relationships. Emphasis in the study is being given to the evaluation of the Vollenweider models for correlating nutrient load with eutrophication response. The U.S. part of this study included the investigation of about 40 waterbodies or parts thereof. The results of the U.S. and the other studies all show a strong correlation between the phosphorus load to a water body as normalized by mean depth and hydraulic residence time, and the planktonic algal chlorophyll, Secchi depth (water clarity), and the hypolimnetic oxygen depletion rate. These relationships have been developed to a sufficient degree of sophistication so that they are the method of choice for estimating the impact of altering the phosphorus load to a P-limited waterbody on eutrophication-related water quality.

While the U.S. OECD waterbodies were primarily lakes and impoundments, included as part of the data base upon which the statistical correlations were made were three parts of the Potomac estuary. It appears that in general, the OECD eutrophication modeling approach is applicable to estuarine systems as well as lakes and impoundments. In addition to reviewing the U.S. OECD study results, this paper presents a discussion of the modifications that may need to be made in the OECD-Vollenweider eutrophication modeling approach in order to apply it to some estuarine systems.

INTRODUCTION

With increasing emphasis being placed on management of excessive fertility-eutrophication within United States and other nations' tidal freshwater, estuarine, and marine waters, increased attention is being given to methods of formulating nutrient load-eutrophication response relationships for these waters. A prime example of this interest within the U.S. is the Chesapeake Bay Program, which is the focal point of this conference. Similar programs of this type need to be conducted for the New York Bight and other areas of the U.S. In the case of the New York Bight, in the summer of 1976 a massive area off the New York-New Jersey coast (8 by 100 miles - 13 by 160 km) of hypolimnetic waters became deoxygenated, which resulted in massive destruction of benthic fish and shellfish in this region (7, 15).

A similar type of problem of even greater magnitude occurred over the past several years in the Sato Inland Sea in Japan, where large populations of yellowtail and other fish have been killed because of excessive growths of toxic planktonic algae. Many of the prime coastal recreational areas of the world's waters, such as the Emilia Romagna Italian waters of the northwest Adriatic Sea and the coastal waters of the Costa del Sol (Spain) are showing significant water quality deterioration due to eutrophication which has or will soon adversely affect the use of and recreation-based revenue from these regions.

The OECD eutrophication modeling approach has been found to be a useful tool to predict changes in eutrophication-related water quality that will result from altering the phosphorus load to a waterbody by a given degree. This paper discusses the applicability of these results to nutrient load-eutrophication response modeling for estuarine, tidal freshwater, and marine systems.

CHARACTERISTICS OF THE OECD EUTROPHICATION STUDY PROGRAM

The Organization for Economic Cooperation and Development Eutrophication Study Program was initiated approximately half-a-dozen years ago to develop nutrient load-waterbody response relationships that could be used for management of water quality in excessively fertile waterbodies. It included detailed nutrient load-eutrophication response studies on approximately 200 waterbodies located in 22 OECD member nations in Western Europe, Japan, Australia, and North America. Within the U.S., approximately 40 waterbodies or parts thereof were included within this program, including the Potomac estuary. Two reports have been generated within the U.S. covering these studies. One (21) consisted of a series of individual waterbody reports in which the investigator who had conducted intensive studies of that waterbody prepared a review of these studies and provided data in a standardized form on the characteristics of the waterbody needed to make a load-response assessment. An overall summary report was prepared by Rast and Lee (16) in which the standardized data from each of the individual studies were examined collectively to determine what trends and in particular what nutrient load-eutrophication response relationships could be generated from these data. Lee et al. (12, 13) prepared summary papers discussing the results of the U.S. OECD eutrophication study.

The principal output from the Rast and Lee report are three relationships which are shown in Figure 1. The abscissa in this figure is the areal phosphorus load to the waterbody normalized by the waterbody's mean depth and hydraulic residence time. Vollenweider (24) proposed that this normalization factor would be suitable for use in relating phosphorus load to eutrophication response as measured by planktonic algal chlorophyll. Rast and Lee have shown that his proposals were appropriate and that there is a remarkably good correlation between the normalized phosphorus load and the planktonic algal chlorophyll for approximately 40 U.S. waterbodies. They extended Vollenweider's concept to include the Secchi depth (water clarity) and hypolimnetic

FIGURE 1. U.S. OECD data applied to P load – summer mean chlorophyll, summer mean Secchi depth, and hypolimnetic oxygen depletion rate relationships.

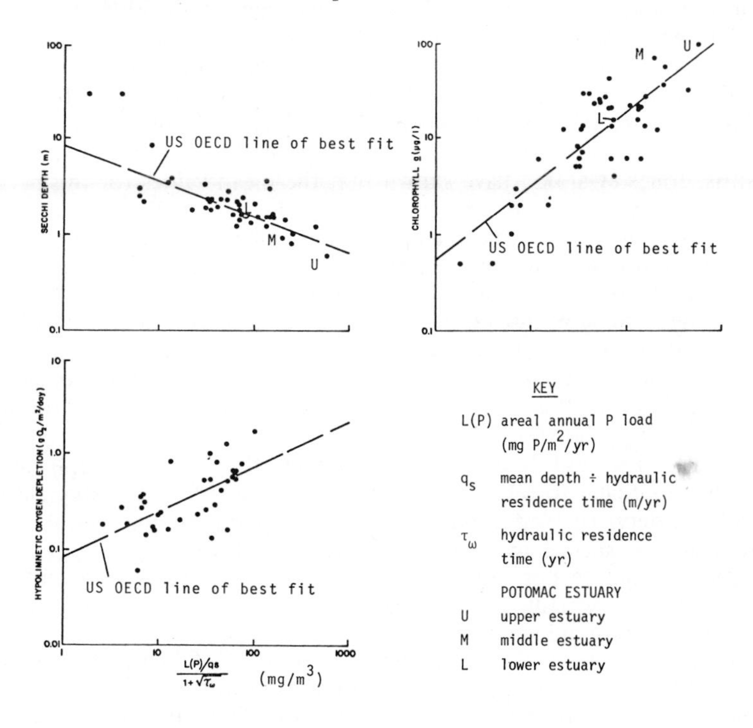

oxygen depletion rate (Figure 1). It should be noted that while the overall OECD eutrophication study results have not yet been published, they will be released in the near future. These results have been compared to the Rast and Lee relationships developed based on the U.S. OECD eutrophication study data base, and it has been found that waterbodies located throughout Western Europe, North America, Japan, and Australia fit these relationships well. As discussed by Vollenweider (25) the normalized P load to a waterbody can also be correlated to the primary productivity in the waterbody.

Since the completion of the Rast and Lee report in the summer of 1977, Lee and his associates have continued to study nutrient load-eutrophication response relationships for a variety of waterbodies in the U.S. and abroad. They have found that the additional 100 or so waterbodies they have studies obey the same load-response relationships that were developed based on the 40 U.S. OECD waterbodies, so that at this time the relationships depicted in Figure 1 have been found to appropriately represent approximately 300 waterbodies. It is now clear that Vollenweider's original concept of relating a normalized phosphorus load to eutrophication-related water quality has general applicability to many waterbodies located throughout the world. In fact, if waterbodies do not fit these relationships, there is a very good indication that either the data which were used as a basis for judging the appropriateness of fit were incorrect or there is something peculiar about the way in which the waterbody utilizes its phosphorus load in developing planktonic algae.

While the relationships shown in Figure 1 for chlorophyll and Secchi depth are based on mean concentrations generally through the summer months, Jones et al. (5) have shown that the mean values for this period can be readily translated into single maximum values. Figure 2 presents the relationship between mean and maximum planktonic algal chlorophyll concentration based on about 90 waterbodies or parts thereof located in various parts of the world. It is evident that below about 50 μg/l chlorophyll which would be generally classified as the lower end of the hypereutrophic range, there is a remarkably simple relationship between mean and maximum chlorophyll concentrations. The same approach can be extrapolated to the water clarity as measured by Secchi depth through the relationships shown in Figure 1.

Recently, Lee and Jones (9) have extended the OECD eutrophication modeling approach to relate normalized phosphorus load to fish yield for a wide variety of waterbodies in various parts of the world. This relationship is shown in Figure 3. It is evident from this figure that phosphorus load as normalized based on the Vollenweider concept is a tremendously powerful predictor of the overall functioning of aquatic ecosystems in producing phytoplankton.

The fact that over 300 waterbodies fit the U.S. OECD load-response relationships imparts considerable credibility to this modeling approach. However, in order to determine the ability of this modeling approach to track changes in water quality with changes in P load, Rast et al. (17) have

FIGURE 2. Relationship between mean and maximum planktonic algal chlorophyll concentration for 90 waterbodies.

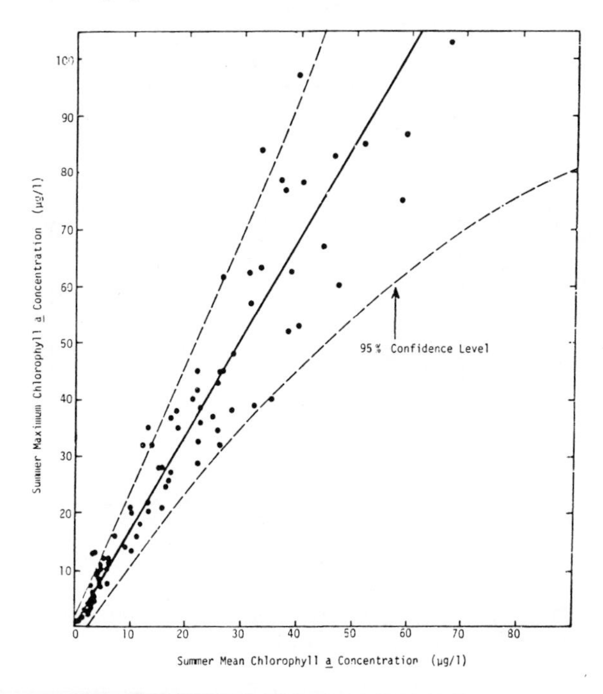

examined "before and after" data for 9 waterbodies which have had changes in their phosphorus loads and for which there were data on the water quality characteristics before and after these changes occurred. The results of this study follow the predictions made by Rast and Lee (16) as to how waterbodies would respond to changes in water quality as a function of changes in phosphorus load. In general, Rast et al. found for a variety of waterbodies located throughout the world that when the P loading term (P load normalized by mean depth and hydraulic residence time) is altered as with a P load reduction, the load-response relationship for that waterbody tracks down parallel to the U.S. OECD eutrophication study lines of best fit shown in Figure 1. It appears that whatever caused a waterbody's load-response relationship to deviate from the U.S. OECD line of best fit was a constant factor in both the "before" and "after" phosphorus loads for the waterbody.

FIGURE 3. Relationship between P load and fish yield. Line of best fit: Log Fish yield = 0.7 log $\{L(P)/q_s)/(1+\sqrt{\tau_\omega})\}$ -1.86. (r^2=0.86)

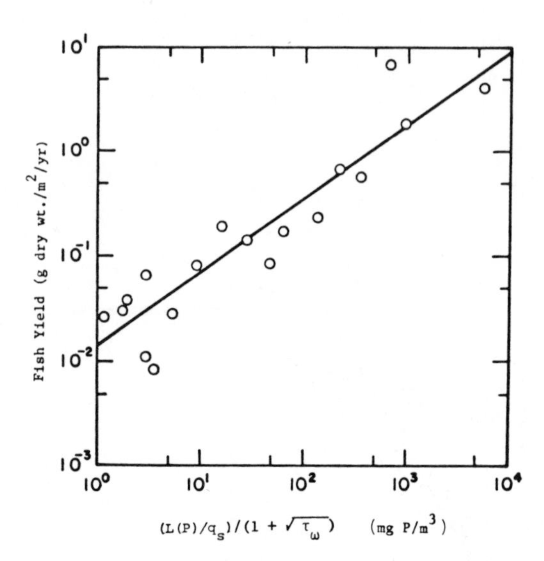

APPLICATION OF OECD EUTROPHICATION STUDY RESULTS TO MANAGEMENT OF EUTROPHICATION-RELATED WATER QUALITY IN ESTUARIES

While the primary focal point of the OECD eutrophication study program is freshwater lakes and impoundments, the U.S. part of this program included three sections of the Potomac estuary. Jaworski (2) has summarized the U.S. EPA studies on the Potomac estuary. He presented a review of these studies at this symposium. Rast and Lee (16) have shown that the data for each part of the Potomac estuary, marked on Figure 1, fall in the same family of points in the U.S. OECD eutrophication studies as freshwater lakes and impoundments.

Table 1 presents various characteristics of the three reaches of the Potomac estuary shown in Figure 4 as determined by Jaworski (2) and reported in Rast and Lee (16). The typical ranges of chlorophyll concentrations and Secchi depth found in sections of each are presented as "measured" levels; those predicted based on the U.S. OECD lines of best fit shown in Figure 1 are presented as "compouted" levels. The computed chlorophyll levels are, for all three reaches, within the range of values typically found; for the upper and lower reaches, the computed value is approximately equal to the average of the high and low measured values. When the computed mean concentration values are converted to maximum values using the relationship shown in Figure 2, maximum

TABLE 1. Characteristics of the Potomac Estuary[1]

Parameter	Reach of Potomac Estuary[2]		
	Upper	Middle	Lower
Mean depth (m)	4.8	5.1	7.2
Hydraulic residence time (yr)	0.04	0.18	0.85
Areal P loading ($gP/m^2/yr$)	85	8	1.2
Mean chlorophyll (μg/l)			
measured	30-150	30-100	10-20
computed[3]	80	31	13
Maximum chlorophyll - computed (μg/l)	136	53	22
Mean Secchi depth (m)			
measured	0.4-0.8	0.5-1.3	1.0-2.3
computed[3]	0.8	1.3	1.6

[1]After Jaworski[2] - Rast and Lee[16].

[2]See Figure 4.

[3]Using US OECD line of best fit.

chlorophyll levels on the order to the high end of the measured range are found. Also as seen in Table 1, the U.S. OECD load-response relationship provided a good approximation of the Secchi depth values on all three reaches of the Potomac estuary.

While at first glance it may seem to be somewhat extraordinary that lakes and estuaries behave in the same manner with respect to algal use of phosphorus, actually when one examines the nature of the OECD eutrophication modeling, approach, it is evident that this approach is based on the fact that a certain amount of phosphorus is needed in a waterbody to produce a certain crop of phytoplankton. It is important to emphasize that the OECD eutrophication modeling approach is currently limited in its applicability to phosphorus-limited systems where the peak of the phytoplankton biomass is governed by the amount of available phosphorus present in the water column. The Vollenweider approach is simply employing what are sometimes called the "Redfield numbers," which indicate that phytoplankton are composed of approximately 106 carbon atoms for every 16 nitrogen and every one phosphorus atom and that algae take up these three elements in roughly that proportion. While individual species of algae show differences in the C:N:P stoichiometry, freshwater, estuarine, and marine algae all have on the order of the same ratios. It makes little difference whether the phytoplankton are marine or freshwater. They have essentially the same composition and nutrient requirements.

FIGURE 4. Potomac estuary.

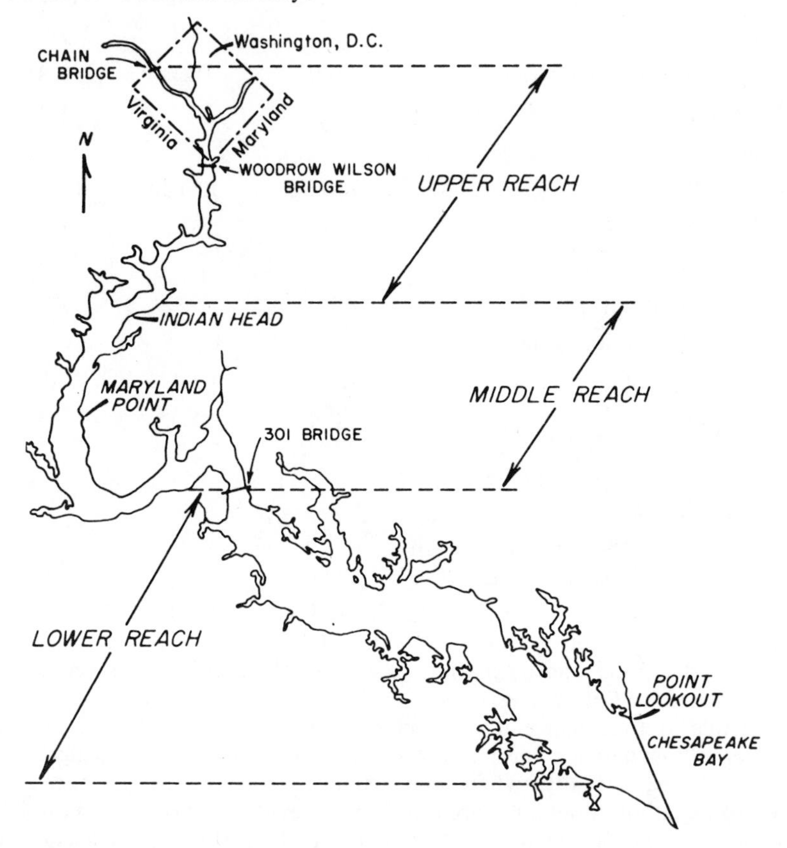

One of the most significant questions about the application of the OECD eutrophication modeling approach to estuaries is: What is an appropriate mean depth and hydraulic residence time for the estuarine system? Jaworski, in developing the data for the Potomac estuary, has assumed a simple plug flow model based on water displacement. While this approach worked well for the Potomac estuary, there may be situations, especially where there is a well developed salt wedge, where the plug flow assumption is not applicable or where the approach needs to be modified to consider mixing between surface freshwater and bottom salt water. Also, the appropriate mean depth to use may not be the morphological mean depth, but rather the depth of the surface freshwater layer. Lee and Jones (6) suggested that this modification in mean depth may be necessary for applying the OECD eutrophication modeling approach to the Emilia Romagna Italian coastal waters of the Adriatic Sea.

Another factor that can influence the applicability of the U.S. OECD eutrophication modeling approach to estuarine systems is the fact that many estuaries have sufficient amounts of inorganic and organic turbidity to limit phytoplankton growth. This is especially true along the south Atlantic and the northern Florida through Texas Gulf Coast. In these systems, the non-phytoplankton related turbidity may prevent the biomass from developing to its maximum potential based on the nutrient concentration of the water. Under these conditions, the chlorophyll that is observed during the summer growing season would average somewhat less than that predicted based on the OECD eutrophication modeling results because of the fact that the phytoplankton growth is more severely light-limited than it would be if the turbidity were due only to phytoplankton.

The first step in establishing appropriate phosphorus loads to a waterbody is to be certain that phosphorus does limit or can be made to limit maximum phytoplankton production during the time of year of water quality concern. This type of determination can best be made by examining the soluble orthophosphate, nitrate, and ammonia concentrations during the period of water quality concern. If phosphorus is limiting peak algal biomass production, then the available P concentration found during peak biomass will be on the order of a few μgP/l. Similarly for nitrogen to be limiting, the sum of the concentrations of nitrite, nitrate, and ammonia will be on the order 15 to 50 μgN/l during this period. If neither available N nor P is reduced to these levels during the period of maximum biomass, then the concentrations of N and/or P are not likely significantly limiting algal growth. Light is generally always a limiting factor in governing phytoplankton development within a waterbody, but given sufficient time, the importance of nutrients as a limiting factor may become dominant for any particular light conditions. It is this condition that is of importance in managing eutrophication-related water quality.

In many marine waters, phytoplankton growth is limited by nitrogen rather than phosphorus. Therefore, there may be a small region in the upper part of the estuary where phytoplankton growth is phosphorus limited during the summer period but as the marine waters are mixed with fresh water, the system should shift over to nitrogen limitation. Since many estuarine systems have nearby population centers and large amounts of domestic wastewaters are discharged into the estuary near the head end, and since the nitrogen to phosphorus ratio in domestic wastewaters is strongly in favor of nitrogen limitation for phytoplankton growth, it is conceivable that some estuaries will have few areas where phosphorus is the element limiting maximum phytoplankton biomass production. If the situation occurs that the estuary of concern is not P limited, then the OECD eutrophication modeling approach will not, without modification, accurately predict biomass as measured by chlorophyll, since these relationships were derived primarily based on waterbodies in which maximum biomass production was limited by phosphorus content of the water. When nitrogen limits peak algal biomass, the relationship between algal biomass and P concentration or load is on the plateau of a plot of algal biomass as a function of P load (or concentration); additional P input does not result in additional algal biomass. In order to apply the Vollenweider-OECD eutrophication modeling approach, a determination must be made of how much P load reduction must take place to cause phosphorus to limit algal biomass production. When that P load is reached, then further reductions in P load will result in decreased algal growth, i.e., the biomass-P concentration relationship is on the steep portion of the plot. In order to make this assessment, the mean available nitrogen concentration during the growing season of water quality concern must be divided by 7.5 to determine the mean P concentration necessary to attain the "optimum" algal stoichiometric ratio of 16:1 N:P atomic ratio discussed previously. At P concentrations below this level, P becomes the potentially limiting element. The P load corresponding to the calculated P concentration can be estimated through the use of the $(L(P)/q_s)/(1 + \tau_\omega)$ term (where L(P) is the areal annual P load, and q_s is the mean depth ÷ hydraulic residence time (τ_ω)) since it is equivalent to the average P concentration in the waterbody.

It is important that all determinations of limiting nutrients be conducted during the period of water quality concern. Algal bioassays such as those developed by the U.S. EPA (27, 28) for determining limiting nutrients should be performed on water samples collected during the period of concern. It has been found that N:P ratios measured and algal bioassays conducted at other times of the year may yield unreliable estimates of limiting nutrients during the critical water quality period. This is a result of the fact that the N:P ratio is a function of the relative rates of supply of available N and P from external as well as internal sources.

In the freshwater part of the estuary, there is potential for blue-green algal development, some of which are nitrogen fixers at times. Under these conditions, of course, the nitrogen load would have to be adjusted for nitrogen fixation since much higher biomasses are possible when nitrogen fixation is progressing at a rapid rate. This does not appear to be a significant problem in the marine part of the estuary where the nitrogen-fixing blue-greens do not seem to be predominant.

The OECD eutrophication modeling approach is patterned from the developments of Vollenweider (22, 23) in which he attempted to relate phosphorus load to a mean depth/hydraulic residence time quotient where, as shown in Figure 5, two curved lines represent "excessive" and "permissible" loading lines. It is important to emphasize, as pointed out by Jaworski (2), that this approach may not be valid for some estuarine systems. As discussed by Rast and Lee (16), the original concept of

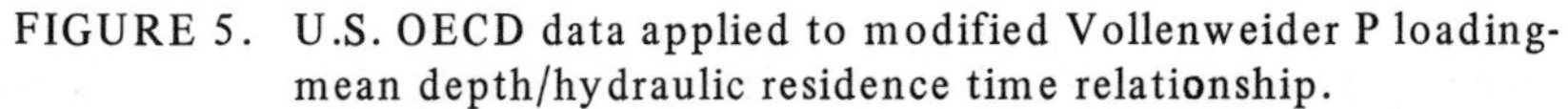

FIGURE 5. U.S. OECD data applied to modified Vollenweider P loading-mean depth/hydraulic residence time relationship.

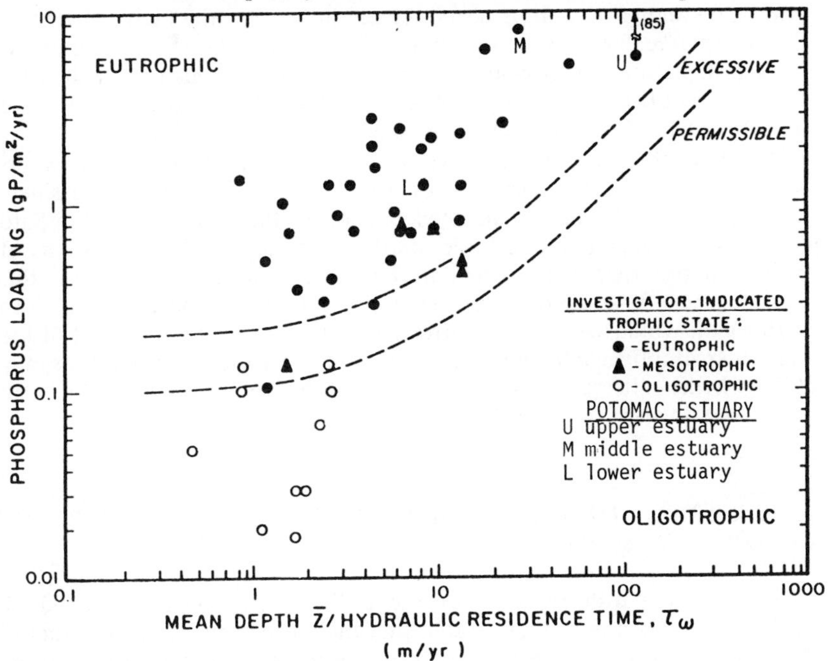

excessive and permissible loadings by Vollenweider was based on the work of Sawyer (20) who correlated nitrogen and phosphorus concentrations in 22 southern Wisconsin waterbodies with his subjective assessment of water quality in these waterbodies in terms of their recreational use. It is important that the concept of excessive and permissible nutrient loads be established on a case-by-case evaluation since what might be deemed "excessive" in tropical waters might be satisfactory in many other estuarine waters. It is far better to work out an approach whereby a certain phytoplankton chlorophyll concentration is established as a desirable goal and then the appropriate corresponding phosphorus load needed to attain this goal is determined based on the waterbody's mean depth and hydraulic residence time. The greeness of a waterbody is an important parameter by which water quality is assessed. The public's perception of greenness varies widely, depending primarily on the background turbidity in the system. At low levels of organic color and detritus and inorganic materials suspended in the water column, a relatively small increase in phytoplankton chlorophyll is readily discernible and objectionable to many individuals. In more turbid systems, however, the chlorophyll content can increase significantly and still not significantly alter a beneficial use of the water. It is important to set water quality objectives based on impairment of beneficial use and not some arbitrarily established phytoplankton chlorophyll value. For some estuarine systems as well as nearshore marine waters, the factor governing the acceptable level of chlorophyll - P load may be the hypolimnetic oxygen depletion which would in some years tend to destroy the benthic fish and shellfish below the thermocline-chemocline in the area, such as occurred in 1976 off the coast of New Jersey.

Another effect of excessive fertilization of estuarine and marine waters which may alter the positions of the "excessive" and "permissible" loading lines is the one mentioned previously that occurred in the Sato Inland Sea in Japan, where toxic dinoflagellate blooms killed millions of yellowtail fish that were being cultured for commercial purposes. If this were the beneficial use of a water, the nitrogen and/or phosphorus load to that region would have to be correlated to the chlorophyll content of the particular dinoflagellate which would be toxic to fish. These numbers may be markedly different than what is typically found to interfere with recreational use of the waterbody.

APPLICATION TO WATER QUALITY MANAGEMENT FOR THE POTOMAC ESTUARY

Jaworski (2) utilized some of the initial relationships developed by Vollenweider as a means of establishing the phosphorus loads that should be obtained in order to achieve a certain water quality in various parts of the Potomac estuary. Since Jaworski's efforts along this line, the OECD Eutrophication Study Program has been completed with the result that a much larger data base is now available for establishing lines of best fit between load and response for various types of waterbodies. This section of this paper uses the data obtained by Jaworski to determine the eutrophication-related water quality that will result from altering the P load to the upper Potomac estuary.

Jaworski (2) reported that phytoplankton biomass production in the upper and middle portions of the Potomac estuary becomes nitrogen limited in the summer months. Examination of Jaworski's data (4) on N and P concentrations shows that N limitation during the summers when the data used herein were collected, is unlikely. If a nutrient is limiting, it is more likely to be P. With a small P load reduction, the phytoplankton growth in the upper Potomac would become P limited; therefore the OECD eutrophication modeling approach should be applicable to predicting the planktonic algal chlorophyll that will develop from a certain P load reduction to the upper Potomac estuary.

Jaworski (2) estimated that in 1970, 10,900 kg/day of phosphorus were added to the upper reach of the Potomac estuary from domestic wastewater sources. Since phosphorus in domestic wastewaters can be readily reduced by about 90 percent by chemical precipitation with iron or aluminum salts during normal secondary wastewater treatment for a cost of less than a quarter of a cent per person per day, the phosphorus load from direct discharge of wastewater treatment plant effluent to the upper estuary could therefore be readily reduced to approximately 1,000 kg/day. Using the relationships developed in the U.S. OECD eutrophication study, achieving this load (assuming that 50 percent of the upper Potomac basin load would also be removed as a result of wastewater P removal (2)) would result, under low flow conditions of 40 m^3/sec, in an average summer planktonic algal chlorophyll of about 45 μg/l, and under median flow of 185 m^3/sec, about 23 μg/l. Jones et al. (5) developed a statistical correlation which showed that the maximum summer planktonic algal chlorophyll concentration is about 1.7 times the summer mean. Therefore, summer maxima would be predicted to be on the orders of 77 and 39 μg/l (for low and median flows, respectively) for the upper parts of the Potomac estuary if 90 percent P removal took place in the basin. Jaworski (2) reported that a 1,000 kg P/day domestic wastewater treatment plant P load (with an assumed 50 percent reduction in upper basin P load) would result in a 25 μg/l chlorophyll concentration in the upper estuary. This was estimated using a dynamic modeling approach and agrees with what is predicted based on the OECD modeling approach. A 25 μg/l mean chlorophyll concentration does not, however, generally represent good water quality for most recreational purposes. Such a water would experience excursions of chlorophyll concentration about 40 μg/l during the summer, according to the relationship developed by Jones et al. (5). It is evident that much smaller loads will have to be attained in order to achieve a generally acceptable eutrophication-related water quality in the upper parts of the Potomac estuary.

Jaworski (2) indicated using another approach, that it would be necessary for the nutrient loads to be reduced to 1913-1920 conditions (about 600 kg P/day) to have no major aquatic plant nuisances (in the upper estuary?). Using the U.S. OECD eutrophication model, this load would result in mean chlorophyll concentrations in the upper estuary of 8.3 and 25 μg/l during median and low flows, respectively. While achieving this level would significantly improve the eutrophication-related water quality, it would not reduce aquatic plant nuisances in this area to levels which are generally considered acceptable for recreational uses. It should be noted that according to Jaworski (Personal Communication, 1980), substantial P load reductions have been attained at the wastewater treatment plants in the upper estuary watershed and a noticeable improvement in eutrophication-related water quality has resulted.

It is possible through advanced wastewater treatment processes to reduce the phosphorus content of domestic wastewaters to a few tenths of a mg/l P. This would represent a significant increase in cost over the 1 mg/l effluent phosphorus that is readily achievable by iron or aluminum hydrous oxide coprecipitation. There are also some questions about the potential benefits of the removal of phosphorus to this extent from domestic wastewaters because of the fact that the additional costs are largely associated with removal of particulate forms of organic and inorganic phosphorus, some of which may not be available to support phytoplankton growth in the receiving waters. Lee et al. (11) have recently completed a comprehensive review on the current state of knowledge on the measurement and environmental significance of various forms of phosphorus that are important in developing a phosphorus management strategy for a waterbody. They pointed out that before any large scale significantly improved sedimentation or filtration is initiated on domestic wastewaters for phosphorus control, studies should be conducted to determine how much of the phosphorus that will be removed by increased particle removal in the treatment plant, would be available to support phytoplankton growth in the receiving waters. These studies must be done using algal bioassays. As discussed by Lee et al. (11), at this time there are no reliable chemical procedures to determine whether phosphorus in a wastewater sample is available to support algal growth in receiving waters. The chlorophyll concentration that would be expected to result from achieving a 0.1 mg P/l effluent concentration would be about 25 μg/l provided that the additional P removed below the 1 mg P/l effluent level was largely available to support algal growth. A 25 μg/l average summer chlorophyll would generally cause the water quality to be considered somewhat degraded.

It is important to note that the various loadings that have been used in developing these chlorophyll estimates are based on Jaworski's early 1970 work. While the average and maximum chlorophyll concentration predicted by the OECD eutrophication model for the P loads would be applicable to the situation today, the phosphorus loads that could be achieved by reducing phosphorus in domestic wastewaters by various amounts has changed since 1970, due to increased population in the Potomac estuary area and voluntary reduction of the P content in household laundry detergent formulations. A comparison should be made between current phosphorus loadings to the upper Potomac from domestic wastewater treatment plants achieving various degrees of phosphorus removal and those used in these computations. If there are significant differences, then adjustments should be made in these values to correct for these differences.

One of the proposed solutions for minimizing eutrophication in the Potomac estuary area is the passage of a detergent phosphate limitation. At the time that Jaworski conducted his studies of the phosphorus content of domestic wastewaters across the U.S., approximately 50 percent of the P in domestic wastewaters was derived from phosphorus use as a builder in household laundry detergents. Today this value has been voluntarily decreased in all parts of the U.S. to about 25 percent, which means that the benefit of improved water quality arising from the removal of phosphorus from detergents is much smaller today than it was ten years ago. In the case of the upper reaches of the Potomac estuary, a detergent P ban would not drop the P concentrations sufficiently to have a discernible effect on water quality.

APPLICATION OF OECD MODELING TO UPPER CHESAPEAKE BAY

Jaworski (3) has provided sufficient information on the upper Chesapeake Bay to enable application of the OECD eutrophication modeling approach to this waterbody. For the main part of the Bay, Jaworski estimated for the period 1969-1971 a mean depth of 6.5 m, average hydraulic residence time of 1.2 years, and a P load of 1.3 g/m^2/yr. Using these values and the U.S. OECD eutrophication study line of best fit relating normalized P load to chlorophyll, a summer average planktonic algal chlorophyll concentration of 26 μg/l is predicted. This estimate is in agreement with the data reported by Salas and Thomann (19) of approximately 20 to 30 μg/l average chlorophyll for the summer of 1965. This close match provides additional verification of the applicability of the OECD modeling approach to estuarine systems.

Jaworski (3) estimated that approximately 70 percent of the P load to the upper part of the Chesapeake Bay is from domestic wastewater sources. Since approximately 90 percent of the P in domestic wastewaters can be readily removed using chemical precipitation techniques in normal secondary wastewater treatment, it is of interest to estimate the potential water quality benefits that could be derived from initiating this removal in the upper Chesapeake Bay watershed. Attainment of a 0.5 gP/m^2/yr load will result in a mean summer chlorophyll concentration of 11 μg/l. This P loading can be achieved through initiating P removal at the domestic wastewater treatment plants in the upper Chesapeake Bay basin at a cost to the residents of less than a half a cent per person per day. Following this approach would result in a significant increase in the overall eutrophication-related water quality in the upper Chesapeake Bay. This part of the Bay could be changed from hypereutrophic to the transition area between mesotrophic and eutrophic based on the results of the OECD eutrophication study program. As with the upper Potomac, passage of a detergent P limitation would not significantly change the water quality of the upper Chesapeake Bay.

COMPARISON OF OECD AND DYNAMIC MODELING APPROACHES FOR EUTROPHICATION-RELATED WATER QUALITY MANAGEMENT

Two basic approaches have developed for estimating the benefits in eutrophication-related water quality that can accrue from controlling phosphorus inputs to a waterbody by a certain amount. One of these, the dynamic modeling approach, formulates the various processes that govern the transformation of phosphorus, nitrogen, carbon, sunlight, and other inputs into phytoplankton biomass, into a series of differential equations which, when solved simultaneously for a certain phosphorus load, provide an estimate of the planktonic algal biomass that will be present in a waterbody. The other approach is that originally developed by Vollenweider in which a relatively simple correlation is established between a normalized phosphorus load and the planktonic algal chlorophyll. It is important to note that while the dynamic modeling approach can be simplified to express the same general relationships as the OECD eutrophication modeling approach, this does not mean that they are both equally valid as management tools. There are very significant differences in the validity of these approaches in predicting P load-eutrophication response relationships for waterbodies. The dynamic modeling approach in general does not cover all of the physical, chemical, and biological processes that occur in waterbodies adequately to be system independent. Therefore, dynamic models must be tuned to each particular system. This tuning process is by far the greatest deficiency of these models and frequently limits their applicability for predicting planktonic algal chlorophyll under significantly altered phosphorus loading conditions. As discussed by O'Connor at this symposium, attempts have been made to use the dynamic modeling approach on a number of estuaries. It has failed to properly predict the biomass that developed in the system after conditions were altered. This failure would be expected of any dynamic model that must be tuned to each particular system. These system-dependent models should not be used for management purposes until this deficiency has been corrected.

An example of the lack of appropriate predictability of dynamic models for management of water quality in the Great Lakes was a model developed for prediction of hypolimnetic oxygen depletion in Lake Erie. A task group of the U.S.-Canadian Great Lakes Water Quality Agreement drafting committee, using a dynamic model, predicted that oxygen could be maintained in the hypolimnion of Lake Erie with a P load to this lake of 11,000 mt/yr. Lee and Jones (8) and Lee et al. (14) using the OECD eutrophication modeling approach, Vollenweider (26) using an independent modeling appraoch, and Chapra (1) have shown that this number is far in excess of the P load that can be allowed and still maintain oxygen in the hypolimnion of this lake under all conditions. According to the calculations of Vollenweider and of Lee and his associates, this lake can receive no more than on the order of 2,000 to 4,000 mt P/yr if oxygen is to be maintained in its hypolimnion under all conditions. For further discussion of this topic consult Lee et al. (14).

The OECD eutrophication modeling approach is also tuned; however, the tuning process is based on a statistical correlation involving load-response relationships for approximately 300 waterbodies in Western Europe, North America, Japan, and Australia. Because of the very large data base available in support of the OECD eutrophication modeling, it is the modeling approach of choice in predicting load-response relationships for waterbodies. Verification of the reliability of this approach has been accomplished in two ways. First, the original OECD load-response relationships, based on the 40 U.S. OECD waterbodies, have been found to be applicable to about 300 waterbodies located in the U.S., Canada, Western Europe, Japan, and Australia. Further, Rast et al. (17) have reviewed data on 9 waterbodies to which there has been a significant change in the P load and for which there was sufficient information on the characteristics before and after the P load change to apply the OECD eutrophication modeling approach. Their review has shown that these waterbodies track properly, i.e., parallel to the U.S. OECD line of best fit for the P load-planktonic algal chlorophyll relationship. The new chlorophyll level in these waterbodies would be reliably predicted based on the altered P load.

One of the most significant advantages of the OECD eutrophication modeling approach over the dynamic modeling approach is the marked difference in the amount of data needed to implement the two approaches. The OECD approach requires a minimal amount of data that can be readily obtained at a low cost. The dynamic modeling approach, because it must be tuned to each particular waterbody, requires a substantial data base; even after a seemingly sufficient data base is obtained it may be found that the model cannot be applied without undertaking a substantial additional research effort. Lee et al. (10) found this to be true in connection with applying the dynamic modeling approach to Lake Ray Hubbard near Dallas, Texas, where much more than simply adjusting the coefficients used in the model for the increased temperature was needed to make this approach fit. Even after a $300,000 15-month study of this waterbody, insufficient information was available to properly tune the model. One of the biggest difficulties in tuning dynamic models is their failure to properly model the exchange between water and sediments for nitrogen and/or phosphorus. Failure of these models to properly handle this exchange was another reason the dynamic model evaluated by Lee et al. (10) could not be fit to Lake Ray Hubbard.

DISCUSSION

It has been found through the OECD Eutrophication Study Program that the nutrient load-eutrophication response relationships originally developed by Vollenweider are a viable tool for establishing management programs for eutrophication-related water quality. While originally developed for lakes and impoundments, they have demonstrated applicability to the three reaches of the Potomac estuary. From a fundamental point of view, based on the nature of the relationships developed by Vollenweider, they should be equally applicable to all estuarine systems. The important parameters that must be evaluated for a particular estuary in the application of the OECD eutrophication model are the appropriate mean depth and hydraulic residence time. It is suggested that initially the approach developed by Jaworski of assuming a plug flow and morphological mean depth be tried. If chlorophyll concentrations measured during the summer and those predicted using these assumptions in the OECD eutrophication models are in reasonable agreement, then the simple assumptions for mean depth and hydraulic residence time would likely be appropriate for use with the system. If, however, there are major discrepancies between measured and predicted levels, then it would be necessary to modify the methods used to determine the mean depth and hydraulic residence time to consider the position and dynamics of the salt wedge. It is important in this tuning process to keep the influence of the salt wedge realistic in terms of expected hydrodynamic behavior. If assuming potentially reasonable mean depths and mixing between the salt wedge and the overlying water does not result in the chlorophyll falling within the population of data obtained in the U.S. OECD study, then this approach should not be used and further studies should be conducted to determine why this discrepancy exists. Until such time as sufficient experience has been gained in applying this approach to water quality management of estuarine systems, it will be necessary to check the validity of the load-response relationships for each particular estuary by comparing the predicted and measured chlorophyll concentrations.

ACKNOWLEDGEMENTS

We wish to acknowledge the U.S. EPA-Corvallis for the support of the U.S. OECD Eutrophication Study which is the basis for this paper. The assistance of Walter Rast of the International Joint Commission, Washington, D.C. in the U.S. OECD study and in subsequent work which aided in the preparation of this paper is also greatly appreciated. The authors express their appreciation to N. Jaworski for his review of this paper and his assistance in providing additional data. The preparation of this paper was supported by the Department of Civil Engineering, Colorado State University, Fort Collins, Colorado, and EnviroQual Consultants and Laboratories, Fort Collins, Colorado.

REFERENCES

1. Chapra, S., NOAA, Ann Arbor, Michigan. Personal Communication to W. Rast, IJC, Washington, D.C. May, 1979.
2. Jaworski, N.A. 1977. Section VIII - Multiple-State Lakes and Special Topics Limnological Characteristics of the Potomac Estuary. *In* L.

Seyb and K. Randolph, North American Project-A Study of U.S. Water Bodies, EPA-600/3-77-086, U.S. EPA-Corvallis.
3. Jaworski, N.A. 1979. Sources of nutrients and the scale of eutrophication problems in estuaries. Proc. International Symposium on Effects of Nutrient Enrichment in Estuaries, Williamsburg, Va. May 1979.
4. Jaworski, N.A., L.J. Clark, and K.D. Feigner. 1971. A water resources - water supply study of the Potomac estuary. Technical Report 35. Chesapeake Technical Support Lab, Middle Atlantic Region, U.S. EPA, Annapolis, Md.
5. Jones, R.A., W. Rast, and G.F. Lee. 1979. Relationship between summer mean and maximum chlorophyll *a* concentrations in lakes. *Environ. Sci. & Technol.* 13:869-870.
6. Lee, G.F., and R.A. Jones. 1977. An approach for development of a eutrophication control program for the Emilia Romagna coastal waters of the Adriatic Sea. Occasional Paper No. 27, Environmental Engineering, Colorado State University, Fort Collins, Col.
7. Lee, G.F., and R.A. Jones. 1977. An assessment of the environmental significance of chemical contaminants present in dredged sediments dumped in the New York Bight. New York District Corps of Engineers, Contract No. DACW 51-77-C-0011, Ocassional Paper No. 28, Environmental Engineering, Colorado State University, Fort Collins, Col.
8. Lee, G.F., and R.A. Jones. 1979. Water quality characteristics of the U.S. waters of Lake Ontario during the IFYGL and modeling contaminant load-water quality response relationships in the nearshore waters of the Great Lakes. Report to National Oceanic and Atmospheric Adminstration Great Lakes Research Laboratory, Ann Arbor, Mich.
9. Lee, G.F., and R.A. Jones. 1979. (In press). Effect of eutrophication on fisheries. American Fisheries Society, Bethesda, Md.
10. Lee, G.F., M. Abdul-Rahman, and E. Meckel. 1978. A study of eutrophication, Lake Ray Hubbard, Dallas, Texas. Occasional Paper No. 15, Environmental Engineering, Colorado State University, Fort Collins, Col.
11. Lee, G.F., R.A. Jones, and W. Rast. 1980. Availability of phosphorus to phytoplankton and its implications for phosphorus management strategies, *In*: Phosphorus Management Strategies for Lakes, Ann Arbor Science, Ann Arbor, MI.
12. Lee, G.F., W. Rast, and R.A. Jones. 1977. Recent advances in assessing aquatic plant nutrient load-eutrophication response for lakes and impoundments. Occasional Paper No. 14, Environmental Engineering, Colorado State University, Fort Collins, Col.
13. Lee, G.F., W. Rast, and R.A. Jones. 1978. Eutrophication of waterbodies: insights for an age-old problem. *Environ. Sci. & Technol.* 12:900-908.
14. Lee, G.F., W. Rast, and R.A. Jones. 1979. Use of the OECD eutrophication modeling approach for assessing Great Lakes' water quality. Occasional Paper No. 42, Environmental Engineering,

Colorado State University, Fort Collins, Col. *In* G.F. Lee and R.A. Jones. 1979. Water quality characteristics of the U.S. waters of Lake Ontario during the IFYGL and modeling contaminant load-water quality response relationships in the nearshore waters of the Great Lakes. Report to National Oceanic and Atmospheric Administration Great Lakes Research Laboratory, Ann Arbor, Mich.

15. Paulson, G. 1976. Report on the fish kill off the New Jersey coast and its potential causes and consequences. State of New Jersey Department of Environmental Protection, Trenton, N.J.
16. Rast, W., and G.F. Lee. 1978. Summary analysis of the North American (US portion) OECD Eutrophication Project: Nutrient loading-lake response relationships and trophic state indices. EPA-600/3-78-008.
17. Rast, W., G.F. Lee, and R.A. Jones. 1981. Predictive capability of the U.S. OECD phosphorus loading eutrophication response models. (In press).
18. Ryther, J.H., and W.M. Dunstan. 1971. Nitrogen, phosphorus and eutrophication in the coastal marine environment. *Science* 171:1006-1013.
19. Salas, H.J., and R.V. Thomann. 1978. A steady-state phytoplankton model of Chesapeake Bay. *Jour. Water Poll. Control Fed.* 50:2752-2770.
20. Sawyer, C.N. 1947. Fertilization of lakes by agricultural and urban drainage. *Jour. New Engl. Water Works Assoc.* 61:109-127.
21. Seyb, L., and K. Randolph. 1977. North American Project-A Study of U.S. Water Bodies, EPA-600/3-77-086, U.S. EPA-Corvallis.
22. Vollenweider, R.A. 1968. Scientific fundamentals of the eutrophication of lakes and flowing waters with particular reference to nitrogen and phosphorus as factors in eutrophication. Technical Report DAS/CSI/68, OECD, Paris.
23. Vollenweider, R.A. 1975. Input-output models, with special reference to the phosphorus loading concept in limnology. *Schweiz. Z. Hydrol.* 37:53-84.
24. Vollenweider, R.A. 1976. Advances in defining critical loading levels for phosphorus in lake eutrophication. *Mem. Ist. Ital. Idrobiol.* 33:53-83.
25. Vollenweider, R.A. 1979. Das Nahrstoffbelastungs-konzept als Grundlage fur den externen Eingriff in den EutrophierungsporzeB stehender Gewasser und Talsperren, Symposium des Wahnbachtalsperrenver bandes Veroffentlichung der Vortrage, Z. F. Wasser - und Abwasser-Forschung, 12, Jahrgang Nr2/79.
26. Vollenweider, R.A. 1979. Canada Centre for Inland Waters, Burlington, Ontario. Personal Communication to G. Fred Lee, Environmental Engineering, Colorado State University, Fort Collins, Col.
27. U.S. EPA. 1971. Algal Assay Procedure Bottle Test. U.S. EPA-Corvallis, Ore.
28. U.S. EPA. 1974. Marine Algal Assay Procedure Bottle Test. U.S. EPA-Corvallis, Ore.

PHOSPHORUS AND NITROGEN LIMITED PHYTOPLANKTON PRODUCTIVITY IN NORTHEASTERN GULF OF MEXICO COASTAL ESTUARIES

Vernon B. Myers* and Richard I. Iverson**
**Florida Department of Environmental Regulation*
2600 Blairstone Rd.
Tallahassee, Florida 32301

***Department of Oceanography*
Florida State University
Tallahassee, Florida 32306

ABSTRACT: An understanding of nutrient limitation of estuarine phytoplankton growth is important in making environmentally sound decisions concerning watershed development and the use of aquatic environments for waste disposal. Experiments to determine nutrient limitation of phytoplankton productivity were conducted monthly during the summers of 1975 and 1976 in several shallow north Florida coastal and estuarine systems by inorganic carbon-14 uptake and phosphorus-32 bioassays. The results of these nutrient enrichment experiments suggest that phosphorus is frequently more important than nitrogen in limiting phytoplankton productivity in nearshore northeastern Gulf of Mexico. A multiple regression model was constructed to determine which combinations of environmental and nutrient variables could explain the most variation of phytoplankton productivity in these coastal systems. The final regression model was:

$$P.P. = 32.1 + 48.4\ S.R.P. = 0.54\ \text{Salinity}$$

where: P.P. is phytoplankton productivity in $\mu gC\ l^{-1} hr^{-1}$, S.R.P. is soluble reactive phosphate in μg-atom PO_4-P l^{-1}, and salinity in part-per thousand. This model explained 64 percent of the variation in phytoplankton productivity.

Nitrogen has been identified as the primary limiting nutrient for phytoplankton in coastal waters and it has been proposed that the removal of phosphorus from marine waste discharges will have little impact on the control of eutrophication. The observation that phosphorus is important in limiting phytoplankton productivity in these coastal and estuarine areas suggests that water quality planning for the coastal zone is best done on a regional basis, with consideration given to local nutrient cycling processes.

INTRODUCTION

Primary productivity in any estuary depends upon complex chemical, physical, and biological processes which control nutrient cycling. Phytoplankton, macroalgae, seagrasses, epiphytes (10, 16, 19), protozoans (12), detritus-bacteria (1, 29), zooplankton (20), sediment-water interactions (21), and watershed drainage (5, 32) are all known to play dynamic roles in phosphorus and nitrogen cycles of shallow estuarine systems. Any process that alters these complex nutrient cycling mechanisms can drastically affect the productivity of an estuary. Watershed urbanization and concomitant nutrient loading have adversely affected several Florida estuaries (31). Currently in Florida, pressure continues for the development of coastal areas, generating increased demand for waste allocation to adjacent estuaries. An understanding of the nutrient limitation of phytoplankton growth in these estuaries is necessary before environmentally sound decisions concerning development can be made.

Study Area

The nearshore environment of the northeastern Gulf of Mexico between the Apalachicola and Econfina Rivers (Figure 1) is characterized by a broad, shallow continental shelf. Water depth at the stations was shallow and varied between 1.4 m and 3.6 m. The region has a low wave energy coast (27), with maximum tidal amplitudes of 1 m. The coastline is fringed with extensive *Juncus* and *Spartina* marshes. The area receives discharge from several rivers: the Econfina, the Aucilla, the St. Marks, the Ochlockonee, the Carrabelle, and the Apalachicola. The eastern section of the study area receives relatively little river discharge and has extensive stands of seagrasses and associated macroalgae (2). The western section of this study area receives large amounts of river discharge from the Ochlokonee and the Apalachicola rivers and has shallow lagoon-type estuaries, bounded by a series of barrier islands.

The estuaries of the northeastern Gulf of Mexico contain commercial fisheries (14). The estuaries and associated marsh systems serve as nursery and feeding areas for oysters, shrimp, blue crabs, mullet, and other commercially important species. The annual oyster harvest from Apalachicola Bay alone accounts for 80 percent of the Florida annual harvest. These estuarine systems have been called one of the least-polluted, least man-influenced subtropical regions in North America (25). However, pressure to develop this region has increased in the past few years. Residential development on the barrier islands has rapidly increased and is relatively unplanned. Modifications to the Apalachicola River have been proposed to increase navigation and aid in industrialization of the area. Port facilities have also been planned for several coastal towns in the region. All of this development has the potential for increasing nutrient mobilization into these estuaries.

The nutrient enrichment experiments discussed in this paper were designed to determine phytoplankton nutrient limitation patterns in these estuaries as part of a basis for estimating the effect of nutrient enrichment on these estuaries.

FIGURE 1. Location of sampling stations in the northeastern Gulf of Mexico. Apal-1A and Apal-7 are in the Apalachicola estuary; ML is between the Apalachicola and Ochlockonee estuaries; Ock-1 and Ock-2 are in the Ochlockonee estuary; and E-12 in the Econfina estuary.

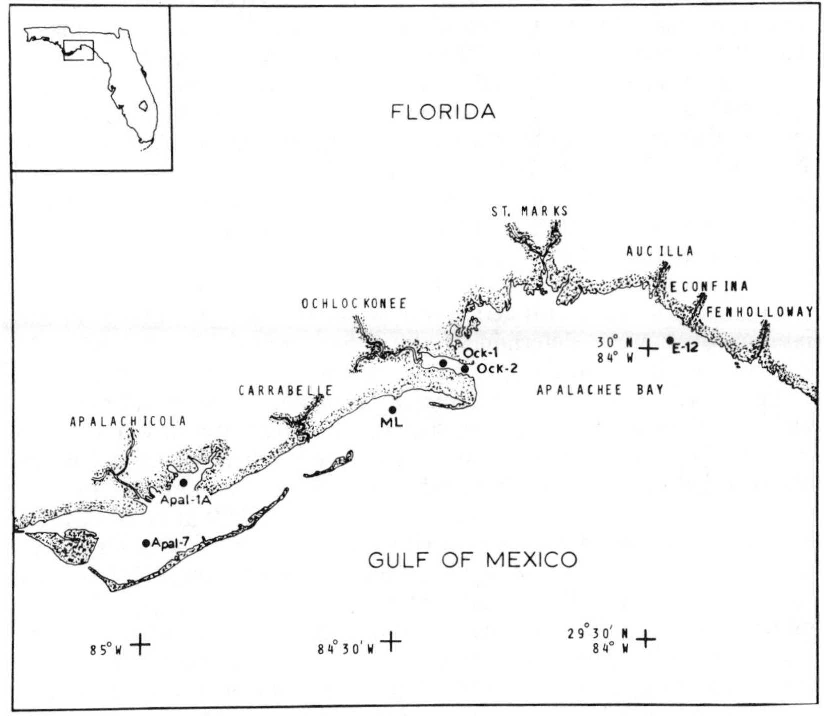

MATERIALS AND METHODS

Water temperature and salinity were measured with a Beckman RS 5-3 in situ portable salinometer equipped with a thermistor. Solar radiation was measured with a Weathermeasure Corporation R401 Mechanical Pyranograph.

Five hundred ml water samples were collected at the surface (z = 0.5m) at the specified stations for nutrient analysis. Samples were immediately filtered through Whatman GF/A glass fiber filters upon collection. One ml of 2 percent $HgCl_2$ solution was added to eliminate microbial activity, after which the samples were placed on ice. All nutrients were analyzed within 48 hours. Soluble reactive phosphate was analyzed by the method of Murphy and Rily as outlined in Strickland and Parsons (24). Briefly, the samples were allowed to react with a combined reagent containing molybdic acid, ascorbic acid, and trivalent antimony. The resulting complex heteropoly acid was reduced and analyzed spectrophotometrically. Nitrate determinations were based on the method of Morris and Riley with modifications given in Strickland and Parsons (24). The samples were passed through a column containing cadmium filings coated with metallic copper, thereby reducing nitrate to nitrite. Nitrite concentration was determined by the method of Randschneider and Robinson given in Strickland and Parsons (24). The method determined nitrite by diazotizing the samples with sulphanilamide and coupling with N-(l-napthyl)-ethylene diamine to form a highly colored azo dye which was analyzed spectrophotometrically. Ammonium was analyzed according to the method of Solorzano given in Strickland and Parsons (24). The samples were treated with an alkaline citrate medium with sodium hypochlorite and phenol in the presence of sodium nitroprusside. The resulting indophenol was measured spectrophotometrically.

Chlorophyll-a was determined by a fluorometric method (24) after the phytoplankton were filtered onto Whatman GF/A filters. Primary productivity was determined by the carbon-14 method (24). Phytoplankton taxonomy was determined by the method of Holmes (8).

Two-factorial nutrient enrichment experiments with nitrogen and phosphorus were conducted on surface phytoplankton in water samples from the Apalachicola Bay system. Water was collected (z = 0.5 m) and placed in 500 ml glass incubation bottles. Nutrient concentrations of 0.0 or 5.0 μg-atom NO_3-N l^{-1} and 0.00, 0.25, 0.50, or 2.00 μg-atom PO_4-P l^{-1} were obtained by adding appropriate amounts of $NaNO_3$ or Na_3PO_4 to each bottle. Phytoplankton were acclimated in situ to the added nutrients for four hours and then incubated in situ with 4 μ Ci of ^{14}C-labeled bicarbonate. Two 100 ml aliquots from each bottle were filtered through Whatman GF/C glass fiber filters. The filters were placed in 5 ml of Aquasol and the activity was determined by liquid scintillation counting.

Phosphate uptake experiments were one-way designs. Surface water samples (z = 0.5 m) were collected and placed in 500 ml glass incubation bottles after Na_3PO_4 was added to obtain concentrations of 0.00, 0.25, 0.50, 2.00 μg-atom PO_4-P l^{-1}. Half the samples were poisoned with 1 ml of 2 percent $HgCl_2$ solution. One ml of approximately 1,000,000 dpm/ml of carrier free ^{32}P-phosphoric acid was added to the bottles, after which samples were incubated in situ. Fifteen ml sub-samples were removed from all bottles at 2, 15, 30, 60, and 120 minutes and filtered through Whatman GF/A glass filters. Ten ml of filtrate were pipeted into LSC vials and the ^{32}P activity was determined by Cherenkov radiation measurements with a liquid scintillation spectrometer (3, 7). Plankton phosphate uptake rates were estimated from linear regression slopes of total phosphate uptake minus $HgCl_2$ poisoned uptake versus time.

RESULTS AND DISCUSSION

Annual Nutrient and Productivity Cycles

The seasonal variability of nutrient and phytoplankton factors in the Econfina estuary are presented in Figure 2. Nutrient fluctuations of this area were trimodal, with distinct seasonal cycles. Nitrate and phosphate seasonal variability were not closely coupled except during the summer and fall. Nitrate concentrations peaked during March-April, October, and December; phosphate peaks occurred during January-February, May-June, and September-October. This suggests that similar mechanisms may be responsible for cycling these two nutrients during the summer-fall period, however, the definitive research has not been completed. Phytoplankton chlorophyll-a concentrations exhibited a bimodal seasonal cycle, peaking in April-May and September-October. This bimodal seasonal pattern is well known for many temperate neritic environments (6). The phytoplankton fluctuations followed water column nutrient changes, especially changes in phosphate concentration, throughout most of the year (March-November). It should be noted that macrophytes dominate the productivity in this area with maximum macrophyte biomass and productivity occurring during the summer (2).

The temporal variability of nutrient and phytoplankton factors for the Apalachicola Bay estuary are presented in Figure 3. The nutrient and phytoplankton levels were an order of magnitude higher in the Apalachicola Bay estuary than in the Econfina estuary. Maximum nitrate and phosphate concentrations occurred during the spring (March-May). Nutrient concentration peaks occurred during the period of maximum river discharge. Several investigators have found strong correlation between nutrient concentration and river discharge (11, 28). Phytoplankton chlorophyll-a concentration was maximum during the summer-fall period (July-October). This is typical for well-mixed south temperate and subtropical estuaries where seasonal productivity cycles closely parallel water temperature changes (12, 23, 28, 32).

Nutrient Limitation Patterns

The following discussion of nutrient limited phytoplankton productivity in the estuarine waters of the northeastern Gulf of Mexico is

FIGURE 2. Seasonal cycles in the Econifina estuary of surface dissolved nitrate (μg-atom NO_3 N L^{-1}), soluble reactive phosphate (μg-atom PO_4 P L^{-1}), and phytoplankton chlorophyll-a (μg L^{-1}) concentrations. Data were collected between 1973 and 1976. Approximate seasonal nutrient cycles are outlined with solid lines.

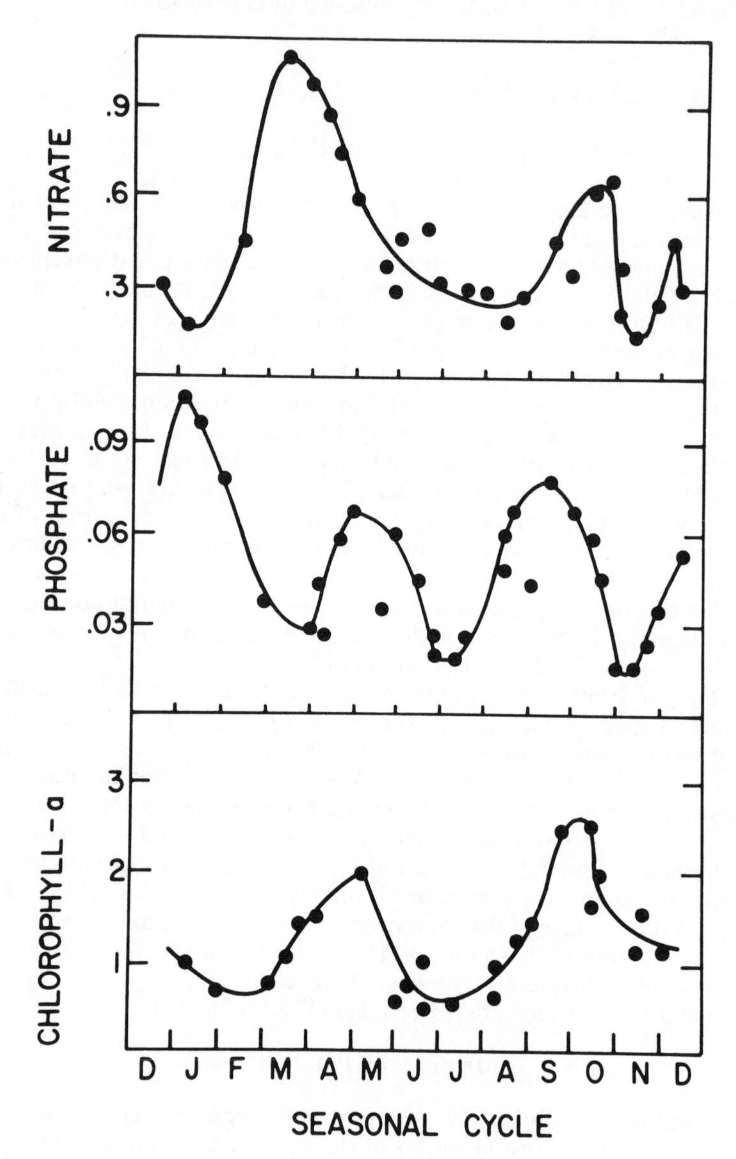

FIGURE 3. Seasonal cycles in the Apalachicola Bay estuary of surface dissolved nitrate (μg-atom NO_3 N L^{-1}), soluble reactive phosphate (μg-atom PO_4 P L^{-1}), and phytoplankton chlorophyll-a (μg L^{-1}) concentrations. Data were collected between 1974 and 1976. Approximate seasonal nutrient cycles are outlined with solid lines.

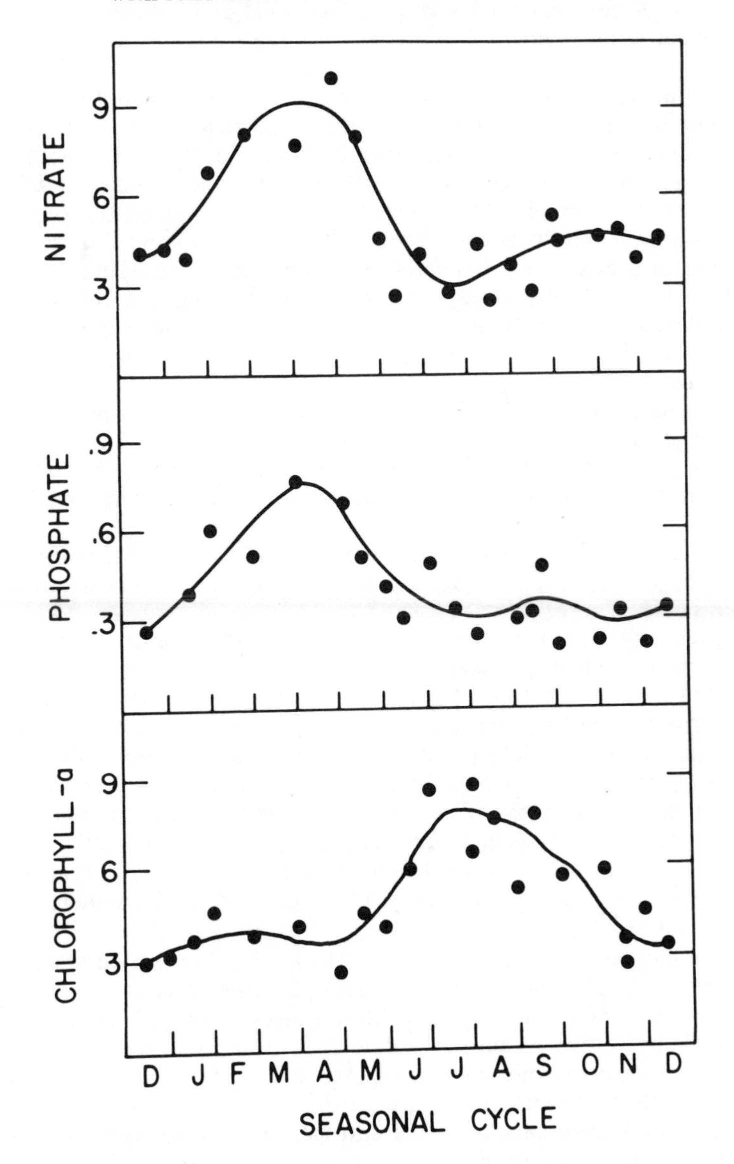

limited to the period between May and September when water temperatures are greater than 25°C. Phytoplankton nutrient limitation bioassays were consistently unresponsive during the rest of the year. Previous investigations of the phytoplankton ecology of these systems suggest that during the rest of the year nutrient concentrations are either high enough to meet phytoplankton demands or temperatures are low enough to be the primary factor limiting phytoplankton productivity (4, 18).

A summary of environmental, nutrient, and phytoplankton data for this period is presented in Table 1. Light and temperature did not vary widely between the various sampling dates and locations, a consequence of data restrictions to summer. Significant differences between stations were observed in salinity, turbidity, nutrient concentrations, and phytoplankton productivity. The relatively low mean salinity values observed at the Apalachicola and Ochlockonee estuarine stations were the result of high river discharge into these areas. The higher turbidity values at the Apalachicola stations relative to the other stations were attributed to high river discharge and mixing processes that suspend bottom sediments (17).

The Apalachicola and Ochlockonee stations exhibited higher nutrient concentrations and higher phytoplankton productivity and chlorophyll-a levels than the other stations. These high nutrient levels are probably the result of watershed runoff and concomitant river discharge into these areas.

Greater variability in the nitrogen values than in the phosphorus values is probably due to the more complex cycling of nitrogen in estuarine systems (15).

Higher chlorophyll-a levels and phytoplankton productivity are a reflection of the higher nutrient concentrations in these systems. It should be noted that the phytoplankton and nutrient levels for the Apalachicola and Ochlockonee area are indicative of productive but not eutrophic estuaries, while the levels for the other estuaries are reflective of oligotrophic areas. Phytoplankton species differences were observed between the high and low salinity stations. Phytoplankton communities at the high salinity stations had a greater proportion of large individuals with 35 percent of the total number having log cell volumes greater than $4.0\mu^3$. The low salinity stations had only 15 percent of total numbers in this larger size fraction.

One indication of phytoplankton nutrient limitation in natural waters is the nitrogen to phosphorus ratio. On a weight basis, total nitrogen to total phosphorus content in phytoplankton ranges between 5:1 and 10:1. Theoretically, ratios in natural waters that are greater than 10:1 would indicate a surplus of nitrogen and possible phosphorus limitation, while ratios less than 5:1 would indicate a surplus of phosphorus and possible nitrogen limitation. Much of the nitrogen and phosphorus in natural waters may not be in forms readily available to phytoplankton, which can

TABLE 1. Summary of an environmental, nutrient, and phytoplankton data.[1]

Station	Temp °C	Salin PPT	Turb FTU	Light ly hr^{-1}	NO_3 µg atomN l^{-1}	NO_2 µg atomN l^{-1}	NH_4 µg atomN l^{-1}	PO_4 µg atomP l^{-1}	Pri Prod µgC l^{-1} hr^{-1}	Chl-a µg l^{-1}
E-12	28.40	26.20	3.15	26.50	0.32	0.01	0.31	0.04	6.00	0.61
	1.01	2.48	0.35	5.60	0.14	0.03	0.20	0.01	1.25	0.17
M.L.	27.80	29.70	3.15	37.80	0.55	0.02	0.61	0.19	9.20	0.52
	1.78	3.53	0.49	3.73	0.10	0.02	0.49	0.04	0.58	0.21
Ock-1	28.20	4.20	4.97	37.90	1.83	0.05	0.79	0.37	30.80	2.14
	0.90	1.60	0.78	7.22	0.37	0.01	0.44	0.07	2.57	0.41
Ock-2	28.20	10.30	4.93	37.90	2.24	0.12	0.91	0.36	26.40	3.00
	0.80	0.70	0.61	7.22	0.83	0.05	1.04	0.09	4.74	0.51
Apal-1A	27.50	3.74	16.50	33.90	3.08	0.15	0.60	0.34	40.30	5.13
	1.19	2.58	8.96	9.17	2.63	0.16	0.83	0.08	10.70	1.12
Apal-7	27.50	11.70	11.70	36.90	3.55	0.21	0.81	0.40	36.70	4.11
	1.34	8.26	6.88	3.50	3.69	0.16	0.89	0.09	5.81	0.84

[1]Data are from surface samples collected between May and September, 1974-1976. The first value under each factor is the mean value of that factor for a given station and the second value is the standard deviation of the values. *Temp* is temperature, *Salin* is salinity, *Turb* is turbidity, and *Pri Prod* is phytoplankton primary production, and *Chl-a* is chlorophyll-a.

TABLE 2. Summary of analysis of variance of carbon uptake and phosphate uptake nutrient enrichment bioassays.[1]

Station	Date	Carbon NO_3	Uptake PO_4	Phosphate Uptake PO_4	Station	Date	Carbon NO_3	Uptake PO_4	Phosphate Uptake PO_4
E-12	6/03/75	*	*	N	Apal-1A	9/02/74	*	*	N
E-12	7/18/75	-	*	N	Apal-1A	5/29/75	-	-	N
E-12	7/12/76	-	*	*	Apal-1A	7/11/75	-	*	N
E-12	9/10/76	-	-	-	Apal-1A	9/11/75	-	*	*
					Apal-1A	9/15/75	-	-	-
M. L.	6/13/76	-	*	*	Apal-1A	6/10/76	-	*	*
M. L.	7/03/76	-	*	*	Apal-1A	6/24/76	-	*	*
M. L.	8/30/76	-	-	-	Apal-1A	7/05/76	-	*	*
M. L.	9/22/76	-	*	-	Apal-1A	8/15/76	-	*	*
					Apal-1A	8/26/76	-	-	-
Ock-1	6/17/76	-	-	-					
Ock-1	7/28/76	-	*	*	Apal-7	9/02/74	*	-	N
Ock-1	8/30/76	*	-	-	Apal-7	5/29/75	-	-	N
					Apal-7	7/11/75	*	*	N
Ock-2	6/17/76	*	-	-	Apal-7	7/11/75	*	-	-
Ock-2	7/28/76	-	*	*	Apal-7	9/15/77	-	-	-
Ock-2	8/30/76	-	-	-	Apal-7	6/10/76	-	-	-
					Apal-7	6/24/76	-	-	-
					Apal-7	7/05/76	*	*	*
					Apal-7	8/15/76	-	*	*
					Apal-7	8/26/76	-	-	-

[1] The symbols under PO_4 and NO_3 indicate the statistical significance (F test) of the effect of that nutrient on either carbon uptake or phosphate uptake. N indicates no data * indicates $\propto < 0.05$, and - indicates ≥ 0.05. When < 0.05 the effect of the nutrient addition was always stimulatory to the physiological process measured.

result in considerable variation to the expected range where either nitrogen or phosphorus is limiting. Therefore, it is best to compare ratios of available nitrogen and phosphorus forms. The inorganic nitrogen forms readily taken up by phytoplankton are ammonia, nitrate, and nitrite (15), while soluble reactive phosphate is the form of phosphorus preferred by phytoplankton. The mean ratio of the inorganic nitrogen forms to soluble reactive phosphate for these coastal stations varied between 6.2 and 16.0. The ratios at stations M.L., Ock-1, and Ock-2 were between 6.2 and 9.0; therefore, no concrete inferences about nutrient limitation at these stations could be made. The nitrogen to phosphorus ratios at stations Apal 1A, Apal 7, and E-12 were greater than 10 and suggested a surplus of available nitrogen and possible phosphorus limitation in these areas.

The phytoplankton and nutrient data were treated with linear regression techniques to gain a better perspective on relationships. Linear correlations coefficients were determined between phytoplankton productivity and soluble nutrient concentrations. Phytoplankton productivity was more strongly correlated with soluble reactive phosphate concentrations ($r = +0.73$) than with either total dissolved inorganic nitrogen concentrations ($r = +0.44$) or with the individual nitrogen forms, soluble nitrate ($r = +0.41$), soluble nitrite ($r = +0.23$), or soluble ammonium ($r = +0.31$). This suggests that soluble reactive phosphate concentration was more important than the concentrations of dissolved inorganic nitrogen forms in explaining summer phytoplankton productivity in the estuaries of the northeastern Gulf of Mexico.

A multiple regression model was constructed to determine which combinations of environmental and nutrient variables could explain the most variation of phytoplankton productivity in these coastal systems. The model was designed with phytoplankton primary productivity as the dependent variable and with temperature, salinity, turbidity, surface light intensity, soluble inorganic nitrogen, and soluble reactive phosphate as possible independent variables. A stepwise regression method was used to enter independent variables into the model. The lower limit of the change of R^2 for addition of a variable to the model was set at 0.05. Soluble reactive phosphate and salinity were the only variables that met the model constraints. The final regression model was:

$$\text{P.P.} = 32.1 + 48.4 \text{ S.R.P.} - 0.54 \text{ Salinity}$$

where: P.P. of phytoplankton productivity in $\mu gC\ l^{-1}\ hr^{-1}$, S.R.P. is soluble reactive phosphate in μg-atom PO_4-P l^{-1}; and salinity in part per thosand. This final model was significant at $a = 0.001$ and explained 64 percent of the variation in phytoplankton productivity in these coastal systems.

Results of both the carbon uptake and phosphate nutrient enrichment bioassays also indicated that phosphorus was more important than nitrogen in limiting phytoplankton productivity during the summer months in these coastal estuaries (Table 2). Phosphate additions stimulated phytoplankton carbon fixation more frequently than did nitrate additions. Phosphate additions significantly stimulated carbon fixation 18 times, while added nitrate stimulated carbon fixation only 8 times. No response to the nutrient additions was recorded 12 times. When phosphate additions stimulated carbon uptake, phytoplankton phosphate uptake was also stimulated. The high positive correlation coefficients between phosphate uptake and both chlorophyll-a ($r = +0.83$) and phytoplankton primary production ($r = +0.77$) during the summer suggest that phytoplankton are the primary source of phosphate uptake of the plankton in these coastal systems, which is consistent with results from other estuaries (26).

The nearshore northeastern Gulf of Mexico environments investigated in this study receive runoff which does not contain high dissolved nutrient concentrations compared to other areas. Shallow, clear waters overlie sandy sediments which remain in the water column for only short periods after suspension in contrast to the silty, turbid waters of the nearshore Georgia coast. Pomeroy (21) suggested that the phosphorus is not a limiting nutrient for phytoplankton growth in any southeastern coastal system; except in some of the clearest, sediment-free estuaries. The Apalachicola Bay water column contains high turbidity for several days following periods of high winds; phytoplankton are not phosphate limited under these conditions, but become phosphate limited after sediments settle to the bottom (17).

Recently, it has been proposed that the removal of phosphorus from marine waste discharges will have little impact on the control of eutrophication in coastal areas because nitrogen is the nutrient which limits phytoplankton productivity in coastal waters (9, 22, 30). In coastal waters, most investigations of phytoplankton nutrient limitation have been performed in eutrophic areas which receive high loads of nutrients from domestic waste discharges. The low nitrogen to phosphorus ratios of domestic wastes compared with the receiving coastal waters has caused an abundance of phosphorus in many eutrophic coastal areas such that immediate improvements in water quality will occur by initially controlling nitrogen discharges (22). However, the nearshore northeastern Gulf of Mexico environments investigated in this study receive runoff which is relatively low in dissolved phosphate (0.6 μg-atom PO_4-P l^{-1}). The watersheds of these estuaries are relatively undeveloped. Any alterations in land use that increase phosphorus mobility should be discouraged as they could lead to eutrophication of these systems.

ACKNOWLEDGEMENTS

Financial support was provided by the Florida Sea Grant Program under NOAA contract 04-3-158-43. H. Bittaker provided data from the Econfina area and helped with field sampling along with D. DiDomenico and D. Connally.

REFERENCES

1. Barsdate, R.J., R.T. Prentki, and T. Fenchel. 1974. Phosphorus cycle of model ecosystems: significance for decomposer food chains and effect of bacterial grazers. *Oikos* 25:239-251.
2. Bittaker, H.F. 1975. A comparative study of the phytoplankton and benthic macrophyte primary productivity in a polluted versus an unpolluted coastal area. 174 pp. M.S. thesis, Florida State University, Tallahassee.
3. Curtis, E.J.C. and I.P. Toms. 1972. Techniques for counting carbon-14 and phosphorus-32 labeled samples of polluted natural waters. 167-179. *In* M.A. Crook, P. Johnson, and D. Scales (eds.), Liquid Scintillation Counting, v. 2. Heyden and Sons Ltd., London.
4. Estabrook, R.H. 1973. Phytoplankton ecology and hydrography of Apalachicola Bay. 166 pp. M.S. thesis, Florida State University, Tallahassee.
5. Flemer, D.A. 1970. Primary production in the Chesapeake Bay. *Chesapeake Science* 11:117-129.
6. Fogg, G.E. 1975. Algal cultures and phytoplankton ecology. University of Wisconsin Press, Madison.
7. Fric, F. and V. Palovickova. 1975. Automatic liquid scintillation counting of ^{32}P in plant extracts by measuring Cherenkov radiation in aqueous solutions. *Int. J. App. Rad. Iso.* 26:305-321.
8. Holmes, R.W. 1962. The preparation of marine phytoplankton for microscopic examination and enumeration on molecular filters. U.S. Fish and Wildlife Service Spec. Sci. Rep. Fish. 433.
9. Goldman, J.C. 1976. Identification of nitrogen as a growth-limiting nutrient in waste-waters and coastal marine waters through continuous culture algal assays. *Water Res.* 10:97-104.
10. Goering, J.J. and P.L. Parker. 1972. Nitrogen fixation by epiphytes on sea grasses. *Limnol. Oceanogr.* 17:320-323.
11. Jefferies, H.P. 1962. Environmental characteristics of Raritan Bay, a polluted estuary. *Limnol. Oceanogr.* 7:21-31.
12. Johannes, R.E. 1965. Influence of marine protozoa on nutrient regeneration. *Limnol. Oceanogr.* 10:434-442.
13. Johansson, J.O.R. 1975. Phytoplankton productivity and standing crop in the Anclote estuary, Florida. 75 pp. M.S. thesis, University of South Florida, Tampa.
14. Livingston, R.J., R.L. Iverson. R.H. Estabrook, V.E. Keyes, and J. Taylor. 1974. Major features of the Apalachicola Bay system: physiography, biota, and resource management. *Florida Sci.* 37:245-271.
15. McCarthy, J.J., N.R. Taylor, and J.L. Taft. 1975. The dynamics of nitrogen and phosphorus cycling in the open waters of the Chesapeake Bay, pp 664-681. *In*: T.M. Church (ed.), Marine Chemistry in the Coastal Environment. Am. Chem. Soc., New York.
16. McRoy, C.P., and J.J. Goering. 1974. Nutrient transfer between the seagrass *Zostera marina* and its epiphytes. *Nature* 248:173-174.
17. Myers, V.B. 1977. Nutrient limitation of phytoplankton productivity in north Florida coastal systems: technical considerations, spatial

patterns, and wind mixing effects. 63 pp. Ph.D. thesis. Florida State University, Tallahassee.
18. ―――――――― and R.L. Iverson. 1977. Aspects of nutrient limitation of phytoplankton productivity in the Apalachicola Bay System. Florida Dept. Nat. Res. Pub. 26:68-74.
19. Penhale, P.A. 1976. Primary productivity, dissolved organic carbon excretion, and nutrient transport in an epiphyte-eelgrass (*Zostera marina*) system. 82 pp. Ph.D. thesis, North Carolina State University, Raleigh.
20. Pomeroy, L.R., H.M. Matthews, and H.S. Min. 1963. Excretion of phosphate and soluble organic phosphorus compounds by zooplankton. *Limnol. Oceanogr.* 8:50-55.
21. ―――――――― L.R. Shenton, R.D.H. Jones, and R.J. Reimhold. 1972. Nutrient flux in estuaries, 274-291. *In*: G.E. Likens (ed.), Nutrients and Eutrophication. Am. Soc. Limnol. Oceanogr. Spec. Symp. V.I.
22. Ryther, J.H. and W.M. Dunstan. 1971. Nitrogen, phosphorus, and eutrophication in the coastal marine environments. *Sci.* 171:1198-1203.
23. Steidinger, K.A. 1973. Phytoplankton ecology: a conceptual review based on eastern Gulf of Mexico research. CRC Critical Reviews in Microbiology 3:49-68.
24. Strickland, J.D.H. and T.R. Parsons. 1972. A practical handbook of seawater analysis, 2nd ed. Bull. Fish. Res. Bd. Can. 167.
25. Swanson, V.E., A.H. Love, and I.C. Frost. 1972. Geochemistry and diagenesis of tidal-marsh sediment, northeastern Gulf of Mexico. Bull. U.S. Geol. Surv. 1360:1-83.
26. Taft, J.L., W.R. Taylor, and J.J. McCarthy. 1975. Uptake and release of phosphorus by phytoplankton in the Chesapeake Bay estuary. *Mar. Biol.* 33:21-32.
27. Tanner, W.F. 1960. Florida coastal classification. Trans. Gulf Coast Assoc. *Geol. Soc.* 10:259-266.
28. Thayer, G.W. 1971. Phytoplankton production and the distribution of nutrients in a shallow unstratified estuarine system near Beaufort, N.C. *Chesapeake Science*. 12:240-253.
29. ―――――――― 1974. Identity and regulation of nutrients limiting phytoplankton production in the shallow estuaries near Beaufort, N.C. *Oecologia* 14:75-92.
30. Thomas, W.H., D.L.R. Seibert, and A.N. Dodson. 1974. Enrichment experiments and bioassays in natural coastal seawater and in sewage outfall receiving waters off southern California. *Estuarine Coast. Mar. Sci.* 2:191-206.
31. Thorhaug, A. 1976. Biscayne Bay - Past/Present/Future. Florida Sea Grant Special Report No. 5.
32. Williams, R.B. and M.B. Murdoch. 1966. Phytoplankton production and chlorophyll concentrations in the Beaufort Channel, North Carolina. *Limnol. Oceanogr.* 11:73-82.

CONTRIBUTED COMPLEMENTARY STUDIES
IN THE YORK RIVER

SHORT TERM CHANGES IN THE VERTICAL SALINITY DISTRIBUTION OF THE YORK RIVER ESTUARY ASSOCIATED WITH THE NEAP-SPRING TIDAL CYCLE

Leonard W. Haas, Fredrick J. Holden and Christopher S. Welch
Virginia Institute of Marine Science and School of Marine Science
College of William and Mary
Gloucester Point, Virginia 23062

ABSTRACT: A multidisciplinary investigation of hydrographic-nutrient-phytoplankton interactions was undertaken in the lower York River estuary of Virginia during August, 1978. The study centered on a spring tide-associated water column destratification event predicted to occur on or soon after August 19, the date of the maximum monthly spring tide. A station in the lower York River (depth 19 m) was occupied during four different periods, August 7-10, 16-17, 21-24 and 28-30, and temperature and salinity were measured periodically at 1 m depth intervals. During August 16-20 salinities were measured through the water column at seven stations in the York River extending from the mouth to 35 km upriver. During the first two sampling periods the water column was moderately to strongly stratified. Destratification was first observed 15-20 km upriver on August 18 and the lower river was destratified by August 21. Destratification persisted in the lower river for four days at which time increasing bottom salinities indicated the beginning of the restratification process. By August 28 restratification, resulting primarily from an increase in bottom salinities, was complete. The results illustrate the highly dynamic hydrographic nature of this estuarine system and the predictability of the stratification-destratification sequence. The effects of this hydrographic cycle on nutrient distributions, phytoplankton dynamics and benthic nutrient fluxes in this estuary are discussed in accompanying papers.

INTRODUCTION

Eutrophication studies historically have emphasized the distribution and concentration of nutrients and plants in bodies of water, and spatial and temporal variations in the abundance of these constituents are often used to infer dynamic processes relating to their interaction and rates of utilization and production. However, hydrodynamic processes may also affect spatial and temporal distribution of nutrients and plants and therefore may complicate eutrophication studies. Thus hydrographic studies must be included with biological and chemical studies of eutrophication, for without an understanding of the fundamental dispersive processes, interpretation of the biological and chemical data is incomplete. Unfortunately, hydrographic processes, which are expensive to measure and difficult to interpret, are often ignored by all but physical oceanographers, who even then are often interested in processes which occur on time scales much longer than those of interest to biologists, chemists and water quality managers.

Estuaries are often characterized on the basis of their morphometry and salinity structure (6). Pritchard (12), for example, has produced the classical conceptual models based on density stratification types, i.e. highly stratified (type A), moderately stratified (type B) and vertically homogeneous (type C). Vertically homogeneous estuaries are generally considered to be shallow, wide and dominated by tidal currents, while stratified estuaries are considered to be deeper, narrower and dominated by freshwater flow (12). It is also recognized that the degree of stratification in some estuaries may change on a time scale ranging from hours to months, a result, respectively, of the semidiurnal tidal cycle (3) and seasonal variation in freshwater flow (9).

The Chesapeake Bay and its tributaries have been regarded as good examples of moderately stratified or type B estuaries, and over the long term (months or years) their salt distribution is adequately explained by such a model. Recently, however, Haas (7) observed that variation in the neap-spring tidal amplitude played a major role in regulating vertical density stratification in the James, York and Rappahannock rivers, all subestuaries of the Chesapeake Bay. In these estuaries, salinity stratification was periodically disrupted by short periods of destratification which were closely correlated with the occurrence of sufficiently strong spring tides. Neither variation in the magnitude of freshwater flow nor meteorological events appeared to affect the occurrence of destratification. This phenomenon, although perhaps predictable on the basis of past physical oceanographic modeling (9), is not well understood in the mathematical sense and is beyond the scope of current physical models of estuarine circulation used in water quality assessment.

However, the time scale of the stratification-destratification cycle suggested to us that hydrographic processes played an integral role in the dynamics of non-conservative chemical and biological constituents in this estuary. To investigate this role, a multidisciplinary study of a predicted destratification event in the York River was undertaken in August, 1978. Personnel and budgetary restrictions limited the observations that could be taken and consequently the physical processes that occurred during the study are still not completely understood. Nevertheless, we have obtained important information about the relationship between biological-chemical estuarine processes and a specific hydrographic cycle which may have wider applicability to nutrient-plankton interactions of the Chesapeake Bay and perhaps other estuaries.

This paper contains a description of the study area, sampling methods and materials used in the study as well as a summary of the hydrographic aspects of the investigation. Results relating to the nutrient dynamics, phytoplankton dynamics and benthic nutrient flux aspects of the study are treated separately in accompanying sections of this volume (5, 8, 11) and elsewhere (16).

FIGURE 1. Station locations in York River during August, 1978.

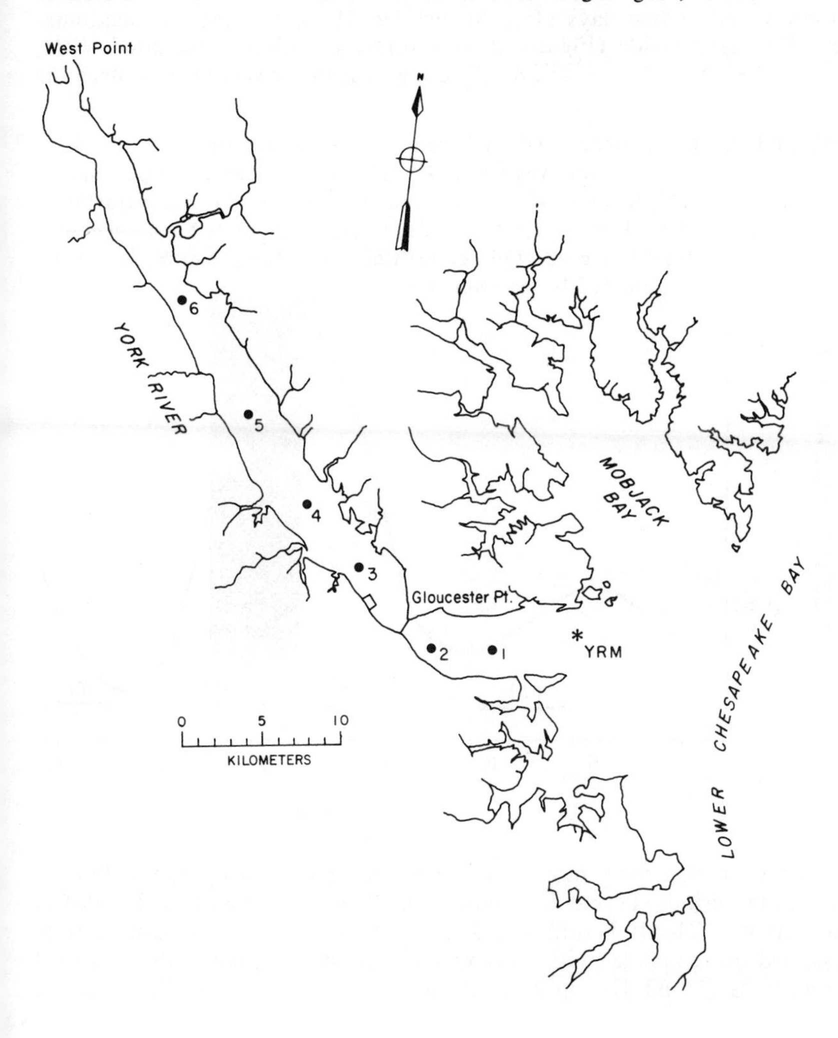

METHODS AND MATERIALS

The York River (Figure 1) is formed by the confluence of the Pamunkey and Mattaponi rivers at a point about 50 km from its point of entry into the Chesapeake Bay. The river is brackish and tidal throughout its entire length. Mean tidal range at the mouth is 0.7 m and salinities at this point range from 15 to 30 $^{o}/oo$ depending on depth and time of year. The 1.0 $^{o}/oo$ isohaline is normally found 65-90 km upriver in the tributaries, and the lower York River contains a deep channel (ca. 18-20 m) which merges with the main channel of the Chesapeake Bay.

Based on previous work in this estuary (7) and calculations of predicted tide heights, destratification in the lower York River was predicted to occur within a few days after August 19, 1978, the date of maximum monthly spring tide (Figure 2). Our primary station (designated YRM, 37°15’40” N. Lat., 76°23’28” W. Long., Figure 1) was located near the

FIGURE 2. Daily mean predicted high tide heights for the lower York River during August, 1978. Duration of sampling periods I, II, III and IV are indicated as (├────┤), and duration of York River slack water sampling is indicated as (├ ‒ ‒ ┤). Period of predicted destratification in the lower York River is indicated by hatched area.

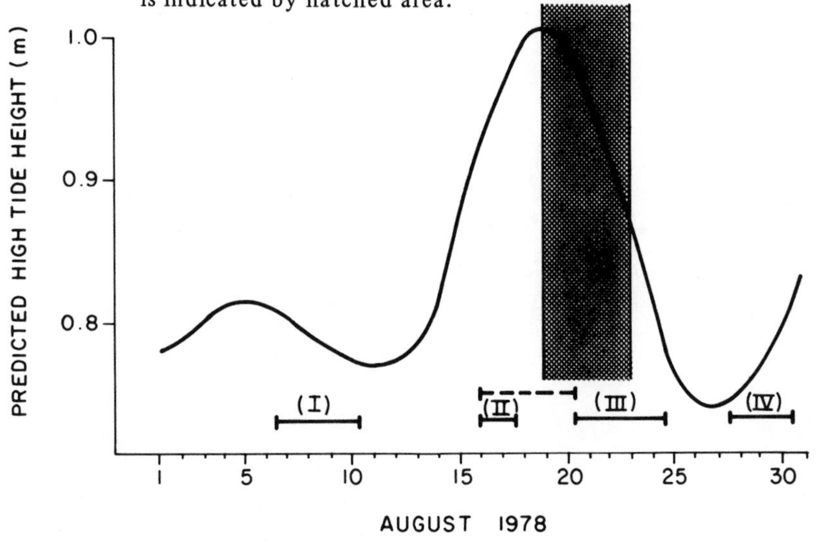

channel at the river mouth and was occupied during the following sampling periods: Period I, August 7-10; Period II, August 16-17; Period III, August 21-24; Period IV, August 28-30. These observations were planned to coincide with anticipated periods of neap tide stratified conditions (I and II), spring tide destratified conditions (III), and the subsequent neap tide restratified conditions (IV) (Figure 2). The R/V *Retriever* from the Virginia Institute of Marine Science was used during periods I, III and IV and the R/V *Ridgely Warfield* was used during period III.

During each sampling period, hydrocasts were made at three-hour intervals using a submersible pump (Silent Giant) to provide water at 1 m depth intervals to an array of shipboard sensors. Measurements included temperature, salinity and conductivity (YSI model 33 S-C-T meter) and dissolved oxygen (YSI model 51A salinity compensated oxygen meter). Water samples from one profile during each s ampling period were analyzed for salinity with a Beckman RS-7A induction conductivity salinometer to serve as calibration controls. Nutrient analyses were performed as often as possible on hydrocast samples. During sampling periods I, III and IV, nutrients were measured at least at every depth, every three hours over a 24-hour period. During sampling period II, nutrients were measured at every depth at successive dawn, dusk and dawn profiles.

In addition to the hydrographic profiles at station YRM, seven stations in the York River (YRM and stations 1 through 6, Figure 1) were sampled on daily slack water runs during August 16-20. At each of these stations, temperature and conductivity were measured at surface (bucket sample), 1m, 2m and two meter depth intervals thereafter to the bottom with a submersible CTD. Salinities were calculated from the temperature-conductivity data. Water samples from the surface and bottom were analyzed for salinity with the Beckman RS-7A salinometer to serve as checks on the CTD performance.

RESULTS

Since salinity is by far the dominant density determinant in this estuary during the summer months, it is used as the stratification index. During the study, thermocline and halocline depths always coincided, i.e. there was no secondary thermal stratification. Salinity isopleths plotted against time for the four sampling periods (Figure 3) illustrate the hydrographic variability observed at YRM during the study. The extent of the variability is summarized in selected salinity profiles from periods II, III and IV (Figure 4). The water column was moderately stratified during period I and more strongly stratified (isohalines more closely spaced at the halocline) during period II. During the interval between periods I and II the bottom salinity increased about 2 $^{o}/oo$. During period II the mean surface and bottom temperatures were 29.0^{o}C and 24.6^{o}C respectively (10 observations).

By period III destratification had already occurred, resulting in nearly homogeneous salinities throughout the water column (Figure 3 and 4). An indication of stratification state is given by Δ $^{o}/oo$, the difference in salinity between bottom and surface samples. The value of Δ $^{o}/oo$ averaged over the period 0200-2300 on August 22 (8 observations) was <0.01 $^{o}/oo$, indicating a destratified condition. Mean surface and bottom temperatures over the same time span were 26.8°C and 26.5°C, respectively.

Salinities from the slack water runs are shown in Table 1. The surface-to-bottom salinity differences for each profile at each station were plotted against time (Figure 5) and indicate that destratification i.e. Δ $^{o}/oo$ <ca. 1.0, was first observed at Stations 3 and 4 (about 15-20 km upriver from YRM) on August 18. Destratification at Station 2 was observed on August 20. As noted previously, station YRM was destratified when initially sampled on August 21.

After three days of destratification, increasing bottom salinities at YRM were first noted at 0900 on August 24, indicating the initiation of the restratification process. A lapse in sampling between 1400 on August 23 and 0900 on August 24 precludes a more exact determination of the onset of restratification at YRM. However, by 1700 on August 24 (the last

FIGURE 3. Salinity profiles at Station YRM.

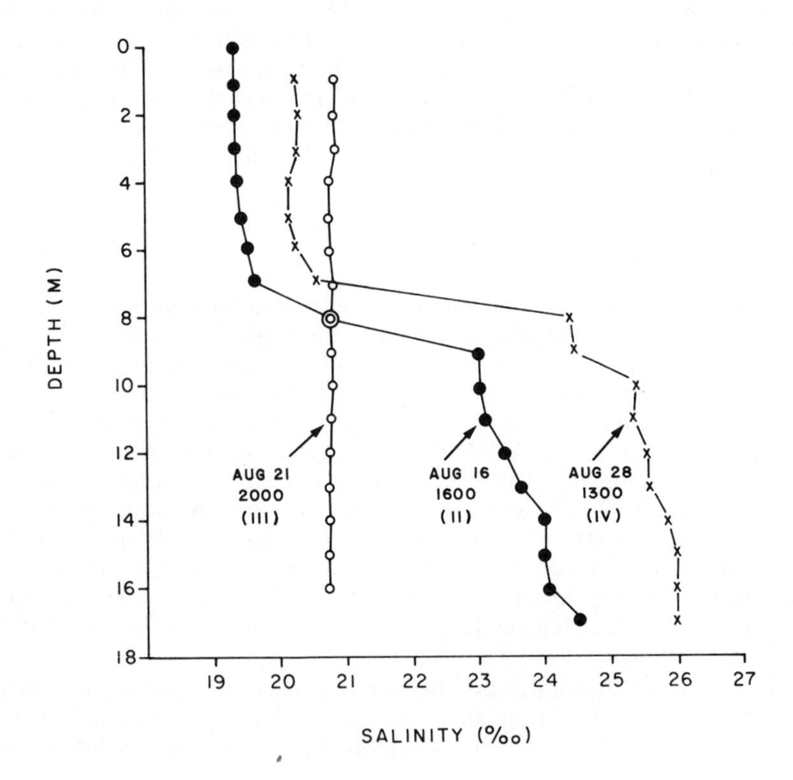

observation of sampling period III) stratification was well advanced, and when YRM was reoccupied on August 28, restratification was essentially complete, as Δ $^{o}/oo$ was ca. 6 and a well developed halocline was observed (Figures 3 and 4). Over the period 0100-2200 on August 29 (8 observations) mean surface and bottom temperatures were 27.5^{o}C and 24.3^{o}C, respectively. Throughout the four sampling periods, the constancy of a given hydrographic state during the semi diurnal tidal cycle (Figure 4) indicates that this short term tidal parameter has little effect on the degree of stratification and that, in the absence of more extensive longitudinal sampling, the conditions observed at YRM existed in the river for at lease the distance of one tidal excursion (7-9 km) from the station. Previous observations in this estuary indicated that the entire river responded essentially uniformly to the neap-spring hydrographic cycle (7).

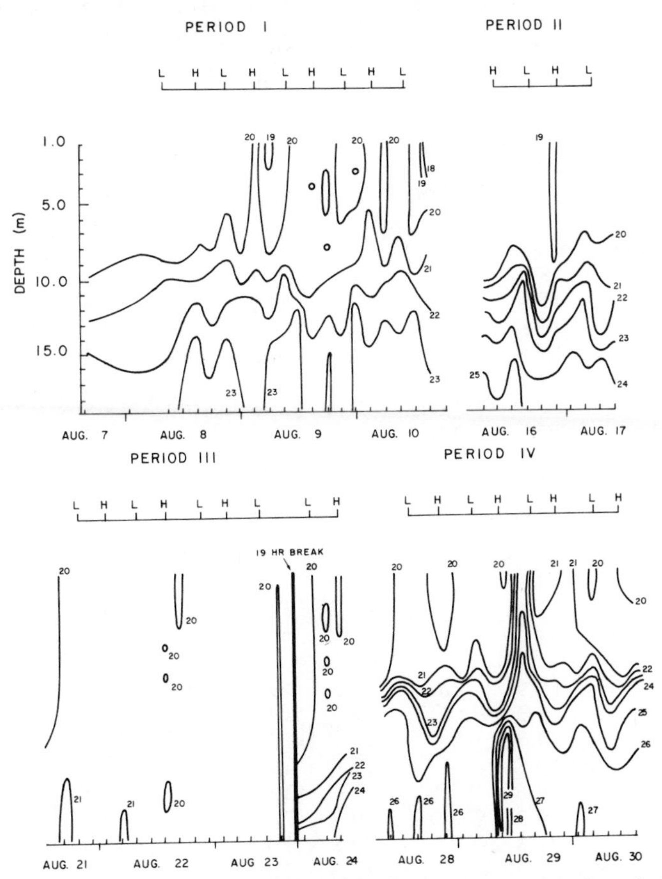

FIGURE 4. Salinity isopleths at YRM plotted against depth and time for the four sampling periods. Values are $^{o}/oo$. No data collected from 1400 August 23 to 0900 August 24. Times of high slack water (H) and low slack water (L) are indicated.

TABLE 1. Slack water salinities in York River. All values $^o/_{oo}$ nd - not determined.

Date: August 16, 1978 — Low Slack at YRM 1650

Station:	YRM	1	2	3	4	5	6
Time:	-	2400	-	1900	1925	1947	2010
Depth (m)							
Surface	nd	18.9	nd	17.2	15.5	13.5	11.0
1	"	18.9	"	17.6	17.3	14.4	11.8
2	"	18.9	"	19.1	18.1	15.2	12.1
4	"	19.0	"	20.0	18.6	16.1	12.9
6	"	19.3	"	20.4	18.7	17.1	15.3
8	"	19.4	"	20.5	18.7	17.5	15.6
10	"	19.9	"	20.6	18.7	17.5	
12	"	20.8	"	20.6			
14	"	22.9	"	20.6			

Date: August 17, 1978 — High Slack at YRM 1037

Station:	YRM	1	2	3	4	5	6
Time:	2209	1020	-	1052	1112	1138	1215
Depth (m)							
Surface	19.4	19.4	nd	19.4	18.7	16.5	13.4
1	19.4	19.4	"	19.5	18.8	17.4	14.0
2	19.4	19.3	"	19.6	19.1	18.1	14.2
4	19.5	19.4	"	20.0	19.9	18.2	14.4
6	19.6	20.0	"	20.5	19.9	18.2	16.3
8	20.2	20.4	"	20.6	19.9	18.3	17.8
10	20.7	21.6	"	21.0		18.3	
12	21.5	23.3	"	21.2			
14	22.9		"				
16	23.3						

Date: August 18, 1978 — High Slack at YRM 1134

Station:	YRM	1	2	3	4	5	6
Time:	1320	-	1347	1415	1440	1525	-
Depth (m)							
Surface	18.9	nd	19.5	19.3	18.3	15.5	nd
1	19.2	"	19.5	19.3	18.9	15.9	"
2	19.5	"	19.5	19.4	19.2	16.8	"
4	19.5	"	19.6	19.8	19.4	17.7	"
6	19.6	"	19.6	20.1	19.5	17.8	"
8	20.3	"	19.7	20.5		18.1	"
10	21.3	"	20.8	20.6		18.1	
12	21.7	"	22.3	20.6			
14	22.1		22.9				
16	22.4						
18	22.7						

TABLE 1. Cont'd.

Date: August 19, 1978					High Slack at YRM 1228		
Station:	YRM	1	2	3	4	5	6
Time:	1312	-	1342	1413	1434	1450	1510
Depth (m)							
Surface	18.9	nd	19.7	19.1	18.7	16.6	13.8
1	19.0	"	19.7	19.2	19.0	17.8	13.9
2	19.2	"	19.7	19.6	19.5	18.1	14.1
4	19.2	"	19.7	19.7	19.7	18.1	15.9
6	19.2	"	19.9	19.7	19.7	18.2	16.1
8	19.3	"	20.1	19.7	19.7	18.4	16.1
10	19.3	"	20.2	19.7	19.7		
12	20.2	"	20.9	20.1			
14	20.6		21.5	20.1			
16	20.9		21.6				
18	20.9						

Date: August 20, 1978					High Slack at YRM 1320		
Station:	YRM	1	2	3	4	5	6
Time:	1618	-	1705	-	-	-	-
Depth (m)							
Surface	19.3	nd	19.7	nd	nd	nd	nd
1	19.3	"	19.7	"	"	"	"
2	19.3	"	19.8	"	"	"	"
4	19.4	"	19.9	"	"	"	"
6	19.5	"	19.9	"	"	"	"
8	19.7	"	19.9	"	"	"	"
10	19.7	"	19.9	"	"	"	"
12	20.0	"	19.9	"			
14	20.1		20.0	"			
16	20.2		20.0				
18	20.4						

DISCUSSION

The results of this study illustrate that significant hydrographic changes occur over relatively short time periods in this estuary. The time scale of the observed stratification-destratification cycle appears to fall between those observed for similar changes reported to occur in other estuaries over a semi diurnal tidal cycle on one hand (3) and over a seasonal cycle on the other hand (9). Although the specific causes of the observed hydrographic sequences are incompletely known, one possible explanation of destratification is that it resulted from greater turbulent mixing accompanying the increased tidal velocities during spring tides. The apparent initiation of destratification upriver from Station YRM confirms earlier interpretations (7) and is consistent with, but not coincident with, increased tidal current speeds encountered upriver.

Restratification following destratification in the lower York River appears to result from the accentuated net upriver transport of deep, higher salinity water as neap tides are approached. The break in observations between sampling periods III and IV precludes a determination of the minimum time necessary for restratification to be completed at YRM, however the process occurred within four days. Periodic deep water replenishment temporally related to the neap-spring tidal cycle has also been observed in Puget Sound (4).

The York River and other tributary estuaries of the lower Chesapeake Bay (James and Rappahannock rivers) are not the only estuarine systems observed to oscillate between conditions of relative stratification and destratification in conjunction with the neap-spring tidal cycle. Similar cycles have been observed in the St. Lawrence estuary (13), the Gironde and Aulne estuaries (1, 2) and Charleston Harbor (15). Indirect evidence

FIGURE 5. Surface-to-bottom salinity differences ($\Delta\,^{o}/oo$) for the York River stations versus time of collection. Station locations are shown in Figure 1. Destratification is defined as $\Delta\,^{o}/oo$ $<$ca. 1.0.

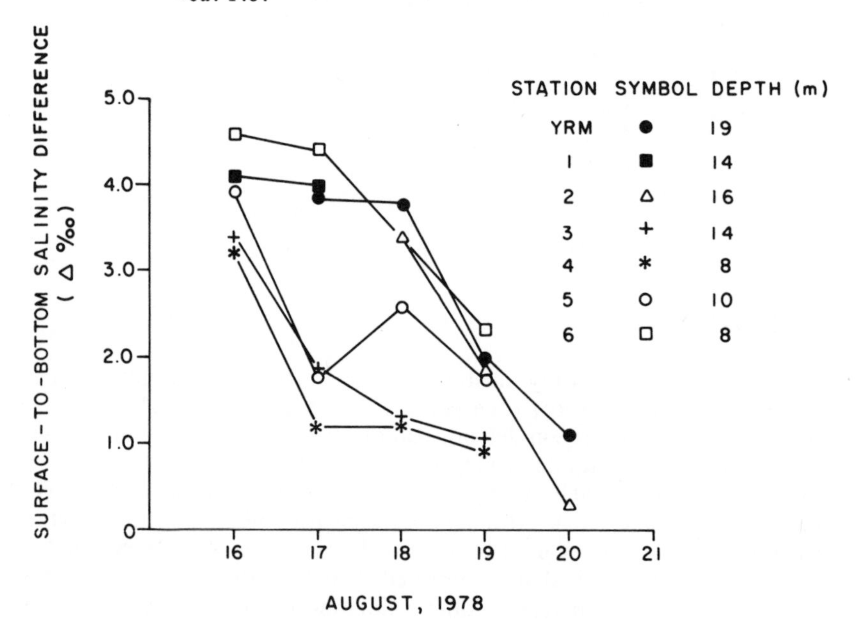

indicates that both Puget Sound (17) and Saanich Inlet (14) may also undergo such cycles. The apparent widespread occurrence of this phenomenon suggests that the potential for spring-neap tidal effects on stratification should be considered in designs for estuarine experiments.

Although this study adds little to our present knowledge of the possible causative mechanisms underlying the stratification-destratification cycle, it does illustrate that, at least for this time of year, destratification events in the York River can be adequately predicted on the basis of predicted tidal heights. Although no actual measurements of vertical or longitudinal transport rates were attempted during this study, the extreme hydrographic conditions encountered suggest that widely varying transfer rates occur. Certainly vertical transport of materials is much greater during destratified periods than during stratified periods, and the observed restratification process suggests that significant short-term longitudinal transport may also occur. Quantifying vertical and horizontal transport rates under these different conditions is necessary for a complete understanding of the biological and chemical processes not only in this estuary but for marine systems in general (10). The demonstrated predictability of the hydrographic processes and the extreme conditions observed in the York River make it an ideal area for further investigations of this type.

ACKNOWLEDGEMENTS

The authors would like to thank S. Hastings, D. Hayward, L. Pastor A. Thomson and S. Earl, Jr. for their help in collecting field data and their participation in the Interdisciplinary York River Study. We would also like to thank the captain and crew of the R/V *Retriever* for their help and cooperation in this project. This research was supported in part by the Virgniia Institute of Marine Science, by the Oceanography Section, National Science Foundation, under NSF grant OCE-77-20228, and by the office of Sea Grant, National Oceanic and Atmospheric Administration, U.S. Department of Commerce grant to the Virginia Institute of Marine Science.

REFERENCES

1. Allen, G.P., 1972. Etude des processus sedimentaires dans l'estuaire de la Gironde. Mem. Inst. Geol. Bassin Aquitaine 5:314 pp.
2. Allen G.P., J.C., Salomon, P., Bassoullet, Y., Du Penhoat, and C. Del Grandpre'. 1980. Effects of tides on mixing and suspended sediment transport in macrotidal estuaries. Sediment. Geol. 26:69-90.

3. Bowden, K.F. and S.H. Sharaf El Din, 1966. Circulation, salinity and river discharge in the Mersey estuary. Geophys. J. Royal Astron. Soc. 10:383-399.
4. Cannon, G.A. and N.P. Laird, 1978. Variability of currents and water properties from year long observations in a fjord estuary. 515-535. *In*: J.C.J. Nihoul (ed.). Hydrodynamics of estuaries and fjords. Elsevier Scientific Publishing Company, New York.
5. D'Elia, C.F., K.L. Webb and R.L. Wetzel. 1980. Impact of hydrographic events on water quality in an estuary. pp. 597-606. *In*: B.J. Neilson (ed.), International symposium on nutrient enrichment in estuaries. The Humana Press, Inc., Clifton, New Jersey.
6. Dyer, K.R. 1973. Estuaries: a physical introduction. J. Wiley & Sons, New York. 140 pp.
7. Haas, L.W. 1977. The effect of the spring-neap tidal cycle on the vertical salinity structure of the James, York and Rappahannock rivers, Virginia, U.S.A. Est. & Coast. Mar. Sci. 5:485-496.
8. Haas, L.W., S.J. Hastings and K.L. Webb. 1980. Phytoplankton response to a tidally induced cycle of stratification and mixing in the York River estuary. *In*: B.J. Neilson (ed.) International symposium on nutrient enrichment in estuaries. The Humana Press Inc., Clifton, New Jersey.
9. Hansen, D.V. and M. Rattray Jr. 1966. New dimensions in estuary classification. Limnol. Oceanogr. 11:319-326.
10. McGowan, J.A. and T.L. Hayward. 1978. Mixing and oceanic productivity. Deep-Sea Res. 25:771-794.
11. Phoel, W.G., K.L. Webb and C.F. D'Elia. 1980. Inorganic nitrogen regeneration and total oxygen consumption by the sediments at the mouth of the York River, Virginia., U.S.A. pp. 607-615. *In*: B.J. Neilson (ed.). International symposium on nutrient enrichment in estuaries. The Humana Press, Inc., Clifton, New Jersey.
12. Pritchard, D.W. 1967. Observations of circulation in coastal plain estuaries, 37-44. *In*: G.H. Lauff (ed.) Estuaries. American Association for the Advancement of Science, Publication 83.
13. Sinclair, M. 1978. Summer phytoplankton variability in the lower St. Lawrence estuary. J. Fish. Res. Bd. Can., 35:1171-1185.
14. Takahashi, M., D.L. Seibert and W.H. Thomas. 1977. Occasional blooms of phytoplankton during summer in Saanich Inlet, B.C., Canada. Deep-Sea Res. 24:775-780.
15. Wastler, T.A. and C.M. Walter. 1968. Statistical approach to estuarine behavior. J. Sanitary Engineering Division, A.S.C.E. SA6, Proc. Paper 6311:1175-1194.
16. Webb, K.L. and C.F. D'Elia. 1980. Nutrient and oxygen redistribution during a spring neap tidal cycle in a temperate estuary. Science 207:983-985.
17. Winter, D.F., K. Banse and G.C. Anderson. 1975. The dynamics of phytoplankton blooms in Puget Sound, a fjord in the Northwestern United States. Mar. Biol. 29:139-176.

TIME VARYING HYDRODYNAMICS AND WATER QUALITY IN AN ESTUARY

C.F. D'Elia*, K.L. Webb and R.L. Wetzel****
**University of Maryland*
Center for Environmental and Estuarine Studies
Chesapeake Biological Laboratory
Solomons, Maryland 20688

***Virginia Institute of Marine Science and*
School of Marine Science of the
College of William and Mary
Gloucester Point, Virginia 23062

INTRODUCTION

This paper is included in this symposium volume to illustrate the effect hydrographic factors such as the spring-neap tidal stratification-destratification cycle (4) can have on the distribution of oxygen and nutrients in the water column of an estuary. All too often scientists, modelers and managers, finding hydrographic processes in estuaries extremely difficult to measure and understand, neglect or ignore key features of hydrodynamics that have substantial bearing on "water quality." As a result, monitoring strategies are often ill-conceived; models are frequently unverifiable; and water quality management objectives may not be cost-effectively met.

We stress in this paper that time-scales involved with physical, biological and chemical processes pertaining to water quality must be considered carefully by all parties contributing to water quality assessment and the management of estuaries. Although excellent and well known conceptual models have been developed to explain longer, usually month-to-month, and more or less steady state features of estuarine hydrodynamics and salt balance (e.g. 12), other conceptual models need also be developed and taken into consideration to explain shorter and perhaps non-steady state hydrographic features occurring on a time scale having greater bearing on the distribution of non-conservative properties. Certainly mathematical models of water quality should include the appropriate features of these conceptual models. The data we present here are intended to convince the reader that water quality in the lower York River is affected by such non-steady state and short-term hydrodynamic processes.

MATERIALS AND METHODS

A description of the site, sampling stations and sampling instrumentation and procedures has been given elsewhere (4). All samples for nutrient analysis were filtered (1.0 μm pore glass filter or 0.45 μm pore cellulose acetate) immediately and processed using an onboard Technicon AutoAnalyzer or stored at 4°C for no longer than 12 hours before analysis. Standard EPA methodologies (6) 353.2, 350.1 and 365.1 were employed for NO_2^-, NO_3^-, NH_4^+, and PO_4^{3-} determinations. Some PO_4^{3-} determinations were done manually using the method described in Strickland and Parsons (15). We were careful to run frequent standard curves and use internal standards to verify that precision and accuracy were within prescribed limits (6).

RESULTS AND DISCUSSION

Representative profiles of oxygen (abbreviated here as "0" to avoid confusion when discussing flux ratios), salinity and nutrients during stratified, destratified and restratified periods are shown in Figure 1. During the period before the destratification event (Figure 1A), 0 and NO_2^- concentrations were high above the halocline and low below it, whereas the reverse was true for NH_4^+ and PO_4^{3-} (not shown). At destratification (Figure 1B), the water column constituents measured became virtually homogeneously distributed with depth; the most important feature of this period was that oxygen was transported to the water overlying the benthos. Another period of stratification followed (Figure 1C), exhibiting all the characteristics of the first except that NO_2 concentration was low in the surface layer. We were surprised that during this entire study, which encompassed most of the month of August, 1978, NO_3^- was virtually undetectable in the water column. The high concentrations of NO_2^- in the absence of NO_3^- observed during our study argue strongly against the frequent assumption that NO_2^- is never environmentally abundant and need not be discriminated from NO_3^- analytically. The high NO_2^- concentration observed also supports the contention (9) that NO_2^- should be included as a separate element in mathematical water quality models.

We have not identified the source of the NO_2^- observed in early August, although we suspect it was produced as the result of ammonia oxidation (see 18, for a review of the N cycle). Review of the historical data base for the Chesapeake and other estuaries has shown that a characteristic late

FIGURE 1. Profiles of salinity in parts per thousand (—·—·—), oxygen in mg· L^{-1} (·····), nitrite in μg at N· L^{-1} (– – –), and ammonia in μ g at N·L^{-1} (——). Changes in bottom depth are related to tidal fluctuations and slight changes in vessel position at anchor.

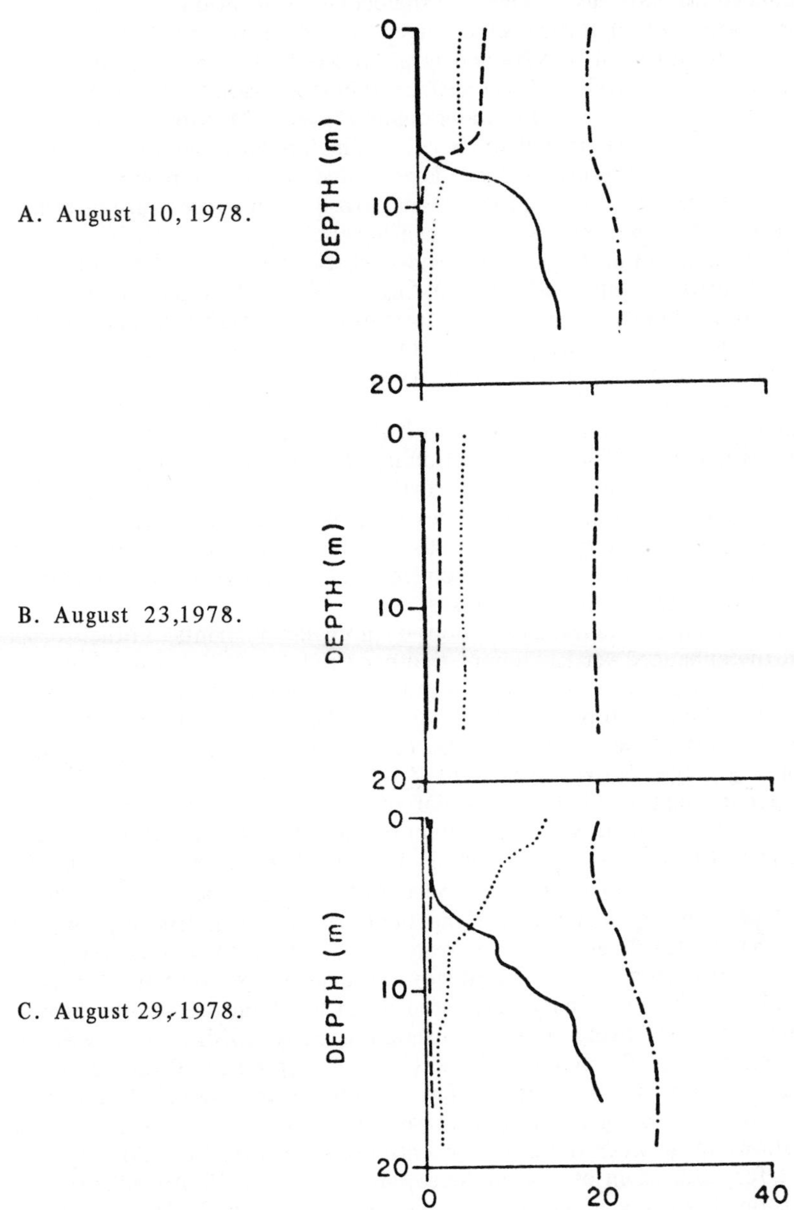

summer NO_2^- maximum has been observed often (19). Biological NO_2^- production most likely occurred in nearby salt marshes (21) and shallow benthic areas, with NO_2^- advected to our sampling station, or NO_2^- was produced in situ just above the halocline, a location in which high nitrification potential has been observed in another estuary (5).

Concentrations of 0, NH_4^+ and PO_4^{3-} in the bottom meter of the water column strongly reflected the surface-to-bottom salinity differences and thus the homogeneity of the water column (Figure 2). Although advective and diffusive transport and water column metabolism undoubtedly affect nutrient concentrations, evidence is presented below and in another paper (11) that the high NH_4^+ and PO_4^{3-} and low oxygen concentrations in this bottom water are primarily due to benthic exchange.

Webb and D'Elia (20) report evidence elsewhere that at the onset of destratification, sometime between August 17 and 21 when we were not on station, there was a rapid transformation of inorganic N and P in the water column into forms we did not measure. This coincided with an enhancement of primary productivity during the period (3) and did not appear to be the result of dilution by low nutrient water as proportional changes in salinity did not occur with the decrease in inorganic nutrients. Periods like this, of sudden reduction in inorganic nutrient and increase in 0 content, are certainly worthy of further study. Our observations support the view that radical changes in water quality parameters can occur in estuaries during very short periods of time (days) and cause one to question the fidelity of using longer term, steady state mathematical models of water quality for such systems.

As a result of the transformations of inorganic nutrients to other forms and the enhanced vertical mixing that occurred during the destratification period, we were provided with low nutrient, oxygenated water in which to observe changes brought about during the subsequent stratification period. This approach is analogous to one used by oceanographers who look at changes in oxgyen content of water masses to determine apparent oxygen utilization (13). Substantial changes occurred in 0, PO_4^{3-}, and total inorganic nitroten (Σ N) content of water below the halocline during the restratification period (Figure 3). Oxygen decreased asymptotically to about 50 percent of its original concentration below the halocline. Water in the boundary meter above the benthos contained less oxygen (not shown). Virtually all of the increase in Σ N observed was as NH_4^+, this increase occurred at a somewhat slower rate as oxygen content decreased. PO_4^{3-}, on the other hand, increased at an accelerating rate--the highest concentrations being observed simultaneously with the lowest 0 concentrations. Webb and D'Elia (20), in expressing PO_4^{3-} and Σ N concentrations in the bottom few meters of the water column as a function of oxygen concentration, demonstrated a nearly linear relationship between 0 and N concentrations. This relationship suggested that for each atom of Σ N produced, approximately 18 atoms of 0 were consumed, somewhat higher than the "Redfield" ratio (14). In contrast, as is suggested by Figure 3, PO_4^{3-} concentration increased rapidly at lower oxygen concentration in a non-linear fashion. Webb and D'Elia (20), by expressing PO_4^{3-} against 0 concentration of deep water samples, found that the ratio of apparent PO_4^{3-} production to oxygen consumption was 1:74 below 125 mg at O_2 liter^{-1}, and 1:869 above 125 mg at O_2liter^{-1}. Since it is well known that oxygen tension can affect the rate of PO_4^{3-} release from sediments and soils (8, 10) and may affect its solubility in Chesapeake waters (16), it is apparent that "water quality" as expressed in terms of PO_4^{3-} concentration, will be dependent on oxygen concentration. Clearly, a simple linear mathematical expression relating the two in water quality models would be inadequate.

Phoel et al. (11), who performed in situ benthic chamber incubations concurrently with our water column sampling, found that ammonia was released from the sediments at our sampling station at up to approximately 0.25 μg at N. m^{-2}. sec^{-1}. This value corresponds very closely to the value of 0.24 μg at N. m^{-2}. sec^{-1} calculated from our data in Figure 3, if we assume that benthic release of N is primarily responsible for the increase in N concentration below the halocline after restratification, and that dispersive and advective processes remove or add little N to the water column during that period. We realize that during the restratification period there was likely to have been some net upstream movement of deep waters and consequently that estimates of benthic nutrient regeneration determined in this way also represent to some extent the effect of downstream benthic processes.

There is a growing body of evidence (e.g. 1) in addition to what we present above which indicates that benthic processes are more significant sources of nutrients to the Chesapeake Bay than was believed previously by Carpenter et al. (2) and McCarthy et al. (7). Taft et al. (17) suggested that a large reservoir of nutrients, presumably of benthic origin, is isolated below the pycnocline in the bay under low flow conditions, but could not account for adequate vertical transport of these nutrients south of 38° 53' N. Lat. to supply appreciable N to surface phytoplankton. Our study shows that at times (i.e. during destratification events), vertical transport is sufficient to transfer nutrients regenerated by the benthos to surface waters in the southern end of the Chesapeake estuary.

We cannot emphasize strongly enough the need for those designing sampling programs and numerical models to give adequate attention to time scales appropriate to key biological, chemical and hydrographic events affecting water quality. Consider the difficulty that might have been encountered in extrapolating York River water quality to a longer time scale were our sampling period limited to a single day in August 1978–had it been on August 21 or 22 we might have mistakenly concluded that the lower York River was relatively unenriched with nutrients and was well oxygenated. Such results would be difficult to reconcile in steady state water quality models based on the hydrographic features of the estuary which affect salt balance.

FIGURE 2. A. Absolute value of surface to bottom salinity differences; B. Bottom meter oxygen concentrations. Curves are fitted by eye.

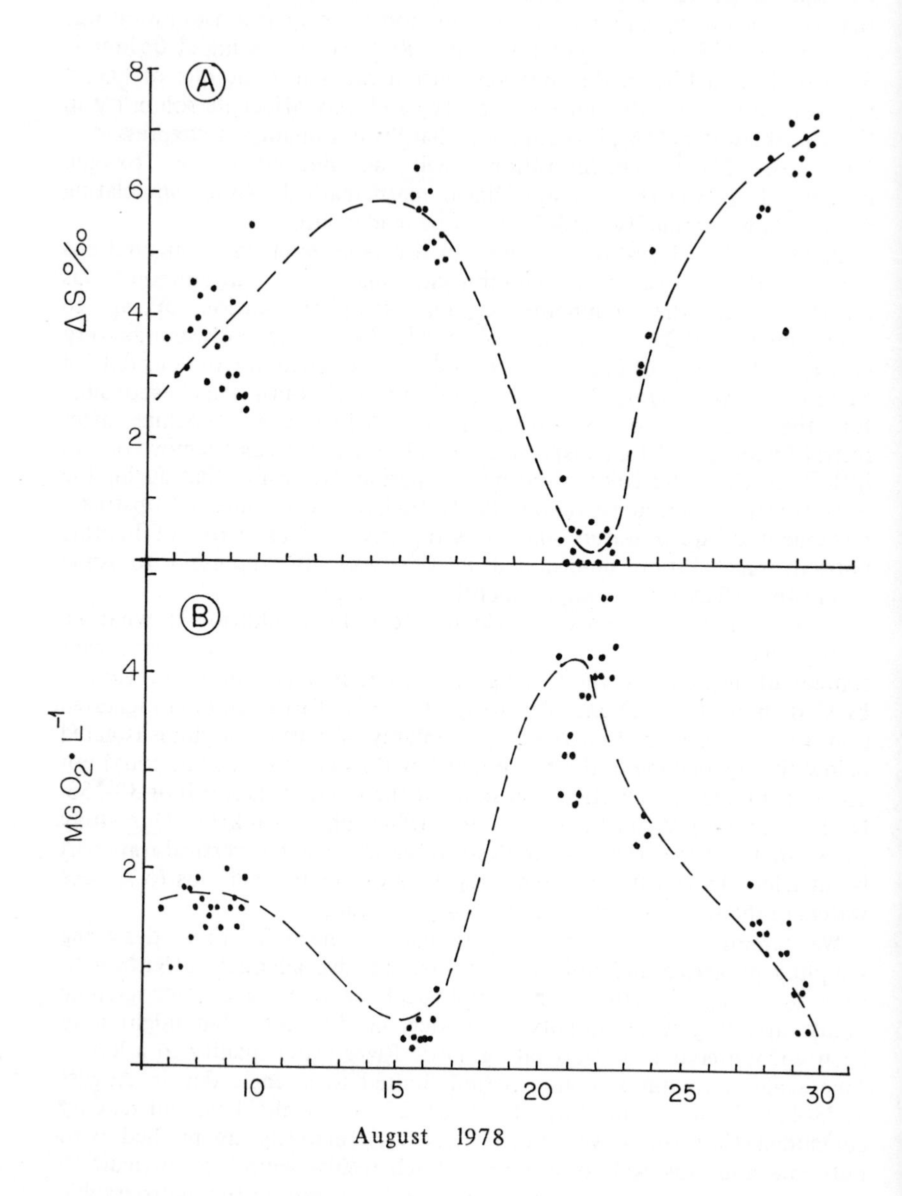

FIGURE 2. (Cont'd.) C. Bottom meter ammonium concentrations; and D. Bottom meter phosphate concentrations during the study period. Curves are fitted by eye.

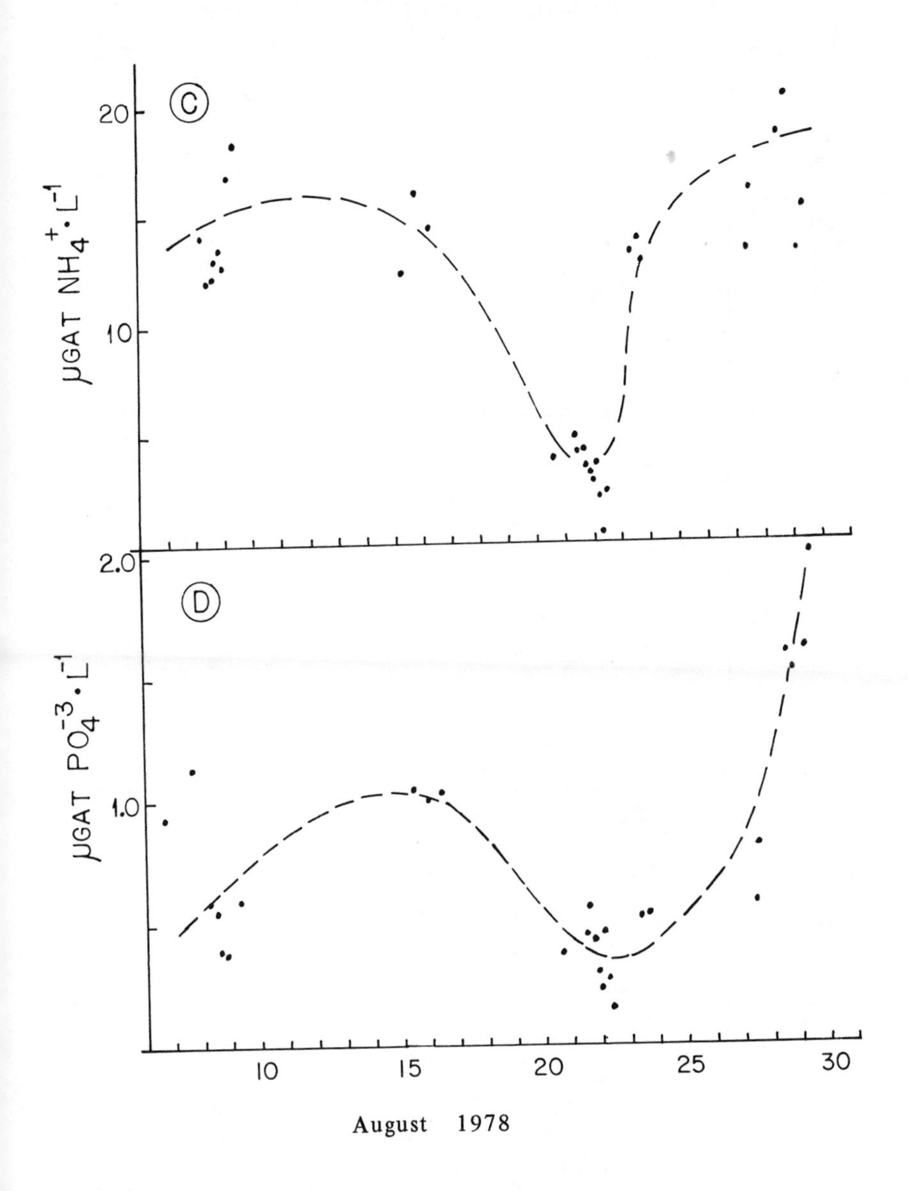

FIGURE 3. Nutrient and oxygen concentrations in the water column below the halocline from the end of the destratification period to the end of our study in August, 1978. A. Dissolved inorganic nitrogen (ΣN), here, primarily NH_4^+, B. Phosphate, and C. Oxygen.

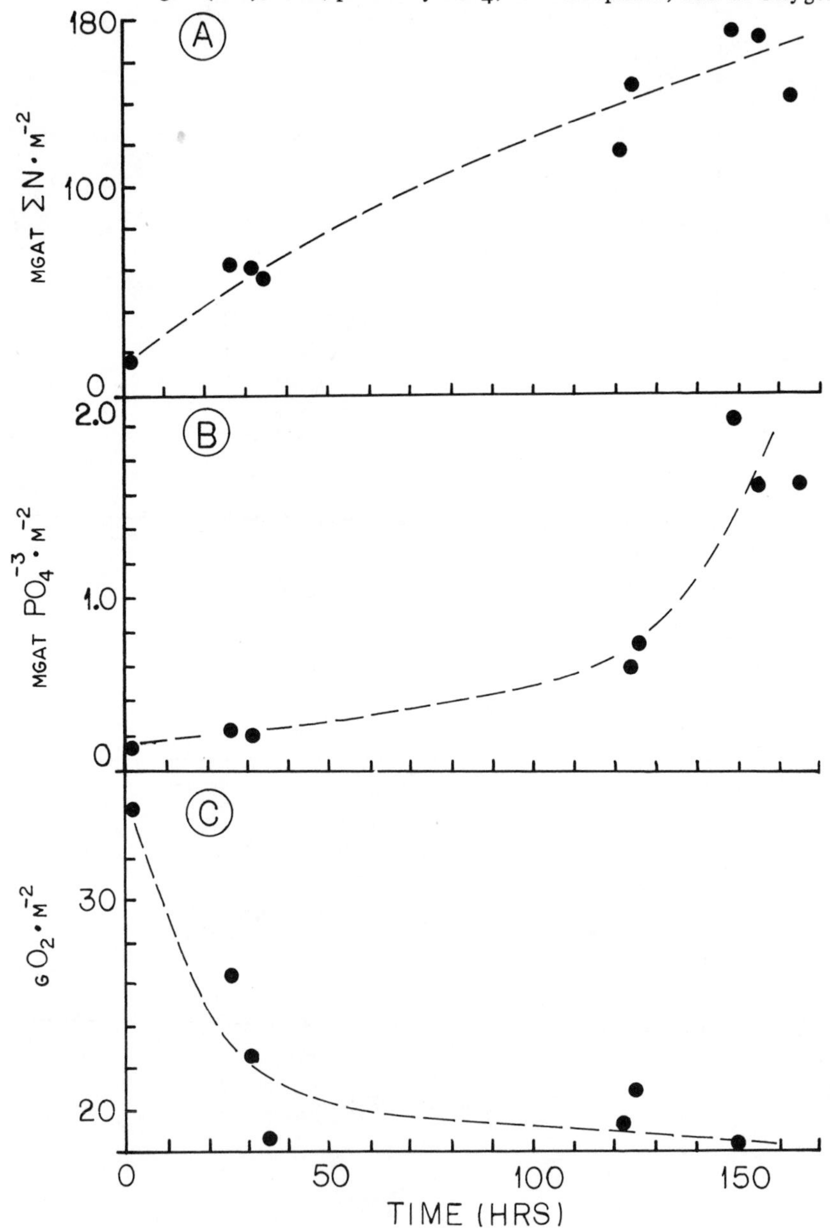

As a final note we point out that although in this study we focused our attention on spring tidal destratification events which are predictable and allow enhanced surface to bottom water column mixing to occur, other destratification events may occur in an irregular or unpredictable fashion as a result of meteorological effects. Any processes accelerating mixing in a stratified or partially stratified estuary are likely to have profound effects on nutrient and oxygen fluxes and distribution as we have shown here. Some accommodation for these events will also be necessary in assessing and managing estuarine water quality.

REFERENCES

1. Boynton, W.R., W.M. Kemp, and C. Osborne. 1980. Benthic nutrient fluxes in the sediment trap portion of the Patuxent estuary, 93-109. *In* V.S. Kennedy (ed.) Estuarine Perspectives, Academic Press, N.Y.
2. Carpenter, J.H., D.W. Pritchard, and R.C. Whaley. 1969. Observations on eutrophication and nutrient cycles in some coastal plain estuaries, 210-221. *In* Eutrophication, Causes, Consequences, Correctives. Natl. Acad. Sci. Publ. 1700.
3. Haas, L.W., S.J. Hastings, and K.L. Webb. 1980. Phytoplankton response to a tidally induced cycle of stratification and mixing in the York River Estuary. pp. 585-596. *In* B.J. Neilson and L.E. Cronin (eds.). International symposium on nutrient enrichment in estuaries. Humana Press, Inc. Clifton, N.J.
4. Haas, L.W., F.J. Holden and C.S. Welch. 1980. Short-term changes in the vertical salinity distribution of the York River estuary associated with the neap-spring tidal cycle. pp. 619-636. *In* B.J. Neilson and L.E. Cronin (eds.). International symposium on nutrient enrichment in estuaries. Humana Press, Inc. Clifton, N.J.
5. Indreb, G., B. Pengerud and I. Dundas. 1979. Microbial activities in a permanently stratified estuary. II. Microbial activities at the oxic-anoxic interface. Mar. Biol. 51:305-309.
6. Kopp, J.F. and G.D. McKee. 1979. Methods for chemical analysis of water and waste. U.S. E.P.A. Publication EPA-600/4-79-020. 460 p.
7. McCarthy, J.J., W.R. Taylor and J.L. Taft. 1977. Nitrogenous nutrition of the plankton in the Chesapeake Bay. 1. Nutrient availability and phytoplankton preferences. Limnol. Oceanogr. 22:996-1011.
8. Mortimer, C.H. 1971. Chemical exchanges between sediments and water in the Great Lakes - speculations on probable regulatory mechanisms. Limnol. Oceanogr., 16:387-404.
9. Najarian, T. 1979. Personal communication.
10. Patrick H.H., Jr. and E.A. Khalid. 1971. Phosphate release and sorption by soils and sediments: effect of aerobic and anaerobic conditions. Science 186:53-55.

11. Phoel, W.C., K.L. Webb and C.F. D'Elia. 1980. Inorganic nitrogen regeneration and total oxygen consumption by the sediments at the mouth of the York River, Virginia, U.S.A. pp. 607-618. *In* B.J. Neilson and L.E. Cronin (eds.). International symposium on nutrient enrichment in estuaries. Humana Press, Inc., Clifton, N.J.
12. Pritchard, D.W. 1955. Estuarine circulation patterns. Proc. Am. Soc. Civil Engrs. 81:1-11.
13. Redfield, A.C. 1942. The processes determining the concentration of oxygen, phosphate and other organic derivatives within the depths of the Atlantic Ocean. Pap. Phys. Oceanogr. Meterol. 9:1-22.
14. Redfield, A.C. 1958. The biological control of chemical factors in the environment. Amer. Scientist 46:205-221.
15. Strickland, J.D., and T.R. Parsons. 1968. A practical handbook of seawater analysis. Bull. Fish. Res. Bd. Can. 167.
16. Taft, J.L. and T. Callender, unpublished data.
17. Taft, J.L., A.J. Elliot, and W.R. Taylor. 1978. Box model analysis of Chesapeake Bay ammonium and nitrate fluxes, 115-130. *In* M.L. Wiley, (ed.) Estuarine Interactions, Academic Press, N.Y.
18. Vaccaro, R.F. 1965. Inorganic nitrogen in sea water, 365-407. *In* J.P. Riley and G. Skirrow, (eds.). Chemical Oceanography. Academic Press, N.Y.
19. Webb, K.L., unpublished manuscript.
20. Webb, K.L. and C.F. D'Elia. 1980. Nutrient and oxygen redistribution during a spring neap tidal cycle in a temperate estuary. Science 207:983-985.
21. Wolaver, T., R.L. Wetzel, J.C. Zieman and K.L. Webb. 1980. Nutrient interactions between salt marsh, intertidal mudflats and estuarine water. 123-133. *In* V.S. Kennedy (ed.). Estuarine Perspectives. Academic Press, N.Y.

INORGANIC NITROGEN REGENERATION AND TOTAL OXYGEN CONSUMPTION BY THE SEDIMENTS AT THE MOUTH OF THE YORK RIVER, VIRGINIA

William C. Phoel*, K.L. Webb and C.F. D'Elia*****

**U.S. Department of Commerce*
National Oceanic and Atmospheric Administration
National Marine Fisheries Service
Northeast Fisheries Center
Sandy Hook Laboratory
Highlands, New Jersey 07732

***Virginia Institute of Marine Science*
School of Marine Science
College of William and Mary
Gloucester Point, Virginia 23062

****University of Maryland*
Center for Environmental and Estuarine Studies
Chesapeake Biological Laboratory
Solomons, Maryland 20688

ABSTRACT: In situ measurements of inorganic nitrogen fluxes and riverbed oxygen consumption were made on sediments in 3, 9, and 16 m of water at the mouth of the York River during stratified and destratified water conditions. Ammonia was regenerated, the rate of which increased with depth and oxygen concentration in the overlying water. Nitrate and nitrite fluxes from the sediment were minimal or non-existent during stratification at the 16-m station but increased and the nutrients were taken up by the sediments under destratified conditions. At the 3-m station, which is above the halocline when developed, nitrate and nitrite appeared to be the major forms of nitrogen being released by the sediments. Oxygen consumption by the riverbed at the 16 and 9-m stations was higher during the increased oxygen tensions associated with vertical destratification. The 3-m station maintained the highest rates of oxygen consumption throughout the sampling period. The in situ incubation of bottom water alone at all three stations indicated negligible rates of oxygen uptake.

INTRODUCTION

As part of the interdisciplinary York River Study (6), we investigated the effects of the cyclic stratification-destratification-restratification of the water column (4, 6, 7, 8) upon inorganic nitrogen regeneration and oxygen consumption by the sediments. To complement the York River mouth 16-m station, which was of primary concern, we expanded the in situ investigation to include sediments at depths of 9 m and 3 m (Figure 1).

DESCRIPTION OF STUDY AREA

The hydrographic and physical properties and the nutrient and oxygen profiles found in the study area have been described elsewhere (4, 6). The sediments at the 16-m station consisted of soft black mud. Bottom water temperatures ranged between 25.0° and 26.0°C and dissolved oxygen (D.O.) concentrations between 1.8 ml 1^{-1} during destratified conditions and 0.8 ml 1^{-1} during stratification. About 9 m down in the water column, just below the halocline, ambient light rapidly decreased to zero, and the turbidity caused by suspended, extremely fine flocculent material, present only during destratified water conditions, limited bottom visibility with artificial light to approximately 20 cm. This flocculent material may be inorganic precipitates due to reoxygenation or be derived from the surface waters through mixing. Turbidity decreased dramatically under stratified conditions, and permitted one of us (WCP) to see, with artificial light, slightly more than 1 m. Sluggish currents occurred in the bottom 3 m of the water column while above this a moderate to strong current prevailed.

At the 9 m station the sediment was very compacted, fine grained, dark-brown sand. During stratified conditions a halocline existed at about 1.5 m off the bottom. As a consequence, this sediment was exposed to the stratification-destratification-restratification phenomenon. That the alternation of exposure occurred was confirmed by a decrease in bottom water oxygen concentration from an average of 4.1 ml 1^{-1} during destratified conditions, to an average of 1.4 ml 1^{-1} during stratification. The temperature of the bottom water ranged from 26.0° to 26.5°C. Bottom currents were moderate and visibility, with aid of artificial light, increased during stratification as turbidity decreased.

The sediments at the 3-m station (hard-packed, medium grained, yellow-brown sand) were influenced by moderate currents and were above the halocline when it developed; consequently, this water remained well oxygenated (>4.0 ml 1^{-1}). Temperature ranged between 27.0° and 27.5°C. Sunlight penetrated to the bottom during all daylight incubations and visibility was usually between 0.5 and 1.0 m without artificial illumination.

FIGURE 1. Station locations and depth contours at the York River mouth, Virginia. Shaded circles with diamonds are navigational buoys.

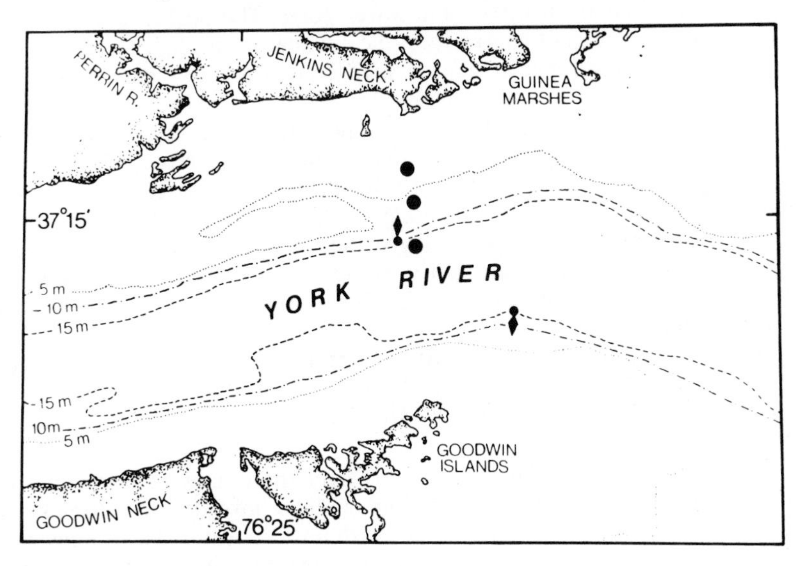

MATERIALS AND METHODS

The incubators were 1 m diameter hemispheres constructed of clear plexiglas with a 10 cm wide horizontal collar around the equator circumscribing the dome to prevent excessive settling into the softer sediments. With the collar on the sediment the volume of incubated water was accurately maintained at 262 ℓ after insertion of the dissolved oxygen probe. A plexiglas flange extended vertically from the collar and penetrated the sediment to a depth of 10 cm to prevent the intrusion of outside bottom water into the incubator. Approximately 5 kg of lead weights were distributed around the collar to aid in stabilizing the incubator. Upper and lower sampling ports, fitted with rubber serum bottle stoppers, were located 5 cm off top center and approximately 20 cm above the collar, respectively.

International Biophysics Corporation "Deep Diver" polarographic oxygen electrodes 1/ were used to determine dissolved oxygen concentrations within the incubator. According to the manufacturers specifications, these electrodes have a precision of ± 0.1 ppm on the 0-5 ppm scale and ± 0.2 ppm on the 0-20 ppm scale. The temperature precision is ± 1.0°C. These electrodes incorporate a battery operated submersible stirrer to maintain a flow over the electrode. The stirrer has been shown to gently circulate the water in domes of this size based on

1/ Does not constitute an endorsement statement.

dye studies (J.M. Wells, personal communication); studies by one of us (WCP) showed no average differences in dissolved oxygen concentrations in samples taken from upper and lower ports. The probes were lowered approximately one-half meter into the incubator through a hole at top center. A rubber stopper cemented with nontoxic epoxy glue to the wire of the probe was used to seal the hole.

To determine the rate of oxygen consumption by the bottom water without the influence of the sediments, a smaller, black plexiglas cylinder with a volume of 9 ℓ was used to incubate the bottom water alone at each station. The cylinder had a diameter of 28 cm and was 15 cm high. No stirring device was utilized with incubations of water only. A second black cylinder of similar dimensions but open bottom was used to determine the riverbed oxygen consumption at the 3-m station where sunlight reached the sediments.

All diving was done using a surface supplied system in accordance with the diving regulations promulgated by both the National Oceanic and Atmospheric Administration and the Occupational Safety and Health Administration.

At the beginning of an experiment, after the domes were lowered by line from a boat moored on station, a diver inverted the domes and replaced the water contained in them with bottom water. The domes were then righted and gently placed on undisturbed sediment. Initial, midpoint and final samples for D.O. and nutrient analyses were drawn with 50 ml plastic hypodermic syringes. Three to five replicate samples of ambient bottom water were taken in 35 ml glass bottles for calibrating the D.O. probe. No substantial differences between the ambient D.O. concentrations and initial syringe concentrations were observed. The dissolved oxygen concentrations were measured using the azide modification of the iodometric method (1) except that amylose was used in place of starch and 0.025 N phenylarsine oxide (PAO) was used in place of sodium thiosulfate (10, 19). Nutrient samples, were immediately filtered and refrefrigerated until delivered to the R/V *Retriever* for analysis as described elsewhere (4).

To determine the effect that increasing D.O. concentration had on the rates of oxygen consumption and nutrient regeneration during stratified conditions, one incubator on the sediment was aerated in situ for one hour with compressed air from a SCUBA bottle and diffuser, increasing the D.O. concentration from 0.8 ml l^{-1} to 3.5 ml l^{-1}. A 9 ℓ cylinder containing bottom water only was similarly aerated for three minutes, which increased the D.O. concentration from 0.8 ml l^{-1} to 6.6 ml^{-1}. No nutrient analysis was performed on the water only incubation. Readings from the dissolved oxygen meters were recorded about every 15 minutes and whenever a sample was drawn. Incubations lasted from three to 19 hours depending upon the oxygen consumption rate and did not consume more than 30 percent of the initial dissolved oxygen, except for the aerated dome which utilized 40 percent. The lowest final D.O.

concentration was 0.7 ml l^{-1}, down from an initial concentration of 1.1 ml l^{-1} during stratified conditions at the 9-m station.

RESULTS AND DISCUSSION

At the 16-m station, ammonia was the predominant source of inorganic nitrogen released by the sediment, with flux rates during periods of destratification, when the water was relatively well oxygenated, at 0.142 μgat N. m^{-2} sec^{-1} (511 μgat N. m^{-2} hr^{-1}), almost twice those attained during stratification (0.080 μgat N. m^{-2} sec^{-1} or 288 μgat N. m^{-2} hr^{-1} under less oxygenated conditions. Nitrate fluxes out of the sediment during stratification were negligible (0.003 μgat N. m^{-2} sec^{-1} or 11 μgat N. m^{-2} hr^{-1}) and those of nitrite were undetectable. During destratified conditions, however, nitrate and nitrite were taken up by the sediments, at the respective rates of 0.013 μgat N. m^{-2} sec^{-1} (47 μgat N. m^{-2} hr^{-1}) and 0.004 μgat N. m^{-2} sec^{-1} (14 μgat N. m^{-2} hr^{-1}). Ammonia was also the largest component of inorganic nitrogen flux at the 9-m station (0.045 μgat N. m^{-2} sec^{-1} or 162 μgat N. m^{-2} hr^{-1}) during destratified conditions but was one-third the rate regenerated at the 16 m station. The nitrate and nitrite fluxes into the sediments at the 9-m station were quite small (0.005 and 0.001 μgat N. m^{-2} sec^{-1} or 18 and 4 μgat N. m^{-2} hr^{-1}, respectively).

The river bed oxygen consumption rates at both the 16-m and 9-m stations were highest during periods of destratification (49 ml m^{-2} hr^{-1} or 1.2 μgat 0. m^{-2} sec^{-1} and 33 ml m^{-2} hr^{-1} or 0.8 μgat 0. m^{-2} sec^{-1}, respectively) and decreased substantially during restratification to rates of 20 ml m^{-2} hr^{-1} (0.5 μgat 0. m^{-2} sec^{-1}) at the 16-m station and 6 ml m^{-2} hr^{-1} (0.1 μgat 0. m^{-2} sec^{-1}) at the 9-m station. The ambient bottom D.O. concentrations for the 16-m and 9-m stations at the time sampling measurements were made during destratification were 1.8 ml l^{-1} and 4.5 ml l^{-1}, respectively. During restratification the concentrations dropped to 0.8 ml l^{-1} (16 m) and 1.0 ml l^{-1} (9 m).

The 3-m station, not under the influence of the stratification-destratification-restratification phenomenon, was well oxygenated at all times with concentrations ranging from 4.4 to 4.9 ml l^{-1} during the duration of the experiments. The regeneration of inorganic nitrogen at this station, however, is complicated by the availability of sunlight to the sediments. The rate of ammonia regeneration from two incubations was lower during the hours of maximum sunlight (0.006 and 0.027 μgat N. m^{-2} sec^{-1} or 22 and 97 μgat N. m^{-2} hr^{-1}) than during the hours of little or no light when the ammonia flux was 0.053 μgat N. m^{-2} sec^{-1} (191 μgat N. m^{-2} hr^{-1}). Conversely, both nitrite and nitrate regeneration increased dramatically from a combined low of 0.002 μgat N. m^{-2} sec^{-1} (7 μgat N. m^{-2} hr^{-1}) under primarily dark conditions to a combined two incubation average of 0.466 μgat N. m^{-2} hr^{-1}; (1678 μgat N. m^{-2} hr^{-1}) during maximum sunlight (nitrate = 0.260 and 0.204 μgat N. m^{-2} sec^{-1} or 936 and 734 μgat N. m^{-2} hr^{-1}; nitrite = 0.305 and 0.158 μgat

N. m^{-2} sec^{-1} or 1098 and 569 μgat N. m^{-2} hr^{-1}). The apparent decrease in ammonia regeneration during a period of normally high photosynthetic activity may have been caused by benthic algae and phytoplankton utilizing the ammonia. The opposite trend for nitrite and nitrate is not understood and the data are too limited to provide an adequate explanation. The riverbed oxygen consumption was high ($\overline{X}$ = 72 ml m^{-2} hr^{-1} or 1.8 μgat 0. m^{-2} sec^{-1}, n = 4, ± 16 S.D.) for the 3-m station. This average includes one high value of 96 ml m^{-2} hr^{-1} (2.4 μgat 0. m^{-2} sec^{-1}) from a mid to late afternoon incubation, a value of 65 ml m^{-2} hr^{-1} (1.6 μgat 0. m^{-2} sec^{-1}) from a night incubation, a maximum sunlight value of 65 ml m^{-2} hr^{-1} and a value of 60 ml m^{-2} hr^{-1} (1.5 μgat 0. m^{-2} sec^{-1}) from a black dome incubated during the early afternoon.

Incubations of bottom water alone at each station exhibited negligible rates of oxygen consumption except after artificial aeration at the 16-m station. At this station, high rates of oxygen consumption were observed for about two hours after aeration during stratified conditions and decreased the D.O. concentration in the dome from 6.6 ml l^{-1} to 5.9 ml l^{-1}. These rates decreased and became negligible for the remaining 16 hours of the incubation producing a final D.O. concentration of 5.8 ml l^{-1}.

The strong associations between both ammonia flux and riverbed oxygen consumption with the stratification-destratification-restratification phenomenon is illustrated in Figure 2. As previously stated, the water at the 16 m and 9 m stations is affected by this phenomenon which is associated with neap-spring tidal cycle. A consequence of such stratification is the isolation of the bottom water and an associated depletion of D.O. by biological metabolism (4, 17). As the D.O. concentration is depleted to a point, generally between 2.0 and 1.0 ml l^{-1} or lower, the rate of oxygen consumption decreases rapidly (5, 12). During stratification, when oxygen consumption is depressed by low D.O., a metabolic potential for the rapid utilization of oxygen or "oxygen debt" is established. AS the system switches to anaerobic conditions reduced substrates are produced which add to this potential for accelerated oxygen consumption.

When oxygen is reintroduced into the bottom water during mixing, this biological and chemical potential for oxygen utilization responds with high rates of oxygen consumption and metabolically associated high rates of ammonia regeneration. This has been demonstrated by Thomas et al. (18) for the New York Bight during an anoxic event in 1976.

The 16-m station, during stratified conditions (Figure 2), maintained a relatively low oxygen consumption rate of 23 ml m^{-2} hr^{-1} (0.6 μgat 0. m^{-2} sec^{-1}) and regenerated ammonia at 288 μgat N. m^{-2} hr^{-1} (0.080 μgat N. m^{-2} sec^{-1}). During mixing, and natural oxygenation of the bottom waters, the oxygen consumption rate approximately doubled to 49 ml m^{-2} hr^{-1} (1.2 μgat 0. m^{-2} sec^{-1}) as did the ammonia flux (511 μgat N. m^{-2} hr^{-1} or 0.142 μgat N. m^{-2} sec^{-1}).

FIGURE 2. Riverbed oxygen consumption and ammonia regeneration as a function of time and condition of stratification of the water column at the York River mouth.

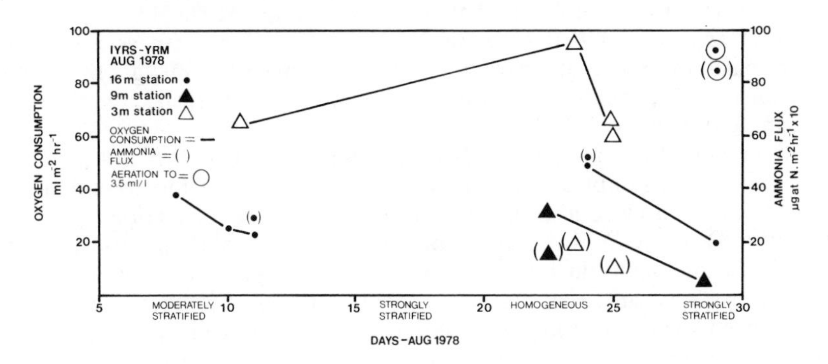

To determine if this association persisted at oxygen consumption rates not normally occurring at this station, we aerated the bottom water in an incubator during stratified conditions for one hour, to increase the D.O. from 0.8 ml l^{-1} to 3.5 ml l^{-1}. Data from this aerated dome are denoted by circled points on Figures 2 and 3. The sediments responded to this aeration with a corresponding increase in oxygen consumption, from 20 ml m^{-2} hr^{-1} (0.5 µgat 0. m^{-2} sec^{-1}) to 91 ml m^{-2} hr^{-1} (2.3 µgat 0. m^{-2} sec^{-1}). This increased rate is almost twice the natural oxygen consumption rate during destratified water conditions present after mixing. The ammonia regeneration rate of 853 µgat N. m^{-2} hr^{-1} (0.237 µgat N. m^{-2} sec^{-1}) is a 60 percent increase over that of natural mixed conditions. This high rate of ammonia release remained stable throughout the incubation period of 19.1 hours.

Nitrite and nitrate were produced at about equal rates (0.051 and 0.059 µgat N. m^{-2} sec^{-1} or 184 and 212 µgat N. m^{-2} hr^{-1}, respectively) for the first 2.5 hours after aeration, but these rates declined and eventually the nitrite and nitrate began to be taken up by the sediments at fairly low rates (0.006 and 0.010 µgat N. m^{-2} sec^{-1} or 22 and 36 µgat N. m^{-2} hr^{-1}, respectively) during the remainder of the incubation. To preclude the influence of low D.O. concentration on the rate of oxygen consumption, the experiment was terminated when the dissolved oxygen concentration of the incubated water reached 2.1 ml l^{-1}.

To ascertain that D.O. concentrations did, in fact, exert a controlling influence on the rate of riverbed oxygen consumption, we plotted oxygen consumption rates as a function of D.O. concentration for the three stations (Figure 3) and found that the correlation at the 16-m station is very strong (r = 0.95). Although higher ratios of oxygen consumption were also observed at increased oxygen concentrations at the other stations, not enough data points were available at either to obtain a meaningful regression line.

Regressing total inorganic nitrogen flux as a function flux as a function of oxygen consumption rates at the 16-m station, during stratified and destratified water conditions, produced an excellent correlation (r = 0.99) (Figure 4). The O/N flux ratios, noted at each data point, are lower during stratification than during destratification, except when the incubated water was aerated. In all cases the O/N flux ratios were lower than would have been expected using the theoretical Redfield ratio of 212 atoms of oxygen required to oxidize 15 atoms of organic nitrogen (9, 13) to ammonia and would indicate that more inorganic nitrogen (predominantely in the form of ammonia) is being regenerated than can be accounted for by benthic aerobic metabolism. This is contrary to studies in other estuaries and coastal areas where the O/N flux ratio is substantially higher than the theoretical Redfield ratio of 13.25:1, indicating less inorganic nitrogen is being regenerated than expected on the basis of benthic oxygen consumption (3, 11, 15, 16). Rowe et al. (14), however, have reported ammonia flux from the sediments off Spanish Sahara to be about three times that which would have been predicted from an oxygen-nitrogen relationship.

The 9-m station, while still affected by stratification and mixing, has significantly higher dissolved oxygen concentrations than the 16-m station during periods of destratification. However, during stratification oxygen concentrations go as low as 1.0 ml l^{-1}. We have no O/N flux data for stratified conditions at the 9-m station. However, for mixed conditions the O/N flux ratio was 20.1:1. This was similar to that reported by Nixon et al. (1976) for the west passage of Narragansett Bay.

The O/N flux ratios for the 3-m station, which was always well oxygenated, were 43.3:1 (n = 1) during the hours of little or no light and an average of 3.9:1 (n = 2) for the period of maximum sunlight the following day. The extremely high O/N flux ratio of 43.3:1 appeared to be caused by a combination of relatively high ammonia regeneration (0.053 μgat N. m^{-2} sec^{-1} or 191 μgat N. m^{-2} hr^{-1}) and negligibly low nitrite and nitrate fluxes (combined flux of 0.002 μgat N. m^{-2} sec^{-1} or 7 μgat N. m^{-2} hr^{-1}). Rowe et al. (14) suggest that high O/N flux ratios may be explained by the denitrification of nitrate and the subsequent loss of N as N_2 gas. The anomalously low O/N flux ratio of 3.9:1 was caused by surprisingly high rates of nitrite and nitrate fluxes from the sediment (combined average = 0.466 μgat N. m^{-2} sec^{-1} or 1678 μgat N. m^{-2} hr^{-1}, n = 2). The similarity in time and environmental conditions, except for the

FIGURE 3. Riverbed oxygen consumption rates plotted as a function of dissolved oxygen concentration for the York River mouth stations.

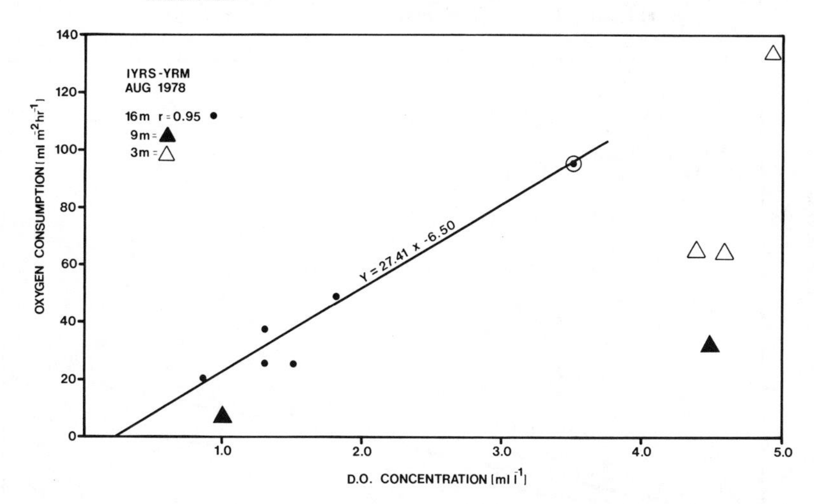

amount of sunlight reaching the incubators, for these two exceedingly disparate values indicates the need for further investigation of euphotic sediments before attempts are made to predict nitrogen regeneration flux based on oxygen consumption rates alone.

The abiotic and biotic conditions at each of the three stations, located within 1.5 km of each other, are quite different, especially with regard to sediment type, infauna and flora, quantity of sunlight, and stratification-destratification-restratification phenomenon. Similarly there are substantial differences among the stations with regard to benthic oxygen consumption, fluxes of the various nutrients and correlation between the two, usually described as O/N flux ratios.

There are complementary indications that the ammonia flux from the benthos is the primary factor influencing the ammonia concentration of the overlying water. First, the ammonia regeneration rates in dome incubations of the benthos were substantial, while the regeneration rates in control water only incubations were negligible. Second, there is good general agreement between these benthic ammonia regeneration rates which were directly measured and regeneration rates calculated by ammonia accumulation in the water column alone (4).

The results of this study, when compared with a similar one (2) from the Chesapeake estuary, indicate that this estuary exhibits some of the highest recorded rates of benthic metabolism. This will undoubtedly have to be considered by those constructing water quality models of this estuary. It will be difficult to reduce the benthic flux components of these

FIGURE 4. Total inorganic nitrogen flux plotted as a function of oxygen consumption for the 16-m station. O/N ratios and condition of stratification are presented as well as the slope of the theoretical 13.25:1 O/N ratio.

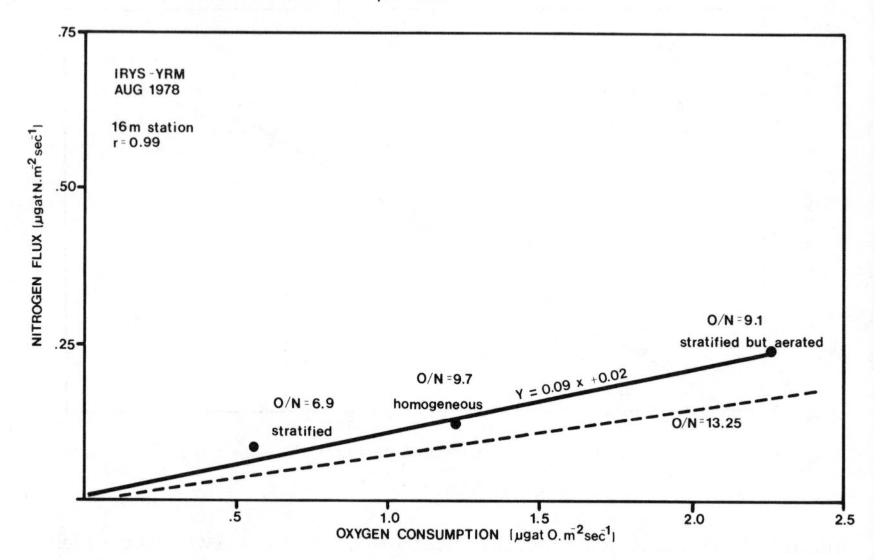

models to a few simple coefficients given the considerable temporal and spatial variations we and others (2) have observed in these fluxes. In this study alone, we have seen variations directly or indirectly due to depth, D.O. concentrations, sediment type, light, infauna and flora, and the spring-neap tidal cycle. More work seems to be needed to assess temporal and spatial variations in benthic fluxes as measured in situ by domes and by indirect mass balance calculations.

SUMMARY

1. Observed ammonia concentrations in the bottom water during stratified conditions were the result of regenerative processes in the sediments. The rate of ammonia regeneration was directly related to the rate of riverbed oxygen consumption at the 16-m station.
2. Riverbed metabolism decreased the D.O. concentration in the bottom water when the bottom water was isolated by stratification. The resulting hypoxia was relieved by reoxygenation during the destratification associated with spring tidal mixing.
3. O/N flux ratios at the 16-m station were lower than would be expected using the Redfield ratio. The 9-m station maintained a ratio which was higher, while ratios at the 3-m station were higher and lower in the extreme.

4. The results of this and other studies suggest that O/N flux ratios for different bottom environments are highly variable and calculations based on data from only one or two of the environments cannot be extrapolated to the entire ecosystem under consideration.

ACKNOWLEDGEMENTS

Special thanks are due to A. Thomson and L. Pastor for their assistance in obtaining the data and for serving as emergency medical technician and stand-by diver, respectively.

REFERENCES

1. American Public Health Assoc. 1975. Standard methods for the examination of water and wastewater, Fourteenth ed. 1193 p.
2. Boynton, W.R., W.M. Kemp and C. Osborne. 1980. Benthic nutrient fluxes in the sediment trap portion of the Patuxent estuary, (In press). *In* V.S. Kennedy (ed.), Estuarine Perspectives. Academic Press, New York.
3. Davies, J.M. 1975. Energy flow through the benthos in a Scottish sea lock. Marine Biology 31, 353-362.
4. D'Elia, C.F., K.L. Webb and R.L. Wetzel. 1980. Inputs of hydrographic events on water quality in an estuary. *In* B.J. Neilson, and L.E. Cronin (eds.), International Symposium on Nutrient Enrichment in Estuaries. The Humana Press, Clifton, New Jersey.
5. Edberg, N. and B.V. Hofsten. 1973. Oxygen uptake of bottom sediments studied in situ and in the laboratory. Water Research 7:1285-1294.
6. Haas, L.W., F.J. Holden and C.S. Welch. 1980. Short term changes in the vertical salinity distribution of the York River estuary associated with the neap-spring tidal cycle. *In* B.J. Neilson and L.E. Cronin (eds.), International Symposium on Nutrient Enrichment in Estuaries. The Humana Press, Inc., Clifton, New Jersey.
7. Haas L.W., S.J. Hastings and K.L. Webb. 1980. Phytoplankton responses to a stratification-mixing cycle in the York River estuary during late summer. *In* B.J. Neilson and L.E. Cronin (eds.), International Symposium on Nutrient Enrichment in Estuaries. The Humana Press, Inc., Clifton, New Jersey.
8. Haas, L.W. 1977. The effect of the spring-neap tidal cycle on the vertical salinity structure of the James, York and Rappahannock Rivers, Virginia, U.S.A. Estuarine and Coastal Marine Science (5):485-496.
9. Hale, S.S. 1975. The role of benthic communities in the nitrogen and phosphorus cycles of an estuary, 291-308. *In* F.G. Howell, J.B. Gentry and M.H. Smith (eds.), Mineral Cycling in Southeastern Ecosystems. ERDA Symposium Series (CONF-740513).

10. Kroner, R.C., J.E. Longbottom and R. Gorman. 1964. A comparison of various reagents proposed for use in the Winkler procedure for dissolved oxygen. PHS Water Pollut. Surveillance System Appl. Develop. Rep. 12. Public Health Serv., DEP. HEW. 18 p.
11. Nixon, S.W., C.A. Oviatt and S.S. Hale. 1976. Nitrogen regeneration and the metabolism of coastal marine bottom communities, 269-283. *In* J.M. Anderson and A. Macfadyen (eds.), The Role of Terrestial and Aquatic Organisms in Decomposition Processes. Blackwell Scientific Publications, Oxford.
12. Pamatmat, M.M. 1971. Oxygen consumption by the seabed. IV. Shipboard and laboratory experiments. Limnol. and Oceanogr. 16(3):536-550.
13. Redfield, A.C., B.H. Ketchum and F.A. Richards. 1963. The influence of organisms on the composition of seawater, 26-77. *In* M.N. Hill (ed.), The Sea, Vol. 2. Wiley-Interscienc , New York.
14. Rowe, G.T., C.H. Clifford and K.L. Smith, Jr. 1977. Nutrient regeneration in sediments off Cap Blanc, Spanish Sahara. Deep-Sea Research 24:57-63.
15. Rowe, G.T., K.L. Smith, Jr., and C.H. Clifford. 1976. Benthic-pelagic coupling in the New York Bight, 370-376. *In* G. Gross (ed.), Middle Atlantic Continental Shelf and the New York Bight. Am. Soc. Limnol. Oceanogr. Spec. Symp. 2.
16. Rowe, G.T., C.H. Clifford, K.L. Smith and P.L. Hamilton. 1975. Benthic nutrient regeneration and its coupling to primary productivity in coastal waters. Nature. 255:215-217.
17. Thomas, J.P., W.C. Phoel, F.W. Steimle, J.E. O'Reilly and C.A. Evans. 1976. Seabed oxygen consumption - New York Bight Apex. *In* G. Gross (ed.), Middle Atlantic Continental Shelf and the New York Bight, 354-369. Am. Soc. Limnol. Oceanogr. Spec. Symp. 2.
18. Thomas, J.P., J.E. O'Reilly, A. Draxler, J.A. Babinchak, C.N. Robertson, W.C. Phoel, R. Waldhauer, C.A. Evans, A. Matte, M. Cohn, M. Nitkowski and S. Dudley. Biological processes: Productivity and Respiration. *In* C. Sindermann and L. Swanson (eds.), Oxygen Depletion and Associated Benthic Mortalities in the New York Bight, 1976. NOAA Professional Paper. (In press).
19. U.S. Environmental Protection Agency. 1974. Methods for chemical analysis of water and wastes. Methods Develop. Qual. Assurance Res. Lab. NERC EPA-625/6-74/003. 298 p.

PHYTOPLANKTON RESPONSE TO A STRATIFICATION-MIXING CYCLE IN THE YORK RIVER ESTUARY DURING LATE SUMMER

Leonard W. Haas, Steven J. Hastings and Kenneth L. Webb
Virginia Institute of Marine Science and
School of Marine Science
College of William and Mary
Gloucester Point, Virginia 23062

ABSTRACT: As part of a larger multidisciplinary study of the lower York River estuary, phytoplankton response to a tidally related cycle of stratification-destratification was examined during August 1978. A "red water bloom" dominated by the dinoflagellate *Cocchlodinium heterolobatum* was initially observed in the lower York River coincident with the spring tide-induced water column destratification event. It is proposed that the dinoflagellates initiating the red tide were advected into the estuary in deep water during the preceding period of stratification or were derived from cysts in the sediments and that destratification provided access to the surface waters. The extent of the red water increased during the ensuing restratified period in the York River, and several lines of evidence indicated that *C. heterolobatum* migrated diurnally between ammonium enriched waters below the halocline (8-10 m) and the relatively nutrient-poor surface waters. Other estuarine systems in which phytoplankton blooms associated with alternating periods of stratification-destratification have been observed are noted. The results illustrate the close relationship between phytoplankton and hydrographic dynamics in this estuarine system and emphasize the necessity to include the study of hydrographic processes in the study of phytoplankton dynamics.

INTRODUCTION

Fundamental relationships between hydrographic and phytoplankton processes in coastal and open ocean areas are well known, e.g. upwellings enhance phytoplankton production (18). In estuaries, however, the inherent complexity of hydrographic processes has hindered our understanding of their effect on phytoplankton dynamics, although progress is being made (25, 29). Hydrography may have direct as well as indirect effects on the temporal and spatial variability of phytoplankton abundance and production. For instance, phytoplankton may be advected into or out of an area by currents or may become concentrated at discontinuities or boundary layers. Hydrography may indirectly affect phytoplankton growth by regulating the availability of light and/or inorganic nutrients. Thus, an understanding of phytoplankton dynamics in estuaries, or other aquatic systems, requires an understanding of the fundamental hydrographic processes occurring therein, and the logistics of phytoplankton sampling must take into consideration the time and space scales of the relevant hydrographic processes.

In this paper we report the results of phytoplankton observations made during an interdisciplinary study of the neap-spring tidally related cycle of stratification-destratification in the lower York River during August, 1978. The results of the hydrographic, nutrient dynamics and benthic nutrient flux aspects of the study are found in accompanying papers in this volume (6, 10, 23).

METHODS AND MATERIALS

A description of the site, sampling stations and sampling instrumentation is provided elsewhere (10). In vivo chlorophyll *a* fluorescence was measured in conjunction with hydrographic sampling, i.e. 1 m depth intervals every three hours for the duration of each sampling period. Water was provided by a submersible pump lowered through the water column and fluorescence was measured with a Turner Designs Model 10 Fluorometer. Chlorophyll *a* was also measured in vitro, using the method of Strickland and Parsons (26) in water samples collected with a Van Dohrn bottle from near surface, mid-depth and near bottom every three hours over a 24-hour span during each sampling period. A paired water sample filtered through a 15 μm Nitex sieve prior to the chlorophyll *a* analysis permitted a determination of the distribution of total chlorophyll *a* between the nannophytoplankton (< 15 μm) and the net phytoplankton (> 15 μm).

The attenuation of photosynthetically active radiation (PAR) through the water column was determined by measuring PAR at 0.5 m intervals through the water column with a Li-Cor LI-185A quantum meter equipped with a LI-192A underwater quantum sensor. Attenuation was calculated as the extinction coefficient, k m^{-1}.

Two different methods were used to enumerate phytoplankton. In the first, a 250 ml water sample was fixed with 2.0 ml lugols preservative and stored in the dark. For enumeration the phytoplankton contained in a 2.0 ml aliquot of this sample were allowed to settle onto a settling chamber for 24 hours which was then observed under an inverted microscope using 200x and 400x magnification. In the second technique, a 2.0 ml water sample was stained with 0.02 ml 0.033 percent proflavin, fixed with 0.20 ml 6 percent gluteraldehyde and gently filtered onto a 0.22 μm Nuclepore filter prestained with irgalen black. The filter was then observed under epifluorescent illumination using a Zeiss Standard Microscope equipped with a 12 V, 100 watt halogen lamp, BP 450-490 exciter filter, FT 510 chromatic beam splitter, an LP 528 barrier filter and a 63X Planapochromat objective.

Water samples for "settled" phytoplankton counts were collected with a Van Dohrn bottle from 1 m below the surface and 1 m above the bottom every six hours for 24 hours during sampling periods I and III, and from 1 m, mid-depth above and below the halocline and 1 m above the bottom every six hours for 24 hours during period IV. Samples for epifluorescent observation were collected from the near surface waters with a hand-held submerged bottle.

Simulated in situ rates of primary production were measured in water samples collected from 1 m and mid-water depths during sampling periods I, III and IV. All incubations were performed between 1300-1700 hours to nullify, insofar as possible, the effect of diel variation on the production rates. Twenty ml aliquots were inoculated with 0.25-1.0 μCi ^{14}C sodium bicarbonate and were placed in an incubation chamber. Light at six different intensities was provided by Vita Lite fluorescent tubes. Attenuation was obtained with combinations of gray plexiglass and plastic window screen. Temperature control was maintained with surface water circulated through the incubator. After two hour incubations, the samples were filtered through Gelman 0.2 μm porosity membrane filters which were subsequently placed in scintillation vials to which 0.3 ml NCS was added. After standing overnight, a toluene based cocktail was added to the vials and the radioactivity counted in a Beckman liquid scintillation counter. Carbon fixation rates were calculated taking into account the total inorganic carbon content of the water, the latter determined according to the method of Strickland and Parsons (26).

RESULTS

The salinity stratification conditions encountered during the study are summarized in salinity profiles from station YRM (Figure 1). During sampling periods I and II the lower river was moderately to strongly stratified. Destratification at YRM occurred on August 20 and the water column remained virtually homogeneous for most of sampling period III. Restratification at YRM commenced on August 24 and was completed by sampling period IV. A more detailed analysis of the hydrography can be found elsewhere in this volume (10).

The most striking observation regarding the phytoplankton aspects of this study was the initial appearance of a "red tide" in the lower York

FIGURE 1. Selected salinity profiles at Station YRM illustrating transition from a stratified to a destratified to a restratified water column encountered during the study.

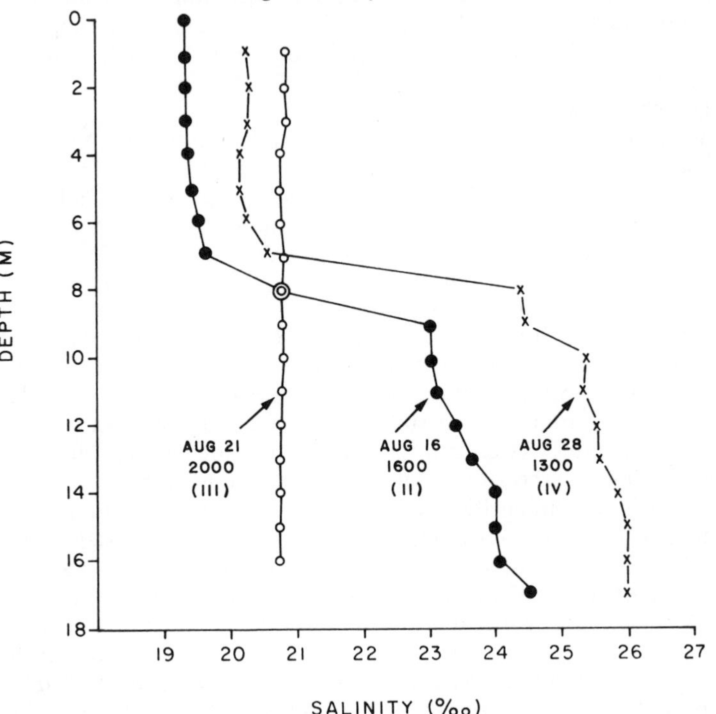

River on August 20, coincident with the onset of water column destratification. Red water had not been observed in the York River for at least a month prior to this observation. From our vantage point at station YRM, red water appeared as small (10-20 meter) patches moving down and up the estuary with the ebb and flood of the semi-diurnal tide. During period III the species composition of the dominant dinoflagellates associated with the red water changed (Table 1). Surface red water samples collected from a small boat in the lower river early in period III were dominated by *Ceratium furca* and *Prorocentrum micans*, while those collected later in period III were dominated by *Cocchlodinium heterolobatum* with some *C. furca* present and no. *P. micans* observed.

During period IV red water continued to be observed in the lower York River. The surface patches were larger (50-100 m) and typically were first observed in the morning as long streaks paralleling the main channel and overlying the area of rapid depth change between the shallows and the main channel on the north side of the river. These patches were probably associated with lateral shear fronts which typically occur in this zone in

TABLE 1. Dinoflagellates in red water samples collected in lower York River and observed with epifluorescent microscopy.

Date (1978)	*Ceratium furca* cells·ml-1	*Prorocentrum micans* cells·l-1	*Cocchlodinium heterolobatum* cells·ml-1
Aug. 21	258	58	0
Aug. 22	573	56	1232
Aug. 23	3	0	220
Aug. 24(a)	229	0	4041
Aug. 24(b)	172	0	2379

the estuary (3). During the morning these patches or streaks moved downriver with the ebb tide and during the afternoon the patches, somewhat dispersed and more evenly distributed over the main channel, moved back upriver with the flood tide. Red water samples during period IV continued to be dominated by *C. heterolobatum* with few if any *C. furca* or *P. micans* observed. During periods III and IV, the color of the red water patches appeared to intensify during the afternoon and the patches typically disappeared during the early evening. Our sampling also indicated that red water at the surface was generally restricted to approximately the top 1-1.5 m of the water column. The incidence of red water in the York River continued to increase into September, until by the middle of that month the entire surface of the river, a distance of about 50 km, appeared red during the afternoon.

The concentration of total chlorophyll *a* averaged over all depths and for all times of collection was not significantly different among the four sampling periods (means ranged from 7.9 to 9.6 μg l^{-1}; Figure 2). The chlorophyll *a* values for period III do not reflect the presence of red water in the river during this period. For example, the August 24 (b) red water sample (Table 1) contained 109 μg chlorophyll *a* l^{-1}. This omission reflects the extreme "patchiness" of the red tide event at that time and illustrates the inadequacy of attempting to monitor such an event by sampling on a predetermined schedule at a fixed point in the estuary. By the time of sampling period IV, however, the increased spatial distribution of the red water is reflected in the chlorophyll *a* concentrations observed at YRM. For example, a concentration of 39 μg chlorophyll *a* l^{-1} was measured in the 1 m sample at 1700 on August 29 when the water surface at YRM was brownish-red in color.

Although no difference was found in mean chlorophyll *a* concentrations among the four sampling periods, a difference in the size distribution of chlorophyll *a* was observed. During period I the < 15 μm size fraction dominated at all chlorophyll *a* concentrations (Figure 3a). During sampling period IV however, there was a positive linear correlation between total chlorophyll *a* concentration and the percent contribution of

FIGURE 2. Total chlorophyll *a* from surface, mid and bottom depths for sampling periods I, II, III and IV. Means for each period are given in parentheses and shown as crosses with horizontal bars indicating ± 1.0 standard error.

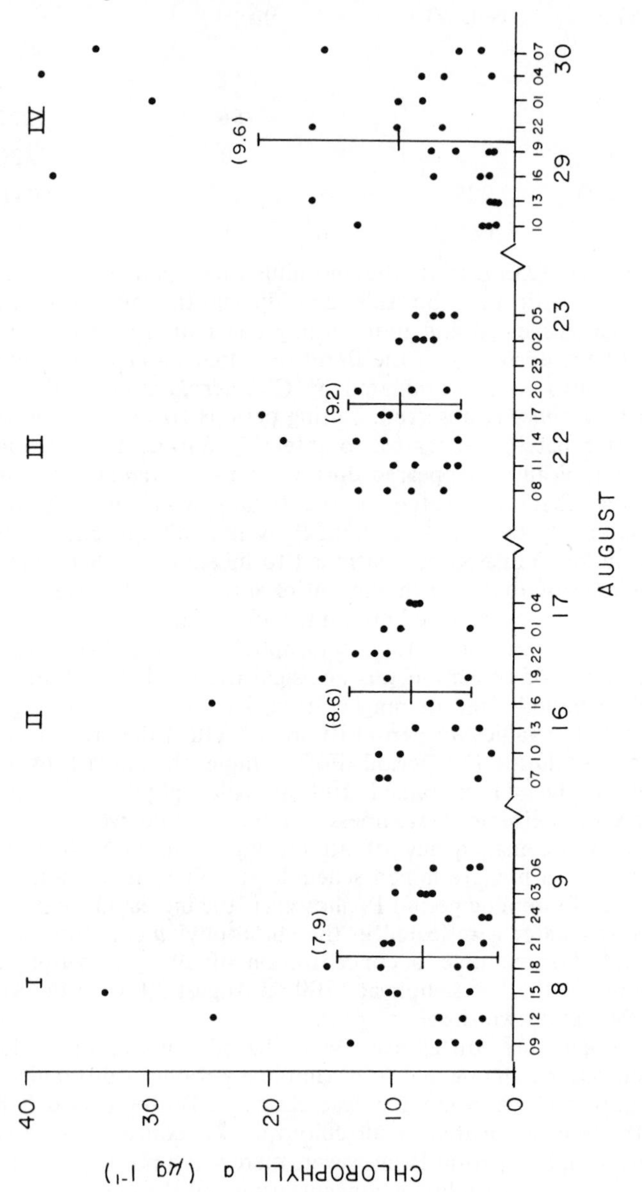

FIGURE 3a. Percent chlorophyll *a* in the < 15 μm size fraction versus total chlorophyll *a* for sampling period I.

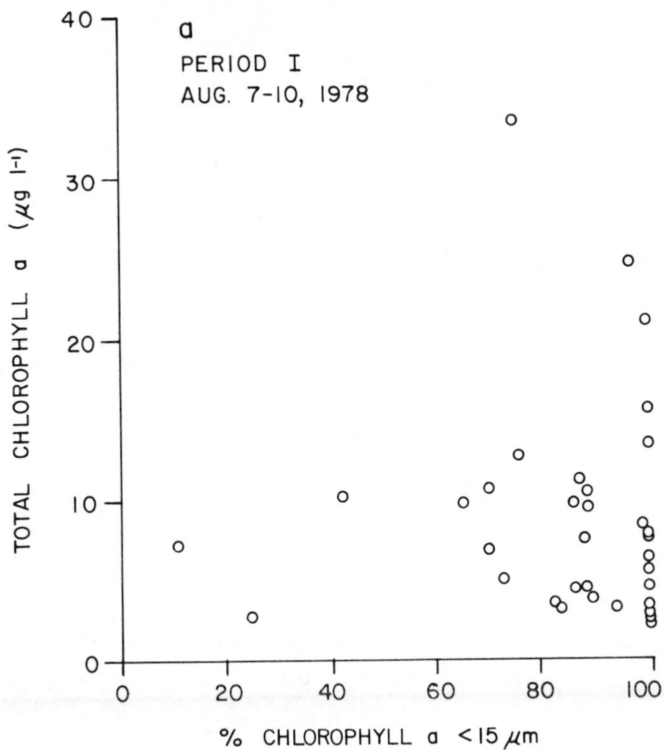

the > 15 μm size fraction (Figure 3b), indicating that the greater chlorophyll *a* concentrations during this period were the result primarily of large-sized phytoplankton.

A positive linear correlation was also observed between in vivo fluorescence and chlorophyll *a* concentration for all sampling periods.

$$\text{chlorophyll } a = 0.031 \text{ fluorescence in vivo} - 1.01$$

$$r^2 = 0.71 \qquad N = 145$$

This suggests that variation in fluorescence was primarily a result of variation in chlorophyll *a* concentration rather than the fluorescent efficiency of chlorophyll *a*. A distinct diurnal variation in mean euphotic zone (1-4 m) fluorescence was observed during all four sampling periods, with highest values at mid to late afternoon and the lowest values after midnight (Figure 4). During periods I, II and III, deep water (> 10 m) mean fluorescence values were always less than euphotic zone values and

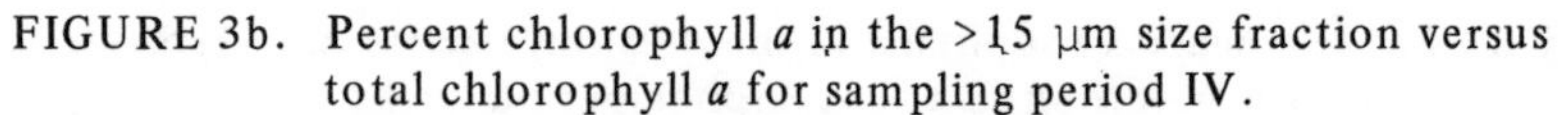
FIGURE 3b. Percent chlorophyll *a* in the >15 µm size fraction versus total chlorophyll *a* for sampling period IV.

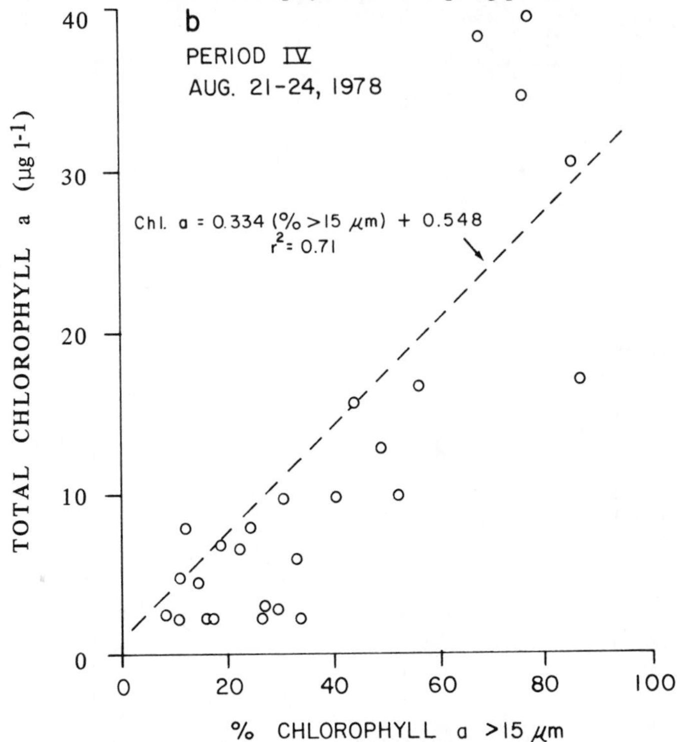

did not change markedly with time. During period IV, however, fluorescence values below 10 m increased sharply at night to levels above those in the euphotic zone. A comparison of individual fluorescence profiles for a 24-hour span during period IV indicates that the increase in deep water fluorescence at night appears to result from the diurnal vertical migration of a fluorescence maximum (Figure 5).

Extinction coefficients from periods I, II and III varied only slightly (range 0.31-0.35) and indicated a euphotic zone depth, i.e. 1 percent light depth, of about 4 m (Table 2). During sampling period IV, light attenuation varied depending on the water mass measured. On August 29 in mid afternoon when the surface water was brownish-red in color, an extinction coefficient of 0.08 m^{-1} (1 percent light depth 1.7 m) was measured, while on August 30, a "clear" water mass had an extinction coefficient of 0.24 m^{-1}. In all cases, the 1 percent light depth was considerably less than the depth of the halocline, when the latter was present.

The phytoplankton cell counts (all values x 10^3 ml^{-1}) indicate that during period I the surface waters were dominated by small (< 15 µm) flagellates with cryptophytes, *Pyramimonas sp.*, a small (3-5 μm) unidentified flagellate and *Katodinium rotundatum* the dominant forms (mean total concentration was 7.0). Large dinoflagellates were generally scarce (0.001-0.005), with *C. furca* and *C. heterolobatum* observed in only one water sample and *P. micans* observed at low concentrations (< 0.005) in several samples and at a concentration of 0.039 in one deep (17m)

FIGURE 4. Surface (1 - 4 m averaged) and deep water (greater than 10 m averaged) in vivo fluorescence values for each hydrocast plotted against time of collection for each sampling period. Horizontal bars indicate range of values. No deep water values for 0100, 0400 and 0700 hours on August 29 as vessel moved slightly off station. No hydrocasts between 1400, August 23 and 0900, August 24.

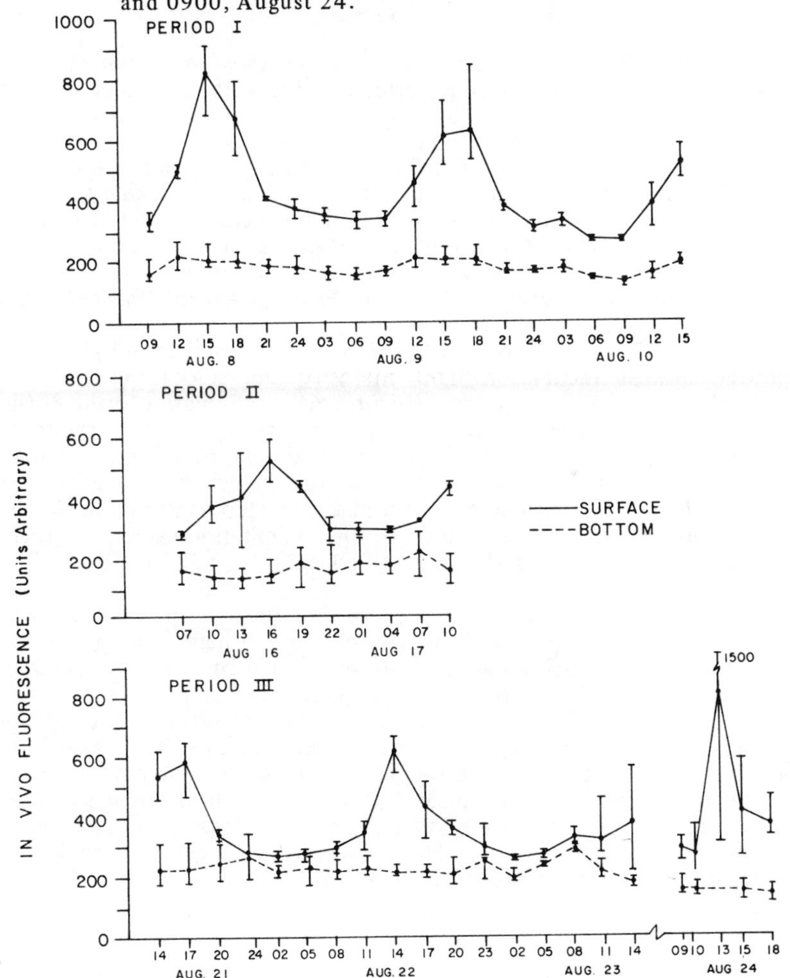

FIGURE 4. Cont'd.

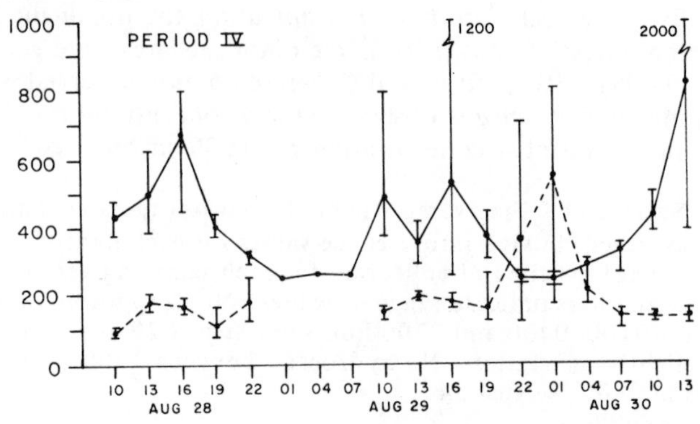

sample. The dominant diatoms were *Skeletonema costatum* (0.7) and *Thalassiosira* sp. (0.3). During period III the surface waters were still dominated by small flagellates (mean total 4.2). However, larger dinoflagellates were more abundant with *C. furca* (0.05), *Gyrodinium sp.* (0.01) and *P. micans* (0.03) the dominant forms. The dominant diatoms were *S. costatum* (0.44), *Cyclotella sp.* (0.15) and *Nitzschia sp.* (0.35). During sampling period IV, small flagellates were less abundant (mean total 1.1) while larger dinoflagellates had increased in abundance with *C. heterolobatum* (0.23) the dominant form. Diatoms were also more abundant with *S. costatum* (2.2), *Nitzschia sp.* (0.45) and *Thalassiosira sp.* (0.3) the dominant forms.

Photosynthesis rates were normalized on a per chlorophyll *a* basis, plotted against incubation light intensity and smooth curves drawn through the resulting points (Figure 6). Assimilation values (light saturated photosynthesis in units of μg C hr^{-1} per μg Chl a) for the surface waters during periods I and IV (i.e. stratified conditions) were 18 and 16 respectively, while values from below the halocline during stratified conditions and for both surface and midwater depths during destratified conditions ranged between 8 and 10. The assimilation value for a surface red water sample during the destratified period (Aug. 24b, Table I) was 20.

The results of the nutrient aspects of this study are provided elsewhere this volume (6); only a brief description is given here. During the initial stratified period, nitrite was high above the halocline (ca. 9 μg-at N l^{-1}) and absent below the halocline. Ammonium was high below the halocline (ca. 15 μg-at N l^{-1}) and absent above. During destratified conditions, ammonium was absent from the water column and nitrite was uniformly low (ca. 3 μg-at N l^{-1}) throughout the water column. Following restratification ammonium concentrations below the halocline increased to 15-20 μg-at N l^{-1} while nitrite remained low ($\leq$ 3 μg-at N l^{-1}) throughout the water column. Nitrate was never observed above 0.5 μg-at N l^{-1} in the water column during this study.

FIGURE 5. In vivo fluorescence profiles from hydrocasts at Station YRM taken at three-hour intervals over a 24-hour period. Date and time for each profile are indicated.

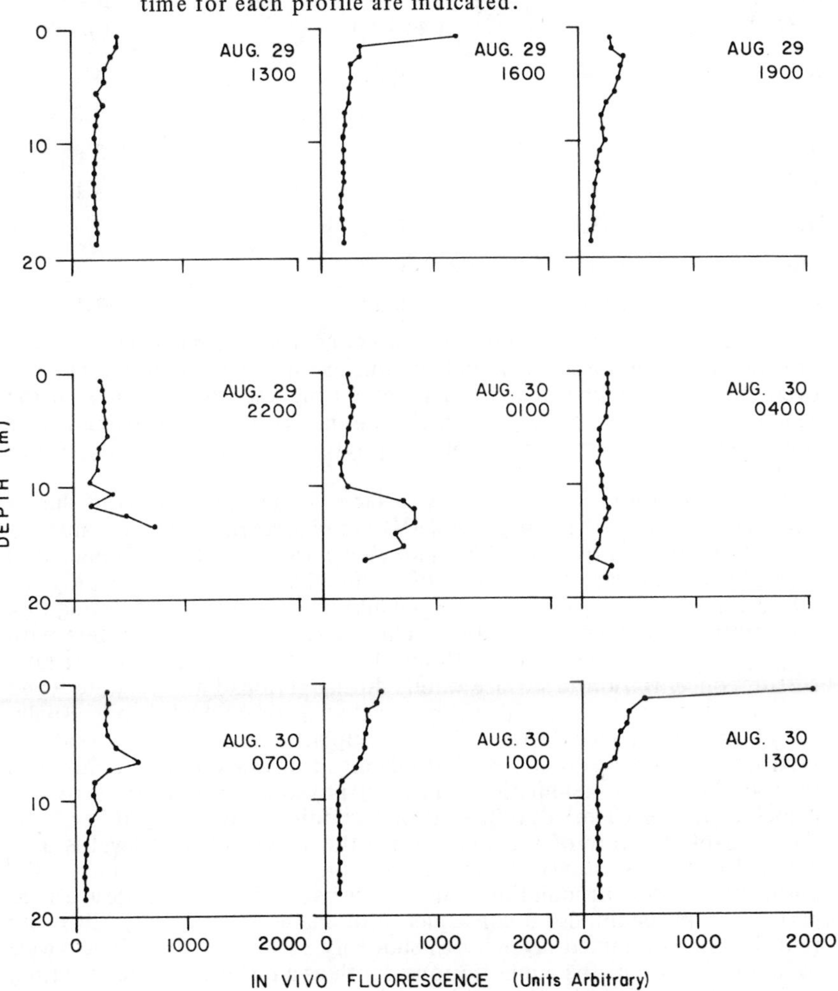

DISCUSSION

The values of plankton parameters observed during periods I and II are typical for this estuary and time of year. Chlorophyll *a* concentrations were generally within the range of 5-25 μg l^{-1} that is characteristic of the lower Chesapeake Bay estuary (16, 22), and the domination of the phytoplankton community by the nannophytoplankton is typical of temperate estuaries during warmer months (5, 15, 19, 20). The diel

TABLE 2. Extinction coefficients and 1 percent light depths at Station YRM.

Date (1978)	Time	Extinction Coefficient (k; m^{-1})	1% light depth (m)
Aug. 9	1200	0.33	4.1
Aug. 16	1100	0.35	4.3
Aug. 22	0930	0.32	3.9
Aug. 22	1340	0.33	4.1
Aug. 22	1630	0.31	3.9
Aug. 29	1534	0.07	1.7
Aug. 30	1600	0.24	3.1

variation in near surface chlorophyll *a* concentrations has been observed previously at this station (9), and the uniformity of the diel fluorescence cycle despite sampling at a fixed point during the ebb and flow of the semi-diurnal tide indicates that, in the absence of red water, patchiness is not a significant characteristic of the phytoplankton community in this estuary.

The assimilation values for the surface (1 m) phytoplankton during stratified conditions (periods I and IV) are characteristic of non-nutrient limited phytoplankton at 27°C and above (5, 20, 21). Although it is recognized that standing stock of a nutrient is not necessarily an indication of its utilization by phytoplankton (17), it is worth noting that the stratified surface assimilation values were observed in waters with varying nitrite content, but without detectable ammonium or nitrate. Nitrite concentration in surface waters during periods I and IV were 9 and 3 μg-at l^{-1}, respectively, which suggests that nitrite, which is not usually found in high concentrations in the euphotic zone, is a potentially significant nitrogen source for phytoplankton in this estuary at this time of year. The lower assimilation values for phytoplankton from below the halocline (i.e. 8-10 m) during stratified conditions probably reflects the shade adapted nature of these cells since the 1 percent light level is only about 4 m in this estuary. Likewise, excluding the red water sample, the uniformly lower assimilation values observed during the destratified period are probably also a consequence of shade adaptation, in this case resulting from a mixed layer depth suddenly increased from about twice to about five times the euphotic zone depth as a result of destratification. Reduced rates of production observed in water columns with a high ratio of mixed layer depth:euphotic zone depth and attributed to shade adaption have been reported in coastal oceanic areas (13) and lakes (28).

Red tides have been reported to occur sporadically throughout most of the year in the York River (33). In the present study the initial appearance of red water simultaneous with destratification suggests that the dinoflagellates initiating the event were derived from below the halocline. Tyler and Seliger (29) report that dinoflagellates causing red tides in the upper Chesapeake Bay traverse the length of the Bay in deep, upestuary flowing water on a seasonal cycle before appearing at the

FIGURE 6. Photosynthetic rates for total phytoplankton, normalized per unit chlorophyll *a* and plotted against light level of incubation. Ten meters is below halocline for August 10 and eight meters is below halocline for August 29. No halocline on August 22. Roman numerals in parentheses indicate sampling period.

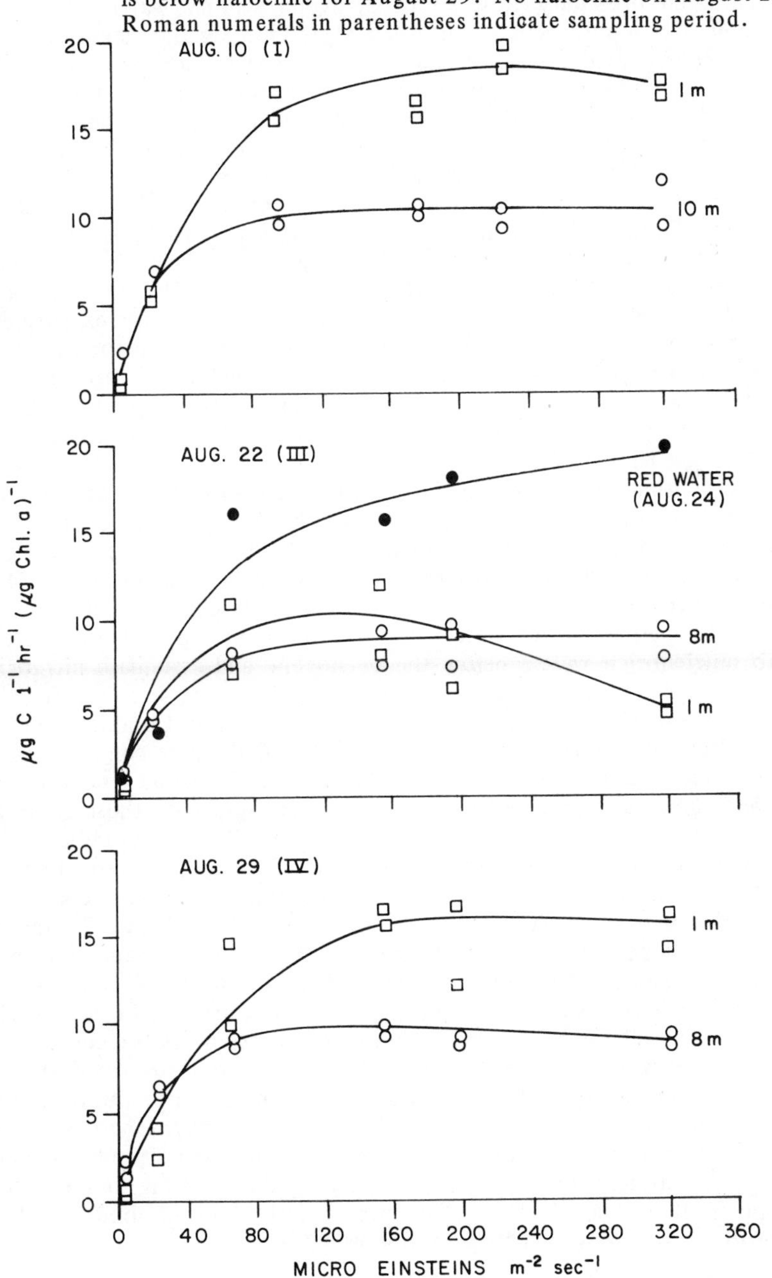

surface. We propose that a similar transport mechanism, albeit reduced in space and time, may be functioning in the York River. The process of restratification in the lower York River appears to result from the accentuated upestuary flow of deep, higher salinity water (10). It is possible that dinoflagellates or other algae entering the estuary in this deep water contribute to, or dominate, the species composition of the surface phytoplankton following a destratification event. In the present study, one species of dinoflagellate observed in red water samples was found in the deep water prior to destratification. However, the small number of samples taken below the halocline prior to destratification does not, in our estimation, warrant conclusions concerning the presence or absence of other dinoflagellate species. Another potential source of cells to initiate a red water event is dinoflagellate cysts in the sediments which may act as a "seed bed" for the surface waters (1). In this case, destratification may function as an environmental cue for excystment (e.g. temperature or oxygen shock) as well as a mechanism for transport to the surface. The proposed mechanisms for providing cells to initiate a red tide should not be considered mutually exclusive, e.g. phytoplankton cells entering the York River in deep water may be derived from "seed beds" in the Chesapeake Bay.

The continuous increase in the extent of red water that was observed during late August-early September as the river remained stratified suggests that the stratified state of the estuary enhanced the development of the red water phenomenon. Eppley and Harrison (7) and Harrison (12) report that the propensity for red tide dinoflagellates to flourish in nutrient-depleted surface waters derives from their capability to migrate into nutrient-rich waters below the pycnocline and assimilate nitrate in the dark. We propose that a similar mechanism may account for the development of the bloom observed in the York River. Conditions in the York River during period IV, i.e. a well-stratified water column with a high concentration of ammonium below the halocline and low or undetectable nutrient concentrations above, are similar to those cited by these authors, and the fluorescence profiles for this period indicate the diurnal vertical migration of a fluorescence maximum. Several lines of evidence indicate that this fluorescence maximum was composed primarily of the dinoflagellate *C. heterolobatum*. This alga was by far the dominant large dinoflagellate observed in the cell count samples during period IV, and the temporal and spatial distribution of *C. heterolobatum* cell counts is consistent with a diurnal migration capability (Table 3). The positive correlations observed between in vivo fluorescence and chlorophyll *a* and between high chlorophyll *a* concentrations and an increased percentage of large-sized phytoplankton also support this contention. The diurnal vertical migration of *C. heterolobatum* is also consistent with our observation of the appearance and disappearance of red water at the surface of the estuary in the morning and evening, respectively, and a calculated minimum migration rate of about 1 m hr^{-1} is within the reported capabilities of dinoflagellates (2).

TABLE 3. *Cocchlodinium heterolobatum* cell counts at Station YRM, nd-not determined.

Date (1978)	Time	Surface (1m)	Mid depth above halocline (6-8m)	Mid depth below halocline (10-14m)	Bottom (18m)
Aug. 29	1000	42*	1	8	nd
Aug. 29	1600	870	2	1	16
Aug. 29	2200	1	4	270	nd
Aug. 30	0400	1	88	19	10

*All values cells·ml^{-1}

In addition to permitting access to high nutrient levels in deep water, a diel migration capability may have additional adaptive significance for *C. heterolobatum*. The relatively high assimilation value for the red water sample compared to other algae during period III is consistent with the hypothesis that by maintaining themselves near the surface during the day these dinoflagellates overcome the shade adapted state imposed on other, presumably less migratory, phytoplankton by the increased mixed layer depth accompanying destratification. It is also noted that the accumulation of dinoflagellates at the surface extended only as deep as the much reduced 1 percent light depth resulting from these accumulations. Thus, migration allows the dinoflagellates to maximize resource utilization by "filling" the available light space in the water column.

The results of this study suggest that the debate concerning whether red tides are more likely to be associated with upwelling-vertically mixed water masses (2, 4, 11) or with stable, well-stratified water columns (12, 24, 32) may be primarily a temporal question. In the York River estuary we propose that at least some red tides are initiated by spring-tide associated destratification and enhanced by ensuing conditions of stratification. However, the implications of hydrographic regulation of phytoplankton dynamics in estuaries need not be limited to red tide dinoflagellates. For example, the marked increase in diatom numbers observed in the York River subsequent to destratification may have resulted from nutrient enrichment to the surface waters occasioned by the destratification event. Periodic phytoplankton blooms associated with alternating periods of stratification and destratification have been observed in a variety of estuaries including Auke Bay, Alaska (14), a British Columbia fjord (8), St. Lawrence estuary (25), Saanich Inlet (27), Puget Sound (31) and the Duwamish estuary (30). In the first two estuaries, the relative stratification state was controlled by wind events while in the latter four estuaries, neap-spring tides apparently were the controlling factor. Thus the fortnightly cycle of neap-spring tides appears to play a significant role in phytoplankton dynamics in certain estuaries.

Although it is apparent that hydrographic and phytoplankton processes may be tightly coupled in the York River estuary, many questions remain. Among the more interesting are the rates and nature of nutrient assimilation by phytoplankton at various depths during different periods of the neap-spring hydrographic cycle. Only after it is ascertained how phytoplankton respond to the varied nutrient regimes which occur during the neap-spring tidal cycle can meaningful predictions concerning phytoplankton response to additional, anthropogenic nutrient inputs be attempted.

ACKNOWLEDGEMENTS

The authors would like to thank M. Petty and A. Thomson for their assistance on this project and R. Jordan and P. Goodwin for providing cell count data. This research was supported in part by the Virginia Institute of Marine Science, by the Oceanography Section, National Science Foundation, under NSF grant OCE-77-20228, and by the Office of Sea Grant, National Oceanic and Atmospheric Administration, U.S. Department of Commerce grant to the Virginia Institute of Marine Science. Contribution number 948 of the Virginia Institute of Marine Science.

REFERENCES

1. Anderson, D.M. and D. Wall. 1978. Potential importance of benthic cysts of *Gonyaulax tamarensis* and *G. excavata* in initiating dinoflagellate blooms. J. Phycol. 14:224-234.
2. Blasco, D. 1979. Changes of the surface distribution of a dinoflagellate bloom off the Peru coast related to time of day. pp. 209-214. *In*: D.L. Taylor and H.H. Seliger (eds.) Toxic Dinoflagellate Blooms. Elsevier North Holland Inc. New York.
3. Bowman, M.J. and R.L. Iverson. 1978. Estuarine and plume fronts. pp. 87-104. *In*: M.J. Bowman and W.E. Esaias (eds.), Oceanic Fronts in Coastal Processes. Springer-Verlag, New York.
4. Dugdale, R.C. 1979. Primary nutrients and red tides in upwelling areas. P. 257-262. *In*: D.L. Taylor and H.H. Seliger (eds.) Toxic Dinoflagellate Blooms. Elsevier North Holland Inc., New York.
5. Durbin, E.G., R.W. Krawiec and T.J. Smayda. 1975. Seasonal studies on the relative importance of different size fractions of phytoplankton in Narragansett Bay (USA). Mar. Biol. 32:271-287.
6. D'Elia, C.F., K.L. Webb, and R.L. Wetzel. 1981. Time Varying hydrodynamics and water quality in an estuary. pp. 597-606. *In*: B. Neilson and L.E. Cronin (eds.). Enrichment of Estuaries. The Humana Press, Inc., Clifton, N.J.
7. Eppley, R.W. and W.G. Harrison. 1975. Physiological ecology by *Gonyaulax polyedra*, a red water dinoflagellate off Southern California. pp. 11-22. *In* V.R. LoCicero (ed.). The first international

conference on toxic dinoflagellate blooms. M. S. T. F., Wakefield, Mass.

8. Gilmartin, M. 1964. The primary production of a British Columbia Fjord. J. Fish. Res. Bd. Can. 21:505-538.
9. Haas, L.W. 1975. Plankton dynamics in a temperate estuary with observations on a variable hydrographic condition. Ph.D. dissertation, College of William and Mary. 202 p.
10. Haas, L.W., F.J. Holden, and C.S. Welch. 1981. Short term changes in the vertical salinity distribution of the York River estuary associated with the neap-spring tidal cycle pp. 585-596. *In*: B. Neilson and L.E. Cronin (eds.), Enrichment of Estuaries. The Humana Press, Inc., Clifton, New Jersey.
11. Haddad, K.D. and K.L. Carder. 1979. Oceanic intrusions: one possible initiation mechanism of red tide blooms on the west coast of Florida. pp. 269-274. *In*: D.L. Taylor and H.H. Seliger (eds.). Toxic Dinoflagellate Blooms. Elsevier North Holland, Inc., New York.
12. Harrison, W.G. 1976. Nitrate metabolism of the red tide dinoflagellate *Gonyaulax polyedra*. J. Exp. Mar. Biol. Ecol. 21:199-209.
13. Huntsman, S.A. and R.T. Barber. 1977. Primary production off northwest Africa: the relationship to wind and nutrient conditions. Deep-Sea Res. 24:25-33.
14. Iverson, R.L., H.C. Curl, Jr., H.B. O'Conners, Jr., D. Kirk and K. Zakar. 1974. Summer phytoplankton blooms in Auke Bay, Alaska, driven by wind mixing of the water column. Limnol. Oceanogr. 19:271-278.
15. McCarthy, J.J., W.R. Taylor and M.E. Loftus. 1975. Significance of nannoplankton in the Chesapeake Bay estuary and problems associated with measurement of nannoplankton productivity. Mar. Biol. 24:7-16.
16. McCarthy, J.J., W.R. Taylor, and J.L. Taft. 1975. The dynamics of nitrogen and phosphorus cycling in the open waters of the Chesapeake Bay, p. 664-681. *In*: T. Church (ed.), Marine hemistry. A.C.S. Symposium Series 18. Washington, D.C.
17. McCarthy, J.J., W.R. Taylor and J.L. Taft. 1977. Nitrogenous nutrition of the plankton in the Chesapeake Bay. 1. Nutrient availability and phytoplankton preferences. Limnol. Oceanogr. 22:996-1011.
18. McGowan, J.A. and T.L. Hayward. 1978. Mixing and oceanic productivity. Deep-Sea Res. 25:771-794.
19. Malone, T.C. 1976. Phytoplankton productivity in the apex of the New York bight: environmental regulation of productivity/chlorophyll *a*. pp. 260-272. *In*: G. Gross (ed.), Middle Atlantic Continental Shelf and the New York Bight. Special Symposia, Vol. 2. Amer. Soc. Limnol. Oceanogr., Inc.
20. Malone, T.C. 1972. Light saturated photosynthesis by phytoplankton size fractions in the New York Bight USA. Mar. Biol. 42:281-292.
21. Malone, T.C. 1977 Environmental regulation of phytoplankton productivity in the lower Hudson estuary. Est. Coastal Mar. Sci. 5:157-171.

22. Patten, B.C., R.A. Mulford, and J.E. Warinner. 1963. An annual phytoplankton cycle in the lower Chesapeake Bay. Ches. Sci. 4:1-20.
23. Phoel, W.C., K.L. Webb and C.F. D'Elia. 1981. Inorganic nitrogen regeneration and total oxygen consumption by the sediments at the mouth of the York River, Virginia, U.S.A. pp. 607-618. *In*: B. Neilson and L.E. Cronin (eds.). Enrichment of Estuaries. The Humana Press, Inc., Clifton, New Jersey.
24. Pingree, R.D., P.R. Pugh, P.M. Holligan and G.R. Forster. 1975. Summer phytoplankton blooms and red tides along tidal fronts in the approaches to the English Channel. Nature, London, 258:672-677.
25. Sinclair, M. 1978. Summer Phytoplankton Variability in the Lower St. Lawrence Estuary. J. Fish. Res. Bd. Can., 35:1171-1185.
26. Strickland, J.D.H. and T.R. Parsons. 1968. A practical handbook of seawater analysis. Bull. Fish. Res. Bd. Can. 167, 311 p.
27. Takahashi, M., D.L. Seibert and W.H. Thomas. 1977. Occasional blooms of phytoplankton during summer in Saanich Inlet, B.C., Canada. Deep-Sea Res. 24:775-780.
28. Tilzer, M.M. and C.R. Goldman. 1978. Importance of mixing, thermal stratification and light adaptation for phytoplankton productivity in Lake Tahoe (California-Nevada). Ecology 59:810-821.
29. Tyler, M.A. and H.M. Seliger. 1978. Annual subsurface transport of a red tide dinoflagellate to its bloom area: water circulation patterns and organism distributions in the Chesapeake Bay. Limnol. Oceanogr. 23:227-246.
30. Welch, E.B. 1969. Factors initiating phytoplankton blooms and resulting effects on dissolved oxygen in Duwamish River Estuary, Seattle, Washington, Geol. Sur. Water-Supply Paper 1873-A.
31. Winter, D.F., K. Banse and G.C. Anderson. 1975. The dynamics of phytoplankton blooms in Puget Sound, a fjord in the Northwestern United States. Mar. Biol. 29:139-176.
32. Wyatt, T. and J. Horwood. 1973. Model which generates red tides. Nature 244:238-240.
33. Zubkoff, P.L., J.C. Munday, R.G. Rhodes and J.E. Warinner. 1979. Mesoscale features of summer (1975-1977) dinoflagellate blooms in the York River, Virginia (Chesapeake Bay estuary). pp. 279-286. *In*: D.L. Taylor and H.H. Seliger (eds.) Toxic Dinoflagellate Blooms. Elsevier North Holland Inc., New York.

INDEXES

Animal Index

Plant Index

G

H

J

K

L

M

N

P

R

S

T

U

Z

Waterbody Index

Subject Index